W0106999

Ultrafast Processes in Spectroscopy

Ultrafast Processes in Spectroscopy

Edited by

O. Svelto
S. De Silvestri
Politecnico di Milano
Milano, Italy

and

G. Denardo
International Centre for Theoretical Physics
Trieste, Italy

Springer Science+Business Media, LLC

Library of Congress Cataloging-in-Publication Data

Ultrafast processes in spectroscopy / edited by O. Svelto, S. De
Silvestri and G. Denardo.
    p.   cm.
    "Ninth International Symposium on Ultrafast Processes in
Spectroscopy, UPS'95, Trieste, Italy, October 30-November 3, 1995"-
-Pref.
    Includes bibliographical references and index.
    ISBN 978-1-4613-7705-4   ISBN 978-1-4615-5897-2 (eBook)
    DOI 10.1007/978-1-4615-5897-2
    1. Laser spectroscopy--Congresses. 2. Picosecond pulses-
-Congresses. I. Svelto, Orazio. II. De Silvestri, Sandro.
III. Denardo, G. (Gallieno), 1935-   . IV. International Symposium
on Ultrafast Processes in Spectroscopy (9th : 1995 : Trieste, Italy)
QC454.L3U433 1996
543'.0858--dc21   96-43801
                                                          CIP

Proceedings of the Ninth International Conference on Ultrafast Processes in Spectroscopy,
held October 30 – November 3, 1995, in Trieste, Italy

ISBN 978-1-4613-7705-4

© 1996 Springer Science+Business Media New York
Originally published by Plenum Press, New York in 1996
Softcover reprint of the hardcover 1st edition 1996

All rights reserved

10 9 8 7 6 5 4 3 2 1

No part of this book may be reproduced, stored in a retrieval system, or transmitted in any form
or by any means, electronic, mechanical, photocopying, microfilming, recording, or otherwise,
without written permission from the Publisher

PREFACE

This volume is a collection of papers presented at the Ninth International Symposium on "Ultrafast Processes in Spectroscopy" (UPS'95) held at the International Centre for Theoretical Physics (ICTP), Trieste (Italy), October 30 - November 3, 1995. These meetings have become recognized as the major forum in Europe for discussion of new work in this rapidly moving field. The UPS'95 Conference in Trieste brought together a multidisciplinary group of researchers sharing common interests in the generation of ultrashort optical pulses and their application to studies of ultrafast phenomena in physics, chemistry, material science, electronics, and biology. It was attended by approximately 250 participants from 20 countries and the five-day program comprises more than 200 papers.

The progress of both technology and applications in the field of ultrafast processes during these last years is truly remarkable. The advent of all solid state femtosecond lasers and the extension of laser wavelengths by frequency conversion techniques provide a large variety of high-performance light sources for ultrashort pulses. With these sources ultrafast phenomena in physical, chemical and biological systems and in electronic devices are now studied extensively. Ultrafast technology is becoming one of the basic and common tools presently entering a wide variety of scientific fields not only for basic research but also for promoting new applications in various areas. We feel that these proceedings vividly reflect the present status of the field.

We would like to acknowledge numerous contributions which helped to make the UPS'95 Conference so successful. We are indebted to the European Community, the ICTP, the Italian Research Council (CEQSE-CNR), the Istituto Nazionale Fisica della Materia (INFM), the Politecnico of Milan and a few industrial companies for their financial support. We would like to thank the members of the International Program Committee for their help in setting up the conference program. Special thanks are also due to the secretariat of UPS'95 Mrs. F. Masserano (CEQSE-CNR), Mrs. P. Passarella (ICTP) and Mr. A. Caporali (INFM) for their work in implementing the meeting arrangements.

Conference Chair	O. Svelto Politecnico, Milan, Italy
Scientific Program Chair	S. De Silvestri Politecnico, Milan, Italy
Local Organization Chair	G. Denardo International Centre for Theoretical Physics, Trieste, Italy

CONTENTS

ULTRAFAST SPECTROSCOPY OF ATOMS AND MOLECULES

ULTRAFAST NONLINEAR OPTICAL PHENOMENA

ULTRAFAST SPECTROSCOPY OF SEMICONDUCTORS

GENERATION AND APPLICATIONS OF INTENSE ULTRASHORT PULSES

FREQUENCY CONVERSION

ULTRAFAST NONLINEAR OPTICS IN ORGANICS

APPLICATIONS OF ULTRAFAST LASERS IN MEDICINE AND ULTRAFAST PROCESSES IN BIOPHYSICS

ULTRAFAST SPECTROSCOPY OF METALS, INSULATORS, AND CONFINED SYSTEMS

NEW ULTRAFAST MEASUREMENT TECHNIQUES

SUPERCONDUCTORS AND TERAHERTZ SPECTROSCOPY

ULTRAFAST OPTOELECTRONICS

ALL-SOLID-STATE TUNABLE DIODE-PUMPED ULTRAFAST LASER OSCILLATORS AND AMPLIFIERS FOR REAL-WORLD APPLICATIONS INCLUDING MEDICAL IMAGING

R. Mellish,[1] N. P. Barry,[1] P. M. W. French,[1*] J. C. Dainty,[1] K. Dowling,[1] S. C. W. Hyde,[1] R. Jones,[1] J. Sutherland,[1] J. R. Taylor,[1] Y. P. Tong,[1] C. J. van der Poel[2] and A. Valster[2]

[1] Optics Section
Physics Department
Imperial College
London SW7 2BZ, United Kingdom
[2] Philips Optoelectronics Centre
Prof. Holstlaan 4, 5656 AA Eindhoven, The Netherlands

INTRODUCTION

The development of Kerr Lens mode-locked (KLM) solid-state lasers [1, 2] has demonstrated the potential for convenient user friendly ultrafast laser systems which may be commercialised and applied in a wide range of scientific and technological areas. Self-starting KLM has recently been shown to be straightforward in Ti:sapphire lasers [3] and may be extended to any other solid-state laser medium to yield a relatively simple picosecond or femtosecond laser with no active components to increase the complexity of the system and no resonant saturable absorbers to compromise the tunability.

Figure 1 is a schematic of the approximate tuning ranges of several solid-state laser media providing coverage from < 300 nm in the ultraviolet spectral region to ~ 3000 nm in the near infra-red (See reference [4] for a recent overview). In the near future we expect to see most of these laser systems Kerr Lens mode-locked: it is, in principle, not significantly more difficult to build and ultrafast KLM laser than a c.w. laser. The further development of appropriate semiconductor diode pump lasers for these media will produce a

* Tel. : 44–171–594 7706 Fax. : 44–171–589 9463 email: paul.french@ic.ac.uk

Ultrafast Processes in Spectroscopy, edited by Svelto et al.
Plenum Press, New York, 1996

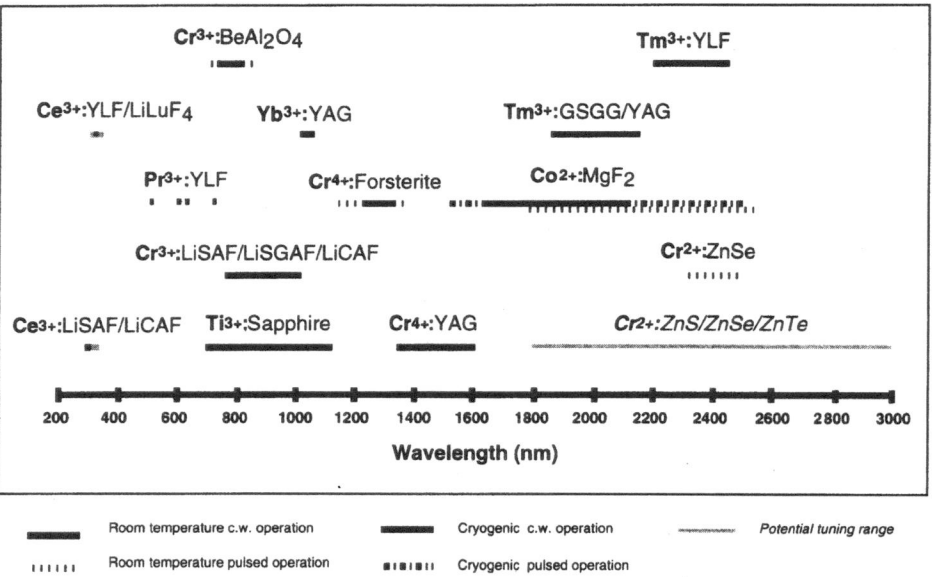

Figure 1. Tuning range of solid-state lasers.

range of versatile and low-cost ultrafast lasers which could revolutionise the range of applications of ultrafast technology.

ALL-SOLID-STATE DIODE-PUMPED ULTRAFAST Cr:LiSAF LASERS

Ti:sapphire was the first laser medium to be employed in a commercial KLM laser and Cr:LiSAF is likely to be the first medium to be employed in an all-solid-state diode-pumped ultrafast laser. All-solid-state Cr:LiSAF lasers can replace Titanium-doped sapphire (Ti:Al$_2$O$_3$) lasers for many applications, delivering reasonably high c.w. average power levels, ultrashort picosecond and femtosecond pulse operation and narrow line-width operation. We report here the first all-solid-state tunable Cr:LiSAF femtosecond laser oscillator/amplifier system which is pumped by AlGaInP semiconductor diodes. The oscillator generates pulses of less than 55 fs, tunable from 835–910 nm, with up to 70 mW average output power in a single beam. The shortest pulsewidths measured to date are 24 fs. This diode-pumped oscillator is used to seed a diode-pumped regenerative amplifier which delivers pulses energies as high as 2.7 μJ (measured before the grating compressor) at a repetition rates up to 25 kHz. The compressed pulses were of ~ 1 μJ energy and ~ 200 fs duration.

Previously we had employed AlGaInP diodes of 160 μm stripe-width to pump the first tunable picosecond all-solid-state Cr:LiSAF laser [5] and the first femtosecond all-solid-state Cr:LiSAF laser oscillator [6]. The latter device employed an intracavity MQW semiconductor saturable to achieve femtosecond pulse generation with the relatively low intracavity power levels available with these pump diodes. We now report an improved oscillator which is pumped by much brighter diodes of 50 μm stripe-width. The resulting intracavity power lev-

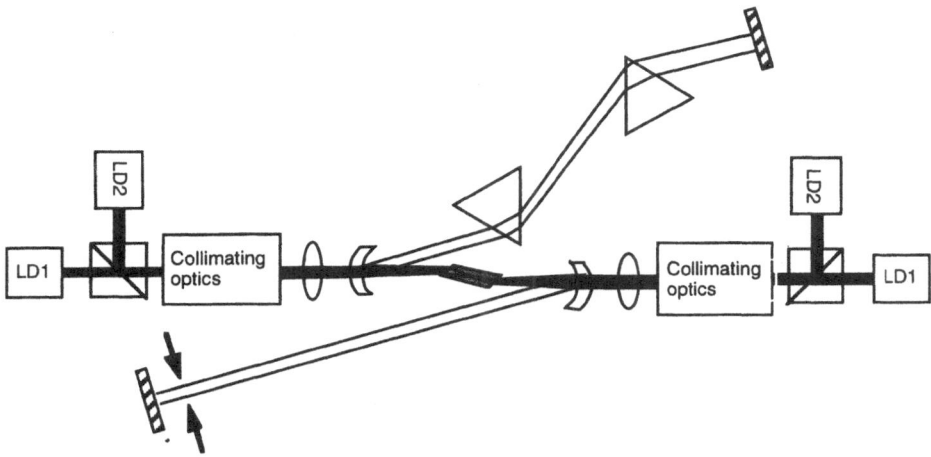

Figure 2. Schematic of diode-pumped KLM Cr:LiSAF femtosecond laser oscillator.

els were sufficient to achieve Kerr Lens Mode-locking (KLM) in the cavity shown in Figure 2. Each pump diode provided 300 mW which after polarisation-combined and beam shaping, resulted in a total of 900 mW of pump radiation being absorbed in the 1.7 % doped laser rod. The cavity mirrors were all of 99.9 % reflectivity except the output coupler. With an output coupler of 0.12 % transmission, the laser delivered pulses of less than 40 fs duration with up to 70 mW average output power. This mirror operated near the edge of its reflectivity curve and this limited the available bandwidth. Using a 0.05 % transmission mirror with a flat reflectivity profile centred on 840 nm, this laser tuned from 835–910 nm generating pulses shorter than 55 fs [6]. KLM was initiated by either tapping a mirror or using an acousto-optic modulator on one of the quartz prisms. The amplitude stability was better than 1 %. Replacing the output coupler with a high reflector produced pulses as short as 24 fs with a time-band-width product of 0.39. This measured duration was limited by the autocorrelator used. We note that slightly shorter pulses have been reported from a similar laser, although with less output power [7].

Figure 3 shows the configuration of the diode-pumped regenerative amplifier. This was considerably improved compared to our earlier system [8] which was previously demonstrated to amplify the pulses from an argon ion-pumped Spectra Physics Tsunami laser oscillator and tuned from 805–875 nm. Improving the mirror reflectivity's to 99.9 % and pumping with four full power (300 mW) diodes with improved beam-shaping permitted us to amplify 100 fs pulses of 200 pJ from the diode-pumped Cr:LiSAF oscillator to energies as high as 2.65 µJ (measured before the compressor). The output energy of this regenerative amplifier remained constant as a function of pulse repetition rate up to 25 kHz, the maximum possible with our pockels cell driver.

APPLICATIONS OF ULTRAFAST LASERS

This all-solid-state diode-pumped laser technology has the potential to provide low-cost compact devices for ultrafast instrumentation. We have already used the femtosecond

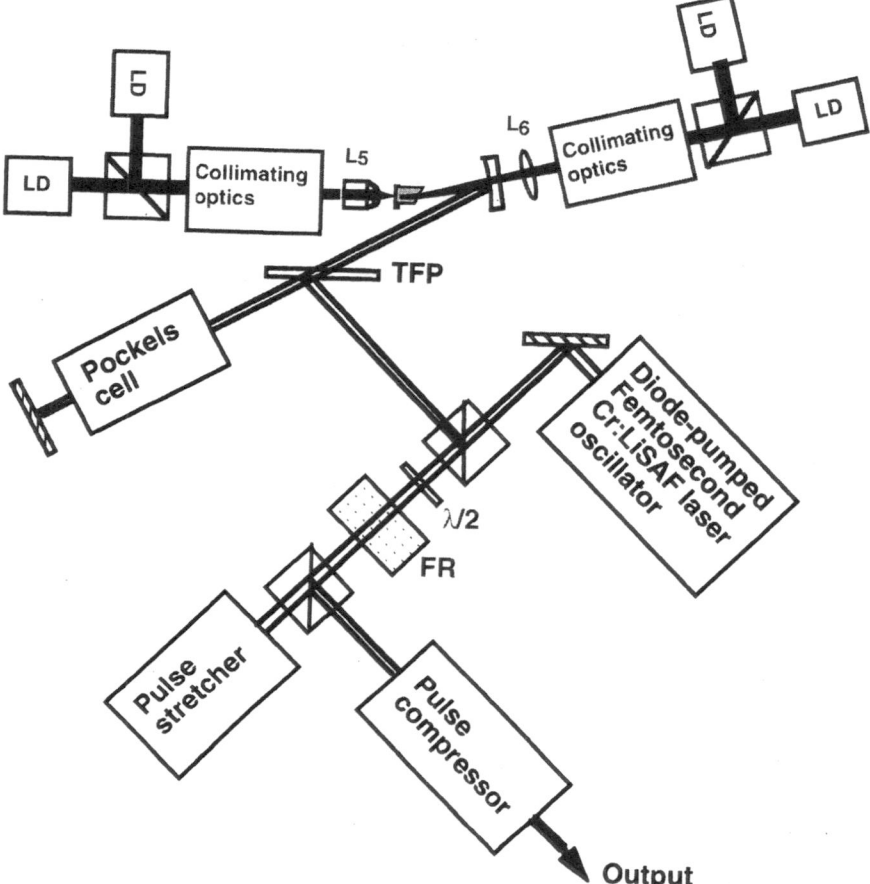

Figure 3. Schematic of diode-pumped Cr:LiSAF femtosecond laser oscillator/amplifier system

oscillator described above in a two photon microscope developed by Carl Zeiss, Oberkochen [9]. Other applications include medical imaging e.g. using time-gated holography [10], the measurement of surface micro-roughness using second harmonic generation and electro-optic sampling. The amplified pulses are sufficiently intense for continuum generation and parametric generation e.g. [11]. In the near future, the demonstration of low threshold and/or self-starting KLM diode-pumped lasers will greatly simplify the oscillator design. Furthermore, the use of higher power pump diode arrays will permit significantly higher powers to be obtained from both oscillators and amplifiers. We anticipate many new applications for versatile, low-cost, ultrafast all-solid-state lasers.

REFERENCES

1. D. E. Spence, P. N. Kean, W. Sibbett, Opt. Lett. **16**, 42 (1991)
2. D. K. Negus, L. Spinelli, N. Goldblatt, G. Feugnet, in *Advanced Solid State Lasers* , G. Dube and L. L. Chase, eds., Vol. 10 of OSA Proceedings Series (Optical Society of America, Washington, D.C., 1991), pp. 120–124

3. G. Cerullo, S. De Silvestri and V. Magni, Opt. Lett., **19**, (1994) 1040

4. P. M. W. French, Laser Focus World, pp. 93—100, September 1995

5. P. M. W. French, R. Mellish, J. R. Taylor, P. J. Delfyett and L. T. Florez, Electron. Lett., **29**, (1993) 1263

6. R. Mellish, P. M. W. French, J. R. Taylor, P. J. Delfyett and L. T. Florez, Electron. Lett., **30**, (1994) 223

7. M. J. P. Dymott and A. I. Ferguson, Opt. Lett., **20**, (1995) 1157

8. S. C. W. Hyde, N. P. Barry, R. Mellish, P. M. W. French, J. R. Taylor, C. J. van der Poel and A. Valster, Opt. Lett., **20**, (1995) 160

9. H. Sakowski, M. Kaschke, R. Mellish, R. Jones, P. M. W. French, J. R. Taylor, V. Petrov and F. Noack, To be published.

10. S. C. W. Hyde, N. P. Barry, R. Jones, J. C. Dainty, P. M. W. French, M. B. Klein and B. A. Wechsler. Optics Lett. **20**, 1331 (1995)

11. M. K. Reed, M. K. Steiner-Shepard and D. K. Negus, Opt. Lett., **19**, (1994) 1855\

2

POWERFUL ULTRASHORT PULSE GENERATION IN Ti:SAPPHIRE

M. Lenzner, Ch. Spielmann, R. Szipöcs,[1] and F. Krausz

Abt. Quantenelektronik und Lasertechnik
Technische Universität Wien
Gusshausstrasse 27-29, A–1040 Wien, Austria
[1]Research Institute for Solid State Physics
H–1525 Budapest, POB 49, Hungary

Kilohertz repetition rate amplifiers of femtosecond pulses turned out to be a powerful tool in spectroscopy during the last years. A number of schemes based on the concept of chirped pulse amplification (CPA)[1] have been devised. Recent advances in the CPA technique resulted in amplification of optical pulses of ≈ 30 fs beyond the TW level at a repetition rate of 10 Hz.[2,3] Amplifier systems with kHz repetition rates produce femtosecond pulses with energies up to 1 mJ.[4,5] Gain narrowing and residual high-order dispersion due to an imperfect matching between the pulse stretcher and compressor have been identified as the major performance limiting effects.

In this paper we report on a kilohertz-repetition-rate multipass Ti:S amplifier seeded by ≈ 12 fs, 100-nm-bandwidth pulses from a mirror dispersion controlled Ti:S oscillator.[6] The material dispersion of the pulse slicing and isolation system leads to a substantial *natural* pulse broadening, and hence to an "automatic" implementation of the CPA concept. This eliminates the need for an additional stretcher. The achieved broadening is sufficient to avoid self-focusing and related degradation of the optical beam in the amplifier.

The amplifier part of the system is shown in Fig. 1. The pulses from the MDC oscillator pass a special broadband Faraday Isolator which prevents the reverse amplified spontaneous emission (ASE) from being reflected back from the output coupler into the amplifier.[7] After two passes of the 80-MHz pulse train through the amplifier single pulses are selected by means of a Pockels cell including polarizers. A further improved ASE suppression is achievable with this setup compared to the one with the pulse selection in front of the amplifier. A 6-mm-long, highly-doped (0.20%) Brewster-cut Ti:sapphire slab is used at the focus of an astigmatically compensated confocal four-pass amplifier formed by two dichroic mirrors (ROC=300 mm, 40 mm diameter) and retroreflectors. A multiple aperture in the beam path prevents the onset of laser action in the high-gain amplifier.

The Ti:sapphire crystal is pumped by a cw-pumped, frequency-doubled, Q-switched Nd:YLF laser, which delivers 300-ns pulses at a repetition rate variable between 1 kHz and 5 kHz with a maximum average power of 12 W. The crystal absorbs 64% of the incident

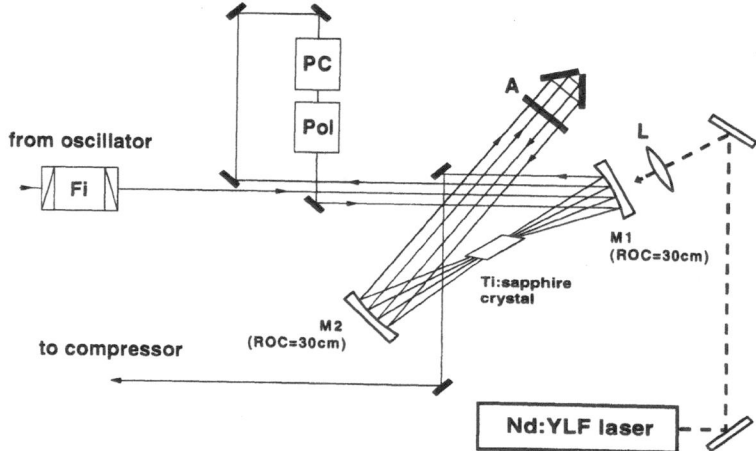

Figure 1. The confocal multi-pass amplifier. M — mirrors, PC — Pockels cell, Pol — polarization compensator, Fi — Faraday Isolator, A — multiple aperture, L — lens ($f = 150$ mm)

pump energy in the first pass. With the transmitted portion backreflected into the crystal an overall absorption efficiency of $> 80\%$ is achieved. An effective single-pass gain of ≈ 18 is attained, yielding a total gain of $\approx 10^5$, by pumping the amplifier with an energy of 6 mJ at a 2 kHz repetition rate.

After amplification the chirped pulses are recompressed in a setup, shown in Fig. 2, comprising of dispersive mirrors and a low-dispersion FK5 (Schott) prism pair. Specially designed dispersive mirrors are employed to compensate for the remaining third order dispersion.[8] The remaining fourth order dispersion was calculated to be -2×10^4 fs.[4]

After amplification and recompression the system yielded pulses with a duration of

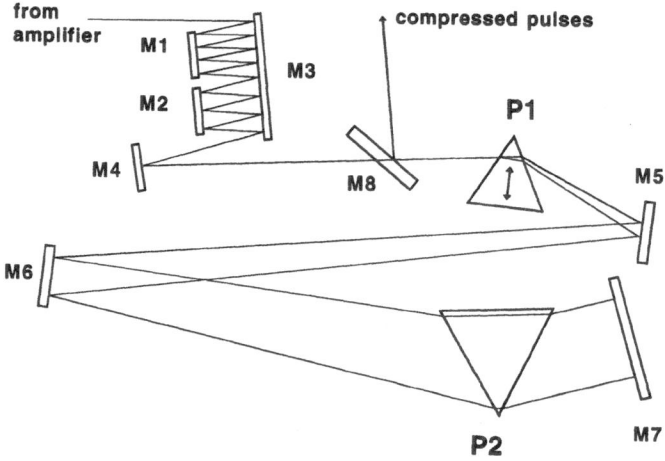

Figure 2. The mirror-prism-compressor. M1...M3 — chirped mirrors, M4..M8 — standard mirrors, P — prisms

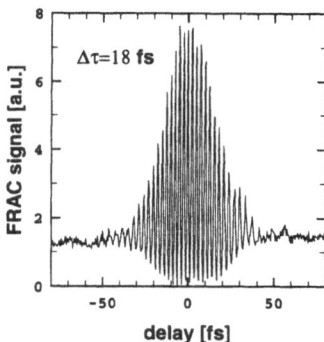

Figure 3. Fringe resolved autocorrelation trace of the amplified pulses.

17–18 fs as evaluated from the fringe-resolved autocorrelation trace shown in Fig. 3. This is a consequence of gain narrowing in the amplifier which reduces the spectrum to ≈ 60 nm (Fig. 4). These measurements have been carried out at a repetition rate of 2 kHz and with a pump energy of ≈ 6 mJ, leading to compressed pulses with an energy of ≈ 50 μJ. This implies a compressor throughput of > 80%. At a repetition rate of 1 kHz the pulse energy rose to approximately 100 μJ (at a pump energy of ≈ 7 mJ, a further increase of the pump pulse energy only lead to an increase of the ASE energy content). The pulse-to-pulse energy fluctuation is < 10%, which can mainly be attributed to the 1-2% energy fluctuation of the pump source No evidence of nonlinear effects was observed which is in agreement with the estimated value of the B-integral of 0.2.

In conclusion, we have demonstrated 18-fs optical pulses amplified to the 100-μJ energy level, yielding a peak power in excess of 5 GW. Due to its stable performance,

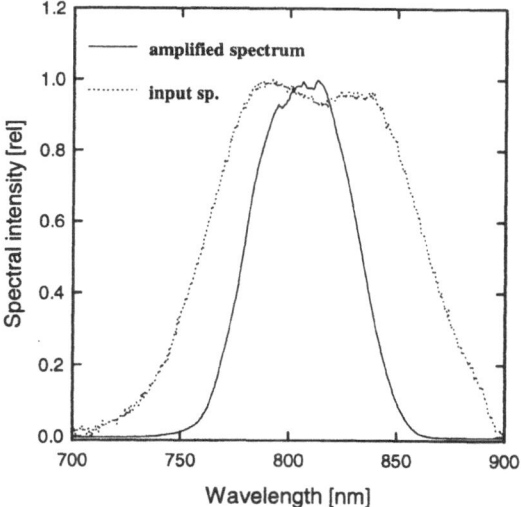

Figure 4. Spectrum of the amplified pulses (solid line) compared to the spectrum at the input of the amplifier (dotted line).

compactness and reliability the presented kHz amplifier promises to become an important tool for high-resolution ultrafast spectroscopy.

This work was supported by the Fonds zur Förderung der Wissenschaftlichen Forschung in Österreich, grants P9710 and P10409 and the Österreichische Nationalbank, grants 5335 and 4915. M. L. acknowledges a grant by the Deutsche Forschungsgemeinschaft.

REFERENCES

1. D. Strickland and G. Mourou, Opt. Comm **56**, 219 (1985).
2. C. P. J. Barty, C. L. Gordon III, and B. E. Lemoff, Opt. Lett. **19**, 1442 (1994).
3. J. Zhou, C. P. Huang, M. M. Murnane, and H. C. Kapteyn, Opt. Lett. **20**, 64 (1995).
4. K. Wynne, G. D. Reid, and R. M. Hochstrasser, Opt. Lett. **20**, 895 (1994).
5. S. Backus, J. Peatross, C. P. Huang, M. M. Murnane, and H. C. Kapteyn, Opt. Lett. **20**, 2000 (1995).
6. A. Stingl, M. Lenzner, Ch. Spielmann, F. Krausz, R. Szipöcs, Opt. Lett. **20**, 602 (1995).
7. M. Lenzner, Ch. Spielmann, E. Wintner, F. Krausz, and A. J. Schmidt, Opt. Lett. **20**, 1397 (1995).
8. Ch. Spielmann, M. Lenzner, R. Szipöcs, and F. Krausz, to be published in Opt. Comm.

3

THEORY OF SOLITON-MODELOCKED FEMTOSECOND SOLID-STATE LASERS

F. X. Kärtner

Institute of Quantum Electronics
Swiss Federal Institute of Technology,
ETH Hönggerberg - HPT, CH - 8093 Zürich, Switzerland

Since the early work of Martinez et al.,[1] it is well known that modelocked lasers will yield pulses that are both shorter and more stable if solitonlike pulse shaping is used, i.e. self-phase modulation (SPM) and negative group velocity dispersion (GVD) is properly adjusted. This work extends these results. Based on soliton perturbation theory we derive stability ranges for actively and passively modelocked femtosecond solid-state lasers. The theory shows that the response time of the saturable absorber in a passively modelocked laser that operates in the soliton regime can be much longer than the pulse width of the generated solitonlike pulses. This is in contrast to the traditional concepts of ultrashort pulse generation that rely either on an artificial fast saturable absorber as is the case for APM or KLM systems or on the interplay between a slow saturable absorber and gain saturation as is the case with dye lasers. From this finding we can conclude that semiconductor absorbers with typical absorption recovery times of 100 fs due to intraband thermalization processes can be used directly to generate 10 fs pulses. The theoretical results are compared with experiments. Thus far we achieved pulses as short as 16 fs from a Ti:sapphire laser modelocked only with a semiconductor absorber[2].

In the existing fast saturable absorber modelocking techniques a short net gain window in time is created that shapes and supports only the pulse while suppressing the background radiation. In contrast, we consider the case were the pulse shaping is solely done by negative GVD and SPM, i.e. soliton formation, when the modelocked laser reaches steady state. Therefore, we call it soliton modelocking. The absorber does not shape the pulse substantially when strongly saturated and only has to stabilize the solitonlike pulse against the background radiation that is called the continuum in soliton perturbation theory. This is possible because for the solitonlike pulse the non-linear effects due to SPM and the linear effects due to GVD are in balance. In contrast, the noise or instabilities which tend to grow, i.e. the continuum, are not intense enough to experience the non-linearity and are therefore spread in time by GVD. When they are spread in time they experience the higher absorption due to the slowly recovering absorber after passage of the soliton and the strong non saturated absorption in front of the soliton. We describe this

Ultrafast Processes in Spectroscopy, edited by Svelto et al.
Plenum Press, New York, 1996

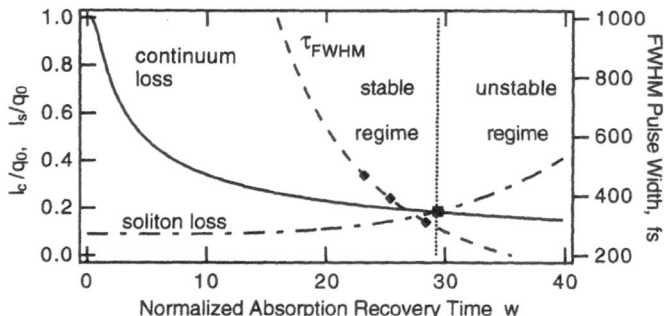

Figure 1. Stability diagram for a soliton-modelocked laser with a slow saturable absorber.

clean-up process of the soliton by deriving an evolution equation for the continuum accompanying the steady state soliton[3,4]. This equation allows to study the stability of the soliton. The soliton is stable if the continuum is below threshold. This is the case when the continuum experiences more loss per roundtrip due to the finite gain bandwidth, intracavity filter bandwidth and the saturable absorber than the soliton. Fig. 1 shows the losses of continuum and soliton normalized to q_0, which is the absorption depth of the absorber bleached by the soliton, as a function of the normalized recovery time of the absorber $w = T_A \sqrt{q_0/\Phi_0}/\tau$, where T_A is the response time of a broadband model two-level absorber and τ is the FWHM pulsewidth of the soliton divided by 1.76. Φ_0 is the phase shift of the soliton per roundtrip in the laser cavity which depends on the amount of negative GVD, $|D| = \Phi_0 \cdot \tau^2$.

In the region where the soliton loss is smaller than the continuum loss the soliton will saturate the gain to the soliton loss so that the continuum does experience a net loss per roundtrip and the pulse formation is stable. We verified this theoretical prediction using a standard Ti:sapphire laser with an additional intracavity etalon that controls the laser bandwidth[5]. As a slow saturable absorber we used a low temperature grown GaAs bulk layer that is excited high in the conduction band at around 800 nm. Since the intraband thermalization and cooling processes at that wavelength are shorter than 100 fs the absorber shows only a response of $T_A = 10\,ps$ due to carrier recombination for pulses longer than 100 fs. We also verified that we operate the cavity in a regime where Kerr lensing is absent. Fig. 2 shows the measured autocorrelation traces and spectra. We extracted the laser parameters and intracavity GVD and marked the operating conditions in the stability diagram in Fig.1 for $|D| = 800, 1000, 1200\,fs^2$. The experiment confirms the theoretical predictions (see Fig.2). For a negative intracavity GVD larger than $|D| = 750\,fs^2$ the laser generates stable perfectly transform limited sech-shaped pulses which preserve their shape over six orders of magnitude in a high dynamic range autocorrelation measurement[6]. Due to the slow saturable absorber the solitonlike pulse undergoes an efficient pulse cleaning mechanism. In each roundtrip the front part of the soliton is absorbed that retards the soliton with respect to the continuum. Therefore after many roundtrips the continuum is separated from the soliton and gets absorbed in the nonsaturated absorber. Reduction of the negative GVD below the critical value results in soliton plus continuum generation which is recognized by a jittering autocorrelation trace due to the relatively long continuum pulses and the appearance of additional spectral components. Further reduction of the negative GVD below $|D| = 300\,fs^2$ leads to a new regime where higher order solitons and continuum are generated. The measured details in the laser dynamics can be reproduced

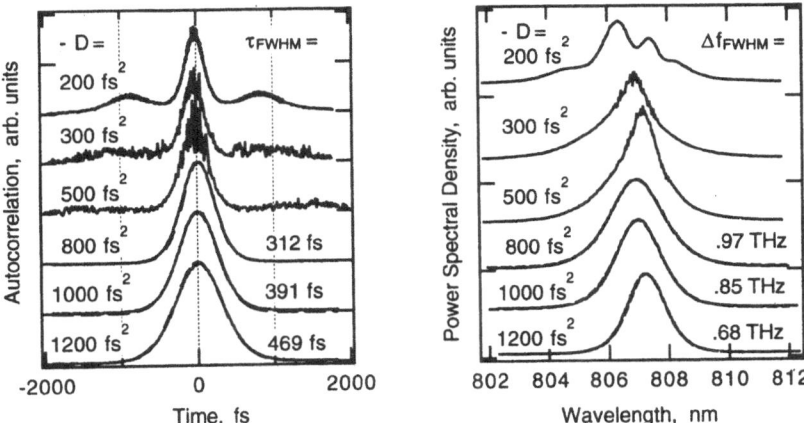

Figure 2. Measured autocorrelation traces and power spectral densities near the critical point. The dashed lines in the lower three traces are the fits to an ideal soliton-like pulse.

by numerical simulation of the full master equation of the modelocked laser including the slow saturable absorber.

The generated solitonlike pulses are up to 35 times shorter than the recovery time of the absorber. With a stronger absorption and by using the fast response time of 200 fs due to intraband carrier thermalization in a GaAs - Quantum Well we achieved pulses as short as 35 fs solely by soliton modelocking[7]. The pulse width was only limited by the Bragg mirror supporting the absorber and not by the modelocking mechanism. Replacement of the Bragg mirror by a silver mirror with a similar absorber on top resulted in pulses as short as 16 fs by pure soliton modelocking[2], see Fig. 3, and 10 fs self starting pulses when Kerr Lens modelocked is additionally employed[8].

The experimental verification of the theoretical predictions of soliton-modelocking with a slow saturable absorber allow the following conclusions. Semiconductor absorbers with a fast recovery time due to intraband relaxation processes of 100 fs or even below are fast enough to allow for sub-10 fs pulse generation without the need of Kerr-Lens-Mode-

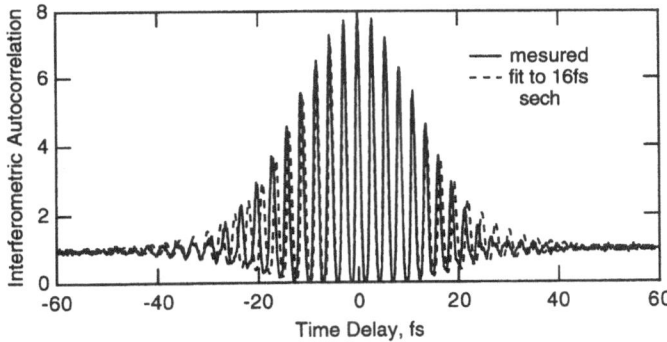

Figure 3. Measured autocorrelation trace for the 16 fs soliton modelocked pulse. The dashed line is the autocorrelation trace computed with a 16-fs sech-shaped pulse.

locking where strong self-focusing of the laser beam is used to produce an equivalent fast saturable absorber action. The strong self-focusing imposes limitations on the cavity design and leads to strong space-time coupling of the pulses in the laser crystal[9] that results in a complex laser dynamics. The soliton modelocking demonstrated here allows for a decoupling of the nonlinearity responsible for self-phase modulation and therefore soliton formation from the saturable absorber nonlinearity which is necessary for stable pulsed operation and self-starting. This approach allows for a design of the laser so that it reliably self-starts without overdriving the nonlinearities when the laser reaches the 10 fs regime. This opens new possibilities for an improved cavity design to obtain even shorter pulses and to create a compact and reliable 10 fs laser technology.

ACKNOWLEDGMENT

The author appreciates very much the collaboration with I.D. Jung, D. Kopf, R. Fluck, R. L. Brovelli, M. Kamp, G. Zhang and U. Keller which made the experimental verification of the theoretical results possible. Thanks go also to M. Moser from the Paul Scherrer Institute Zürich for supplying the AlAs / AlGaAs Bragg mirrors for the saturable absorbers. This work is supported by the Swiss National Fund under contract number 21–43474.95.

REFERENCES

1. O. E. Martinez, R. L. Fork and J. P. Gordon, Opt. Lett. **9**, 156 (1984).
2. R. Fluck, I. D. Jung, G. Zhang, F. X. Kärtner and U. Keller, to be published.
3. F. X. Kärtner, D. Kopf, U. Keller, JOSA B **12**, 486 (1995).
4. F. X. Kärtner, U. Keller, Opt. Lett. **20**, 16 (1995).
5. I. D. Jung, F. X. Kärtner, L. R. Brovelli, M. Kamp and U. Keller, Opt. Lett. **20**, 1892 (1995).
6. A. Braun, J. V. Rudd, H. Cheng, G. Mourou, K. Kopf, I. D. Jung, K. J. Weingarten and U. Keller, Opt. Lett. **20**, 1889 (1995).
7. L. R. Brovelli, I. D. Jung, D. Kopf, M. Kamp, F. X. Kärtner and U. Keller, Electr. Lett. **31**, 287 (1995).
8. U. Keller, same proceedings, paper MC3.
9. I. P. Christov, H. C. Kapteyn, M. M. Murnane, C. P. Huang, J. Zhou, Opt. Lett. **20**, 309 (1995).

4

MODE-LOCKING OF SOLID STATE LASERS BY CASCADED SECOND ORDER NONLINEARITIES

V. Magni, G. Cerullo, S. De Silvestri, A. Monguzzi, and M. Zavelani-Rossi

Centro di Elettronica Quantistica e Strumentazione Elettronica del CNR
Dipartimento di Fisica del Politecnico
P.za L. da Vinci 32, 20133 Milano, Italy

In the last years considerable efforts have been devoted to the development of new passive mode-locking (ML) techniques for cw picosecond solid-state lasers exploiting fast nonlinearities. For Nd:YAG and Nd:YLF the most successful techniques up to now have been Additive Pulse ML[1], Kerr Lens ML[2] and Antiresonant Fabry-Perot Saturable Absorber ML[3] based on absorption saturation in semiconductors. Pulses with duration below 10 ps have been obtained from Nd:YAG and Nd:YLF lasers mode-locked using these techniques. Together with their well known advantages these methods present some drawbacks: additive pulse ML requires a very accurate interferometric control of the cavity length, Kerr-lens ML shows weak tendency to self-starting, semiconductor saturable absorbers are rather difficult to fabricate.

In this article we report on two recently developed passive ML techniques for the generation of picosecond pulses from cw Neodymium lasers based on the real or imaginary part of the equivalent third order nonlinearity originated by second order processes[4–9]. The two ML techniques have been applied to various Neodymium lasers pumped by lamps, Ti:sapphire laser and laser diodes. Intracavity second order nonlinearities are attractive for ML of cw lasers because they provide large nonlinear loss modulations, thus ensuring self-starting operation; furthermore they do not require interferometric cavity length control. These advantages are particularly significant with low peak power lasers, generating ps pulses. The main limitation of these techniques is due to the group velocity mismatch (GVM) between the FW and the SH in the second harmonic crystal[9,10], which tends to broaden and distort the pulses. A compensation technique is applied to overcome this limitation.

The basic scheme used to exploit the $\chi^{(2)}$ cascading effects is shown in Fig. 1. It consists of a nonlinear crystal (NLC) for second harmonic generation, a tilted glass plate, a birefringent crystal for GVM compensation and a mirror that totally reflects the second harmonic (SH) and has a suitable reflectivity, R, for the fundamental wavelength (FW). If the glass plate is suitably tilted to provide a proper phase shift between SH and FW between the two passes in the NLC, the SH power is almost completely converted back into

Ultrafast Processes in Spectroscopy, edited by Svelto et al.
Plenum Press, New York, 1996

15

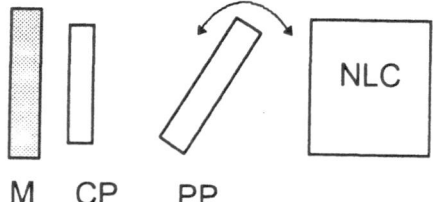

M   CP   PP

Figure 1. Basic scheme used to exploit the $\chi^{(2)}$ cascading for mode-locking. NLC, second harmonic nonlinear crystal; PP glass plate; CP, birefringent crystal for GVM compensation; M, mirror.

the FW during the second pass due to the parametric interaction. In these conditions, if R < 1 the reflectivity for the FW increases with power and the system behaves as a fast saturable absorber with a negative equivalent Im[$\chi^{(3)}$]. This ML technique, known as the non-linear mirror[4], has been demonstrated with pulsed[5] and with cw lasers[6,7,9]. If R = 1 a double pass through the NLC, although not directly changing the cavity losses, results in a nonlinear phase shift, corresponding to an equivalent Re[$\chi^{(3)}$], analogous to that provided by a Kerr effect[11]. The phase shift causes power-dependent mode variations which, in combination with a suitably positioned aperture, provide nonlinear loss modulations similar to those of Kerr lens ML. This technique, which we call CSM (Cascaded Second order nonlinearity Mode-locking), was recently demonstrated with a cw Nd:YAG laser[8,9]. With these ML schemes the difference between the group velocity of FW and SH in the NLC plays a crucial role. The effect of GVM on the pulse duration can be understood as follows: since the group velocity of the SH, $v_{g2\omega}$, is smaller than that of the FW, $v_{g\omega}$, the SH pulse is delayed with respect to the FW one after the first pass through the NLC and it is not completely converted back during the second pass. Furthermore, the trailing edge of the FW pulse, which in the second pass through the NLC better overlaps the SH, is shortened more effectively than the leading one. This effect becomes more pronounced for shorter pulses ultimately limiting the pulse duration.

The GVM compensation technique[12,9] we applied in the experiments, consists in introducing between the NLC and the dichroic mirror a transparent plate that delays the FW pulse with respect to the SH one, like in a material with anomalous dispersion. Since such materials are not available, a suitable equivalent system must be used. If the FW and SH pulses are orthogonally polarized, a plate of a birefringent crystal with the fast axis parallel to the polarization of the SH and the slow axis parallel to that of the FW can delay the FW. Compensation is therefore possible if the SH is generated in a NLC with type I phase matching, since in this case the FW and the SH are orthogonally polarized. The SH pulse after the first pass through a NLC of length L_c is delayed by approximately $\tau_c = \delta_c L_c$, where $\delta_c = 1/v_{g2\omega} - 1/v_{g\omega}$ is the GVM parameter of the crystal. If a compensating birefringent plate of thickness L_p is inserted between the NLC and the dichroic mirror, the SH pulse will experience a further delay $\tau_p = \delta_p L_p$, where δ_p is the GVM parameter of the plate, $\delta_p = 1/v_{g2\omega p} - 1/v_{g\omega p}$ ($v_{g2\omega p}$ being the group velocity of the SH and $v_{g\omega p}$ that of the FW). To obtain efficient GVM compensation, the SH and FW pulses must overlap temporally during the second pass in the NLC, therefore $\tau_p \approx -0.5\,\tau_c$. Our numerical analysis of the steady state ML regime has shown that, for laser parameters relevant to our experimental conditions, $c = -\tau_p/\tau_c$ should be within 0.3 - 0.4 to achieve the shortest self-consistent pulse.

The NLC used in our experiments is a 15 mm long lithium triborate crystal (LBO) cut for type I temperature tuned noncritical phase matching. The GVM parameter of the crystal can be calculated to be $\delta_c = 55$ fs/mm at a FW of 1050 nm. For a crystal length of

15 mm, this corresponds to a delay τ_c = 825 fs. The compensating plate must provide a negative delay of the order of -400 fs. To this purpose we used plates of β-barium borate (BBO), which at the FW of 1050 nm has δ_p = -280 fs/mm. We used two plates wedged by 0.5°, of thickness 1.5 and 2.5 mm respectively, with the optical axis lying in the plane of the plate. The plates were inserted in the cavity with the ordinary axis parallel to the polarization of the FW. To change the compensation parameter, we also inserted, between the NLC and the dichroic mirror, plates of BK7 (which has a GVM parameter δ = 100 fs/mm at 1050 nm) of varying thickness.

A first set of experiments was performed with the lamp-pumped Nd:YLF laser[6,9] shown in Fig. 2(a). The laser crystal has a diameter of 6.35 mm and a length of 104 mm. The TEM$_{00}$ mode, selected with a diaphragm, has a spot size in the rod of 0.7 mm and of 35 μm inside the LBO. In this condition the output power is 2 W at 5.6 kW pump power. The reflectivity of M_3 is R = 0.78 at the FW and 1 at the SH wavelength. We inserted between mirrors M_2 and M_3 compensating plates and BK7 plates, so as to achieve different values of c, both positive and negative. The ML regime was self-starting. With positive compensation the pulses shorten significantly, reaching the minimum value for c ≈ 0.6; the pulse duration is not very sensitive to the exact value of c in a range 0.3–0.8 but increases for c > 1. Note that these values of c correspond to the delays introduced by the compensating and glass plates, while the delays due to the air path and to the dichroic mirrors (estimated to be ≈ 0.2 ps), are not taken into account. The background-free autocorrelations measured for c = 0.6 and corresponding to the minimum pulse duration, is shown in Fig. 3(a). The autocorrelation is fitted almost perfectly by a sech^2 pulse shape, giving a FWHM pulse duration of 5.1 ps.

The laser configuration used for the experiments[7,9] with the Nd:YAG is shown in Fig. 2(b). The 10 mm long Nd:YAG rod is longitudinally pumped, by a Ti:sapphire laser. The mode spot size is 150 μm in the active material and 30 μm in the LBO. The laser has an output power of 0.8 W for a pump power of 2.5 W. The laser is mode-locked using the nonlinear mirror scheme, in which mirror M_4 reflects totally the SH and only partially (R = 0.78) the FW. The ML regime is self-starting. With a compensation parameter c = 0.6, we measured a pulse duration of 6.4 ps as shown by the autocorrelation, reported in Fig. 3(b), almost perfectly fitted by a sech^2 pulse shape.

The third set of experiments concerns the CSM of a Nd:YAG laser[8,9]. The resonator, shown in Fig. 1(c), is basically that used in the previous case, but mirror M_4 is a total reflector for both the FW and the SH, and a slit is placed in a suitable intracavity plane. Self-starting ML is obtained when the LBO position around the focus and the slit width are properly adjusted to exploit the nonlinear spot size variations[13]. In this configuration power is extracted from the laser using a beam splitter. With a GVM compensation parameter c ≈ 0.6 we measured a pulse duration of 5.9 ps, as shown in Fig. 3(c), together with its best fit by sech^2 pulse shape. The pulses obtained in this configuration are shorter than those measured with Nd:YAG and the nonlinear mirror technique, because during the first pass through the NLC, SH generation lengthens the FW pulse, so that the pulse transmitted from the dichroic mirror is longer than that extracted after the second pass through the NLC.

In conclusion, we have demonstrated ML of a cw laser using two different techniques based on the effects resulting from cascaded second order nonlinear processes. Both techniques combine the advantages typical of passive ML with the very large nonlinear loss modulations obtainable from second order nonlinearity. With respect to the nonlinear mirror technique, the CSM has the advantage that the nonlinear loss modulation does not depend on the reflectivity of the output coupler, which can thus be separately op-

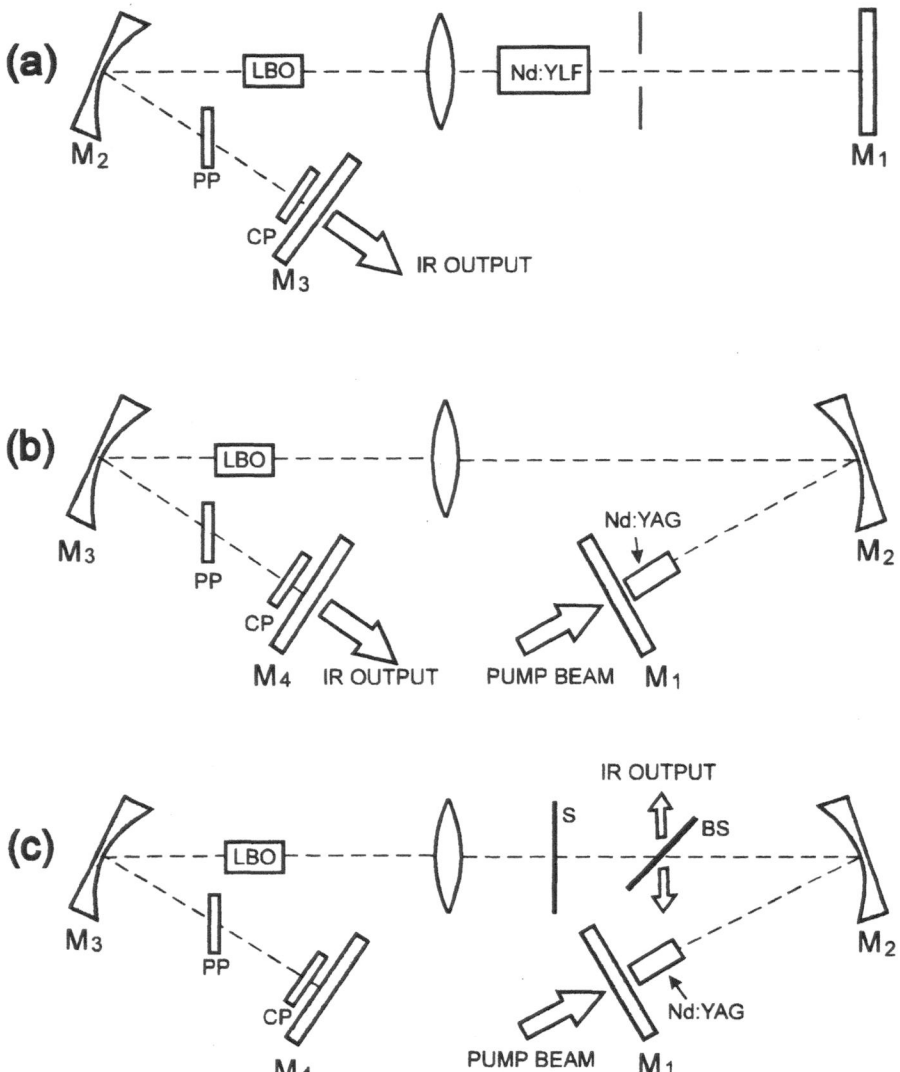

Figure 2. (a) Lamp-pumped Nd:YLF laser and (b) longitudinally pumped Nd:YAG laser, mode locked by the nonlinear mirror technique. (c) Longitudinally pumped Nd:YAG laser mode locked by the CSM technique. PP, glass plate; CP, birefringent compensating plate; S, slit; BS, beam splitter.

timized, and that the nonlinear loss modulation is about two times as large resulting in a two times lower ML pump power threshold. Using the GVM compensation pulses with FWHM duration of 5–6 ps have been obtained at a power level of 0.5–1.5 W. We believe that these ML methods could find significant applications in the picosecond domain where they present significant advantages over other passive ML techniques. Moreover it should be also possible to extend these methods to other active materials and, if the group velocity mismatch in the SH crystal is properly compensated, to the femtosecond domain.

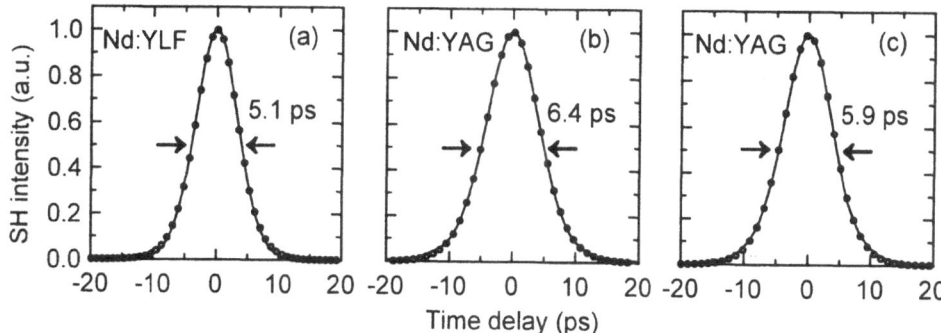

Figure 3. Autocorrelations (solid line) and best fit with sech^2 pulse (circles) of the pulses from: (a) Nd:YLF laser, nonlinear mirror ML; (b) Nd:YAG laser, nonlinear mirror ML; (c) Nd:YAG laser, CSM. The pulse widths in this figure are the autocorrelation FWHM's divided by 1.55.

REFERENCES

1. Goodberlet, J. Jacobson, J.G. Fujimoto, P.A. Schulz, and T.Y. Fan, Opt. Lett. **15**, 504 (1990).
2. G. P. A. Malcom, and A. I. Ferguson, Opt. Lett. **16**, 1967 (1991).
3. U. Keller, Appl. Phys. B **58**, 347 (1994).
4. K. A. Stankov, Appl. Phys. B **45**, 191 (1988).
5. K. A. Stankov, and J. Jethwa, Opt. Commun. **66**, 41 (1988).
6. M. B. Danailov, G. Cerullo, V. Magni, D. Segala, and S. De Silvestri, Opt. Lett. **19**, 792 (1994).
7. G. Cerullo, M. B. Danailov, S. De Silvestri, P. Laporta, V. Magni, D. Segala, and S. Taccheo, Appl. Phys. Lett. **65**, 2392 (1994).
8. G. Cerullo, S. De Silvestri, A. Monguzzi, D. Segala, and V. Magni, Opt. Lett. **20**, 746 (1995).
9. G. Cerullo, V. Magni, and A. Monguzzi, Opt. Lett. **20**, 1785 (1995).
10. K. A. Stankov, V. P. Tzolov, and M. G. Mirkov, Opt. Lett. **16**, 1119 (1991).
11. R. DeSalvo, D. J. Hagan, M. Sheik-Bahae, G. Stegeman, and E. W. Van Stryland, Opt. Lett. **17**, 28 (1992).
12. K. A. Stankov, V. P. Tzolov, and M. G. Mirkov, Appl. Phys. B **54**, 303 (1992).
13. V. Magni, G. Cerullo, and S. De Silvestri, Opt. Commun. **101**, 365 (1993).

5

ADVANCES IN ULTRAFAST ALL-SOLID-STATE LASERS USING SEMICONDUCTOR SATURABLE ABSORBERS

Ursula Keller

Institute of Quantum Electronics
Swiss Federal Institute of Technology (ETH)
ETH Hönggerberg
CH-8093 Zurich, Switzerland

We review recent progress in "real-world" ultrafast lasers which are possible for the first time due to the recent progress in novel passive modelocking techniques and diode-pumped solid-state lasers. The requirements for "real-world" include reliable self-starting of the modelocking process, no critical cavity alignment, and diode-pumping. Such turn-key, hands-off laser system, requiring little or no laser expertise, will be important for system applications in medicine, lidar, and optoelectronic switching, for example.

Semiconductor saturable absorbers have been successfully used to modelock semiconductor diode lasers [1] and color center lasers [2]. However, early examples of passive modelocking with solid-state lasers with long upper state lifetimes (> 100 µs) were always accompanied by self-Q-switching, where the modelocked pulse train was strongly modulated by much longer macro pulses, typically of ≈µs duration and at ≈kHz repetition rate. Therefore, initially there were no suitable intracavity saturable absorbers available, and stable cw modelocking was only achieved with nonlinear coupled cavities such as additive pulse modelocking (APM) [3, 4, 5, 6] and resonant passive modelocking (RPM) [7, 8]. This was the first time that an artificial fast saturable absorber was tailored through the coupling to a nonlinear coupled cavity. A major breakthrough came with the invention of Kerr lens modelocking (KLM) [9, 10, 11, 12], which quickly resulted in sub-10 fs pulse generation [13, 14]. However, KLM has several disadvantages for "real-world" lasers: it relies on critical cavity stability regimes to generate sufficiently large effective saturable absorption through Kerr lensing and to self-start modelocking. We demonstrate solutions to both of these limitations with our semiconductor saturable absorber devices and soliton modelocking [15, 16].

The antiresonant Fabry-Perot saturable absorber (A-FPSA) [17, 18] was the first intracavity saturable absorber that started and sustained stable cw modelocking of Nd:YLF and Nd:YAG lasers. This was mainly because the parameters such as saturation intensity, losses, and saturation fluence could be adjusted to prevent self-Q-switching. Further work

Ultrafast Processes in Spectroscopy, edited by Svelto et al.
Plenum Press, New York, 1996

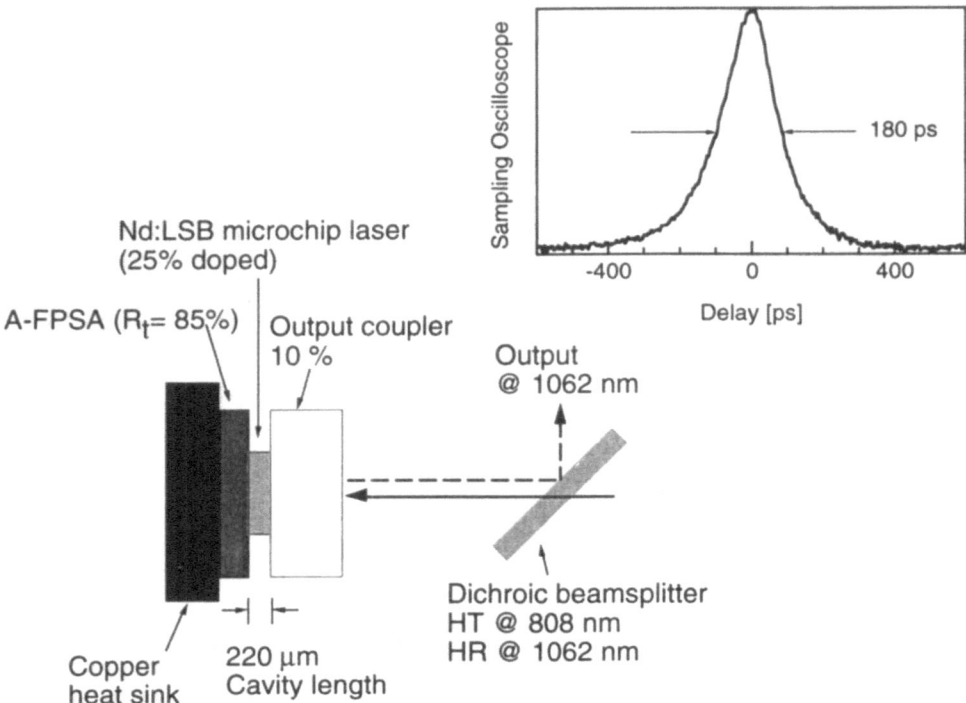

Figure 1. Passively Q-switched Nd:LSB microchip laser producing pulses as short as 180 ps, at a pulse repetition rate of ≈ 100 kHz and pulse energy of 0.1 μJ. This are the shortest passively Q-switched pulses generated with a solid-state laser.

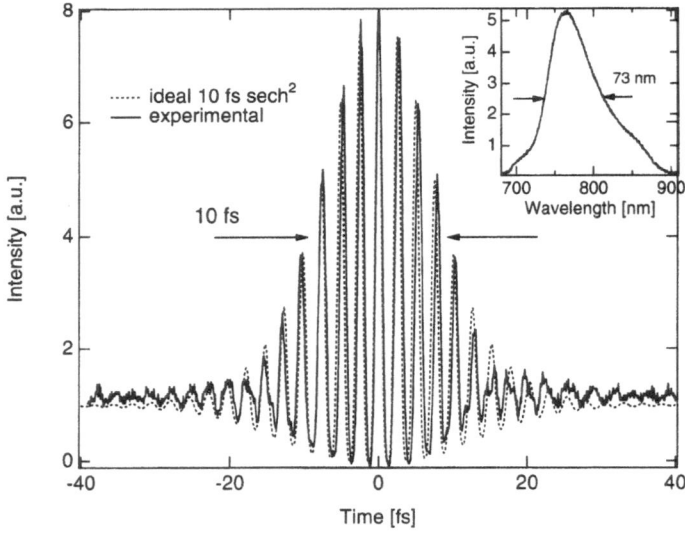

Figure 2. Self-starting 10 fs pulses from a KLM Ti:sapphire laser using a broadband low-finesse A-FPSA.

has allowed us to explore the stability regime for stable modelocking by varying those absorber parameters. The main limitation for stable modelocking is due to modelocked Q-switching which occurs earlier than pure Q-switching [19]. Therefore, the capability to custom-design the saturable absorber parameters has turned out to be essential. Today we can adapt the A-FPSA design for both passive Q-switching, generating pulses as short as 180 ps with a Nd:LSB microchip laser (Fig. 1) [20], and for passive cw modelocking, supporting pulses as short as 10 fs with a Ti:sapphire laser (Fig. 2) [21].

The A-FPSA offers more than enough adjustable parameters such as absorber thickness, top reflector, and impulse response to fulfill the stability requirements of solid-state

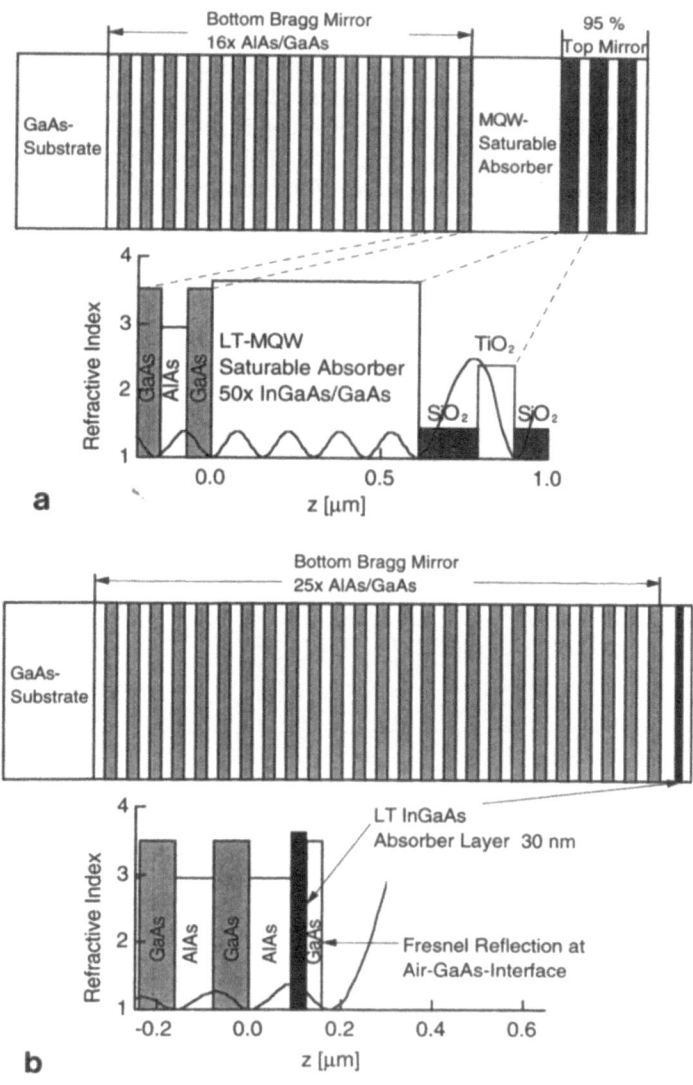

Figure 3. Typical A-FPSA design for an operation wavelength of 1 μm. (a) High-finesse A-FPSA, (b) Low-finesse A-FPSA.

lasers [22]. We distinguish between different design regimes of the A-FPSA such as the high-finesse and low-finesse A-FPSA [23, 24] which can be scaled by adjusting the incident laser mode area and absorber thickness (Fig. 3). The ultimate scaling limit of the high-finesse to the low-finesse A-FPSA is an AR-coated thin saturable absorber which has been used to passively modelock a Ti:sapphire laser [25]. The low-finesse A-FPSA design puts more stringent requirements on the quality of the bottom mirror such as bandwidth, reflectivity, and group velocity dispersion. Meanwhile, other solid-state lasers such as Yb:YAG [26], Nd:LSB [27], Nd:YLF at 1.3 μm, and Nd:YVO$_4$ at 1.3 μm [28] have been passively modelocked in the picosecond regime. Additionally, a p-i-n modulator integrated within an A-FPSA allows for possible synchronization of the pulse arrival time to an external reference [29].

In the femtosecond regime, negative group velocity dispersion (GVD) and self-phase modulation (SPM) allows for soliton modelocking. In this case, soliton formation can be adjusted to be the dominant pulse forming mechanism. The saturable absorber is only required to stabilize the soliton, e.g. to introduce enough loss so that the continuum, which is the energy that the soliton looses due to gain dispersion and losses in the cavity, does not reach threshold. This lost energy is initially contained in a low intensity background pulse which spreads in time due to GVD, but negligible SPM, and experiences increased losses due to the saturable absorber window. Therefore, a resonant saturable absorber can potentially stabilize an ultrashort pulse with a pulse spectrum beyond the bandwidth of the absorption edge. In contrast, if the semiconductor saturable absorber is used as a fast saturable absorber alone, the maximum supported bandwidth would be limited by the bandgap. Femtosecond pulses have been generated with Yb:YAG (Å 500 fs pulses) [26], diode-pumped Nd:glass (Å 100 fs pulses) [30] and Cr:LiSAF (45 fs pulses) [31, 32] lasers. A compact Cr:LiSAF laser has been demonstrated with a device that combines both saturable absorption and negative GVD [33]. For scaling towards higher output powers of "real-world" ultrafast lasers, we have demonstrated 400 mW average output

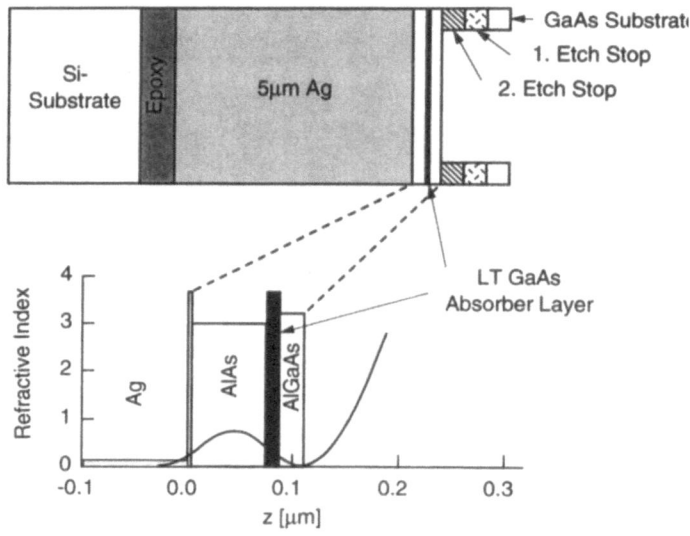

Figure 4. Broadband low-finesse A-FPSA which self-started and stabilized 10 fs pulses from a Ti:sapphire laser (Fig. 2).

power from a diode-pumped Cr:LiSAF laser [34]. We improved this result just recently to 1 W average output power. Using a Ti:sapphire laser, pulses as short as 34 fs [25], and more recently 10 fs [21] have been generated. To obtain a larger bandwidth of the A-FPSA, we replaced the bandwidth limiting lower AlGaAs/AlAs Bragg mirror with a broadband silver mirror (Fig. 4).

ACKNOWLEDGMENT

The author gratefully acknowledges contributions from D. Kopf, I. Jung, R. Fluck, B. Braun, C. Hönninger, G. Zhang, and F. X. Kärtner at ETH, K. J. Weingarten at Time Bandwidth Products AG, M. Moser at the Paul Scherrer Institute in Zürich, G. Huber at University of Hamburg, A. Giesen at University of Stuttgart, G. L. Bona and P. Roentgen at IBM Rüschlikon and R. J. Beach, M. A. Emanuel, and J. A. Skidmore at Lawrence Livermore National Laboratory.

REFERENCES

1. Y. Silberberg, P. W. Smith, D. J. Eilenberger, D. A. B. Miller, A. C. Gossard, W. Wiegmann, *Optics Letters* **9**, 507 (1984)
2. M. N. Islam, E. R. Sunderman, C. E. Soccolich, I. Bar-Joseph, N. Sauer, T. Y. Chang, B. I. Miller, *IEEE J. Quantum Electronics* **25**, 2454 (1989)
3. K. J. Blow, D. Wood, *J. Opt. Soc. Am B* **5**, 629 (1988)
4. K. J. Blow, B. P. Nelson, *Optics Letters* **13**, 1026 (1988)
5. P. N. Kean, X. Zhu, D. W. Crust, R. S. Grant, N. Landford, W. Sibbett, *Optics Letters* **14**, 39 (1989)
6. E. P. Ippen, H. A. Haus, L. Y. Liu, *J. Opt. Soc. Am. B* **6**, 1736 (1989)
7. U. Keller, W. H. Knox, H. Roskos, *Optics Letters* **15**, 1377 (1990)
8. H. A. Haus, U. Keller, W. H. Knox, *J. Opt. Soc. Am. B* **8**, 1252 (1991)
9. D. E. Spence, P. N. Kean, W. Sibbett, *Optics Letters* **16**, 42 (1991)
10. U. Keller, G. W. 'tHooft, W. H. Knox, J. E. Cunningham, *Optics Letters* **16**, 1022 (1991)
11. D. K. Negus, L. Spinelli, N. Goldblatt, G. Feugnet, in *Advanced Solid-State Lasers* G. Dubé, L. Chase, Eds. (Optical Society of America, Washington, D.C., 1991), **10**, 120
12. F. Salin, J. Squier, M. Piché, *Optics Letters* **16**, 1674 (1991)
13. J. Zhou, G. Taft, C.-P. Huang, M. M. Murnane, H. C. Kapteyn, I. P. Christov, *Optics Letters* **19**, 1149 (1994)
14. A. Stingl, M. Lenzner, Ch. Spielmann, F. Krausz, *Optics Letters* **20**, 602 (1995)
15. F. X. Kärtner, U. Keller, *Optics Letters* **20**, 16 (1995)
16. I. D. Jung, F. X. Kärtner, L. R. Brovelli, M. Kamp, U. Keller, *Optics Lett.* , to be published Sept. 15 (1995)
17. U. Keller, D. A. B. Miller, G. D. Boyd, T. H. Chiu, J. F. Ferguson, M. T. Asom, *Optics Letters* **17**, 505 (1992)
18. U. Keller, *Applied Phys. B* **58**, 347 (1994)
19. F. X. Kärtner, L. R. Brovelli, D. Kopf, M. Kamp, I. Calasso, U. Keller, *Optical Engineering* **34**, 2024 (1995)
20. B. Braun, F. X. Kärtner, U. Keller, J.-P. Meyn, G. Huber, *Optics Letters* **21**, submitted Oct. 1995
21. R. Fluck, I. D. Jung, G. Zhang, F. X. Kärtner, U. Keller, *Optics Lett.* **21**, submitted Nov. 1995
22. L. R. Brovelli, U. Keller, T. H. Chiu, *Journal of the Optical Society of America B* **12**, 311 (1995)
23. I. D. Jung, L. R. Brovelli, M. Kamp, U. Keller, M. Moser, *Optics Letters* **20**, 1559 (1995)
24. S. Tsuda, W. H. Knox, E. A. d. Souza, W. Y. Jan, J. E. Cunningham, *Optics Letters* **20**, 1406 (1995)
25. L. R. Brovelli, I. D. Jung, D. Kopf, M. Kamp, M. Moser, F. X. Kärtner, U. Keller, *Electronics Letters* **31**, 287 (1995)
26. C. Hönninger, G. Zhang, U. Keller, A. Giesen, *Optics Lett.* **20**, Dec. 1 (1995)
27. B. Braun, F. X. Kärtner, U. Keller, J.-P. Meyn, G. Huber, T. H. Chiu, *Advanced Solid-State Lasers '95*, paper ThB1

28. R. Fluck, K. J. Weingarten, G. Zhang, U. Keller, *Advanced Solid-State Lasers '96*, paper submitted
29. L. R. Brovelli, M. Lanker, U. Keller, K. W. Goossen, J. A. Walker, J. E. Cunningham, *Electronics Lett.* **31**, 381 (1995)
30. D. Kopf, F. X. Kärtner, K. J. Weingarten, U. Keller, *Optics Lett.* **20**, 1169 (1995)
31. D. Kopf, K. J. Weingarten, L. Brovelli, M. Kamp, U. Keller, *Optics Lett.* **19**, 2143 (1994)
32. D. Kopf, K. J. Weingarten, L. R. Brovelli, M. Kamp, U. Keller, *Conference on Lasers and Electro-Optics, CLEO 1995*, paper CWM2
33. D. Kopf, G. Zhang, M. Moser, U. Keller, *Conference on Lasers and Electro-Optics, CLEO 1995*, paper postdeadline CPD41
34. D. Kopf, J. Aus der Au, U. Keller, G. L. Bona, P. Roentgen, *Optics Letters* **20**, 1782 (1995)

6

RECENT PROGRESS IN FEMTOSECOND FIBER LASERS

Eiji Yoshida and Masataka Nakazawa

NTT Access Network Systems Laboratories
Ibaraki, Japan

The generation of femtosecond pulses in the GHz region is important in terms of realizing Tera bit/s communication. A method for compressing the output pulse from a stable mode-locked fiber laser is very attractive for the generation of femtosecond pulse trains in the GHz region.

In this paper, we report the successful operation of a harmonically and regeneratively mode-locked erbium-doped fiber laser at 10 and 20 GHz, from which we obtained stable pulse trains of 2.7 ps and 1.9 ps, respectively[1,2]. These pulse trains at 10 GHz and 20 GHz were compressed to as short as 170 fs by using a dispersion-decreasing erbium-doped fiber amplifier (DD-EDFA)[2,3].

The experimental setup is shown in Fig.1. The laser had a polarization-maintaining erbium-doped fiber (PM-EDF), a WDM coupler, a polarization-maintaining dispersion-shifted fiber (PM-DSF), a 15 % output coupler, a polarization dependent isolator, a $LiNbO_3$ intensity modulator and an optical bandpass filter with a bandwidth of 2.5 nm. The pumping source was 1.48 µm laser diodes. The dispersion-shifted fiber was 190 m long and the group velocity dispersion (GVD) was -3.3 ps/km/nm. All fibers in the cavity were polarization-maintaining fibers to prevent polarization fluctuation. The clock extraction circuit was composed of a high speed photodetector, a high Q filter (Q~1000) and a high gain amplifier. The center frequency of the electric filter was set at 10 or 20 GHz. A part of laser output beam was coupled into the clock extraction circuit. A longitudinal self-beat signal near 10 or 20 GHz, which was one of the harmonic longitudinal modes of the cavity, was detected with the clock extraction circuit. This harmonic beat signal was amplified through an electrical amplifier and fed back to the $LiNbO_3$ intensity modulator. This technique enabled us to achieve regenerative mode-locking. The phase between the pulse and the modulation signal was adjusted by a phase controller. The output pulse was amplified through a high-power erbium-doped fiber amplifier (EDFA) and coupled into the DD-EDFA. The dispersion-decreasing erbium-doped fiber was 1 km long and the GVD was changed from -9.3 ps/km/nm at the input to -0.5 ps/km/nm at the output. In the DD-EDFA two mechanisms of adiabatic soliton compression reduce the pulse width. One is adiabatic soliton compression resulting from decreasing dispersion[4] and the other is adiabatic soliton narrowing via optical amplification.

Ultrafast Processes in Spectroscopy, edited by Svelto et al.
Plenum Press, New York, 1996

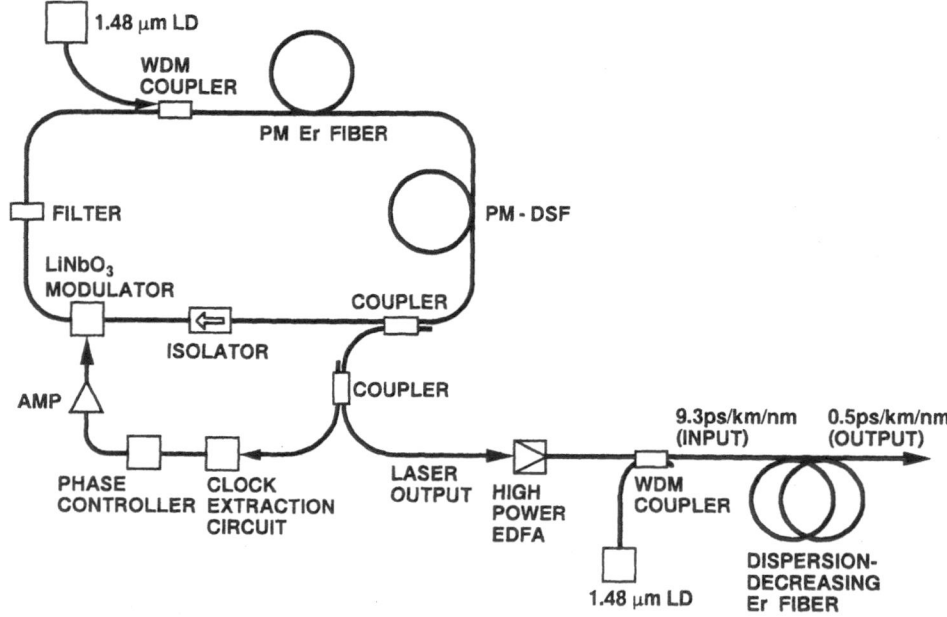

Figure 1. Experimental setup.

Regenerative mode-locking was accomplished automatically by adjusting the phase between the pulse and the clock signal. The laser was very stable because the clock signal always followed the changes in cavity length. When the center frequency of the electric filter was set at 10 GHz, a stable 2.7 ps pulse train was obtained. The central wavelength was 1.55 μm and the spectral width was 1.0 nm. The bandwidth-pulse width product was 0.34, indicating that the output pulse was a nearly transform-limited sech pulse. The repetition rate was increased to as high as 20 GHz. Figure 2 shows the oscillation characteristics at 20 GHz. Figure 2(a) is an autocorrelation waveform. The pulse width was as short as 1.8 ps. This result agrees with active mode-locking theory[5]. Figure 2(b) is an autocorrelation waveform showing a repetitive feature. The interval between the two pulses was 50 ps, which corresponded to a repetition rate of 20 GHz. Figure 2(c) shows the optical spectral profile, where the central wavelength is 1.551 μm and the spectral width is 1.6 nm. The fine structure with a spectral separation of 0.16 nm corresponds to a longitudinal mode separation of 20 GHz. The bandwidth-pulse width product was 0.36, indicating that the output pulse was a nearly transform-limited sech pulse. Figure 2(d) shows the extracted clock spectrum. It is important to note that no supermodes or parasitic noise was present.

The compressed pulse at 20 GHz is shown in Fig. 3. Fig. 3(a) shows an autocorrelation waveform for a DD-EDFA pump power of 90 mW, and the corresponding spectrum is shown in Fig. 3(b). The pulse width was as short as 170 fs and had no pedestal. The central wavelength shifted from 1.551 μm to 1.560 μm due to the soliton self frequency shift. A spectral oscillation also developed from near the original central wavelength of 1.551 μm toward the longer wavelength region. The spectral width was approximately 15 nm and the bandwidth-pulse width product was 0.32, which indicated that the compressed pulse was still a transform-limited sech pulse. The output pulse at 10 GHz was also compressed to 170 fs.

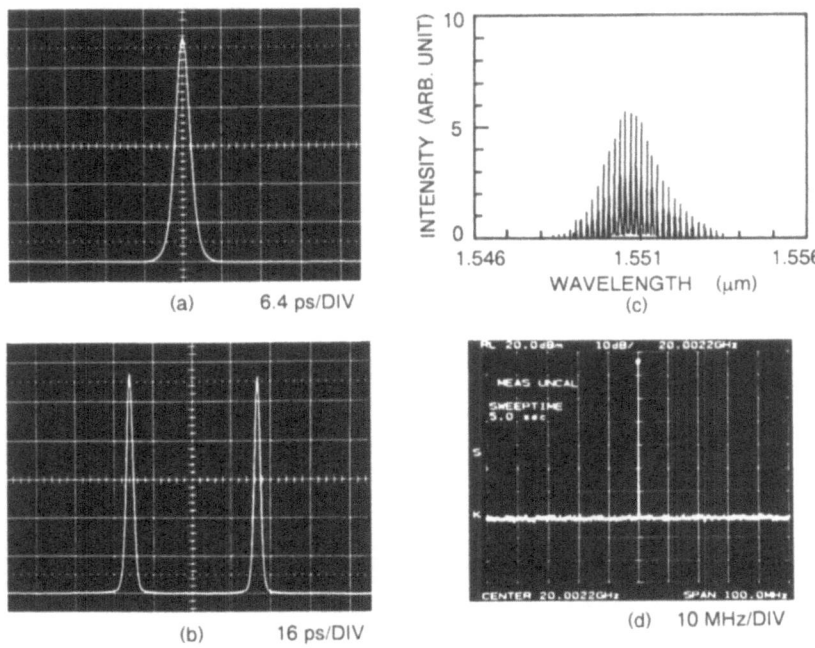

Figure 2. Oscillation characteristics at 20 GHz. (a) Autocorrelation waveform. (b) Autocorrelation waveform showing the repetition rate. (c) Spectral profile. (d) Extracted clock spectrum at 20 GHz.

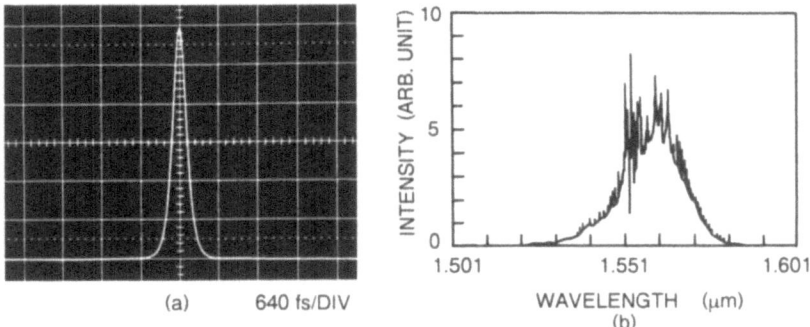

Figure 3. Compressed pulse at 20 GHz. (a) Autocorrelation waveform. (b) Spectral profile.

In conclusion, we obtained a 2.7 ps pulse train at 10 GHz and a 1.8 ps pulse train at 20 GHz. These stable pulses were compressed to 170 fs through the use of a DD-EDFA external compressor.

REFERENCES

1. M. Nakazawa, E. Yoshida and Y. Kimura, Electron. Lett. **30**, 1603 (1994)
2. E. Yoshida, Y. Kimura and M. Nakazawa, Electron. Lett. **31**, 377 (1995)

3. M. Nakazawa, E. Yoshida and Y. Kimura, Electron. Lett. **30**, 2038 (1994)
4. K. Tajima, Opt. Lett. **12,** 54 (1987)
5. H. A. Haus, *Waves and fields in optoelectronics* (Prentice-Hall series in solid state physical electronics, New Jersey, 1984)

FORMATION OF FREQUENCY-SHIFTED QUASI-SOLITONS IN SOLID-STATE LASERS

I. G. Poloyko, V. L. Kalashnikov, V. P. Kalosha, and V. P. Mikhailov

International Laser Center
7 Kurchatov Str., Minsk, 220064, Belarus

Recently, new quasi-soliton mechanism of ultra-short pulse formation in solid-state lasers different from Schrödinger-type has become known [1–6]. Its distinctive feature is the generation field carrier frequency walk-off from the gain band center due movement of cavity mirror, cavity length modulation [2, 3], splitting of gain band [4], or use of the band-limiting filter with unilateral spectral cut-off [5, 6].

Here we perform the analysis of this ultra-short pulse formation mechanism in solid-state lasers within the framework of the self-consistent approach to continuous-wave generation. Basic significance of the saturation of the gain medium within one pulse and influence of intensity and power saturation of the cavity loss are shown. The stability of the obtained quasi-solitons against perturbations of their parameters and against noise perturbations of generation field is investigated.

Based on the self-consistent approach to the analysis of continuous-wave generation of ultra-short pulses by the solid-state lasers with four-level active medium let us consider an analytical representation for such pulse. For this approach a periodic recurrence not only of pulse temporal shape $A(k, \eta)$, but also gain temporal shape $\alpha(k, \eta)$ with each passage k through the cavity is assumed, where η is the local time. The solution for single isolated ultra-short pulse generation with the duration τ_p is:

$$\alpha(k, \eta) = \alpha_0 \, exp\left[-\sigma \int_0^\eta |A(k, \eta')|^2 \, d\eta' \right],$$

(1)

where

$$\alpha_0 = \alpha_m \frac{1 - exp(-U_p)}{1 - exp(-U_p - E)} exp(-E/2)$$

is the gain at pulse peak, α_m is the maximum gain of the active medium, $U_p = \sigma_{abs} T_{cav} I_p$ is the normalized pump energy,

Ultrafast Processes in Spectroscopy, edited by Svelto et al.
Plenum Press, New York, 1996

$$E = \sigma \int_{-\infty}^{\infty} |A|^2 \, d\eta$$

is the normalized total pulse energy, σ_{abs} is the absorption cross-section of the active medium, I_p is the pump intensity.

Generation field evolution as a function of round-trip number k is described by the following dynamic equation

$$\frac{\partial A}{\partial k} = \left(\alpha - \gamma - \frac{\partial}{\partial \eta} + \frac{\partial^2}{\partial \eta^2} - i|A|^2 \right) A, \tag{2}$$

where γ is the linear loss in the cavity. The terms with temporal derivatives in the right-hand side of Eq. 2 describe spectral filtering due to finite width of the gain band, the last term accounts for the self-phase modulation. Here the local time and field intensity are normalized on τ_g and p, respectively.

Expansion of the gain in power series of local energy

$$\varepsilon(\eta) = \int_0^\eta |A(k, \eta')|^2 \, d\eta'$$

where $\tau = \sigma \tau_g / p$, makes an Eq. 2 a generalized Landau-Ginsburg equation. Being limited by the linear term on ε, one can convinced by the direct substitution, that its solution as a quasi-soliton with the temporal shape unchanged from passage to passage is as follows

$$A = A_0 \, sech^{1+i\psi} \left[(\eta - \delta k) / \tau_p \right] exp \left[i\omega(\eta - \delta k) \right], \tag{4}$$

where A_0 is the amplitude, ψ is the chirp, ω is the frequency shift from the gain band center, δ is the time delay at cavity period, and the pulse phase delay is omitted.

It should be emphasized that the solution 4 exists only when the gain saturation is accounted for. It justifies the qualitative interpretation of the mechanism of ultra-short pulse formation mentioned above.

Now we perform the analysis of the quasi-soliton stability against laser continuum. Let us assume that perturbed solution of the master dynamic equation 2 is $A(k, \eta) = A^0(k, \eta) + a(k, \eta)$, where $A^0(k, \eta)$ is the unperturbed solution 4, and perturbation is $a(k, \eta) = \varsigma(\eta) exp(\lambda \eta)$ where $\varsigma(\eta)$ is the perturbation temporal envelope, λ is the perturbation increment which in the general case is complex. Then in the linear approximation the Eq. 2 is reduced to a stationary Schrödinger equation for the perturbation

$$\left[-\frac{d^2}{d\eta^2} + q(\eta) \right] \varsigma = \mu \varsigma, \tag{3}$$

where the complex potential is

$$q(\eta) = \alpha_0 A_0^2 \tau_p th(\eta / \tau_p) \left[1 - \frac{1}{2} \tau_p A_0^2 th(\eta / \tau_p) \right] + i A_0^2 \, sech^2(\eta / \tau_p),$$

$\mu = \alpha_0 - \gamma - \lambda$ is eigenvalue.

Quasi-soliton 4 is stable against perturbations, when for all solutions of Eq. 3 real parts of eigenvalues are positive. As the real part of $q(\eta)$ is a barrier with a negative left-hand wing, Eq. 3 has a continuous spectrum of eigenfunctions with $Re\lambda > 0$. Therefore quasi-soliton 4 is unstable when Eq. (2) includes only the term taking into account the gain saturation.

For quasi-soliton stabilization it is possible to use an instantaneous saturable absorber, or additional cavity with period matched to main cavity period, or to introduce a spectral cut-off of pulse high-frequency component by optical "knife" between a prism pair. The latter causes loss reduction with intensity growth because of anti-Stokes shift of high-frequency components as a result of interaction between self-phase modulation and gain saturation.

In the first case a term $-\beta_0\sigma_b A_0^2 sech^2(\eta/\tau_p)$ must be added to the potential in Eq. 3 and linear loss must be replaced with total initial loss. To obtain eigenvalue spectrum of Eq. 3 it is convenient to do the variable substitution $y = 2\sqrt{2}\left[1+exp(2\eta/\tau_p)\right]$ which reduces Eq. (13) to Whittaker equation with the following solution:

$$\varsigma = y^{m+1/2} exp(-y/2)\,_1F_1(-n,2m+1,y),$$

where

$$m = \sqrt{16\tau_p^2 A_0^2(\beta_0\sigma_b - i)+\frac{1}{4}}, \quad n = l-m-\frac{1}{2}, \quad l = \frac{\alpha_0 E + 4\beta_0\sigma_b A_0^2 - 2iA_0^2}{\sqrt{\alpha_0 E - \mu}},$$

$_1F_1$ is the degenerate hypergeometric function.

The stability diagram of quasi-soliton with a fast saturable absorber in the system is shown in Fig. 1. Inside the range I there is a continuous spectrum of eigenvalues with

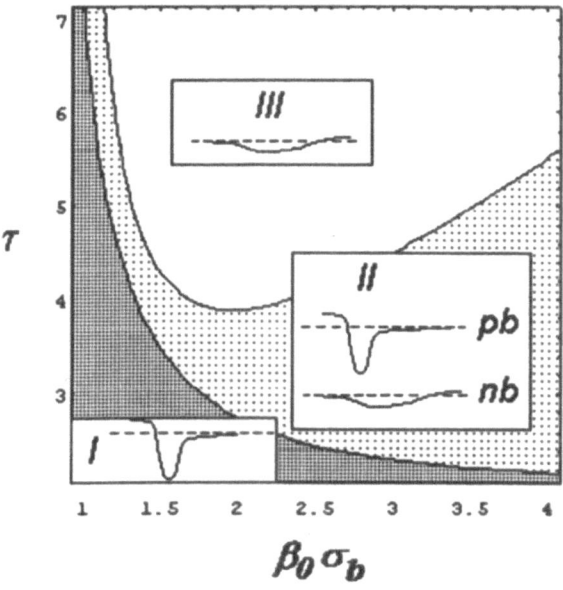

Figure 1.

*Re*λ > *0* for the both negatively and positively frequency-shifted pulses (unstable solutions). There is no unstable continuum in this case, however Eq. 3 may have a discrete spectrum of eigenvalues for integer *n*. In the range *II* the potential well for the negativly frequency-shifted soliton (labeled by "*nb*" at the Fig. 1) is not deep enough to "contain" a descrite set of undecayng perturbations, so the quasi-soliton is stable. For the solution with positive frequency shift the potential well (labeled by "*pb*") is essentially deeper and it is unstable against the perturbations with discrete spectrum of λ.

At last, the both negatively and positively frequency-shifted pulses are stable inside the range *III*. This is due to the reduction of the potential well depth.

Thus, the combined action of self-phase modulation, gain saturation and instantaneous saturable absorption of the intracavity loss results in generation of the stable frequency-shifted quasi-soliton (4).

ACKNOWLEDGMENTS

This work was partially supported by the Belorussian Soros Foundation.

REFERENCES

1. C. Cutler, "Why does linear phase shift cause mode locking," *IEEE J. Quantum Electron.*, vol. 28, pp. 282–288, 1992.
2. M.J. Kelly and J.R. Taylor, "Mode locking of a continuous-wave titanium-doped sapphire laser using a linear external cavity," *Opt. Lett.,* vol. 15, pp. 378–380, 1990.
3. Wu, J.Y. Zhou, X.G. Huang, Z.X. Li and Q.X. Li, "Mode locking with linear and nonlinear phase shifts," *J. Opt. Soc. Am. B*, vol. 10, 2080–2084, 1993.
4. Bai, S. Chen, Zh. Wang and G. Zhang, "Novel self-mode locking mechanism in narrow-band lasers," *Appl. Phys. Lett.*, vol. 63, pp. 2597–2599, 1993.
5. Kopf, F.X. Kärtner, K.J. Weingarten and U. Keller, "Pulse shortening in a Nd: glass laser by gain reshaping," in *Conf. Lasers Electro-Opt.* Washington, DC: Opt. Soc. Amer., 1994, paper CWA7, pp. 183–184.
6. Kopf, F.X. Kärtner, K.J. Weingarten, and U. Keller, "Pulse shortening in a Nd: glass laser by gain reshaping and soliton formation," *Opt. Lett.*, vol. 19, pp. 2146–2148, 1994.

8

THEORY OF APM LASERS USING SECOND-ORDER NONLINEARITY IN AN EXTERNAL CAVITY

I. V. Mel'nikov[1*] and A. V. Shipulin[2]

[1] Optoelectronics Research Centre
University of Southampton
Hampshire SO17 1BJ, United Kingdom
[2] Institut für Angewandte Physik
F. Schiller - Universität Jena
D-07745 Jena, Germany

The steady-state passive additive-pulse modelocking of a laser incorporating a quasiphase-matched frequency converter in an external cavity is analysed. This model includes a main cavity with a slowly-bleaching lasing transition, and a fibre with $\chi^{(2)}$ - grating, bent as a Fabry-Perot interferometer. Based on the phase-plane analysis, the dependence of the parameters of the spectral-limited pulse on the laser control parameters is investigated.

INTRODUCTION

One of the main constraints on the design of all-solid state lasers with additive-pulse modelocking (APM) is a relatively weak value of the intensity-dependent refractive index of materials used in an external cavity. In these lasers, to reach the threshold of self-starting the nonlinear phase of approximately $\pi/10$, acquired in the external cavity, is needed.[1] Since the nonlinearity is weak, this demands an initiating kick and leads to a phase distortion of the output. Alternative way of approaching desirable values of the phase shift may be based on a cascaded second-order nonlinearity. This causes a pulse of fundamental frequency to experience a nonlinear phase shift while passing through a phasemismatched frequency doubler.[2,3] The value of such an effective refractive index varies from 10^{-10} cm^2/W for polymer waveguides to 10^{-13} cm^2/W for poled silica fibres,[3,4] the later value is three orders of magnitude larger than n_2 produced by the third-order nonlinearity. This im-

---

* Permanent address: General Physics Institute RAS, Moscow 117942, Russian Federation.

Ultrafast Processes in Spectroscopy, edited by Svelto et al.
Plenum Press, New York, 1996

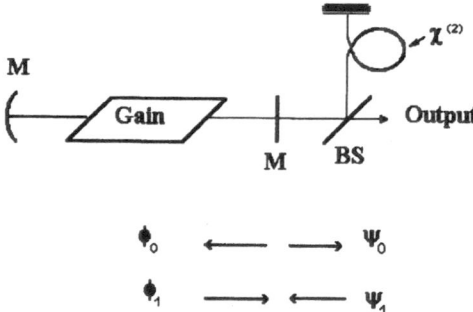

Figure 1. Cascading additive-pulse modelocked laser containing $\chi^{(2)}$-structure in external cavity.

plies that such third-order phenomenon as an APM laser can be implemented with mWatt power level in a fibre geometry on the basis of cascading. In this Report we applied the APM treatment developed earlier by Ippen et al.[5] for a soliton-like output and small phase shifts in the external cavity.

MODEL

We consider a Nd^{3+}:YAG laser with an external cavity configured as a Fabry-Perot interferometer and containing a length of a zero-dispersion fibre in which a $\chi^{(2)}$-grating is drawn (Fig. 1). We restrict our analysis to a steady-state regime of operation of the novel modelocking technique which we denote cascading additive-pulse modelocking (CAPM).

The pulse ϕ_0 reflected by the coupler M passes through the main cavity where it undergoes an amplification and widening due to a bandwidth-limited amplification. After passing the active medium the pulse ϕ_1 interferes with pulse ψ_1, experienced the cascaded phase shift, on the coupler M. Thus the output pulse ψ_0, having the intensity $T_{BS}|\psi_0|^2$, is produced.

Following the analytical approach of Ref. 5 we set the spectrum of the pulse ϕ_0, circulating inside the main cavity, narrower than the bandwidth Ω_g and assume that the saturated gain g_s does not change in time. In addition, we incorporate into the model the phase modulation of the pulse ψ that is due to the cascaded nonlinearity $\chi^{(2)}{:}\chi^{(2)}$ and the losses ε which are due to the second harmonic generation. Then the amplitude of the output pulse ψ_0 obeys the following equation,

$$\frac{d^2\psi_0}{dt^2} + \left[\frac{(g_s R_c - 1)(g_s - R_c)}{g_s T_c^2} - \frac{(g_s - R_c)^2}{g_s T_c^2}\hat{L}_e(\psi_0)\right]\psi_0 = 0,$$

(1)

where the time normalized to the bandwidth $t = t \, \Omega_g$ is used, R_c and T_c is the reflectivity and transmitivity of the coupling mirror, respectively. The recursion operator L_e, describing the pulse evolution on passing through the external cavity, is defined by

$$\hat{L}_e = (\varepsilon R_{BS})^2 \, exp(i\gamma |\psi_0|^2 + iF_0) \psi_0, \tag{2}$$

with the reflectivity of the beamsplitter R_{BS}, $\gamma = n_2 \, L \, (\varepsilon R_{BS})^2 \, \Gamma$, where n_2 is the cascaded nonlinearity, L is the fibre length, Γ is the mode confinement factor, and F_0 is the linear phase shift effectively describing both the phasemismatch and detuning between the cavities.

Limiting the consideration by the case of spectral-limited pulses, the set of Eqs. (1) and (2) can be replaced by a classical equation of motion of the particle with a unit mass,

$$\tfrac{1}{2} \left(\frac{dE}{dt} \right)^2 + U(E, \beta) = 0, \tag{3a}$$

$$U(E, \beta) = \tfrac{1}{2} \, (\beta E^2 - sin(E^2 + F_0) + sin F_0), \tag{3b}$$

with $\beta = (g_s R_c - 1)(\varepsilon R_{BS})^{-2}(g_s - R_c)^{-1}$. Fig. 2 shows the shape of the potential U as a function of E for different β. The trajectory corresponding to the solitary-pulse generation is a separatrix outcoming from the saddle point (Fig. 2b) what limits the value of β or, physically, the saturated gain, $g_{min} < g_s < g_{max}$.

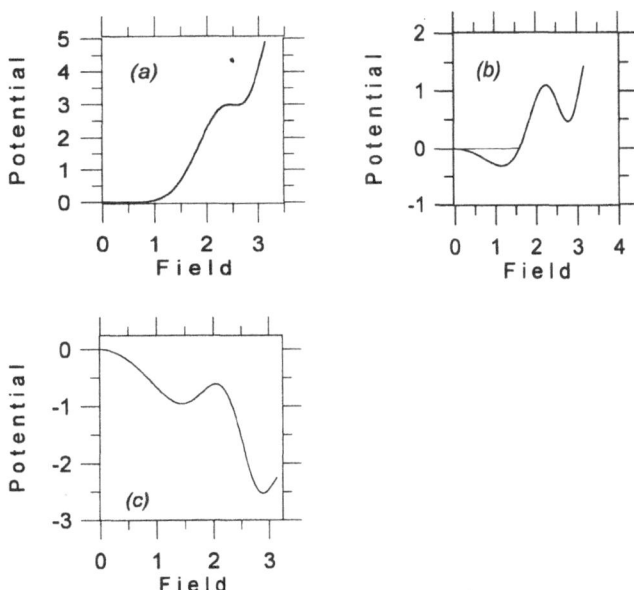

Figure 2. Different regimes of the CAPM laser represented by different forms of the potential $U(E, \beta)$: train of pulses (a); solitary pulse generation (b); spike generation (c).

The origin of the lower boundary for the gain is that at the smaller g_s the cascaded phase modulation is small as well, and the interference on the coupling mirror can not compensate temporal broadening of the pulse in the main cavity. Alternatively, if g_s is too high, it results in the development of wings of the output pulse; in other words, the trajectory does not return back into the saddle point.

The zero point of the potential U gives us the maximum value of the field E by means of solving the equation,

$$\beta E_{max}^2 = sin(E_{max} + F_0) - sin F_0.$$

Then Eq. (3) defining of parameters of the output pulse can be reduced to the Cauchy problem,

$$\frac{dE}{dt} = -\sqrt{sin(E^2 + F_0) - sin F_0 - \beta E^2},$$

$$E(t = 0) = E_{max}. \tag{4}$$

It is of extremely importance to emphasize here that the net result of the phase-plane analysis is the replacement of the complex nonlinear equation (3) by a first-order differential equation. This enables us to take into account a non-small value of the nonlinear phase shift what has been lacking in other models in the literature.

RESULTS AND DISCUSSION

Eq. (3) is now solved numerically to characterize the steady-state regime of the CAPM laser. First, the cascaded phase modulation is plotted against F_0 for a device with R = 0.95 and $k = 0.4$ what, to our knowledge, could be applicable for Nd^{3+} -based CAPM. The generation of the spectral-limited pulse is accessible only within a certain gap of F_0, - $0.25\pi < F_0 < 0.3\pi$, the highest modulation occuring at the lowest F_0 and being equal to 1.5π (Fig. 3a). The process of the interference on the coupler is characterized by the total phase shift (TPS) between the pulses at maximum intensity. This is also explored in Fig. 3a where the TPS being the sum of the cascaded phase shift, cavity detuning and phasemismatch, plus $\pi/2$ added by the second pass through the coupling mirror is plotted against F_0. It may be concluded that as the TPS is approximately $3\pi/2$ the interference has a destructive character throughout all pulse width, except for a small interval around the maximum intensity. In Fig. 3b the full-width at a half-maximum (FWHM) given in the units of the gain bandwidth and the time-bandwidth product (TBP) are plotted to assess the temporal parameters of the pulse. It is seen that for the maximum value of the cascaded phase modulation there is a shortest achievable value of FWHM. An obvious question to ask is whether a fundamental soliton can be generated as the output. This is quite amazing for the dispersionless main cavity and external fibre, but the chirpless soliton with TBP = 0.314 may be generated by the CAPM laser, and what is more the width of such a soliton is close to the shortest one. But it is worth to bear in mind that this belongs to a variety of different solitary pulses producable by the configuration.

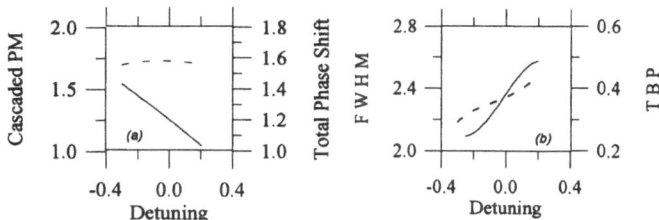

Figure 3. Characterization of the CAPM laser: intrafibre cascaded phase modulation (solid) and total phase shift between colliding pulses (dashed) vs F_0 (a), all in the units of π; FWHM in the units of of the gain bandwidth (solid) and TBP (dashed) vs F_0 given in the units of π (b).

CONCLUSION

To summarize, a simple model for the steady-state regime of a solid-state laser passively modelocked by a cascaded nonlinearity has been presented. The pulse shaping mechanism represents the balance between pulse stretching in the gain medium and compression on the coupling mirror. Assuming the chirpless output, the resulting characteristics can be estimated without expensive computer methods. This method has the benefit that it may be applied to many of the available APM systems with a non-small single-pass nonlinear phase shift, using a third-order nonlinearity.[6]

ACKNOWLEDGMENTS

One of the authors, IVM, acknowledges The Royal Society, London, for the financial support and University of Southampton for hospitality; he would also like to thank P.St.J.Russell for his kind and insistent suggestions to consider this problem. The final stage of this work was supported by Deutsche Forschungsgemeinschaft via Innovationkolleg "Optische Informationstechnik".

REFERENCES

1. D.J.Richardson, A.B.Grudinin, and D.N.Payne, *Electron.Lett.* **28**, 778 (1992).
2. A.Armstrong, N.Bloembergen, J.Ducuing, and P.S.Pershan, *Phys.Rev.* **127**, 1918 (1962).
3. G.I.Stegeman, M.Sheik-Bahae, E.W.Van Stryland, and G.Assanto, *Opt.Lett.* **18**, 13 (1993).
4. P.G.Kazansky, V.Pruneri, and P.St.J.Russell, *Opt.Lett.* **20**, 843 (1995).
5. H.A.Haus, J.G.Fujimoto, and E.P.Ippen, *JOSA* **B8**, 2069 (1991).
6. G.Sucha, S.R.Bolton, S.Weiss, and D.S.Chemla, *Opt.Lett.* **20**, 1794 (1995).

9

HIGH CONTRAST RATIO SUB-PICOSECOND PULSE GENERATION AND ITS CHARACTERIZATION IN Nd:GLASS LASER SYSTEM BASED ON CHIRPED PULSE AMPLIFICATION

A. B. Van'kov, A. A. Kozlov, S. A. Chizhov, V. E. Yashin, V. A. Gorbunov, and N. G. Gogoleva

Institute for Laser Physics
Research Center "S.I.Vavilov SOI"
St. Petersburg, Russia

1. INTRODUCTION

Currently the interest to the study of interaction of ultrashort light pulses with matter is strongly increasing[1]. For these experiments pulses are required with high contrast ratio.

In this paper different factors are discussed for the contrast ratio improvement in CPA laser system.

2. LASER SYSTEM DESCRIPTION

Blocks of CPA laser system are presented in Fig.1.

Picosecond pulse is generated in the master oscillator on the phosphate Nd:glass GLS-22 with passive mode-locking and negative feedback. Implementation of the latter gives the possibility to improve significantly pulse energy and duration stability in comparison with a simple passive mode-locking[2]. Autocorrelation trace gives a pulse duration 1.2 ps.

The short pulse is sent into the stretcher, consisting of the pair of two holographic diffraction gratings with the spatial frequency 1800 lines/mm and the telescope between them. Output pulse duration is 450 ps. Then the pulse was injected into the regenerative amplifier on the Nd:phosphate glass rod with the dimension 5x100 mm. After the sufficient level of the energy (2–5mJ) is achieved, the pulse is extracted from the amplifier by the Pockels cell.

Ultrafast Processes in Spectroscopy, edited by Svelto et al.
Plenum Press, New York, 1996

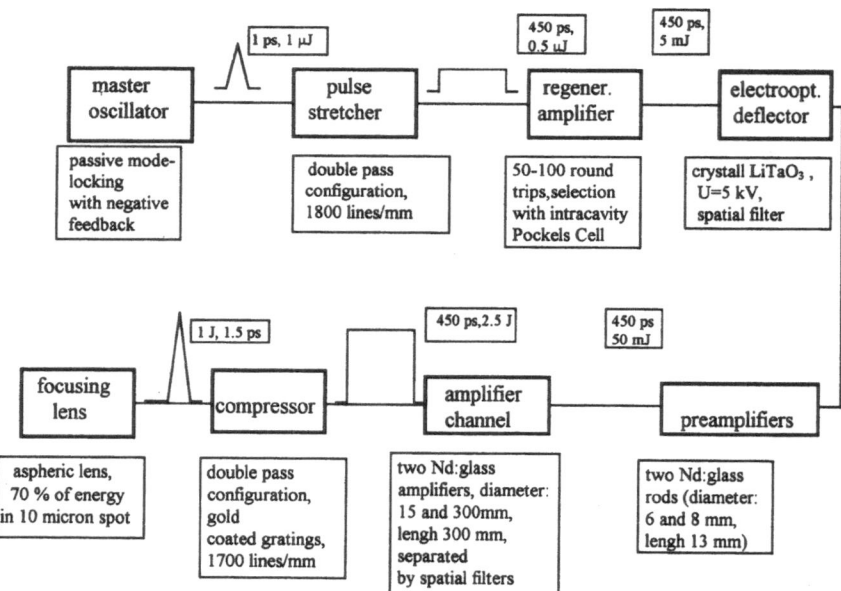

Figure 1. Start laser complex, system for improving pulse contrast ratio, relay-imaging amplifier system and diffraction grating compressor.

Then the stretched pulse passes through the electrooptical deflector system for elimination of the prepulses arising from the regenerative amplifier. This system provides the intensity contrast ratio better than 10^7 To increase the pulse energy we use two preamplifiers with rods 6 and 8 mm in diameter. At the output the beam has gaussian spatial distribution and to provide a flat-topped one we use two hard apertures with relay-imaging telescopes also serving as spatial filters.

Main amplifier channel consist of two GLS-22 Nd:phosphate glass rods 15 mm and 30 mm with length 300 mm separated by spatial filters.

The 1–5 J output pulse is relayed into the compressor system. Compressor consists of two gold coated gratings (1700 lines/mm, 110x110 mm) mounted near Littrow angle with corner cube in double pass. Output energy is limited by the damage threshold of diffraction gratings and for our condition was 0.5–1 J. Gratings (N=1700 lines/mm) were mounted in the standard double pass geometry. It was obtained the compressed picosecond 1 TW laser pulse with contrast ratio 10^8.

3. REASONS FOR PULSE CONTRAST RATIO DEGRADATION IN CPA LASER SYSTEMS

- pulse satellites from MO
- prepulses arising due to RA intracavity Pockels cell imperfection
- parasitic optical connection of MO and RA
- temporal aberration due to misalignment of a stretcher-compressor system
- spectrum-modulation due to optical elements, which can act as Fabry-Perot etalone

- self-phase modulation
- amplified spontaneous emission (ASE)

Autocorrelation trace after MO shows no satellites with contrast ratio more than 10^4.

Initial contrast of main pulse relatively to prepulses from regen is about 10^2 because of finite contrast of output Pockels cell. To improve contrast ratio the electrooptic deflector system[3] has been used. Prepulses were removed by spatial filter placed after the electrooptic deflector. The contrast ratio is provided up to 10^8.

Parasitic optical connection of M.O. and regenerative amplifier R.A. leads to multipass amplification of M.O. prepulses in R.A. and its time interference with main pulse. As a result the strong spectrum modulation and prepulses is arised. To suppress this parasitic connection we have used Pockels cells between M.O. and R.A.

Stretcher adjustment was performed to obtain spectrum homogeneous transverse beam distribution. Under weak stretcher misalignment the spectrum modulation with depth up to 30% was observed. On the basis of measurements and calculations[4] the general requirements for stretcher alignment are stated.

Intracavity optical elements can introduce selective spectrum losses as a Fabry-Perot etalone[5]. All possible reflections should be eliminated. We use new immersion for optical contact between crystal surface of Pockels cells and protective glass.

4. NUMERICAL MODELLING

To predict a laser system characteristics we have developed a numerical model which accounts for the effects of gain narrowing, gain saturation and self-phase modulation on final pulse recompensation.

Further we have investigated the possibilities to increase the output power by shortening of the starting pulse down to 100 fs and by increasing of the gratings in size. In this case the all above-listed distorting factors have an influence on the output pulse shape and background pedestal enhancement.

5. SUMMARY

Nd:glass laser system with energy up to 1 J and pulse length 1 ps have been developed and constructed. The contrast ratio up to 10^8 have been obtained. The amplitude and phase distortions of the chirped pulse in the Nd:phosphate glass power amplifier were investigated and numerical model may be used for optimization of the laser scheme. The most severe limitations of output pulse power and contrast are caused by the gain narrowing and self-phase- modulation in the amplifiers.

REFERENCES

1. M.D.Perry, G.Mourou, Science, **264**, 917 (1994).
2. S.Burdulis et al., Experimentelle Technik der Physik, **39,** 342 (1991)
3. V.I.Krizhanovskyi et. al. Kvantovaya elektronika, **12,** 372 (1985)
4. C.Fiorini et al., Journal of Quantum Electronics, **30,** 1062 (1994).
5. M.D.Perry, F.G.Patterson, J.Weston, Opt.Lett., **15,** No.7 (1990)

10

SOLID-STATE LASERS MODE-LOCKED BY LINEAR AND NONLINEAR FREQUENCY SHIFT

V. L. Kalashnikov, V. P. Kalosha, V. P. Mikhailov, and I. G. Poloyko

International Laser Center
7 Kurchatov St., Minsk, 220064, Belarus

An effective pulse formation in solid-state lasers is available through the wide variety of resonant and non-resonant optical nonlinearities, e.g. self-phase modulation,[1] nonlinear optical Kerr effect[2] and self-focusing.[3] An extremely short pulses have been achieved due to the femtosecond response of these third-order nonlinearities. It was shown that the start of ultra-short pulse generation may be provided by external disturbances of a laser system, for example, by modulation of cavity period,[4] also without the traditional amplitude modulation.[5] It seems that mode locking can be obtained "easier" than one used to think before.

Here we have shown that the laser system, with only self-phase modulation and energy saturation of gain in active medium with finite gain band width, potentially possesses a nonlinear mode, which can be excited under certain conditions, that is displayed in generation of ultra-short pulse train.

It is well-known, that under the assumption of small field change for a pass through the cavity, the generation evolution in a laser system can be described by a generalized Landau-Ginsburg equation:

$$\frac{\partial}{\partial k}a(k,t) = Q\langle t|a\rangle + q\langle t|a\rangle, \quad Q\langle t|a\rangle = \left[\alpha(t) - \gamma - \frac{\partial}{\partial t} + \frac{\partial^2}{\partial t^2} - i|a(k,t)|^2\right]a(k,t), \tag{1}$$

where $a(k,t)$ is the generation field temporal shape, k is the cavity round-trip number, t is the local time, $a(t)$ is the active medium gain, γ is the linear loss, $q\langle t|a\rangle$ is the nonlinear operator accounting for active or passive modulation. The last term in "laser" operator Q is connected with the self-phase modulation, where the field intensity is normalized on the self-phase modulation factor β. The other terms describe the dissipative effects by gain, linear loss and Lorenzian spectral selective element with a half-width $\Delta\omega = 2\pi/t_f$, where t_f defines the group-velocity delay. In Eq. (1) all times are normalized on t_f.

Taking into account the energy saturation of the active medium the gain can be written as

Ultrafast Processes in Spectroscopy, edited by Svelto et al.
Plenum Press, New York, 1996

$$\alpha(t) = \alpha_0 (1 - \varepsilon + \varepsilon^2 / 2 - \ldots), \tag{2}$$

where

$$\varepsilon = \tau \int_0^t |a(k, t')|^2 \, dt'$$

is proportional to pulse local energy, $\tau = \sigma t_f / \beta$, σ is the radiation cross-section of the active medium, α_0 is the gain at the pulse peak determining by the energy of continuous pump.

Direct substitution convinces that for first two or three terms in Eq. (2) the unperturbed Eq. (1) (without modulation operator $q\langle t|a\rangle$) has the quasi-soliton solution

$$a(k, t) = a_0 \, sec \, h^{1+i\psi}\left[(t - \delta k)/\tau_p\right] exp\left[i\omega(t - \delta k)\right], \tag{3}$$

where a_0 is the amplitude, ψ is the chirp, τ_p is the pulse duration, δ is the cavity round-trip group delay and ω can be interpreted as a carrier frequency shift of the pulse from the gain band center.

The physical treatment of the nonlinear mode (3) is the following. A pulse in nonlinear medium receives a positive chirp due to the self-phase modulation, so its instant frequency grows with time. At the same time, the gain is removed along a pulse temporal profile. Thus the low-frequency components located on the pulse front amplify in a greater degree, than the high-frequency ones located on pulse tail. It causes displacement of the pulse spectrum to the longer-wavelength side. On the other hand, the high-frequency spectral components are closer to center of the gain band and, hence, they amplify predominatingly. The final total gain has a local maximum corresponding to the quasi-soliton (3).

Thus, the balance of the factors, inherent to a solid-state laser system, such as self-phase modulation, energy saturation of the active medium gain and gain band spectral profile, results in formation of ultra-short pulse with Stokes frequency shift.

While limiting by the first two terms in expansion of local gain in Eq. (2), the parameters of the pulse (3) can be found out as follows:

$$a_0^2 = 3\omega^2 / \sqrt{2}, \quad \psi = -\sqrt{2}, \quad \tau_p = -\sqrt{2} / \omega,$$
$$\delta = 1 + \omega\sqrt{2}, \quad \alpha_0 = \gamma + 3\omega^2 / 2 = -\omega / \tau, \tag{4}$$

where

$$\omega = \left(-1 \pm \sqrt{1 - 6\gamma\tau^2}\right) / 3\tau \, .$$

As far as the realization of this generation regime concerned, it is interesting to consider the stability of the quasi-soliton (3) against laser noise continuum. To do this, the contribution of the operator $q\langle t|a\rangle$ in Eq. (1) was assumed to be small in comparison with the contributions of the gain and self-phase modulation, but decisive factor for laser continuum. The same assumption was used in analysis of the stability of the Shrödinger laser

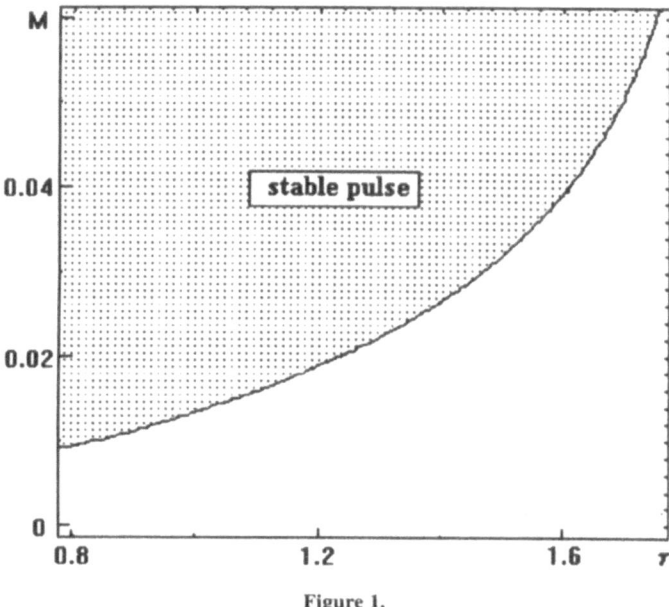

Figure 1.

soliton in Ref. [6]. Let the solution of complete Eq. (1) be $A(k,t) = a(k,t) + \zeta$, where $a(k,t)$ is the solution (3), $\zeta = \zeta_0 \, exp(\lambda k)$ is the perturbation constant in local time, λ is the increment of the perturbation.

It is obvious that the dynamical behavior of the perturbation before the pulse front, where the gain is maximum, is decisive for its stability. Therefore it should be investigated the influence of $q\langle -\infty | a \rangle$. For definiteness sake we have considered an active mode locking with modulation frequency equal to cavity repetition rate and modulation depth M. The condition $\lambda = \alpha(-\infty) - \gamma - M < 0$ means that the net gain at pulse front is negative and the quasi-soliton (3) is stable against noise continuum. The stability diagram obtained from the condition $\alpha(-\infty) = \gamma + M$ is presented in Fig. The laser parameter range where quasi-soliton (3) is stable (unstable) against laser noise is located above (below) curve.

As is seen, the quasi-soliton (3) is unstable in the absence of the active modulation ($M = 0$) similarly to the Shrödinger soliton in the laser system. Additional modulation permits to stabilize the quasi-soliton. As follows from the analysis of the quasi-soliton parameters, stability of the pulse with smaller duration and greater energy can be achieved by the active modulation with greater depth. For $\tau \leq 1$ typical modulation depths are sufficient for stabilization of the quasi-soliton (3).

In practice, the excitation of quasi-soliton can be carried out not only by introduction of ordinary active or passive modulation (including those based on the resonant nonlinearity), but also by use of Stokes shift due to gain saturation and self-phase modulation responsible for formation of the quasi-soliton (3). To do this, in Ref. [6] cut-off of shorter-wavelength side of the pulse spectrum by optical knife between two intracavity prisms is used and this results in loss reduction for pulses of greater intensity which have larger Stokes shift.[7] Thus, optical knife between prism pair together with the effect of Stokes shift due to the gain saturation and the self-phase modulation acts as an instant saturable absorber. Similarly, in Ref. [5] a moving mirror plays a role of phase modulator, it produces anti-Stokes shift for generation field because of the Doppler effect. Though the shift per single passage through the cavity is small, its accumulation results in reasonably large "running-off" of a field carrier frequency from the gain line maximum. Nonlinear Stokes shift for more intensive component towards the gain band center provides their prevailing amplification.

In conclusion, the generation quasi-soliton (3) is provided only by laser ever-presented factors. This makes this mechanism universal in ultra-short pulse solid-state lasers in contrast to Shrödinger type soliton, considered in Refs. [6].

REFERENCES

1. L.F.Mollenauer, R.H.Stolen, Opt. Lett. **9**, 13 (1984).
2. V.L.Kalashnikov, V.P.Kalosha, and V.P.Mikhailov, Quantum Electron. **21**, 101 (1994).
3. D.E.Spence, P.N.Kean, and W.Sibbett, Opt. Lett. **16**, 42 (1991).
4. P.M.W.French, S.M.J.Kelly, and J.R.Taylor, Opt. Lett. **15**, 378 (1990).
5. Q.Wu, J.Y.Zhou, X.G.Huang, Z.X.Li, and Q.X.Li, J. Opt. Soc. Am. B **10**, 2080 (1993).
6. D.Kopf. F.X.Kartner, K.J.Weingarten, and U.Keller, Opt. Lett. **19**, 2146 (1994).
7. J.Herrmann and M.Muller, Opt. Lett. **20**, 22 (1995).

11

OPTICAL SOLITON PROPAGATION IN ERBIUM DOPED NONLINEAR FIBER MEDIA

K. Nakkeeran and K. Porsezian

Department of Physics
Anna University
Madras, 600 025, India

After the invention of laser, the practical applications of nonlinear optical effects are utilized in many fields. Out of those, one in optical fiber is the most important one. That is the optical solitons, which is useful in optical communication and pulse compression.

An interesting phenomenon occurs when one examines the Kerr effect in the negative dispersion region of a fiber. For a high intensity pulse the Kerr nonlinearity causes the leading edge of the pulse to develop a red shift and the trailing edge to develop a blue shift. This is called the Self Phase Modulation (SPM). In addition, because of the negative group velocity dispersion, the red-shifted light will travel slower than the unshifted center of the pulse and the blue-shifted light will travel faster. By carefully choosing a proper combination of pulse shape, pulse intensity, and pulse width, the effects of the Kerr nonlinearity and the negative dispersion will cancel exactly. Such a pulse, which will travel without any changes, is referred to as a soliton[1].

The governing Nonlinear Partial Differential Equation (NPDE) of the fiber system with the effects of Group Velocity Dispersion (GVD) and Kerr nonlinearity is called the Nonlinear Schrödinger (NLS) equation. NLS equation takes the form[1]

$$E_z = i\left(\frac{1}{2}E_{tt} + |E|^2 E\right),$$ (1)

where E is the slowly varying envelope of electric field, the subscripts z and t denote the spatial and temporal partial derivatives. The solution of the NPDE will determine the conditions, pulse shape, intensity and width. The one soliton solution of Eq.1. is[1]

$$E(z,t) = \operatorname{sech}(t)\exp(iz/2),$$ (2)

The NLS equation reveals that the optical soliton is an secant shape. So if one transmits a pulse of hyperbolic secant shape and width (to have required GVD), in the negative dispersion regime, it will travel as a soliton.

Ultrafast Processes in Spectroscopy, edited by Svelto et al.
Plenum Press, New York, 1996

All the above principle of propagation of soliton is possible only when the fiber has zero optical losses. Towards this aim, erbium atoms are doped in the fibers, as the energy difference between two levels of the erbium atoms are doped in the fibers, as the energy difference between two levels of the erbium atoms are nearly equal to the wavelength at which the soliton pulses are propagated. The presence of erbium atoms cause, the additional nonlinear effect, self induced transparency (SIT). The Maxwell–Bloch (MB) equations govern the SIT system. So, the erbium doped fiber system is governed by the coupled system of the NLS–MB equations. The set NLS–MB equations are,

$$E_z = i\left[\frac{1}{2}E_{tt} + |E|^2 E\right] + \langle p\rangle,$$

$$p_t = 2i\omega p + E\eta,$$

$$\eta_t = -(Ep^* + Ep^*), \tag{3}$$

where p and η are given by $v_1 v_2^*$ and $|v_1|^2 - |v_2|^2$ respectively. Here v_1 and v_2 are the wave functions of the two energy levels of the resonant atoms. The bracketed term $<\cdots>$ means, the averaging over the entire frequency range,

$$\langle p(z,t;\omega)\rangle = \int\limits_{-\infty}^{\infty} p(z,t)h(\omega)d\omega,$$

$$\int\limits_{-\infty}^{\infty} h(\omega)d\omega = 1, \tag{4}$$

where $h(\omega)$ is the uncertainty in the energy level. The NLS–MB system also allows soliton type lossless propagation only for the certain choices of parameters involved in the system like, the GVD parameter, Kerr coefficient and the parameter concerned with the interaction of the field and the erbium atoms[2].

When we propagate an intense pulse, in addition to Kerr nonlinearity, the pulse will suffer from additional effects like self steepening and stimulated inelastic scattering. As the pulse width is small, the higher order dispersion will also comes into picture. Only with the effects of higher order dispersion and self steepening, the wave propagation in the erbium doped nonlinear fiber guide is governed by the coupled system of the Hirota equation and the Maxwell–Bloch (H-MB) equations. We have proved that the H-MB equations posses soliton solutions[3].

With all the additional effects, the governing equation for the erbium doped fiber system is the coupled system of the Higher order NLS (HNLS)–MB equations. The HNLS–MB system of equations are,

$$E_z = i\left[\frac{1}{2}E_{tt} + |E|^2 E\right] - \varepsilon\left[\alpha_1 E_{ttt} + \alpha_2 |E|^2 E_t + \alpha_3 E(|E|^2)_t\right] + \langle p\rangle,$$

$$p_t = 2i\omega p + E\eta,$$

$$\eta_t = -(Ep^* + Ep^*), \tag{5}$$

Here α_1, α_2 and α_3 are parameters related to higher order dispersion, self steepening and stimulated inelastic scattering respectively. When we apply Painlevé singularity structure analysis to Eq.5. we find that the HNLS–MB system allows soliton type pulse propagation only when the relation $3\alpha_1 = \alpha_2 = 2\alpha_3$ is satisfied[4]. The linear eigen value problem (Lax pair) of Eq.5 is,

$$\begin{aligned} \Psi_t &= U_2\Psi, \\ \Psi_z &= V_2\Psi, \end{aligned} \qquad \Psi = (\Psi_1 \ \Psi_2 \ \Psi_3)^T \tag{6}$$

where

$$U = \begin{pmatrix} -i\lambda & E & E^* \\ -E^* & i\lambda & 0 \\ -E & 0 & i\lambda \end{pmatrix},$$

$$V = (-8i\lambda^3 + 2i\lambda^2)\begin{pmatrix} 0 & 0 & 0 \\ 0 & 1 & 0 \\ 0 & 0 & 1 \end{pmatrix} + (-4\varepsilon\lambda^2 + \lambda)\begin{pmatrix} 0 & E & E^* \\ -E^* & 0 & 0 \\ -E & 0 & 0 \end{pmatrix}$$

$$+ \left(-2i\varepsilon\lambda + \frac{i}{2}\right)\begin{pmatrix} 2|E|^2 & E_t & E_t^* \\ E_t^* & -|E|^2 & -(E^*)^2 \\ E_t & -E^2 & -|E|^2 \end{pmatrix} + \varepsilon\begin{pmatrix} 0 & E_{tt} + 4|E|^2E & E_{tt}^* + 4|E|^2E^* \\ -E_{tt}^* - 4|E|^2E^* & -(E^*E_t - EE_t^*) & 0 \\ -E_{tt} - 4|E|^2E & 0 & -(EE_t^* - E^*E_t) \end{pmatrix}$$

where λ is the iso-spectral parameter. The HNLS-MB equations are obtained from the consistency condition $U_z - V_t + [U,V] = 0$. As Eq.5. has linear eigen value problem then it posses soliton solutions. But the construction of soliton solution for Eq.5. is still an open problem.

To conclude, the erbium doped fiber amplifiers with all the higher order effects also allow soliton type pulse propagation only for a particular relation between the physical parameters. We consider our results will be very useful for practical realization of the optical soliton.

ACKNOWLEDGMENTS

Author K.N. wishes to thank Prof. S. De Silvestri and Prof. V. Masilamani for their immense help. K.N. also thank European Community for the financial support to attend UPS'95. He also thank CSIR, Govt. of India for the partial travel grant to attend UPS'95 and for awarding him the Junior Research Fellowship. Author K.P. expresses his thanks to DAAD for offering the German Fellowship and to DST and CSIR, Govt. of India for their financial support through projects.

REFERENCES

1. G.P. Agrawal, *Nonlinear Fiber Optics* (Academic, New York, 1989).
2. K. Porsezian and K. Nakkeeran, J. Mod. Opt. (In Press).
3. K. Porsezian and K. Nakkeeran, Phys. Rev. Lett. **74**, 2941 (1995).
4. K. Nakkeeran and K. Porsezian, J. Phys. **A 28**, 3817 (1995).

12

PARAMETRIC NONLINEAR MIRROR IN ACTIVELY MODE-LOCKED Nd:YAG LASER WITH NEGATIVE FEEDBACK CONTROL

O. V. Chekhlov and V. A. Zaporozhchenko

Institute of Physics Academy of Sciences of Belarus
Minsk

Recent progress in the development of fast-response nonlinear modulating devices is closely associated with their application for mode-locking of solid-state lasers. The intracavity second harmonic generation (SHG) nonlinear mirror (NLM)[1] is one of such devices which has been intensively investigated as a mode- locker both for pulsed[2] and cw lasers[3]. A stable NLM operation providing the reliable formation or shortening of light pulses takes place at the sufficiently high intensity of radiation in laser cavity which can be attained due to the use of additional active mode-locking.

We present the results of experimental investigation of a hybrid mode-locking in Nd:YAG laser with acoustooptic modulator and NLM. The flashlamp pumped Nd:YAG laser was operating in quasi-cw regime provided by the negative feedback loop. The laser cavity was formed by a concave 200cm totally reflecting mirror and flat output coupler with 50% and 100% reflectivities at 1064nm and 532nm wavelengths. The 30mm-long $LiIO_3$ nonlinear crystal was used in NLM configuration.

As known the efficiency of NLM action is dependent on the phase differences acquired by the laser and second harmonic waves in nonlinear crystal due to the phase-matching detuning $\varphi_1 = \Delta k_c l_c$ and in the space between crystal and cavity mirror due to air dispersion $\varphi_2 = 4\pi l_a (n_{2\omega}^{(a)} - n_\omega^{(a)})/\lambda$ and phase shifts on the mirror $\Delta\varphi_{mir}$. At $\Delta\varphi_{mir} = \pi$ the simple expression for the efficiency of the double-path SHG obtained in[4] in the low intensity limit:

$$\eta \propto \left\{ \sin(\varphi_1/2)\sin[(\varphi_1 - \varphi_2)/2]/\varphi_1 \right\}^2 \qquad (1)$$

gives the picture shown in Fig.1. The working point of NLM have to be chosen at $\varphi_1 \approx 0$ to provide the highest SHG efficiency and

$$\varphi_2 + \Delta\varphi_{mir} \approx \pi(2m+1), \, (m = 0,1,2,...) \qquad (2)$$

Ultrafast Processes in Spectroscopy, edited by Svelto et al.
Plenum Press, New York, 1996

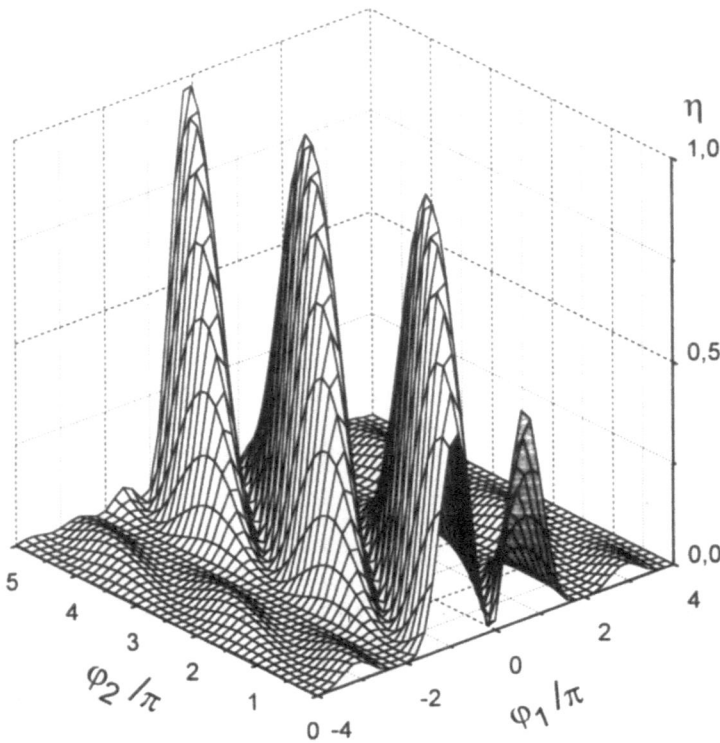

Figure 1. The double-pass SHG efficiency dependence on the phase mismatch in crystal φ_1 and the phase shift in crystal-mirror air space φ_2.

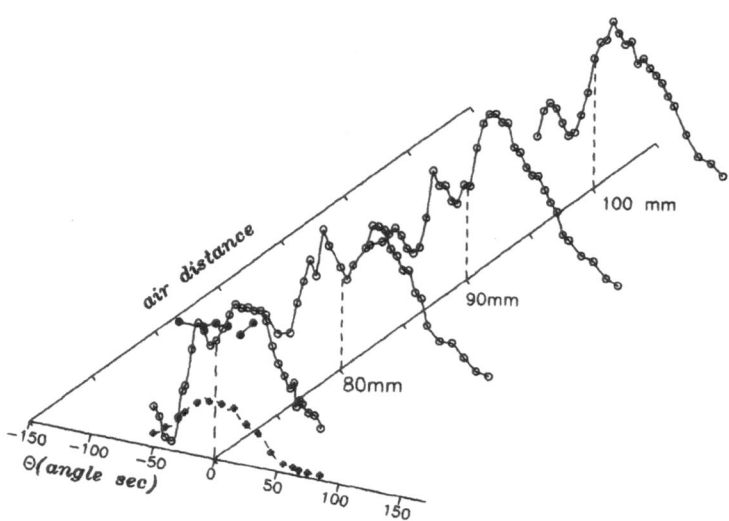

Figure 2. Experimentally measured double-pass SHG efficiency in nonlinear mirror in the vicinity of working point.

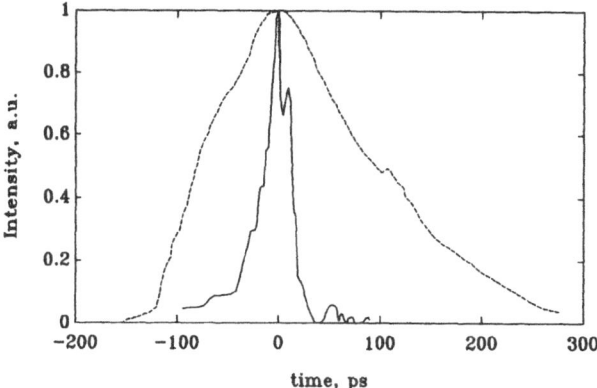

Figure 3. Laser pulse shapes at the properly adjusted nonlinear mirror (solid) and at the large phase mismatch (dashed).

to invert the energy exchange between the interacting waves at their pass through the crystal in opposite direction.

The measurements of the used NLM characteristics have been fulfilled out of laser cavity with the single 100ps-pulse from the actively mode-locked laser. The results obtained in the vicinity of working point are shown in Fig.2. These results repeat the main features of Fig.1 but show also certain asymmetry in φ_2 coordinate (air distance) which can arise from the nonlinear phase modulation inherent in the frequency conversion process.

In actively mode-locked regime laser generated a ~3.2mJ energy long (~80ms) train of pulses with durations 120ps. The hybrid mode-locking was readily observed after setting of the air-space distance between nonlinear crystal and dichroic mirror in the range of 75–100mm and adjustment of the nonlinear crystal near a phase matching angle for SHG. The results of the laser pulse shape streak-camera measurements shown in Fig.3 evidence a more then four-fold pulse shortening due to the NLM action. The obtained pulses have the duration ~(25:33)ps and asymmetrical shape with the sharp trailing edge. A ~70ms pulse train had energy ~2mJ.

As the NLM behaviour is similar to that for fast passive mode- locker[2] we have also faced with abrupt laser operation jumping in periodic self-Q-switching regime observed in[3]. The Q- switching was observed in the vicinity of the highest reflectivity of NLM and interfered in the maintaining of quasi-cw laser operation. Therefore we have been forced to work at the increased crystal-mirror distance accompanied by a turning of the crystal from the phase matching angle.

It should be noted that the detail analysis of NLM action in the mode-locked laser have to take into account the nonlinear phase modulation effects arising both from efficient energy exchange between the interacting waves and nonzero phase detuning in crystal[4-6]. These effects can give rise to frequency chirp and probably explain the observed asymmetric pulse shape and some features of Fig.2.

REFERENCES

1. K.A.Stankov. Appl.Phys. B, **45**, 191 (1988).

2. K.A.Stankov, V.Kubecek, K.Hamal IEEE J.Quant.Electr., **27**, 2135 (1991).
3. M.B.Danailov, G.Cerullo, V.Magni, D.Segala, S.De Silvestri. Opt.Lett., **19**, 792 (1994).
4. P.A.Apanasevich, O.V.Chekhlov, A.V.Kachinskii, I.V.Pilipovich, R.G.Zaporozhchenko, V.A.Zaporozh-chenko. Optics Communs., **51**, 289 (1984).
5. P.A.Apanasevich, V.A.Zaporozhchenko, R.G.Zaporozhchenko, A.V.Kachinskii. Kvantovaja elektronika, **11**, 897 (1984).
6. P.A.Apanasevich, V.A.Zaporozhchenko, R.G.Zaporozhchenko, A.V.Kachinskii, V.A.Mukha, I.V.Pilipovich, O.V.Chekhlov. Kvantovaja elektronika, **13**, 1132 (1986).

13

BROADENING OF THE CAVITY PARAMETER REGION FOR KERR-LENS MODELOCKING BY GAIN GUIDING AND ASTIGMATISM

S. Gatz and J. Herrmann

Max-Born-Institut
Rudower Chaussee 6, D-12474 Berlin

Kerr-lens mode-locking (KLM) is a powerful technique for the generation of femto-second pulses in solid-state lasers. Mode locking with this technique is caused by self-focusing and provides a fast amplitude modulation (SAM) by a suitable intracavity aperture. SAM in KLM lasers can also be caused without aperture by the combined effects of self-focusing and a radially decreasing gain (gain guiding), which yields a saturable diffraction loss (described by a power-dependent resonator magnification) and a power-dependent averaged gain[1].

The aim of this paper is to investigate the effect of a Kerr medium cut at Brewster's angle on KLM performance extending the former results in Ref. 1. Starting from the wave equation for an elliptical Gaussian beam in the gain Kerr medium and taking into account diffraction, the Kerr effect, and a radially decreasing gain by a Gaussian pump beam, we can derive similar equation for the complex-valued ABCD element as given in Ref 1.

Using the matrix elements of the gain Kerr medium, we can calculate the transition matrix $\hat{A}, \hat{B}, \hat{C}, \hat{D}$, for one complete round trip with the matrices of the additional elements and the rules for matrix multiplication. We consider a standard four-mirror configuration as depicted in Fig. 1(a). The two focusing curved m mirrors M_3 and M_4 with a radius of curvature of $2f$ are tilted by the angle. Therefore the effective focal lengths in the x and y planes differ an are given by $f_x = f / \cos \Omega$, $f_y = f \cdot \cos \Omega$.

In a KLM laser with an internal aperture, the aperture loss at the end mirror is given to the first order by $T_a = q_L - (q_{NL} P / w_2)(dw_2 / dP)$, where nonlinear loss coefficient $\gamma_a = -d T_a / dP$ depends mainly on the relative spot size change of the beam size w_2 at the out couple mirror M_2, and q_L, q_{NL} are the aperture-dependent coefficients. An independent mechanism of mode locking is caused by the radially varying gain, which yields a saturable diffraction loss and a power-dependent transversely averaged gain. The averaged power amplitude factor per round trip is given by $T_\infty = exp(2G^+ + 2G^-) / |M_2|^2$, where M_2 is the magnification at the outcoupling mirror and G^+, G^- are the averaged gain in forward and backward direction, respectively[1]. The characteristic parameter for SAM then is given by $\gamma_{gg} = d T_{gg} / dP |_{P=0}$.

Ultrafast Processes in Spectroscopy, edited by Svelto et al.
Plenum Press, New York, 1996

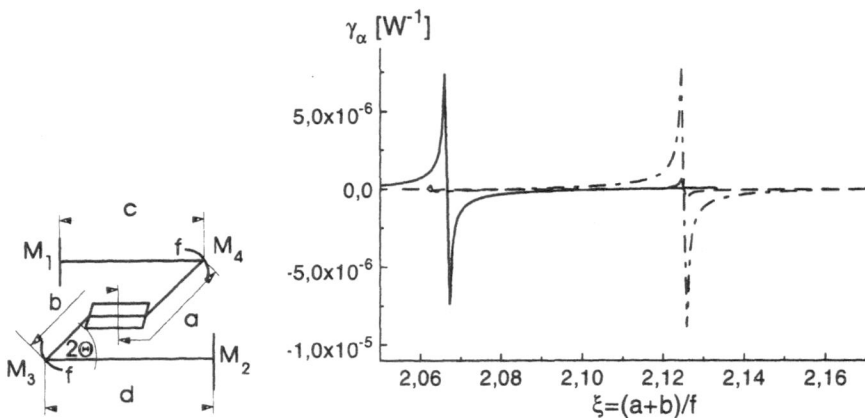

Figure 1. (a) Resonator configuration: the distance a (b) is the optical path in saggittal direction form mirror M4 (M3) to the center of the crystal. (b) Nonlinear loss coefficient γ_a for a radially constant gain (w_p). Parameters are c = 80 cm, f = 5 cm, L = 1.5 cm, g = 0, n = 1.76, $n_2 = 3 \times 10^{-16}$ cm^2 W^{-1}, λ = 785 nm, q_{NL} = 0.5. Solid curve (without astigmatism), b = 4.74 cm; dashed (γ_{ax}) and dotted-dashed (γ_{ay}) curves, b = 5.14 cm, 2 = 26.5.

The main effect of astigmatism and the beam ellipticity results from the change of the stability boarders by the tilted focusing mirrors and the Brewster angled laser crystal. Using as cavity variable in Fig. 1 the normalized optical path of the inner mirror distance in the saggittal direction $\xi = (a+b)/f$ the two stability zones $\left(\xi_{Cx/y}, \xi_{Ax/y} \right)$, $\left(\xi_{Dx/y}, \xi_{Bx/y} \right)$ (defined by ABCD $\leq$ 0) are different in the x and y plane. However, we can choose a proper mirror tilting angle Θ in such a way that all stability boarder nearly coincide in x and y direction.

Let us consider how this astigmatism compensation influences the property of KLM. In Fig. 1(b) the nonlinear loss coefficient γ_a of an aperture is plotted in dependence on the variable $\xi = (a+b)f$ for a cavity with symmetric long arms and without gain guiding (g = 0). In this case both inner stability boarders, ξ_A and ξ_D, degenerate to a singular point with $\xi_A = \xi_D$, The case without astigmatism is calculated by the solid line in Fig. 1(b). The maximum of $\gamma_a = 7.1 \cdot 10^{-6} W^{-1}$ is reached at the degenerated stability boarder $\xi_A = \xi_D = 2.067$, for Ti:sapphire parameters and an optimized mirror-crystal distance b = 4.74 cm Taking into account beam ellipticity with the compensation of astigmatism by a tilting angle 2Θ = 26.5 in Fig. 1(b) we plot the nonlinear loss coefficients γ_a, now is reached at the stability boarder of the astigmatism-compensated cavity, $\xi_{Dx} = \xi_{Ax} = 2.125$.

To consider the effect of gain guiding, we calculated corresponding curves [Fig. 2(a)] for the same cavity configuration but with gain g = 0.1 cm^{-1}, a beam waist w_p = 75 m, and b = 4.92 cm or b = 4.75 cm in the astigmatic or nonastigmatic cavity, respectively. In comparison with that Fig. 1(b), the region of the cavity parameter for possible KLM becomes broader without astigmatism (solid line), but in an astigmatic cavity the broadening is dramatic. The tangential component γ_{ay} with sufficient high values $\gamma_a \geq 10^{-7} W^{-1}$, covers more than the half of the stability region.

In Fig. 2(b) the nonlinear coefficient for the saturable diffraction loss parameter $\gamma_{gg} = d\Gamma_{gg}/dP$ is depicted, and it illustrates the reasons for KLM without aperture. Here the dashed curve refers to the total effect of sagittal and tangential components in an astigmatic cavity. In comparison with those in Fig. 2(a), the curves for γ_{gg} are reflected into the

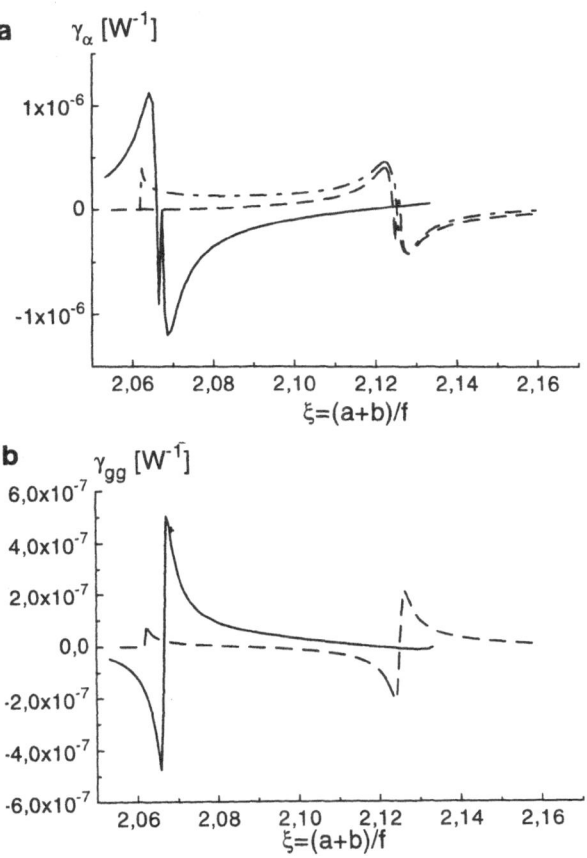

Figure 2. SAM coefficient γ_a (a) and (b) γ_{gg} with a radially decreasing gain. Parameters are $g = 0.1$ cm^{-1} and $w_p =$ 75 m. Solid line (without astigmatism) $b = 4.76$ cm dashed and pointed-dashed line (astigmatic cavity): $b = 4.92$ cm, $2\Theta = 26.5$ (other parameters as in Fig. 1).

reverse direction. Consequently KLM caused by saturable diffraction losses is possible for cavity parameter regions where KLM by the saturable aperture loss is not allowed and viceversa. The allowed region for KLM is in this case also relatively broad without astigmatism and the astigmatism here does not lead to broadening of this region.

REFERENCE

1. J. Hermann, J. Opt. Soc. Am. 11, 498 (1994).

14

HIDDEN MECHANISM FOR MODE LOCKING SOLID-STATE LASERS BY USE OF A SLOW OFF-RESONANT SATURABLE ABSORBER

J. Herrmann and M. Müller

Max-Born-Institut
Berlin, Germany

Recent experiments succeeded in sub-100fs pulse generation from solid-state lasers by use of a saturable single-quantum-well absorber (SQW) yielding self-starting 34fs pulses in a Ti:sapphire laser[1] as well as 90fs pulses in a diode pumped Cr:LiSAF laser[2]. These experimental findings can hardly be explained by the common theoretical understanding of passive mode locking in solid-state lasers which requires a fast saturable loss for stable pulse generation. The relaxation of the absorption bleaching in a GaAs quantum well at room temperature exhibits a bitemporal behavior where the fast component of about 200fs owing to thermal ionization of the exzitons is remarkably slower than the meanwhile achieved pulse width in the experiments. Recently we have proposed a novel mechanism for mode locking solid state lasers by use of a slow off-resonant saturable two-level absorber[3].

Here we analyze the operating conditions for this new mode locking scheme by use of a slow saturable SQW-absorber and calculate the dependence of the pulse parameters on the characteristical laser and absorber parameters. The underlying principle of mode locking by this novel mechanism is that in a first order expansion beyond the rate equation approximation a slow saturable absorber with an arbitrary large response time T_1 produces an additional intensity dependent loss $\Delta\alpha = -\gamma(d(\Delta n)/d\omega)I$, where $\Delta n(\omega)$ is the linear refractive index change owing to the near-resonant transition, γ is an absorber parameter and I is the pulse intensity. The physical origin of the intensity dependent loss $\Delta\alpha$ is that due to the finite phase relaxation time τ_a the coherent medium polarization encounters a group velocity change which leads owing to the slow saturation term under off-resonance $d(\Delta n)/d\omega\big|_{\omega_L} > 0$ to a fast saturable loss contribution $\Delta\alpha < 0$.

The pulse evolution in the laser resonator is governed by a recursion formulae for the complex field amplitude πσι with carrier frequency ω_L which is in the kth round trip given by $\psi_{k+1} = r\hat{D}\hat{Q}^2\hat{D}\psi_k$, where the nonlinear refractive index solid state gain medium with group velocity dispersion compensation is described by $\hat{D}$, $\hat{Q}$ accounts for the transmission through the SQW absorber and r is the reflectivity of the output mirror. To obtain a simple analytical expression for $\hat{Q}$ we consider a 2D free carrier model with band filling

Ultrafast Processes in Spectroscopy, edited by Svelto et al.
Plenum Press, New York, 1996

61

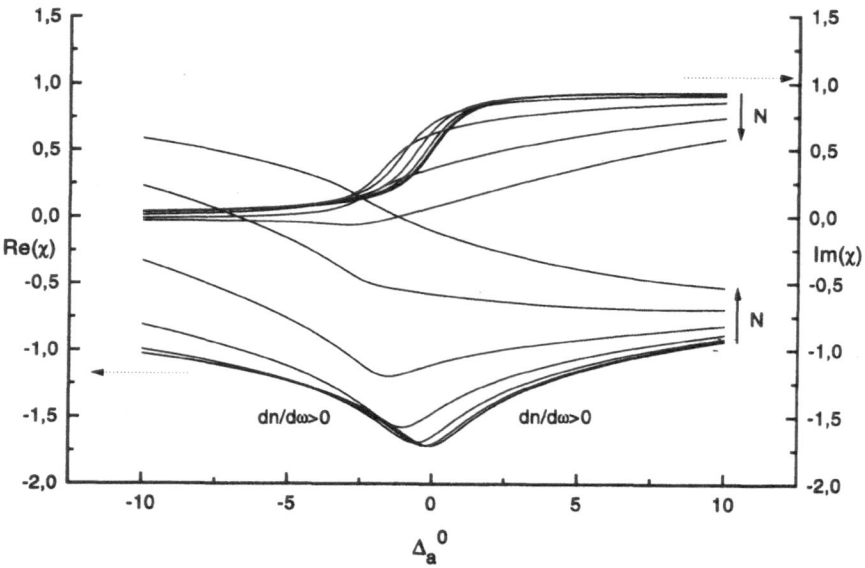

Figure 1. Spectrum of the normalized SQW-susceptibilty χ / χ_0 for different carrier densities $N < N_T$.

and band gap renormalization neglecting excitonic absorption as well as coulomb enhancement of the continuum absorption[4]. Using these simplifications we derive the spectrum of the complex SQW-susceptibiliy $\chi(\omega, N)$ in dependence on the carrier density N. For different carrier densities in Fig.1 we have depicted the real and imaginary part of the normalized free-carrier SQW-susceptibiliy χ / χ_0 in dependence on the normalized spectral detuning parameter $\Delta_a^0 = (\hbar\omega_L - E_g^0) / \Gamma$ with the linear band gap E_g^0 and a phase relaxation time $\tau_a = \hbar / \Gamma$ due to the carrier-carrier scattering rate Γ.

From Fig.1 it is obvious that a refractive index change $\Delta n \propto Re(\chi)$ with a positive spectral slope $d\,Re(\chi(\omega)) / d\omega > 0$ requires a positive spectral detuning $\Delta_a^0 \geq 0$. For an interband recombination time T_1 which is significantly greater than the pulse width τ but shorter than the cavity round trip time we obtain for the optically induced carrier density the expression $N = N_T \left[\int_{-\infty}^{\eta} |\psi|^2 dt / U_a \right]$ where U_a is the effective absorber saturation energy and N_T is a normalization density of the quantum well absorber indicating the crossing from absorption to gain. Expanding the time-domain round trip operators $\hat{D}, \hat{Q}$ with respect to small parameters we obtain with the steady-state condition $\psi_{k+1}(\eta) = e^{\gamma}\psi_k(\eta + h)$ the following stationary mode locking equation[3]

$$\left\{ c_2 \frac{\partial^2}{\partial \eta^2} + c_1 \frac{\partial}{\partial \eta} + c_0 + c_{N1}|\psi|^2 + c_{N2}\int_{-\infty}^{\eta}|\psi|^2 dt + c_{N3}\left[\int_{-\infty}^{\eta}|\psi|^2 dt\right]^2 + c_{N4}\int_{-\infty}^{\eta}|\psi|^2 dt \frac{\partial}{\partial \eta} \right\}\psi(\eta) = 0 \quad (1)$$

Here all coefficients are complex-valued and depend on the absorber, amplifier and pulse parameters. The condition for stable mode locking operation is $Re(c_{N1}) \propto \gamma(d(\Delta n) / d\omega) > 0$ which requires in correspondence to Fig.1 a positive laser-absorber detuning $\Delta_a > 0$. The steady-state mode locking equation is exactly solved by the ansatz

$$\psi(\eta) = v_0 \left[cosh\left(\frac{\eta}{\tau}\right) \right]^{-1+i\beta}$$

yielding algebraic equations for the pulse parameters. In the calculation of the pulse parameters we used a round trip time of 5ns, a nonlinear Kerr-coefficient of $2*10^{-7}W^{-1}$ a pumprate of 10% over threshold and a group velocity dispersion of $D_2/\tau_g^2 = -0.1$ where τ_g is the amplifier phase relaxation time. As SQW absorber we consider a GaAs/ $Al_xGa_{1-x}As$ structure of 10nm well width yielding for $x = 0.09$ a gap at 847nm which is suitable for the Cr:LiSAF center wavelength around 840nm. The other SQW parameter are a phase relaxation time of $\tau_a = 100$ fs, and a saturation energy at the band gap of $U_a^{\Delta_a=0} = 0.5nJ$.

In Fig.2 the normalized pulse width τ/τ_g is plotted versus the gain-absorber detuning parameter $\Delta_{ga} = \tau_a(\omega_g - \omega_a)$ for (a) typical solid-state laser bandwidths ranging from $\tau_g = 40$ fs (Nd:glass) to $\tau_g=2$ fs (ti:sapphire) at a fixed loss parameter of $\alpha_0 = 1\%$ and (b) for different linear loss parameters $\alpha_0 = 0.25\%...5\%$ at a fixed gain bandwidth of $\tau_g = 5$ fs corresponding to a Cr:LiSAF laser. The calculated pulse parameters in dependence on the SQW and laser parameters shows that a near-resonant slow saturable SQW-absorber in a sufficiently extended operating region with $\Delta_a > 0$ always provides an additional fast SAM contribution sufficient for stable passive mode locking in broad band solid state lasers.

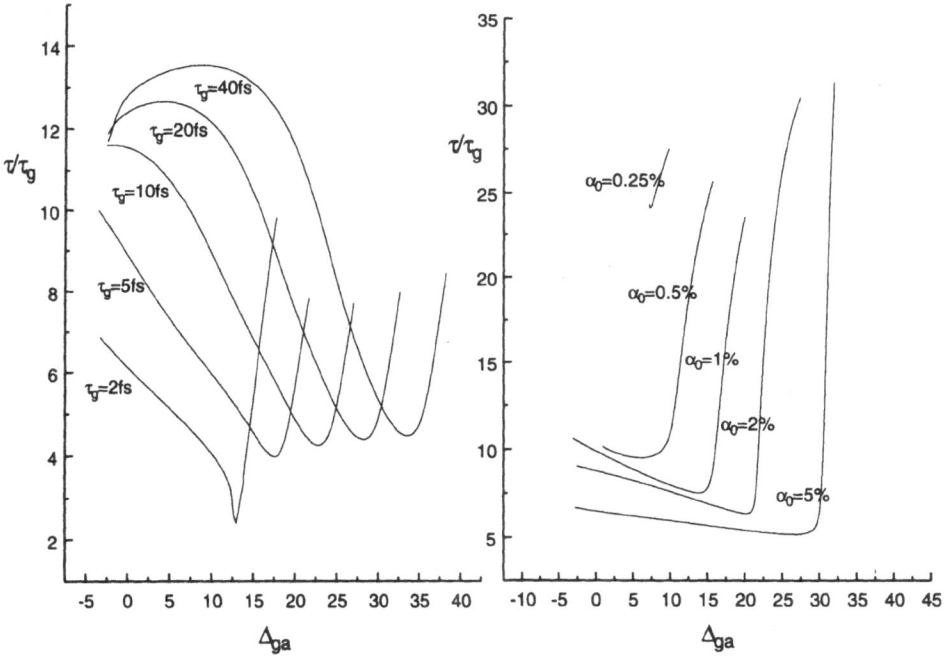

Figure 2. Normalized pulse width τ/τ_g over gain-absorber detuning Δ_{ga} for different gain bandwidths (a) and different loss parameters α_0 (b).

Thus, off-resonant slow passive mode locking (OPSM) with a slow SQW-absorber represents a reliable novel method for fs-pulse generation in solid-state lasers and can explain some of the recent experimental findings[1,2]. This new approach to passive mode locking ensures self-starting, cavity design independence and the possibility of diode-pumping, it seems to be particular attractive for all-solid-state fs-pulse generation.

We gratefully acknowledge support by the Deutsche Forschungsgemeinschaft.

REFERENCES

1. L. R. Brovelli, et al. Electron.Lett. 31, 287 (1995).
2. S. Tsuda, et.al. Opt.Lett. 20, 1406 (1995).
3. J. Herrmann, M. Müller, Opt.Lett. 20, 994 (1995).
4. H. Haug, S.W. Koch, "Quantum Theory of the optical and electronic properties of semiconductors", World Scientific (1994).

15

TUNABLE UV PICOSECOND PULSE GENERATION USING FAST SPECTRAL EVOLUTION OF LOW-Q SHORT DYE LASER OSCILLATORS

Nguyen Dai Hung,[1,2] Y. Segawa,[1] D. V. Trung,[2] P. Long,[2] L. H. Hai,[2] and N. H. Tam[2]

[1] Photodynamic Research Center
Institute of Physical and Chemical Research (RIKEN)
19–1399 Nagamachi Koeji, Aoba, Sendai 980, Japan
[2] Institute of Material Sciences
National Centre for Sciences and Technology of Vietnam
Nghia do, Tu liem, Hanoi, Vietnam

Tunable ultrashort pulsed lasers in UV wavelength region are indispensable to research the ultrafast mechanism of photochemical processes of materials. In recent years numerous approaches have been made to generate UV ultrashort pulses, including sum or difference-frequency mixing in nonlinear optic crystals and noble gases, third-harmonic generation in noble gases and anti-stock Raman shifting[1,2]. The existing commercially available UV ultrashort pulsed laser systems are mainly based on non-linear frequency converting of tunable visible or near infrared laser systems that are complicated and expensive up to now.

Recently we presented the direct production of picosecond and subpicosecond laser pulses tunable from 400 nm to 700 nm using organic dyes in short low-Q cavity and a standard nanosecond pumping laser[3–5]. In this paper, we report: 1) theoretical and experimental results of the study of spectral dynamics of the broadband laser emission from the p-terphenyl dye in short, low-Q oscillators; 2) application of fast spectral evolution in such broadband laser emission to direct production of tunable picosecond UV laser pulses within the 320–400 nm range with a nanosecond pumping laser.

A multiwavelength analysis on spectral and time processes for p-terphenyl, low-Q short-cavity dye laser have been performed with the rate equation system, as given below, in order to expose the spectro-temporal evolution in transient laser emission:

$$\frac{dN_1}{dt} = \left(P + \sum_{i=1}^{n} \sigma_{ai} I_i\right) N_o - \left(K + \sum_{i=1}^{n} \sigma_{ei} I_i\right) N_1 \tag{1}$$

Ultrafast Processes in Spectroscopy, edited by Svelto et al.
Plenum Press, New York, 1996

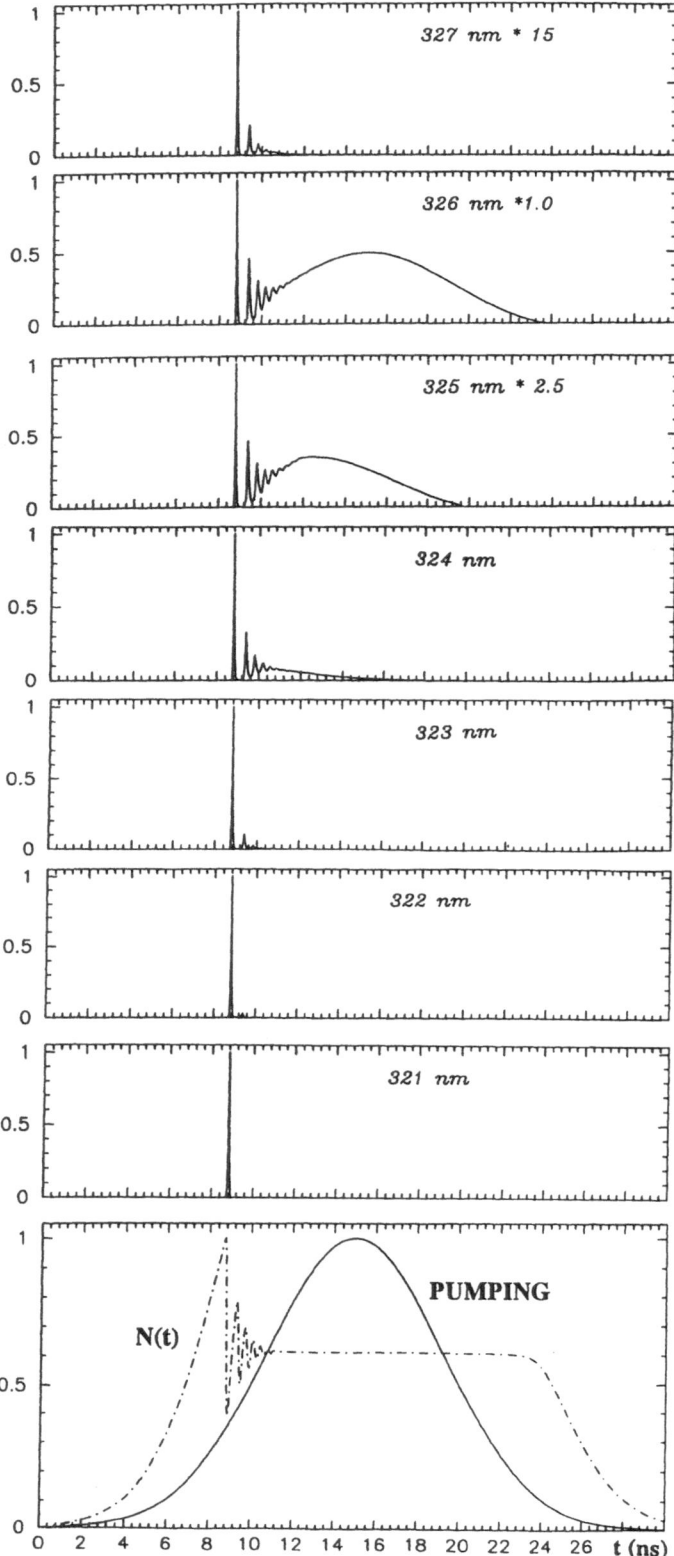

$$\frac{dI_i}{dt} = \frac{I_i}{T}\left[2L\left(\sigma_{gi}N_1 - \sigma_{ai}N_o\right) - \alpha_i\right] + U_iN_1$$

(2)

where I_i is the laser intensity at wavelength λ_i ; P is the pumping rate (s^{-1}) and n is the number of channel with significant intensity; $K(s-1) = 1 / \tau$, τ is the fluorescence decay time; T is the round trip cavity time; L is the active medium length; α_i is the round-trip loss at wavelength λ_i, $\alpha = - \ln R_1R_2$ with R_1 and R_2 - the mirror reflectivity. $U_i = 10^{-10}$ cm s^{-2} is a constant that stimulated the spontaneous emission, its value has very little importance, can be taken equal for all channels. N_o and N_1 are the ground state S_o and the lowest excited single state population, respectively. σ_g is the difference σ_e-σ_u between σ_e, the stimulated emission cross section of the lowest excited single state S_1 and σ_u, the corresponding absorption cross section from the S1 state to upper states. σ_a is the absorption section from the state S_o.

We used 11 coupled equations for P-terphenyl dye (one equation for the excited state and 10 equations for the monochromatic intensities at wavelengths between 319 nm and 328 nm which corresponds to a part of the gain region for P-terphenyl in cyclohexane). We take L=1 cm, T =100 ps, N = 6.10^{17} molecules / cm^3, α = 4.6 with R_1 = 0.25, R_2 = 0.04 , τ = 1 ns and a Gaussian smooth pumping rate P (s^{-1}) of 10 ns - FWHM duration. The molecular emission and absorption cross sections were quoted from Ref.[6,7]. Unfortunately, some spectroscopic parameters of P-terphenyl dye are not well known taking into account reabsorption of the lasing intensity I_i at the wavelength λ_i by the non-excited molecules, absorption from the excited single state S_1 to upper states (σ_u), and particular the triplet absorption at the wavelength range considered. The latter was found to be in the long wavelength wing of the fluorescence spectrum and becomes dramatic at high excitation energy (~150 mJ)[7,8]. In the calculations, we take $\sigma_u \ll \sigma_e$ and so $\sigma_g = \sigma_e$, the triplet state absorption is negligible for the short-wavelength wing of emission spectrum.

The computed results are presented in Fig.1. The intensity of the laser emission is as a function of time and wavelength and shows damped relaxation oscillations as well as a fast spectral evolution. The laser spectrum initiates in the form of a broad band which rapid narrows while its maximum shifts toward longer wavelengths (326 nm). Although the laser gain reaches an oscillating regime, the spectro-temporal evolution nevertheless continues in the same spectral direction and is non oscillating. The emphasized result is that the oscillation damping strongly depends on the wavelength. The single short spike at the short -wavelength side which is of high interest for short pulse generation, can be isolated by a spectral selection. The pulse duration of the single short spike at 321 nm is 85 ps (FWHM) for the pumping level at 6 times above threshold.

The experiments were done with the p-terphenyl dye dissolved at a concentration of 10^{-3} M / l. The solvent was cyclohexane. The low-Q short cavity was formed by a quartz cell with a length of 10 mm and two parallel windows normal to the cell axis. Both the windows are wedged to avoid parasitic reflections. One of them was aluminum coated with 25–30 % reflectivity, the other is an uncoated quartz plate as the output coupler. A conventional transverse pumping configuration was used for excitation of the dye laser oscillator. The pumping beam (~1.5 mj at 266 nm and 10 ns- FWHM) was transformed by a

Figure 1. Computed result of the spectro-temporal evolution in the low-Q, p-terphenyl dye, 1-cm oscillator emission (10 times above threshold). The dashed curve presents the calculated excited-state population $N_1(t)$. The continuous curve presents the Gaussian-shaped pump pulse with 10 ns FWHM duration.

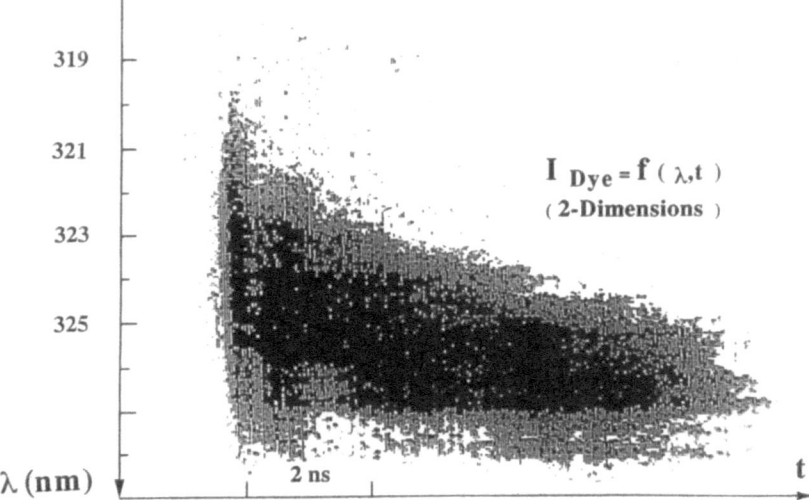

Figure 2. Two-dimensions analysis of the spectro-temporal evolution in the low-Q, p-terphenyl, 1-cm oscillator emission. Pumping level 6 times the threshold.

cylindrical lens in to a rectangular spot with dimensions of 0.4×10 mm^2 on the surface of the dye cell. The active volume has the shape of a narrow cylinder, the axis of which is perpendicular to the two windows. At the pumping level of 6 times threshold, we observed a laser broadband emission having the maximum wavelength at 326 nm. The bandwidth was ~3.5 nm (FWHM). A conversion efficiency was 8.6% for broadband laser operation.

The spectral evolution in the laser emission from the low-Q, p-terphenyl, 1- cm oscillator was directly measured with a two-dimension streak camera, as shown in Fig. 2. Damped relaxation oscillations and a fast non-oscillatory spectral evolution are clearly observed at high pumping level. The damping time depends strongly on wavelength and is faster in the short wavelength wing (321 nm).

The pulse shortening effect resulting from the spectral selection is presented in Fig. 3. A single short pulse of <100 ps can be obtained by a spectral filtering at ~322 nm. Clearly, these allow one to produce of tunable UV picosecond dye laser pulses using spectro-temporal selection (STS) method for UV laser dyes.

The tunable UV picosecond STS dye laser configuration is shown in Fig. 4. At first experiments, we used an excimer laser as nanosecond pumping laser. However, the pumping from the fourth harmonic of a Nd:YAG oscillator was also employed. The pump beam was divided 1: 5 on the dye oscillator (O) and the multipass amplifier (MPA). Pumping energy at 266 nm was 6 mJ in a nominal 10 ns duration and at repetitive rate adjustable between 1- 20 Hz. The dye oscillator output was spectrally selected by a grating-based filter (G) and then directed to the MPA. The path from the oscillator to the MPA is adjusted with the optic delay (DL) for amplifying all six passes by the maximum of the pumping pulse at the MPA cell. The amplifier pump pulse was delay by ~ 2 ns relative to the dye oscillator pump pulse. The grating-based filter consists a grazing incidence grating arranged in the folded dispersion delay line[9]. This arrangement simultaneously provides a spectrally high-selectivity and a temporal compensation for elimination of pulse broadening during the spectral filtering. Wavelength tuning is conveniently obtained by translat-

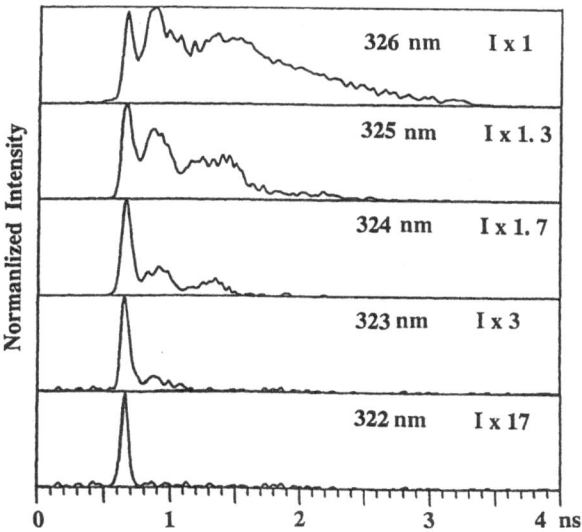

Figure 3. Pulse shortening effect of spectral selection in the STS picosecond dye laser. Single picosecond pulse was obtained by spectral filtering.

ing the stripe mirror (TM) horizontally, in the focal plan of the lens (L4), with the respect to the grating groove. Output UV laser linewidth can be controlled by stripe mirror width or grating incident angle.

Direct production of UV picosecond ~90 ps dye laser pulses spectrally adjustable between 320 nm and 400 nm is demonstrated using some UV dyes. Corresponding a pumping pulse energy of 6 mJ , output UV picosecond pulse energy is typically 8 µJ

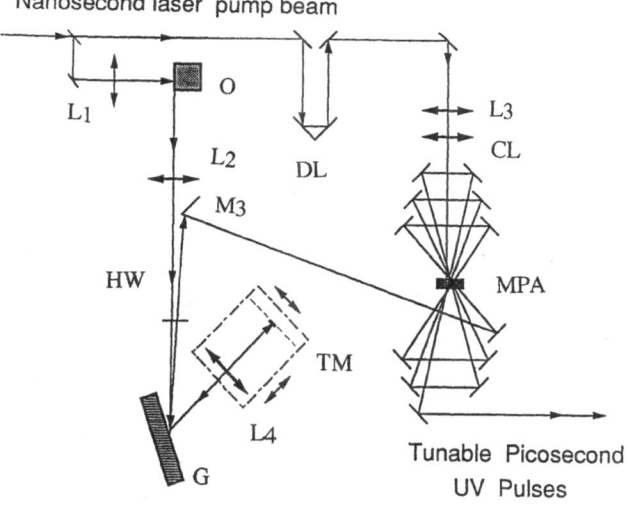

Figure 4. The STS dye laser for generation of single picosecond UV pulses from nanosecond pump laser pulses.

which is straightforward to further amplification to high energy picosecond UV pulses. For the first time, the spectro-temporal selection (STS) picosecond dye lasers have been successfully operated in the ultraviolet spectral range.

The dye solutions in the amplifier and oscillator were contained in 250 ml volumes and circulated at a flow rate of about 1.2 l / min. A slow deterioration of the dye solution was observed in the output pulsed energy. Satisfactory laser operation (9% rms.in pulse energy and duration stability) of the UV picosecond laser is indicated for period in excess of 54 h.

In conclusion, a fast spectral evolution in the broadband laser emission of the P-ter-phenyl dye (and some ultraviolet dyes) in short, low-Q laser oscillators is studied and for the first time, used to produce tunable picosecond UV laser pulses between 320 nm and 400 nm with a nanosecond laser pumping. This direct approach to the generation of picosecond UV pulses is more efficient and simpler than multi-step approaches.

ACKNOWLEDGMENTS

This research was partially supported by the fundamental research Program (KT-04) of the Vietnam MSTE and Japan JRDC.

REFERENCES

1. *Ultrafast Phenomena VIII*. Ed. J.L Martin, A. Migus, G.A. Mourou and A.Zewai. (Spring Series in Chemical Physics, 1992)
2. *Ultrashort Laser Pulses Generation and Applications* (2nd edition). W. Kaiser. (Topics in Applied Physics Vol. 60. Springer- Verlag Berlin Heidelberg, 1993)
3. Nguyen dai Hung and Y. H. Meyer. Appl. Phys. B **53**, 226 (1991).
4. Nguyen Dai Hung, P. Plaza, M. Martin, Y. Meyer. Appl. Opt. **31** (1992) 7046.
5. Nguyen Dai Hung, P.Long, Y. Meyer. et al. SPIE proceeding **2321**, 286 (1994)
6. I.B. Berlman. *Handbook of Fluorescence Spectra of Aromatic Molecules* (Academic Press). p.220 (1971)
7. V.I. Tomin, A.J.Alcock, W.Sarjeant and K.E. Leopold. Opt. Comm. **28**, 336 (1979). V.I. Tomin, A.J.Alcock, W.Sarjeant and K.E. Leopold. Opt. Comm. **26**, 396 (1978).
8. T.G. Pavlopoulos. Opt. Comm. **24**, 170 (1978).
9. Nguyen Dai Hung, Y. Segawa, P. Long and et al. Proceeding of the 1995 International Conference on Opto-Electronics and Lasers (ICOEL'95). P.289 (1995).

16

PICOSECOND STUDY OF ENERGY AND ELECTRON TRANSFER PROCESSES IN SELF-ORGANIZED PORPHYRIN COMPLEXES

C. von Borczyskowski,[1] U. Rempel,[1] E. Zenkevich,[2] A. Shul'ga,[2] and A. Chernook[2]

[1] University of Technology Chemnitz-Zwickau
Chemnitz, Germany
[2] Institute of Molecular and Atomic Physics
Belarus Acad. Sci.
Minsk, Belarus

Natural and synthetic complexes of tetrapyrrolic compounds are known to be used in wide areas of investigations and applications: vectorial energy and electron transfer[1,2], photochemical molecular devices[3], light-harvesting and -transforming systems[4], sensitizers for photodynamic therapy[5]. The main aim of our study is focused on the constriction of supramolecular systems containing more than two π-conjugated porphyrin-like macrocycles and investigation of electronic excitation energy deactivation processes in such arrays. With respect to studies performed so far[2,6,7] in our case porphyrin or chlorin subunits are connected in two fundamentally different ways[8]: i) by chemical spacers of various nature (-CH_2-CH_2- bond or phenyl ring) that link together monomeric compounds in Zn-porphyrin and Zn-chlorin chemical dimers and, additionally ii) by self-assembling non-covalent binding interactions of these dimers with pyridyl-mesosubstituted tetrapyrroles (two-fold extra-ligation effect with "mix and match" principle) which lead to the formation of trimeric and pentameric complexes stable in methylcyclohexane solutions at 295 K. More than 40 combinations of interacting moeties have been obtained and investigated. Computer simulated (HyperChem) structures of some systems under consideration are presented on Fig.1.

Here we present the results of picosecond time-resolved experiments and analyze the mechanisms of interchromophoric interactions as well as excitation energy deactivation processes taking place in the assemblies of interest. Fluorescence lifetimes and time-resolved spectra were measured using a laser fluorescent setup with 2-D (wavelength-lifetime) registration. The excitation system consisting of a dye laser (Spectra Physics Model 375 including cavity dumper Spectra Physics Model 344) synchronously pumped by an Ar-ion laser (Spectra Physics Model 171) produces pulses of app. 10 ps at a repetition rate of 10 MHz. The detection system (Streak-Scope Hamamatsu Model C4334

Ultrafast Processes in Spectroscopy, edited by Svelto et al.
Plenum Press, New York, 1996

71

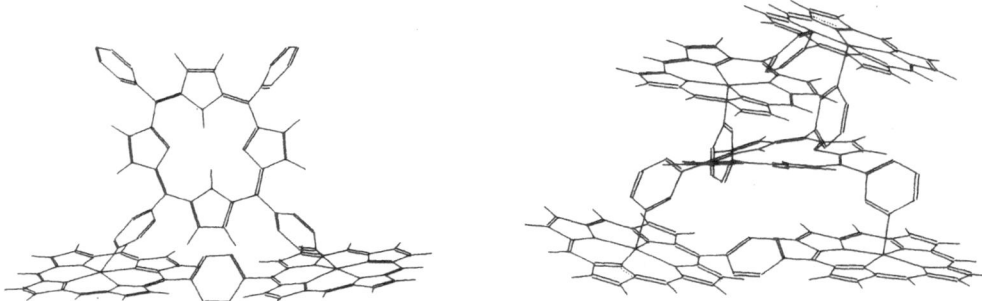

Figure 1. Structures of triadic and pentadic arrays of tetrapyrrolic compounds: A -(ZnHTPP)$_2$+H$_2$P-(p^Pyr)$_2$, extra-ligand with 2 nitrogens in para-positions of adjacent pyridyl rings; B - (ZnHTPP)$_2$+H$_2$P-(m-Pyr)$_4$, extra-ligand with 4 nitrogens in meta-positions of pyridyl rings.

with Trigger Unit Hamamatsu Model C4792) response time is 30 ps in the 1-ns range of the Streak-Scope. One of the typical results is depicted in Fig.2.

It has been shown that two-fold coordination plays an essential role in the formation of supramolecular ensembles and complexation constants (K_C=6·10^5-5·10^7 M^{-1}) as well as activation energies (E_a=0.5 - 1.2 eV) depend on the nature and structure of interacting components. Optical properties of triads and pentads are explained by interchromophoric

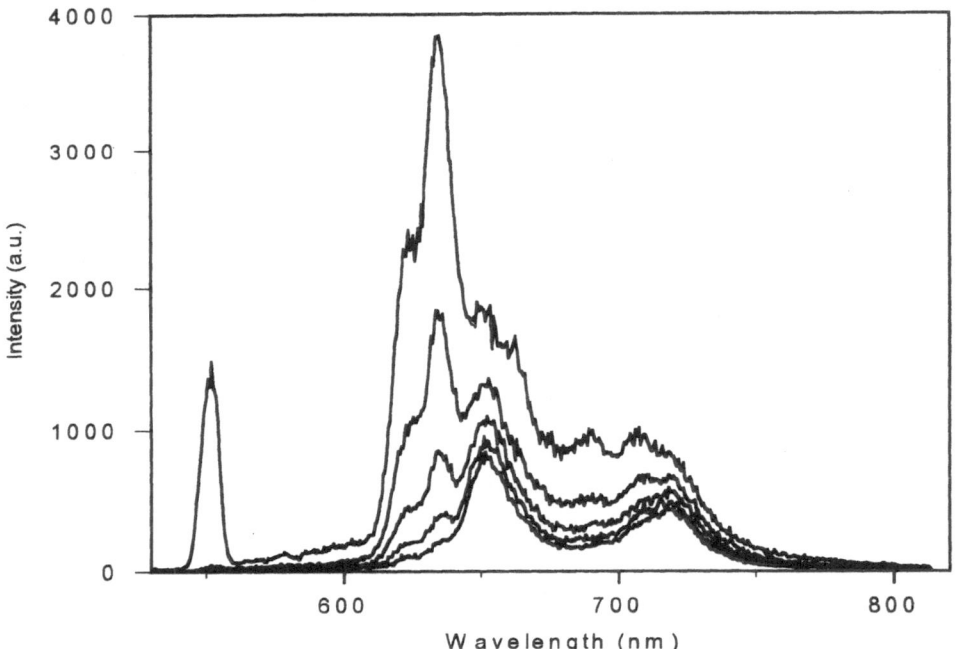

Figure 2. Time-resolved fluorescence spectra of the (ZnOEChl)2+H2P-(m^Pyr)2 triad in methylcyclohexane at 273 K (λex=550 nm, the integration range is 1ns, the time delay step is 1ns).

interactions of different nature. The red shift of Zn-porphyrin dimer electronic Q- and B-bands ($\Delta\nu \approx 500$ cm^{-1}) upon complexation with pyridyl containing extra-ligands is due to ligation effect which influences the relative position of HOMO's a_{1u} and a_{2u} according to the four-orbital model. The splitting of Zn-porphyrin dimer Soret B-bands ($\Delta E \geq 600$ cm^{-1}) as well as the appearance of additional red shifted spectral components ($\Delta\nu \approx 1900$ cm^{-1}) upon the formation of pentads depending on the spatial arrangement of interacting macrocycles are connected with excitonic coupling of resonant strong B-transitions of the dimer subunits. Observed experimental findings of splitting are in a good quantitative correspondence with theoretical values calculated using the point dipole approximation, quantum-chemical CNDO/S approach and a computer-simulated structures of the complexes under investigation.

The fluorescence spectra of triads and pentads of various composition and structure are dominated by an extra-ligand emission, while the Zn-dimer fluorescence is substantially quenched. This clearly indicates the non-radiative energy transfer (ET) from the Zn-dimer moiety to the free base. The dynamics of ET (see Fig.2) as well as quenching efficiency depend on the overlapping of donor emission and acceptor absorption spectra being the smallest one for pyridyl-substituted tetrahydroporphyrin as acceptor. Time-resolved fluorescence studies show an unresolved fast Zn-dimer emission decay, which means that singlet-singlet ET to free base subunits takes place within < 10 ps. The analysis of the fluorescence decay dynamics at various wavelengths of excitation and detection in the temperature range of 140-360 K as well as the examination of amplitude distribution for different decay components over the total range of solutions emission have shown that the backward energy transfer (within ~ 100 ps) activated thermally (activation energy ΔE = 950-1050 cm^{-1}) takes place in these complexes but does not influence essentially on the high efficiency of the direct ET from the Zn-dimer to an extra-ligand subunit.

For trimeric and pentameric arrays containing Cu-porphyrin as an extra-ligand the shortening fluorescence lifetime of the Zn-porphyrin or Zn-chlorin dimers from 1.2-1.7 ns to < 5 ps is due to two reasons: i) an exchange d-π effect of ligand central Cu ion (rising of intersystem crossing probabilities $F_{ISC} > 10^{12}$ s^{-1} in the subunits of the dimer) and ii) a thermally activated ($\Delta E = 460$ cm^{-1}) energy transfer to Cu-porphyrin.

It has been found that in complexes with (ZnHTPP)$_2$ no significant change of the decay times of extra-ligand components as compared to uncomplexed free bases is observed, while in complexes with (ZnOEP)$_2$ this fluorescence lifetime is reduced from 9 ns to 4... 6 ns (depending on the kind of free base used). We apt to believe that this effect is caused by a charge transfer ZnP-H$_2$P$^* \rightarrow$ ZnP$^+$-H$_2$P$^-$. This is in accordance with the fact that the oxidation potential of ZnTPP is 0.23 eV higher than that of ZnOEP which leads to estimations of the energy of the charge separated states of 1.90 eV and 1.66 eV in the case of (ZnTPP)$_2$ and (ZnOEP)$_2$ complexes respectively, whereas the H$_2$P excited state energy is 1.73 eV.

Financial support from Volkswagen-Stiftung (Nr. I/68 941) is gratefully acknowledged. Certain of the experiments have been partly supported by National Foundation for Basic Research of Belarus under Grant Nr. Ph 5-111 and by International Science Foundation (Grant Nr. RW 1300).

REFERENCES

1. G.I.Karvanos, *Fundamentals of Photoinduced Electron Transfer* (VCH Publishers, New York/Weinheim/Cambridge, 1993).

2. J.L.Sessler, B.Wang and A.Harriman, J. Amer. Chem. Soc. **117**, 704 (1995).

3. V.Balzani and F.Scandola, *Supramolecular Photochemistry* (Elis Horwood, New York/London/Toronto/ Sydney/Tokyo/Singapore, 1991).

4. V.S.Chirvonii, E.I.Zenkevich, R.Gadonas, V.Krasauskas and A.Pelakauskas, in *Laser Applications in Life Sciences*, SPIE vol. **1403**, part II, 638 (1991).

5. R.Bonnett, Chem. Soc. Rev. **24**, 19 (1995).

6. I.Salabert, T.-H.Tran-Thi, H.Ali, J.van-Lier, D.Houde and E.Keszei, Chem. Phys. Lett. **223**, 313 (1994).

7. A.K.Barrel, D.L.Officer and D.C.Reid, Angewandte Chem. Internat. Edition in English **33**, 900 (1995).

8. C. von Borczyskowski, U.Rempel, E.Zenkevich, A.Shul'ga and A.Chernook, J. of Mol. Struct. **348**, 441 (1994).

17

VIBRATIONAL COHERENCE IN PHOTOISOMERIZATION REACTION OF CIS-STILBENE IN SOLUTION

Dipak K. Palit,[1] Arpad Z. Szarka,[2] Nick Pugliano,[2] and Robin M. Hochstrasser[2]

[1] Chemistry Division
Bhabha Atomic Research Centre
Bombay 400 085, India
[2] Chemistry Department
University of Pennsylvania
Philadelphia, Pennsylvania 19104

An important goal of the study of chemical reaction dynamics in solution is to characterize reactive coordinates and understand how the solvent influences the reaction pathways and energy flow. The Photoisomerization of cis-stilbene, in which a significant structual change occurs without severing any chemical bond, is an ideal one for this purpose. The isomerization of cis-stilbene in the excited singlet state involves motion on a surface that is close to being barrierless along the reaction coordinate[1]. The time dependence of the disappearance of the electronically excited state has been shown to be sensitive to solvent and temperature[2]. Vibrational dephasing times longer than 200 fs, as can be estimated from the full Raman line width of 50 cm^{-1} for the lowest frequency (165 cm^{-1}) mode of the ground state[3], dictates the possibility of probing vibrational wave-packet dynamics on the reactive surface of photoisomerization reaction of cis-stilbene using femtosecond (fs) laser pulses. Such an experiment is made possible by the existence of a strong excited state absorption band of cis-stilbene in the 650 nm region which is spectroscopically isolated from any ground state absorption as well as the gain spectra[4].

In this communication we demonstrate that wave packet motion on the cis-stilbene reactive surface can be detected with ultrafast time resolution. The wave packet has been initially prepared by a fs laser pulse (60 fs, 330 nm) and the temporal evolution of the wave-packet has been probed by transient absorption measurements. The experimental set up has been described in detail else- where[5]. In brief, the 70 fs laser output from a colliding pulsed mode locked (CPM) dye laser was preamplified in a two stage dye amplifier pumped by a 20 Hz Nd:YAG laser and was launched into a single mode fiber. The resulting white light continuum was used to seed a three stage dye amplifier pumped by the same Nd:YAG laser. The amplifying medium which was a mixture of DCM and rho-

Ultrafast Processes in Spectroscopy, edited by Svelto et al.
Plenum Press, New York, 1996

damine B in methanol, produced laser pulses having bandwidth (FWHM) of 22 nm centered at 660 nm, which were compressed to 33 fs using a grating pair in double pass configuration. A portion of the 660 nm light was frequency doubled to create the 330 nm pump and the remaining part was used to generate continuum probe in ethylene glycol jet.

In fig. 1 a typical excited state absorption kinetics has been shown for the probe wavelength of 650 nm in hexadecane solvent. The transient absorption decay is clearly non-exponential and has been fitted with the following molecular response function (equation 1),

$$M(t) = A \exp(-t/\tau) + B \exp(-t/\tau_d) \cos(\omega t + \phi) \tag{1}$$

in which the first term describes the excited state population with initial amplitude (A) and decay (τ) and the second term characterises the oscillatory component with the initial coherence amplitude (B), dephasing time (τ_d), the oscillation frequency (ω) and the phase ϕ. The oscillatory part of the signal have been obtained from the experimental data subtracting off the portion of the signal due to cis* population dynamics. The fitted wave-packet frequencies were verified by Fourier Transforming the oscillatory portion of the signal. Transient kinetics have been studied for the two additional solvents, hexane and methanol and also two other probe wavelengths, 600 and 690 nm. The results of the fits are summerised in Table 1. The data were fitted with other molecular response functions (e.g. inclusion of more than one frequency or a time dependent frequency[6]) but none of them yielded better fits than those presnted here.

The life time of cis* in methanol (474 fs) is shorter than those in hexane (1.4 ps) and hexadecane (1.6 ps). The wave packet oscillation period is also shorter in methanol (270 fs) than in hexane (310 fs) or hexadecane (290 fs). This suggests that the coherently excited mode is other than the reactive mode of cis-trans isomerisation, i.e. the C = C bond

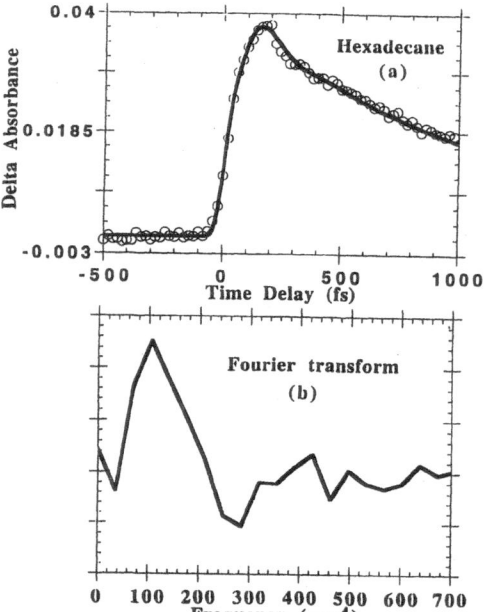

Figure 1. (a) Transient absorption signal of cis* in hexadecane probed at 650 nm. (b) Fourier Transform of the oscillatory portion of the transient absorption shown in (a).

Table 1. Fitting parameters for the experimental data set using the molecular response function defined in eq. 1

	Hexane			Hexadecane	Methanol
	600 nm	650 nm	690 nm	650 nm	650 nm
τ (fs)	1363	1363	1363	1632	474
τ_d (fs)	91	91	109	90	140
ω (cm^{-1})	107	107	107	113	125
ϕ (rad)	1.5	2.4	3.8	1.8	0.6

twist. The observed coherence involves a mode with frequency 107 - 125 cm^{-1}, depending on the solvent. The lowest frequency mode seen in resonance Raman is at 165 cm^{-1}, which is associated largely with the phenyl torsion in the ground state[3]. It therefore seems reasonable to associate the observed mode with this torsional motion. The lowering of the frequency by 20 - 40% on excitation could be a result of decoupling of the phenyl torsion from the skeletal motions making the motion closer to being a pure phenyl torsion in a bound potential.

The phase of the oscillation is also quite different in methanol compared to that in hexane, being close to zero (0.19 π) in methanol and close to π (0.76 π) in hexane for the 650 nm probe. This implies that a significant shift of the combining surfaces is introduced by solvent effects. The results of the probe wavelength dependent study show that by tuning the probe to lower frequencies, the wave packet moves into resonance with the probe frequency at later times as revealed by the gradual increase in value of phase from 600 to 690 nm (Table 1).

The vibrational dephasing times (τ_d) of the wave-packet on the excited cis-surface (~ 100 fs in hydrocarbons and 140 fs in methanol) are somewhat shorter than that (212 fs) for the 165 cm^{-1} mode of the ground state estimated from the full line width of the Raman line (50 cm^{-1} in cyclohexane)[3]. The more than two fold increase in the dephasing rate in the excited cis-stilbene is consistent with there being a new relaxation channel on excited state surface, most evidently due to coupling of the isomerization coordinate to the superposition state through intramolecular vibrational distribution. The time periods of oscillation (e.g. 290 fs in hexadecane) are only 2 - 3 times longer than the dephasing time for this coherent mode. Hence the vibrational energy relaxation occurs quickly on the excited cis surface which near the Franck-Condon maximum is steeply sloping towards the twisted intermediate[4].

Although the data presented here provide a remarkable amount of new informations about the excited cis surface, no definitive surface could be deduced from these limited informations. Further experiments involning the temperature dependence of the wave packet dynamics should be able to provide better insight into the potential energy surface of this photoisomerization reaction.

REFERENCES

1. D. C. Todd and G. R. Fleming, J. Chem. Phys. 98, 269 (1993).
2. S. Abrash, S. Repinec and R. M. Hochstrasser, J. Chem. Phys. 93, 1041 (1990).
3. A. B. Myers and R. A. Mathies, J. Chem. Phys. 81, 1552 (1984) .
4. R. J. Sension, S. T. Repinec, A. Z. Szarka and R. M. Hochstrasser, J. Chem. Phys. 98, 6291 (1993).
5. A. Z. Szarka, N. Pugliano, D. K. Palit and R. M. Hochstrasser, IQEC, Technical Digest, 9, 13 (1994).
6. S. Peddersen, L. Banares and A. H. Zewail, J. Chem. Phys. 97, 8801 (1992).

18

PROBING THE INTERMOLECULAR DYNAMICS OF LIQUIDS BY FEMTOSECOND SPECTROSCOPY

John D. Simon, Peijun Cong, and Yong Joon Chang

Department of Chemistry and Institute for Nonlinear Science
University of California at San Diego
La Jolla, California 92093-0341

Solvent-solvent and solvent-solute intermolecular forces play an important role in determining solution chemistry. We outline below some of our recent experimental efforts using femtosecond nonlinear spectroscopy to probe the intermolecular dynamics in liquids. Our approaches center on four-wave mixing techniques and thus the experimental signals probe the intermolecular potential through the third-order nonlinear susceptibility tensor. First, we examine results of femtosecond optically-heterodyne-detected Raman-induced Kerr Effect spectroscopy (OHD-RIKES) for a series of aromatic liquids. Second, we show that the OHD-RIKES spectrum is identical to that obtained from the frequency domain technique of stimulated gain spectroscopy. Finally, we present data for benzene collected using the technique of optically-heterodyne detected impulsive-stimulated Raman scattering (OHD-ISRS).

OHD-RIKES STUDIES OF AROMATIC LIQUIDS

Based on previous studies of several series of related liquids,[1] we have found that the shape of the spectral densities derived from OHD-RIKES measurements depend on both the anisotropic polarizability of the molecule as well as the moments of inertia about the rotational axes for the molecular motion that contribute to the experimental signal.[2] In particular, for a contributing rotational motion, an increase in the anisotropic polarizability leads to a broadening of the spectral density while an increase in the moment of inertia leads to a narrowing of the spectral density.

Here we discuss OHD-RIKES data collected for room temperature samples of benzene, benzonitrile, and o-methyl-benzonitrile. We focus on the spectral densities, which are determined from the time domain data using published Fourier transform procedures. In Figure 1(a), the librational spectral densities of benzene and benzonitrile are compared. Despite the dramatic differences between the polarity, viscosity, and various thermodynamic properties of these two liquids, the librational spectral densities are essentially iden-

Ultrafast Processes in Spectroscopy, edited by Svelto et al.
Plenum Press, New York, 1996

79

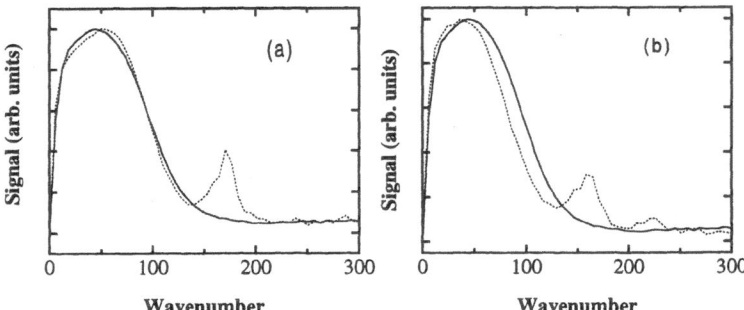

Figure 1. (a) The librational spectral densities of benzene (solid line) and benzonitrile (dotted line); (b) The librational spectral densities of benzene (solid line) and o-methyl-benzonitrile (dotted line).

tical. This observation can be explained by careful consideration of the rotational modes that contribute to the experimental signal. Because of the D_{6h} symmetry of benzene, only one rotational motion contributes to the OHD-RIKES measurement, namely, rotation around the in-plane axis. Benzonitrile has lower symmetry and now rotation around the in-plane axis that contains the CN moiety as well as rotation around the axis that passes through the center of the aromatic ring but is perpendicular to CN bond contribute to the experimental signal. Consider the axis along the CN bond. The moment of inertia for rotation around this axis is similar to that of benzene. We also expect the anisotropic polarizability about this axis to be nearly the same for these two aromatic molecules. For the axis perpendicular to the CN group, the anisotropic polarizability increases relative to benzene and should therefore be reflected by a broader spectrum. However, the moment of inertia about this axis is smaller than that for benzene and this should cause a narrowing of the spectral density. These compensating effects result in a spectral density that is similar to that of benzene.

In Figure 1(b), the librational spectral densities of benzene and o-methyl benzonitrile are compared. We see here that placing a methyl group at the ortho position of the benzonitrile results in a narrowing of the spectral density in comparison to benzene or benzonitrile. This is the expected result based on the above arguments. The major effect of the methyl group is to increase the moment of inertia, thereby causing a narrowing of the spectral density.

COMPARISON OF THE OHD-RIKES AND SGS DATA FOR BENZENE

The time-domain technique of OHD-RIKES is equivalent to the frequency domain technique of stimulated gain spectroscopy (SGS).[3] Therefore it is worth examining whether or not the spectral densities obtained by the Fourier transform of OHD-RIKES data quantitatively agree with that obtained from SGS. We first consider the connection between these two experiments (see Figure 2).

For SGS, the signal is given by

$$S_{SGS}(\Omega_1 - \Omega_2) \propto \mathrm{Im}_{xxyy}^{(3)}(-\Omega_2, \Omega_2, \Omega_1, -\Omega_1) \tag{1}$$

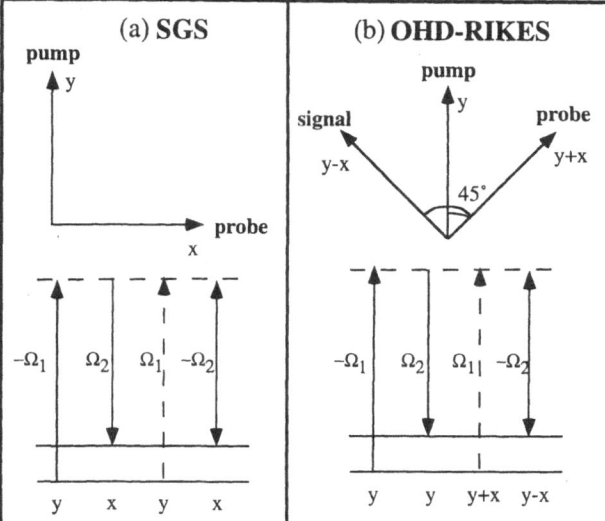

Figure 2. (a) The polarization directions of the pump and probe fields and the density matrix time evolution pathways for (a) SGS and (b) OHD-RIKES. Solid and dashed arrows correspond to the bra and ket side of the time evolution. Time increases from left to right. The polarization directions are labeled under each field arrow with respect to the coordinate system given for the particular experiment.

The frequency arguments are arranged in accord with the Maker-Terhune notation[4] ($\chi^{(3)}(-\Omega, \Omega_1, \Omega_2, -\Omega_3)$, $\Omega = \Omega_1 + \Omega_2 + \Omega_3$). The time-evolution pathway in Figure 2(a) shows that the vibrational excitation in SGS is accomplished by a sequential beating process by the pump and probe light fields.

For OHD-RIKES, the excitation is impulsive. The vibrational population is prepared coherently through the spectral width of the femtosecond pump light pulse. Thus, the first two fields and the last two fields shown in the time-evolution pathway of Figure 2(b) are derived from the pump and probe pulse, respectively. The signal is given by

$$S_{\text{OHD-RIKES}}(\Omega_1 - \Omega_2) \propto \chi^{(3)}_{yyyy}(-\Omega_2, \Omega_2, \Omega_1, -\Omega_1) - \chi^{(3)}_{xyxy}(-\Omega_2, \Omega_2, \Omega_1, -\Omega_1)$$
$$= \chi^{(3)}_{xyyx}(-\Omega_2, \Omega_2, \Omega_1, -\Omega_1) + \chi^{(3)}_{xxyy}(-\Omega_2, \Omega_2, \Omega_1, -\Omega_1) \tag{2}$$

The second equality is obtained by using the relationships between the three non-zero elements of the third-order susceptibility tensor for isotropic media. Because only the nuclear motions contribute to the time-resolved signal (except at t=0 where electronic contributions are observed),

$$\chi^{(3)}_{xxyy} = \chi^{(3)}_{xyyx} = \left(\chi^{(3)}_{yyyy} - \chi^{(3)}_{xyxy}\right) / 2 \tag{3}$$

and we see that OHD-RIKES and SGS measure the same component of $\chi^{(3)}$, namely, $\chi^{(3)}_{xxyy}$.

Figure 3 compares the librational spectral density of benzene obtained by the Fourier transform of OHD-RIKES data with that obtained from SGS. The agreement is quantitative. We have discussed the relative merits of these two techniques elsewhere and have shown that care must be exercised in calculating spectral densities from the time-domain OHD-RIKES data.[3] In many cases, the OHD-RIKES data underestimates the contributions from diffusive reorientational motions because of the limited time-delay commonly used in recording these data.

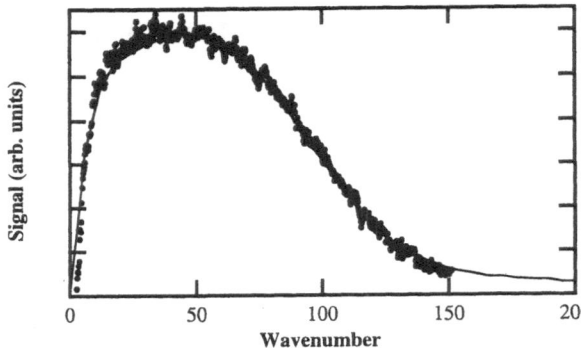

Figure 3. Comparison of the librational spectral density of benzene obtained from SGS (filled circles) and the Fourier-transform of the OHD-RIKES data (solid line).

OHD-ISRS STUDIES OF BENZENE

Equation (3) shows that for isotropic media there are two independent tensor elements of $\chi^{(3)}$. The experiments discussed so far only measure the difference between these two tensor elements, not the values of each individual element. In addition, the elements of $\chi^{(3)}$ are complex functions, and so we desire to separately measure the real and imaginary parts. To accomplish this goal we have implemented an optically-heterodyne detection scheme into a transient grating impulsive-stimulated Raman scattering experiment (OHD-ISRS).[5] By selecting the polarization of the pump and probe light fields used to generate the transient grating, a specific tensor element of $\chi^{(3)}$ can be probed. In the absence of heterodyne detection, the time-resolved scattering signal from this grating is proportional to the squared modulus of the tensor element (the homodyne signal). Introduction of a local oscillator that is in-phase or in-quadrature with the signal beam enables the selective measurement of the real and imaginary parts of a single tensor element, respectively.

Figure 4 shows the experimental data obtained for the $\chi^{(3)}_{yyyy}$ tensor element for room temperature benzene. Both the homodyne- and in-phase heterodyne-detected signals are

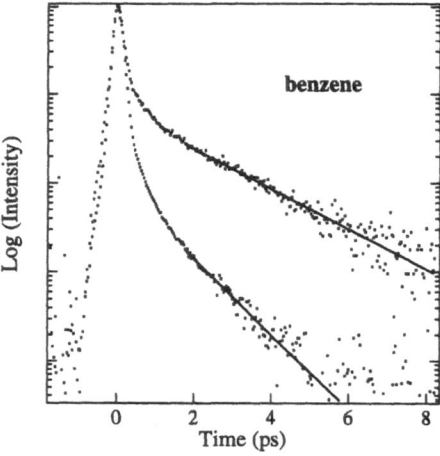

Figure 4. Comparison of the heterodyne- and homodyne-detected transient grating ISRS response for benzene.

plotted. The introduction of the local oscillator, the signal is linearly related to $\chi^{(3)}_{yyyy}$ and selectively measures the real part of this tensor element. This can be confirmed by an analysis of the time constant for the decay of the long-time tail of the experimental data. The signal at long times is determined by the rotational diffusion time of the solvent molecules. The tensor element at long time is then described by an exponential function, with a decay time equal to the rotational relaxation time. Because the homodyne experiment measures the square of the tensor element and the heterodyne experiment is linearly related to the tensor element, the decay times observed in these two experiments is expected to differ by a factor of two. The fits to the two data sets give decay times of 1.9 ± 0.1 ps and 1.0 ± 0.1 ps, which differ by the expected factor of two.

ACKNOWLEDGMENTS

We thank Professor C. Y. She for valuable discussions concerning the relationship between OHD-RIKES and SGS and Dr. Hans Deuel for assistance in the collection of the OHD-RIKES data. Financial support from the Chemistry Division of the National Science Foundation is greatly appreciated.

REFERENCES

1. (a) D. McMorrow, W. T. Lotshaw, and G. Kenney-Wallace, *IEEE J. Quantum Electron.* **QE-24**, 443 (1988); (b) Y. J. Chang and E. W. Castner, Jr., *J. Chem. Phys.* **99**, 7289 (1993); (c) Y. J. Chang and E. W. Castner, Jr., *J. Chem. Phys.* **99**, 113 (1993); (d) E. W. Castner, Jr., Y. J. Chang, Y. C. Chu, and G. E. Walrafen, *J. Chem. Phys.* **102**, 653 (1995); (e) D. McMorrow and W. T. Lotshaw, *J. Phys. Chem.* **95**, 10395 (1991); (f) M. Cho, M. Du, N. F. Scherer, G. R. Fleming, and S. Mukamel, *J. Chem. Phys.* **99**, 2410 (1993).
2. (a) H. P. Deuel, P. Cong, and J. D. Simon, *J. Phys. Chem.* **98**, 12600 (1994); (b) P. Cong, H. P. Deuel, and J. D. Simon, *Chem. Phys. Lett.* **240**, 72 (1995)
3. P. Cong, J. D. Simon, and C. Y. She, *J. Chem. Phys.*, in press.
4. M. D. Levenson and S. S. Kano, *Introduction to Nonlinear Laser Spectroscopy* (Academic Press, San Diego, 1988).
5. Y. J. Chang, P. Cong, and J. D. Simon, *J. Phys. Chem.* **99**, 7857 (1995).

19

PHOTODISSOCIATION DYNAMICS OF I_3^-

Comparison of Ultrafast Pump Probe Experiments to Raman Scattering

Guy Ashkenazi, Ronnie Kosloff, and Sandy Ruhman

Department of Physical Chemistry and the Fritz Haber Research Center
The Hebrew University
Jerusalem, Israel

When a UV light pulse is applied to I_3^- in solution, it either absorbs the light leading to fragmentation to $I_2^- + I$, or scatters the light usually as a Raman process. Since both the fragmentation and the Raman processes are governed by the same ground and excited potential energy surfaces, crossing the information gained from the two independent measurements leads to enhanced insight on the photodissociation event. A unified quantum computational scheme addresses simultaneously the dynamics of photodissociation and the absorption and Raman cross sections.

The photoinduced processes of the I_3^- system in different solvents has been the subject of extensive experimental and theoretical studies[1-4]. Of importance to this study are the coherent vibrations of the I_2^- probed by two different wavelengths, exhibiting antiphased modulations[1,2,5]. In conjunction with this process the leftover I_3^- reactant is also coherently excited due to the dynamical "hole" left on the ground surface[3,4]. This vibration can be correlated to the Raman spectra of I_3^- [6].

To allow numerical simulation of this complex experimental system, a simplified two dimensional model was used. The two degrees of freedom are the symmetric and antisymmetric stretch modes of I_3^-. The dynamics take place on a ground harmonic electronic potential, and on an excited LEPS potential[7]. The solvent effects are incorporated phenomenologicaly, through the use of the Liouville-Von Neuman equation which includes energy relaxation and dephasing processes[8].

The basic algorithem for simulating these processes is the Newtonian interpolation scheme with Leja interpolation points[9]. Given a function f, an operator $\hat{O}$ and an initial wavefunction ψ, the algorithem produces $f(\hat{O})\psi$. By casting the dynamics as well as the Raman and absorption spectra into such templates, all experimental observable can be extracted using this single algorithem.

All the processes under study have a common starting point i.e. the initial I_3^- wavefunction positioned on the ground electronic potential surface. These eigenstates are produced by operating the relaxation function

Ultrafast Processes in Spectroscopy, edited by Svelto et al.
Plenum Press, New York, 1996

$$f(\hat{\mathbf{H}}_{gr}) = e^{-\hat{H}_{gr}t} \tag{1}$$

where $\hat{\mathbf{H}}_{gr}$ is the ground surface Hamiltonian, on an arbitrary wavefunction[10].

To simulate the dynamics, the evolution function

$$f(\hat{\mathbf{O}}) = e^{-i\hat{O}\Delta t} \tag{2}$$

is applied to an initial state function. Different operators can be used for different propagation schemes, such as propagation under the influence of an electromagnetic field[11], free Hamiltonian evolution on the excited surface[12], and dissipative evolution using the Liouville - Von Neuman equation[8,13].

The excitation dynamics is demonstrated for a vibrationally excited initial state (with one quanta in the anti-symmetric stretch mode) induced by a Gaussian 60 fsec pump pulse.

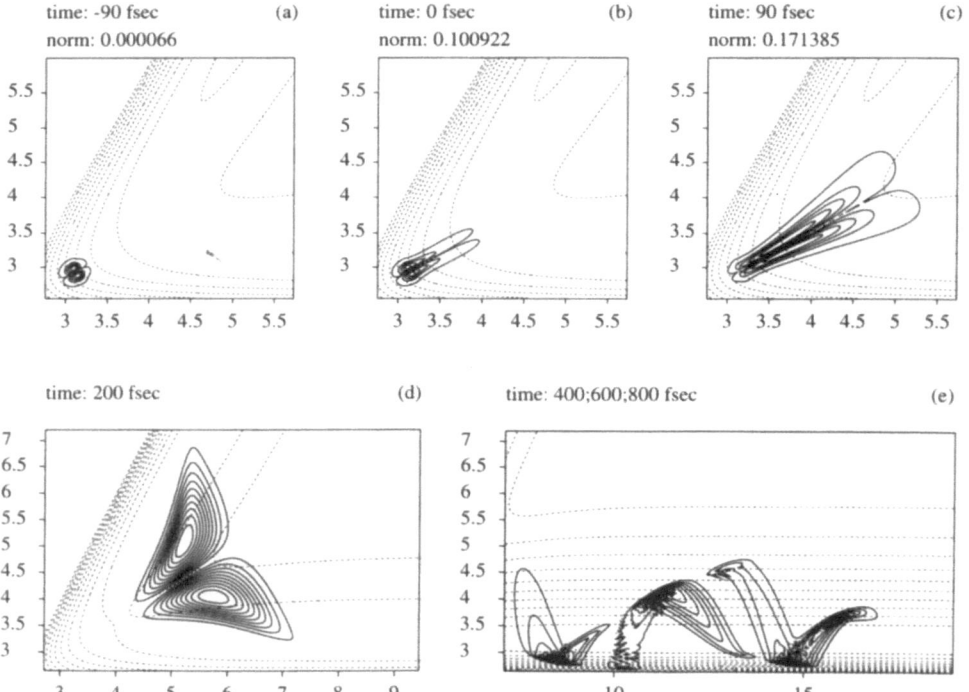

Figure 1. Snapshots of wavefunctions as contour maps on the excited electronic potential of I_3^-. (Distances are measured in angstroms in mass scaled coordinates.) **(a)** shows the wavefunction 90 fsec before the peak of the pump pulse. At this short time, the wavefunction is just a Frank - Condon projection of the ground surface eigenstate. **(b)** (0 fsec) shows the wavefunction at the instant of maximum intensity and **(c)** (90 fsec) just before the pulse is over. The wavefunction evolves predominantly along symmetric stretch direction throughout the pulse, while dynamics along the anti-symmetric direction is insignificant. After the end of the pulse, free Hamiltonian evolution is applied to the excited state wavefunction.. In **(d)** the wavefunction starts to branch among the exit channels, and in **(e)**. half of the wavefunction is vibrating coherently in one of the exit channels.

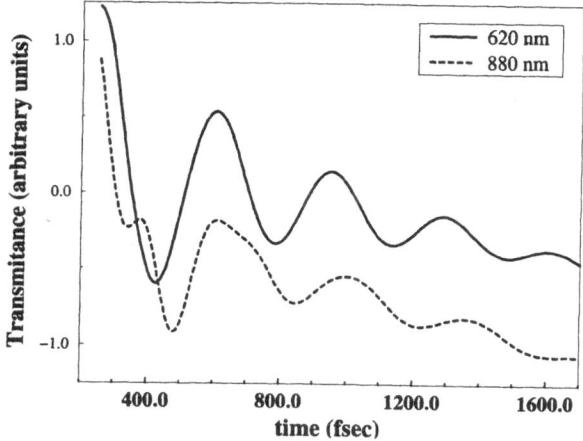

Figure 2. Transient absorption spectrum for I$_2^-$ for two probe wavelengths.

By combining the final wavefunctions with Boltzman weights a density operator is constructed. Applying dissipative evolution dynamics enables to simulate the observed transient spectrum[14].

The calculated spectrum clearly reproduces the decaying modulations seen in the experiment, and at the later times the anti-phased behavior of the two wavelengths.

The absorption and Raman spectrum are calculated using the absorption and Raman functions respectivly[15]:

$$f(\hat{\mathbf{O}}) = \int_{-\infty}^{\infty} e^{-i(\hat{O}-\omega)t} dt \tag{3}$$

$$f(\hat{\mathbf{O}}) = \int_{0}^{\infty} e^{-\left[i(\hat{O}-\omega)+G\right]t} dt \tag{4}$$

where g(t) is the decay factor of a brownian oscilator[6] for t < 75 fsec and g(t) = ∞ after that time.

The Raman function operator was applied to the first 16 ground surface eigenstates. Fig.3a. shows an example of such a Raman wavefunction originated from the ground state of I$_3^-$. The Raman spectrum was calculated by overlapping the resulting Raman wavefunctions with the first 60 eigenstates, finding the Raman cross section for each transition and than averaging for a temperature of 300K. Fig.3b. is the spectrum calculated from the Raman wavefunctions.

The peaks in the spectrum correlate with the fundamental frequency of the symmetric stretch mode (111 cm^{-1}) and its overtones. The fundamental frequency of the anti-symmetric stretch mode (143 cm^{-1}) is symmetry forbidden, but the absence of its overtones indicate that no significant dynamics take place along the anti-symmetric direction. This is clearly seen from the Raman wavefunction (Fig.3a.), as well as from the dynamical picture in Fig.1b.

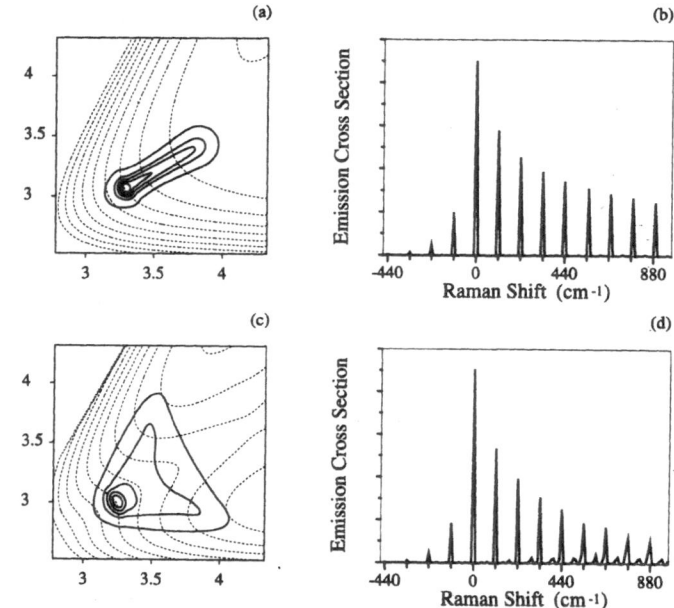

Figure 3. (a) Raman wavefunction and **(b)** Raman spectrum at 300K for the LEPS potential. **(c)** and **(d)** are the same for an ab-initio potential.

Recently, an ab-initio calculation of the I_3^- system was conducted by Yamashita[16], and preliminary simulations were made on the calculated ground and excited potential surfaces. The ab-initio potential differs greatly from the LEPS potential used in previous studies, and the early separation to the exit channels in this potential induce significant dynamics along the anti-symmetric direction. The change in dynamics is reflected in both the Raman wavefunction (Fig.3c.) and Raman spectrum (Fig.3d.). The appearance of the second overtone of the anti-symmetric stretch mode, and the rapid decay of the Raman cross section in higher shifts are both indicative of this type of dynamics.

To conclude, the unified time-dependent quantum computational tools are able to supply insight into the ultrafast dissociation dynamics of a complicated molecular system. Propagating the initial wavefunctios supplies the absorption spectrum determined by the very short time dynamics, the Raman spectrum sensitive to slightly longer times and transient spectra of the products. The phase of the transient spectra is found to be strongly influenced by the bifurcation dynamics of the excited wavefunction into the product channel. This combination of experiment and theory is therefore able to illustrate a nascent chemical event.

REFERENCES

1. U. Banin, A. Waldheim and S. Ruhman, J. Chem. Phys., **96**, 2416 (1992).
2. U. Banin, R. Kosloff and S. Ruhman, Israel. J. Chem., **33**, 141 (1993).
3. U. Banin and S. Ruhman, J. Chem. Phys., **98**, 4391 (1993).
4. A. Bartana, U. Banin, S. Ruhman and R. Kosloff, Chem. Phys. Lett., **229**, 211 (1994).
5. N. Pugliano, D. K. Palit, A. Z. Szarka and R. M. Hochstarsser, J. Chem. Phys., **99**, 7273 (1993).

6. A. E. Jhonson and A. B. Myers, J. Chem Phys., **102**, 3519 (1995).
7. I. Benjamin, U. Banin and S. Ruhman, J. Chem. Phys., **98**, 8337 (1993).
8. M. Berman, R. Kosloff and H. Tal-Ezer, J. Phys. A, **25**, 1283 (1992).
9. G. Ashkenazi, R. Kosloff, S. Ruhman and H. Tal-Ezer, J. Chem. Phys., **103** (1995).
10. R. Kosloff and H. Tal-Ezer, Chem. Phys. Lett., **127**, 223 (1986).
11. U. Banin, A. Bartana, S. Ruhman and R. Kosloff, J. Chem. Phys., **101**, 8461 (1994).
12. R. Kosloff, Annu. Rev. Phys. Chem., **45**, 145 (1994).
13. A. Bartana R. Kosloff and D. Tannor, J. Chem. Phys., **99**, 196 (1993).
14. R. Kosloff, G. Ashkenazi, S. Ruhman, *Femthochemistry, The Lussane Conference,* ed. by M. Chergui (1995).
15. R. Kosloff, J. Phys. Chem., **92**, 2087 (1988).
16. K. Yamashita, G. Ashkenazi, R. Kosloff and S. Ruhman, in preperation.

20

INTRALIGAND CHARGE-TRANSFER AS A TRIGGER FOR CATION PHOTORELEASE

Monique M. Martin, Pascal Plaza, and Yves H. Meyer

Laboratoire de Photophysique Moléculaire du CNRS
Bâtiment 213, Université de Paris-Sud
91405 Orsay, France

INTRODUCTION

Many physiological functions are controlled by fluctuations of intracellular free Ca^{2+} concentration. Photochemical systems able to produce concentration jumps of cations would be of interest for the study of intracellular reaction dynamics. The possibility of using DCM-crown for photorelease of cations was investigated in acetonitrile solutions[1]. Key points of the work are reported here.

DCM-crown was designed by J. Bourson and B. Valeur[2] for optical detection of cations. This compound consists of the laser dye DCM where the amino group has been replaced by an aza-ether-crown group able to interact with metal ions (M^{n+}) as shown on the scheme.

DCM-crown

In polar solvents DCM exhibits an intense and strongly red-shifted fluorescence band attributed to a highly polar excited state[3,4]. The photophysics of DCM-crown is very similar to that of DCM but upon complexation with a cation its absorption spectrum is blue-shifted and its fluorescence is quenched although spectrally quasi-unshifted[2]. Photoinduced intraligand charge-transfer[5,6] is expected to occur in DCM-crown-cation complexes and lead to the repulsion of the cation. The results obtained by picosecond transient spectroscopy suggests cation release in the locally excited state of the complex, followed by further intraligand charge transfer most likely accompanied by ejection. Similar studies were carried out for some crowned stilbene derivatives[7].

Ultrafast Processes in Spectroscopy, edited by Svelto et al.
Plenum Press, New York, 1996

EXPERIMENTAL

Transient absorption and gain spectra were measured by the pump-probe technique using a 0.7ps, 15–30μJ pump pulse at 425nm and a continuum probe. The white-light probe was produced by focusing a 0.3ps, up to 200μJ, 710nm pulse in a 1cm H_2O cell. The 425nm and 710nm subpicosecond beams were generated by a cheap all-dye laser system, according to unconventional methods[8]. The whole system is driven by a standard seeded 10Hz Q-switched Nd:YAG laser delivering smooth 6 ns pulses at 532 and 355nm.

Photoinduced Intramolecular Charge-Transfer in DCM

The time-resolved transient absorption and gain spectra (ΔD) of DCM in tetrahydrofuran (THF) are given in Fig. 1 (left). Within 5ps after excitation, the spectra show a temporary isosbestic point at 580nm while new transient absorption maxima rise at 420 and 540nm, suggesting the fast formation of a transient state. The observation of a temporary isosbestic point for a negative ΔD where the ground state (D_0) does not absorb indicates that at this wavelength both the locally excited state and the transient state have a dominant stimulated emission cross section, i.e. are both emissive.

In a very polar solvent such as acetonitrile (Fig. 1, right), the 420nm and 540nm bands are already observed during the excitation pulse showing that the transient state is populated in less than 1ps. Furthermore the gain maximum is shifted to the red indicating that the transient state is polar. We thus attribute the photoinduced process to the rapid formation of an emissive intramolecular charge-transfer (CT) state from the emissive locally excited (LE) state of DCM. In femtosecond up-conversion experiments, H. Zhang et al observed[7] that 50% of the integrated fluorescence intensity of DCM in methanol is lost in a few picoseconds and concluded in favor of a model involving strong adiabatic coupling between the LE and CT states, the charge separation dynamics being controlled by the solvation dynamics.

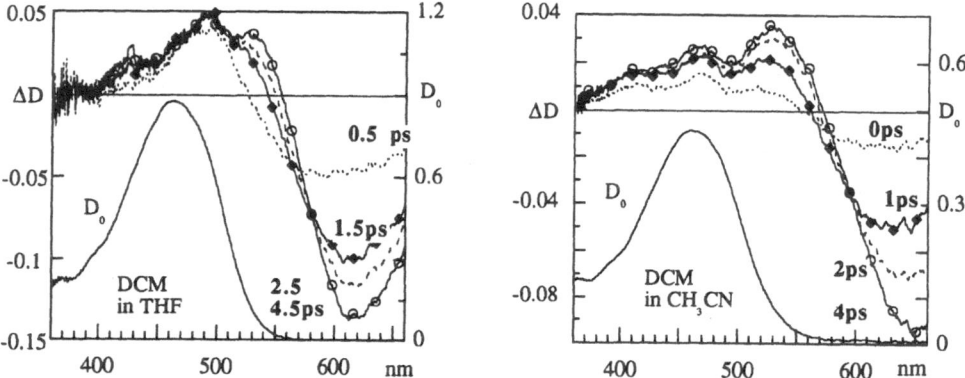

Figure 1. Transient absorption and gain spectra (ΔD) of DCM in tetrahydrofuran (left) and in acetonitrile (right). Unexcited sample absorption (D_0).

Cation Photorelease from DCM-crown-Ca^{2+} Complexes

In acetonitrile, the time-resolved transient spectra of DCM-crown are very similar to those of DCM indicating that efficient photoinduced intramolecular charge transfer occurs also in the crowned derivative[1].

The transient absorption and gain spectra of DCM-crown fully complexed with Ca^{2+} (0.05M) in acetonitrile are given in Fig. 2 (left). It is seen that the initial ΔD band shows a broad absorption peak around 470nm, a gain band above 550nm and some bleaching around 390nm. Then, in the same way as for DCM or DCM-crown but at a slower rate, transient absorption increases below 430nm and around 540nm, and gain increases in the red. The whole process takes about 30ps and temporary isosbestic points are observed at 510nm and 570nm. We thus attribute the delayed growth of the 540nm differential band and the disappearance of the bleaching below 430nm to the formation of the CT state of the free ligand.

Excited-State Mechanism

At very short delays, the time-resolved differential spectra of DCM in THF (Fig. 1 left, 0.5ps) and of DCM-crown-Ca^{2+} complexes in acetonitrile (Fig. 2 left, 2ps) show a broad transient absorption band in the 450–520nm wavelength range although their ground state absorption spectra are much shifted from each other. We checked whether the same initially excited state spectrum $\varepsilon_g (\lambda)$ could be extracted from these early transient spectra by assuming that in both cases only the initially excited state and the ground state are present in the solution. The extraction was done by using Eq. 1:

$$\varepsilon_g (\lambda) = \{\Delta D (\lambda) . C/C^* + D_0\} / CL \tag{1}$$

where eg represents the difference between the excited state extinction and stimulated emission coefficients, D0 the optical density of the unexcited sample, L the cell length and C/C* the ratio of the total solute concentration to that of the excited solute. This could not be done for DCM or DCM-crown in acetonitrile due to the ultrafast formation of the CT state.

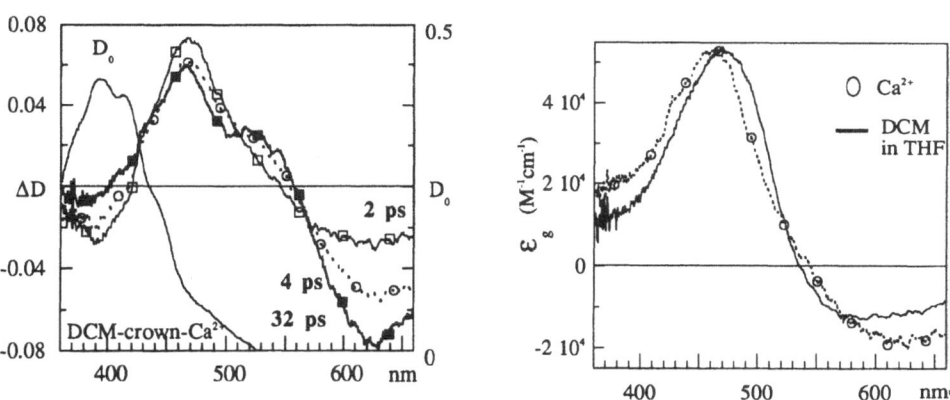

Figure 2. Left: transient absorption and gain spectra (ΔD) of DCM-crown-Ca^{2+} complexes in acetonitrile and un-excited sample absorption (D_0). - Right: Extinction coefficients ε_g of the initially probed excited state of DCM and DCM-crown-Ca^{2+} complex, calculated from Eq. 1.

Very similar ε_g spectra, with a broad maximum of ~47000 $M^{-1} cm^{-1}$ around 470 nm, could be found (Fig. 2, right) for both DCM in THF and DCM-crown-Ca^{2+} in acetonitrile, for values of C^*/C of 0.2 and 0.15 respectively. The spectra are just slightly red-shifted from each other. We obtained very similar results with DCM-crown-Li^+ [1].

This analysis suggests that the species initially probed is the same in all cases, namely the LE state of DCM-crown or DCM. One thus proposes that photodisruption of the binding between the cation and the crown nitrogen occurs in the LE state of the complex, leading to a loose excited complex with an absorption spectrum close to that of the free ligand. This process is followed by intraligand charge transfer but at a slower rate than in the free ligand. This mechanism is supported by Marguet et al.'s theoretical investigation[9] which foresees that the charge on the nitrogen atom becomes slightly positive in the LE state of the trans-planar DCM. On the other hand, Zhang et al[6] reported for DCM (in methanol) that the transition from the LE state to the CT state corresponds to an increase of the excited state CT character from 70% to 100%, indicating the strong CT character of the LE state. One may think that cation ejection to the bulk accompanies the transition to the highly polar CT state since the formation rate of the free ligand CT state rate was found to be slower for the Ca^{2+} complex than for the Li^+ complex[1], by a factor 15 at most, which can be understood in terms of higher charge density of Ca^{2+} and a better fit to the cavity size of the crown. Unfortunately, we could not evidence the excess of free ligand resulting from the photoejection process on the transient spectra measured 20ns after excitation. Bleaching was found to dominate the 20ns transient spectra of solutions containing Ca^{2+} concentration low enough to prevent fast diffusion-controlled recombination with the free ligand ground state after excited state relaxation (2ns lifetime). This bleaching is most likely due to trans-cis photoisomerization which is known to compete with intramolecular charge transfer in DCM[4].

ACKNOWLEDGMENT

This research was supported by the GDR 1017 of CNRS, France.

REFERENCES

1. a) M. M. Martin, P. Plaza, N. Dai Hung, Y. H. Meyer, J. Bourson, B. Valeur, Chem. Phys. Lett., **202**, 425 (1993). -b). M. M. Martin, P. Plaza, Y. H. Meyer, L. Bégin, J. Bourson, B.Valeur, J. of fluor., **4**, 271 (1994). - c) M. M. Martin, P. Plaza, Y. H. Meyer, F. Badaoui, J. Bourson, J. P. Lefevre, B. Valeur, J. Phys. Chem. submitted.

2. J. Bourson, B. Valeur, J. Phys. Chem. **93**, 3871.(1989).

3. M. Meyer, J. C. Mialocq, Opt. Commun. **64**, 264 (1987).

4. W. Rettig, W. Majenz, Chem. Phys. Lett. **154**, 335 (1989).

5. M. M. Martin, P. Plaza, Y. H. Meyer, Chem. Phys. **192**, 367 (1995)

6. H. Zhang, A. M. Jonkman, P. Van der Meulen, M. Glasbeek, Chem. Phys. Lett. **224**, 551 (1994).

7. P. Dumon, G. Jonusauskas, F. Dupuy, Ph. Pée, C. Rullière, J. F. Létard, R. Lapouyade, J. Phys. Chem. **98**, 10391 (1994).

8. N. Dai Hung, P. Plaza, M. M. Martin, Y. H. Meyer, Appl. Opt. **31**, 7046 (1992).

9. S. Marguet, J. C. Mialocq, P. Millie, G. Berthier, F. Momicchioli, Chem. Phys. **160**, 265 (1992).

21

COHERENT PHONONS IN FULLERITES UNDER FEMTOSECOND LASER EXCITATION

A. L. Dobryakov,[1] N. P. Ernsting,[2] V. M. Farztdinov,[1] S. A. Kovalenko,[2]
V. S. Letokhov,[1] Yu. E. Lozovik,[1] G. Marowsky,[3] and Yu. A. Matveets[1]

[1] Institute of Spectroscopy
Russian Academy of Science
Troitsk, Moscow Region, Russia
[2] Max - Planck - Institut für Biophysikalische Chemie
Göttingen, Germany
[3] Laser- Laboratorium
Göttingen, Germany

1. INTRODUCTION

The C_{60} molecules form in the solid state a semiconductor crystal with fcc Bravais lattice at room temperature. The upper group of valence bands is formed by the h_u bands and the lower group of conduction bands, by the t_{1u} bands. Experimental studies of the optical properties of C_{60} crystals have shown that the energy gap is in the range 1.6–2 eV. The optical transition $h_u \rightarrow t_{1u}$ (HOMO - LUMO transition) is dipole-forbidden in the C_{60} molecule. Weak absorption in this region results from the vibronic mixing of states. In solid phase mixing of states even increases at the expense of interaction of charge carriers with a crystal field.

A series of studies [1–7] of the temporal characteristics of relaxation processes in fullerites has been published. It was revealed that optical response of solid C_{60} films has complex temporal dynamics on femtosecond and longer time scales. The data on relaxation times and the mechanisms, suggested to contribute to the dynamics of relaxation process, presented in this works differ greatly. The investigation of the spectral dependence of relaxation properties is important for revealing the physical origin of relaxation processes.

In the present work the technique of femtosecond laser spectroscopy has been used to study the temporal variation of the optical density of C_{60} film over the spectral range 1.78–2.34 eV. The data obtained were used to investigate the dependences of ultrafast relaxation time and of the spectrum of excited coherent oscillations on the probing wavelength, to reveal the charge-carrier relaxation mechanisms and the role played by intra- and intermolecular vibrations in the electron-phonon interaction at the various stages of the relaxation process.

Ultrafast Processes in Spectroscopy, edited by Svelto et al.
Plenum Press, New York, 1996

2. EXPERIMENTAL TECHNIQUE AND EXPERIMENTAL RESULTS

The sample under study was C_{60} film on a quartz substrate. It was excited with optical pulses with the duration of ≈ 50 fs and a photon energy of $\hbar\omega_{pu} = 2.58$ eV. The exciting pulse intensity was $\sim 2.6 \times 10^{12}$ W/cm^2. The exciting beam spot on the sample was 100 µm across. Probing was effected in the range 1.78–2.34 eV with the help of ≈ 50 fs probing pulses with wide spectrum. The probing beam spot was 50 µm across. The optical delay between the exciting and probing pulses was varied with constant step of 27 fs. The maximum delay time reached 4 ps. The repetition frequency of the exciting and probing pulses amounted to some 1 Hz. Throughout the experiment, we have controlled the optical density of the samples and have not found any irreversible changes in it.

The absorption of pumping photons (in C_{60} film it corresponds to $h_u \rightarrow t_{1u}$ and $h_u \rightarrow t_{1g}$ transitions) caused the darkening of the sample in the entire spectral range 1.78–2.34 eV. The amount of darkening at a fixed positive time delay was found to depend on the probing photon energy, reaching its maximum near $\hbar\omega \approx 1.8$ eV. Approximating the experimental dependencies $\Delta D_{exp}(t)$ by the two-exponential fitting functions $\Delta D_{fit}(t)$ we obtained that the shortest relaxation time τ_1 depends on probing photon energy: its value is $\tau_1 = 370\pm70$ fs in the region $1.78 < \hbar\omega < 2.1$ eV and $\tau_1 = 540\pm90$ fs in the region $2.1 < \hbar\omega < 2.35$ eV (see Fig.1). The longer relaxation time τ_2 was estimated as $\tau_2 = 30\pm10$ ps.

The Fourier-analysis of temporal dependence of the optical density variation $[\Delta D_{exp}(t) - \Delta D_{fit}(t)]$ was used for the determination of excited lattice and intramolecular coherent vibrations in the field of wavenumbers from 10 to 400 cm^{-1} with the resolution of about 10 cm^{-1}. It is found out, that the excitation of coherent phonons occurs in the entire spectral area of probing $\hbar\omega_{pr} = 1.78 \div 2.34$ eV. The Fourier-intensity distribution of coherent phonons depended on the energy of probing. For example, in the low-energy region the vibrations of modes with wavenumbers ≈ 27 and 56 cm^{-1} are dominant, while in the high-energy region the vibration with frequency of 105 cm^{-1} is observed with maximum intensity. For the determination of coherent phonons, which are effectively excited through the whole spectrum region under study, the mean-geometrical Fourier- spectrum of vibrations (see Fig. 2) was calculated. As a result it was found out, that in the sample the vibrations with wavenumbers ~ 27, 56, 90, 125, 194, 290, and 357 cm^{-1} (dominant in the given spectral area) are excited. Vibrations of 145, 252, 325, 386 cm^{-1} and a number of other modes are observed with lower amplitude. In the Fourier-spectra we observed not only known[8] lattice and intramolecular phonons (27, 56, 90, 145 cm^{-1} - lattice modes; 252,

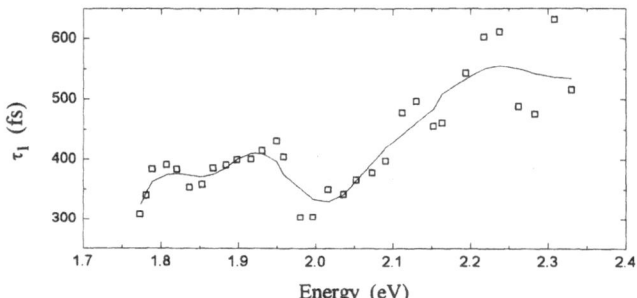

Figure 1. The spectral dependence of relaxation time τ_1.

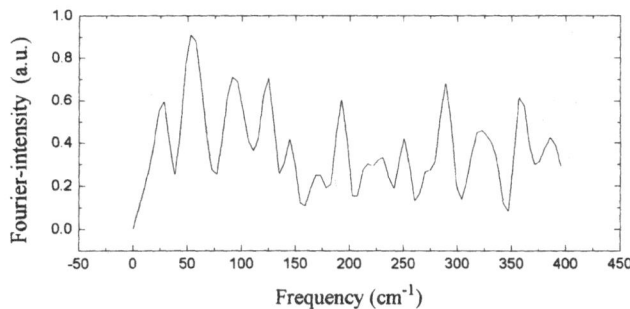

Figure 2. The mean-geometrical spectrum of vibrations.

290 cm^{-1} - splitted $H_g(1)$ intramolecular mode), but also modes[9,10], associated with dimerized and/or polimerized state of C_{60} : 125, 325 and 386 cm^{-1}.

3. DISCUSSION

The dependence of the spectrum of relaxation time τ_1 as well as of the spectrum of coherent phonons from the probing photon energy specifies that in these processes participate electrons of various groups (in the region under investigation there are the $h_u \rightarrow t_{1u}$, $h_g + g_g \rightarrow h_u$, and $t_{1u} \rightarrow h_g$ probing photon transitions). The main contribution to the photo-darkening in the spectral region $\hbar\omega_{pr} \approx 1.78 \div 2$ eV comes from the $h_g + g_g \rightarrow h_u$ transitions and in the spectral region $\hbar\omega_{pr} \approx 2 \div 2.34$ eV comes from the $t_{1u} \rightarrow h_g$ transitions. Therefore in the first region the relaxation of holes in the h_u band and coherent vibrations strongly interacting with these carriers is observed. In the second region the relaxation of electrons in the t_{1u} band and coherent vibrations strongly interacting with electrons is observed. The important role is played also by the dependence of relaxation rate on the position of carriers in the particular band relatively to its edge. The mechanisms responsible for this dependence could be the decrease of phase volume of final states and switching off the relaxation channel with emitting one of high frequency intramolecular oscillations with decrease of excitation energy.

We consider that the temporal evolution on the picosecond time scale results from the relaxation of charge carriers involving intra- and intermolecular vibrations. At first, both high-frequency a_g and h_g intramolecular vibrations and low-frequency lattice phonons would be emitted. The cooling of carriers will cause the filling of states near the bottom of corresponding band. Some part of the carriers will form singlet excitons and some other part will be trapped by structural defects, whose concentration may be rather high. When the energies of the charge carriers are insufficient for the emission of intramolecular vibrations, the emission of lattice optical phonons would become the primary relaxation mechanism. Correspondingly, the relaxation time would increase significantly.

The process of absorption on $h_u \rightarrow t_{1u}$ transition (for the removal of the forbiddeness on parity) occurs with radiation of lattice and intramolecular oscillations of odd symmetry. These oscillations will be coherent, if for excitation an optical pulse with a duration shorter than half-cycle of lattice or intramolecular oscillations was used[11]. Besides, the final state is unstable in relation to Jahn - Teller transitions with radiation of intramolecular oscillations with H_g symmetry. Hence, in given experiment the coherent vibrations of H_g modes can be observed also.

The increase of population of excited states results in an increase of molecular susceptibility, that, in turn, results in strengthening of intermolecular attraction. The increase of attraction decreases the intermolecular distance. As a result the formation of a new nonequilibrium crystal phase with smaller lattice parameter, the formation of dimers (lattice of dimers), and even the nonequilibrium polymerization[9] of lattice becomes possible in the excited state. During transition into this state radiation of low-frequency phonons will occur: optical lattice phonons and librons in the first case, as well as vibrations of dimers (polymers) in the second case. The nonequilibrum state can exist only during a rather limited period of time, as far as it will be destroyed[12] by acoustic phonons emitted during the process of charge carriers energy relaxation.

The density of absorbed photons in a film corresponds about to two photons per a molecule. Hence, the low-frequency phonons observed by us can be connected to both abovementioned phases of a fullerite in an excited state. The approximately three times decrease of the density of absorbed photons leads to the suppression[13] of the line at 125 cm^{-1} — the suppression of the transition into new nonequilibrium dimerized/polymerized state.

4. CONCLUSION

We have studied experimentally the temporal variation of the optical density of a C_{60} film over a wide spectral range - from 1.78 eV to 2.34 eV. We have investigated a nature of excited state and a spectrum of excited coherent phonon oscillations and its dependence on the probing photon energy. The temporal dynamic on the time scale of several picoseconds and shorter is due to the relaxation of the charge carriers interacting with intra- and intermolecular vibrations. The discovered spectral dependence of the ultrafast relaxation time (and of the excited coherent phonon oscillations) has demonstrated the possibility of simultaneous observation of the relaxation of different groups of excited carriers (created by the same exciting pulse). The formation of new nonequilibrium phase in strongly excited solid C_{60} was observed.

5. ACKNOWLEDGMENT

This work was supported by the Russian Foundation of Basic Research, International Science and Technology Center, Scientific Program "Fullerenes and Atomic Clusters".

6. REFERENCES

1. M.J. Rosker, H.O. Marcy, T.Y. Chang, J.T. Khoury, K. Hansen, and R.L. Whetten, Chem.Phys.Lett. **196**, 427 (1992).
2. R.A. Cheville, N.J. Halas, Phys.Rev. B45, N 8, 4548- 4550, 1992.
3. S.L. Dexheimer, M.D. Mittlemann, R.W. Schoenlien, W. Vareka, X.-D. Xiang, A. Zettl, C.V. Shank, in *Ultrafast Phenomena VIII*, ed's J.-L.Martin, A.Migus, G.A.Mourou and A.H.Zewail, (Springer-Verlag, Berlin, 1992); S.L. Dexheimer, W. Vareka, M.D. Mittlemann, A. Zettl, C.V. Shank, Chem.Phys.Lett. **235**, 552 (1995).

4. S.D. Brorson, M.K. Kelly, U. Wenschuh, R.Buhleier, J.Kuhl, Phys.Rev. **B46**, 7329 (1992).

5. S.B. Fleischer, E.P. Ippen, G Dresselhaus, M.S. Dresselhaus, A.O. Rao, P. Zhou, P.C. Eklund, Appl. Phys. Letters, **62**, 3241 (1993).

6. I.E. Kardash, V.S. Letokhov, Yu.E. Lozovik, Yu.A. Matveets, A.G. Stepanov, V.M. Farztdinov, Pis'ma ZhETF, **58**, 134 (1993); V.M. Farztdinov, Yu.E. Lozovik, Yu.A. Matveets, A.G. Stepanov, and V.S. Letokhov, J. Phys. Chem., **98**, 3290 (1994).

7. T.N. Thomas, R.A. Tyalor, J.F. Ryan, D. Mihailovic, and R. Zamboni, Europhys. Lett. **25**, 403 (1994).

8. P.H.M. van Loodsdrecht, P.J.M. van Bentum, M.A.Verheijen, and G.Meijer, Chem. Phys. Lett. **198**, 587 (1992); L.Pintschovius, B.Renker, F.Gompf, R,Heid, S.L.Chaplot, M.Haluska and H.Kuzmany, Phys.Rev.Lett. **69**, 2662 (1992). C.Coulombeau, H.Jobic, P.Bernier, C.Fabre, D.Schutz, and A.Rassat, J.Phys.Chem. **96**, 22 (1992).

9. A.M.Rao, P.Zhou, K.-A.Wang, G.T.Hager, J.M.Holden, Y.Wang, W.-T.Lee, X.-X. Bi, P.C.Eklund, D.S.Cornett, M.A.Duncan, and I.J.Amster, Science, **259**, 955 (1993); W.S.Bacsa, J.S.Lannin, Phys.Rev. 49, 14750 (1994).

10. M.Menon, K.R.Sabbaswamy, M.Sawtarie, Phys.Rev. **B49**, 13966 (1994).

11. W.Kütt, W.Albreht, and H.Kurz, IEEE Journ. of QE, **28**, 2434 (1992).

12. Y.Wang, J.M.Holden, X.-x.Bi, and P.C.Eklund, Chem.Phys.Lett. **217**, 413 (1994).

13. A.L.Dobryakov, V.M.Farztdinov, S.A.Kovalenko, Yu.E.Lozovik, Yu.A.Matveets, et. al., to be published.

22

SPONTANEOUS RAMAN SCATTERING WITH PICOSECOND PULSES

H. Graener,[1] R. Zürl,[2] and M. Hofmann[2]

[1] Martin-Luther-Universität Halle
Fachbereich Physik
06108 Halle, Germany
[2] Universität Bayreuth
Physikalisches Institut
95440 Bayreuth, Germany

The first information about vibrational relaxation processes in liquids was obtained by observing the time evolution of the spontaneous anti-Stokes Raman scattering from vibrational states with a non equilibrium population[1]. Due to the small Raman scattering cross sections and the involved relaxation time constants these experiments are rather inefficient and require a picosecond time resolution. With the increasing performance of ultrashort laser systems and detection setups it is now possible to use the whole potential of this technique especially to monitor the time evolution of complete (anti-Stokes) Raman spectra with sufficient resolution[2]. This allows to observe all Raman active vibrational states which carry a transient excess population. Exciting one single vibrational mode via resonant IR absorption rather complex relaxation pathways can be identified and analysed.

A theoretical description of such an experiment, which takes into account different polarization states of the involved optical pulses, shows, that the reorientational motion influences the observed Raman transients. Such an analysis starts with a separation of the contributions of population and orientation induced by the IR pump pulse by a development of the angular distribution in Legendre polynomials $P_l(\cos\vartheta)$, where ϑ is the angle between the IR polarization and the molecular induced dipole moment[3]:

$$N_i(\Omega, t) = \frac{N_i(t)}{4\pi}\left(1 + \sum_{l=2}^{\infty} c_l(t) P_l(\cos\vartheta)\right)$$

(1)

$N_i(t)$ describes the time evolution of the vibrational population of a state i and $c_l(t)$ contains the information about the reorientational processes. Probing such an anisotropic distribution by anti-Stokes Raman scattering the result will depend on the polarization conditions chosen for the experiment and the direction of the induced dipole with respect to the Raman polarizability tensor. (In the following it is assumed, that the dipole coin-

Ultrafast Processes in Spectroscopy, edited by Svelto et al.
Plenum Press, New York, 1996

101

cides with one principal axis.) Defining the observation direction of the Raman light as y with the monochromator slit direction as z, and using pulses propagating in the xy plane two important polarization configurations are discussed: for pump and probe pulse parallel in z - direction one calculates:

$$I^{par}(t) \propto N_i(t)\left[\left(45a^2+7\gamma^2\right)-2c_2(t)\left(6a\gamma-\frac{11}{14}\gamma^2\right)+c_4(t)\frac{4}{7}\gamma^2\right]$$

(2)

where a and γ are the invariants of the Raman polarizability tensor. Choosing the IR excitation polarization in the xy - plane, one obtains:

$$I^{per}(t) \propto N_i(t)\left[\left(45a^2+7\gamma^2\right)+c_2(t)\left(6a\gamma-\frac{11}{14}\gamma^2\right)+c_4(t)\frac{3}{7}\gamma^2\right]$$

(3)

For both situations an unpolarized detection is assumed. The leading term in both cases is well known from conventional Raman spectroscopy. A quick look at Eq. 2 and 3 shows that for highly polarized Raman lines ($\gamma = 0$) the observable anti-Stokes Raman intensity of a specific vibration is only influenced by the population but not by the orientation of the excited state. For moderate excitation intensities the fourth order term $c_4(t)$ can be neglected and the linear combination $I^{rf} = I^{par} + 2I^{per}$ gives the pure population contribution $N_i(t)$. On the other hand one can define an induced dichroism D (in analogy to fluorescence or transmission studies) as:

$$D = \left(I^{par}-I^{per}\right)\Big/I^{rf} \propto -\frac{3c_2(t)\left(6a/\gamma-11/14\right)}{45a^2/\gamma^2+7}$$

(4)

which contains the reorientational information: It is interesting to note that the sign of D is determined by the sign of γ (at least for intermediate values of a/γ); thus it should be possible to obtain direct experimental information about the shape of the Raman tensor which cannot be deduced from conventional Raman spectroscopy which only gives a^2/γ^2.

Experimentally we work with a APM-FCM Nd:YLF laser system[4] running with a repetition rate of 50 Hz. By help of an optical parametric oscillator pumped by the second harmonic of the Nd:YLF pulse train and a single pulse driven down conversion stage the desired infrared excitation pulses are generated (tuning range 2.5 - 4 μm; energy 20 μJ, duration 4 ps). A minor part of the laser pulse is frequency doubled and time delayed; it serves as probe pulse (energy 20 μJ, duration 4 ps). The IR and green pulse are noncollinearly focused into the sample. The scattered Anti-Stokes light is collimated with a f/1.4 objective, passes a notch filter to suppress the Rayleigh light and is focused with an achromatic lens into the entrance slit of a f = 64 cm monochromator (150 l/mm grating). The scattered Raman light is detected by a 1100×330 pixel slow scan, back illuminated, LN cooled CCD array. With a typical detection rate of 1 count per channel and second the summation over app. 1000 laser shots (20 s) gives anti-Stokes Raman spectra of sufficient quality.

Results for $CHCl_3$ will be discussed in the following. Exciting the CH - stretching vibration at 3020 cm^{-1} the time evolution of the anti-Stokes Raman spectra as shown in Fig. 1 is observed.

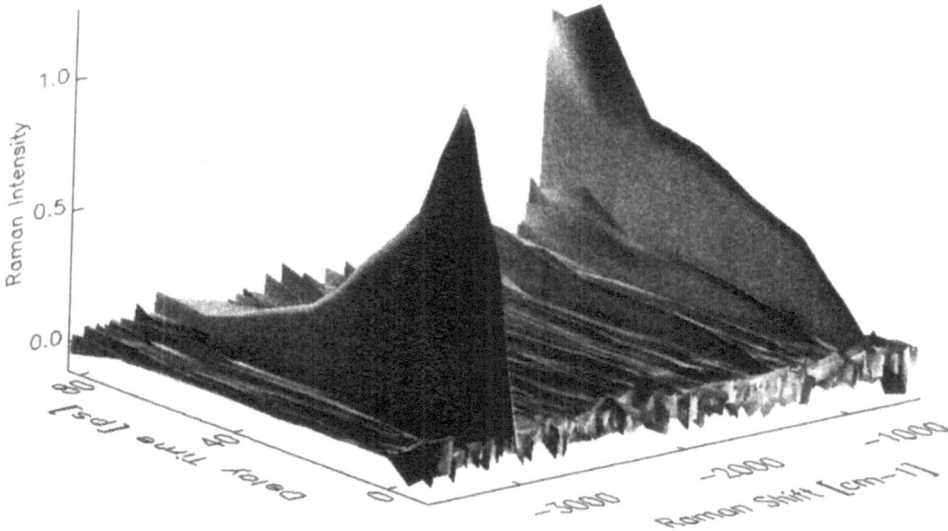

Figure 1. 3-D plot of the observed anti-Stokes Raman scattering of $CHCl_3$ versus frequency and delay time for an excitation frequency of 3020 cm^{-1}; negative frequencies denote anti-Stokes side of the spectrum.

Clearly visible is the rapid rise of the Raman scattering around the pump frequency, which decays on a 20 ps time scale. It is interesting to observe that with increasing delay time the low frequency vibrations are populated. They decay on a much longer time scale (not shown in Fig. 1). To get more precise information about the relaxation time constants, one can integrate over the vibrational bands of interest and normalize the result with respect to the input energies. Fig. 2 shows the corresponding result for the CH-stretching vibration. A rate equation calculation (solid line in Fig. 2) gives an effective lifetime of 23 ± 1 ps. This value is in excellent agreement with previous results from IR studies[5].

To obtain the reorientational contribution the anisotropy D has the be considered. The results for the CH stretching vibration of $CHCl_3$ is shown in Fig. 3.

Figure 2. Time evolution of the CH - anti-Stokes scattering; the pure population signal transient I^{rf} is plotted versus delay time; measured points and calculated curve.

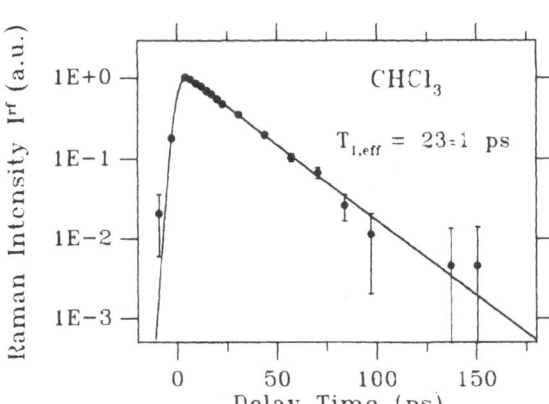

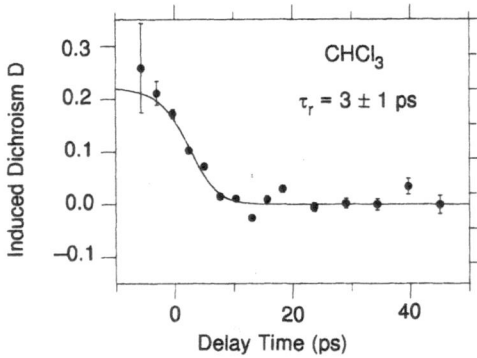

Figure 3. Induced dichroism (calculated from experimental data as $D = (I^{par}-I^{per})/I^{rf})$ for the CH -stretching vibration of $CHCl_3$ as a function of delay time; measured point and calculated curve.

From Fig. 3 two important conclusions can be drawn: the induced dichroism is positive and decays very rapidly. From a calculation with angle dependent rate equations[3] (result shown as solid line in Fig. 3) the reorientational time constant is determined to 3 ± 1 ps in agreement with Raman line shape studies[6]. From the positive sign if D the sign of γ can be determined. Conventional Raman spectroscopy gives a value of 0.5 for a^2/γ^2. Inserting this in Eq. 4 it is clear, that γ has to be negative $(a/\gamma \approx -0.7)$ to obtain the observed positive induced dichroism.

In summary we point out, that incoherent anti-Stokes Raman Spectroscopy is an extreme useful tool for the study of relaxation processes in liquids. Besides the vibrational population dynamics detailed information about reorientational processes and even about structural parameters as the shape of the Raman tensor can be obtained.

REFERENCES

1. A. Laubereau, D. von der Linde, A. Kaiser, Phys. Rev. Lett. **28**, 1162 (1972), A. Laubereau, W. Kaiser, Rev. Mod. Phys. **50**,607 (1978)
2. A. Tokmakoff, B. Sauter, A.S. Kwok, M.D. Fayer, Chem. Phys. Lett. **221**, 412 (1994)
3. Y.B. Band, R. Bavli, Phys. Rev. A **36**, 3203 (1987), H. Graener, G. Seifert, A. Laubereau, Chem. Phys., **175**, 193 (1993)
4. K. Wolfrum, P. Heinz, Opt. Comm. **84**, 290 (1991)
5. H. J. Bakker, J. Chem. Phys. **98**, 8496 (1993)
6. J.E. Griffiths, in *Vibrational Spectra and Structure 6*, ed. J.R. Durig (Elsevier, Amsterdam 1977)

23

FEMTOSECOND INTRAMOLECULAR CHARGE SEPARATION IN DCM

H. Zhang, P. van der Meulen, A. M. Jonkman, and M. Glasbeek

Laboratory for Physical Chemistry
University of Amsterdam
Nieuwe Achtergracht 127
1018 WS Amsterdam, The Netherlands

Upon photo-excitation of the laser dye molecule DCM (4-(dicyanomethylene)-2-methyl-6-(p-dimethylaminostyryl)-4-H-pyran), an intramolecular electron transfer (ET) process is known to occur. The charge separation causes an electronic polarization of the DCM molecules which in turn results in a Stokes shift of the emission band. Following ultrashort pulsed photoexcitation of DCM in liquid solution, the temporal evolution of the solvation energy of the system to the minimum value characteristic of the charge transfer state is determined by the reorientational motions of the polar solvent molecules. The effect is manifested as a time dependent red shift in the DCM fluorescence spectrum. The shift of the maximum, ν_{max}, of the fluorescence band in time is represented by C(t) = $(\nu_{max}(t)-\nu_{max}(\infty))/(\nu_{max}(0)-\nu_{max}(\infty))$, where $\nu_{max}(t)$ corresponds to the frequency of the fluorescence band maximum of dissolved DCM. In linear response theory the time-dependence of the Stokes shift is determined by the time autocorrelation function of the free energy fluctuations. Thus, as is well known for the solvents commonly used, the time scale for the Stokes shift dynamics will be on the order of tens of femtoseconds up to tens of picoseconds.

We have performed time-resolved fluorescence up-conversion experiments (with a time-resolution of ~ 100 fs) on DCM dissolved in a number of different polar solvents. Following the method of Maroncelli and Fleming[1], the emission spectra at different times after the excitation pulse were reconstructed from the observed fluorescence transients. The reconstructed spectra were fitted to a log-normal shape function; the frequency of the band maximum in the log-normal shaped spectrum was taken as $\nu_{max}(t)$ in the C(t) function mentioned above. Typically, C(t) as extracted from the experimental data for DCM dissolved in the different polar solvents methanol, ethylene glycol and ethylacetate, showed a bimodal decay behavior,

$$C(t) = A_G \, exp(-0.5 \, \omega_G^2 t^2) + A_1 exp(-t \, / \, \tau_1) + A_2 exp(-t \, / \, \tau_2), \qquad (1)$$

Ultrafast Processes in Spectroscopy, edited by Svelto et al.
Plenum Press, New York, 1996

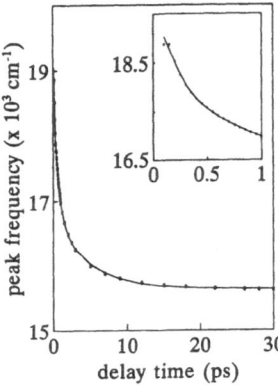

Figure 1. Time evolution of the fluorescence peak frequency of DCM in methanol. Squares represent experimental data, whereas the result of the fit according to Eq. 1 is drawn as a solid line.

where the fastest component is a Gaussian decay function and the remainder of C(t) (as represented by the second and third decay components) is on a time scale of a few picoseconds. In figure 1 we present the results obtained for DCM dissolved in methanol.

The best fit values are $A_G = 0.2$, $A_1 = 0.5$, $A_2 = 0.3$, $\omega_G = 6$ ps^{-1}, $\tau_1 = 0.7$ ps, and $\tau_2 = 5$ ps. The Gaussian component of the function, C(t), with a characteristic decay rate constant of 6 ps^{-1}, is most likely to a large extent due to the "inertial free streaming" motion recently reported in the literature[2]. This free streaming motion is characteristic of the rapid collisionless rotational motions of the solvent molecules in the solvent cage. Our data for DCM in methanol indicate an appreciable contribution of about 30% of the approximately 100 fs Gaussian decay component to the experimentally resolved dynamic Stokes shift. On the other hand, from molecular dynamics calculations it is estimated that free streaming motions might contribute as much as 70% of the solvation relaxation energy. It thus could be that an appreciable part of the inertial solvation relaxation process is missed in our experiments, most likely due to the presence of relaxation components that are too fast to be resolved with our equipment. These very fast contributions to the solvation relaxation dynamics could arise for example in contributions from small angle fluctuations in the fast O-H librational motions (~ 50 fs), rapid electronic dephasing and also from dispersive interactions[3] between the solute and the solvent molecules.

The decay components on the picosecond time scale (characterized by τ_1 and τ_2) are representative of the rotational diffusion motions of the solvent molecules. The solvation times we obtained for the various solvents (e.g., in methanol, $\tau_1 = 0.7$ ps and $\tau_2 = 5$ ps) are compatible with results of MD simulations for such rotational diffusion motions[2].

We now turn to the ultrafast transient behavior of the integrated intensity of the DCM emission after the pulsed photoexcitation. As illustrated in figure 2, the total emission intensity of DCM in ethylene glycol initially increases up to times of about 1 ps and then decreases. The decrease on the longer time scale is as expected because, as time progresses, on the one hand a Stokes red shift is observed (viz., figure 1) and on the other hand the total intensity for spontaneous emission is proportional to ν_{max}^3. Thus a ν_{max}^3-dependence of the integrated emission intensity is expected on a time scale characteristic for the Stokes shift.

To explain the initial increase of the integrated fluorescence intensity we first remark that apparently the population of the charge transfer state is still increasing while the solvation of the DCM molecules in the vibrationally relaxed charge transfer state has already been initiated. The fast replenishing of the probed fluorescent state is thought to

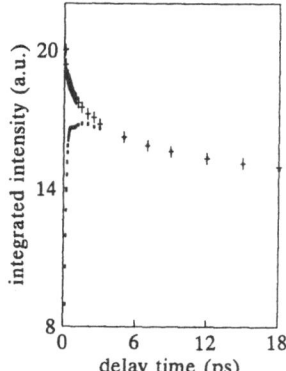

Figure 2. Time evolution of the integrated emission intensity of DCM dissolved in ethylene glycol. Experimental points are represented by squares. The ν_{max}^3 dependence of the integrated emission intensity is given by the plus symbols.

arise from the relaxation of one or more higher lying levels that are formed concurrently with the lowest charge transfer state during the charge separation process, but at a larger rate. Recently it has been discussed elsewhere[4] that the formation of such higher lying states is more likely when these states have a lower barrier for electron transfer than the lowest excited charge transfer state. In particular, in the event of electron transfer in the inverted regime this mechanism may be important in determining the electron transfer rate.

Starting from a Smoluchowsky-type diffusion equation[5] we have performed numerical calculations of the evolution with time of the population of the probed charge transfer state taking into account both solvation relaxation and additional feeding of this state from just one higher lying dark state[6]. Figure 3 illustrates a few typical results. Fig. 3a shows

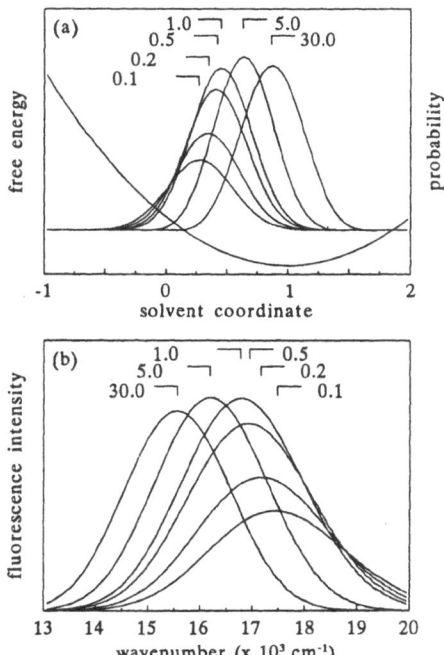

Figure 3. Probability distributions (a) and fluorescence spectra (b) obtained from a simulation of the fluorescent behaviour of DCM dissolved in ethylene glycol, as described in the text. The drawn curves correspond to the situation at 0.1 ps, 0.2 ps, 0.5 ps, 1.0 ps, 5.0 ps and 30.0 ps, respectively, following impulsive laser excitation. The free energy surface of the emissive charge transfer state is depicted in figure 3a.

the variation of the electronic potential energy of the solute as a function of a generalized solvation coordinate, X. The calculated probability distribution in the probed charge transfer state as a function of X at a few different times after the pulsed excitation are depicted in figure 3a also. Figure 3b displays (i) the time development of the fluorescence spectra of DCM in ethylene glycol as reconstructed from the experimental data as discussed above, and (ii) the emission spectra as simulated taking into account the solvation relaxation plus the intramolecular relaxation dynamics. A comparison of the experimental and simulated results shows satisfactory agreement for the peak position and the integrated emission intensity. On the other hand, the band width and the peak height parameters match only qualitatively. Currently, calculations taking into account anharmonic terms in the potential of figure 3a are in progress[7].

REFERENCES

1. M. Maroncelli and G.R. Fleming, J. Chem. Phys. **86**, 6221 (1987).
2. S.J. Rosenthal, R. Jimenez, G.R. Fleming, P.V. Kumar and M. Maroncelli, J. Mol. Liq. **60**, 25 (1994).
3. R. Brown, J.Chem.Phys. **102**, 9059 (1995).
4. I. Rips and J. Jortner, J.Chem.Phys. **87**, 2090 (1987); J.Jortner and M. Bixon, J.Chem.Phys. **88**, 167 (1988).
5. K. Tominaga, G.C. Walker, W. Jarzeba, and P.F. Barbara, J. Phys. Chem. **95**, 10475 (1991).
6. P. van der Meulen, H. Zhang, A.M. Jonkman, and M. Glasbeek, J.Phys.Chem., to be published.
7. P. van der Meulen, H. Zhang, and M. Glasbeek, to be published.

24

VIBRATIONAL AND VIBRONIC DYNAMICS OF LARGE MOLECULES IN SOLUTION STUDIED ON A 20 FEMTOSECOND TIME SCALE

E. Riedle, T. Hasche, S. H. Ashworth, M. Woerner, and T. Elsaesser

Max-Born-Institut für Nichtlineare Optik und Kurzzeitspektroskopie
12489 Berlin, Germany

Recent experiments with the shortest available light pulses have shown that wave packet motion is even detectable for molecules in solution and persistent on the timescale of typically 1 ps[1,2]. However, in these one color experiments it was generally not possible to uniquely decide whether the wave packet propagates on the upper or the lower potential surface. In the experiments reported here, we performed a set of two color measurements with dye molecules which unambiguously show vibrational wave packet motion in the electronic ground state[3]. The electronic relaxation of highly excited singlet states observed simultaneously proceeds on a surprisingly slow timescale and this points to a complex relaxation scheme.

IR125 was used as a prototypical large organic molecule in solution. The absorption spectrum of the dye dissolved in ethylene glycol has a first strong maximum around 794 nm, corresponding to the S_0-S_1 transition, and further absorption bands below 500 nm. We pumped the molecules with blue pulses to vibronic singlet states above S_1, hereafter referred to as S_n. The blue pump pulses of 40 mW average power were centered at 425 nm and were of about 30 fs duration[4]. The dynamics were then probed with weak 20 fs pulses centered at 850 nm or with a small fraction of the blue pulses. The red probe spectrum overlapped with both the absorption spectrum of IR125 and its fluorescence spectrum with a maximum at 822 nm. The large bandwidth of the probe pulses (55 nm) allowed us to measure the change in transmission either integrally or spectrally resolved (detection bandwidth 10 nm).

The fully synchronized and pre-compensated pulses at second harmonic and fundamental wavelengths of a home built Ti:sapphire laser were used as pump and probe. After chopping, the pump beam was combined collinearly with the probe beam and focused into a free jet of IR125 (5×10^{-4} mol dm^{-3}) dissolved in ethylene glycol. The pump beam was filtered out with colored glass filters and the probe beam then detected using a photodiode. The photodiode was situated downstream of a monochromator and the signal was recorded either integrally (zeroth order) or spectrally resolved (first order).

Ultrafast Processes in Spectroscopy, edited by Svelto et al.
Plenum Press, New York, 1996

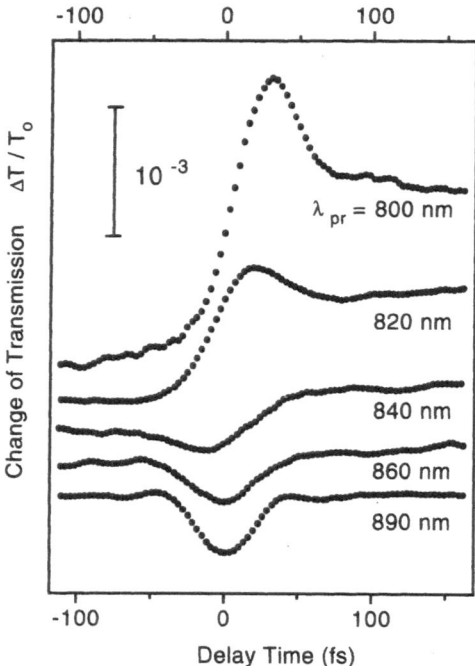

Figure 1. Spectrally resolved two color pump-probe signals of IR 125 dissolved in ethylene glycol.

The spectrally integrated pump-probe signal shows an initial steep rise of transmission within about 50 fs. This is comparable to the length of the cross-correlation. Spectrally resolved recordings are shown in Fig. 1. On short timescales, that is to say within the first 80 fs, these show an additional oscillatory component. At the shorter wavelengths the transmission first increases, whereas at the longer wavelengths a decrease in transmission is observed. On a longer timescale of about 1 ps both the integral and the spectrally resolved recordings show a slow increase in transmission.

Initially the S_n and S_0 states are coherently coupled by the ultrafast excitation pulse. A wave packet created on the upper potential surface propagates before being projected back to the ground state. Thus a vibrational wave packet in the ground state is produced through Resonant Impulsive Stimulated Raman Scattering (RISRS) which can be monitored with the probe pulses. This is possible because the dephasing time of the vibronic transition is comparable to the duration of the pump pulses. Unlike earlier one color experiments we are here in a position to uniquely identify the wave packet motion as occurring in the ground state. This is due to the fact that on an ultrafast timescale the S_0 state is the only common electronic state which can mediate the coherent coupling between the two vibronic transitions.

The rapid, incoherent, step-like rise in transmission on the femtosecond timescale is due to bleaching of the ground state. This is confirmed by the close quantitative agreement of the signal level at 80 fs with the stationary absorption spectrum (see Fig. 2; left part). From this agreement between the bleaching and the absorption cross section we can also conclude that transient absorption or stimulated emission from the initially excited S_n state is not a significant contribution to the observed transient signals.

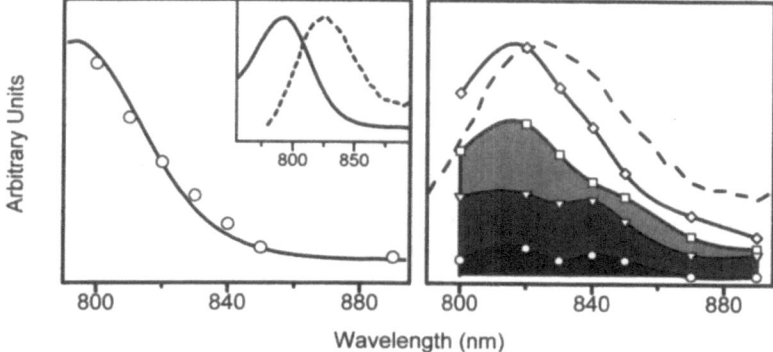

Figure 2. *Left*: Comparison of the bleaching signal (o) with the absorption spectrum of IR125. The scaling of the bleaching signal is quantitatively correct. *Right*: Comparison of stimulated emission signal at 200 fs -O-, 800 fs - ▽-, 1.3 ps -□-, and 10 ps -◇- with the stimulated emission spectrum calculated from the measured fluorescence spectrum. *Inset*: The relative positions of the absorption (solid line) and stimulated emission spectra (broken line).

The coherent contributions shown in Fig. 1 are strongly damped. This is in contrast to observations on molecules of similar size in solution where dephasing times of individual modes of typically one picosecond are found[1,2,5]. The one color femtosecond pump probe experiments that rendered this information produced, however, only a small oscillatory contribution on top of the bleaching signal[5]. If we investigate our spectrally resolved transients closely, we also find persistent modulations. These are clearest at the extremes of the probe spectrum, in agreement with theoretical predictions[6]. To obtain an estimate of the number of Raman active modes in IR125 we measured a spontaneous Raman spectrum with excitation at 1064 nm. This spectrum is shown in Fig. 3 and contains a large number

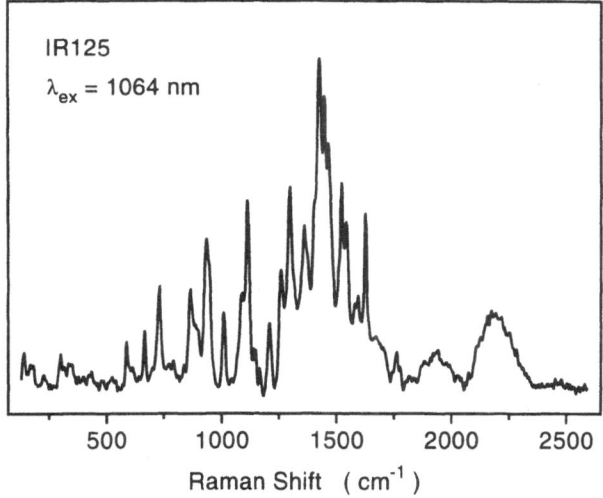

Figure 3. Spontaneous Raman spectrum of IR125 with excitation at 1064 nm.

of lines. From this we can conclude that the wave packet produced by RISRS is multidimensional. Similar to previous investigations in the picosecond regime, where wave packets of many coupled eigenstates in S_1 were found to dephase on the timescale of the coherence width of the laser pulses[7], the ground state wave packet only gives rise to a small modulated contribution to the transient absorption signals after the initial strong peak.

Having been excited to the S_n state the molecules decay to the S_1 state by internal conversion. After vibrational redistribution and relaxation due to intramolecular processes and the solvent environment, they collect at the bottom of the S_1 state and are detected in our experiment through stimulated emission. This is observed as the picosecond rise of transmission in both the integral and the spectrally resolved transients.

The slow rise in transmission corresponds closely to the stimulated emission spectrum of IR125. In the right-hand portion of Fig. 2 the signal level at discrete time intervals is shown, having first subtracted the bleaching component at that wavelength. The spectrum can be seen to grow in at a position shifted slightly to the blue of the stationary stimulated emission spectrum. It is likely that this shift is due to the solvent environment not having fully relaxed at delay times as short as 10 ps. Solvation in ethylene glycol was observed for molecules of similar size to occur in about 100 ps[8].

A one color pump-probe experiment performed with the 30 fs blue pulses results in a transient absorption curve that shows a very strong coherent peak with pulse limited width on top of a step like increase in transmission. This step like increase is again due to the ground state bleaching. The dominant peak is believed to contain contributions from degenerate four wave mixing (DFWM). In addition, the multidimensional vibronic wave packet created by the pump pulse will most likely disappear from the Franck-Condon region of the S_0-S_n transition on a time scale comparable to the pulse length. A classical decay time of the S_n state that would be appropriate in an incoherent description of the nonradiative decay following excitation is therefore not meaningful in the present situation of broadband ultrafast excitation.

Regardless of the details, we find that the excited molecules disappear from the optically accessible region within less than 80 fs. The stimulated emission from the S_1 state, however, has a rise time with a picosecond time constant. From the present experiments we cannot judge whether the wave packet propagates for an extended period on the S_n potential surface before it finds its way directly to the S_1 state. Alternatively, internal conversion to the S_1 state could be quite rapid, however, to regions of the potential surface with small Franck-Condon overlap in the spectral regime of the probe pulses. An increase in transmission would then require subsequent vibrational relaxation which is known to proceed on the picosecond timescale. Last but not least we can infer from the absorption spectrum that we have initially excited an electronic state S_n with n at least 4. Therefore the nonradiative decay could proceed sequentially through all intermediate states. Since these will, in general, not be optically active, it would follow that the stimulated emission from the S_1 state appears with considerable delay.

The differentiation between these scenarios will rely on future experiments with independently tunable probe pulses. In the present experiments we have, however, already shown that there is vibrational wave packet motion on the ground state electronic surface and ultrafast relaxation from a highly excited electronic state of a large molecule in solution. These observations were possible through two color experiments of the vibrational and vibronic dynamics of large molecules in solution with an unprecedented time resolution of 20 fs.

REFERENCES

1. H. L. Fragnito, J.-Y.Bigot, P. C. Becker, and C. V. Shank, *Chem. Phys. Lett.* **160**, 101 (1989); S. L. Dexheimer, Q. Wang, L. A. Peteanu, W. T. Pollard, R. A. Mathies, and C. V. Shank, *Chem. Phys. Lett.* **188**, 61 (1992)
2. T. Joo and A. C. Albrecht, *Chem. Phys.* **173**, 17 (1993)
3. T. Hasche, S. H. Ashworth, E. Riedle, M. Woerner, and T. Elsaesser, *Chem. Phys. Lett.* **244**, 164 (1995)
4. S. H. Ashworth, M. Joschko, M. Woerner, E. Riedle, and T. Elsaesser, *Opt. Lett.* 20, in print
5. P. Vöhringer, R. A. Westervelt, T.-S. Yang, D. C. Arnett, M. J. Feldstein, and N. F. Scherer, *J. Raman Spectrosc.* **26**, 535 (1995)
6. W. T. Pollard, S. L. Dexheimer, Q. Wang, L. A. Peteanu, C. V. Shank, and R. A. Mathies, *J. Phys. Chem.* **96**, 6147 (1992)
7. P. M. Felker and A. H. Zewail, in *Advances in Chemical Physics* (Wiley, New York, 1988), Vol. LXX, p. 265
8. J. D. Simon, *Acc. Chem. Res.* **21**, 128 (1988)

25

VIBRATIONAL POPULATION DYNAMICS IN LIQUIDS AND GLASSES

IR Pump-Probe Experiments from 10 K to 300 K

C. Ferrante, A. Tokmakoff, C. Taiti, A. S. Kwok, R. S. Francis, K. D. Rector, and M. D. Fayer

Department of Chemistry
Stanford University
Stanford, California 94305

The temperature dependent vibrational relaxation of the CO stretching mode of rhodium dicarbonyl acetylacetonate ($Rh(CO)_2(acac)$) in dibutyl phthalate (DBP), and tungsten hexacarbonyl (W(CO)6) in 2-methylpentane (2-MP) were measured with IR pump and probe (PP) experiments. The experiments were performed with ~1.5 ps pulses (spectral width $\Delta v \approx 12$ cm^{-1}) at $\lambda \approx 5$ μm generated by the Stanford superconducting-accelerator-pumped free electron laser (FEL)[1] and 10–15 ps pulses ($\Delta v \approx 2$ cm^{-1}) at the same wavelength generated by an $LiIO_3$ optical parametric amplifier (OPA)[2]. Typical pulse energies used in the experiments were 10 nJ for the probe and 50–100 nJ for the pump. Both beams were focused to a ~150 μm spot on a liquid cell with 400 μm optical path. The concentrations of the sample solutions were 6×10^{-3} M for $Rh(CO)_2(acac)$ and 4×10^{-3} M for $W(CO)_6$. The differential absorption of the probe beam was measured as a function of the time delay between the pump and the probe pulses. The decay curves of the probe signal were fitted using a single or a double exponential function.

For $W(CO)_6$ in 2-MP, measurements were performed with both laser sources on the triply degenerate asymmetric CO stretching mode ($\lambda=5.05$ μm) in the temperature range 10–300 K with parallel and magic angle polarization. This mode is characterized by an absorption linewidth ranging from 3.7 cm^{-1} at 300 K to 10.5 cm^{-1} full width at half maximum (FWHM).

The decay curves measured at parallel polarization show a biexponential behavior over the whole temperature range with both laser sources. The decay time constants and zero time amplitudes for the slow and fast component are reported in Figure 1a. At magic angle polarization, the decay curves measured with the FEL show a monoexponential behavior over the whole temperature range, while those taken with the OPA are monoexponential from 300 K to 200 K, and then become biexponential at lower temperatures[2]. The zero time amplitude and decay time constant of the slow and fast component of the OPA data are reported in Figure 1b.

Ultrafast Processes in Spectroscopy, edited by Svelto et al.
Plenum Press, New York, 1996

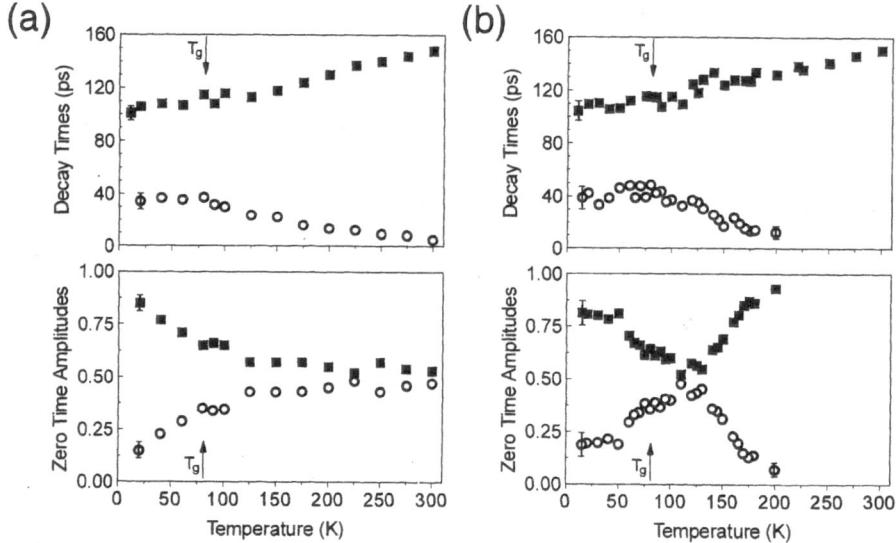

Figure 1. Temperature-dependence of the decay time constants and zero time amplitudes of the slow (squares) and fast (circles) components determined by fitting the PP decay for the CO mode of the $W(CO)_6$ in 2-MP. a) results obtained with the FEL for parallel polarization of the probe beam, b) results obtained with the OPA at magic angle polarization (see text)[2].

The slow component is attributed to the population relaxation of the vibrational mode, T_1. The fast component is instead attributed redistribution of population in between the three degenerate levels of the CO asymmetric stretching mode. This process causes orientational relaxation of the molecule and thus gives rise to the fast component shown in the decay curved taken at parallel polarization. The appearance of this component in the magic angle data taken with the OPA is due to a spectral diffusion caused by same physical process since the OPA source has a spectral width smaller than the FWHM of the CO mode. This appears at temperatures ≤200K when the absorption line of the CO mode becomes inhomogeneously broadened as confirmed by photon echo measurements[3]. Since the FEL pulse has a spectral width that is bigger than the FWHM of the CO mode, this component does not appear in the magic angle data taken with the FEL.

In the case of $Rh(CO)_2(acac)$, measurements using the FEL were performed on the CO asymmetric stretching mode (λ = 4.98 μm) over the temperature range 10–300 K with parallel and magic angle polarization. The decay curves of the probe signal both at parallel and magic angle are well fitted by a biexponential function over the entire temperature range. The decay time constants and zero time amplitudes for the fast and slow components are shown in Figure 2 as a function of temperature.

The slow component is attributed to the population relaxation of the CO asymmetric stretching mode. This is in agreement with previous work of Heilweil and co-workers on the CO symmetric stretching mode of the same compound in n-hexane and chloroform performed only at room temperature[4]. For the fast component, Heilweil and co-workers proposed a fast v-v coupling mechanism that redistributes population between the two CO stretching modes. A simulation based on this model is also presented in Figure 2 (solid line) where the lifetime for the asymmetric and symmetric CO modes are 45 ps and 35 ps

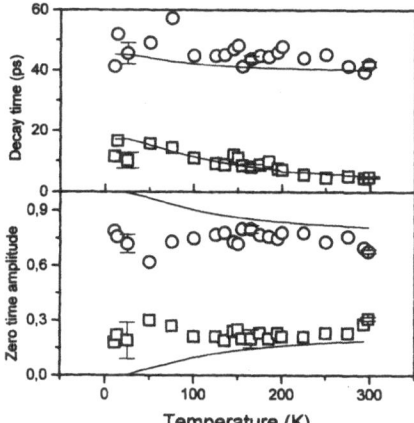

Figure 2. Temperature-dependence of the decay time constants and zero time amplitudes obtained by fitting the PP decay of the CO mode of Rh(CO)$_2$(acac) in DBP. All data points represent the average of the data taken at magic angle and parallel polarization for the probe beam with the FEL .The solid line is the result of a simulation (see text).

respectively and the rate of redistribution of population is properly chosen in order to fit the room temperature value of the fast component decay constant. As is clear from the Figure this process satisfactorily describes our data only for temperature $\geq$ 150 K. A mechanism of orientational relaxation is excluded because the biexponential behavior in the decays is present even with magic angle probing, and the hydrodynamic relaxation time is far too long to be observable in the decay. We tentatively attribute it to spectral diffusion, since the laser bandwidth is smaller (12 cm^{-1}) than the absorption linewidth of the CO mode (17 cm^{-1}). Photon echo experiments on W(CO)$_6$ have shown that the absorption line of the CO mode is inhomogeneously broadened even at room temperature in DBP, and thus the line of the Rhodium compound is expected to be inhomogeneously broadened in this solvent[1]. On the other hand, the PP data on the W(CO)$_6$ show that the spectral diffusion mechanism vanishes at very low temperature, while in our case it remains practically unchanged in the whole temperature range. It is possible that in this system a static distribution of relaxation rates exists for the CO asymmetric stretching mode. Such behavior would give rise to a decay described by a stretched exponential function, which would also be consistent with this data.

REFERENCES

1. A. Tokmakoff, A.S. Kwok, R.S. Urdahl, R.S. Francis, and M.D. Fayer, Chem. Phys. Lett., **234**, 289 (1995).
2. A. Tokmakoff, R.S. Urdahl, D. Zimdars, R.S. Francis, A.S. Kwok, and M.D.Fayer, J.Chem Phys. **102**, 3919 (1995).
3. A. Tokmakoff and M.D. Fayer, J. Chem. Phys. **103**, 2810 (1995).
4. J.D. Beckerle, M.P. Casassa, R.R. Cavanagh, E.J. Heilweil, and J.C. Stephenson, Chem. Phys., **160**, 487 (1992).

26

ULTRAFAST INTRAMOLECULAR PROTON TRANSFER IN THE CONDENSED PHASE STUDIED BY RESONANCE RAMAN AND FEMTOSECOND SPECTROSCOPY

T. Elsaesser, M. Pfeiffer, K. Lenz, A. Lau, C. Chudoba, S. Lutgen, T. Jentzsch, E. Riedle, and M. Woerner

Max-Born-Institut für Nichtlineare Optik und Kurzzeitspektroskopie
D-12489 Berlin, Germany

Intramolecular proton transfer after electronic excitation of aromatic molecules proceeds frequently on ultrafast time scales. Femtosecond studies of the transient electronic spectra occurring with and after the formation of a new molecular geometry provide insight into the reaction dynamics, whereas resonance Raman experiments give information on the vibrational modes involved in the initial phase of excited state reactions and - thus - on the microscopic mechanisms of the structural changes. In this paper, we report new results on proton transfer reactions and nonequilibrium vibrational excitations of 2-(2'-hydroxy-5'- methylphenyl)benzotriazole (trade name: TINUVIN P, TIN), an ultraviolet stabilizer of polymers.

enol   keto-type

In the time-resolved experiments, a femtosecond pump pulse promotes enol tautomers to the S_1 state and initiates the proton transfer cycle. The transient absorption and emission of the keto-type product molecules is recorded with weak probe pulses in the range between 400 and 1000 nm (time resolution 100 fs). Independently tunable pump and probe pulses were derived by spectral selection and - for the pump - by frequency doubling from two femtosecond white-light continua generated with pulses from a regenera-

Ultrafast Processes in Spectroscopy, edited by Svelto et al.
Plenum Press, New York, 1996

tively amplified Ti:sapphire laser. The measurements were performed under conditions of small signal excitation. The femtosecond experiments were complemented by continuous-wave resonance Raman studies in the range of the S_0-S1 absorption band of enol-TIN.

The femtosecond data demonstrate that TIN undergoes a closed reaction cycle with (i) excited state proton transfer, (ii) radiationless deactivation of the keto-type reaction product, and (iii) keto-enol proton back-transfer in the electronic ground state.[1-3]

(i) The excited state reaction forming keto-TIN proceeds on a 100 fs time scale, as is evident from the build-up dynamics of the keto-type emission. In Fig. 1, we present the transient spectrum of stimulated emission measured after excitation at λ_{ex} = 370 nm (27030 cm^{-1}), nearly resonant to the purely electronic S_0-S_1 transition of the enol tautomers.[4,5] The spectrum extends from about 600 nm to well beyond 800 nm. The time evolution of emission at 700 nm is shown in Fig. 2. We observe a delayed rise with a time constant of 100 fs which is due to the finite formation time of the product, i.e. the time needed to transfer the proton. The formation kinetics of keto-TIN remain unchanged if the enol species is excited at other wavelengths well above or below its 00-transition. This fact demonstrates that the proton transfer dynamics do not depend on the initial amount of vibronic excess energy in enol-TIN, pointing to a barrierless pathway of the excited state reaction[6].

(ii) The S'_1 state of the keto-type molecules has a very short lifetime of 150 fs, as is obvious from the rapid decay of stimulated emission in Fig. 2. The fast depopulation is caused by the highly efficient radiationless deactivation and results in the very low quantum yield of spontaneous emission of 10^{-5}. After depopulation of the excited state, a broad ground state absorption of keto-TIN is found between 400 and 700 nm. The transient spec-

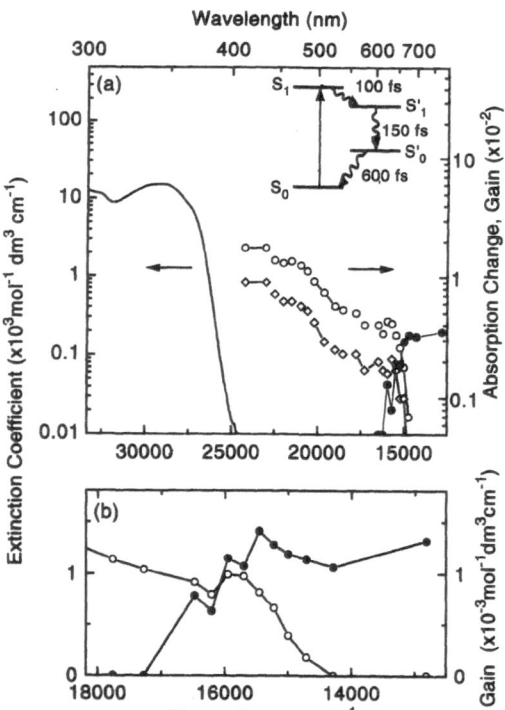

Figure 1. (a) Steady state absorption band of enol-TIN dissolved in cyclohexane (solid line), and spectra of transient absorption ΔA=-ln(T/T$_0$) (open symbols) and gain G=ln(T/T$_0$) (solid circles) of keto-TIN (logarithmic ordinate scales; T$_0$,T: transmission of the sample before and after excitation at λ_{ex}=370 nm). The emission spectrum was recorded at a delay time of 200 fs, the transient absorption was measured with delays of 750 fs (circles) and 1.25 ps (diamonds). Insert : Schematic of the ultrafast reaction cycle of TIN. (b) Molar coefficients of absorption and stimulated emission of keto-TIN plotted on an extended energy scale (linear ordinates).

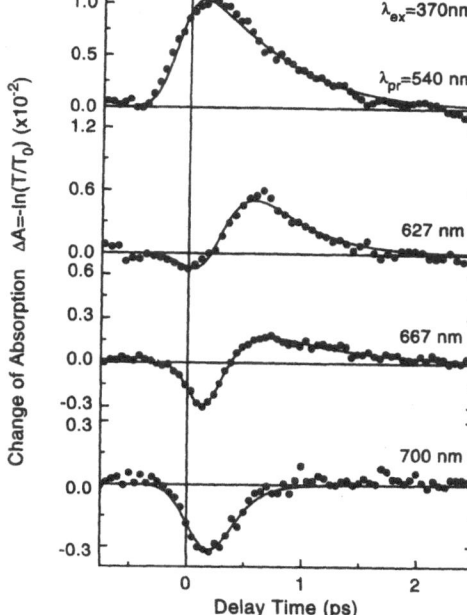

Figure 2. Time resolved transmission changes of TIN after excitation at λ_{ex} = 370 nm. The change of absorption $\Delta A=-\ln(T/T_0)$ is plotted as a function of delay time between pump and probe (T_0,T: transmission before and after excitation) for probe wavelengths λ_{pr} of 540 nm, 627 nm, 667 nm, and 700 nm (points). Positive signals correspond to a transiently enhanced absorption, whereas negative components are due to stimulated emission. At 700 nm, exclusively the gain component is found. Solid lines : calculated kinetics.

trum is plotted in Fig. 1 for two different delay times and shows spectral overlap with the emission band over more than 2000 cm⁻¹. This is due to the strong vibrational excitation of the keto-type molecules. The excited state reaction and the subsequent internal conversion process depopulating the keto S'_1 state transfer a large amount of excess energy into the vibrational system. In the excited vibrational manifold, substantial populations occur in high lying vibrational levels some of which couple to the electronic transitions and give rise to an enhancement of the S'_1-S'_0 emission at short wavelengths and of the S'_0-S'_1 absorption at long wavelengths. The more quantitative analysis of the spectral overlap in Ref. 3 shows that such modes exhibit strong non-equilibrium populations, much higher than in a thermal distribution with the same content of excess energy.

(iii) The keto-type molecules transform back to the enol geometry by ground state proton transfer. This process occurs with a time constant of 600 fs which is found both in the decay of the keto-type ground state absorption plotted in Fig. 2 and in the repopulation dynamics of the enol ground state.[2] The fast reaction cycle generates enol molecules with a highly excited vibrational system as is evident from the transient broadening and reshaping of the enol S_0-S_1 absorption band shown in Fig. 3. The transient absorption band evolves towards the steady-state S_0-S_1 absorption on a slower time scale of 50 ps due to cooling of the hot molecules by collisional interaction with the solvent.

Our resonance Raman data give direct information on the vibronic structure of the enol S_0-S_1 band.[4,5] About 15 vibrational modes in the range between 300 and 1650 cm⁻¹ show a pronounced coupling to the electronic transition. Theoretical calculations based on the time-propagator method allow a simulation of the initial propagation of vibronic wavepackets in the S_1 state and of the steady state and transient absorption spectra. The simulations suggest that a 469 cm⁻¹ mode providing a distinct shift of the proton along the O-H...N direction is essential for the ultrafast proton transfer in the excited state. Further-

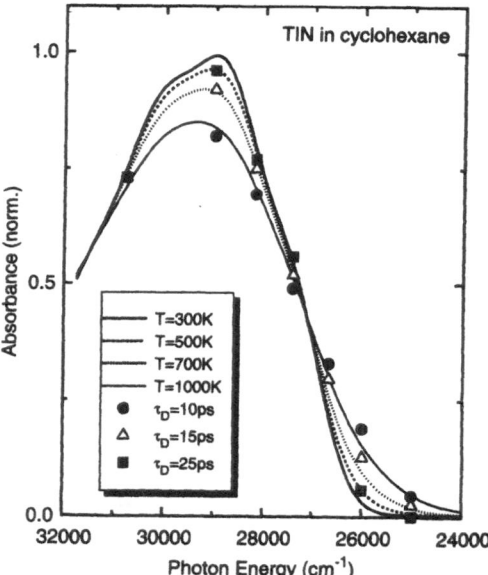

Figure 3. Transient absorption spectra of enol-TIN. The absorbance A=-ln(T) derived from femtosecond experiments is plotted for different delay times after excitation (symbols, T: transmission). The lines are calculated for different vibrational temperatures from a theoretical analysis using parameters from resonance Raman experiments. The 300 K spectrum agrees very well with the steady-state absorption of TIN.

more, our analysis reproduces the steady state absorption band quantitatively, demonstrating progressions in the 1400 cm^{-1} range which are attributed to modes involving O-H bending and C-O stretching motions. In Fig. 3, the enol spectra observed in the time-resolved experiments are compared to calculations for elevated vibrational temperatures, i.e. assuming an equilibrated vibrational system of enol-TIN. For delay times longer than 10 ps, we find an excellent agreement of experiment and theory whereas substantial deviations occur at earlier times. This points to nonthermal vibrational distributions created by the proton-back-transfer which will be the subject of future experiments.

REFERENCES

1. M. Wiechmann, H. Port, W. Frey, F. Laermer, and T. Elsaesser, J. Phys. Chem. **95**, 1918 (1991)
2. W. Frey, and T. Elsaesser, Chem. Phys. Lett. **189**, 565 (1992)
3. C. Chudoba, S. Lutgen, T. Jentzsch, E. Riedle, M. Woerner, and T. Elsaesser, Chem. Phys. Lett. **240**, 35 (1995)
4. K. Lenz, M. Pfeiffer, A. Lau, and T. Elsaesser, Chem. Phys. Lett. **229**, 340 (1994)
5. M. Pfeiffer, K. Lenz, A. Lau, and T. Elsaesser, J. Raman Spectrosc. **26**, 607 (1995)
6. A.L. Sobolewski, and W. Domcke, Chem. Phys. Lett. **211**, 82 (1993)

27

TEMPERATURE-DEPENDENT VIBRATIONAL DYNAMICS AND INHOMOGENEOUS BROADENING IN GLASS-FORMING LIQUIDS STUDIED WITH INFRARED PHOTON ECHOES

A. Tokmakoff and M. D. Fayer

Department of Chemistry
Stanford University
Stanford, California 94305

Picosecond infrared (IR) photon echo and pump-probe experiments have been used for a detailed study of the temperature-dependence of the homogeneous vibrational lineshape and inhomogeneous broadening of a high-frequency vibrational transition in three molecular glass-forming liquids[1]. The homogeneous vibrational linewidth of the T_{1u} CO stretch of tungsten hexacarbonyl ($\sim$1980 cm^{-1}) was measured in 2-methylpentane (2-MP), 2-methyltetrahydrofuran (2-MTHF), and dibutylphthalate (DBP) from 300 K to 10 K. The dynamics cover temperature regions from the glass, through the glass transition and super-cooled liquid, to the room temperature liquid. These experiments are the first to directly follow the temperature-dependent vibrational dynamics in amorphous condensed phases, and quantify the degree of inhomogeneous broadening for these systems.

The difficulty with determining microscopic dynamics from linear spectroscopic techniques is that they have no means for separating inhomogeneous contributions from the homogeneous vibrational lineshape[2]. Nonlinear vibrational spectroscopy such as the IR photon echo[1,3] and Raman echo[2,4] can determine the homogeneous vibrational dynamics, even when this lineshape is masked by inhomogenous broadening. Yet, to determine the relative contribution of various dynamics to the vibrational lineshape, such as lifetime and orientational relaxation, traditional pump-probe spectroscopies are also required.

Two aspects of the vibrational lineshape in these systems have been studied in detail. Initially, ps IR photon echo experiments were used to compare the homogeneous linewidths in the glasses and the transition to the room temperature liquids for the three systems. The temperature dependence of the homogeneous vibrational linewidth in each of the three glassy solvents is approximately T^2, but the behavior is distinct in each of the liquids. A T^2 temperature dependence is indicative of dephasing through a two-phonon Raman scattering process, which is expected above the effective Debye temperature of the glass.

Ultrafast Processes in Spectroscopy, edited by Svelto et al.
Plenum Press, New York, 1996

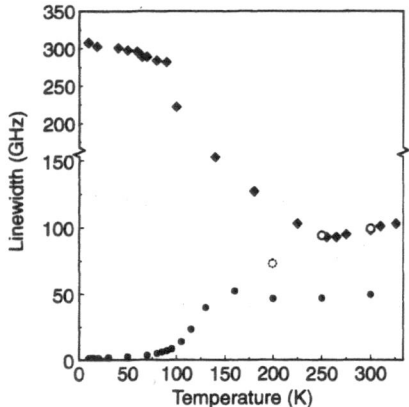

Figure 1. The homogeneous (circles) and absorption linewidths (diamonds) for the T_{1u} mode of $W(CO)_6$ in 2-MP. The open circles represent the interpretation of the echo signal as a free induction decay. The point at 200 K is only an interpolation between the two limiting cases.

In the low temperature glass, all of the absorption linewidths are seen to be massively inhomogeneously broadened. While in 2-MP the vibrational line becomes homogeneously broadened at room temperature, the line in DBP is still massively inhomogeneous in the room temperature liquid. The data in DBP represent a direct observation of massive inhomogeneous broadening in a room temperature liquid. The transition to a homogeneously broadened line in 2-MP is shown in Fig. 1. A rapid increase in the rate of dephasing is observed near the glass transition temperature T_g. At temperatures below 150 K, the homogeneous linewidth obtained from the photon echo experiments ($\Gamma = 1/\pi T_2$) is substantially narrower than the absorption linewidth. In this limit, the photon echo decays with an exponential rate of $4/T_2$, shown by the solid circles in Fig. 1. In the limit of homogeneous broadening, the echo will become a free induction decay, which relaxes at a rate of $2/T_2$. This limit, shown by the open circles at 250 K and 300 K, matches the absorption linewidth, indicating a homogeneously broadened line at these temperatures. The temperatures near 200 K represent the transition between these two limits.

The homogeneous vibrational lineshape for $W(CO)_6$ in 2-MP has been analyzed in greater detail. The contributions to the vibrational lineshape from different dynamic processes have been delineated by combining the results of photon echo measurements of the homogeneous lineshape[1] with pump-probe measurements of the lifetime and reorientational dynamics[5]. This combination of measurements allows the decomposition of the total homogeneous vibrational lineshape into the individual components of pure-dephasing (T_2^*), population relaxation (T_1), and orientational relaxation:

$$\Gamma = 1/\pi T_2^* + 1/2\pi T_2^* + 2D_{or} \tag{1}$$

The contribution of orientational relaxation to the infrared linewidth is related to the isotropic orientational diffusion constant D_{or}. The results, shown in Fig. 2, demonstrate that each of these processes contributes significantly, but to varying degrees at different temperatures. The homogeneous linewidth is dominated by the vibrational lifetime at low temperatures and by pure dephasing in the liquid. The contribution from pure dephasing is observed to increase rapidly near the glass transition temperature. The temperature dependence of pure dephasing is well described by the sum of a T^2 power law and a Vogel-Tammann-Fulcher type equation

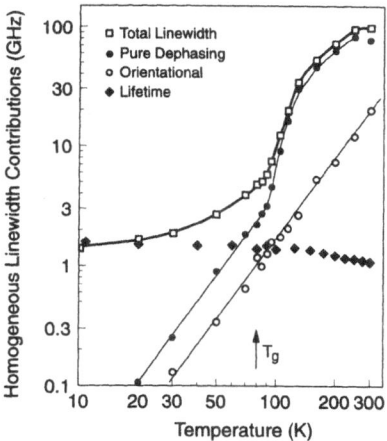

Figure 2. The temperature-dependent dynamic contributions to the homogeneous vibrational linewidth of the T_{1u} mode of $W(CO)_6$ in 2-MP.

$$\Gamma^*(T) = A_1 T^2 + A_2\, exp(-B\,/\,(T - T_g))$$

(2)

Here the divergence temperature for the activated process is seen to be the laboratory glass transition temperature. The orientational relaxation contribution to the line is significant, and is described by a continuous T^2 power law over all temperatures. Although T_1 varies from 100 ps at 10 K to 150 ps at 300 K, the temperature dependence of the contribution from the lifetime is negligible in comparison to the other processes.

REFERENCES

1. A. Tokmakoff and M. D. Fayer, J. Chem. Phys. **103**, 2810 (1995).
2. R. F. Loring and S. Mukamel, J. Chem. Phys. **83**, 2116 (1985).
3. (a) D. Zimdars, A. Tokmakoff, S. Chen, S. R. Greenfield and M. D. Fayer, Phys. Rev. Lett. **70**, 2718 (1993). (b) A. Tokmakoff, D. Zimdars, B. Sauter, R. S. Francis, A. S. Kwok and M. D. Fayer, J. Chem. Phys. **101**, 1741 (1994). (c) A. Tokmakoff, A. S. Kwok, R. S. Urdahl, R. S. Francis and M. D. Fayer, Chem. Phys. Lett. **234**, 289 (1995).
4. (a) D. Vanden Bout, L. J. Muller and M. Berg, Phys. Rev. Lett. **67**, 3700 (1991). (b) L. J. Muller, D. Vanden Bout and M. Berg, J. Chem. Phys. **99**, 810 (1993). (c) R. Inaba, K. Tominaga, M. Tasumi, K. A. Nelson and K. Yoshihara, Chem. Phys. Lett. **211**, 183 (1993).
5. A. Tokmakoff, R. S. Urdahl, D. Zimdars, A. S. Kwok, R. S. Francis and M. D. Fayer, J. Chem. Phys. **102**, 3919 (1995).

28

FEMTOSECOND PUMP&PROBE SPECTROSCOPY ON THE K₂ MOLECULE

Perturbations in different Isotopomeres

Soeren Rutz, Elmar Schreiber, and Ludger Wöste

Institut für Experimentalphysik
Freie Universität Berlin
14195 Berlin, Germany

With femtosecond three-photon ionization spectroscopy we studied the wave packet dynamics in the spin-orbit perturbed K_2 $A^1\Sigma_u^+$ state. Mass-selective detection enabled us to distinguish between the two isotopomeres $^{39,39}K_2$ and $^{39,41}K_2$. We directly measured the oscillation time of K_2 excited in the A state. Besides this, we detected revival structures totally differing for the two studied isotopomeres. We could as well determine the vibrational periods and revival times for the isotopomere $^{39,41}K_2$ and the energy spacings between the involved vibrational levels.

In femtosecond vibrational spectroscopy the first molecules to be studied had been the relatively simple dimer systems[1,2]. By use of a variety of femtosecond pump&probe techniques the dynamics of these systems was investigated. These techniques opened the possibility to directly observe the ro-vibrational dynamics of the excited molecules. Furthermore spectral information could be deduced from time-resolved measurements although the spectral width of ultrashort laser pulses in the sub-100 fs regime is fairly broad. In case of a pump&probe experiment with a subsequent fluorescence detection oscillation periods could be directly measured and vibrational and rotational revivals were observed for I_2[1,3]. It was possible to study the wave packet dynamics in different excited electronic states combined with different ionization pathways in Na_2 with femtosecond multi photon ionization spectroscopy[1]. In Cs_2, the phase of laser-induced wavepackets could be controlled by a pump-control&probe experiment[4].

The $A^1\Sigma_u^+$ state of alkali dimers is one of the best known systems from the viewpoint of energy resolved spectroscopy. Vibrational and rotational energy levels could be observed by many experimental techniques e.g. in the case of Na_2, Li_2, NaK. a RKR analysis was performed to extract the spectroscopic constants and the potential energy curves[5,6,7,8,9,10]. In case of $^{39,41}K_2$ some fluorescence lines could be identified[11]. Several authors studied the influence of triplet states to the energy levels in electronic states of the alkali dimers Li_2, Na_2, NaK and K_2[5,6,8,11,12,13]. Especially the spin-orbit coupling between

Ultrafast Processes in Spectroscopy, edited by Svelto et al.
Plenum Press, New York, 1996

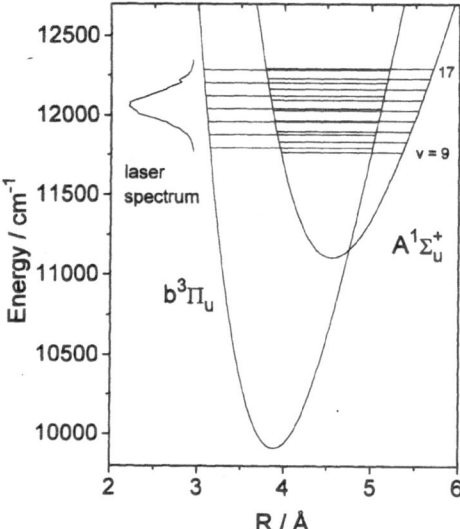

Figure 1. Potential energy curves of the K_2 $A^1\Sigma_u^+$ and the $b\,^3\Pi_u$ states[11]. The applied laser spectrum is indicated as well as the involved vibrational levels v.

the $b^3\Pi_u$ state and the $A^1\Sigma_u^+$ state (fig. 1) in these molecules with the resulting energy shifts of vibrational energy levels was a main point of interest.

We present investigations on the wave packet dynamics in the spin-orbit perturbed $A^1\Sigma_u^+$ state of K_2. (see fig. 1). Pump&probe experiments were performed on the two isotopomeres $^{39,39}K_2$ and $^{39,41}K_2$. One ultrashort laser pulse of 12000 wavenumbers was used to prepare a wave packet on the potential energy surface of the A state around the vibrational levels $v = 12,13$ in case of the lighter isotopomere. A subsequent laser pulse of the same energy with a defined time delay was applied to probe the induced wave packet with two photons being required to ionize the dimer. Hence, a three-photon ionization (3PI) process is used to detect the dimer's wave packet propagation.

We applied an argon ion laser-pumped regeneratively modelocked titanium:sapphire laser (Spectra Physics model 2080 and 3960) to produce ultrashort laser pulses of about 90 fs duration (FHWM, assuming sech2 pulse shape)[14]. In a common pump&probe setup the laser beam was split and realigned collinearly with the same polarization.

A molecular beam of high stability and a large amount of dimers was produced by evaporating pure potassium at a temperature of 550 K and coexpanding the vapor with argon through a 70 µm nozzle. The rotational and vibrational temperatures in this continuous supersonic beam source amounted about 10 K and 50 K, respectively. The laser beams were focussed on the interaction area with the cluster beam by means of a 400 mm lens where each pulse reached a peak power of about 0.5 GWcm^{-2}.

Photoionized potassium dimers were mass-selectively detected by means of a quadrupole mass spectrometer with a resolution of m/Δm > 240. The amount of K_2 ions was continuously sampled as a function of the delay time between the pump and the probe pulse.

We recorded pump&probe spectra for the two isotopomeres of K_2 for delay times up to more than 200 ps. The first parts of the transients for both isotopomeres are shown in fig. 2. A fine oscillatory structure with an oscillation period $T_A \approx 500$ fs can be clearly

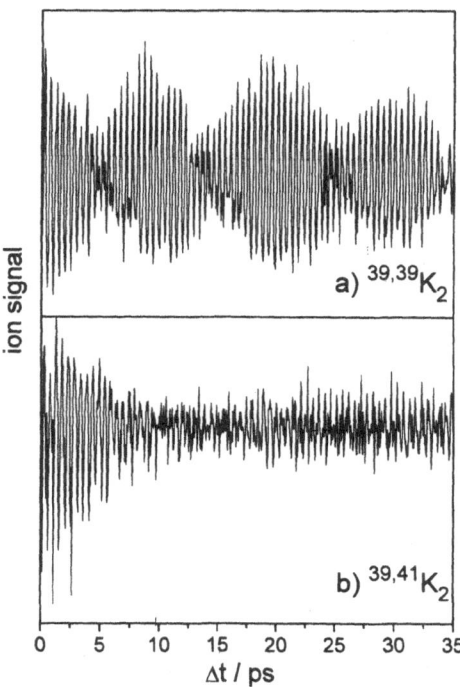

Figure 2. Transient evolution of the 3PI signal for a) ^{39,39}K$_2$ and b) ^{39,41}K$_2$.

seen in case of both isotopomeres. A 250 fs shift of the oscillation's first maximum with respect to the temporal origin is directly observed. The fast oscillatory structure of the ion signal's intensity is superimposed on a characteristic coarse structure being totally different for the two isotopomeres: for ^{39,39}K$_2$ a beat structure with a 10 ps period is dominant, whereas for the ^{39,41}K$_2$ a time-dependent structure with a fast decrease of the wave packet's amplitude and some fractional revivals[15] are observed (fig. 2b).

The frequencies involved in the transients are calculated by a Fourier analysis. The Fourier spectra are dominated by a group of frequencies around $\omega_0^{(1)} \approx 65$ cm^{-1}. Two additional frequency groups with lower amplitudes appear at $\omega_0^{(2)} \approx 130$ cm^{-1} and at $\omega_0^{(3)} \approx 195$ cm^{-1}. An additional peak is observed at $\omega_x \approx 90$ cm^{-1}.

The frequency group around $\omega_0^{(1)}$ is illustrated in fig. 3 for the case of the studied isotopomeres of K$_2$. Part a) of fig. 3 shows the fourier components in the pump&probe spectrum of ^{39,39}K$_2$: Here three main Frequencies at 63.8 cm^{-1}, 67 cm^{-1} and 67.2 cm^{-1} dominate this spectrum and another fairly strong frequency is present at 65.6 cm^{-1}. The spectrum for the isotopomere ^{39,41}K$_2$ is presented in fig 3b. There, we observe five distinct components in the frequency group at 64.3 cm^{-1}, 64.7 cm^{-1}, 65.5 cm^{-1}, 65.9 cm^{-1} and 66.4 cm^{-1}.

The classical oscillation period T with the energy level spacing between the vibrational levels of a diatomic molecule is given by $T = 1/(\omega c)$[16], where c is the speed of light in appropriate units. One immediately sees that the oscillation period $T_A \approx 500$ fs can be principally assigned to the classical oscillation period of K$_2$ in the A state in the energy region around the vibrational states with $v = 12,13$[10]. Frequencies around $\omega_0^{(1)} \approx 65$ cm^{-1}

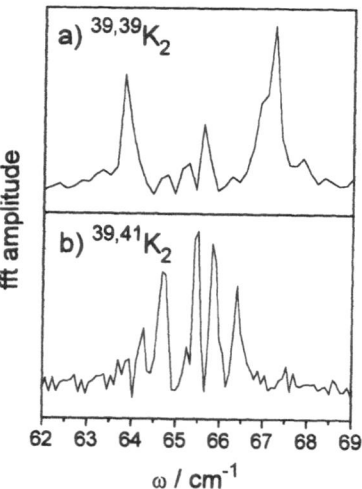

Figure 3. Frequency spectrum obtained by Fourier analysis for a) $^{39,39}K_2$ and b) $^{39,41}K_2$.

correspond to the energy level spacing in that energy region. Frequencies around $\omega_0^{(2)} \approx$ 130 cm^{-1} and $\omega_0^{(3)} \approx 195$ cm^{-1} are present in the Fourier spectra due to coherent excitation of energy levels with a difference in the vibrational quantum number of $\Delta v = 2$ and $\Delta v = 3$, respectively. The observed weak frequency component around 90 cm^{-1} can be assigned to a ground state wave packet being induced by impulsive stimulated Raman scattering, a process which becomes more dominant at higher laser peak powers[17].

The measured 250 fs phase shift of the first maximum (180° with respect to T_A) in the ion signal gives the information about the favored geometry of the K$_2$ molecules at the moment of ionization. The creation site of the wave packet in the A state is located near the inner turning point of the dimer's vibration. Therefore, it is evident that the ionization from the A state appears at the outer turning point. A deeper analysis indicates that the $(2)^1\Pi_g$ state can act as a Franck-Condon window to favor the two-photon ionization step at the internuclear distance corresponding to the outer turning point of the K$_2$ vibration[18].

From the RKR data[10] one would expect in case of $^{39,39}K_2$ that the applied laser pulses with a spectral width of 190 cm^{-1} would allow to excite the frequencies $\omega\{v,v+1\}$ for values of v between 10 and 16. Therefore, in the Fourier analysis of the transient data the frequencies 67.3 cm^{-1}, 66.9 cm^{-1}, 66.6 cm^{-1}, 66.3 cm^{-1}, 65.9 cm^{-1} and 65.6 cm^{-1} should appear. The discrepancy between the measured frequencies in fig. 3b and these expected values can be explained by assuming an energy level shift of the vibrational level $v = 13$ of the $A^1\Sigma_u^+$ state due to spin-orbit perturbations of that level by the $b^3\Pi_u$ state. An up-shift of this level will result in a smaller value for the spacing between $v = 13$ and $v = 14$ and a higher value for that of $v = 12$ and $v = 13$. A shift of 2.5 cm^{-1} would then give the frequencies 69.1 cm^{-1} and 63.8 cm^{-1} instead of 66.6 cm^{-1} and 66.3 cm^{-1}. Due to the de-pendence of the perturbation on the rotational quantum number and due to the fact that a variety a rotational levels can be excited the unperturbed frequency components should appear in the frequency spectrum, too. With this knowledge we assign the frequency 67.2 cm^{-1} to the levels $v = 10,11$, 67 cm^{-1} to $v = 11,12$, 65.6 cm^{-1} to $v = 15,16$ and 63.8 cm^{-1} to $v = 13,14$ where v=13 is shifted in energy. At 69.1 cm^{-1} and at the expected values for v=14,15 only weak frequency components are present. For a deeper interpretation the appropriate spin-orbit coupling and the proper Franck-Condon factors will be included in the analysis[19]. The appearance of the strong frequency at 63.8 cm^{-1} and the strong frequencies around 67 cm^{-1} is the reason for the beat structure observed in the transient for ^{39,39}K$_2$.

For the larger isotopomere, ^{39,41}K$_2$ only few spectroscopic data are available up to know, especially for the b$^3\Pi_u$ state. Hence, a comparison with the frequencies in fig. 3 and spectroscopic energy values and an assumption of eventually perturbed levels has not yet been performed. Although a regular frequency spectrum appears, we cannot totally exclude the perturbation of some of the vibrational levels. Only slight energy shifts would have an influence on fractional revival times of such a system. Further work will include the calculation of the vibrational energies of this isotopomere for the A$^1\Sigma_u^+$ state and the b$^3\Pi_u$ state[19].

Three photon ionization pump&probe spectroscopy enabled us to follow the evolution of a wave packet created in the K$_2$ A$^1\Sigma_u^+$ state. The vibration period of dimers excited by the pump laser pulse amounts about 500 fs which corresponds to frequencies in the range of 65 cm^{-1}. For the isotopomere ^{39,39}K$_2$ we measured a strong perturbation of the vibrational level v=13 which results in unexpected frequency components and a dominant beat structure with a period of 20 ps. The appearance times of the revivals for ^{39,41}K$_2$ will give hints to perturbations in this rarely studied dimer system. Hence, we think we could show that femtosecond spectroscopy enables to observe perturbations of excited vibrational levels with high sensitivity. While the revival times allow to estimate a first rough value of perturbations in comparison to RKR data, the Fourier analysis gives information on the energy level spacing of vibrational states with an accuracy of 0.1 cm^{-1}. Applying higher peak powers to the system opens an ionization pathway where impulsive stimulated Raman scattering (ISRS) becomes stronger. Then the oscillation of a ground state wave packet with T$_X$ = 360fs dominates the ion signal[17].

This work has been financially supported by the Deutsche Forschungsgemeinschaft within the Sonderforschungsbereich 337 (TP A8).

REFERENCES

1. M. Gruebele, G. Roberts, M. Dantus, R:M: Bowman,, and A.H. Zewail, Chem. Phys. Lett. **166**, 459 (1990).
2. T. Baumert, M. Grosser, R. Thalweiser, and G. Gerber, Phys. Rev. Lett. **67**, 3753 (1991).
3. M. Gruebele, and A.H. Zewail, J. Chem. Phys **98**, 883 (1993).
4. V. Blanchet, M.A. Bouchene, O. Cabrol, and B. Girard, Chem. Phys. Lett. **233**, 491 (1995).
5. P. Kusch, and M.M. Hessel, J. Chem. Phys **63**, 4087 (1975).
6. C. Effantin, O. Babaky, K. Hussein, J. d'Incan, and R.F. Barrow, J. Phys. B **18**, 4077 (1985).
7. G. Gerber, and R. Möller, Chem. Phys. Lett. **13**, 546 (1984).
8. X. Xie, and R.W. Field, Chem. Phys. **99**, 337 (1985).
9. A.J. Ross, R.M. Clements, and R.F. Barrow, J. Mol. Spectry. **127**, 546 (1988).
10. A.M. Lyyra, W.T. Luh, L. Li, H. Wang, and W.C. Stwalley, J. Chem. Phys. **92**, 43 (1990).
11. A.J. Ross, P. Crozet, C. Effantin, J. d'Incan, and R.F. Barrow, J. Phys. B **20**, 6225 (1987).
12. J.B. Atkinson, J. Becher, and W. Demtröder, Chem. Phys. Lett. **87**, 92 (1982).
13. A.J. Ross, C. Effantin, J. d'Incan, and R.F. Barrow, J. Phys B **19**, 1449 (1986).
14. S. Rutz, E. Schreiber, and L. Wöste, Surf. Rev. and Lett., in press.
15. I.Sh. Averbukh, and N.F. Perelmann, Phys. Lett. A **139**, 449 (1989).
16. L.D. Landau, and E.M. Lifshitz, *Quantenmechanik* (Akademie-Verlag, Berlin, 1988).
17. R. de Vivie-Riedle, K. Kobe, J. Manz, W. Meyer, B. Reischl, S. Rutz, E. Schreiber, and L. Wöste, J. Phys. Chem, submitted.
18. R. de Vivie-Riedle, B. Reischl, S. Rutz, and E. Schreiber, J. Phys. Chem, in press.
19. S. Rutz, R. de Vivie-Riedle, and E. Schreiber, to be published.

29

FEMTOSECOND DYNAMICS OF THE GROUND STATE OF Ag_3

A New Approach to Study the Ultrafast Dynamics of Mass-Selected Neutral Clusters

E. Schreiber,[1] S. Berry,[2] T. Leisner,[1] S. Rutz,[1] S. Wolf,[1] and L. Wöste[1]

[1] Freie Universität Berlin
Institut für Experimentalphysik
Arnimallee 14, 14195 Berlin, Germany
[2] The University of Chicago
Chicago, Illinois, 60637

We describe a new approach which we have used to investigate small silver clusters, Ag_n, (n=3,5,7 and 9). The method starts with a beam of mass-filtered, negatively-charged clusters. These anions are subjected to photodetachment. Subsequently, after a variable but selected temporal delay they are photoionized. The produced cations are then mass-analyzed and collected. Their intensity as a function of the delay interval between the ultrashort pulses is a measure of the Franck-Condon factor for photoionization of a neutral prepared by a vertical detachment process from a low-lying vibrational state of the anion. Hence, the vibrational motion of the neutral is probed. For convenience, we shall refer to the process as NeNePo, Negative-to-Neutral-to-Positive (see also[1]). In this letter, the NeNePo results for the Ag_3 will be presented, only. Fig. 1 displays the principle of the NeNePo process.

The silver cluster anions are generated in a sputtering ion source by bombarding targets of elemental silver with fast Xe^+ ions.[2] Vaporizing cesium metal onto the targets lowers the work function of the silver surface and enables to increase the yield of cluster anions drastically. An ion lens collects the generated cluster anions and leads them into a first quadrupole ion guide of large cross-section. Here the cluster ions are cooled and moderated by collisions with a background gas of He ($p_{He} \approx 10^{-2}$ mbar).[3] The cations then approximately have room temperature. Next the anions pass a quadrupole mass filter, where the cluster size of interest can be selected. A beam of well thermalized mass selected metal cluster ions is created by this method. To increase the density of the anions, they are stored in a linear quadrupole ion guide which is filled with gas and operates as an ion trap.[4] Therefore, the entrance lens of the ion guide is kept at a potential being slightly

Ultrafast Processes in Spectroscopy, edited by Svelto et al.
Plenum Press, New York, 1996

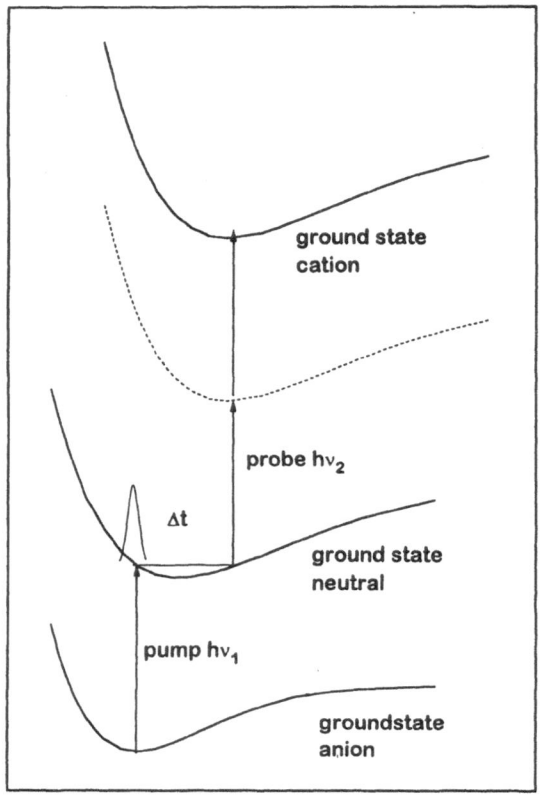

ground state cation

probe hν_2

Δt

ground state neutral

pump hν_1

groundstate anion

molecular coordinate (arb. units)

Figure 1. Principle of the NeNePo process: Beginning from the anion's potential energy surface an ultrashort pump pulse detaches an electron and prepares a wave packet in the neutral. After a certain delay time Δt a probe pulse photoionizes the neutral. The time-dependent signal of the cation's intensity is detected.

below the kinetic energy of the clusters, while the exit lens is on a higher potential. Hence, the anions can enter the ion guide, but are reflected back at the exit lens. Traveling through the gas cell, the ions loose kinetic energy by collisions and are no longer able to escape via the entrance lens. The number of stored ions can be monitored by pulsing the exit lens open and recording the magnitude of the ion-current pulse. Without the detachment laser beam, more than 10^8 ions can be stored in the trap. With the detachment laser, an equilibrium between the continuous filling and the depletion of the trap by the detachment process is reached. In this case, an ensemble of about 10^6 mass selected cations is stored and can interact with the subsequent detachment and ionization pulses. The cations created by the laser pulses are extracted by the exit lens. Having passed a second quadrupole mass filter the cations are detected by means of a secondary electron multiplier detector. This allows to detect possible fragmentation of the clusters during or after the interaction with the light pulses.

To analyze the temporal evolution of the neutral trimer we employed a laser system consisting of a Titanium sapphire oscillator (Spectra Physics Tsunami) which is pumped by a 12 W argon ion laser and of a Nd:YLF-pumped regenerative amplifier (Quantronix 4800). This configuration enables to produce ultrashort ($\tau < 100$ fs, 500 μJ/pulse) laser pulses at a repetition rate of 1 kHz. The pulses are efficiently (~40%) frequency doubled in a BBO crystal. The second harmonic is split into pump and probe pulses, with the probe pulse retarded with respect to the pump pulse by a computer controlled translation stage. Pump and probe laser beams are imaged into the trap collinear with the ion trajectories and overlap throughout the whole length of the trap. The electrons of the stored cluster anions are detached by the pump pulse, and after a certain delay time Δt, the just created neutrals are ionized by the probe pulse. This is achieved by nonresonant two-photon ionization.[5,6] Direct one-photon ionization experiments, however, are under work. The detected ion signal gives the yield of mass selected cluster cations produced from mass selected cluster anions as a function of the delay time between pump and probe pulse at fixed wavelength and fluence of these pulses.

In our first experiments we studied the silver trimer. Mass selected Ag$_3^-$ ions were produced with an intensity of about 2 nA, and stored in the ion trap. For the detachment process laser pulses with a central wavelength of 420 nm, 415 nm, 400 nm, and 390 nm were applied. This allowed one-photon detachment of the anions. The ionization was performed nonresonantly using two photons of the same wavelength. Since the energy of two photons of 420 nm is just slightly above the ionization potential of the silver trimer a very soft ionization is realized. Employing a wavelength of 415 nm, positive ions are mainly detected when there is a nonzero time delay between pump and probe laser pulses. This confirms that sequential processes of detachment and ionization are involved in the creation of the cations.

In Fig. 2 the yield of Ag$_3^+$ is displayed as a function of the delay time Δt for various wavelengths of the detachment and ionization laser. At $\Delta t=0$, pump and probe laser pulses exchange their role. However, the traces are not symmetrical since the respective fluences of pump and probe pulse are different. For the longer wavelengths, the ion yield rises from almost zero to a maximum around $\Delta t=750$ fs, and then decays to a constant value at longer time delays. There it stays constant for more than 100 ps which is the longest time delay used in our experiment. If we use light of shorter wavelength this phenomenon is progressively washed out. Using pulses at a central wavelength of 390 nm the ionization efficiency is almost independent of the delay time Δt. The time required to reach the maximum grows with increasing wavelength from 500 fs to about 800 fs. In our opinion this is consistent with the notion that the extra energy goes at least in part into the bending vibration. The dependence of the cation yield on the power of the pump and probe laser pulses shows that the detachment process depends linearly, but the ionization process quadratically on the respective light intensity. This is in good agreement with the creation process of the cations as discussed above.

As theoretical predictions[7,8] indicate the most stable geometry of the trimer is linear for the anion, obtuse isosceles for the neutral and equilateral for the cation (see Fig.3). The neutral trimer is presumably generated in a linear and therefore highly vibrationally excited configuration, at a saddle point, from which it bends slowly at first and then faster, comes through the geometry of the obtuse isosceles minimum and then decelerates until it approaches equilateral geometry, where its overlap with the positive ion is greatest.

In fact the results obtained with the silver trimers do not show the multiply-periodic behavior of a simple vibrational spectrum measured in the time domain. Rather, some of the results reveal less of the vibrational spectra but more of the dynamics of the internal rearrangements of these species, as shown by the time-dependent currents of positive ions in Fig. 2. Hence, we use the following hypothesis to interpret the behavior of the trimer signal: initially, the neutral is produced in a linear configuration by the vertical Franck-Condon detachment process. The Franck-Condon overlap factor of the linear neutral with the equilateral positive ion is so low that virtually no positive ions are generated. However, the neutral starts to bend and after passing through the obtuse equilibrium geometry

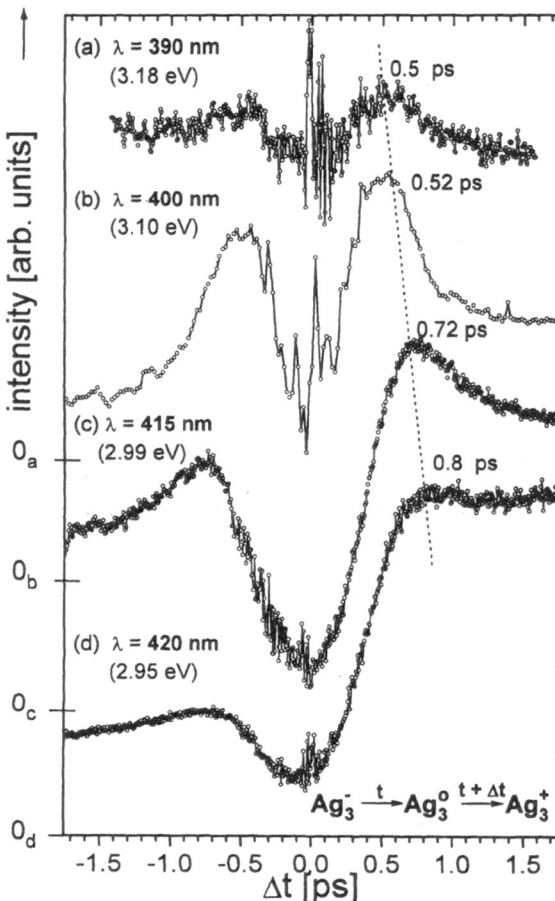

Figure 2. NeNePo spectra of the silver trimer taken with wavelength of $\lambda = 390$ nm, 400 nm, 415 nm, and 420 nm. Note that each curve has its own axis of zero signal, and that the time-independent background increases steadily with decreasing wavelength. The fine structure around $\Delta t = 0$ is due to interference of pump and probe pulses.

of the neutral it approaches the equilateral equilibrium geometry of the cation after a certain delay time Δt. During this time the cation signal grows continuously reaching a maximum when the system is close to the classical turning point near the equilateral triangular geometry. Then the system rebounds and the signal decreases. However the vibrational excitation is high enough that the modes mix, and after the rebound, the still-unionized neutral trimers are left with enough energy to pseudorotate through their three equivalent obtuse-triangular equilibrium structures, going around the trough of their "Mexican hat" potential energy surface[9]. In so doing, they remain at a roughly constant distance from the equilateral geometry, so that the Franck-Condon factor also remains nearly constant, and therefore so does the positive ion signal. The signal at short times is more pronounced at

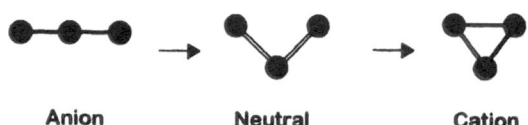

Anion   Neutral   Cation

Figure 3. Geometric configurations of the silver trimer as anion, neutral, and cation.

relatively long wavelength of the ionization laser, as then the ionization probability is strongly dependent on the vertical ionization potential in the momentary configuration of the neutral.

To conclude we like to emphasize that NeNePo is a quite general scheme for investigating the time evolution of a coherent nonequilibrium state in neutral clusters. Further investigations will be carried out using different cluster systems and different ionization pathways. NeNePo might also be used to investigate reactive compounds (i. e. cluster ligand systems) where a chemical reaction or molecular rearrangement starts after the neutralization. The associated dynamics could eventually deduced by detecting and energy analyzing the photoelectrons of the probe process as a function of Δt.

ACKNOWLEDGMENTS

The authors want to thank Professor V. Bonacic Koutecky for valuable discussions. Support from the Deutsche Forschungsgesellschaft (SFB337) is gratefully acknowledged.

REFERENCES

1. S. Wolf, G. Sommerer, S. Rutz, E. Schreiber, T. Leisner, L. Wöste , and S. Berry, Phys. Rev. Lett. **74**, 4177 (1995)
2. T. Leisner, S.Wolf, and L.Wöste, in preparation
3. L. Hanley, S. A. Ruatta, and S. L Anderson, J. Chem. Phys. **87**, 260, (1987)
4. G. G. Dolnikowski, M. J. Kristo, C. G. Enke, and J. T. Watson, Int. J. Mass Spectr. Ion Proc. **82** 1 (1988)
5. P. Y. Cheng and M. A. Duncan Chem. Phys. Lett. **152**, 341, (1988)
6. A. M. Ellis, E. S. J. Robles, and T. A. Miller, Chem. Phys. Lett. **201**, 132, (1993)
7. V. Bonacic-Koutecky, L. Cespiva, P. Fantucci and J. Koutecky, J. Chem. Phys. **98**, 7981 (1993).
8. V. Bonacic-Koutecky, L. Cespiva, P. Fantucci, J. Pittner and J. Koutecky, J. Chem. Phys. **100**, 490 (1994).
9. M. Broyer, G. Delacretaz, P. Labastie, J. P. Wolf und L. Wöste, J. Phys. Chem. **91**, 2626 (1987)

30

FEMTOSECOND DYNAMICS OF MOLECULAR IODINE IN RARE GAS SOLIDS

R. Zadoyan, M. Sterling, and V. A. Apkarian

Department of Chemistry
University of California
Irvine, California 92717

A molecular level of understanding the many-body interactions that govern chemical dynamics in condensed media remains one of the main challenges to modern chemical physics. The major impediment to the development of such an understanding in the past has been the lack of detailed experimental observables against which theories can be tested and refined. Ultrafast spectroscopies, such as femtosecond pump-probe studies, promise to change this situation. This has been demonstrated most dramatically in recent studies of the model system of molecular iodine isolated in cryogenic rare gas solids, where by taking advantage of the reduction of initial phase space distribution, it has been possible to follow the most elementary of chemical processes, namely: the breaking and remaking of a chemical bond, in its entirety, in real-time.[1] In contrast with frequency domain spectroscopies, where the available time window for observations is limited by dephasing times of states prepared in a bath, in the case of ultrafast pump-probe measurements, the observation times are limited by the duration of the classical population coherences created by the pump pulse. Such times are typically many picoseconds, and therefore long enough to be meaningful on time scales relevant to chemical change. The information content of such measurements is clearly determined by the extent of the created coherences. Understanding the system parameters that determine the extent of coherences, and developing the tools necessary to extract the information content from observables, are the necessary developments in *dynamical spectroscopy* which is uniquely suited for characterization of condensed media on an atomistic level of detail. We exemplify this aspect through an exposition of the vibrational coherences and predissociation in the bound $B(^3\Pi_{0u})$ state of I_2 in solid Kr. While real-time observations of condensed phase dynamics is an important development, its successful execution suggests as a natural second step the possibilities of *controlling dynamics* by the coherence of the laser field. We have demonstrated this possibility using chirped laser pulses.[2,3] In particular, we consider the effect of chirped pulses on both preparation and interrogation of the highly nonlinear dynamics involved in the caging of photoproducts after photodissociation in a matrix.

Ultrafast Processes in Spectroscopy, edited by Svelto et al.
Plenum Press, New York, 1996

The well known structured B←X absorption spectrum of gas phase I_2 is commonly used as a wavelength calibration standard. When isolated in rare gas matrices, this spectrum collapses into a broad continuum due to electronic dephasing of the excited state on a time scale of <50 fs. Information on a somewhat longer time scale is afforded through Resonance Raman spectra of this system in condensed media, solids[4] and liquids,[5] where long overtone progressions are observed. The information content of these spectra, dictated by the projection of the correlation function of a prepared superposition state on the I-I coordinate, is limited to ~100 fs. In contrast, when prepared by a pulse of ~100 fs, the population coherence can be probed for many piscoseconds as illustrated by an example in figure 1. The analysis of such a transient requires a full treatment of the system dynamics. To this end, we carry out molecular dynamics simulations, by computing the classical equations of motion of an I_2 molecule embedded as a doubly substituted impurity in an assembly of 500 lattice atoms. The generated trajectory ensemble is then inverted, using the classical Franck principle for absorption, as a convolution between the probe window function W(r, t) with the evolving trajectories $\{R_i(t)\}$. The simulated signal, which is indistinguishable from the experiment, is also shown in figure 1. The inputs to this simulation were the I-Kr, Kr-Kr, and I_2(B) pair potentials. In generating the match we iterate the two window function parameters, width and position, and an exponential decay function to represent the rate of predissociation. The window parameters define the gradient and position of the difference potential, and pinpoint the $E(0_g^+)$ ion-pair potential, the location of which is otherwise unknown in the solid state. The exponential decay function determines the predissociation probability at this energy to be 5% per period, and yields the off diagonal matrix element that couples the B state to the repulsive surface that crosses it, presumably the a(1_g) state. Using the gas phase curvatures of a and B potentials, a coupling element $V_{12} = 1000$ cm^{-1} is extracted for the predissociation in solid Kr. The data illustrated in figure 1 is only one of a set conducted as a function of pump and probe wavelengths. Analysis of the full set characterizes the diagonal and off-diagonal interactions among various solvated electronic states, both for covalent and ionic configurations of the molecule. Such information is simply not at accessible by frequency domain spectroscopies.

To investigate the controllability of dynamics by the coherence of the laser, we focus on the elementary step of caging: the collision of photodissociation products with the immediate host atoms and their recoil to reform a molecular bond. We prepare the system by dissociative excitation on the A($^3\Pi_{1u}$) surface, while we tune the probe near the minimum of the β-A difference potential. The latter choice guarantees the observation to be limited to only two dynamical resonances: the pre-collision resonance and the post-colli-

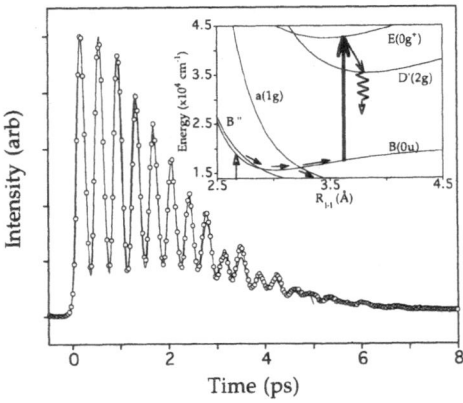

Figure 1. Experiment and simulation of pump probe signal for I_2 isolated in solid Kr. The amplitude decay, t = 5ps, is due to predissociation assumed to be via the a(1_g) state. The relevant potentials are illustrated in the inset. In the example λ(pump) = 549 nm, λ(probe) = 400 nm, while the LIF is collected on the D'→A' transition at 420 nm.

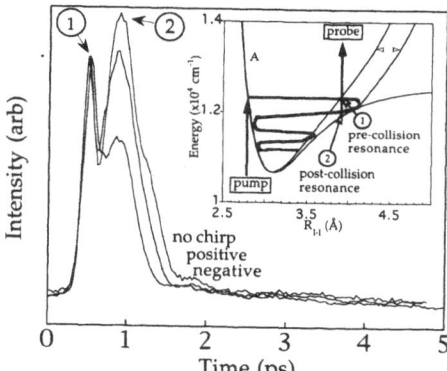

Figure 2. Pump-probe signals of caging with chirped pulses. The system is prepared on the A state, above dissociation, and probed such that only two resonances are observed as indicated in the inset.

sion resonance, as indicated in the inset to figure 2. Accordingly, the signal is composed of two peaks (see fig. 2). The data presented are recorded by chirping both pulses by ~5cm⁻¹/fs. Note, the pre-collision resonance is laser response limited and as such can be used to measure the cross correlation between pump and probe pulses. It can be seen that while the time response of the system is virtually unchanged, the post collision resonance shows a strong chirp dependence. We conclude that the highly nonlinear dynamics of: collision with the wall, energy loss, and recoil, proceeds with retention of memory of the initial preparation of the wavepacket. A detailed theoretical analysis of these effects – the effect of the probe chirp on the observable signal, and the effect of the pump chirp on the coherence of the dynamics – have already been given.[2] Probing with chirped pulses is subject to a Doppler effect which arises from the fact that the signal corresponds to convolution of a traveling wavepacket with a traveling window function, and therefore provides vectorial information about the evolving dynamics. Pumping with a chirped pulse has the effect of timed launches of trajectories. In the case of a negatively chirped pulse, the higher energy trajectories are launched first. Since these trajectories travel a longer distance in the strongly anharmonic potential of the cage, they will catch-up with the lower energy trajectories which are launched last. The result is a focusing during recoil, which in turn leads to a significantly more coherent dynamics – a more compact distribution evolving in phase space – for several picoseconds. In effect, the particular dynamics under consideration is controllable.

REFERENCES

1. R. Zadoyan, Z. Li, P. Ashjian, C. C. Martens, and V. A. Apkarian, *Chem. Phys. Lett.* **218** (1994) 504; R. Zadoyan, Z. Li, C. C. Martens, and V. A. Apkarian, *Proceedings of SPIE*, **2124** (1994) 233; R. Zadoyan, Z. Li, C. C. Martens, and V. A. Apkarian, in *Ultrafast Phenomena IX*, edited by G. A. Mourou and A. H. Zewail, (Springer, New York, 1994) p.79; R. Zadoyan, Z. Li, C. C. Martens, and V. A. Apkarian, *J. Chem. Phys.* **101** (1994) 6648; Z. Li, R. Zadoyan, V. A. Apkarian, and C.C. Martens, *J. Phys. Chem.* **99** (1995) 7453.
2. M. Sterling, R. Zadoyan, and V. A. Apkarian, J. Chem. Phys. (submitted, 1995); V. A. Apkarian, Proceedings of "Femtochemistry: The Lausanne Conference", (in press).
3. R. Zadoyan, B. Kohler, K. R. Wilson and V. A. Apkarian, (to be published).
4. W. F. Howard and L. Andrews, J. Raman Spec. **2** (1974) 447; ibid. **4** (1975) 99.
5. J. Xu, N. Schwentner, M. Chergui, J. Chem. Phys. **101** (1994) 7381; J. Xu, N. Schwentner, S. Hennig, M. Chergui, J. Chim. Phys. Phys-Chim Biol., **92** (1995) 541.

31

TIME-RESOLVED SPECTROSCOPY OF HIGH EXCITED STATES OF ALKALI DIMERS

K. Ahmed,[1] T. Weyh,[2] E. Mehdizadeh,[3] and W. Demtröder[2]

[1] Physics Department
University of Karachi-75270, Pakistan
[2] Fachbereich Physik Der Universitat Kaiserslautern
67657, Germany
[3] Physics Department
Shahid-Bahonar University of Kerman, 76175, Iran

Alkali dimers are also interesting and important since they are thought to undergo primarily excimer like triplet-triplet electronic transitions in which the lower triplet state is repulsive [1,2]. Because of inherent population inversion and relatively broad spectral ranges, they are potentially important continuous tunable laser sources.

The present paper reports on such time-resolved fluorescence measurements which yields radiative lifetime and quenching cross-section of excited states of Li_2 and Na_2.

A mode-locked argon ion laser at $\lambda = 488$ nm is used for synchronous pumping of a jet stream dye laser with a DCM dye. The pulse duration of the mode-locked output pulses was below 2 ns, whereas dye laser pulse widths of less than 15 psec could be reached. However, in order to obtain spectral narrowing of the dye laser pulse, which is necessary for selective excitation of molecular levels, a three-plate bire-fringent filter and two solid etalons were placed inside the dye laser resonator. This decreases the spectral width down to 1.5 GHz but increases the time duration of the pulse to 0.5 nsec. For coarse wavelength tuning the birefringent filter was tuned while for the fine tuning the etalons are simultaneously tilted [3].

The life time of the levels to be measured are in the range of 5–70 nsec. In order to be observed the fluorescence decay of an excited level over a time scale sufficiently long for a statistical analysis, the interval between two successive laser pulse should be greater than approximately 4 mean lifetimes. This is achieved by the cavity dumping technique [4], where all mirror of the dye laser cavity are high reflectors and the mode-locked pulses circulation inside the resonator are coupled out of the cavity by synchronized acousto-optic switch (Bragg cell) at a repetition frequency of 4 MHz.

A small fraction of the output beam is coupled by a beam splitter into a Computer-controlled Fabry-Perot wavemeter [5,6]. Since the excitation energies of the investigated upper sate in both alkali molecules are around 30,000 cm^{-1} the mode-locked optical pulses

Ultrafast Processes in Spectroscopy, edited by Svelto et al.
Plenum Press, New York, 1996

143

with a wavelength around 650 nm have to be frequency doubled. This is realized by focusing the output pulses with a lens of f = 15 cm into a 1 cm long LiIO$_3$ crystal that can be rotated around a vertical axis in order to achieve optimum phase-matching for all wanted wavelengths.

The UV-pulses with a peak powers of about 18 mW and pulse energies of 8 PJ are focused by a quartz lens with f = 15 cm into the heat-pipe containing alkali vapors and argon buffer gas. The laser induced fluorescence is imaged by two quartz lenses onto the entrance slit of a monochromator. A fast photomultiplier behind the exit slit detects single photon. Its output pulses start a time-to-amplitude converter (TAC) which is stopped by the next laser pulse. This modified delayed coincidence single photon counting techniques [7,8] allows one to operate the TAC at a high stop pulse rate (4MHz) but a much lower start pulse rate (10 - 100 KHz) which minimizes the dead time of the TAC. The output pulses of a TAC are stored in a multichannel analyzer and delivered to Computer which performs a least square fit to the experimental data.

The frequency doubled laser was tuned to wavelength around the 325 nm in order to excite the rovibrational levels (v' , J') in the C$^1\Pi_u$ state of Li$_2$ and Na$_2$ [9–11]. The laser induced fluorescence spectra are recorded between 310 to 620 nm. We have also recorded collisionally populated states in violet and visible regions.

Keeping the dye laser on the selected excitation line and the monochromator on a selected fluorescence transition of C$^1\Pi_u$ state, the time-resolved decay curves were measured as a function of a heat-pipe temperature (which determine the Alkali dimer pressure) [12].

A least square fit of the experimental data to the function.

$$I(t) = a + b \, \exp(-t \, / \, \tau_{eff}) \qquad (1)$$

where t = 0 is chosen at the end of the exciting laser pulse, gives an effective lifetime. With the deconvolution programmme [13] the time t = 0 can also be chosen at the beginning of the excitation pulse if the laser pulse profile I$_L$(t) has been measured using laser scattered from the cold heat-pipe.

A Stern-Volmer plot of the inverse effective lifetimes

$$1 \, / \, \tau_{eff} = (1 \, / \, \tau_{rad}) + n\sigma\bar{v} = (1 \, / \, \tau_{rad}) + \sigma \sqrt{(8 \, / \, \pi\mu kT)} \cdot P_s \qquad (2)$$

measured at different atomic densities at a temperature "T" with a mean relative velocity $\bar{v}$ = (8kT / $\pi\mu$)$^{1/2}$ of the collisional partners with reduced mass μ. The radiation lifetime τ_{rad} and the quenching cross-section of optically excited states in both molecules are compiled in Table 1.

The lifetime of the collisionally populated states were measured in a similar way with the monochromator tuned to the maximum intensity, corresponding to strong fluorescence. The time-dependent fluorescence intensity I$_2$(t) Fig. 1. of the collisionally populated states (CPS) which are proportional to the population density N$_2$(t) = N(CPS) of its upper state. It can be described by the rate equation

$$dN_2 \, / \, dt = N_1(t) \cdot C_{12} - N_2(t) \, (A_2 + C_2) \qquad (3)$$

where |1> is the optically excited level in the C$^1\Pi_u$ state, A$_2$ and C$_2$ are the probabilities of radiative and collisionally induced transitions from |2> and C$_{12}$ = n σ_{12} $\bar{v}$ in the probability

Table 1. Lifetime and Quenching Cross-section of Optically and Collisionally excited states of Li_2 and Na_2

Molecule	State	Excited level (v', J')	Wavelength [nm]	Lifetime τ_{rad} [nsec]	σ_{tot} [alkali dimmer-alkali atom] [A^2]	σ_{tot} [alkali dimmer-Ar atom] [A^2]
Li_2						
	$C^1\Pi_u$	(2,7) f	321.44	60.60 ± 2.80	383.70 ± 15	73.30
	$C^1\Pi_u$	(2,7) f	323.20	62.40 ± 5.40	303.50 ± 13	—
	$2^3\Pi_g$	—	436.00	15.30 ± 0.50	093.50 ± 13	23.00 ± 3
	$C^1\Pi_u$	(2,7) f	323.20	60.50 ± 5.40	342.90 ± 11	—
	1^3A_g	(v' = 2)	516.00	19.40 ± 0.60	222.00 ± 16	62.00 ± 3
Na_2						
	$C^1\Pi_u$	(9,34)e	327.49	07.90 ± 0.50	850.00 ± 50	—
	$C^1\Pi_u$	(9,34)e	327.49	07.60 ± 0.70	875.00 ± 80	—
	$2^3\Pi_g$	—	436.00	24.00 ± 2.00	320.00 ± 30	220 ± 50
	$C^1\Pi_u$	(9,34)e	327.49	07.96 ± 0.70	828.00 ± 70	—
	$2^1\Sigma_u^+$	—	452.00	52.50 ± 5.00	325.00 ± 30	—

of collision-induced transitions $|1>\!\!-\!\!-\!\!-\!\!-\!\!-\!\!|2>$ at a density n of collisional partners with reduced mass μ. The integration of eq. 3. yields with $N_1(t) = N_{10} \exp(-t / \tau_1)$

$$N_2(t) = N_{10} (\tau_2 / (\tau_2 - \tau_1)) [\exp(-t / \tau_1) - \exp(-t / \tau_2)] \qquad (4)$$

A least squares fit of the experimental data to eq. 4. yields the effective lifetimes τ_1^{eff} and τ_2^{eff}. Measurements at different pressure allow one to make Stern-Volmer plots from which the radiative lifetime τ_1 and τ_2, collisional quenching cross-sections σ_1 of level $|1>$ and σ_2 of level $|2>$ are evaluated [14] and presented in Table 1.

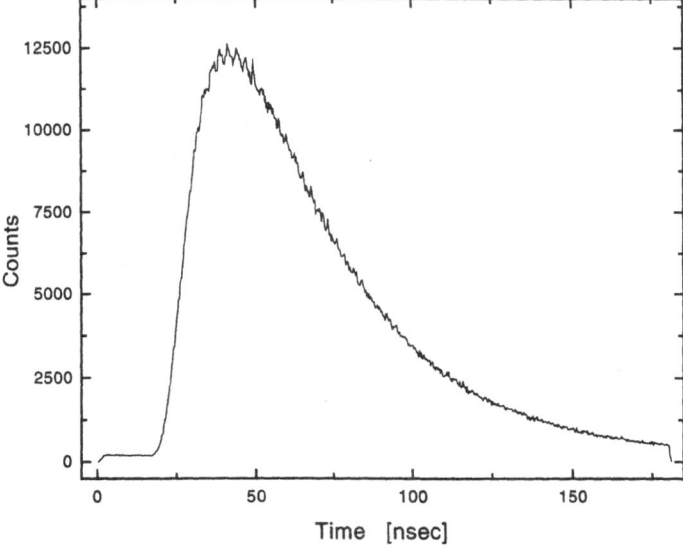

Figure 1. Time dependence fluorescence decay curve of the $1^3\Delta_g$ state of Li_2 (Excitation wavelength $\lambda = 321.44$ nm in the $C^1\Pi_u$).

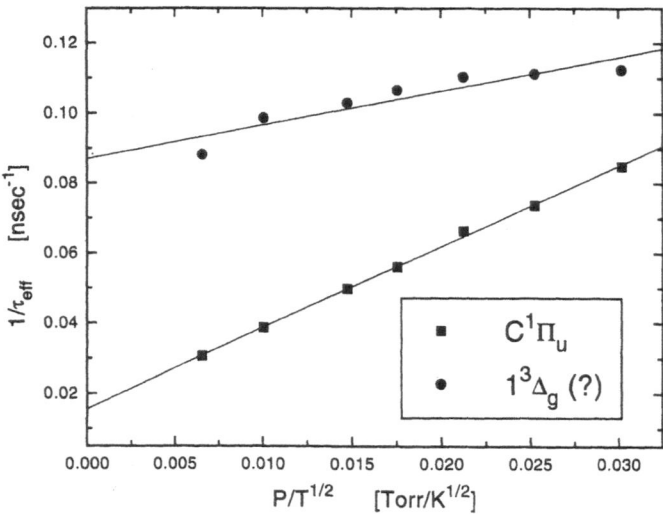

Figure 2. Stern-Volmer plot of $(\tau_1^{eff})^{-1}$ and $(\tau_2^{eff})^{-1}$ for the $C^1\Pi_u$ ($v'=3$, $J'=14$) and $1^3\Delta_g$ state of Li_2 as a function of atomic density, expressed by the product $PT^{-1/2}$ of pressure and temperature measured in heatpipe.

The combination of pulse modulation laser technique and single photon counting method is a very powerful tool for cascade-free measurements of excited state lifetime. By employing a deconvolution procedure considerably shorter lifetime can also be measured with good accuracy.

REFERENCES

1. T.S.Vih and C.Y. Robert Wu, Opt. Commun. **70**, 319 (1989).
2. S. Shahdin, B. Wellegehausen and Z. G. Ma, Appl. Phys. B **29**, 195 (1982).
3. E. Mehdizadeh, K. Ahmed and W. Demtröder, Appl. Phys. B **59**, 509 (1994).
4. R. H. Johnson, IEEE Jour. Quant. Elect. **QE-9**, 255 (1973).
5. A. Fischer, R. Kullmer and W. Demtröder, Opt. Commun. **39**, 277 (1981).
6. R.Castell, W. Demtröder, A. Fischer, R. Kullmer, H.Weickenmeier and K. Wickert, Appl. Phys. **38**, 1 (1985).
7. D.V.O. Connor, D. Phillips, "Time Correlated single photon counting method" (Academic Press, NewYork 1984).
8. W.Demtröder, "Laser Spectroscopy" (Third Edition Springer-Verlag 1992).
9. G.-Y. Yan, B. W. Sterling, T. Kalka and A. L. Schawlow, Jour. Opt. Soc. Am. B **6**, 1975 (1989).
10. K. Ishikawa, S. Kubo and H. Kato, Jour. Chem. Phys. **95**, 8803 (1991).
11. D. K. Hsu, Ph. D. Thesis, Fordham University (1974).
12. M.Nesmeyanov, "Vapor Pressure of the elements" (Academic Press, NewYork 1963).
13. B.Bieniak, Private Communication.
14. K. Ahmed, Ph. D. Thesis, Karachi University (1994).

32

ULTRAFAST TEMPORAL DYNAMICS IN AN OPTICAL MICROSCOPIC CAVITY

P. Mataloni, A. Aiello, F. De Martini, and M. Giangrasso

Physics Department and I. N. F. N.
University of Rome "La Sapienza"
Rome, Italy

Optical plane dye microcavities with dimensions approaching a wavelength of the emitted light have been investigated in several experiments. Although the particular geometry of a Fabry-Perot microcavity closed by two multilayer mirrors and the large gain bandwidth of the active medium limit the value of the confinement on the forward cavity mode, microlaser action has been demonstrated in these systems[1]. High gain and low threshold behaviour have been recognized to be distinctive properties of the dye microlaser[2]. In this paper we report some important results related to the ultrafast temporal dynamics of the microlaser. The basic experimental set-up is shown in Fig.1. The microcavity, filled with a solution of Oxazine 725 in ethylene glycol, was pumped by the amplified pulse (50 fs) of a CPM dye laser. The pump laser was focused to a spot of 15 μm of diameter, well inside the transverse coherence length of the microlaser[3]. The microcavity with finesse f $\cong$ 1000, was terminated by two plane multilayer dielectric mirrors reflecting at the emission wavelength ($\lambda \cong 700$ nm) and transparent to the pump wavelength ($\lambda_p = 615$ nm). A silver deposition of .25 μm, partially evaporated on one of the two mirrors determined the effective mirrors spacing. The microlaser temporal dynamics could be analyzed by adopting a conventional time-resolved frequency mixing technique in a non - linear crystal (KDP). The ultrafast response capability is a direct consequence of two peculiar characteristics of the microlaser: the extremely short photon lifetime and the large enhancement of the spontaneous emission rate in the forward cavity mode[4]. The relevant parameter of the temporal evolution of the microlaser output pulse, following the femtosecond excitation, is the build-up time, τ_{bu}, which corresponds to the time elapsing between the pump excitation and the onset of the pulse due to stimulated emission. Its expression is: $\tau_{bu} \propto 2n^{\cdot}(1+2n)/(N_i \, \Gamma)$, where n is the cavity order, N_i is the initial upper state population and Γ represents the spontaneous emission rate. For a thin microcavity the build-up time is very short in appropriate pumping conditions. In Fig. 2 the shape of the microlaser output pulse, compared with the theoretical fit, is reported. In this case the microlaser was excited by only one pump beam. A 3 ps build-up time, together with a rise-time of nearly 1 ps have been measured for a pumping rate of nearly 30 times above

Ultrafast Processes in Spectroscopy, edited by Svelto et al.
Plenum Press, New York, 1996

147

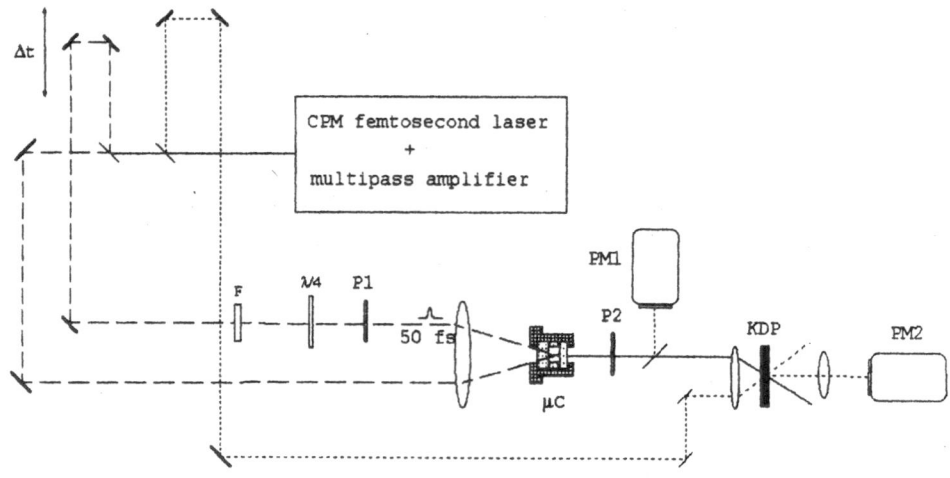

Figure 1. Experimental apparatus.

the threshold. The trailing edge of the output pulse corresponds to the exponential decay of the stored photon energy with a time constant, 5.7 ps, equal to the cavity photon life-time.

   Another important feature of the microlaser dynamics is given by its polarization be-haviour in the time domain which is influenced by the angular diffusion of the dye mole-cules[5]. In this experiment we measured threshold and build-up time for different cavity lengths and varying the polarization angle. We could change the pump beam polarization by means of the $\lambda/4$ plate and of the polarizer P_1 and selected the output polarization with

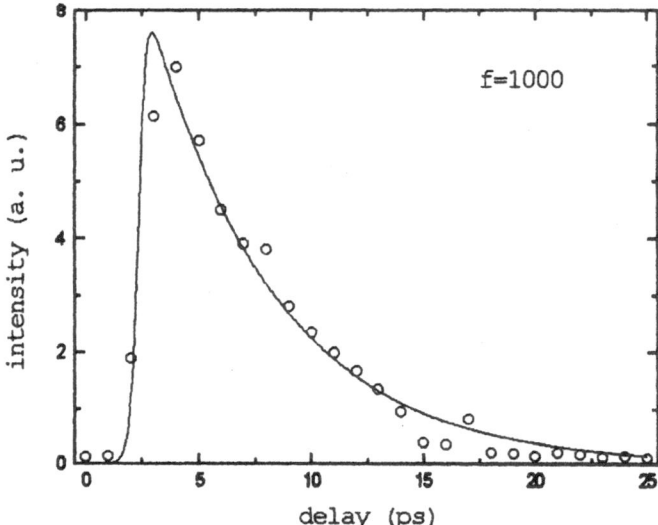

Figure 2. Microlaser output pulse. The continuous line represents the numerical simulation.

the analyzer P_2. Let's consider a set of dye molecules excited by a linearly polarized femtosecond pulse. For random orientation, the excitation probability is larger for the molecules whose dipole moment μ is parallel to the pump polarization. After the excitation, the orientation of the dipoles can be modified because of the angular diffusion, whose characteristic time constant τ_D is a function of the solvent viscosity, at a given temperature. For a short cavity, for which the diffusion time is much longer than the microcavity photon lifetime, the dipoles appear "frozen" in their position during the microlaser emission, and the degree of polarization is equal to 1. For larger values of the cavity length the role of the angular diffusion increases more and more and the polarization component, perpendicular to that of the pump beam, increases.

The third process we investigated is related to the onset of "gain coupling" between two microlasers in the time domain[6]. In this experiment two identical microlasers, confined in the active region of the same microcavity and excited by two ultrashort laser pulses (see Fig.1), are brought into a mutual optical correlation within the process of stimulated emission, by setting the mutual distance R at a value which is is of the order of the transverse coherence length of the cavity. In the case of the experiment described in this paper the two microlasers can cooperate in the emission process with a degree of correlation ≤ 1, provided they emit in the same transverse mode. If the excitation of the two microlasers is induced by two identical short laser pulses with delay Δt, it is possible, by an appropriate experiment, to investigate the space-time dynamics of the transverse interaction among the two microlasers over R. Let's define a "correlation time" T_c, as the "travelling" time of the transverse interaction over the intermicrolaser distance. Once the first microlaser has been excited, the second one is dynamically coupled to the first if the total time, $\Delta t + T_c$ does not exceed the pulse duration. This effect of coupling causes an increase of the gain of the second microlaser, with a consequent reduction of its build-up time. In this experiment the shape of the overall output pulse is given by two pulses in temporal sequence $\Delta\tau$. The pulse shape and the delay between the two output pulses are modified as a function of the excitation delay Δt. The results relative to the measured values of $\Delta\tau$ vs. Δt, for R = 50 μm, are shown in Fig. 3. Note that the experimental points

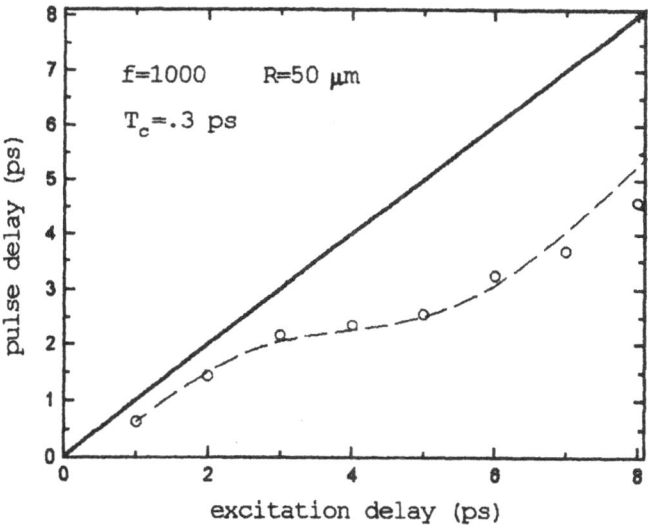

Figure 3. Temporal correlation curve of the microlasers.

are below the 45° straight line corresponding to the limit of absence of correlation between the two microlasers ($\Delta t = \Delta\tau$). The dashed line reported in the Figure is the result of a theoretical simulation obtained by following a dynamic rate equations model written for the two coupled microlasers. The results are consistent with a correlation time $T_c \cong .3$ ps, corresponding to the direct interaction between the microlasers through the distance $\dot{R}$. Absence of correlation ($\Delta t = \Delta\tau$) is obtained when R is much larger than the transverse coherence length.

REFERENCES

1. F. De Martini and G. R. Jacobovitz, Phis. Rev. Lett. **60**, 1711 (1988).
2. F. De Martini, F. Cairo, P. Mataloni and F. Verzegnassi, Phys. Rev. **A 46**, 4220 (1992).
3. A. Aiello, F. De Martini, M. Marrocco and P. Mataloni, Opt. Lett., **20**, 1492 (1995).
4. P. Mataloni, A. Aiello, D. Murra and F. De Martini, Appl. Phys. Lett. **65**, 1891 (1994).
5. A. Aiello, F. De Martini and P. Mataloni, "Polarization temporal dynamics in a dye microlaser", Opt. Lett., submitted.
6. A. Aiello, F. De Martini, M. Giangrasso and P. Mataloni, Quantum and Semiclassical Optics, **7**, 677 (1995).

33

SPATIALLY AND TEMPORALLY NONDIFFRACTING ULTRASHORT PULSES

Peeter Saari

Institute of Physics
Tartu, Estonia

During the last decade a number of novel exact solutions to the three-dimensional wave equation have been discovered, whose names in themselves – "splash modes," "self-focus wave modes," "directed-energy pulse trains," "electromagnetic bullets," "nondiffracting Bessel beams," "nondiffracting X waves," "slingshot pulses," etc. – give an idea of localized or even particle-like properties of these exotic wave packets (see Refs.[1-4] and references therein). Unfortunately, so far the majority of them exists in formulae only, even if approximations of finite aperture and of finite invariant-propagation ("diffraction-free") distance are considered. A serious obstacles in the way toward *optical* implementation of the most interesting and forward-propagating versions of these solutions is an ultra-wide-band spectral content needed for the temporal and spatial localization. This is why the experimental results in this emerging prospective field have been so far obtained in radio-frequency domain and in acoustic imaging, the latter being very intensively studied in the context of medical ultrasound diagnostics. The monochromatic Bessel beam is a remarkable exception. The field amplitude of an axisymmetric Bessel beam can be expressed as $\Phi_B \propto J_0(r\ k\ sin\theta)\ exp[i\ (z\ k\ cos\theta - \omega\ t)]$, where J_0 stands for the zeroth-order Bessel function, r is the transversal distance from the propagation axis z, $k = \omega\ /c$ is the wave number of the monochromatic light and θ is the tilt of the plane wave components with respect to the z axis. As a cylindrical counterpart of the ideal plane-wave solution, it has been rather well known in mathematical physics already since the last century. However, as recently as in 1987 it was demonstrated that the beam essentially maintains its propagation invariance property even in its physically realizable approximate versions: the beam possessing a central bright spot, e.g. of 70 μm FWHM, in the cross-section, shows no spread of this small spot over a distance of 70 cm. Since that discovery the number of papers on optical studies of Bessel beams has been growing constantly. The beam can be generated by a conical lens (Axicon) or by an annular slit and collimator (see Ref.[3] or Fig.2 below with the transparency and the 1st lens replaced simply by a telescope-expanded cw laser beam). The second method of generation is to use computer-generated holograms[5]. Although promising optical applications such as phase-matched second-harmonic generation have been accomplished already[6], to our best knowledge *optical pulsed*

Ultrafast Processes in Spectroscopy, edited by Svelto et al.
Plenum Press, New York, 1996

versions of Bessel beams have not been considered so far (of course, one can use nanosecond or even shorter pulses instead of cw laser for convenience, but in the present context qualitatively new picture opens only if pulse duration or the coherence length is shorter than the aperture diameter weighted by $sin\theta$, i.e. in the sub-picosecond region). A physically clear and straightforward method to obtain the pulsed versions of the Bessel field is to use an ultrashort pulse as an input to a Bessel beam generator, which corresponds to integration of Φ_B over the pulse spectral amplitude profile. In distinction to the novel ultra-wide-band pulsed solutions listed above, to our best knowledge nobody has found any analytical expressions of the integral for the *optical* region. The difficulty is that a spectral profile of an optical pulse cannot start from $\omega = 0$ but should be up-shifted to a carrier frequency ω_0 instead, while in cylindrical coordinates the well-known shift-modulation relations of Fourier- or Laplace transforms does not hold.

The motivation of the present paper is to show that nondiffracting pulses are realizable in optics and to demonstrate their possible applications. The limits for the articles in the Proceedings allow only to list our main results and reproduce here in black-and-white only some of the computed pictures discussed in the oral version of the paper.

We have found that with a quite physical pulse spectral amplitude

$$S(k) \propto (k/k_0)^{1/2} \exp\{-[d_0(k-k_0)]^2/2\} ,\tag{1}$$

which is a Gaussian profile slightly modified to suppress the non-optical low-frequency tail, centered at optical carrier wavenumber k_0 and corresponding to initial laser pulse duration $c^{-1}d_0$, it is possible to get an approximate analytical result for the integral. i.e. to obtain the resulting field amplitude

$$\Phi_{BX}(r,z,t) \propto \sqrt{Z(d)} \cdot exp\left[-\frac{1}{2d_0^2}\left(r^2 \, sin^2\,\theta + d^2\right)\right] \cdot J_0\left[Z(d) \cdot r\,k_0\,sin\theta\right] \cdot e^{ik_0d},$$

$$d \equiv z\,cos\,\theta - ct, \qquad\qquad Z(d) \equiv 1 + i \cdot d/k_0 d_0^2 \tag{2}$$

We have shown by various ways, that this expression is practically exact if $d_0 k_0 > 2$, which means for $\lambda_0 \equiv 2\pi/k_0 = 0.6$ μm that FWHM of the laser pulse should exceed 1 femtosec.

We call the pulse as Bessel-X nondiffracting pulse due to the characteristic shape of its intensity profile in a plane of the propagation axis z , see Fig.1.

We see that the pulse intensity possesses the revolved-X-like interference pattern resulted from the conical angular spectrum, propagating without any change with a superluminal phase and group velocities $c/cos\theta$ (all this known from the studies[1,2,4] of slingshot pulses or X-waves) and, simultaneously, a rudiment of the Bessel-function radial modulations suppressed by Gaussian envelope. For longer than ~ 50 fsec pulses this peculiar profile is replaced by the usual Bessel beam modulated longitudinally by the pulse temporal Gaussian profile, while the X-branching appears uninterestingly far away from the propagation axis (if the beam apeture is sufficiently large for that at all). However, the most remarkable feature of the pulses with durations of the order of ≤ 10 fsec is the high-intensity highly-localized (within few μm) central spot appearing as spike in Fig.1.a and enclosed into the most central $1/e^2$ surface in Fig.1.b. It should be stressed that the properties of the pulse are not bound to the particular spectrum used for obtaining the analytical result, they are common for any pulse-shape (double-exponential, etc.) of the comparable duration.

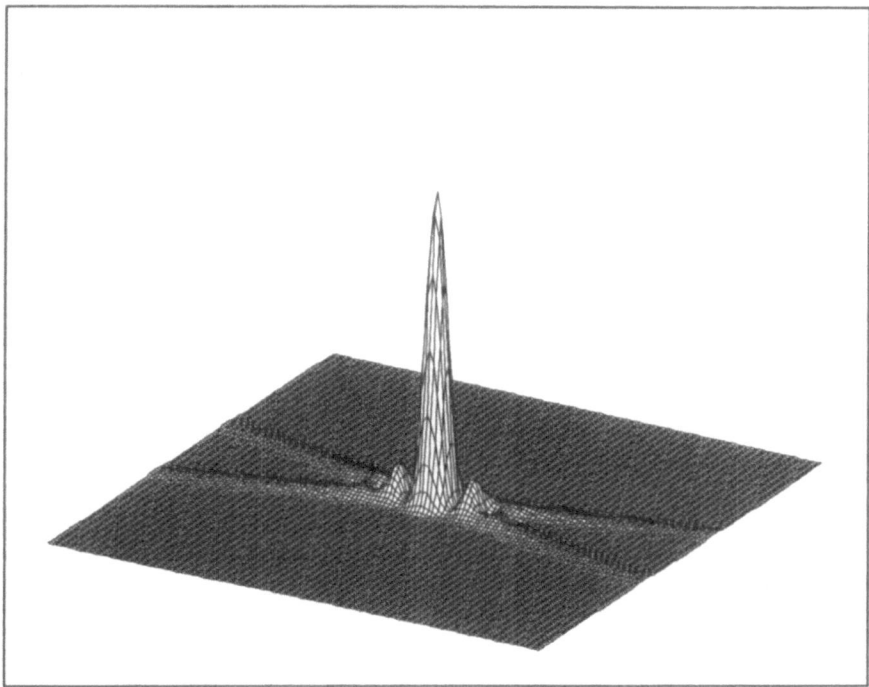

INTENSITY

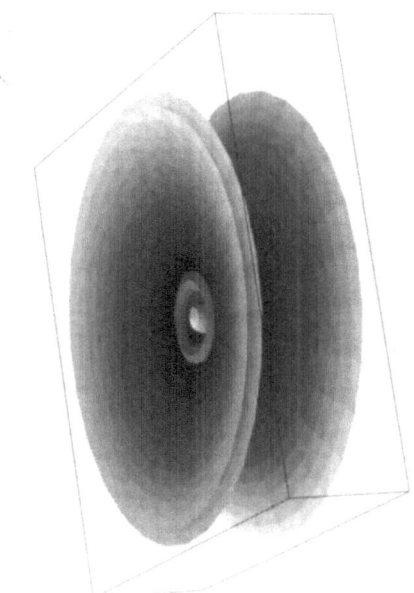

Figure 1. (a) Intensity (upwards) of the Bessel-X pulse versus propagation distance $z - ct/cos\theta$ (from foreground to the right) and a lateral axis on a 20x20 μm square of these coordinates. (b) Bessel-X pulse flying to the right in the figure plane projection of physical 3-D space, shown as surfaces on which the intensity of the field is equal to $(1/e^2 r)$-fraction of its maximum value in the central point; the box corresponds to the X-pattern-carrying strip of the square in Fig.1.a. The plots are computed from Eq.2 for 3-femtosec (FWHM) pulse and parameters $\lambda_0 = 0.6$ μm and $\theta = 14^0$.

Although it is intuitively obvious, we have proved it by direct numerical integration of corresponding plane wave pulses over the conical angular spectrum.

The discussion is so far concerned with propagation in vacuum or dispersionless medium only.

The problem of temporal spread of a light pulse propagating in a dispersive medium is well known in ultrafast spectroscopy. The effect is treatable in much the same way as spatial Fresnel diffraction and such a formal but deep analogy justifies the using of the term 'temporal diffraction'. One interesting consequence of this analogy is the concept of 'time-lens' which has been intensively studied in recent years (see Ref.[7] and references therein). Another consequence is that the temporal spread is as inevitable as the common diffraction. However, it would be very advantageous not only in spectroscopy but even more in the fields like Fourier optics, imaging, image processing, light energy transmission, etc., if one could generate freely-traveling wave packets remaining spatially and temporally highly localized over extensive distances in the propagation medium. Very recently a solution to the problem for a model 1-dimensional case has been proposed[8]. We have shown, that optical elements generating dispersive Bessel beams, i.e. the beams with frequency-dependent phase and group velocities do not give propagation-invariant Bessel-X pulses in vacuum but can be used for suppression/compensation the pulse spread in a given dispersive medium. Computer-generated holograms fabricated for Bessel beam generation[5] and 4–D holograms (see Refs.[9,10] and references therein) are promising examples of such optical elements.

The pulses under discussion are very interesting in the context of optical image propagation and processing. As the intensity in the cross-section plane at $z\cos\theta - ct$ decays with increasing of the lateral distance r essentially more strongly than in the case of single-frequency Bessel beam, an optical filter/processor of 2-D images, the point-spread function of which is the Bessel-X pulse, transmits the image through space propagation-invariantly, i.e. without diffractional spread. The simplest optical scheme for that is depicted in Fig.2.

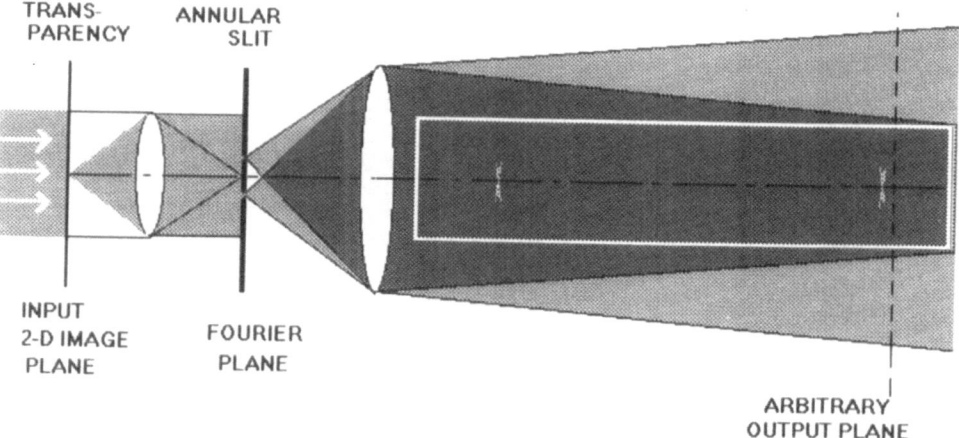

Figure 2. Optical filter for producing non-diffracting images with femtosecond-duration pulses (expanded beam of femtosecond laser pulses enter the scheme from the left; the white box indicates approximate boundaries of diffraction-free propagation area according to the criterion for limited-aperture Bessel beams[3], the recording in an arbitrary plane within this area should be accomplished with femtosecond temporal resolution).

In the terms of optical coherent filtering of 2-D images on amplitude/phase-modulated transparencies the 4-F scheme shown passes only single radial frequency of the image due to the infinitely small width of the annular slit in the spatial filtering plane. This is why in the case of monochromatic illumination the transmitted image, although diffraction-free as convolution of Bessel beam cross sections, maintains not much more information than the symmetry of the input image (see Fig.3.b). However, in the case of pulsed

SIGNAL $_{IN}$

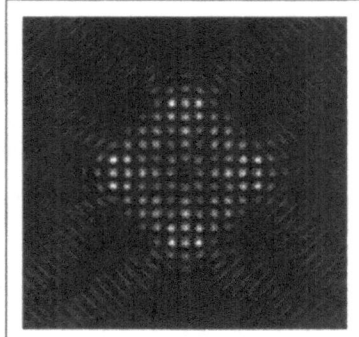

FIELD $_B$

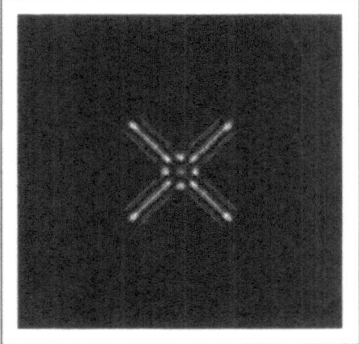

FIELD $_{BX}$

Figure 3. (a) An example binary input image for the scheme in Fig.2. (b) Computed propagation-invariant intensity in the case of monochromatic illumination. (c) Propagation-invariant intensity in arbitrary output plane (see Fig.2) at moments $t = z\cos\theta \, / \, c$, which is computed by using the Bessel-X field from Eq.2.

illumination, i.e. by Bessel-X pulses, a wide band of the radial frequencies of the image is passed through the annular slit due to the different scaling of the Fourier plane for different temporal frequency components of the illuminating laser pulse. As a matter of fact the scheme carries out 2-D axisymmetric wavelet transform (well-known in the image processing) the carrier wavelength being in the role of the scaling variable. It should be pointed out that all this can work also without femtosecond pulses and -resolution − the corresponding correlation times are essential, which means that laser noise-burst pulses and correlation recording techniques can be exploited (see also Ref.[10]).

Our computer simulations demonstrate that with exception of the some halo effect caused by the next-to-zeroth maximum of the Bessel profile and contour-enhancing caused by suppression of the lowest spatial frequencies of the image, the diffraction-freely flying image is identical to the input one.

In conclusion, as less-than-ten-femtosecond pulse generation is becoming a routine task thanks to the new techniques (see, e.g. review[11]), it is high time to start studies of optical versions of localized waves and their possible applications. The research described in this publication was made possible in part by Grant No. LL8100 from the Joint Fund Program of Estonia.

REFERENCES

1. R.W.Ziolkowski, I.M.Besieris, and A.M. Shaarawi, JOSA A **10**, 75 (1993).
2. P.L.Overfeldt, Phys. Rev. A **44**, 3941 (1991).
3. J.Durnin, J.J.Miceli, Jr., and J.H.Eberly, Phys. Rev. Lett. **58**, 1499 (1987).
4. J.-Y.Lu and J.F.Greenleaf, IEEE Trans. Ultrasonics, Ferroel. & Freq. Contr. **39**, 19 (1992).
5. J.Turunen, A.Vasara, and A.T.Friberg, Appl. Opt. **27**,3959 (1988).
6. T.Wulle and S.Herminghaus, Phys. Rev. Lett. **70**, 1401 (1993).
7. A.A.Godil, B.A.Auld, and D.M.Bloom, IEEE J.Quantum Electron. **30**, 827 (1994).
8. J.Rozen, B.Salik, A.Yariv, H.Liu, Optics Lett. 20, (1995), 423.
9. P.M.Saari, R.K.Kaarli, R.V.Sarapuu, and H.R.Sõnajalg, IEEE J. Quant. Electr. **25**, 339 (1989).
10. P.Saari, R.Kaarli, and M.Rätsep, J.Lumin. **56**, 175 (1993).
11. R.Szipöcs, A.Stingi, C.Spielmann, and F.Krausz, Optics & Photonics News No.6, 16 (1995).

34

HIGH FREQUENCY BRIGHT AND DARK SOLITON SOURCES BASED ON DISPERSION PROFILED FIBRE CIRCUITRY AND THEIR APPLICATIONS

D. J. Richardson,[1] R. P. Chamberlin,[1] L. Dong,[1] D. N. Payne,[1] A. D. Ellis,[2] T. Widdowson,[2] W. A. Pender,[2] and D. M. Spirit[2]

[1] Optoelectronics Research Centre
Southampton University, United Kingdom
[2] BT Laboratories
Martlesham Heath, Ipswich, United Kingdom

There has been considerable recent interest in the development of all-optical techniques for the generation of high frequency soliton trains based on nonlinear beat signal to soliton train conversion in dispersion profiled fibre circuits (see e.g. Refs.[1,2,3,4,5]). These schemes offer advantages of ultra-high repetition rates, high pulse quality, and broad wavelength and repetition rate tunability. Moreover, it is a non-resonant technique offering increased environmental stability relative to other more conventional fibre based short pulse generation schemes. However, although impressive source demonstrations have been made, their practical applications to date have been limited, due primarily to issues relating to timing jitter [4,5], Brillouin scattering [3,4] and difficulties in applying the techniques to repetition rate ranges <40 GHz to allow compatibility with state of the art, high data rate electronic signals [1]. In this paper we report on the development and performance of a diode-driven, ultra-low jitter, 30–40 GHz soliton transmitter with potential for telecommunication applications [5,6]. The results illustrate that synchronisation, modulation and electrical detection issues specific to the use of such high frequency pulse sources can be overcome to give stable, error free operation.

The source configuration (figure 1) consists of three principal components: an optical beat-signal source, an Er^{3+}/Yb^{3+} optical power amplifier and a dispersion varying fibre section. In order to obtain a low timing jitter beat signal we used a 20 GHz amplitude modulator tuned to a transmission null and driven at 17.5 GHz to obtain 35 GHz sinusoidal modulation of the output from a DFB laser diode. Two equal amplitude frequency components separated by 35 GHz are obtained with almost no component at the carrier wavelength. The dispersion decreasing fibre (DDF) section consisted of 2 km of dispersion-shifted fibre to spectrally enrich the input beat signal [3], and an 8 km DDF. The

Ultrafast Processes in Spectroscopy, edited by Svelto et al.
Plenum Press, New York, 1996

157

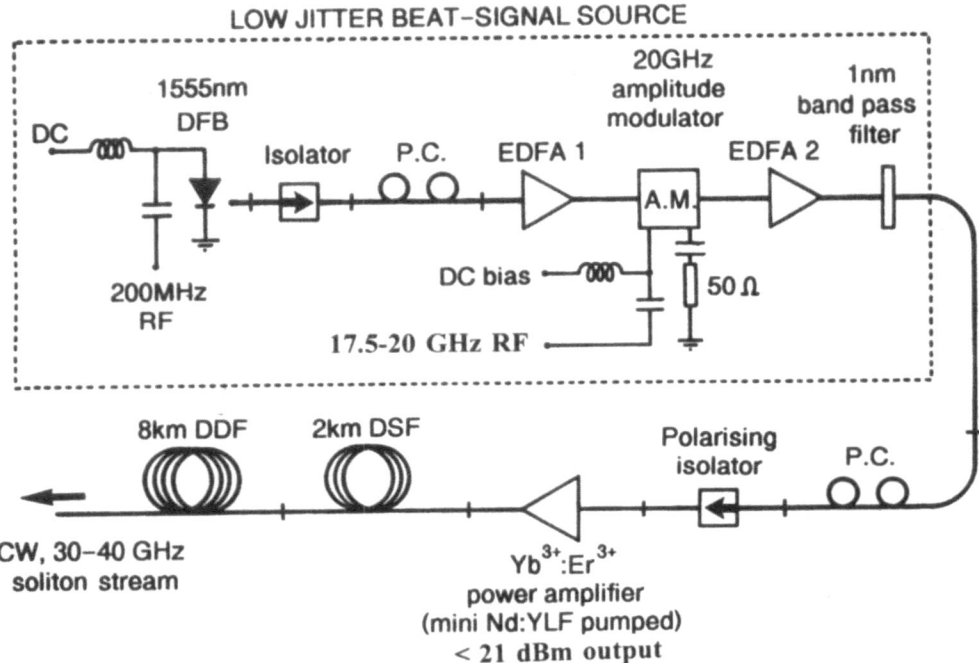

Figure 1. Schematic of diode-driven, stabilised 40 GHz bright soliton source.

DDF dispersion followed a hyperbolic profile at 1550 nm tapering along the 8 km length from 13.75 to 2.75 ps/(nm.km). The profile and output dispersion were chosen so as to reduce the absolute physical length of fibre required to obtain high quality, adiabatic 40 GHz pulse generation at an MSR of 5:1, whilst maintaining a practical optical power requirement on the input beat signal (80 mW). (Note that we have recently made significant developments in dispersion decreasing fibre fabrication extending the technology to fibre lengths of 40 km and to fibres with dispersion variation in both the anomalous and normal dispersion regimes [7,8]).

The source was tested under a wide range of input powers and beat frequencies in the range 32–40 GHz. Transform-limited, soliton pulses of durations 4.5–6.5 ps were obtained for input beat signal powers <20 dBm within the wavelength range of the available diodes 1547–1563nm. A typical autocorrelation function (ACF) and spectrum of a 35 GHz pulse train at the source output are shown in figure 2. In addition, 35 GHz, 4.5ps pulse propagation experiments using the source were performed over a transmission line incorporating four ≈50 km spans of dispersion shifted fibre and 3 EDFAs illustrating for the first time the stability of the pulse trains over prpoagation distances of a terrestrial scale [5].

Due to the low jitter beat signal seed source we were able to make direct electrical domain measurements on the pulse trains for the first time. The output pulse timing jitter was determined to be defined entirely by the phase noise of the modulator drive synthesizer (<300 fs). Furthermore, using electrical clock recovery circuits we have recently been able to synchronize a 6.4ps, 40 GBit/s optical data stream to the output of the soliton

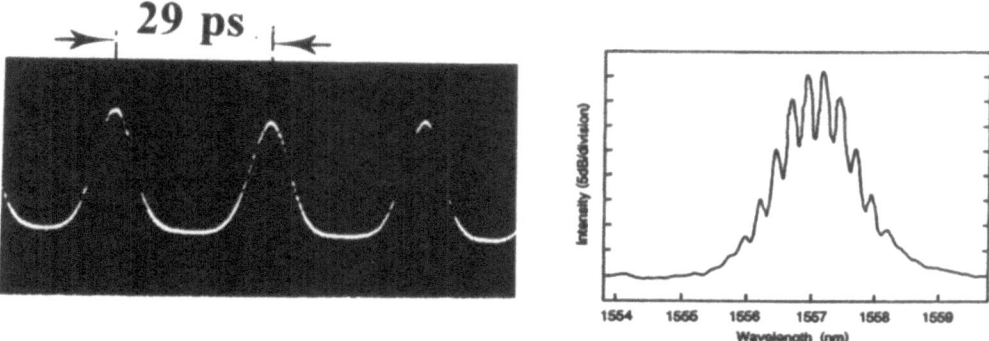

Figure 2. Autocorrelation and optical spectrum of 35 GHz pulse trains at source output.

source. (The 6.4ps pulses were generated using a gain-switched diode operating at 1545nm. The diode output was externally modulated at 10 GBit/s, and the resulting signal multiplexed to 40 GBit/s. We could then employ an all optical modulator (Kerr Gate) to encode data all-optically directly onto the beat signal soliton stream. Demultiplexing using a second clock recovery unit and EA modulator could then be performed enabling BER measurements on the four 10Gbit/s soliton data channels to be made. Error free operation, without a noise floor was readily obtainable with a power penalty of <1.5 dB (see figure 3.) [6].

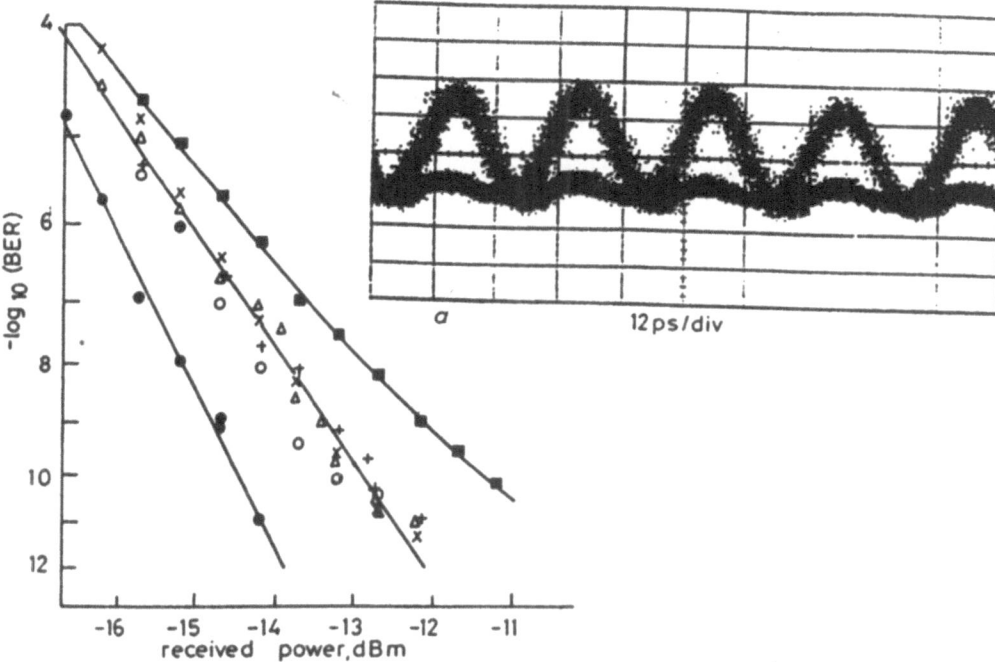

Figure 3. Eye diagram and BER rate curves for data encoded 40 Gbit/s data stream demultiplexed down to 10 Gbit/s illustrating error free all-optical modulation. ●) Gbit/s, 1545 nm back-to-back; ■) demultiplexed 40 Gbit/s, 1545 nm; O, Δ, +, X) all four 10 Gbit/s channels from 40 Gbit/s beat frequency derived signal.

In conclusion, we have demonstrated a practical low-timing jitter, diode-driven, soliton source capable of operating in the range 30–40 GHz and confirmed the pulse train stability during propagation over terrestrial distances for the first time. Transmission over considerably greater distances can be expected from an optimised system, with reduced dispersion variation and/or amplifier spacing. In addition, we have now demonstrated 40 Gbit/s, error free all optical data encoding onto the output of such sources illustrating for the first time that timing jitter and synchronisation issues associated with such sources can readily be overcome. Such techniques are also of relevance to dark soliton sources based on dispersion decreasing fibre [7]. The soliton transmitter should enable a true assessment of the suitability of such sources for future high speed communication applications.

REFERENCES

1. P.V. Mamyshev, S.V. Chernikov, E.M. Dianov: IEEE J. Quantum Electron., 27, 2347, (1991).
2. V.A. Bogatyrev et al.: IEEE J. of Lightwave Technol., 9, 561–565, (1991).
3. S.V. Chernikov, D.J. Richardson, R.I. Laming, E.M. Dianov, D.N.Payne: Electron. Lett., 28, 1220–1221, (1992).
4. E.A. Swanson, S.R. Chinn: IEEE Photonics Technol. Lett., 7, 116, (1995).
5. D.J. Richardson, R.P. Chamberlin, L. Dong, D.N. Payne, A.D. Ellis, T. Widdowson and D.M. Spirit: Electron. Lett., 31, 470, (1995).
6. A.D. Ellis, W.A. Pender, T. Widdowson, D.J. Richardson, L. Dong, R.P. Chamberlin: Electron. Lett., 31, 1362, (1995).
7. D.J. Richardson, R.P. Chamberlin, L. Dong, D.N. Payne: Electron. Lett., 30, 1322, (1994).
8. D.J. Richardson, R.P. Chamberlin, L. Dong, D.N. Payne: Electron. Lett., 31, 1681, (1995).

35

TEMPORAL BEHAVIORS OF STIMULATED DYNAMIC LIGHT SCATTERING

H. Z. Wang, X. G. Zheng, Z. X. Yu, and Z. L. Gao

Institute for Laser and Spectroscopy
Zhongshan University
Guangzhou, China

Dynamic light scattering (DLS), known as depolarized light scattering, is related to intermolecular motion and interaction, and is a rich source of dynamical and structural information on molecules[1]. Sponteneous DLS is a broadband light scattering around Rayleigh line or Raman line, and therefore also called Rayleigh-wing scattering or Raman-wing scattering. When the pump density is high enough, stimulated dynamic light scattering (SDLS) can be observed[2]. In recent years, large Stokes broadening of SDLS in waveguide has been investigated by several groups[3–5], and their experimental results show that the broadband SDLS is applicable to transient spectroscopy and investigation of molecular dynamics. SDLS of molecules in liquid consists of two primary components, and it is pump density dependent[5]. At high pump density, SDLS is related to several kinds of intermolecular motion and interaction[5,6], whereas SDLS mainly results from molecular orientation at an average pump density lower than 1 GW/cm^2. In this communication, the temporal behaviors of SDLS in CS_2 liquid-core waveguide at an average pump density lower than 1 GW/cm^2 are studied. The experimental results demonstrate that the temporal behaviors and the threshold of SDLS are pump pulse duration dependent, and implies that the relaxation time of molecular orientation in high alignment is longer than 80 ps.

The experiments are performed with a frequency-doubled passively mode-locked Nd:YAG laser and a frequency-doubled saturable absorber Q-switch Nd:YAG laser as pump sources. The pulse durations of the Q-switch Nd:YAG laser and the passively mode-locked Nd:YAG laser are 6 ns and 40 ps, respectively. The scattering is detected by a single pulse sweep streak camera (Hamamatsu Model C1587, 2 ps time resolution) connected to a polychromator. Thus the spectral and the temporal behaviors can be recorded simultaneously. A glass capillary, 30 cm in length and 0.5 mm in internal diameter, is employed as a waveguide. There is not any colour filter between the waveguide and the streak camera for cutting off pump light.

The temporal behavior of SDLS in a CS_2 liquid-core waveguide at a 0.2 GW/cm^2 average pump density for a frequency doubled Q-switch Nd:YAG laser is shown in Figure 1. It is demonstrated in Figure 1 that SDLS is effective only during the risetime of the

Ultrafast Processes in Spectroscopy, edited by Svelto et al.
Plenum Press, New York, 1996

161

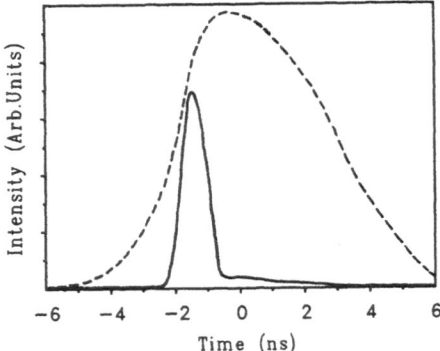

Figure 1. Temporal behavior of SDLS (solid line) in a CS_2 liquid-core waveguide at a 0.2 GW/cm^2 average pump power density for a frequency-doubled Q-switch Nd:YAG laser. Dashed line denotes the pump pulse.

nanosecond pump pulse. In Figure 2 is shown the temporal behavior of SDLS in a CS_2 liquid-core waveguide at a 0.2 GW/cm^2 average pump density for a frequency-doubled passively mode-locked Nd:YAG laser. It is revealed in Figure 2 that not only an effective SDLS appears during the risetime of the picosecond pump pulse, but also a gradually weakening SDLS appears during the falltime of the picosecond pump pulse.

The difference between the temporal behaviors of SDLS in Figure 1 and in Figure 2 concerns condition of molecular orientation distribution equilibrium with the light field. At nanosecond pulse laser pumping, the duration of the pump laser pulse is much longer than that of the molecular orientation distribution tending to equilibrium with the light field against thermal randomization and repulsive interaction between molecules in high alignment. The degree of molecular alignment increases with the transient pump power during the risetime of the nanosecond pump pulse, and therefore SDLS mainly appears in this duration. At picosecond pulse laser pumping, the risetime of the pump pulse is shorter than the duration that the molecular orientation distribution tends to equilibrate with the light field. Although the light field has decreased in the first part of the falltime of the picosecond pump pulse, the molecular orientation distribution has not yet reached equilibrium with the light field. Therefore SDLS not only is effective during the risetime but also appears during the falltime of the picosecond pump pulse.

The experimental results above may be explained as follows. The mixing of the incident light field $\mathbf{E}_1$ and the scattering light field $\mathbf{E}_2$ reorient the molecules, and the molecu-

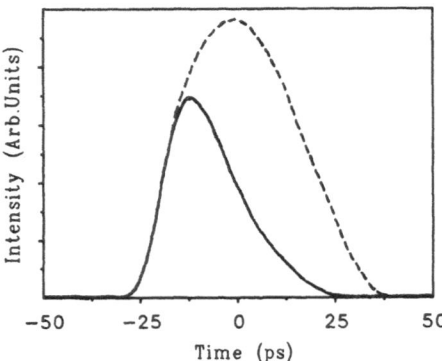

Figure 2. Temporal behavior of SDLS (solid line) in a CS_2 liquid-core waveguide at a 0.2 GW/cm^2 average pump power density for a frequency-doubled mode-locked Nd:YAG laser. Dashed line denotes the pump pulse.

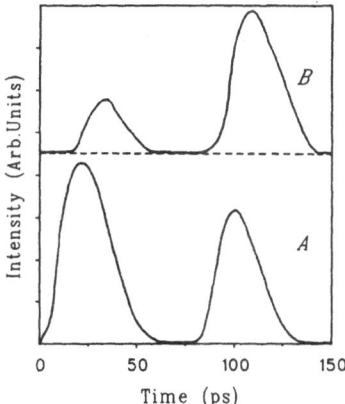

Figure 3. Temporal behaviors of SDLS (*A*) and SRS (*B*) under double pulse pumping.

lar reorientation in turn beats with $\mathbf{E}_1$ to enforce $\mathbf{E}_2$. When the molecular orientation distribution reaches an equilibrium with the light field, little molecular reorientation beats with $\mathbf{E}_1$. Therefore $\mathbf{E}_2$ is weakened. There is also a quantum description similar to the description of stimulated Raman scattering (SRS). One incident photon at $\omega_1(\mathbf{k}_1)$ is annihilated and one photon at $\omega_2(\mathbf{k}_2)$ is scattered, and simultaneously a molecule transits from its initial orientation state to its final orientation state. When the molecular orientation distribution equilibrates with the light field, most of the molecules are in their final states. SDLS occurs only when most molecules are in their initial states. Therefore it only occurs under non-equilibrium condition, or occurs during the risetime of a long duration pump pulse, in which the molecular orientation degree is increasing.

The temporal behaviors of SDLS shown in Figure 1 and Figure 2 are detected at the same average pump density of 0.2 GW/cm^2. As shown in Figure 1 and Figure 2, SDLS appears even at the start of the picosecond pump pulse, whereas it appears a quarter of the pump pulse risetime later in case of nanosecond pulse pumping. This demonstrates that the threshold of SDLS, which relates to condition of molecular orientation distribution equilibrium with the light field, is pump pulse duration dependent, and that the shorter the pump pulse, the lower the threshold of SDLS.

When the power supply of Nd:YAG laser is adjusted, doubled train laser is obtained, and double pulses with an interval of 80 ps and the same power intensity can be selected. Under pumping of these double pulses, SDLS is more effective during the first pulse than the second pulse, as shown in Figure 3. On the contrary, SRS is much weaker during the first pulse than the second pulse. The fact that SDLS dominates during the first pulse and is weaker than SRS during the second pulse implies that the molecular orientation in high alignment has not completely relaxed in 80 ps.

In summary, at low density nanosecond pulse laser pumping, SDLS mainly appears during the risetime of the pump pulse; at picosecond pulse laser pumping, SDLS not only is effective during the risetime but also appears during the falltime of the pump pulse. The threshold of SDLS at picosecond pulse pumping is lower than that at nanosecond pulse laser pumping. It is demonstrated that SDLS relates to condition of molecular orientation distribution equilibrium with the light field. Moreover, competition of SDLS and SRS under pumping of double laser pulses implies that the relaxation time of molecular orientation in high alignment is longer than 80 ps.

ACKNOWLEDGMENT

This work is supported by an NNSFC grant and a Frontier Sciences grant of Zhong-shan University.

REFERENCES

1. R. Pecora, *Dynamic light scattering* (Plenum Press, New York, 1985).
2. Y. R. Shen, *The Principles of Nonlinear Optics* (John Wiley & Sons, New York, 1984).
3. G. S. He and P. N. Prasad, Opt. Comm. **73**, 161 (1989).
4. J. Y. Zhou, H. Z. Wang and Z. X. Yu, Appl. Phys. Lett. **57**, 643 (1990).
5. H. Z. Wang, X. G. Zheng, W. D. Mao, Z. X. Yu, Z. L. Gao, Phys. Rev. A **52**, 1740 (1995).
6. D. Wang and G. Rivoire, J. Chem. Phys. **98**, 9279 (1993).

36

STIMULATED MULTIPHOTON BREMSSTRAHLUNG OF ELECTRONS ON HIGHLY CHARGED IONS IN A STRONG LASER FIELD

V. Astapenko[1] and A. Kukushkin[2]

[1] Moscow Institute of Physics and Technology
Dolgoprudnyi, Russia
[2] Institute of Nuclear Fusion
Russian Research Center "Kurchatov Institute"
Moscow, Russia

The theory of stimulated multiphoton Bremsstrahlung in the presence of a strong laser field (not exceeding the Coulomb field in the spatial region responsible for the emission) has been developed in calculable limited cases, for Born domain energies ($\xi = Z/v \ll 1$) in[1] for classical current approximations ($\xi \gg 1$) in[2,3] (atomic units are used).

For the probability of stimulated multiphoton emission the result[1], $\xi \ll 1$, has the form:

$$W(n) = J_n^2(\mathbf{aq}) \tag{1}$$

where J_n is Bessel function $\mathbf{a} = \mathbf{E}_0 / \omega^2$, $\mathbf{E}_0$, and ω are laser field amplitude and frequency, respectively, $\mathbf{q}$ is momentum transfer.

In alternative limit, $\xi \gg 1$, the result has similar structure:

$$W(n) = J_n^2(\mathbf{av}_\omega) \tag{2}$$

where $\mathbf{v}_\omega$ is the Fourier transform of particle's velocity at classical trajectory.

The relation between the case of quasiclassical motion of incident electron and classical current approximations has been traced at a qualitative level in[2].

In the limit of *weak* inelasticity of the process both the above results may be presented by a universal formula (cf. [2]):

$$W(n) = J_n^2(\mathbf{aq}) \tag{3}$$

Ultrafast Processes in Spectroscopy, edited by Svelto et al.
Plenum Press, New York, 1996

165

where **q** is a momentum transfer in the case of Born approximation and the change of particle momentum caused by elastic scattering in classical limit.

It appears that formula (3) may be extended to the case of *strong* inelasticity (in terms of ratio between total energy of emitted photons and particle's initial energy). Here, the only restriction is the smallness of the perturbation of particle's trajectory in the spatial region responsible for the emission process. In fact, this covers, in additional to formula (3), the case of strong inelasticity for quasiclassically moving incident particle. In the latter case, the actual criterion of *weak* distortion of electron trajectory appears to be described by the smallness of the ratio between emitted energy $\Delta E = n\omega$ and the local kinetic energy, $E_{kin}(r_\omega)$, where r_ω [4,5] is the characteristic distance between electron and highly charged ion in the region of the emission of quantum ω. In quasiclassical case, $\xi \gg 1$, which is peculiar to the electron Bremsstrahlung on highly charged ions, the smallness of $\mu = \Delta E / E_{kin}(r_\omega)$ appears to be feasible and being of practical interest just for highly charged ions, $\mu \sim n\omega^{1/3}/Z^{2/3} \sim n/\xi \leq 1$.

The following results are presented:

1. further extension of formula (3) to the case of strong inelasticity and arbitrary ξ substantiated in the frame of the formalism for the matrix element of multiphoton transition for the exact wave functions of an electron in the Coulomb/central potential;
2. straightforward transition from quasiclassical description of multiphoton Bremsstrahlung to classical current approximation, which extends the Correspondence Principle [5,6] for the matrix element of inelastic free-free transition in a central field to the multiphoton case.

REFERENCES

1. V.F. Bunkin, M.V. Fedorov , Zh.Exp.Teor.Fiz. (Sov.Phys.JETP), **49**, 1215 (1965).
2. I.Ya. Berson , Sov.Phys.JETP, **53**, (1981), 891; **56**, 731(1982).
3. V.S. Lisitsa, Yu.A. Saveliev , Sov.Phys.JETP, **65**, 273 (1987).
4. V.I. Kogan, A.B. Kukushkin, Sov.Phys.JETP, **60**, 665 (1984).
5. V.I. Kogan, A.B. Kukushkin, V.S. Lisitsa , Phys. Reports, **213**, 1–116 (1992).
6. V.I. Kogan, A.B. Kukushkin , Preprint of Kurchatov Institute, IAE-3660/6, Moscow, 1982.

37

NONRESONANT SELF-ACTION OF PULSES
WITH A FEW LIGHT OSCILLATION DURATION

N. R. Belashenkov,[1] V. V. Bezzubik,[1] S. A. Kozlov,[1] and S. V. Sazonov[2]

[1] State Institute of Fine Mechanics and Optics (Technical University)
Saint-Petersburg, Russia
[2] State Technical University
Astrakhan, Russia

The development of laser systems producing the ultrashort laser pulses resulted finally in a creation of laser sources providing the pulses with a duration of several light wave oscillation periods[1]. Actually the most important question is how do such pulses propagate in optical media? It is evident that the nonlinear modes of their propagation are of great interest because according to well-known relations[1] the low intensive SSP broadens rapidly due to dispersion.

The nonresonant self-action of light pulses with a duration up to femtosecond range is studied usually solving the wave equation[1]:

$$i\frac{\partial \mathscr{E}}{\partial z} + \alpha_1 \frac{\partial^2 \mathscr{E}}{\partial \tau^2} + \alpha_2 \frac{\partial^3 \mathscr{E}}{\partial \tau^3} + ... + \beta_0 |\mathscr{E}|^2 \mathscr{E} + \beta_1 \frac{\partial}{\partial \tau} |\mathscr{E}|^2 \mathscr{E} + ... = 0 \tag{1}$$

where $\mathscr{E}(z,\tau)$ is slowly varying envelope of light pulse

$$E = \frac{1}{2}\mathscr{E}\exp(i\omega\tau) + c.c. \tag{2}$$

ω is the light wave frequency, $\tau = t - z/v$ is the time in the accompanying coordinate system, v is the group velocity. The dispersion of linear and nonlinear refraction indices of a medium is taken into account in different approximations of dispersion theory by introducing the phenomenological coefficients α_1, α_2 ... and β_0, β_1, ... respectively. However the analysis of propagation of pulses with a duration of few light wave oscillation periods becomes qualitatively more complex. The term of slowly varying envelope (see Eq. 2) hardly could be applied to such pulses and wave equation (Eq. 1) derived in this approximation is not valid.

In our previous works[2-4] the mathematical model of interaction of pulses which duration is about few light wave oscillation periods (named supremely short pulses (SSP))

Ultrafast Processes in Spectroscopy, edited by Svelto et al.
Plenum Press, New York, 1996

167

with dielectrics has been proposed. This model is based on density matrix formalism and is interpreted in the form of classic dispersion theory for high-intensive light. It has been shown that polarization response $P = P_e^L + P_n^L + P^{NL}$ of dielectrics:

$$\begin{cases} \dfrac{\partial^2 P_e^L}{\partial t^2} + \dfrac{2}{T_e}\dfrac{\partial P_e^L}{\partial t} + \omega_e^2 P_e^L = \alpha_e E \\[2mm] \dfrac{\partial^2 P_n^L}{\partial t^2} + \dfrac{2}{T_n}\dfrac{\partial P_n^L}{\partial t} + \omega_n^2 P_n^L = \alpha_n E \\[2mm] \dfrac{\partial^2 P^{NL}}{\partial t^2} + \dfrac{2}{T_e}\dfrac{\partial P^{NL}}{\partial t} + \omega_e^2 P^{NL} = \beta(R_e + R_v)E \\[2mm] \dfrac{\partial^2 R_e}{\partial t^2} + \dfrac{2}{T_{e'}}\dfrac{\partial R_e}{\partial t} + \omega_{e'}^2 R_e = \gamma_e P^L E \\[2mm] \dfrac{\partial^2 R_v}{\partial t^2} + \dfrac{2}{T_v}\dfrac{\partial R_v}{\partial t} + \omega_v^2 R_v = \gamma_v P^L E \end{cases} \tag{3}$$

where the phenomenological parameters of a medium α_e, ω_e, T_e and α_n, ω_n, T_n characterise the dispersion of linear polarization response of electron and nuclear (ion, vibration) nature; the coefficients $\beta, \gamma_e, \omega_{e'}, T_{e'}$ and γ_v, ω_v, T_v characterise the dispersion of nonlinear polarization response of electron and electron-nuclear (electron-vibration, Raman) nature, respectively, describes adequately the dispersion of both linear and nonlinear refraction indices of dielectrics in near whole range of transparency. So, for example, under the values of $\omega_e = 2.096 \cdot 10^{16} s^{-1}$, $\alpha_e = 3.862 \cdot 10^{31} s^{-2}$, $\omega_n = 2.154 \cdot 10^{14} s^{-1}$, $\alpha_n = 2.534 \cdot 10^{27} s^{-2}$ the dispersion of linear refraction index of fused silica in visible and near IR spectral range is described by Eq. 3 with accuracy 0.0002 and under the values of $\omega_{e'} = 3.0 \cdot 10^{16} s^{-1}$, $\omega_v = 8.3 \cdot 10^{13} s^{-1}$, $\beta = 2.0 \cdot 10^{43} ESU$, $\gamma_e = 2.9 \cdot 10^9 ESU$, $\gamma_v = 8.0 \cdot 10^3 ESU$ the dispersion of fast nonlinear refraction index n_2 is in good agreement with earlier published experimental data[3].

Since the spectrum of pulses with a duration even of 1.5–2 periods of light wave oscillations covering the considerable part of visible and IR range is within the transparency range of most glasses and crystals the interaction of SSP with dielectrics can be considered as nonresonant one. Taking into account that electron-nuclear mechanism of nonlinearity is "frozen" in the field of SSP[3] in the first order of nonresonant approximation the material equations (Eq. 3) may be reduced to:

$$\begin{cases} \dfrac{\partial^2 P_n}{\partial t^2} \approx \alpha_n E \\[3mm] \dfrac{\partial^2 P_e}{\partial t^2} \approx \dfrac{\alpha_e}{\omega_e^2}\dfrac{\partial^2 E}{\partial t^2} + \dfrac{\partial^2}{\partial t^2}\left(\dfrac{\alpha_e \beta \gamma_e}{\omega_e^4 \omega_{e'}^2} E^3 - \dfrac{\alpha_e}{\omega_e^4}\dfrac{\partial^2 E}{\partial t^2} \right) \end{cases} \tag{4}$$

The wave equation describing the propagation of SSP in a medium with polarization response (Eq. 4) may be written as follows:

$$\dfrac{\partial^2 E}{\partial z^2} - \dfrac{n_0(0)}{c^2}\dfrac{\partial^2 E}{\partial t^2} = \dfrac{2n_0}{c}A\dfrac{\partial^4 E}{\partial t^4} + \dfrac{2n_0}{c}BE + \dfrac{2n_0}{3c}C\dfrac{\partial^2 E^3}{\partial t^2} \tag{5}$$

where

$$n_0^2(0) = 1 + \frac{4\pi\alpha_e}{\omega_e^2}, \quad A = -\frac{2\pi\alpha_e}{n_0 c\omega_e}, \quad B = \frac{2\pi\alpha_n}{n_0 c}, \quad C = -\frac{6\pi\alpha_e\beta\gamma_e}{n_0 c\omega_e^4\omega_{e'}^2}$$

In the one-way running wave approximation[5] the wave equation (Eq. 5) may be represented as:

$$\frac{\partial E}{\partial z} + A\frac{\partial^3 E}{\partial \tau^3} + B\int_{-\infty}^{\tau} E d\tau' + CE^2\frac{\partial E}{\partial \tau} = 0 \tag{6}$$

Kozlov and Sazonov first derived and published[6] this nonlinear wave equation. It is important to notice that in the case of "long" pulses (Eq. 2) with a duration of more than ten periods of light wave oscillations Eq. 6 could be transformed to Eq. 1 while $\alpha_1 = 3A(\omega_c^4 - \omega^4)/\omega^3$, $\alpha_2 = iA(1 + 3\omega_c^4/\omega^4)$, $\beta_0 = -C\omega$, $\beta_1 = -iC$, $\omega_c^4 = -B/3A$.

Let's consider the self-action of SSP which spectrum is within the range of normal dispersion of group velocity. In this case the third term in Eq. 6 could be neglected and Eq. 6 becomes the modified Korteweg de Vries's equation. For glasses and crystals the coefficients A<0 and C>0 so SSP self-action results in the time defocusing adding to linear dispersive pulse broadening. The soliton mode of pulse propagation is impossible.

The formation of solitons from SSP is possible if the considerable part of SSP's spectrum is within the range of abnormal dispersion of group velocity.

The breather with 1.5 light wave period duration obtained during the numerical analysis of Eq. 6 is represented in Fig.1.

This work is supported by Grant of State Committee for Higher Education of Russian Federation.

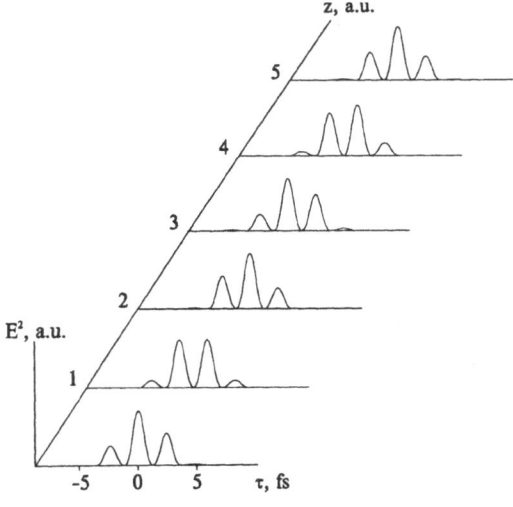

Figure 1. The propagation of breather with duration of only 1.5 period of light wave oscillations.

REFERENCES

1. S.A. Akhmanov, V.A.Vysloukh and A.S. Chirkin, *Optics of Femtosecond Laser Pulses* (Science, Moscow, 1988)
2. A.N. Azarenkov, G.B. Altshuler and S.A.Kozlov, *Huygens'Principle 1690–1990: Theory and Applications.* North-Holland, Studies in Mathematical Physics, **3**, 429 (1992)
3. A.N. Azarenkov, G.B. Altshuler, N.R. Belashenkov and S.A. Kozlov, Quant. Electron. (Rus), **20**, 733 (1993)
4. S.A. Kozlov, Opt. & Spectr. (Rus), **79**, 200 (1995)
5. S.V. Sazonov, Laser Physics, **2**, 795 (1992)
6. S.A. Kozlov and S.V. Sazonov, in *Technical Digest of 15-th International Conference on Coherent and Nonlinear Optics*, St.-Petersburg, **1**, 370 (1995)

38

SUPERLUMINAL PULSE PROPAGATION CAUSED BY GOUY PHASE SHIFT

Z. L. Horváth and Zs. Bor

Department of Optics and Quantum Electronics
JATE University
Szeged, Hungary

The propagation of ultrashort light pulses in optical media is one of the most interesting problems of optics. The group velocity is generally used as a quantity which gives the speed of a wave packet. In dispersive media the group velocity can behave abnormally exceeding c, the velocity of light in vacuum, or even become negative[1]. Superluminous pulse propagation has also been observed upon propagation of pulses in deeply saturated amplifiers[2], in a resonant medium with inverted population[3-4] and in quantum-mechanical tunneling of single photons through dielectric mirrors[5]. It should be noted that the effect discussed below has a completely different origin.

Because of the close analogy between the pulse propagation problems and the diffraction theory[6] superluminal pulse propagation caused by diffraction is expected. The intensity distribution of a focused beam near focus can be obtained by the calculation of the diffraction integral[7]. It is known that the phase behaviour of the focused beam differs from that of a converging spherical wave. This phase difference observed by Gouy is called phase anomaly[7-9]. Below we show that this phase behaviour leads to superluminal pulse propagation in the vicinity of the focus.

For reasons of simplicity and applicability to laser beams we assume that the incoming beam is a spatially Gaussian beam. A lens having focal length f converts the incoming Gaussian beam into a Gaussian beam having beam waist w_f at a distance d from the lens given by[11]

$$w_f^2 = \frac{w_0^2}{1 + \left(\dfrac{z_0}{f}\right)^2}, \qquad d = f - \frac{f}{1 + \left(\dfrac{z_0}{f}\right)^2},$$

$$(1)$$

where w_0 is the beam waist on the lens and $z_0 = \pi w_0^2/\lambda$ is the Rayleigh length of the input beam and λ is the wavelength. Then $E(r, z^*) = E_0 \exp\left[-r^2 / w^2(z^*)\right] \exp[-i\Phi(r, z^*)] / w(z^*)$ describes the space dependent part of the electric field behind the lens, where z^* is a coor-

Ultrafast Processes in Spectroscopy, edited by Svelto et al.
Plenum Press, New York, 1996

171

dinate along the optical axis measured from the beam waist, and $w^2(z^*) = w_f^2[1 + (z^*/z_f)^2]$ gives the spot size at a point z^* on the optical axis, and

$$\Phi(r, z^*) = k z^* + \frac{k r^2}{2R(z^*)} - \arctan\frac{z^*}{z_f} \tag{2}$$

is the phase of the electric field ($k = 2\pi/\lambda$ is the wave number). Here $z_f = \pi w_f^2/\lambda$ is the Rayleigh length of the focused beam and $R(z^*) = z^*[1 + (z_f/z^*)^2]$ is the curvature of the phase front at a point z^* on the optical axis. Eq. 2 shows that the phase of a Gaussian beam differs from a converging spherical beam. The phase difference on the optical axis is $\arctan(z^*/z_f)$. Since z_f depends on λ the phase anomaly also depends on λ. Because of this dispersion a difference between the phase and the group velocities is expected.

In order to study the pure effects caused by the Gouy phase shift we assume that the lens is free of chromatic and any other aberration, that is $f(\lambda) = f_0$. For practical cases $(z_0/f)^2 \gg 1$. Then the small shift in position between the geometrical focus and the beam waist is negligible (eq. 1), that is $d = f_0$ and so $z^* = z$ where z is a coordinate along the optical axis measured from the focus. In this approximation the phase of the focused beam along the optical axis is given by

$$\Phi(\omega, z) = \frac{\omega}{c} z - \arctan(K\omega z), \quad \text{where } K = \frac{w_0^2}{2c f_0^2} . \tag{3}$$

The phase velocity on the optical axis is defined by $v_p = \omega/(\partial\Phi/\partial z)$ [7], which results in

$$v_p(u) = c \frac{1 + (\vartheta^2 u)^2}{1 + (\vartheta^2 u)^2 - \frac{\vartheta^2}{2}}, \quad \text{where } u = \pi\frac{z}{\lambda} = \frac{\omega z}{2c} \tag{4}$$

is a dimensionless variable and $\vartheta = w_0/f_0$ is the divergence of the focused beam. Fig. 1a shows the phase velocity on the optical axis for different values of divergence. It is easy to see from eq. 4 that the phase velocity on the optical axis is larger than c and it reaches its maximum in the focus given by $v_{p,\max} = c/[1 - \vartheta^2/2] \approx c/\cos\vartheta$, which is the sweep speed of a pulse on the optical axis falling under the angle of divergence ϑ.

The position of the phase front on the optical axis at a moment t is determined by $t = \Phi(\omega, z)/\omega = z/c - \arctan(K\omega z)/\omega$. Of course one can obtain this equation by integration using eq. 4, too . For $|z| \gg z_f$ it has an asymptotic form: $z = ct + \text{sign}(z)\lambda/4$. This means that far from the focus if $z < 0$, the phase front lags behind if $z > 0$, it precedes the phase front of a spherical wave with a distance $\lambda/4$. This is why the phase velocity is larger than c. The group velocity on the optical axis is defined by [7]

$$v_g = \left(\frac{\partial}{\partial z}\frac{\partial\Phi}{\partial\omega}\bigg|_{\omega_0}\right)^{-1} \tag{5}$$

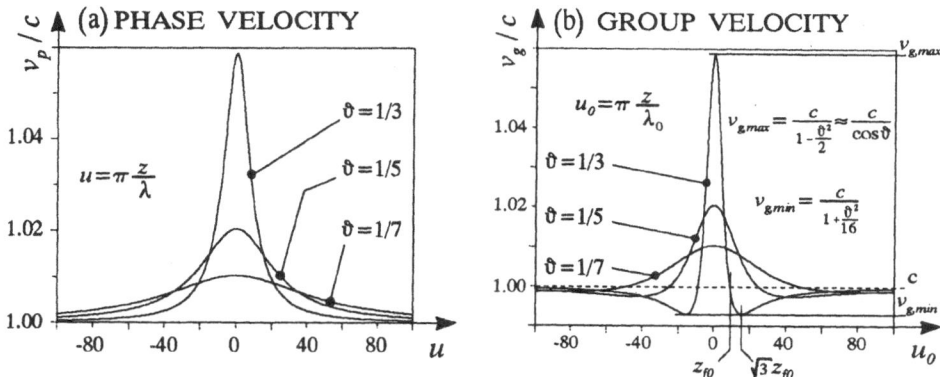

Figure 1. Phase (a) and group (b) velocity on the optical axis for different values of the divergence of the focused beam.

where ω_0 is the central frequency of the pulse. Inserting eq. 3 into 5 the group velocity on the optical axis is given by

$$v_g(u_0) = c \frac{\left[1+\left(\vartheta^2 u_0\right)^2\right]^2}{\left[1+\left(\vartheta^2 u_0\right)^2\right]^2 - \dfrac{\vartheta^2}{2}\left[1-\left(\vartheta^2 u_0\right)^2\right]}, \qquad \text{where } u_0 = u(\omega_0) \tag{6}$$

Eq. 6 is plotted in fig. 1b for different values of divergence ϑ. Analyzing eq. 6 one can conclude that the group velocity on the optical axis equals to c if $u_0 = \pm 1/\vartheta^2$. According to eq. 4 $u_0 = \pm 1/\vartheta^2$ corresponds to $z = \pm z_{f0}$ where $z_{f0} = z_f(\omega_0)$ that is the Rayleigh length at ω_0. The group velocity on the optical axis is larger than c inside the interval $(-z_{f0}, z_{f0})$ and less than c outside. It reaches its maximum in the focus $(u_0 = 0)$ given by $v_{g.max} = c / [1 - \vartheta^2 / 2] \approx c / \cos\vartheta$. The minimum is achieved in $z = \pm \sqrt{3}\, z_{f0}$ and it can be calculated by $v_{g.min} = c / [1 + (\vartheta / 4)^2]$. The position of the pulse front on the optical axis at a moment t is determined by $t = \partial\Phi(\omega_0, z)/\partial\omega = z/c - Kz/[1+(K\omega_0 z)^2]$. The asymptotic form of this equation is $z = ct$, which means that far from the focus the pulse front moves on the optical axis as the phase front of a spherical wave does . This is why the areas above and below the dashed line (corresponding to $v_g = c$) enclosed by the dashed line and the solid line (graph of v_g /c) in fig. 1b are equal.

Since the Gouy phase shift is caused by the diffraction, it does not depend on the dispersion of the lens material. The phase anomaly in the vicinity of the focus appears in case of homogeneous illumination, but the phase behaviour is more complicated. So similar abnormal pulse propagation can occur for that case as well.

REFERENCES

1. L. Brillouin, *Wave propagation and group velocity*, Academic Press, New York, 1960
2. P. G. Kryukov and V. S. Letokhov, Soviet Physics Uspekhi **12**, 641 (1970)
3. R. Y. Chiao and J. Boyce, Phys. Rev. Lett. **73**, 3383 (1994)

4. R. Y. Chiao, J. Boyce, M. W. Mitchell, Appl. Phys. **B 60**, 259 (1995)
5. A. M. Steinbergm P. G. Kwiat and R. Y. Chiao, Phys. Rev. Lett. **72**, 708 (1993)
6. J. Jones, Am. J. Phys. **42**, 43 (1974)
7. M. Born and E. Wolf, *Principles of optics* (Pergamon Press, Oxford, 1989)
8. R. W. Boyd, J. Opt. Soc. Am. **70**, 877 (1980)
9. A. E. Siegman, *Lasers* (University Science Books, Mill Valley, California) ch. 17, p. 682
10. H. Kogelnik and T. Li, Appl. Optics **5**, 1550 (1966)

39

ULTRAFAST KERR DEMULTIPLEXING AT 460 GB/S IN A 1.5 CM LONG OPTICAL FIBER

O. Dühr,[1] F. Seifert,[2] V. Petrov,[1] and F. Noack[1]

[1] Max Born Institute for Nonlinear Optics and Ultrafast Spectroscopy
D-12474 Berlin, Germany
[2] Federal Institute of Physics and Technology
D-10587 Berlin, Germany

All-optical polarization switches exploiting the subpicosecond response time of the Kerr effect in glass fibers are promising for future multi-hundred-Gb/s transmission systems. The maximum switching rate of such Kerr shutters can be increased by reduction of the fiber length with higher pump (gate) powers, by using shorter pulse durations, and by closer frequency separation of the pump and probe pulses in the vicinity of the zero group velocity dispersion wavelength. In a previous work using picosecond pulses we demonstrated ultrafast demultiplexing based on induced polarization rotation at 170 Gb/s[1]. We present here results obtained with femtosecond gate and probe pulses which permit demultiplexing at switching rates as high as 460 Gb/s in a single-mode fiber. Although larger fiber lengths reduce the switching power, their application simultaneously increases the latency of a future communication system. This motivated us to investigate relatively short fibers with lengths comparable to the walk-off length.

Synchronized gate and probe pulses are derived from a common femtosecond pulse source through difference-frequency generation[2]. A commercial mode-locked Ti:sapphire laser is pumped by ≈10 W all lines from a cw Ar-ion laser. About 80% of the output power near 800 nm is used as a pump beam at 82 MHz. The remaining 20% are used to generate a multi-hundred-GHz burst consisting of four pulses in a double Michelson interferometer. The temporal separation between the four pulses is adjustable and repetition rates as high as 4 THz (4 Tb/s) can be achieved. Satisfactory resolution of the individual pulses in the cross-correlation functions used to characterize the burst is possible, however, up to repetition rates as high as ≈3 THz (Fig. 1). The GHz-THz burst near 800 nm is frequency mixed with about 300 mW of cw power at 514.5 nm from the Ar-ion laser in a 5-mm thick lithium triborate (LBO) crystal to produce a probe sequence in the near IR region[2]. The crystal is cut along the x-axis for type-I uncritical phase-matching with temperature tuning. In this geometry the generated IR pulses are somewhat broadened because of the group velocity walk-off in LBO (of the order of 18 fs/mm for an IR wavelength of 1442 nm). The average powers available in the near IR pulse train exceed 150 nW.

Ultrafast Processes in Spectroscopy, edited by Svelto et al.
Plenum Press, New York, 1996

175

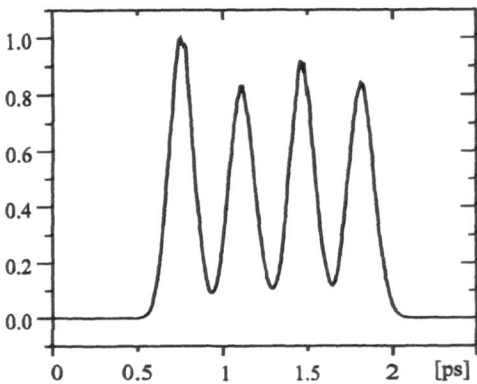

Figure 1. Cross-correlation function between a 2.9 THz burst and single pulses at 82 MHz. Both wavelengths are 800 nm and the noncollinear second harmonic generation was performed in a 1-mm thick β-barium borate (BBO) crystal (type-1 phase-matching). The FWHM's in the figure correspond to a pulse duration of ≈100 fs assuming Gaussian pulse shapes.

The probe and gate pulses are coupled into a commercial single-mode fused silica fiber (6.8 μm mode-field diameter) by a M 10X microscope objective under 45° relative polarization angle. A filter behind the fiber blocks the pump beam, and the polarization rotation of the probe pulses is detected by an analyzer adjusted for minimum transmission in the absence of pump pulses. Demultiplexing is detected by recording the switched signal through the analyzer in dependence on the relative delay between the 4-pulse near IR burst and the pump pulses at 82 MHz[1].

The observed τ_{signal}=450 fs in Fig. 2 has the potential of increasing the demultiplexing rate up to 1 Tb/s. It should be noted that the result in Fig. 2 has been achieved by adjusting the Ti:sapphire laser to produce somewhat longer pulses than normal to limit the pump intensity, preventing in such a way self-phase modulation and the drastic increase of the influence of the group velocity dispersion on the pump pulses that is associated with it. The latter is possible because the fiber length employed in this case (L=1.5 cm) only slightly exceeds the walk-off length of 1.2 cm. At comparable values of the fiber and walk-off lengths a satisfactory agreement with the experimental value of τ_{signal} can be obtained using the simple relation $\tau_{signal}=[(\tau_{probe})^2+(\tau_{pump})^2+(\beta L)^2]^{1/2}$ where β is the inverse group velocity mismatch. For the parameters corresponding to Fig. 2 we calculate using this relation τ_{signal}=415 fs which differs only slightly from the experimental value.

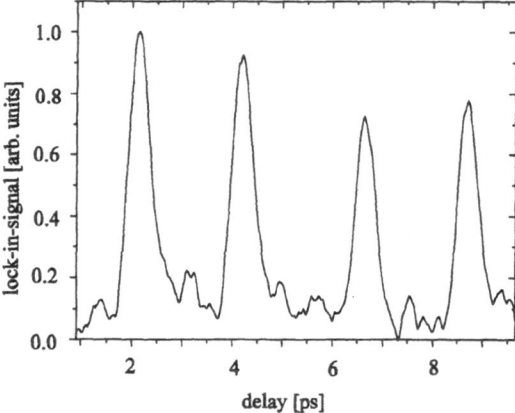

Figure 2. Demultiplexing at 460 Gb/s. λ_{pump}=783 nm, λ_{probe}=1500 nm, τ_{pump}=210 fs, τ_{probe}=235 fs.

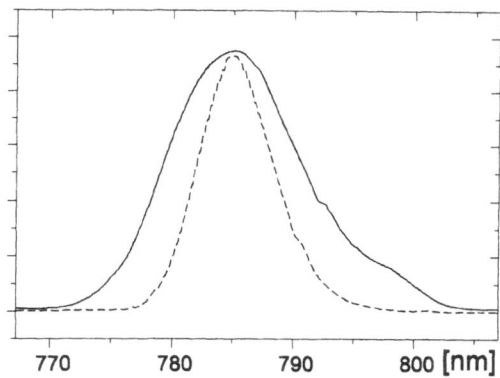

Figure 3. Spectra of the pump pulses recorded before (dashed line) and behind (solid line) the single-mode fiber.

We conclude therefore that the temporal profile of the observed signal, and consequently the maximum switching rates, can be explained and estimated only on the basis of the group walk-off. This means that self-phase modulation and group velocity dispersion are not detrimental for our pump intensities at the short fiber length used. This can be seen also from Fig. 3 which shows the spectra of the pump pulses before and after the fiber where the spectral width has increased by about 80%.

It can be assumed on the basis of our experiment that $\Delta v \tau_{pump}$ values of the order of 1, where Δv is the spectral width, are still acceptable without further limitations on the maximum achievable switching rate. The spectral width Δv due to self-phase modulation is related to the Kerr coefficient n_2 in the absence of group velocity dispersion as $\Delta v = 4 \ln 2 n_2 I_o L / \tau_{pump} \lambda_{pump}$ where I_o is the peak pump intensity. The maximum switching efficiency is obtained at a phase shift (induced phase difference between the two orthogonal components of the probe pulses)[3] of $\Delta \phi_o = (16 \pi n_2 I_o L / 3 \lambda_{probe} \zeta) \mathrm{erf}(\zeta/2)$ where $\zeta = 2(\ln 2)^{1/2} \beta L / \tau_{pump}$. With typical pump intensities of $I_o = 50$ GW/cm^2 this leads to the conclusion that 100% switching (i.e. $\Delta \phi_o = \pi$) at the reported repetition rates (of the order of 500 Gb/s) is realistic if the value of β is decreased approximately 3 times and the pump wavelength λ_{pump} is shifted into the near IR region. In this case one arrives at a necessary fiber length of $L = 4$ cm.

REFERENCES

1. O. Dühr, V. Petrov, F. Noack, and F. Seifert, in *Conference on Lasers and Electro-Optics*, (Optical Society of America, Washington, D.C., 1995), paper CThI27, pp. 305–306.
2. F. Seifert and V. Petrov, *J. Appl. Phys.* **74**, 4798 (1993).
3. T. Morioka and M. Saruwatari, *IEEE J. Select. Area Commun.* **6**, 1186 (1988).

40

GROUP VELOCITY MISMATCH EFFECTS IN ULTRAFAST OPTICAL MODULATORS BASED ON CASCADED SECOND ORDER NONLINEARITIES

G. Toci,[1] R. Pini,[2] R. Salimbeni,[2] and M. Vannini[2]

[1] Dipartimento di Fisica
   Università degli studi di Firenze, Italy
[2] Istituto di Elettronica Quantistica–CNR
   Firenze, Italy

The intensity dependent phase and amplitude modulation arising from cascaded second order optical nonlinear processes in frequency doubling crystals are getting growing attention[1,2,3] for all-optical switching devices and to obtain passive mode-locking techniques complementary to the Kerr-lens mode-locking (KLM)[4,5] . Up to now these effects have been theoretically and experimentally investigated in stationary conditions, i.e. with interacting pulses fundamental and second harmonic at ω and at 2ω respectively, much longer their propagation delay through the nonlinear crystal length, as determined by their Group Velocity Mismatch (GVM). Due to the high nonlinearity of the available frequency doubling crystals, the obtained phase shift can be much higher that with the usual Kerr materials. Nevertheless, in nonstationary conditions arising when the fundamental pulse duration is equal to or shorter than the propagation delay we have to take into account of the GVM effects.

We recently developed[3] a perturbative method that in the limit of weak interaction deals with the propagation problems set by the GVM (nonstationary interaction). Whenever these second order effects are used to obtain a fast saturable absorber action for achieving a mode-locking condition, our method can describe the temporal behavior of the device when the nonstationary interaction limit is reached, showing in particular that the temporal selectivity of the device decreases as the ω pulse shortens, determining the ultimate limit achievable for a given crystal length.

The amplitude and phase modulation in nonstationary condition can be calculated from the time-dependent Bloembergen equations[6]:

$$\left(\frac{\partial}{\partial \xi}+\frac{1}{\upsilon_1}\frac{\partial}{\partial t}\right)\rho_1(\xi,t) = -i\eta_0\rho_1^*(\xi,t)\rho_2(\xi,t)\exp[-i\delta\xi] \tag{1}$$

Ultrafast Processes in Spectroscopy, edited by Svelto et al.
Plenum Press, New York, 1996

$$\left(\frac{\partial}{\partial \xi} + \frac{1}{\upsilon_2}\frac{\partial}{\partial t}\right)\rho_2(\xi,t) = -i\eta_0 \rho_1^2(\xi,t)\exp[i\delta\xi] \tag{2}$$

where $\rho_{1,2} = E_{1,2}/|E_1|_{peak}$ with $E_{1,2}$ the fundamental and harmonic envelopes of the pulses and $|E_1|_{peak}$ the peak amplitude of the fundamental, $\upsilon_{1,2} = v_{1,2}/L$, with $v_{1,2}$ their group velocities, L the crystal length, $\xi = z/L$, $\eta_0 = 4\pi k_1(\chi_{eff}^2/n_1^2)|E_1|_{peak}L$, is the efficiency in stationary condition neglecting the fundamental field depletion with: k_1 and k_2 the wave numbers of the fundamental and of the second harmonic respectively, $\delta = (k_2 - 2k_1)L$ is the phase rotation at the crystal output, χ_{eff}^2 .is the effective second order susceptibility.

We found a solution of the Eq. 1,2 with a perturbative method, expanding the envelopes in series of η_0:

$$\rho_i(\xi,t) = \sum_{n=0}^{\infty}\eta_0{}^n \rho_{i,n}(\xi,t) \qquad (i = 1,2) \tag{3}$$

leading to the following hierarchy of integral equations:

$$\rho_{i,0}(\xi,t) = \rho_{i,in}(t - \xi/\upsilon_i) \qquad (i = 1,2) \cdot \tag{4}$$

$$\rho_{1,n}(\xi,u_1) = -i\int_0^{\xi} \sum_{\substack{i+j=n-1 \\ i\geq 0, j\geq 0}}\rho_{1,i}^*(x,u_1)\rho_{2,j}(x,u_1)\exp(-i\delta x)dx \tag{5}$$

$$\rho_{2,n}(\xi,u_2) = -i\int_0^{\xi} \sum_{\substack{i+j=n-1 \\ i\geq 0, j\geq 0}}\rho_{1,i}(x,u_2)\rho_{1,j}(x,u_2)\exp(i\delta x)dx \tag{6}$$

where $u_1 = t - \xi/\upsilon_1$, $u_2 = t - \xi/\upsilon_2$ and $\rho_{i,in}(t)$ are the pulse shapes injected in the crystal at $\xi = 0, (z = 0)$.

In a double pass configuration, the fundamental pulse is injected into the crystal where it generates the second harmonic pulse (in this case the fundamental and the harmonic pulse at the output are composed respectively only by the even and the odd terms in η_0). Then the two pulses are injected back into the crystal with no alteration but with reflectivities r_1, r_2, a phase shift θ_1, θ_2 respectively and a relative delay Δt.

We verified that for commonly available injected pulse shapes described by $\mathrm{sech}(t/\tau)$ or $\exp(-t^2/\tau^2)$ an analytical solution exists only if we are in phase-matching condition. Otherwise a numerical code is necessary to calculate the integrals in Eq. 5 and 6 or to solve numerically Eq. 1 and 2. Assuming $\rho_{1in}(t) = \mathrm{sech}(t/\tau)$ and to be in phase-matching conditions ($\delta = 0$), the fundamental pulse envelope after the second pass, up to the second order in η_0 is :

$$\rho_1(t) = r_1\,\mathrm{sech}(t/\tau)\left\{1 + r_2\exp(-i\Delta\Theta)\eta_0^2\delta_1(t) - (1 + r_1^2)\eta_0^2\delta_2(t)\right\} \tag{7}$$

where, setting $\sigma = (1/\upsilon_2 - 1/\upsilon_1)/2\tau$ (steadiness parameter[3]) is:

$$\delta_1(t) = \frac{1}{4\sigma^2} \ln\left\{ \frac{\cosh((t+\Delta t)/\tau - 2\sigma)^2}{\cosh((t+\Delta t)/\tau)\cosh((t+\Delta t)/\tau - 4\sigma)} \right\} \tag{8}$$

$$\delta_2(t) = \frac{1}{2\sigma} \tanh\left(\frac{t}{\tau}\right) + \frac{1}{4\sigma^2} \ln\left[\frac{\cosh(t/\tau - 2\sigma)}{(t/\tau)} \right] \tag{9}$$

Depending on the relative phase difference $\Delta\Theta = \theta_2 - 2\theta_1$ at the input of the second pass the fundamental pulse experiences a phase and/or an amplitude modulation proportional to the pulse peak intensity. With various arrangements (i.e. the nonlinear mirror,[4] or exploiting the parametric focusing effect of a gaussian beam through a suitable aperture[1,5,7]) such a modulation induces a time dependent modulation of the transmission of the ω pulse (with respect to its low power level) that is usually proportional to the phase or the amplitude modulation, depending on the arrangement.

Furthermore it is worth to note that part of the modulation profile, arising from the single pass depletion, has a constant phase relationship with the fundamental pulse; this determines an unavoidable loss mechanism.

Considering, as an example, the parametric lens-aperture effect, the Fig. 1 reports the temporal behavior of the transmission of the fundamental pulse through the modulator, at different decreasing pulse durations (increasing σ) for a given peak intensity. The transmission is normalized to its peak value as done in stationary conditions. It can be seen that the transmission peak value decreases, and the relative width of the transmission profile increases when the pulse duration decreases, due to the GVM. Similar results hold also for the nonlinear mirror experiment.

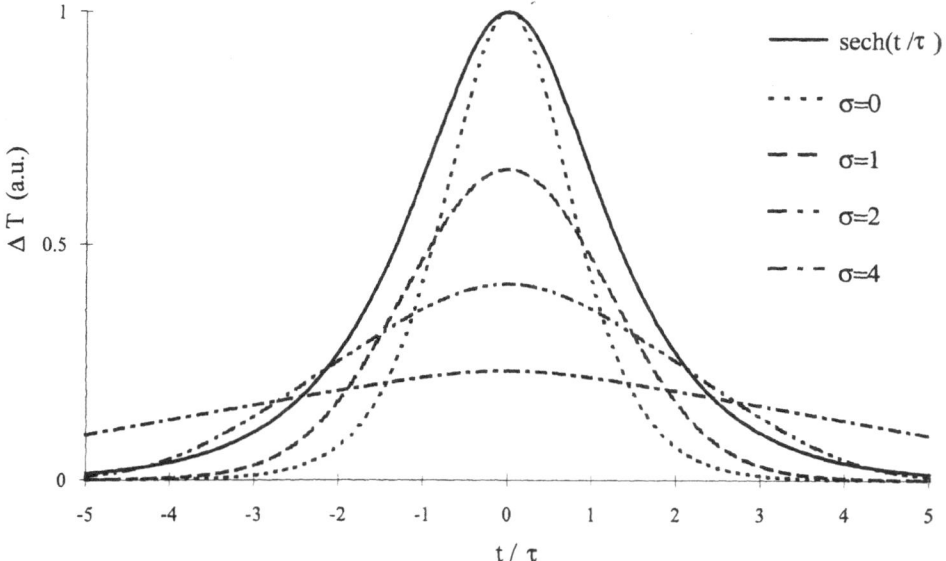

Figure 1. Time dependent transmission through a second order modulator, for increasing values of the steadiness parameter σ.

In conclusion, we propose a mathematical method based on a series expansion that for the first time gets an analytical expression for the fundamental and harmonic pulse envelopes at the output of a cascaded second order nonlinear process when the GVM of the nonlinear crystal is not anymore negligible. We applied this method to the analysis of the temporal behavior of a passive modulator suitable for passive mode-locking and we showed that the temporal selectivity of this class of modulators exhibits an effectiveness limit when the fundamental pulse duration decreases and becomes comparable with the propagation delay through the crystal. Thus when the presence of GVM can not be neglected, an experiment requires a careful design. The proposed method can also be useful to describe the temporal dynamic of the switching properties of the proposed all-optical modulators based on the second-order optical processes. An extended general analysis of the Bloembergen equations in non phase-matched conditions for the non-stationary case requires a computer code that we developed. This all numerical study was validated and compared with all the available analytical solutions showing interesting results to be discussed for the future applications. Work is in progress on these first outputs.

REFERENCES

1. D. Pierrotet, B. Berman, M. Vannini, D. McGraw, Opt. Lett. **18**, 263 (1994)
2. G. Assanto, G. I. Stegeman, M. Sheik-Bahae, E. VanStryland, IEEE J. Quantum Electron. **31**, 673 (1991)
3. G. Toci, D. McGraw, R. Pini, R. Salimbeni, M. Vannini, Opt. Lett. **20**, 1547 (1995)
4. K. A. Stankov, V. P. Tzolov, M. G. Mirkov, Opt. Lett. **16**, 1119 (1991)
5. G. Cerullo, S. De Silvestri, A. Monguzzi, D. Segala, V. Magni, Opt. Lett. **20**, 746 (1995)
6. J. A. Armstrong, N. Bloembergen, J. Ducuing, P. S. Pershan, Phys. Rev. **127**, 1918 (1962)
7. G. Toci, M. Vannini, R. Salimbeni, R. Pini, Optics Commun. **120**, 78 (1995)

41

DYNAMIC OF SOLITONS IN INHOMOGENOUS JOSEPHSON JUNCTIONS

N. Riazi

Department of Physics and Biruni Observatory
Shiraz University, Shiraz 71454, Iran and
Institute for Studies in Theoretical Physics and Mathematics (IPM)
P.O. Box 19395–5746, Tehran, Iran

Superconducting Josephson junctions[2] exhibit soliton behavior which can be described by sine-Gordon equation[4]. Solitons are quantums of magnetic flux, in this case. Their production, transmission and storage as stable objects is quite feasible, and therefore very important information processing systems.

The standard form of sine-Gordon equation is

$$\frac{\partial^2 \varphi}{\partial x^2} - \frac{1}{c^2}\frac{\partial^2 \varphi}{\partial t^2} = a \sin b\varphi \tag{1}$$

in which a and b are constants assumed to have the same sign. By using the following change of variables

$$u = \sqrt{ab}\,x; \quad v = \sqrt{ab}\,ct; \quad \sigma = b\varphi\,; \tag{2}$$

equation (1) becomes

$$\sigma_{uu} - \sigma_{vv} = \sin \sigma\,. \tag{3}$$

Multi-soliton solutions of this equation can be obtained systematically by applying Bäcklund transformations[1].

The time-independent inhomogeneity can be introduced into sine-Gordon equation in different ways. We will mention three ways, although only two cases will be worked out in detail.

The first kind of inhomogeneity is introduced via a varying 'refractive index':

$$\frac{\partial^2 \varphi}{\partial x^2} - \frac{n^2(x)}{c^2}\frac{\partial^2 \varphi}{\partial t^2} = a \sin b\varphi\,. \tag{4}$$

Ultrafast Processes in Spectroscopy, edited by Svelto et al.
Plenum Press, New York, 1996

This looks like a variation in the optical refractive index of a transparent medium in the context of electromagnetic wave propagation. A physical fulfillment of equation (4) can be accomplished by an inhomogeneous Josephson transmission line.

The inhomogeneity of the second kind can be introduced via spatially varying a and b:

$$\frac{\partial^2 \varphi}{\partial x^2} - \frac{1}{c^2} \frac{\partial^2 \varphi}{\partial t^2} = a(x) \sin b(x) \tag{5}$$

The third kind of inhomogeneity is introduced via an external field $\chi(x)$:

$$\frac{\partial^2 \varphi}{\partial x^2} - \frac{1}{c^2} \frac{\partial^2 \varphi}{\partial t^2} = a \sin b\varphi + \chi(x) \tag{6}$$

This case discussed in Reinisch and Fernandez[5] and Kaup[3].

Several interesting questions arise when the kink dynamics in an inhomogeneous medium is concerned. If the kink is considered as a classical particle interacting with a background potential, what would be the characteristics of such a potential in relation with $n(x)$, $a(x)$, $b(x)$, and the kink velocity v_k? In what circumstances the particle aspect fails to be adequate? What are the interesting features of the wave aspect? etc...

REFERENCES

1. A.V. Bäcklund, Mathematische Annalen, **9**, 297 (1876).
2. B.D. Josephson, Phys. Lett., **1**, 251, (1962).
3. D.J. Kaup, Phys. Rev. **B29**, 1072 (1984).
4. N.F. Pederson, Solitons in Josephson Transmission Lines, in *Solitons*, ed. Trullinger, S.E., Zakharov, V.E., and Prokovsky, V.L., North Holand (1986).
5. G. Reinisch, and C. Fernandez, Phys. Rev.**B24**, 835 (1981).
6. N. Riazi Int. J. Theor. Phys., **32** 2155 (1993).

42

INVESTIGATION OF STRUCTURE OF TRANSFORMATION PHASE SPACE AND CHAOTIC INSTABILITIES IN TUNABLE Cr^{4+}:$Y_3Al_5O_{12}$ LASER

J. J. Broslavez, A. A. Fomitchev, V. D. Lokhnygin, and O. O. Silichev

Moscow Institute of Physics and Technology, Laser Center, Russia
Institutsky per., 9, Moscow reg., Dolgoprudny, 141700, Russia

Carried out preliminary investigations of the Cr^{4+}:YAG material have shown very good prospects for using it in the master oscillator to obtain Kerr mode-locked. For this reason we have investigated theoretically and by computer simulation the formation of ultra-short pulses in our laser system using the idea of the description of ultra-short coherent optical pulses as temporal Gaussian beams analogous to complex Ermit-Gaussian beams[4] with displaced center which are well known in the spatial beam optics. The impulse passes trough optical elements of the cavity in our model were presented by the Gaussian function. Each element of the cavity were described by a formulas of transformation of five-dimensional space: A_o, t_o, α_o, R, and ω with parameters of elements:

$$\left\{ A_{o(n+1)}, \tau_{o(n+1)}, \alpha_{o(n+1)}, R_{(n+1)}, \omega_{(n+1)} \right\} = f_i \left\{ A_{o(n)}, \tau_{o(n)}, \alpha_{o(n)}, R_{(n)}, \omega_{(n)} [parameters] \right\} \quad (1)$$

where f_i - is the function relates the output and input of the five-dimensional space for i - element of cavity. The computer simulation have shown the presence of asymptotically stable stationary point in behavior of temporal Gaussian beam similar spatial mode structure in the resonators, when the temporal mode does not change passing through all dispersion element in laser. Our calculations show that the sign of dispersion is very important for formation of phase portrait in our laser system

Thus, the developed theoretical model for formation of structure of radiation in spatial and temporal coordinates of a pulse, has allowed to define the optimal area for obtaining Kerr mode-locking , as well as to simulate process of modification of a pulse at its propagation through various elements of a laser system. Though the results obtained in modeling a real laser system, allowed to receive the most optimal parameters of a system and to define a zone of stability by phase portraits, they nevertheless do not give complete understanding neither of process of pulses formation, nor of the reasons why a system

Ultrafast Processes in Spectroscopy, edited by Svelto et al.
Plenum Press, New York, 1996

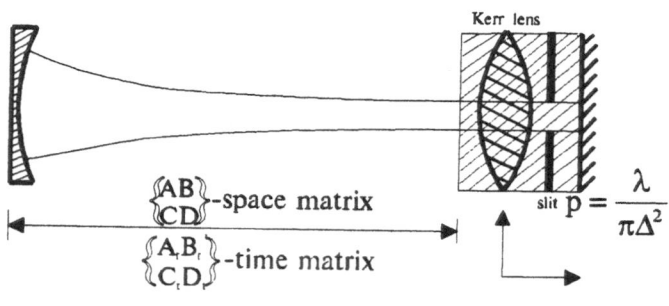

Figure 1.

goes to chaotic mode generation. That is why, we analyzed simplified model of a laser system with Kerr lens, shown on fig. 1.

In a offered system the resonator is formed by two mirrors - one of which is flat. The Kerr lens is formed in a active element located closely to a flat mirror. Let us consider, that irising of radiation takes place in an active element. For a offered configuration, using the matrix description of an optical system, system of equations, describing transformation of complex parameters - both in space and in time is obtained. Let us designate:

$$\mu = \frac{1}{q} = \frac{1}{R} - i\frac{\lambda}{\pi\omega^2} = x - iy\,; \qquad \gamma = \frac{1}{q} = \frac{2}{c}\alpha + i\frac{2}{c\tau^2} = x_t + iy_t$$

(2)

where μ, γ - spatial and temporal complex parameters. Having added to the received system the equation of describing change of pulse energy - we receive a system, where S_n and T_n - small nonlinear additives resulting from Kerr lens and frequency modulation of a pulse:

$$
\begin{cases}
E_o = E_i a_E \left\{ \dfrac{y}{y+p} \right\} \exp\left\{ \dfrac{GL}{1+\dfrac{E_i y}{pE_s}} \right\} \left\{ \dfrac{y+p}{y+2p} \right\} \exp\left\{ \dfrac{GL}{1+\dfrac{E_i(y+p)}{pE_s}} \right\} \\[4mm]
m' = \dfrac{C+Am}{A+Bm} - ip - S_n \\[3mm]
g' = \dfrac{C_t + A_t g}{A_t + B_t g} + ip_t + T_n; \\[3mm]
\text{where -} \\[2mm]
S_n = a_{\text{kerr\_lens}} \left\{ E_i\left[y^2 \sqrt{y_t} \right] + E_o\left[(y+p)^2 \sqrt{y_t + p_t} \right] \right\} \\[3mm]
T_n = a_{\text{shirp}} \left\{ E_i\left[yy^2 \sqrt{y_t} \right] + E_o\left[(y+p)^2 (y_t + p_t) \sqrt{y_t + p_t} \right] \right\}
\end{cases}
$$

(3)

The analysis of a system was conducted on the basis of numerical modeling. Phase portraits with modes both of stable generation of pulses and close to chaotic one are received. Analogy between phase portraits received in modeling of a real laser system and simplified model shows, that the main mechanism of transition of a system to chaotic mode is caused by Kerr formation of a lens and "chirp". In limiting case, when Sn=0 and Tn =0, spatial and temporal equations have the same form and have got simple analytical solution. Therefore, we solve a equation:

$$\mu = \frac{C + A\mu}{A + B\mu} - ip \tag{4}$$

As far as the solution is searched near a stationary point, complex parameter can be presented in a form: $\mu = \mu_o + z(t)$, where $z(t)$ - small function, and μ_o- stationary point. Having substituted in a equation Eq.4 we find the change in $z(t)$ for one: Round trip graduation through the resonator:

$$z' = \frac{A^2 - BC}{\left(A + B\mu_o\right)^2 + zB\left(A + B\mu_o\right)} z = \frac{z}{\left(A + B\mu_o\right)^2 + zB\left(A + B\mu_o\right)} \tag{5}$$

let us designate:

$$\frac{1}{\left(A + B\mu_o\right)^2 + zB\left(A + B\mu_o\right)} = \Theta = \exp\left\{i\xi t^*\right\}$$

We can present the solution in a form: $z' = \Theta \exp\left\{i\xi t_o\right\}$; where $z(t_o) = \exp\left\{i\xi t_o\right\}$ - initial point. Similarly, for a stationary point it is possible to receive:

$$\mu_o = -\frac{ip}{2} \pm (\vartheta - i\frac{pA}{2B\vartheta}) \quad \text{where} \quad \vartheta - \sqrt{\frac{1}{2}\left(\frac{C}{B} - \frac{p^2}{4}\right) \pm \sqrt{\left(\frac{C}{B} - \frac{p^2}{4}\right)^2 + \frac{p^2 A^2}{B^2}}} \tag{6}$$

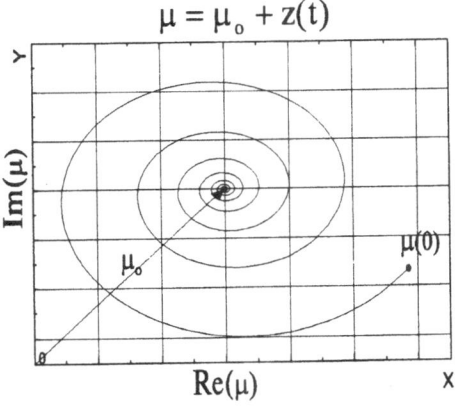

$$\mu = \mu_o + z(t)$$

Figure 2.

Thus, the received solution represents a spiral fig. 2 rounded to a stationary point. It should be noted, that dependence of pre-exponential multiplier on parameter results in distortion of a location of a stationary point. Presence of the nonlinear additives because of the Kerr effect results in deviation of a phase trajectory from the received solution in assumption by their smallness. As they increase the system can have various kinds of attracting centers, and eventually the phase trajectory becomes chaotic.

Thus, the conclusion is, that for obtaining steady Kerr mode-locking the value of the nonlinear additive should be reasonably large to make it possible to obtain a mode-locking, but at the same time it should not exceed the value leading to chaotic generation of pulses. On the other hand, mode of generation close to chaotic one may be used probably for generation of sequences of pulses with some specific parameters.

REFERENCES

1. Howard Nathel, Alphan Sennaroglu, Clifford R.Pollock, Cornell Univ, 'Mode-locked, Cr^{+4} solid-state lasers.', Lawrence Livermore National Laboratory.
2. S.A.Akhmanov, V.A.Visloukh, A.S.Chirkin 'OPTICS OF FEMTOSECOND LASER PULSES', (1988), Moscow.
3. T.Brabec, P.F.Curley, S.M.J.Kelly, E.Wintner,...., 'Ultrashort Pulses from Ti:Sapphire Lasers', (1993), UPS'93.
4. Silichev O.O, Fomichev A.A., 'Matrix Method of Femto-Second Laser Pulses Propagation Calculation', (1993), UPS'93.
5. Zaslavski G.M., Sagdeev R.Z. 'Vvedenie v nelineinyu physics: ...', M.:Nayka, 1988.

43

OPTICAL CROSS-TALK BETWEEN QUANTUM WELLS

D. Weber,[1] R. Hellmann,[1] J. Feldmann,[2] E. O. Göbel,[3] D. S. Citrin,[4] and
K. Köhler[5]

[1] Department of Physics and Materials Sciences Center, University of Marburg
Renthof 5, D-35032 Marburg, Germany
[2] Physics Department, Ludwig-Maximilians University of Munich
Amalienstr. 54, D-80799 München, Germany
[3] Physikalisch-Technische Bundesanstalt
Bundesallee 100, D-38116 Braunschweig, Germany
[4] Department of Physics, Washington State University
Pullman, Washington 99164–2814
[5] Fraunhofer-Institut für Angewandte Festkörperphysik
D-79108 Freiburg, Germany

In recent years, the dynamics of excitons in quantum wells (QWs) has attracted much attention especially under resonant excitation. Many time-resolved photoluminescence experiments have been performed to study the formation and recombination dynamics of QW excitons[1–6]. However, in all these experiments it has been assumed that the number of QWs does not influence the decay dynamics of the excitonic excitation, i.e., the interaction between excitons located in different QWs mediated by emitted and reabsorbed photons has been neglected. But this interaction may lead to a coupling between spatially separated excitons even though electronic coupling is absent and thus may influence the decay dynamics of excitonic polarizations and densities. Recent theoretical investigations showed that such coupling effects may become important for excitonic decay processes in multiple QW stacks[7–9].

In this article we show that the resonantly excited, coherent exciton dynamics indeed depends on the number of QWs. We conjecture that the interaction between excitons in adjacent QWs via photons emitted by the first-order polarization $P^{(1)}$ might be responsible for the modified dynamics.

We have investigated three samples consisting of 1, 2, and 5 periods of GaAs/AlGaAs QWs. Both, the GaAs quantum wells and the $Al_{0.3}Ga_{0.7}As$ barriers have a thickness of 12nm leading to negligible electronic coupling between states in different QWs. The low total linewidth for the lowest n=1 heavy-hole exciton resonance and the vanishing Stokes-Shift observed in photoluminescence spectra reveal the high quality of all three

Ultrafast Processes in Spectroscopy, edited by Svelto et al.
Plenum Press, New York, 1996

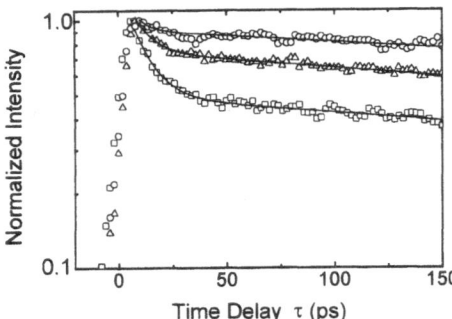

Figure 1. Single-logarithmic plot of the normalized differential reflection as a function of time delay between pump and probe pulses for the SQW (circles), DQW (triangles), and QQW (squares).

samples, i.e., localization effects do not play a major role. All experiments are performed at 10 K and at low excitation densities ($5 * 10^8$ cm^{-2} per QW).

We use pump and probe (P&P) experiments to study the coherent as well as the incoherent population dynamics of the excitons[10]. A synchronously mode-locked dye laser with 7ps pulses is tuned to the resonance of the n=1 heavy-hole exciton transition. A low-intensity laser pulse originating from the same laser source as the pump pulse probes the reflectivity of the respective sample with and without the pump pulse, i.e., the differential reflectivity (ΔR) is measured as a function of time delay τ between pump and probe pulses. The probe beam is cross-polarized with respect to the pump beam to suppress scattered light from the pump pulse. We prefer the reflection geometry in our experiments, since a chemical etching procedure necessary to allow transmission experiments could possibly lead to undesirable inhomogeneous strain effects.

The temporal evolutions of ΔR are shown in Fig. 1 for all three samples. The solid lines represent double exponential fits to the measured P&P transients. For all three samples the long-time behavior is characterized by an exponential decay with a time constant of approximately 650ps. This value is in good agreement with previously reported results[5]. Localization effects[11] and thermalization effects[5] might be responsible for the long excitonic decay time. Altogether, the long-time behavior of the P&P transients reflects the dynamics of the incoherent, probably localized exciton population and thus does not depend on the QW number.

The P&P transients in Fig. 1 clearly show that the initial temporal evolutions of the various curves deviate from each other. For the quintuple QW (QQW) sample we find an initial fast decay of the P&P transient. This fast initial decay is less pronounced for the double QW (DQW) and almost absent for the single QW (SQW). From the double exponential fit a time constant of 9 ps is obtained for the initial fast decay and is approximately

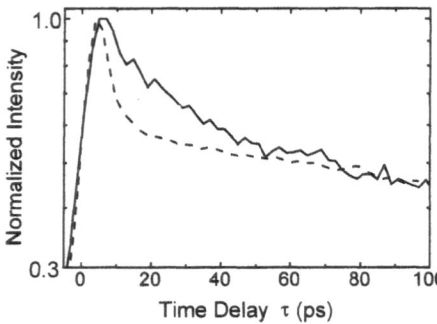

Figure 2. Single-logarithmic plot of the normalized differential reflection as a function of the exciton density n for the QQW. Solid line: $n=5*10^8$cm^{-2}, and dotted line: $n=5*10^9$cm^{-2}.

the same for all three samples. The rise times of all transients are much shorter and limited by the laser pulse width. First of all the question arises, whether the initial excitonic population decay occurs within the temporal window, where the excitations are still coherent. If this is true the initial population decay is expected to become faster when increasing the intensity of the pump pulse, since exciton-exciton scattering shortens the dephasing time T_2. Figure 2 shows P&P transients for the QQW sample for various excitation intensities. We find a shortering of the initial decay time with increasing exciton density. The results of the P&P experiments thus lead us to the conclusion that only the coherent dynamics of excitons depends on the QW number N, whereas the decay dynamics of incoherent excitons is the same for SQWs and for MQWs.

The energy connected with the initial population decay is expected to be found in the light field emitted by $P^{(1)}$. Since this light emission is highly directional, it should follow the reflected and transmitted excitation pulses. The fact that the coherent population decay is most pronounced for the QQW means that the intensity of the coherent light emission should be a superlinear function of the QW number. In order to verify this presumption we have measured the electromagnetic light emission induced by $P^{(1)}$. In order to distinguish this coherent light emission from the reflected laser pulse we have to time-resolve the optical signal in the direction of the reflected laser beam. For this purpose we must use shorter laser pulses, i.e., a mode-locked Ti:sapphire laser providing 110 fs pulses is taken as the excitation source. Again the laser photon energy is tuned into resonance with the lowest n=1 heavy hole exciton transition. The reflected laser beam is up-converted by focusing it onto a 1mm $LiIO_3$ crystal together with a second 110 fs reference pulse. The light intensity at the sum frequency is then monitored as a function of the time delay t between the reference pulse and the reflected pulse. If the intensity of the reference pulse is kept constant, the intensity of the up-converted signal I(t) is proportional to the squared sum of the reflected laser light $E_L(t)$ and the retarded electromagnetic field emitted from the first-order polarization $P^{(1)}_i(t)$ in the i'th QW.

Figure 3 shows the reflected up-converted signal I(t) from the QQW sample as a function of the time delay t. The relatively strong signal at zero delay is caused by the reflected laser pulse from the sample surface. The reflected laser pulse is followed by a modulated optical signal. For time delays larger than approximately 700fs the modulation is periodic with a time period of 485fs. This beat period corresponds to an energy splitting of 8.5meV and thus to the spectral separation of the heavy-hole and light-hole exciton transitions. In this time regime we assign the measured optical signal only to coherent light emission from the QWs. A qualitative similar behavior is found for the DQW- and SQW-sample. In Fig. 4a we have plotted the optical signals emitted from the QWs for all three samples on a linear scale. The initial temporal parts of the reflected laser pulses have

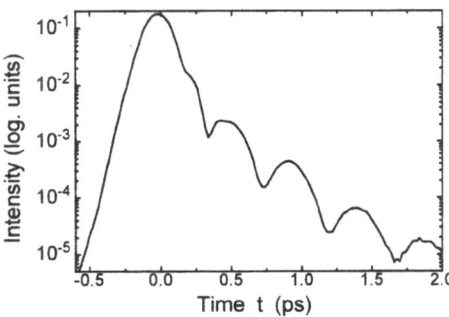

Figure 3. The up-converted optical signal for the QQW on a logarithmic scale in the direction of the reflected laser beam as a function of the time delay between the reflected laser pulse and a second reference pulse.

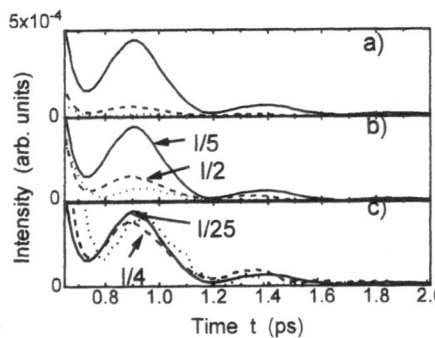

Figure 4. Up-converted optical signals for all three samples on a linear scale (a). In order to illustrate the superlinear intensity dependence on the number of QWs we have divided the measured signal by N (b) and by N^2 (c).

always been omitted. In order to illustrate the superlinear intensity dependence on the number of QWs we have divided the measured signal by N (Fig. 4b) and by N^2 (Fig. 4c). Obviously, the intensity of the light field emitted by $P^{(1)}$ shows a quadratic increase with increasing N.

First of all a possible explanation for this superlinear emission behavior could be that for the reflected beam direction a constructive interference effect may enhance the coherent emission. However, we can exclude that this is the only responsible contribution by considering the sample structure. For a quadratic increase of the intensity of the coherent light emission the excitonic excitations must have exactly the same phases or a multiple of π in the case of the re flected direction. Only under this condition the amplitudes of the light fields emitted from each QW add up constructively, which then would result in a quadratic increase of the light intensity. But this is not the case for the samples under investigation. In addition, without a coupling between the photon and the excitonic resonances the population dynamics would not depend on N, which is in contrast to our experimental results. We can also exclude a stimulated emission process, since all experiments are performed at low intensity, i.e., no gain is reached after excitation.

We conjecture that the interaction between $P^{(1)}$ in a certain QW and the light field emitted by $P^{(1)}$ in a neighbouring QW might lead to the observed effects. Reabsorption and induced emission of the light field should be important processes for the QW structures under investigation. As a result the excitonic population shows the collective behavior as experimentally observed during the coherence lifetime. However, detailed theoretical treatments are needed to clarify the microscopic origin for this superlinear emission process. In particular, it is not clear up to now, why an intensity increase is found without any changes in the decay time of the coherent emission. Probably, many-body Coulomb-effects, disorder induced localization effects, and nonlinear optical effects must all be considered as important ingredients for a theoretical treatment.

ACKNOWLEDGMENT

We thank J. Shah, T. Stroucken A. Knorr, and S.W. Koch for many helpful discussions and M. Preis for expert technical assistance. This work has been financially supported by the Deutsche Forschungsgemeinschaft through the Gerhard Hess-Förderpreis and the SFB 383. One of us (DSC) was supported by the National Science Foundation by grant STC PHY 892018.

REFERENCES

1. J. Feldmann, G. Peter, and E.O. Göbel, Phys. Rev. Lett. **59**, 2337 (1987).
2. T.C. Damen, K. Leo, J. Shah, and J.E. Cunningham, Appl. Phys. Lett. **58,** 1902 (1991).
3. B. Deveaud, F. Clérot, N. Roy, K. Satzke, B. Sermage, and D.S. Katzer, Phys. Rev. Lett **67**, 2355 (1991).
4. R. Eccleston, B.F. Feuerbacher, J. Kuhl, W.W. Rühle, and K. Ploog, Phys. Rev. B **45**, 11403 (1992).
5. A. Vinattieri, J. Shah, T.C. Damen, and D.S. Kim, Solid State Commun. **88**, 189 (1993).
6. H. Wang, J. Shah, and T.C. Damen, Phys. Rev. Lett. **74**, 3065 (1995).
7. D.S. Citrin, Solid State Commun. **89**, 140 (1993).
8. L.C. Andreani, Phys. Stat. Sol. (b) **188**, 29 (1995).
9. T. Stroucken, A. Knorr, C. Anthony, A. Schulze, P.Thomas, S.W. Koch, S.T. Cundiff, J. Feldmann, and E.O. Göbel, Phys. Rev. Lett. **74**, 2391 (1995); T. Stroucken, A.Knorr, P. Thomas, S.W. Koch, "Coherent dynamics of radiatively coupled quantum-well excitons," accepted for publication in Phy. Rev. B.
10. J. Shah, K. Leo, E.O. Göbel, S. Schmitt-Rink, T.C. Damen, W. Schäfer, and K. Köhler, Surface Science **267**, 304 (1992).
11. D.S. Citrin, Phys. Rev. B **47**, 3832 (1993).

44

FEMTOSECOND INVESTIGATION OF THE DOMINANT COLD HOLE SCATTERING PROCESS IN GaAs

R. Tommasi, P. Langot, N. Del Fatti, and F. Vallée

Laboratoire d'Optique Quantique du C.N.R.S.
Ecole Polytechnique
91128 Palaiseau, France

Femtosecond spectroscopy is an important source of information on carrier interaction processes in semiconductors. Most of the investigations have however focused on electron relaxation dynamics and although electron scattering mechanisms are relatively well characterized, little is known on hole thermalization dynamics in intrinsic direct gap semiconductors. It has been recently shown that this problem can be selectively addressed using femtosecond two-color absorption saturation spectroscopy.[1] We have used this technique to study the role of hole-electron and hole-phonon scattering in cold hole heating in GaAs at room temperature.

After photoinjection of a carrier plasma, the semiconductor absorption is modified by both direct filling of the conduction and valence bands and by renormalization of the optic and energetic system properties (many body effects).[2] Band filling is dominated by either the electron or hole contribution depending on the experimental conditions. In particular, because of the higher k-space electron localization, information on the transient hole distribution can be obtained by photoexciting small wave vector (i.e., cold) carriers and probing large wave vector carrier states to monitor the high energy tail of the heavy-hole distribution.[1] The initial hole average energy $\bar{E}_h^o$ (heavy- and light-holes) being lower than the thermal one, hole heating is thus observed. The heating source might be both the optical phonons and the simultaneously created electrons whose initial energy content is larger than the hole one. However, LO phonon absorption being much slower for electrons than for holes, LO phonon mediated heat-up of a cold electron distribution occurs on a much longer time scale. The role of electron-hole energy exchanges can thus be studied by changing the initial electron excess energy, making the transient electron temperature either larger or smaller than the hole one.

Measurements were performed for two pump photon energy: $\hbar\omega_{pp}$=1.43eV and 1.48eV corresponding to an electron initial excess energy of, respectively, ~5meV and ~40meV (close to the thermal energy). In the latter situation, electrons quickly thermalize with the lattice and might act as an energy reservoir for the cold holes ($\bar{E}_h^o$~12meV) as

Ultrafast Processes in Spectroscopy, edited by Svelto et al.
Plenum Press, New York, 1996

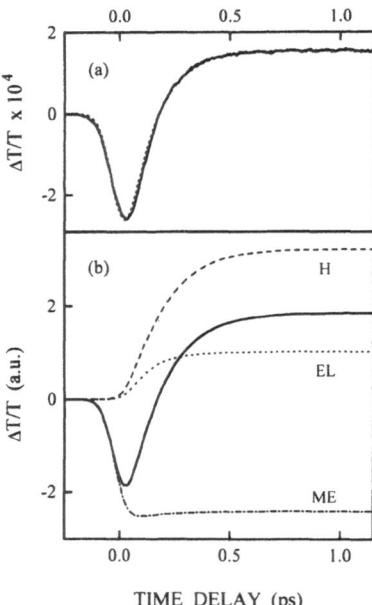

Figure 1. (a) Measured transient transmission change, ΔT/T (full line) for $n_{eh}=8\times10^{16}$ cm^{-3} in GaAs for a pump (probe) photon energy of 1.48eV (1.63eV). The dotted line is a fit using (1). (b) Calculated total ΔT/T (full line) and individual contributions due to filling of the conduction (EL), heavy and light-hole (H) bands, and to many body effects (ME).

has been observed for hot hole cooling in n-doped GaAs at low temperature.[3] For carrier injected closer to the band edge, this possible heating channel is suppressed for the colder holes ($\bar{E}_h^o$ ~2meV).

Experiments were performed using a two-color femtosecond system based on frequency conversion of a Ti:Al$_2$O$_3$ laser.[1] Two synchronized pulses are sent into a standard pump-probe set up and measurements are performed in an optically thin antireflection-coated GaAs sample. A probe photon energy, $\hbar\omega_{pr}$, of 1.63eV has been used corresponding to a region where band filling is dominated by the heavy-hole contribution.[1] The results are shown in Fig. 1 for $\hbar\omega_{pp}$=1.48eV and a carrier density n_{eh}~8×10^{16}cm^{-3}. As the probed states are not directly populated by the pump pulse, the short time delay transmission change, ΔT/T, is dominated by band gap renormalization (BGR) leading to a fast signal decrease. The subsequent rise is mainly due to filling of the heavy hole states as the initial distribution heats up and spreads in k-space.[1] The signal then reaches a plateau indicating that the system is in quasi-equilibrium in less than 1ps.

Our results can be described by a simple phenomenological response function R$_1$:

$$R_1(t) = u(t)\left\{-A + B\left[1 - \exp(-t/\tau_{th})\right]\right\} \tag{1}$$

where u(t) is the unit step function and the first term accounts for the assumed instantaneous BGR. A monoexponential rise with a characteristic time τ_{th} has been assumed to model the hole thermalization. The weaker electron contribution (~3.5 time smaller than the hole one for $\hbar\omega_{pr}$=1.63eV)[1] that quickly rises with the internal electron thermalization time, has been disregarded. This approximation is justified by the fact that τ_{th} has been found to be almost independent of $\hbar\omega_{pr}$ (i.e., of the change of the electron relative contribution) in the range 1.6–1.7eV. A good description of the data is obtained by convoluting the measured pump-probe cross correlation with R$_1$ for τ_{th}~120fs (Fig. 1-a, dotted line).

Photoexciting closer to the band edge ($\hbar\omega_{pp}$=1.43eV), a similar transient transmission behavior is observed with a sharp decrease and a fast rise which is now followed by a slow component that dominates for t≥1ps. As the initial hole distributions are weakly modified (only ~35% more energy has to be absorbed), the fast contribution can be associated to hole thermalization and the slower one to cold electron heating. Taking explicitly into account this process, our results can be reproduced using a modified response function R_2:

$$R_2(t) = u(t)\left\{-A + B_h\left[1 - \exp(-t/\tau_{th})\right] + B_e\left[1 - \exp(-t/\tau_{te})\right]\right\} \tag{2}$$

where B_h/B_e reflects the relative hole and electron contribution. The ratio calculated for $\hbar\omega_{pr}$=1.63eV at thermal equilibrium, B_h/B_e~3.5, has been used.[1] The best fit is obtained for τ_{th}~150fs and τ_{te}~700fs. The effective thermalization time τ_{th} is only slightly larger than the previous one indicating that electrons are not the main energy source for hole heating which is dominated by optical phonon absorption. This larger value is consistent with reduction of the initial hole average energy and, because of the fast optical phonon-hole interaction, with energy transfer from the holes to the cold electrons. The latter process together with electron-LO phonon interaction leads to a slow heating of the cold electrons.[4]

Our conclusions are in good agreement with simulations based on a numerical resolution of the coupled carrier-phonon Boltzmann's equations for electron, light-hole and heavy-hole bands.[2] Polar and nonpolar hole-optical phonon scattering are considered and the nonequilibrium phonon effect is included for LO phonon-electron interactions. The many body effects are computed from the transient distributions.[2] The results show a fast contribution associated to hole thermalization (Fig. 1 and Fig. 2) and a slowly rising com-

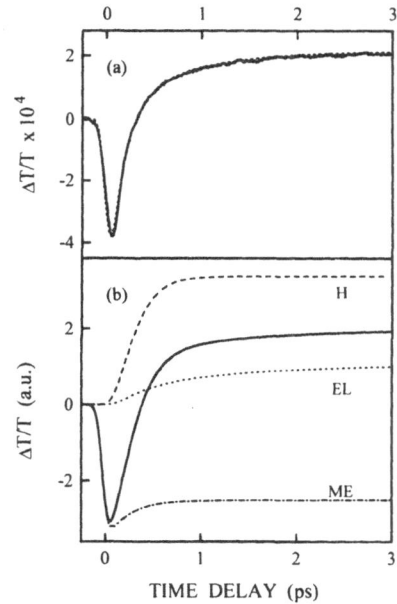

Figure 2. Same as Fig. 1 for $\hbar\omega_{pp}$=1.43eV. The dotted line is a fit using (2).

ponent corresponding to the cold electron heating for $\hbar\omega_{pp}$=1.43eV (Fig. 2). This is consistent with the slow electron-hole energy transfer for hot carriers.[5]

The minor role played by electron-hole energy exchanges is confirmed by carrier density dependent measurements which show a ~40% increase of τ_{th} (from ~120fs to ~170fs) when n_{eh} is decreased by a factor of 35 from $7 \times 10^{17} cm^{-3}$ to 2×10^{16}. This weak increase has been attributed to the reduction of hole-hole interactions which significantly increases the internal hole thermalization time, in agreement with our simulations.

The dominant role of hole-phonon scattering in hole heating is consistent with time resolved luminescence measurements in n-type GaAs which have shown that hot hole cooling is dominated by optical phonon emission at room temperature.[5] In contrast, at low temperature (77K), the energy reservoir for hole cooling has been determined to be the high density thermal electrons injected by doping.[3] The contributions of hole-phonon and hole-electron scattering can be further separated by performing measurements as a function of the lattice temperature. This considerably modifies the optical phonon reservoir, the LO phonon occupation number being for instance reduced from ~0.33 at room temperature to ~0.14 at T=200K, and hence drastically modifies the efficiency of optical phonon absorption by carriers. Preliminary measurements indicate a large slowing down of the hole heating dynamics with decreasing temperature. The characteristic thermalization time τ_{th} is, for instance, measured to increase by a factor of ~2 when the lattice temperature is decreased from 295K to 200K, in very good agreement with our numerical simulations.

These investigations open up many possibilities for the selective analysis of hole scattering in direct gap semiconductors. Our measurements performed as a function of the carrier density and of the initial energy content of the photoexcited carriers demonstrate that the main cold-hole heating-source at room temperature is the lattice through absorption of optical phonons. Temperature dependent measurements should yield additional quantitative information on both hole-phonon and hole-electron interactions.

REFERENCES

1. R. Tommasi, P. Langot, and F. Vallée, Appl. Phys. Lett. **66**, 1361 (1995).
2. J. H. Collet, S. Hunsche, H. Heesel, and H. Kurz, Phys. Rev. B **50**, 10649 (1994).
3. A. Chébira, J. Chesnoy, and G. M. Gale, Phys. Rev. B **46**, 4559 (1992).
4. P. Langot, R. Tommasi, and F. Vallée, to be published.
5. R. D. Joshi, R. O. Grondin, and D. K. Ferry, Phys. Rev. B **42**, 5685 (1990).
6. X. Q. Zhou, K. Leo, and H. Kurz, Phys. Rev. B **45**, 3886 (1992).

45

ULTRAFAST COHERENT EXCITON SPECTROSCOPY AND CARRIER CONTROL IN QUANTUM STRUCTURES

Albert P. Heberle,[1] Jeremy J. Baumberg,[1] Klaus Köhler,[2] and Klaus Ploog[3]

[1] Hitachi Cambridge Laboratory, Cavendish Laboratory
Madingley Road, Cambridge CB3 0HE, UK
[2] Fraunhofer-Institut für Angewandte Festkörperphysik
Tullastraße 72, 79108 Freiburg, Germany
[3] Paul-Drude-Institut für Festkörperelektronik
Hausvoigteiplatz 5–7, 10117 Berlin, Germany

Interference effects between ultrafast laser pulses become important when the relative phase is precisely determined. The response of an optical system then shows strong oscillations when the pump-pulse separation is changed.[1–4] These oscillations appear even for non-overlapping pulses as long as the phase relaxation time of the optical transition exceeds the pump-pulse separation. This interference spectroscopy, in contrast to other coherent optical techniques,[5] gives direct access to the phase evolution in time, which is especially interesting in semiconductors because many-body interactions cause distinct departures from expected behavior of ideal two-level systems.[6,7]

In this paper we report on ultrafast phase spectroscopy of excitons in quantum wells and demonstrate the coherent manipulation of exciton density. We demonstrate the complete destruction of photoexcited excitons within a few hundred femtoseconds of their creation. This technique produces femtosecond optoelectronic nonlinearities which are faster than suggested by the transition line width and shows that carriers are only absorbed irreversibly once phase relaxation has occurred.

Figure 1 shows the phase correlator which measures the differential reflection (ΔR) in a dual-pump-probe geometry. A mode-locked Ti:sapphire laser serves as the source of 100 fs transform-limited pulses. A computer-controlled Michelson interferometer splits the pump beam into two pulses of temporal separation τ_{12}. Active stabilization with a HeNe laser fixes τ_{12} within an accuracy of 0.01 λ/c (0.03 fs) for any phase delay. The pump-induced reflectivity change (ΔR) accesses the exciton population sampled at the probe delay time τ_x. The excitation density was 1 W/cm^2.

Two high quality samples (luminescence line widths of 0.7 and 0.3 meV) were used for the experiments: sample A contains five 120 Å GaAs QWs separated by 120 Å $Al_{0.3}Ga_{0.7}As$ barriers whilst sample B contains ten 250 Å GaAs QWs separated by barriers

Ultrafast Processes in Spectroscopy, edited by Svelto et al.
Plenum Press, New York, 1996

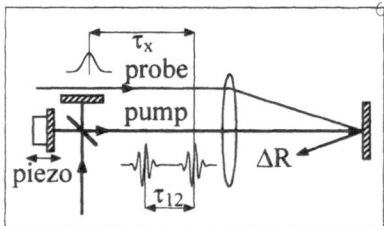

Figure 1. Schematic arrangement of the phase correlator.

which consist of 8.5 Å AlAs, 300 Å $Al_{0.3}Ga_{0.7}As$ and 8.5 Å AlAs. The sample temperature was 4 K.

Figure 2 shows the phase-dependent effect of the second pump pulse on the total carrier density as a function of τ_{12}. Oscillations are visible in ΔR recorded at $\tau_x = 10$ ps at which time most excitons have lost their phase coherence. Pump and probe beam had the same intensity and were perpendicularly polarized. Figure 2(a,b) shows selected sections of the several thousand interference fringes recorded on sample A when the laser was tuned to the heavy-hole exciton energy E_{hh} and a narrow pump spectrum (150 fs pulse width) minimized free-carrier excitation. The reflectivity change (closed circles) directly tracks the total exciton population and oscillates with a period $T_{hh} = h/E_{hh}$. Simultaneously-recorded interference of the pump pulses on a photodiode (open circles) demonstrates the strong influence of their relative phase long after they have ceased to overlap in time. The oscillations originate from the interference of the second pump pulse with the coherent polarization created by the first pulse, and decay for increasing τ_{12} with the polarization.

The oscillations can be characterized by a slowly varying amplitude $A(\tau_{12})$ and phase $\varphi(\tau_{12})$. This technique is ideally suited to explore several current issues in exciton spectroscopy, such as the effect of the local polarization field on the excitons and the microscopic nature of phase scattering outside the impact approximation.[8,9] Figure 2(c,d)

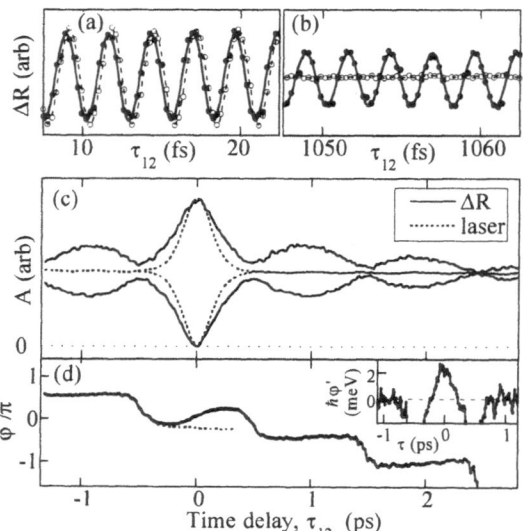

Figure 2. (a,b) Oscillations in ΔR (solid) and the pump interferogram (dashed) for $\tau_{12} \sim 0$ ps and $\tau_{12} \sim 1$ ps (sample A). (c,d) Amplitude A and phase φ of fit to oscillations (sample B). Inset: differential of the phase near zero delay.

shows results for sample B excited 0.9 meV above the heavy-hole exciton energy at 1525 meV. The dashed lines show amplitude and phase of the laser interferogram, and the lack of curvature in φ_{laser} shows chirp is absent. Although the absolute phase value at $\tau_{12} = 0$ is arbitrarily set, the slope of $\varphi(\tau_{12})$ dynamically tracks the evolution of the exciton energy. Strong heavy- light-hole quantum beating is observed with a period of $h/(4.2\ meV)$ in agreement with the separation of heavy- and light-hole exciton lines seen in the linear reflectance spectrum, both of which are covered by the laser spectrum. The quantum beats appear because the interference depends on wavelength; in the anti nodes the hh oscillations are out of phase with the lh oscillations.

The quantum beats appear as steps in $\varphi_{\!ER}$ whose height is expected to be π only when equal electronic polarizations from hh and lh are generated. The bias of the laser towards the lower line and the different oscillator strengths of the two transitions imbalance the modulation and reduce this step height. The step height increases with τ_{12} suggesting that the phase of the light holes relaxes more *slowly* than the phase of the heavy holes. This effect is also visible in the increased modulation with time. Such non-intuitive results would imply that dephasing strongly depends on the angular momentum or the mass of the excitons. Further analysis of these results is hindered by the lack of microscopic models for the dephasing process.

The phase shift around $\tau_{12} = 0$ indicates an instantaneous frequency excursion shown in Figure 2(d,inset) and demonstrates the critical influence of nonlinear contributions. Such an ultrafast transient would be expected from the optical Stark effect (OSE) which cannot be measured in the small detuning, non-adiabatic regime using any other existing technique.[10] In particular there exist many substantial theoretical treatments of the OSE which show the dominating effect of Coulomb interactions at small detunings.

Figure 3 time-resolves the interference of the second pump pulse with the coherent polarization from the first pulse for sample A. The curves show $\Delta R(\tau_{12})$ at fixed $\tau_x = 445$ fs for pulse 1 only (1), pulse 2 only (2), constructive (1+2), and destructive (1–2) interference on the hh exciton line (804 nm). Pump and probe were polarized cross-linearly in the left and co-circularly in the right diagram. The probe intensity was 50 times the pump intensity to avoid coherent probe contributions at the low densities. Curve 1–2 demonstrates the creation of a temporary exciton population by pulse 1 which is coherently destroyed again by pulse 2. The absorbed light is then re-emitted. The possibility of coherent destruction underlines the fact that light is only absorbed irreversibly after phase relaxation has occurred and allows creation of femtosecond exciton pulses which are only limited by the available laser pulse width. In a single pulse experiment the fastest response is limited by the inverse line width of the exciton. The slow rise in ΔR for longer times can be attributed to creation of free carriers by the broad laser spectrum and subsequent cooling and formation of excitons. Complete coherent destruction is only possible when the polarization of the hh and lh excitons are in phase. The inter pump-pulse delay $\tau_{12} = 445$ fs is therefore set to the hh-lh beat period (as shown by the in phase beating of curves 1 and 2) and ensures coherent destruction of hh and lh excitons but cannot avoid incomplete destruction of the broad band free carrier continuum or of higher excitons (2s, 3s, ...). The longer hh-lh beat period of sample B strongly reduces the efficiency of coherent destruction. The linear polarization enhances hh-lh beating but has no influence on coherent destruction as might be expected by the possibility of bi-exciton excitation which is impossible for circular polarization. In the case of constructive interference (curve 1+2) the total exciton density is 4 times the density created by a single pulse. The in-phase carriers that are already excited increase the subsequent probability of absorption if they remain coherent.

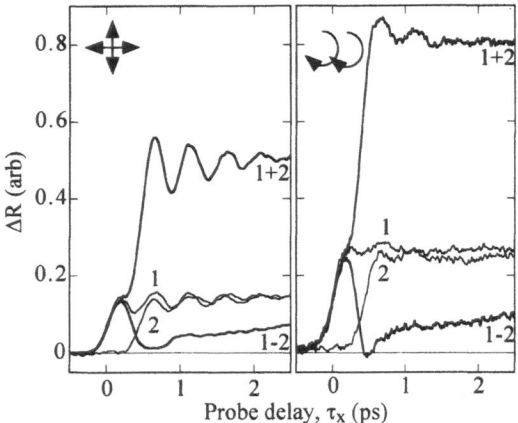

Figure 3. Time-resolved coherent control for a pump pulse separation of $\tau_{12} = 445$ fs. The curves show the differential reflectivity ΔR versus probe delay τ_x for pulse 1 only (1), pulse 2 only (2), constructive (1+2), and destructive (1–2) interference on the hh exciton line. The symbols in the corner show the polarization of the pump and probe beams.

In conclusion, phase-locked femtosecond pulses can coherently manipulate the photoexcited carriers in quantum structures. Analysis of the phase and intensity of the observed oscillations in the carrier density reveal energy shifts underlying the coherent exciton dynamics. Resolving the destruction process of excitons involved in the interference process reveals ultrafast optical dynamics which are not limited by the optical transition line width.

REFERENCES

1. W.S. Warren, H. Rabitz, and M. Dahleh, Science 259, 1581 (1993) and references therein.
2. P.C.M. Planken *et al.*, Phys. Rev. B 48 (1993) 4903; I. Brener *et al.*, J. Opt. Soc. Am. B 11, 2457 (1994).
3. E. Dupont *et al.*, Phys. Rev. Lett. 74, 3596 (1995).
4. A.P. Heberle, J.J. Baumberg, K. Köhler, Phys. Rev. Lett. 75, 2598 (1995).
5. for recent reviews see, phys. stat. sol. (b) 173 (1992) and references therein.
6. S. Weiss *et al.*, Phys. Rev. Lett. 69, 2685 (1992).
7. S.T. Cundiff *et al.*, Phys. Rev. Lett. 73, 1178 (1994).
8. H. Wang *et al.*, Phys. Rev. Lett. 74, 3065 (1995).
9. J.-Y. Bigot *et al.*, Phys. Rev. Lett. 70, 3307 (1993).
10. S. Schmitt-Rink, D.S. Chemla, and H. Haug, Phys. Rev. B 37, 941 (1988).

46

ULTRAFAST DYNAMICS OF EXCITONIC RESONANCES IN InGaAs/InP MQWs

M. Nisoli,[1] S. De Silvestri,[1] O. Svelto,[1] D. Campi,[2] C. Coriasso,[2] and
A. Varanavicius[3]

[1] Centro di Elettronica Quantistica e Strumentazione Elettronica
Politecnico, Milano, Italy
[2] CSELT - Centro Studi e Laboratori Telecomunicazioni
Torino, Italy
[3] Laser Research Center, Vilnius University
Vilnius, Lithuania

Semiconductors with energy gap around 0.8 eV have attracted much attention due to applications in optoelectronics devices compatible with fiber-optics systems. In particular, multiple-quantum well (MQW) structures with InGaAs wells, lattice matched to InP barriers, are very interesting as it is possible to tune the band gap in the near infrared. The growth and the optical characterization of these heterostructures are still topics of extensive research aimed to understand electronic properties and to improve material quality. However up to now the lack of widely tunable infrared femtosecond laser sources did not make it possible a detailed investigation of the time behavior of the non-linear optical properties on time scales interesting for signal processing in optical communication systems operating at ultra-high bit rate.

In this work we report on femtosecond measurements of the room-temperature relaxation dynamics at the excitonic resonances in a InGaAs/InP MQW[1]. The investigated sample has been grown by chemical beam epitaxy on a InP substrate and consisted of 60 $In_{0.55}Ga_{0.45}As$ wells (7.7 nm) separated by InP barriers (7 nm). The data have been obtained by femtosecond pump-probe experiments using a novel widely tunable infrared laser source[2] based on an optical parametric amplifier pumped by a femtosecond Ti:sapphire laser with chirped pulse amplification (150 fs pulses at 780 nm and energy up to 750 μJ at 1 kHz repetition rate). Parametric light conversion was accomplished in a three-pass optical parametric amplifier based on two angle tuned BBO crystals. Parametric conversion at this pump wavelength is particularly efficient in BBO since pump, signal and idler pulses present a long interaction distance. This source can be tuned from 1.1 to 2.7 μm. Almost in all tuning range the pulse energy exceeds 100 μJ, with an average conversion efficiency of ~30%. The intensity autocorrelation functions of signal pulses give pulse duration below 110 fs in the tuning region. A considerable signal pulse shortening

Ultrafast Processes in Spectroscopy, edited by Svelto et al.
Plenum Press, New York, 1996

203

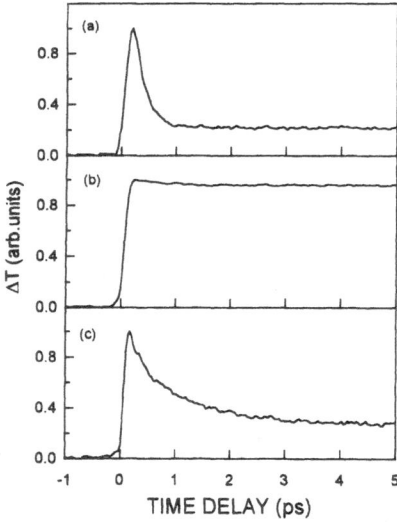

Figure 1. Transmission change ΔT measured as a function of probe delay for different excitation wavelengths: (a) λ=1.55 μm near the n=1 h-h exciton peak; (b) λ=1.45 μm near the n=1 l-h exciton peak; (c) λ=1.27 μm near the n=2 h-h exciton peak.

down to 60 fs was observed near 1.3 μm. The spectral bandwidth of signal and idler pulses was closed to the transform limit in almost all the tuning range.

The room temperature absorption spectrum of the MQW sample clearly shows the resonance peaks of the n=1 and n=2 heavy-hole (h-h) excitons and a broad spectral feature, which corresponds to the n=1 light-hole (l-h) exciton. The time behavior of the n=1 h-h exciton bleaching with an excitation wavelength of 1.55 μm is shown in Fig. 1(a). The best fitting to the data is represented by a single exponential function with time constant of 170 fs followed by a long-lived plateau.

This result is consistent with exciton relaxation dynamics[3]. The resonant pump pulse generates excitons that inhibit further creation of excitons due to exchange and phase space filling effects and, less effectively, to Coulomb screening. After then, in about 170 fs, the excitons are ionized by collisions with the longitudinal optical phonons of the lattice. We find that the bleaching due to photogenerated excitons is stronger than that due to a thermal electron/hole plasma by a factor ~8. Even larger factors (~40) were observed by us for longer excitation wavelengths. Numerical modelling, based on many-body theory, yields qualitative explanation of this unusual behavior[4]. The long lived plateau describes the e-h plasma recombination dynamics (radiative recombination with a time constant of 2.5 ns).

The n=1 l-h exciton bleaching curve, obtained by tuning the pump and probe wavelengths to 1.45 μm, is almost a step function (see Fig. 1(b)). At this wavelength l-h excitons and unbound e-h pairs are created. The lack of a peak is consistent with the large spectral broadening of the n=1 l-h resonance, which should corresponds to an exciton-ionization time constant below the experimental time resolution. Moreover, both the transition selection rules and the relative heights of the density of state favor the formation of unbound electron h-h pairs over l-h excitons.

The bleaching decay of the n=2 h-h exciton, measured at an excitation wavelength of 1.27 μm, is shown in Fig. 1(c). The curve can be fitted by a two exponential function, with time constants of 100 fs and 1 ps, followed a long living plateau. The faster time constant represents the exciton ionization time constant of the n=2 h-h excitons, while the longer one is due to the intersubband relaxation and corresponds to the switching off of

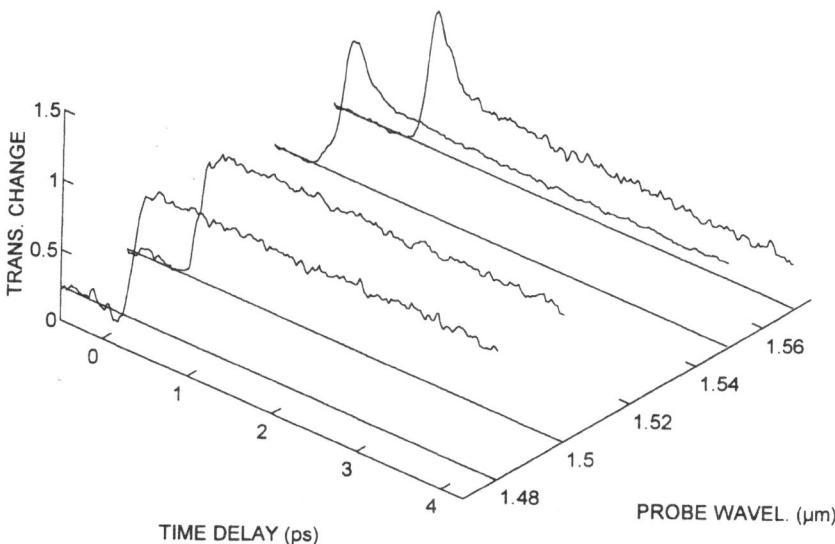

Figure 2. Transmission changes ΔT at different probe wavelengths ranging from 1.48 μm to 1.57 μm. The pump wavelength is 1.55 μm near the n=1 h-h exciton peak.

the exchange and phase-space filling contributions to absorption bleaching, as the carriers thermalize to the n=1 valence and conduction subbands.

We have investigated the saturation behavior of the n=1 h-h and l-h exciton under resonant and non-resonant excitation with frequency continuum pump-probe measurements. By focusing the parametric light pulses in glass continuum generation in the infrared can be obtained. In Fig. 2 we report the results obtained by tuning the pump wavelength at the n=1 hh resonance.

The probe wavelength ranges from the region of the n=1 hh resonance to the region of the n=1 lh resonance. The same pump energy was used for all the traces throughout Fig.2. In the region of the n=1 hh resonance, we observed a peak-and-plateau evolution, consistently with the outcomes of the degenerate experiments illustrated in the above. At shorter wavelengths the response was step-like, the rising edge following the integral of the cross-correlation function. In order to clarify the observed behavior, we calculated the relative strengths of transmittance changes induced by a cold exciton gas, generated by a pump beam tuned to the n=1 hh resonance, or by an e/h plasma with the same density. At positive detuning with respect to the hh resonance (i.e. at shorter wavelengths), there is no region in which the cold exciton gas has substantially larger effect than the carrier plasma. In particular our calculation indicates that the probe transmittance variations induced by the exciton gas are of the same order of those induced by the thermal e-h plasma resulting from its ionization, when the probe is tuned slightly above the n=1 hh resonance and excitation is moderate.

In the nonresonant experiments we tuned the pump at a wavelength corresponding to 38 meV excess energy imparted to the electrons; a value comparable to, or slightly in excess of one optical phonon energy. The pump-probe measurements at probe photon energies below the pump are nearly step-like functions. There is no apparent influence of the initial carrier distribution excited by the pump, or of thermalization of the carrier distribution. This may indicate that the carriers scatter out of their initial states and assume a broad distribution in energy with time constant beyond the resolution of the experiment.

We have also investigated the n=1 h-h exciton dynamics with degenerate pump-probe measurements and repetitive excitation. We used two pump pulses delayed by ~7 ps. This delay permits ionization of the photoexcited excitons, but is short compared to the recombination time and transverse carrier diffusion time. The main result is that the e-h plasma after the ionization of the excitons created by the first excitation pulse does not inhibit further creation of excitons. Moreover the exciton dynamics are completely unchanged in presence of an e-h plasma.

ACKNOWLEDGMENT

A. Varanavicius wants to acknowledge financial support from the NATO Collaborative Research Grant n. 941165.

REFERENCES

1. M.Nisoli, V.Magni, S.De Silvestri, O.Svelto, D.Campi, and C.Coriasso, Appl. Phys. Lett. **66**, 227 (1995).
2. M.Nisoli, S.De Silvestri, V.Magni, O.Svelto, R.Danielius, A.Piskarskas, G.Valiulis, and A.Varanavicius, Opt. Lett. **19**, 1973 (1994).
3. W.H.Knox, R.L.Fork, M.C.Downer, D.A.B.Miller, D.S.Chemla, and C.V.Shank, Phys. Rev. Lett. **54**, 1306 (1985).
4. D.Campi, C.Coriasso, M.Nisoli, and S.De Silvestri, Appl. Phys. Lett. **67**, 953 (1995).

47

SHORTER THAN 100 fs CARRIER LIFETIMES IN AS-GROWN BY LOW TEMPERATURE MBE GaAs

S. Marcinkevičius,[1] A. Krotkus,[2] R. Viselga,[2] and U. Olin[1]

[1] Department of Physics II, Royal Institute of Technology and
Institute of Optical Research
S-10044 Stockholm, Sweden
[2] Semiconductor Physics Institute
2600, A.Goštauto 11, Vilnius, Lithuania

GaAs layers grown by molecular-beam-epitaxy at low (~200^{0}C) substrate temperatures (LT-GaAs) are characterized by very short, subpicosecond carrier lifetimes. Because, at the same time, these layers have reasonably high electron mobilities, large resistivities, and high breakdown fields the number of their applications in ultrafast optoelectronic and microwave electronics devices is steadily growing[1]. Annealed at high temperature LT-GaAs layers are, as a rule, used in these applications. The parameters of as-grown LT-GaAs are by far less documented than such of the annealed layers, mainly as a result of a more complicated experiment in this case. However, the knowledge of as-grown material behaviour is of a crucial relevance to the understanding of the physics lying behind the unique properties of LT-GaAs and other highly nonstoichiometric semiconductors.

In the present work we have investigated the photoexcited carrier relaxation in as-grown LT-GaAs epitaxial layers by a time-resolved photoluminescence (PL) up-conversion technique and have estimated the carrier trapping time in these layers by comparing the experimental results with molecular dynamics calculations.

The LT-GaAs layers were grown on semi-insulating GaAs substrates at the temperature of 200^{0}C under arsenic supersaturation. The growth rate was 2.5 µm/h; as-grown layers contained up to 1% of excess arsenic. Self-mode-locked Ti:sapphire laser with the pulse duration of 100 fs and the central wavelength of 770 nm was employed in the time-resolved up-conversion set-up. time-resolved PL set-up. Photoexcited carrier density was 10^{18} cm^{-3}, PL decay was monitored at different emitted photon energies.

Figure 1 shows PL dynamics in an as-grown LT-GaAs layer measured at three different photon energies. The traces of the transient PL are symmetrical and their width is close to the width of the laser pump and probe pulse cross-correlation trace, which is shown by a dashed line for comparison. This indicates that the characteristic times at

Ultrafast Processes in Spectroscopy, edited by Svelto et al.
Plenum Press, New York, 1996

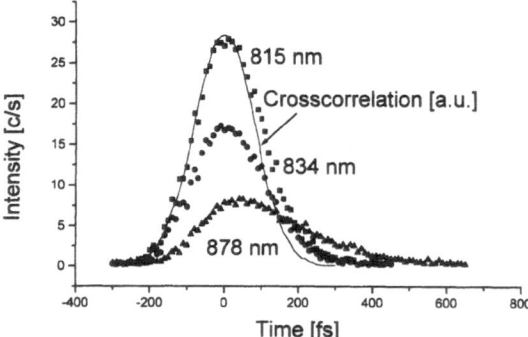

Figure 1. PL transients in as-grown LT-GaAs layer measured at three different emission wavelengths. Full curve shows the cross-correlation between the pump and the probe optical pulses of the time-resolved up-conversion measurement.

which the photoexcited carriers disappear from the conduction and the valence bands should be shorter than the duration of the laser pulses. Ultrafast carrier capture times are evidenced also from the spectral measurements of the PL on as-grown LT-GaAs samples. PL spectra of as-grown and annealed samples, which are presented in Figure 2, show striking differences. The spectra of an annealed sample, which has a carrier trapping time of 400 fs[2], have their maxima at photon energies close to the bandgap energy and are witnessing that the hot, optically excited electron distributions are thermalized even as early as 100 fs after the photoexcitation. On the other hand, the PL spectrum measured on as-grown sample is highly nonthermal and can be rather well approximated by a Lorentian function centred at the energy of the excitation quanta. By comparing the integral PL intensities for annealed and as-grown samples of LT-GaAs we had found that the carrier trapping time should be shorter than 70 fs in the later case.

More exact evaluation of the trapping time can be made by comparing experimental results with the numerical simulation. Semi-classical Monte Carlo with its static-screening approach for carrier- carrier interaction cannot be used on such a short time scale because the assumption of the same screening length for all particles, and because the many-body interaction cannot be treated as binary collisions. Instead, we had calculated initial optical excited carrier dynamics by a molecular dynamics method. The strength of this method lies in not requiring any assumptions on the screening between carriers. The model employed by us was similar to that used previously in Reference[3]. Figure 3 shows the calculated electron, light and heavy hole distributions at 50 and 100 fs after the excitation. At

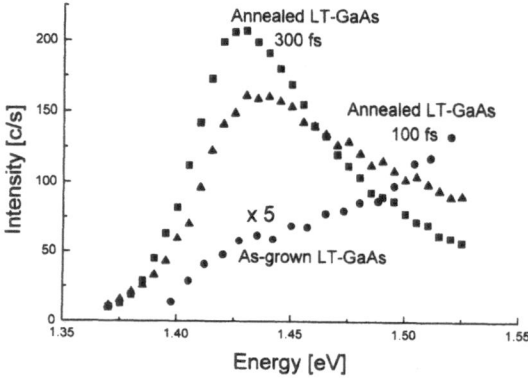

Figure 2. PL spectra for as-grown (1) and annealed (2–4) LT-GaAs layers. The spectrum for as grown layer is measured at the point of a zero delay between pump and probe optical pulses. The spectra for annealed layer are measured at 100 fs (2), 300 fs (3) and 1 ps (4) later than this point.

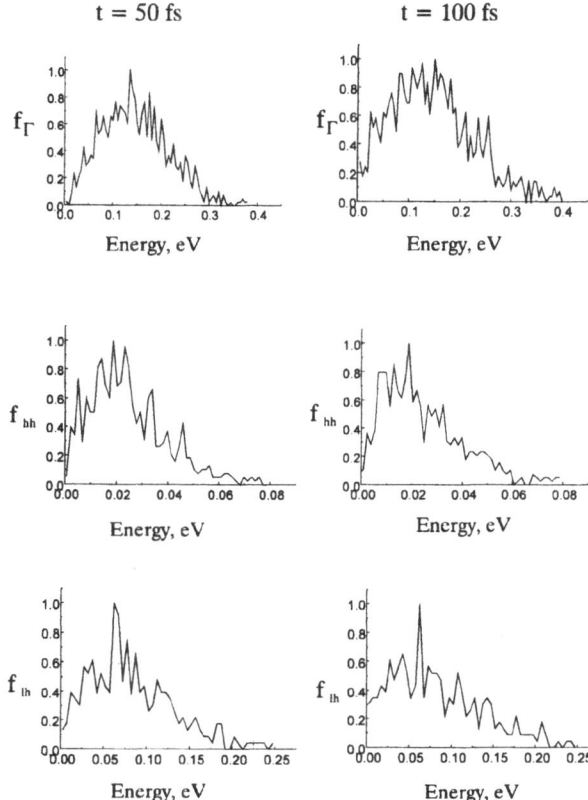

Figure 3. The results of the molecular dynamic calculation for electron, heavy-hole, and light-hole distribution function at two different moments after the photoexcitation.

100 fs all three distributions are close to the thermal ones, whereas the electron and light hole distributions at 50 fs still remain nonthermal. The halfwidth of the PL spectrum, which, essentially, is a product of the electron and hole distribution functions, was found in the later case to be close to 200 meV. This value coincides with the spectral halfwidth found in PL measurements on as-grown LT-GaAs. Thus, the carrier trapping time in this material is in the range between 50 fs and 70 fs.

Besides of being the shortest carrier trapping time ever found in semiconductor materials, this value is also important for the solution of a long-standing controversy on the mechanism of the ultrafast carrier recombination in LT-GaAs. Carrier trapping time increases after annealing by almost one order of magnitude, to the similar extent at which the excess arsenic related point defect density is being reduced during the thermal treatment. On the other side, arsenic precipitates, which, according to some authors[4], are crucial for explaining unique properties of LT-GaAs, are not present in as-grown layers. Therefore our results support the arsenic point defect model of semi-insulating and recombination properties of LT-GaAs[5] rather than the precipitate model.

ACKNOWLEDGMENT

This work has been supported by International Science Foundation (Grant Nr.LH7100) and by the Lithuanian Science and Study Foundation.

REFERENCES

1. G.L.Witt, in *Semi-insulating III-V Materials*, ed. by M.Godlewski (World Scientific, Singapore, 1994).
2. A. Krotkus, R. Viselga, K. Bertulis, V. Jasutis, S. Marcinkevièius, U. Olin, Appl. Phys. Lett. **66**, 1939 (1995).
3. U. Hohenester, P. Supancic, and P. Kocevar, X. Q. Zhou, W. Kütt and H. Kurz, Phys. Rev. B **47**, 13233 (1993).
4. A. C. Warren, N. Katznellenbogen, D. Grischkowsky, J. M. Woodall, M. R. Melloch, N. Otsuka, Appl. Phys. Lett., **58**, 1512 (1991).
5. M. Kaminska, E. R. Weber, Mater. Sci. Forum, **83–87**, 1033 (1992).

48

THE IMPORTANCE OF BIEXCITONS IN FOUR-WAVE MIXING EXPERIMENTS IN QUANTUM WELLS WITH INTERFACE DISORDER

G. Bongiovanni,[1] A. Mura,[1] D. Leadley,[1] F. Quochi,[2] S. Gürtler,[2] and JL. Staehli[2]

[1] Dipartimento di Scienze Fisiche
Università degli Studi
via Ospedale 72, I-09124 Cagliari, Italy
[2] Institut de Physique Appliquée
Ecole Polytechnique Fédérale, PH-Ecublens
CH-1015 Lausanne, Switzerland

ABSTRACT

We present time integrated, time resolved as well as spectrally resolved two-beam four-wave mixing (FWM) experiments. On the low energy side of the X resonance we observe a component in the FWM emission that is emitted as a photon echo and whose decay time does not depend on the polarisations of the input beams. This emission shows strong beatings that can be observed in time integration as well as in time resolution.. The interpretation of these phenomena is based on a five level scheme that includes excitons and biexcitons. A simple model accounts for the different effects the interface roughness has on the exciton and biexciton states.

The Coulomb interaction between the photo excited excitons (Xs) in a semiconductor and its effects on the optical properties has been studied extensively in the past[1]. During the last few years, also the coherence of these Xs has been investigated, mainly through transient four-wave mixing (FWM) experiments. Recently, in high quality samples where the X transition shows practically only homogeneous broadening, there is strong experimental evidence that the formation of biexcitons (biXs, bound two-exciton states) has a strong influence on the polarisation[2]. In most samples, however, the inhomogeneous broadening is comparable to or larger than the effect of the X interactions. Nevertheless, quantum beats due to the presence of biXs have been observed also in quantum wells (QWs) having such a large inhomogeneous broadening[3]. However, several of the observed features attributed to the effect of the biX states on the FWM response are not

Ultrafast Processes in Spectroscopy, edited by Svelto et al.
Plenum Press, New York, 1996

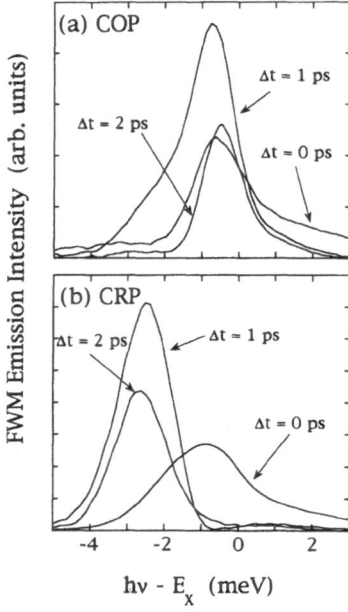

Figure 1. Time integrated FWM emission spectra, measured at different delay times Δt, (a) for co-linearly polarised (COP) and (b) for cross-linearly polarised (CRP) input beams. E_X is the exciton energy.

yet understood. As an example, we mention the influence of disorder on the biX state. While several investigations have been published on the role played by the well width fluctuations on the X energies, as far as we know, very little is known concerning their influence on the two-X state.

In this paper, we present time-integrated (TI), time resolved (TR) and spectrally (SR) resolved femtosecond FWM experiments. They have been performed at a temperature of 2 K. We investigated several samples having an inhomogeneously broadened heavy hole X absorption line width ≥ 2.5 meV. All the considered samples have a quite similar FWM response. In this report, we discuss the results obtained on a sample consisting of twenty 8 nm wide QWs of GaAs/(GaAl)As, with a heavy-hole X absorption line width of about 2.7 meV.

Fig. 1 shows the spectra of the TI FWM emission excited by 110 fs long pulses. The spectra are shown for various delays between the two incident beams. For co-linearly polarised (COP) input fields, the emission is centred at about 0.5 to 1 meV below the X resonance E_X, for all delay times Δt. Most of this red shift is caused by the absorption the diffracted beam experiences close to the X resonance. The FWM spectra also present a weak shoulder, 2 to 3 meV below E_X. For linearly cross polarised (CRP) fields and for Δt ≥ 1 ps, this non-resonant component is the only one that persists, while the contributions emitted at the X resonance and above disappear very quickly with delay time.

The temporal dynamics of the non-linear signals at different emission energies hv is reported in Fig. 2a. All the TI traces show an instantaneous response at Δt = 0, caused by the large detuning between the laser energy and the absorption peak[4]. For hv - E_X ≤ -2 meV (the non-resonant component in Fig. 1), the decay rate of the signals is almost independent of the polarisations of the incident fields. Clear beats with a period T ≈ 1.7 ps are observed for COP beams. A weaker modulation with the same period but shifted by T/2 is observed also in the CRP configuration.

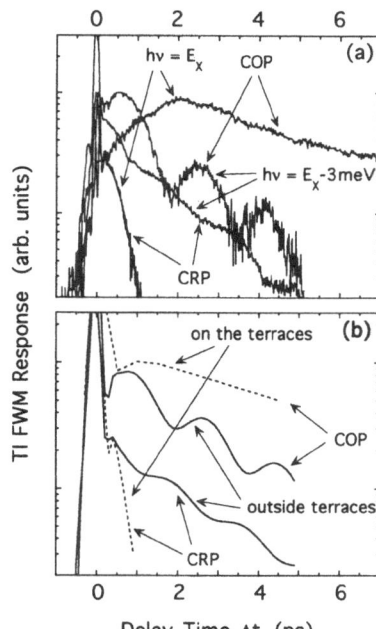

Figure 2. Time integrated FWM response vs. delay time Δt, at different emission energies hv (E_X is the X energy). (a) experimental results, (b) calculated using a five level scheme as described in the text.

The delay times Δt at which the extrema of the oscillations occur do not depend on the emission energy hv. According to Reference 5, this suggests that the modulation of the spectrally resolved TI signal is not caused by polarisation interference, but is a real quantum beat, ie. it results from a quantum mechanical interference of two coupled systems that are coherently excited. The modulation of the signal, however, cannot be attributed to a quantum beating between two transitions (with different energies) with a common ground state. In contrast to our experimental results, such a three level system would give rise to a beating with a maximum at zero delay when the incident pulses are COP[5]. As will be shown later on, the correct dependence of the beats on delay time can be obtained accounting for biX contributions.

Assuming that the modulation of the signal is due to biXs , from the beating period we deduce a binding energy for the biX state of B = 2.3 meV. B is approximately twice the value usually found in samples with a homogeneously broadened X resonance[1]. This enhancement of the binding energy is probably related to the localisation of biXs in regions (*terraces*) of a size comparable to the linear extent of the biX state[6]. Since we observed a distinct modulation frequency in the TI experiments, the spread of the biX binding energies must be relatively small; in other words, a considerable number of terraces must have a similar size.

Moving the detection energy towards smaller detunings, the beats disappear. At resonance, the decay rate of the signal is no longer equal in the two polarisations: The relaxation is much faster for CRP fields, as usually found in QWs[7]. The lack of any beat reveals that biX non-linearities are heavily affected by disorder in those regions of the QWs responsible for the resonant non-linear emission.

The TR FWM response, measured at different fixed delay times, gives further insight on the temporal dynamics of the FWM process in the crystal regions with different

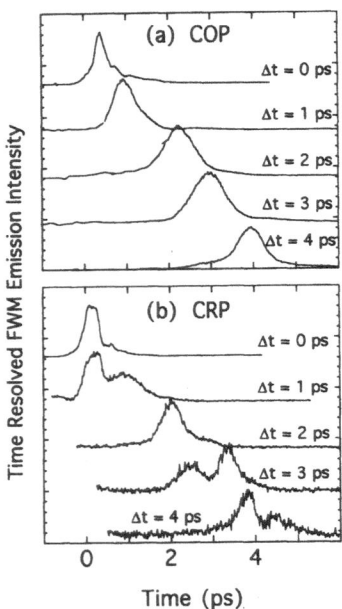

Figure 3. Time resolved FWM response, measured for different time delays Δt, (a) for COP and (b) for CRP input beams.

interface disorder. However, as this response is spectrally integrated, both the resonant (without apparent biX non-linearities) and the non-resonant (with strong biX non-linearities) emissions are detected. Fig. 3 shows that the FWM signal excited by COP fields is a clear photon echo. For CRP beams, we observe a prompt response for Δt ≤ 1 ps. At longer delays, the non-linear signal is emitted as a photon echo whose intensity decays slowly with increasing delay time Δt. A striking distortion of the temporal shape of the echo is also observed. This effect cannot be ascribed to pulse propagation as it appears only for CRP fields. The prompt response disappears very quickly with delay time and is, therefore, associated to the rapidly decaying component of the TI data. The photon echo corresponds to the slowly decaying component in the TI traces and comes from the terraces where the biX non-linearities are quite strong. Thus, the distortion of the echo is directly connected with the quantum beats. Further, the presence of a photon echo shows that also the biX level is inhomogeneously broadened. We note that the modulation of the TR signal and its dependence on Δt is a typical feature of a quantum beat. This has been already noted in Ref. 8, but as far as we know, the modulation of the echo has not yet been observed. Finally, we realise that the surprising absence of pulse distortion of the TR signal excited by COP fields now has a natural explanation: Regions with negligible biX contribution emit an intense photon echo if excited by COP beams. This strong signal hides the weak and (distorted) echo arising from the terraces.

The interpretation of our experimental data is based on a five level scheme consisting of the crystal ground state, left and right polarised X states, a biX state with binding energy B, and a continuum (dissociated) two-X state[2]. The optical Bloch equations for this five level system have been solved numerically up to the third order in the incoming electric fields. Further, our model includes also the effects of interface irregularities in an elementary way, as briefly outlined in the following. In the centre of mass approximation[9], the Xs move in a "random" potential V_X, which is determined by the spatial changes of the single electron and hole energies, averaged over the spatial extension of the X wave function. On the terraces we assume that V_X does not vary appreciably, ie. the correlation length of V_X is comparable to or larger than the biX diameter. In the limit of a weak confinement, the biX can be considered as composed by two X states that undergo weak X-X interactions[6]. The bound and unbound two-X states have the formation energies $E_{X1} + E_{X2}$ - B and $E_{X1} + E_{X2}$, respectively. On the terraces, we therefore assume that $E_{X1} = E_{X2}$ and that B = const. Hence, the average third order polarisation becomes:

$$<P^{(3)}>_T = \int P^{(3)}(E_X, E_X)\, \rho_T(E_X)\, dE_X, \tag{1}$$

Outside the terraces, we assume that the energies of the X and the two-X states (bound and unbound) are completely uncorrelated (the correlation length of V_X less than the biX diameter). Thus, we have calculated the average third order polarisations using the expression:

$$<P^{(3)}> = \int P^{(3)}(E_{X1}, E_{X2})\, \rho(E_{X1})\, \rho(E_{X2})\, dE_{X1}\, dE_{X2}. \tag{2}$$

In the two previous equations, $\rho_T(E_X)$ is the probability that an X is located on a terrace with energy E_X, and $\rho(E_{Xi})$ is the same for an X located outside. In general $\rho_T(E_X) \neq \rho(E_{Xi})$. The spectral widths of $\rho(E_{Xi})$ and $\rho_T(E_X)$ correspond to the temporal widths of the X and biX photon echoes, respectively.

Using Gaussians with an appropriate width for the distribution functions ρ, our numerical results reproduce, in a semi quantitative way, the observed phenomena: In spatial regions where the correlation length of V_X is smaller than the biX extent, ie. where eq. 2 applies, for COP fields the FWM response is emitted as an echo, while in CRP configuration the response is prompt. No beats are observed. On the terraces (ie. where V_X is almost flat over the extension of a biX and eq. 1 applies), the non-linear signal is a photon echo independent of the polarisation configurations. The experimentally observed quantum beats are also obtained (see Fig. 2b), for CRP beams the first oscillation maximum occurs at $\Delta t = 0$, and for COP fields at $\Delta t = T/2$. For CRP beams, the modulation of the TI signal is absent if the biX scattering state is neglected. Finally, it is worth noting that eqs. 1 and 2 represent two limiting cases that have been suggested by our experimental results. Of course, to account for different morphologies of the well/barrier interfaces, other distribution functions ρ could be used.

REFERENCES

1. H. Haug and S. Schmitt-Rink, Prog. Quant. Electr. **9**, 3 (1984).
2. E.J. Mayer et al., Phys. Rev B **51**, 10909 (1995), and references therein.
3. S. Bar-Ad and I. Bar-Joseph, Phys. Rev. Let. **68**, 349 (1992); S. Albrecht et al., 1995 Conference on Quantum Electronics and Laser Science (Baltimore,. 21 - 26 May)
4. G. Bongiovanni et al., Proc. 22nd Int. Conf. Phys. Semicond. (ICPS-22, Vancouver 1994), edited by DG. Lockwood (World Scientific, Singapore 1995), pp 1119
5. V.G. Lyssenko et al., Phys. Rev B **48**, 5720 (1993).
6. G.W. Bryant, Phys. Rev B **41**, 1243 (1990); YZ Hu et al., Phys. Rev. Let. **64**, 1805 (1990)
7. S.T. Cundiff et al., Phys. Rev B **46**, 7248 (1992).
8. M. Koch et al., Phys. Rev. Let. **69**, 3631 (1992).
9. R. Zimmermann, Phys. Status Solidi (b) **173**, 129 (1992).

49

TIME-RESOLVED SPECTROSCOPY IN SEMICONDUCTORS USING INTENSE FAR-INFARED (SUB)PICOSECOND PULSES

P. C. M. Planken,[1] H. P. M. Pellemans,[1] C. J. G. M. Langerak,[2] T. O. Klaassen,[1] and W. Th. Wenckebach[1]

[1] Department of Applied Physics
Delft University of Technology
P.O. Box 5046, 2600 GA Delft, The Netherlands
tel: +31 30–6096816/fax: +31 30–6031204/e-mail: planken@rijnh.nl
[2] FOM-Institute for Plasma Physics 'Rijnhuizen'
3430 BE Nieuwegein, The Netherlands

We have performed far-infrared time-resolved experiments, in semiconductors, using (sub)picosecond pulses from the Free-Electron Laser for Infrared eXperiments FE-LIX at the FOM-institute 'Rijnhuizen', the Netherlands. As a first example we show time-resolved measurement of polarization beats in GaAs:Si, in which we combine the use of ultrashort pulses with photoconductivity measurements[1]. Our sample consists of a 400 µm thick GaAs substrate with a 10 µm thick Si-doped GaAs layer on top. The sample is mounted in a magnet cryostat and kept at a temperature of 8 K. The energy of the Si-impurity transitions can be tuned by application of a magnetic field. We concentrate on the 1s-2p$^+$ donor transition. Two identical collinear far-infrared pulses of 5 ps duration from a Michelson interferometer setup, with a variable time separation τ_d, are weakly focused onto the sample. Electrons, resonantly photo-excited to the 2p$^+$ state thermally decay to the conduction band and will give rise to an increase in the conductivity[2]. In Fig. 1 we plot the measured photo-conductivity of the sample as a function of τ_d when the laser is resonant with the 1s-2p$^+$ transition, for three values of the magnetic field.

The wavelength of the pulses is indicated in the figure. The three curves show rapid oscillations that are still present even when the electric fields of the pulses emerging from the Michelson have *no* temporal overlap. In addition, a node is visible in the three graphs that shifts to longer time delays for decreasing magnetic fields. The oscillations in the photo-conductivity are the result of the interference between the coherence stored in the medium by the first pulse, and the electric field of the second pulse. This gives an increase or decrease in the 2p$^+$ population depending on the phase difference between the two. When the coherence has completely decayed, no interference is possible and the oscilla-

Ultrafast Processes in Spectroscopy, edited by Svelto et al.
Plenum Press, New York, 1996

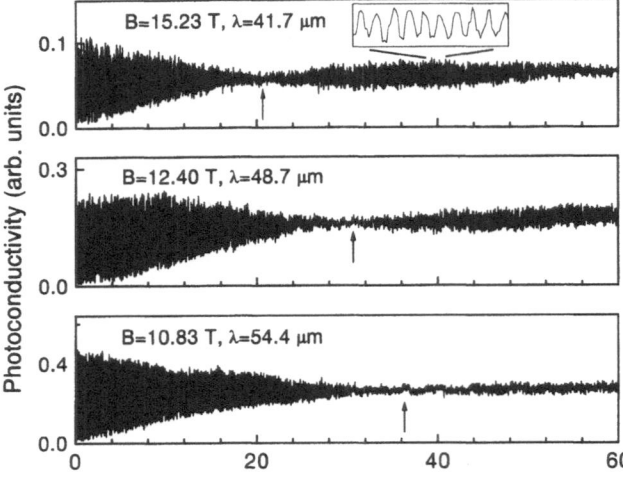

Figure 1. Measured photoconductivity versus time separation, for three different values of the magnetic field. The wavelength of the pulses is adjusted to remain resonant with the 1s-2p$^+$ transition. The arrows indicate the position of the node that results from the beating between the spin-up and the spin-down transitions.

tions disappear. The node observed in the photoconductivity measurements can be explained as follows: At the magnetic-field strengths used in the experiment, the 1s-2p$^+$ transition splits into two separate spin-up and spin-down transitions[3]. This magnetic-field dependent spin splitting gives rise to interference ('beats') in the measurements of the photoconductivity versus time delay as illustrated by the presence of the nodes in Fig. 1. The spinsplitting increases for increasing magnetic field due to band nonparabolicity and consequently, the node shifts to smaller time-delays.

The second example is a far-infrared three pulse transient-grating experiment in n-type InAs. The aim is to study the dynamics of the electron distribution after heating of the electrons with an ultrashort FIR laser pulse. Similar experiments were performed in the past in highly doped InAs (1.5×10^{18} electrons cm^{-3}) using midinfrared pulses of 7 ps duration. There, screening of the electron-LO-phonon interaction and hot phonon effects gave rise to relatively long cooling times with signals lasting tens of picoseconds[4,5]. In contrast, our doping density is roughly thirty times smaller (5×10^{16} electrons cm^{-3}) while our laser pulses are approximately 0.3 ps long. As our results below show, the former decreases the significance of hot-phonon effects and results in fast decay times, the latter allows us to time resolve signals lasting several picoseconds only. Infrared pulses are a necessity in this experiment to avoid the creation of additional electron-hole pairs by interband transitions across the bandgap.

In our experimental setup, shown in Fig. 2, two pump pulses with a wavelength of 18 μm have spatial and temporal overlap in the InAs crystal. The pulses excite the electron gas and transform the intensity grating into an absorption/refractive index grating from which we can diffract a third delayed probe pulse.

In Fig.3, we plot the diffracted probe intensity versus time separation between pump pair and probe for two different temperatures. At T=300 K, the diffracted signal peaks around zero delay times and shows a rapid, non-exponential decay when the time separation increases. At 80 K the results are dramatically different. After the diffraction has peaked around zero delay times, the diffraction decay is initially rapid, after which the decay slows down considerably. In fact at 80 K, the slower part of the decaying signal lasts almost ten times longer than the decay at 300 K.

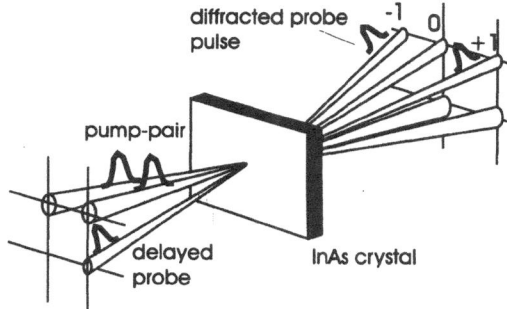

Figure 2. Detail of the experimental configuration used in the three-pulse transient grating experiment.

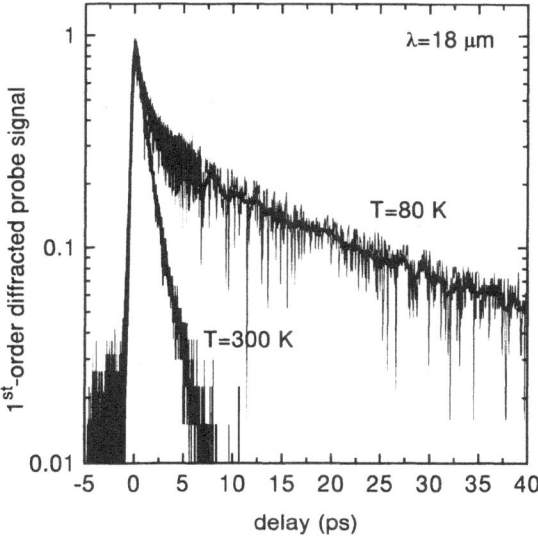

Figure 3. Measured probe diffraction versus time separation between pump pair and probe pulse, at two different temperatures.

Our interpretation of the data is as follows. After excitation of the electrons with the pump-pair, the electron gas rapidly thermalizes ($\approx 10^2$ fs) at a temperature which is higher than the lattice temperature. Cooling of the thermalized distribution then progresses through the emission of LO-phonons by those electrons within the distribution that have enough energy to emit an LO-phonon. As the temperature of the distribution decreases, less and less electrons within the distribution have sufficient energy to emit an LO-phonon. As a result, the cooling speed reduces. At T=300 K, the fraction of electrons with an energy sufficiently high to emit an LO-phonon, is larger than at 80 K and hence the electron gas cools more rapidly than at 80 K.

REFERENCES

1. P. C. M. Planken, P. C. Van Son, J. N. Hovenier, T. O. Klaassen, B. N. Murdin, G. M. H. Knippels, and W. Th. Wenckebach, Phys. Rev. **B 51**, 9643 (1995).
2. A. v. Klarenbosch, T. O. Klaassen, W. Th. Wenckebach, and C. T. Foxon, J. Appl. Phys. **67,** 6323 (1990).
3. H. Sigg, J. A. A. J. Perenboom, P. Pfeffer, and W. Zawadzki, Solid State Commun. **61**, 685 (1987).
4. T. Elsaesser, R. J. Bäuerle, and W. Kaiser, Phys. Rev. **B 40**, 2976 (1989).
5. R. J. Bäuerle, T. Elsaesser, and W. Kaiser, Semicond. Sci Technol **5**, S176 (1990).

50

INFLUENCE OF MANY-BODY EFFECTS ON THE QUANTUM COHERENCE OF 2D EXCITONS IN GaAs QUANTUM WELLS

J. Kuhl,[1] E. J. Mayer,[1] M. Hübner,[1] G. O. Smith,[1] K. Bott,[2] V. Heuckeroth,[2] D. Bennhardt,[2] S. W. Koch,[2] and P. Thomas[2]

[1] Max-Planck-Institut für Festkörperforschung
Heisenbergstr. 1, D-70569 Stuttgart, Germany
[2] Department of Physics
Philipps University
Renthof 5 , D-35032 Marburg, Germany

The polarization dependence of time-integrated (TI) and time-resolved (TR) degenerate four-wave-mixing (DFWM) on the lowest heavy hole (hh) and light hole (lh) exciton transitions in GaAs quantum wells (QW) demonstrates that heavy and light hole excitons cannot be described in the framework of the two non-interacting three-level systems with opposite circular polarization selection rules suggested by Schmitt-Rink et al.[1] In several recent papers,[2-5] we have shown that the discrepancies between experiment and theory can be resolved if many-body Coulomb interactions and disorder-induced coupling of excitonic states are taken into account. Exciton/exciton Coulomb interaction which involves a variation of the local electric fields (LFE), excitation-induced dephasing (EID) and energy shifts of excitonic states, especially the formation of biexcitons (BIF), is phenomenologically implemented into the theory by solving the optical Bloch equations for multi-level model systems which represent single-exciton and two-exciton contributions to the nonlinear optical response by discrete transitions with adjustable oscillator strengths, dephasing rates, and frequencies. Comparison of a large variety of experimental data with theoretical curves calculated with this model demonstrates, in particular, the important contribution of the heavy hole biexciton to the third-order nonlinear optical response of GaAs QW's. Whereas modulation of the signal with a frequency corresponding to the biexciton binding energy are the unique "fingerprint" for the formation of biexcitons, their importance is further proved by the polarization dependence of the signal strength.

We have found a new 3-pulse TI-DFWM configuration which can distinguish between the three different contributions, i.e. LFE, EID, and BIF to the third-order nonlinearity[4]. Experiments are performed on a single 20 nm GaAs/(Al,Ga)As QW using three linearly or circularly polarized pulses (duration 1.1 ps) from a mode-locked Ti:sapphire la-

Ultrafast Processes in Spectroscopy, edited by Svelto et al.
Plenum Press, New York, 1996

221

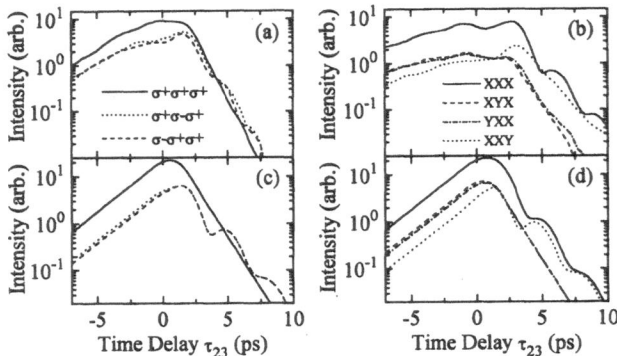

Figure 1. The measured 3-pulse TI-DFWM intensity as a function of the delay time τ_{12} for (a) circular and (b) linear polarizations. The calculated curves for the 3-pulse TI-DFWM intensity as a function of τ_{12} for (c) circular and (d) linear polarizations using the parameters: LFE = 1.2 meV and a biexciton binding energy Δ = 1.2 meV.

ser with wavevectors $\mathbf{k_1}$, $\mathbf{k_2}$, $\mathbf{k_3}$ and delays τ_{12} and τ_{23}. The signal was monitored in the direction $\mathbf{k_1} + \mathbf{k_2} - \mathbf{k_3}$. Figures 1a and b depict the TI DFWM signals recorded for 7 different polarization configurations. Comparison of the experimental data with solutions of the optical Bloch equations for a 5-level model including both LFE and BIF shown in Figs. 1c and d demonstrates that the theory reproduces all the detailed signal features such as the relative amplitudes, the signal rise and decay times and even the presence and absence of pronounced modulations with a frequency corresponding to the biexciton binding energy for the different polarization geometries. The modulations originate from quantum beats between two exciton continuum states and biexcitons as proved by spectral resolution of the DFWM signal.

Further evidence for the role of biexcitons is obtained from TR-DFWM with parallel (PP) or cross-polarized (CP) pulses of 120 fs duration in a two-pulse self-diffraction configuration[5]. The signal diffracted into the direction $2\mathbf{k_2} - \mathbf{k_1}$ is time-resolved via up-conversion with a reference pulse in a 2 mm LiIO$_3$-crystal. The signals shown in Figs. 2a and 2c for the PP and

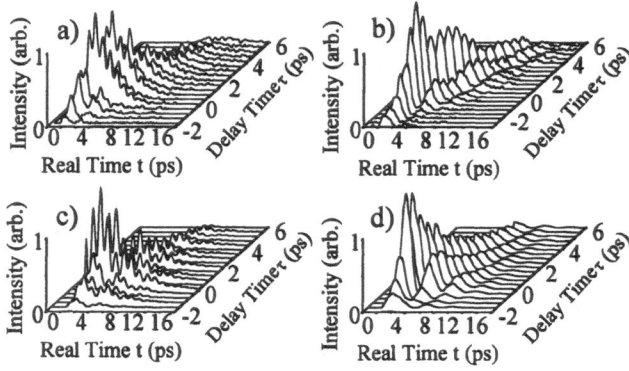

Figure 2. The measured TR-DFWM intensity for (a) parallel and (b) cross-polarized laser fields. The calculated intensity of the TR-DFWM signal at various probe delays based on the 10-level model for (c) parallel and (d) cross-polarized laser fields using the parameters: LFE = 0.1 meV and Δ = 1.2 meV.

CP excitation geometry, respectively, display a diverse and rich quantum beat structure. Surprisingly, the dominant beat frequencies appearing for the PP and CP configuration are completely different and correspond to the hh-lh exciton splitting (3.7 meV) and the biexciton binding energy (ΔE =1.2 meV) of our QW sample, respectively. Comparison with the theoretical curves of Figs. 2 b and d demonstrates that the change of the beat pattern with polarization of the exciting fields is correctly reproduced by solutions of the optical Bloch equations calculated for a 10-level scheme consisting of a ground state, two hh and two lh single-exciton states and 5 two-exciton states. Four of the latter are two-exciton scattering states consisting of either two hh, two lh or a hh and a lh-exciton, the fifth level represents the hh biexciton down-shifted by ΔE from the two-exciton continuum. The occurrence of different beat periods is explained by the fact that identically polarized hh and lh exciton states yield the dominant contribution to the signal for the PP geometry and $\tau > 0$, whereas the signals for $\tau < 0$ and for the CP configuration necessarily involve the polarization of bound and unbound two-exciton states (see also ref. 6).

Furthermore, our simple phenomenological model correctly predicts the polarization rules for quantum beating between lh- and hh-excitons[7] as well as the experimentally observed pronounced dependence of the magnitude I_{sig} ($\tau = 0^+$) and polarization Θ_{sig} of the DFWM signal on the polarization angle Θ_{12} between two linearly polarized excitation pulses used in a two-pulse self-diffraction experiment[2,3]. Both I_{sig} ($\tau = 0^+$) and Θ_{sig} vary with Θ_{12} due to excitation-induced dephasing which provides an additional contribution to the nonlinearity for parallel polarized pulses[8]. Figures 3a and 3c display the calculated and measured peak intensity as a function of Θ_{12}. The data were measured on a high-quality 25 nm single QW with a photoluminescence line width as small as 0.3 meV. The symmetry breaking associated with the Coulomb interaction results in a $\cos^2 \Theta_{12}$ of the signal maximum which is very well reproduced in the experiments at low excitation densities where the dephasing rates γ_μ and γ_ν of the one- and two-exciton states are assumed to be

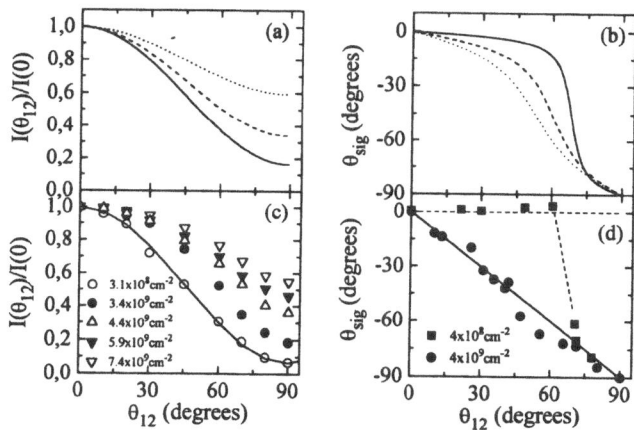

Figure 3. (a) the calculated intensity of the TI-DFWM versus Θ_{12} using the five-level scheme with parameters $\mu=1$, $\nu=1$, $\Delta=0.15$ meV, $\gamma_\mu^{-1} = 12$ ps, $\gamma_\nu^{-1} = 6$ ps (solid line); and $\gamma_\mu^{-1} = 12$ ps and $\gamma_\nu^{-1} = 4.5$ (dashed line); and $\gamma_\mu^{-1} = 4$ ps, $\gamma_\nu^{-1} = 4$ ps (dotted line). (b) Maximum of the polarization angle of the TI-DFWM signal Θ_{sig} at delay $\tau = 0^+$ versus Θ_{12}. The parameters used are the same as in (a). (c) Normalized intensity of the measured TI-DFWM signal versus Θ_{12} at $\tau = 0$. (d) Measured polarization angle of the TI-DFWM signal versus Θ_{12}. The solid line depicts the relation $\Theta_{sig} = - \Theta_{12}$ expected for the two uncoupled two-level systems. The dashed line indicates $\Theta_{sig} = 0$.

significantly different. At higher excitation levels , both γ_μ and γ_ν increase and their values approach each other. The model also successfully describes the complex variation of θ_{sig} with θ_{12} and its dependence on excitation density. This dependence of θ_{sig} on θ_{12} has been impressively confirmed by a complete polarization analysis of the diffracted signal[9].

Finally, it should be mentioned that the exciton/exciton coupling via Coulomb fields considered so far has been confined to excitons excited in the same well since the rapid drop of the dipole-dipole interaction potential with the interparticle distance r prevents interaction if excitons located in different wells even for barrier thicknesses as small as 10–15 nm. Recent dephasing measurements on special MQW structures where the QWs are equidistantly separated by either $\lambda/2$ or $\lambda/4$ (λ =wavelength of the exciton resonance in the material) have revealed the importance of interference effects and inter-well coupling via confined electro-magnetic fields in these structures which result in remarkable variations of the TI-DFWM signal's amplitude and shape.[10,11]

The extension of the theoretical model for the 2D exciton described so far has been confined to the intrinsic excitonic coupling phenomena caused by Coulomb fields. In real samples, short- and long-range disorder lead to local symmetry breaking and inhomogeneous line broadening. If the corresponding mixing of the electron and/or hole states is implemented as an additional coupling mechanism into the optical Bloch equations,[12] the phenomenological model predicts for inhomogeneously broadened samples a distinctly faster decay of the TI-signal for the CP as compared to the PP configuration and a real time shift of the TR-signal from $t = 2\tau$ (photon echo) to $t=\tau$ (free polarization decay) if the excitation is switched from PP to CP pulses. The PP signal's decay is governed only by the dephasing time T_2. Whether the decay of its CP counterpart is equal or possibly faster depends on the width of the Fourier transform of the distribution function for the spatially inhomogeneous, disorder-related coupling strength. Experimental data taken on substantially inhomogeneously broadened samples show excellent agreement with these predictions. In high quality samples, however, the signal always appears as a free polarization decay and its decay time is the same for the PP and CP configuration. Schneider and Ploog[13] have shown that the surprising variations of the temporal evolution of the DFWM signal can be attributed to the presence of short range disorder in the sample.

In summary, we have shown that the pronounced variations of signal amplitudes, temporal evolution and beating phenomena found in TI- and TR-DFWM for different polarization configurations can be consistently explained if the coupling between excitonic states via many body Coulomb interaction and sample disorder are implemented into the theory.

REFERENCES

1. S. Schmitt-Rink et al., Phys. Rev. **B 46**, 10460 (1992).
2. R. Eccleston et al., Sol. State Comm. **86**, 93 (1993).
3. K. Bott et al., Phys. Rev. **B 48**, 17418 (1993).
4. E.J. Mayer et al., Phys. Rev. **B RC 50**, 14730 (1994).
5. E.J. Mayer et al., Phys. Rev. **B 51**, April 15 (1995) .
6. H. Wang et al., Sol. State Comm. **91**, 869 (1994).
7. G.O. Smith et al., Sol. State Comm. **94**, 373 (1995).
8. Y.Z. Hu et al., Phys.Rev. **B 49**, 14382 (1994) and H. Wang et al., Phys. Rev.Lett. **71**, 1261 (1993).
9. S. Patkar et al., Phys. Rev. **B 51**, 10789 (1994).
10. T. Stroucken et al., submitted to Phys. Rev. **A** .
11. M. Hübner et al., see article in this book .
12. D. Bennhardt et al., Phys. Rev. **B 47**, 13485 (1993).
13. H. Schneider and K. Ploog, Phys. Rev. **B 49**, 17050 (1994).

51

EXCITON/EXCITON COUPLING IN MULTIPLE QUANTUM WELL BRAGG STRUCTURES

M. Hübner,[1] J. Kuhl,[1] T. Stroucken,[2] A. Knorr,[2] P. Thomas,[2] S. W. Koch,[2] R. Hey,[3] and K. Ploog[3]

[1] Max-Planck-Institut für Festkörperforschung
Stuttgart, Germany
[2] Department of Physics and Materials Sciences Center
Philipps Universität, Marburg, Germany
[3] Paul Drude Institut für Festkörperelektronik
Berlin, Germany

In the last years, comprehensive theoretical and experimental work has led to rapid progress in the understanding of exciton-exciton interactions, contributing to the nonlinear optical response, such as local field effects, excitation-induced dephasing or the formation of biexcitons in semiconductor quantum wells (QWs). All these interaction mechanisms have their origin in the Coulomb interaction between excitons in the same QW. The purpose of this paper is to show, that in multiple quantum well (MQW) structures with an appropriate geometry radiative coupling of excitons in different QWs can strongly influence the amplitude and the dynamics of the nonlinear optical response in degenerate-four-wave-mixing (DFWM) experiments.

Recent theoretical investigations [1], [2] predict, that in MQWs coupling of the excitons in the different wells via reemitted photons and light propagation effects should play an important role. In special cases, this photon coupling should change the linear and nonlinear optical response of the MQW drastically. These special structures, called Bragg (and anti-Bragg) structures, which have a varying number N of equidistantly separated, identical QW's with a QW spacing equal to a half (or a quarter) of the heavy hole exciton resonance wavelength λ inside the medium are the subject of our investigations.

We present DFWM measurements of GaAs MQW Bragg and anti-Bragg structures grown by MBE on semi-insulating GaAs substrates. The QW thickness was 20 nm and $Al_xGa_{1-x}As$ with x = 0.3 was used as the barrier material. Our samples, matched to the resonance wavelength λ of the hh-exciton, are a single quantum well (SQW), a sample with N = 10, separated by $\lambda/4$, one sample with N = 3, separated by $\lambda/2$, and two samples with N = 10 separated by $\lambda/2$ which have different compositions of the barrier materials.

For the structures, where the barrier thickness is exactly equal to $\lambda/2$, Ref. [1] predicts, that after optical excitation radiative coupling of reemitted photons and light propa-

Ultrafast Processes in Spectroscopy, edited by Svelto et al.
Plenum Press, New York, 1996

gation effects should result in the formation of one superradiant state, whose radiative linewidth is broadened by a factor of N as compared to a SQW. The theory predicts a crucial dependence of this effect on the perfect matching between the barrier width and $\lambda/2$. Even deviations in the range of 1% should give rise to optical coupling to modes which have a very small radiative linewidth that were optically forbidden if the spacing is exactly $\lambda/2$.

In the other interesting case, where the barrier width is equal to $\lambda/4$, the interwell coupling should lead to a splitting of the resonance and hence under coherent excitation to a beating in the dynamical response. In a simple way this can be understood in the picture of coupled oscillators because of the reabsorption of photons between neighbouring wells.

All the features described above should be observable in 2-pulse self-diffraction DFWM experiments. In this geometry, one measures the transversal relaxation time T_2 of the excitons, which is related to the radiative linewidth γ_{rad} by $1/T_2 = \gamma_0 + \gamma_{rad}$, where γ_0 is the nonradiative homogeneous linewidth which is due to scattering with sample imperfections, phonons, electrons or excitons. For these studies, the samples were kept at T = 8 K in a Helium cryostat and excited by two pulses with 770 fs duration and a spectral width $\Delta\omega$ = 3.3 meV from a Kerr-lens mode-locked Ti:Sapphire laser. The pulses with wavevectors k_1 and k_2 were circularly polarised and had equal intensity. The circular polarisation was used to suppress biexcitonic contributions to the DFWM signal. The signal, detected by a photodiode in the backward diffraction geometry in the direction $2k_2-k_1$, was monitored as a function of the time delay τ_{12}. The signal could also be spectrally resolved for a certain time delay.

Figure 1a depicts a normalized DFWM spectrum of the 10 $\lambda/4$ structure at τ_{12} = 0.66 ps. The excitation intensity was 24 W/cm^2. The corresponding time-integrated delay curve is shown in Fig. 1b. Both, the splitting of the DFWM spectrum and the beating in the DFWM delay curve are obvious features, which are qualitatively in good agreement with the theoretically predicted radiative coupling effects for 10 QWs separated by $\lambda/4$. Figure 3a and the inset show a computed time-integrated DFWM delay curve and a DFWM spectrum at τ_{12} = 1.66 ps respectively. The parameters for the exciton transition used for calculation are γ_0 = 12.6 μeV and ΔLT = 0.08 meV.

The Figures 2a and 2b show normalized DFWM curves for the SQW and the three different $\lambda/2$ MQWs recorded for different laser excitation intensities. At low excitation (7 W/cm^2) an enhanced decay of the signals for both 10 $\lambda/2$ samples compared to the SQW is evident during the first 7 ps, where the signal declines to roughly 10% of its maximum

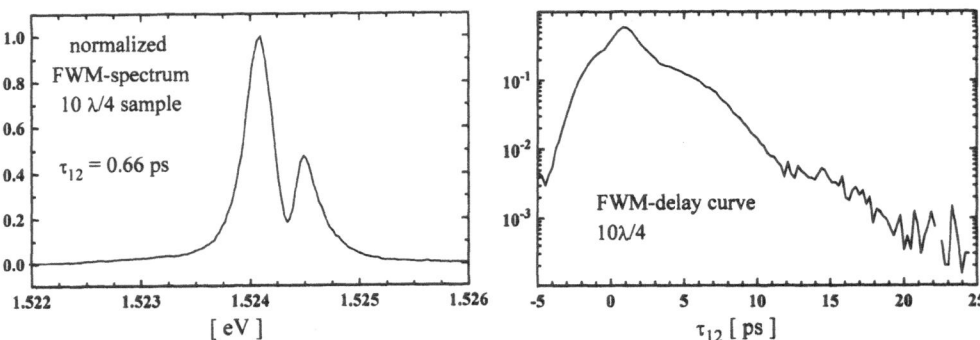

Figure 1. a. Normalized DFWM spectrum of the GaAs 10 $\lambda/4$ anti-Bragg sample at τ_{12} = 0.66 ps. b. Time-integrated DFWM delay curve of the GaAs 10 $\lambda/4$ anti-Bragg sample.

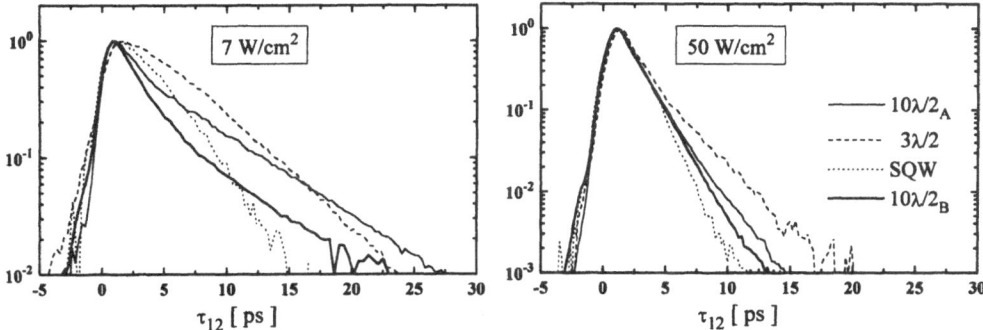

Figure 2. a. Normalized time-integrated DFWM delay curves of the different GaAs $\lambda/2$ Bragg samples at 7 W/cm^2. b. Normalized time-integrated DFWM delay curves of the different GaAs $\lambda/2$ Bragg samples at 50 W/cm^2.

value. After this fast initial decay, which is more pronounced for the sample 10-λ/2-B, both 10-λ/2 samples decay with a dephasing time even distinctly slower than that of the SQW. In contrast the 3-λ/2 sample does not show any enhanced decay compared to the SQW. Its decay is even slower but still faster than the long time decays of the 10-λ/2 samples. For higher laser intensities the difference in the initial decay between the SQW and the 10-λ/2 samples becomes smaller and vanishes almost completely at 50 W/cm^2 (Fig. 2b).

Further evidence for the presence of efficient radiative coupling between the excitons coherently excited in different Qws is provided by the intensity dependence of the signal magnitude. Fig. 3b shows the DFWM signal amplitudes for all MQWs divided by the amplitude of the SQW as a function of the incident laser intensity. For the 3-λ/2 sample this ratio starts for lowest excitation at a value of roughly 7 and remains remarkably constant. For the 10-λ/2 samples the ratio starts at values of 2 and 6 respectively and then increases strongly with excitation intensity. Above 50 W/cm^2 the ratio saturates for both samples at values of 50- 70. In the absence of the enhanced radiative decay of the excitons in the MQW samples, one would expect an increase of the maximum signal by a factor of

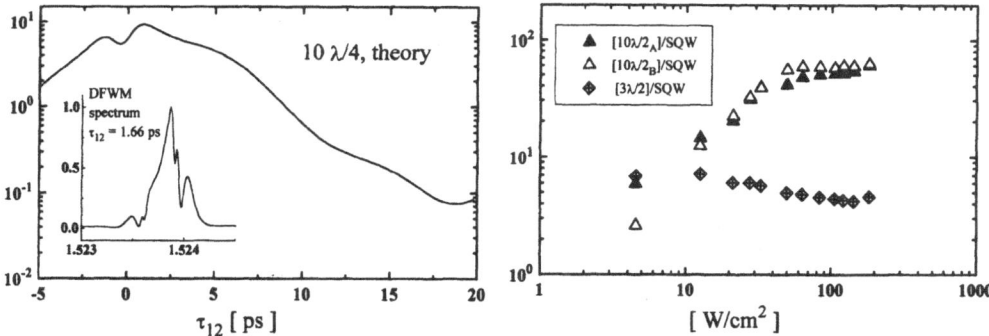

Figure 3. Computed time-integrated DFWM delay curve for a 20 nm GaAs 10 λ/4 anti-Bragg sample. Inset : computed DFWM spectrum at τ_{12} = 1.66 ps. b. Ratios of the DFWM signal amplitudes of the different GaAs λ/2 Bragg samples compared to the SQW in dependence on the excitation intensity.

9 and 100 compared to the SQW for the 3 λ/2 and 10 λ/2 structure, respectively, because of the quadratic growth of the signal with the sample thickness. Thus the signal increase with laser intensity observed for the 10 λ/2 sample reflects an increasing reduction of radiative coupling between excitons excited in different wells which can be easily explained by enhanced dephasing caused by exciton/exciton scattering among the excitons in the same well. On the other hand the formation of a superradiant state favored at lower excitation in a perfect structure should result in a decreased absorption and hence a much smaller DFWM signal. For an ideal high quality sample at very low temperatures where radiative recombination provides the dominating contribution to the loss of phase coherence the signal of the MQWs is expected to be even smaller than that of a SQW.

For all our λ/2 samples we find, however, $R=I(MQW)/I(SQW) > 1$. This result has to be attributed to a slight mismatch between the QW-spacing and the exciton resonance λ in the structures which is hardly avoidable because of the uncertainties of the refractive indices. The largest mismatch obviously occurs in the 3-λ/2 sample. The value of about 7 for R shows, that the formation of the superradiant state is almost completely suppressed.

The fact, that the 3-λ/2 sample always shows a longer dephasing than the SQW could be an indication that there is still a coupling between the wells, but the signal is dominated by the coupled states described in the introduction having a very small radiative linewidth.

For the 10 λ/2 samples but especially for sample B this mismatch in the barrier width seems to be considerably smaller. Both samples show after an enhanced initial decay of the DFWM-signal very long dephasing times T_2 which are at lowest excitation intensities even longer than 15 ps, corresponding to a homogeneous linewidth as small as 0.09 meV.

If all the QWs in the MQW structures are identical with the SQW whose dephasing time at lowest excitation densities is about 6 ps, the results show, that the radiative coupling in λ/2-MQW-Bragg structures with a small mismatch results in the simultaneous formation of a radiatively enhanced excitonic state and excitonic states with a very small radiative linewidth which dominate the DFWM signal at long delays.

The fact, that sample 10 λ/2-B shows the more pronounced fast initial decay in Fig. 1a agrees with the smaller amplitude ratio of its DFWM-signal in Fig. 3b. The better the barriers are matched to the exciton resonance, the more pronounced is the decrease of the DFWM-signal compared to the SQW. The vanishing enhanced initial decay of the 10 λ/2 samples compared to the SQW with increasing excitation intensities described in Figs. 2a-2b also agrees with the strongly increasing ratios of the DFWM signals in Fig. 3b.

In conclusion, we have presented time-integrated DFWM experiments on MQW Bragg and anti-Bragg structures which reveal efficient radiative coupling of excitons excited in different wells. In high-quality Bragg samples, radiative recombination provides the dominant contribution to the dephasing rate at low excitation intensities. Anti-Bragg structures reveal a pronounced beating of the DFWM signal which can be attributed to a splitting of the excitonic resonance due to radiative interaction of neighbouring wells. The experimental data show very good agreement with solutions of the semiconductor Maxwell-Bloch equations for an ultrashort optical pulse interacting with such MQW structures.

Partial funding of this work by the Deutsche Forschungsgemeinschaft is gratefully acknowledged.

REFERENCES

1. T. Stroucken, A. Knorr, P. Thomas and S.W. Koch, to be published in Phys. Rev. B (1995).
2. E. L. Ivchenko, A. I. Nesvizhskii and S. Jorda, in *Phys. Solid State* **36 (7)**, 1156 (1994).

52

CARRIER TRANSPORT IN MULTIPLE QUANTUM WELL REGION OF InGaAsP/InP STRUCTURES

S. Marcinkevicius,[1] N. Tessler,[2] U. Olin,[3] C. Silfvenius,[4] B. Stålnacke,[4] and G. Landgren[4]

[1] Department of Physics II, Royal Institute of Technology
Stockholm, Sweden
[2] Advanced Optoelectronics Center, Department of Electrical Engineering
Technion, Haifa, Israel
[3] Institute of Optical Research
Stockholm, Sweden
[4] Semoconductor Laboratory, Royal Institute of Technology
Stockholm, Sweden

After injection by a current pulse, initial carrier distribution between the quantum wells (QW) of a multiple QW laser is highly nonuniform[1]. For a laser to operate effectively, it is desirable that all the QWs equally contribute to the lasing action. As far as the high frequency modulation is concerned, the rate of the carrier transport and redistribution between the QWs should be considerably faster than the modulation rate.

Here we present a study of transport between the QWs for carriers injected by an optical pulse. The measurements are performed by time-resolved photoluminescence (PL) on a number of InGaAsP/InP multiple QW structures. The experimental results are described by a diffusion model which allows to resolve the influence of structure parameters on the carrier transport process.

Several sets of QW structures were grown by low-pressure MOVPE on n-InP substrates. A schematic picture of a structure is shown in the inset to Fig. 1. The top InP cap layer of the structures was 500 nm thick (or 300 nm for the structures termed 7 to 9), so most of the exciting radiation at 770 nm (1.61 eV) was absorbed in this layer (α^{-1}=300 nm). One or several deep wells were introduced into the structures to detect the carriers which have passed the InP cap layer and the shallow QW region. The structures prepared to study the electron transport were uniformly p-doped ($p = 1^{18}$ cm^{-3}), the ambipolar carrier transport was studied in the undoped structures. The band gap values for the barriers, deep QWs and shallow QWs are, respectively, 0.95 eV, 0.82 eV and 0.76 eV for the structures 1 to 4, and 1.15 eV, 0.95 eV and 0.80 eV for the structures 5 to 9. Other parameters of the structures are given in Table 1.

Ultrafast Processes in Spectroscopy, edited by Svelto et al.
Plenum Press, New York, 1996

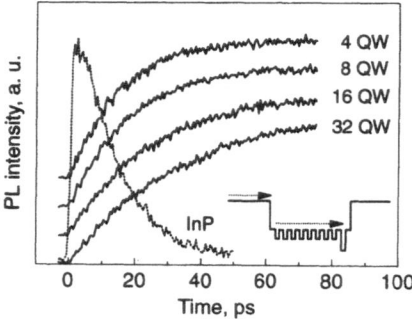

Figure 1. Normalised PL transients measured at the band gap energies of InP and the deep QWs for the structures 1 to 4. The inset shows a schematic profile of the conduction band of a structure.

Time-resolved PL experiments were performed by an upconversion technique. The setup[2] is based on a self-modelocking Ti:sapphire laser (pulse duration 100 fs, repetition frequency 80 MHz). Relatively low excitation intensities, corresponding to the average excited carrier densities in the InP cap layer of $(2 - 8) \times 10^{16}$ cm^{-3}, were used. The measurements were performed at room temperature.

First, the carrier transport in the p-doped structures is discussed. It has been shown[3] that electron transport can be investigated by the time-resolved PL if the excited carrier density is much lower than the doping level of the structure. Then the PL intensity would be proportional to the number of photoexcited electrons, and changes in the PL signal can be associated with changes of the electron distribution. The temporal transients of the PL intensity measured at the band gap energies of the InP layer and the deep QWs, for structures 1 to 4, are shown in Fig. 1. The InP PL signal decays exponentially with a time constant of 14 ps for a photoexcited carrier density of 2×10^{16} cm^{-3}. The rise of the deep QW PL intensity can be approximated by a relation $I_{PL} \sim (1-\exp(-t/\tau))$. Then, the times of carrier transport across the shallow QW region can be simply evaluated as the difference $\tau_2 - \tau_1$.

For a given excitation intensity, the InP PL decay time is the same for all the structures, but the deep QW PL rise times depend on the structure parameters. As can be seen from Table 1, the deep QW PL rise times are very different for the structures which have the same number of shallow QWs, but barriers of different thickness. For the structures which differ only by the number of shallow QWs the deep QW PL rise time increases linearly with the number of QWs (Fig. 2) for a constant excitation intensity.

Table 1. Structural parameters and deep QW PL rise times for the InGaAsP QW structures

Structure No.	Number of shallow QWs	Barrier width, nm	Deep QW PL rise time, ps
1.00	4.00	5.00	16.00
2.00	8.00	5.00	18.00
3.00	16.00	5.00	22.00
4.00	32.00	5.00	29.00
5.00	16.00	8.00	43.00
6.00	4.00	12.00	17.00
7.00	16.00	12.00	45.00
8.00	1.00	12.00	52.00
9.00	2.00	12.00	67.00
10.00	4.00	12.00	88.00

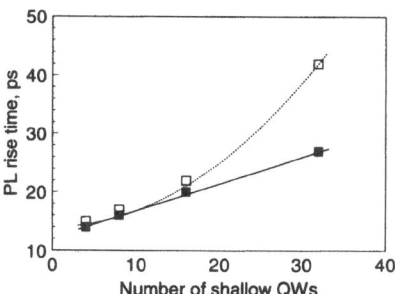

Figure 2. Deep QW PL rise times for the structures 1 to 4 as a function of the number of shallow QWs. Full squares - constant excitation intensity, empty squares - excitation intensity proportional to the number of shallow QWs.

To interpret the experimental results, temporal dependencies of the PL signal at the InP and the deep QW band gap energies were calculated by solving a drift-diffusion equation, with the diffusion coefficient expressed as[4]

$$D(z,t) = \frac{kT}{e} \frac{n(z,t) + p_0 + p(z,t)}{n(z,t)/\mu_p + (p_0 + p(z,t))/\mu_n} \tag{1}$$

Here p_0 is the background hole density, and $n(z,t)$ and $p(z,t)$ ($p = n$) are the photoexcited electron and hole densities, respectively. The electron and hole effective mobilities are denoted as μ_n and μ_p. Under such approach, the diffusion coefficient in the shallow QW region is both, position and time dependent.

The PL rise times measured at the deep QW band gap energy were fitted using the following mobility values: $\mu_e = 3500 \pm 400$ cm^{2}V^{-1}s^{-1} and $\mu_h = 6.5 \pm 0.5$ cm^{2}V^{-1}s^{-1} for the narrow barrier structures 1 to 4, and $\mu_e = 600 \pm 50$ cm^{2}V^{-1}s^{-1} and $\mu_h = 9.5 \pm 0.5$ cm^{2}V^{-1}s^{-1} for the wide barrier structures 6 and 7. The linear dependence of the deep QW PL rise time on the number of shallow QWs, observed in the experiment, is also obtained in the calculations. This linear dependence is due to that, under conditions of constant excitation, the excited carrier density is smaller for structures with a large number of QWs. Thereby, according to Eq. 1, the effective diffusion coefficient in the shallow well region is larger and, hence, a linear instead of a quadratic dependence is obtained. We get a different picture if the carriers in these structures are excited at different excitation intensities proportional to the number of shallow QWs. Under such excitation conditions, the ratio of the shallow QW PL intensity and the number of wells is approximately constant, suggesting that the electron density in this region is about the same for all structures. Then, as shown in Fig. 2, the deep QW PL rise times scale quadratically with the number of shallow QWs.

In the structures with 8 or 12 nm thick barriers the deep QW PL rise time is significantly larger than in the narrow barrier structures (Table 1). This indicates that electron effective mobility is reduced. The influence of the barrier width on the electron transport is evaluated by comparing the effective mobilities with the mobility of the InGaAsP alloy ($\mu_n = 2200$ cm^{2}V^{-1}s^{-1}, $\mu_h = 50$ cm^{2}V^{-1}s^{-1}, Ref. 5). The estimation, based on the values of the parameter [1], shows that in the narrow barrier structures all the electrons are mobile, and most of the holes are confined ($\eta = 15$). However, for the wide barrier structures η is equal to 7 and 12 for the electrons and the holes, respectively.

The fact that all electrons in the narrow barrier structures participate in transport is not surprising, since, according to our calculations, the electron miniband in these structures extends up to the barrier energy. For the wide barrier structures, the miniband width is 16 meV for the 8 nm barriers and 7 meV for the 12 nm barriers. This is too little to make electron transport through the miniband effective at room temperature. For the holes, the minibands are narrow in all cases, thus the holes have rather large

However, even though electrons in the narrow barrier structures can move freely, at the given densities their motion is considerably slowed down by the confined holes, the diffusion of which is limited by the effect of trapping. For the wide barrier structures in which also the electrons are confined, the electron diffusion is mainly limited by trapping into the QWs while the holes further reduce the effective diffusion coefficient by approximately a factor of 2.

In the undoped structures, the photoexcited carrier transport would proceed in the ambipolar manner with the mobility $\mu_{amb} = 2 \, \mu_h$[4]. A slower carrier transport as compared to the p-doped structures is reflected by longer times of the PL rise and decay. The InP PL decay time is 40 ps, the deep QW PL rise times are over 50 ps, their values are given in Table 1. The carrier transport times, evaluated using the effective hole mobilities obtained for the p-doped samples, agree reasonably well with the experimental values for the structures 7 and 8. However, for the structure 9 with 4 shallow QWs, the measured transport time of 44 ps is significantly shorter than the evaluated time which is over a 100 ps. Besides, the linear dependence of the transport times on a number of shallow QWs observed for the undoped structures can not be attributed to the density dependence of the diffusion constant, as it was done for the p-doped samples. Probably, the diffusion model used above is too simplified for the case of the undoped structures. This may be due to several reasons. First, the electrons are much more mobile and have a lower capture probability than the holes. This may lead to different local electron and hole densities and electric fields which would influence the carrier transport. Secondly, the photoexcited carrier gradient, because of an effective carrier trapping into the QWs close to the InP cap layer, should be highly nonuniform across the structure. In any case, the model based on a constant ambipolar diffusion coefficient would not be suitable to describe the carrier transport process. More elaborate modelling of the carrier transport in the undoped structures would be useful to interpret the experimental results.

In conclusion, we have measured carrier transport times across the QW region for a number of InGaAsP structures. The carrier transport times are in the range from several to tens of ps, depending on doping, the number of QWs, and the barrier width. The modelling of the experiment for the p-doped structures revealed that the transport proceeds with mobilities that are dependent on carrier density and, hence, also on position and time.

REFERENCES

1. N. Tessler and G. Eisenstein, IEEE J. Quantum Electron. **29**, 1586 (1993).
2. S. Marcinkevicius, U. Olin, J. Wallin, K. Streubel, and G. Landgren, Appl. Phys. Lett. **65**, 2057 (1994).
3. B. Lambert, B. Deveaud, A. Chomette, A. Regreny, and B. Sermage, Semicond. Sci. Technol. **4**, 513 (1989).
4. S. M. Sze, *Physics of Semiconductor Devices* (Wiley, New York, 1981).
5. R. J. Hayes, A. R. Adams, and P. D. Greene, in *GaInAsP Alloy Semiconductors*, ed. T. P. Pearsall (Wiley, New York, 1982).

53

SPIN RELAXATION OF EXCITONS LOCALIZED IN GaAs/AlGaAs COUPLED QUANTUM WELL STRUCTURES

A. Vinattieri,[1] A. L. C. Triques,[2] M. Colocci,[1] and Ph. Roussignol[3]

[1] Dipartimento di Fisica-Unita' INFM and LENS, Universita' di Firenze
Largo E.Fermi 2, 50125 Firenze, Italy
[2] Departamento de Física do Estado Sólido e Ciênza dos Materiais
Instituto de Física Gleb Wataghin, Universitade Estadual de Campinas
13081–970, Campinas-SP, Brazil
[3] Ecole Normale Supérieure
24 rue Lhomond, 75005 Paris, France

The dynamics of the exciton spin in low-dimensional semiconductor structures is one of the basic processes involved in exciton relaxation and recombination and, therefore, has been widely investigated[1]. Nevertheless, while in high quality samples, where inhomogeneous broadening plays a minor role, it was shown that the exciton spin-relaxation is mainly driven by the exchange interaction[2] and a good agreement has been eventually obtained with theory[3], much less attention has been devoted to samples where excitons are clearly localized, at least at low temperatures[4]. We believe that a deeper understanding of how spin-relaxation is modified in inhomogeneously broadened systems can possibly provide explanation of the dispersion in the experimental results published so far.

In this paper we present experimental results on the spin-polarization dynamics in a GaAs/$Al_{.29}Ga_{.71}$As symmetrical double quantum well structure. The sample consists of two 40 Å thick GaAs quantum wells coupled through a 20 Å thick $Al_{.29}Ga_{.71}$As barrier, grown on a vicinal GaAs surface. Such a thin barrier strongly couples the electronic states of the two wells, producing a splitting of the electronic levels in the conduction and valence bands which originates symmetrical (S) and antisymmetrical (A) states. The energy separation between the S and A excitonic levels is 48 meV for the heavy-hole exciton and 90 meV for the light hole exciton. We performed continuous wave (cw), picosecond time-resolved photoluminescence (PL) and photoluminescence excitation (PLE) spectra at low excitation density ($\leq 1 W/cm^2$), after circularly polarized excitation (σ^+, σ^-), using a cw Ti:Sapphire for the cw experiment and a picosecond Styryl 7 dye laser synchronously pumped by the second harmonic of a Nd:YAG laser for the time-resolved experiment. The overall time-resolution, using a synchroscan streak camera, was 20 ps. Hereafter we indicate with I^+ and I^- the PL intensities in the two different excitation-detection polarization

Ultrafast Processes in Spectroscopy, edited by Svelto et al.
Plenum Press, New York, 1996

233

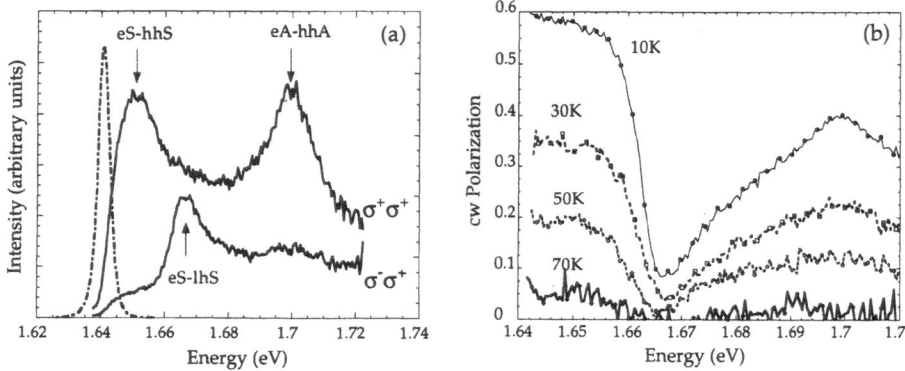

Figure 1. (a) Photoluminescence (dash-dotted line) and photoluminescence excitation (solid lines) spectra at 10 K. $\sigma^+\sigma^+$ and $\sigma^-\sigma^+$ indicate the polarization of the exciting and detected light. (b) cw polarization, detected at the PL maximum, as a function of the excitation energy for various temperatures.

conditions, $\sigma^+\sigma^+$ and $\sigma^-\sigma^+$, respectively; S indicates the total PL signal ($I^+ + I^-$) and D the difference ($I^+ - I^-$), so that the polarization P is given by the ratio D/S.

Fig.1a shows a comparison between the PL and PLE spectra at 10 K after continuous wave excitation with circularly polarized light σ^+ or σ^-; only the positive circularly polarized component of the emitted light was detected. The PL linewidth at 10 K is about 4 meV and an important Stokes shift (10 meV) is measured between the PL and PLE spectra, which disappears when increasing the temperature; therefore at 10 K the luminescence originates from strongly localized excitons. In the PLE spectra very pronounced excitonic resonances are detected, corresponding to the heavy-hole symmetric (eS-hhS) and antisymmetric (eA-hhA) exciton transition energies, when exciting and detecting σ^+ light, while the light hole exciton transition comes out under $\sigma^-\sigma^+$ excitation and detection condition. Not shown in Fig.1a is the antisymmetric light hole exciton state which occurs at the energy of 1.76 eV. In Fig.1b the cw polarization, detected at the PL maximum, is reported as a function of the excitation energy for different temperatures. It turns out that strongly polarized luminescence is measured when the excitation is resonant with the eS-hhS and eA-hhA transition energies, while the polarization goes to zero at the corresponding light-hole exciton transition energies. Moreover, a polarized PL is observed up to 50 K while it disappears at higher temperatures.

The first important consequence of the cw results is that the initial energy relaxation and the subsequent localization do not produce an important loss of polarization, at least at low temperatures, for the heavy-hole exciton states. This result is confirmed by the time-resolved experiment; in fact initial polarization as high as 90% is measured when exciting resonantly with the eS-hhS transition energy, decreasing to 60% when the excitation energy is tuned to the eA-hhA exciton energy (roughly 60 meV above the PL peak).

In Fig.2a a typical decay of I^+ and I^- is shown, when exciting resonantly at the eS-hhS exciton transition energy; in Fig.2b the sum S and the difference D is reported.

The time-resolved data have been analyzed in the framework of a simple three-level model, where the excitons, pumped into the optically active spin-states, are quickly captured by the localized states $|\pm 1\rangle$ where the spin-flip of the exciton (as a whole) takes place. All non-radiative processes, including the spin-flip to the exciton dark states $|\pm 2\rangle$, are described by a non-radiative time τ_{NR}. From the localized $|\pm 1\rangle$ states recombination

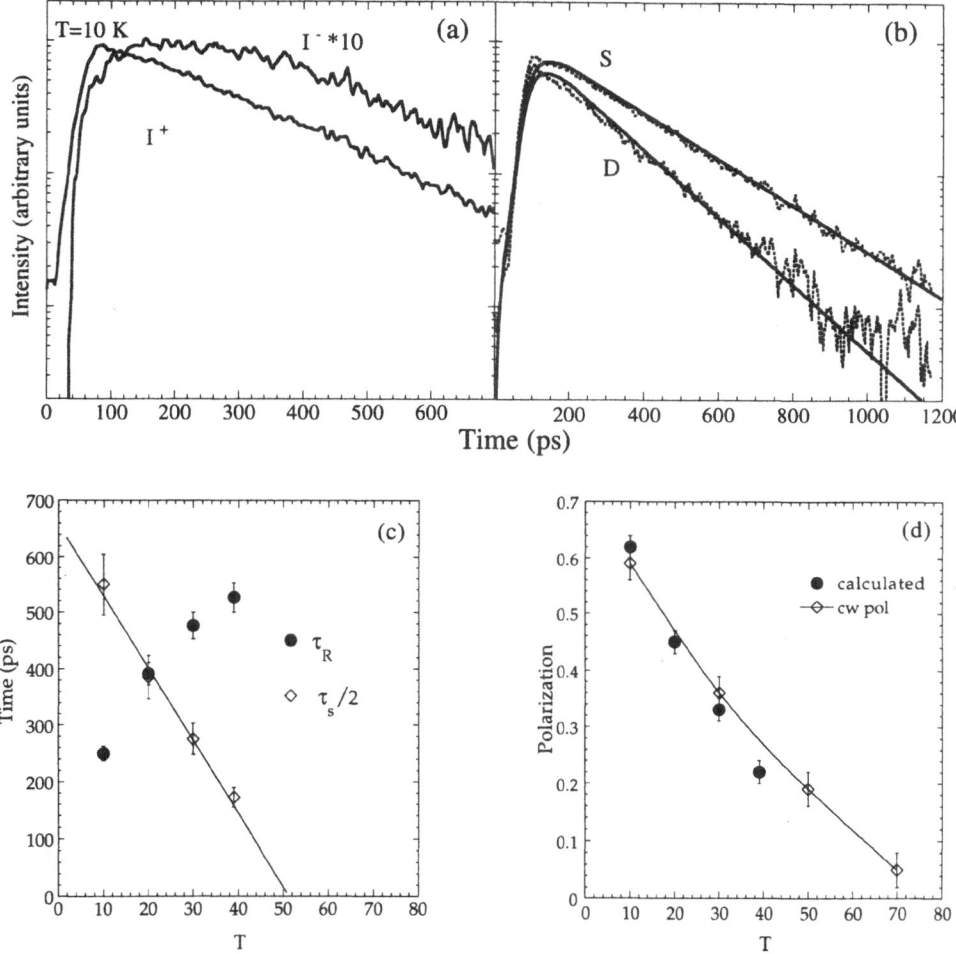

Figure 2. (a) PL decay, detected at 1.641 eV, at 10 K when exciting resonantly at the eS-hhS exciton transition energy. I^+ and I^- indicate the two circularly polarized components of the PL signal. I^- has been multiplied for a factor 10 to evidence the slower risetime with respect to I^+. (b) PL decay of the sum S and the difference D, as obtained from data shown in Fig.2a. Also shown are fits (solid lines) to S and D, as discussed in the text.(c) Recombination time τ_R (full dots) and effective spin relaxation time $\tau_S/2$ (empty rhombs), as deduced from fits to the experimental data for excitation at the eS-hhS exciton transition energy, versus temperature. The solid line is only a guide for eyes. (d) cw polarization (empty rhombs) as a function of temperature for excitation at the eS-hhS exciton transition energy. Also shown are the values calculated from fits to the time-resolved data (full dots). The solid line is only a guide for eyes.

occurs with a characteristic decay time τ_R. This model allows for the independent extraction of τ_R from the temporal behavior of S, while D turns out to decay with a time constant given by $(1/\tau_R + 2/\tau_S)^{-1}$ where $\tau_S/2$ is the effective spin-relaxation time. Typical fits to S and D are shown in Fig.2b, as solid lines. From fits to the experimental data we deduce a PL decay time τ_R of 250 ps at low temperatures which increases, increasing the temperature, as commonly found in quantum well structures[5]. A long spin-relaxation time $\tau_S/2$ of the order of 600 ps is determined at 10 K, roughly independent of the excitation energy, decreasing to 170 ps at 40 K (fig.2c). From the fits to the time-resolved data we can nicely reproduce the polarization values obtained from the cw data (fig.2d).

We want to stress two major results from our analysis of the experimental data. First of all, it comes out that the strong localization of the exciton, as shown by the 10 meV Stokes shift, does not produce an important loss of polarization as observed when exciting resonantly at the eS-hhS exciton transition energy. Nevertheless, if we compare the spin-relaxation time obtained in this case (of the order of a few hundreds of ps, depending on T) with the value of a few tens of ps, previously measured in quantum wells of similar thickness[6], we conclude that the exciton localization strongly modifies the spin-relaxation process. On one hand we could expect, because of such a narrow well, the exchange inter-action as the dominant mechanism; however, exciton localization should produce an increase in the spin-relaxation rate respect to the case of free exciton spin-relaxation. Similarly if we consider the D'yakonov-Perel mechanism, the spin-relaxation rate is pro-portional to the momentum relaxation which is expected to be quite long (few tens of ps or even longer) for localized excitons, once again producing a fast spin-relaxation. On the other hand, we cannot simply invoke some kind of effect due to impurity scattering by the fact that, despite the broad inhomogeneous PL line, no major change has been found both in cw and time-resolved experiments when changing the detection energy over the PL band. Nevertheless the dependence of the polarization on the excitation energy seems to indicate that the exciton flips the spin as a whole particle, suggesting again the exchange interaction as the responsable mechanism. Moreover, it is remarkable that also energy re-laxation does not modify the spin of the exciton, when it is formed at the symmetric and antisymmetric heavy-hole transition energies. Being the energy separation between the eS-hhS and eA-hhA states comparable with the AlAs LO phonon energy, this suggests that whenever a LO phonon is involved in the energy relaxation, the interaction is strictly spin-conserving.

In conclusion we showed that even in samples with strong exciton localization nice excitonic features can be observed in the spin-dynamics. The long spin-relaxation time is possibly explained by the localization mechanism, giving reason of the wide spread in the experimental results concerning the exciton spin-relaxation. Experiments trying to clarify the role of exciton dephasing on the spin-relaxation of localized excitons are in progress.

REFERENCES

1. T.C.Damen, Karl Leo, Jagdeep Shah, and J.E.Cunningham, Appl. Phys. Lett. **58**, 1902 (1991); see also ref 2. and references therein.
2. A.Vinattieri, Jagdeep Shah, T.C.Damen, D.S.Kim, L.N.Pfeiffer, M.Z.Maialle, and L.J.Sham, Phys. Rev. B **50**, 10868 (1994).
3. M.Z.Maialle, E.A.de Andrada, and L.J.Sham, Phys. Rev. B **47**, 15776 (1993).
4. A.Frommer, E.Cohen, Arza Ron, L.N. Pfeiffer, Phys. Rev. B **48**, 2803 (1993); L.Muniz, E.Perez, L.Viña, K.Ploog, Phys. Rev. B **51**, 4247 (1995).
5. J.Feldmann, G.Peter, E.O.Göbel, P.Dawson, K.Moore, G.Foxon and R.J.Elliot, Phys. Rev. Lett. **50**, 2337 (1987).
6. Ph.Roussignol, P.Rolland, R.Ferreira, C.Delalande, G.Bastard, A.Vinattieri, L.Carraresi, M.Colocci, Surf.Sci. **267**, 360 (1992).

54

QUANTUM BEATS IN PHOTOLUMINESCENCE OF EXCITONS IN ZnSe/ZnSeTe SUPERLATTICE QUANTUM WELLS

X. G. Zheng,[1] H. Z. Wang,[1] Z. X. Yu,[1] F. L. Zhao,[2] Z. D. Tang,[2] Z. L. Zhang,[2] C. Xie,[2] W. Z. Lin,[2] and G. Xu[2]

[1] Institute for Laser and Spectroscopy
Zhongshan University
Guangzhou, China
[2] Department of Physics
Zhongshan University
Guangzhou, China

Quantum wells (QWs) of wide and direct band gap II-VI semiconductor materials have aroused intense research interest due to their application to development of green and blue laser diodes[1-4]. Among them, ZnSe/ZnSeTe strained layer QW system, in which Te acts as an isoelectronic trap and a constituent of the alloy as well, has been considerably studied because of its peculiar optical properties. Lee et al.[5] and Chang et al.[6] observed a blue and a green broad bands (S_1 and S_2 bands) in photoluminescence of $ZnSe_{1-x}Te_x$ alloys, and attributed them to extrinsic exciton self-trapping induced respectively by single Te atom and Te_n ($n \geq 2$) clusters. The temperature and Te-concentration dependent characteristics of the alloys have also been inverstigated in details[5-8]. More recently, time-resolved and excitation density dependent photoluminescence properties of ZnSe/ZnSeTe superlattice QWs were demonstrated[9]. In addition to previous studies, photoluminescence dynamics of ZnSe/ZnSeTe superlattice QWs are presented in this work, exhibiting characteristics of quantum beats in photoluminescence of excitons.

The superlattice QW sample studied in this work was grown on GaAs (100) substrates by atomic layer epitaxy method, and its structure is determined as $\{(ZnSe)_{70}(ZnSe:Te)_{20}(ZnSe)_8 + [(CdSe)_1(ZnSe)_3]_7 + (ZnSe)_8(ZnSe:Te)_{20}(ZnSe)_{70}\} \times 1$. The exact concentration of Te in structure of $(ZnSe:Te)_{20}$ has not been determined. We estimate that Te:Se>1:10. Photoluminescence dynamics experiments are performed at a sample temperature of 77 K with a frequency-tripled single pulse mode-locked Nd:YAG laser (~30 ps pulse duration) as an excitation source and a single sweep streak camera (Hamamatsa Model C1587, 2 ps time resolution) connected to a polychromator as a recorder, the spectral and temporal characteristics of the sample thus being simultaneously recorded.

Ultrafast Processes in Spectroscopy, edited by Svelto et al.
Plenum Press, New York, 1996

237

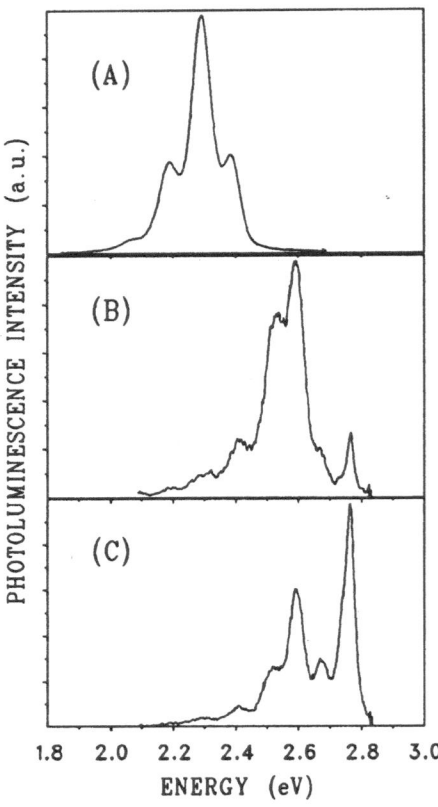

Figure 1. Photoluminescence spectra of ZnSe/ZnSeTe superlattice quantum wells excited by 365-nm line of low density CW Hg lamp (A), 0.4 mJ/cm^2 frequency-tripled Nd:YAG picosecond pulse laser (B), and 4 mJ/cm^2 frequency-tripled Nd:YAG picosecond pulse laser (C), respectively.

In Figure 1 are shown the photoluminescence spectra of ZnSe/ZnSeTe under different excitation. When the excitation source is the 365-nm line of a CW Hg lamp, only a green broad band (S_2) around 541 nm (2.29 eV) is recorded, as shown in Figure 1(A). Similiar spectropic characteristics are obtained when the sample is excited by a rather weak frequency-tripled mode-locked Nd:YAG laser. As the excitation density increases, the emission band blue-shifts. The photoluminescence is dominated by a blue broad band (S_1) around 478 nm (2.59 eV) when the excitation density is 0.4 mJ/cm^2, as shown in Figure 1(B). Meanwhile, a sharp emission line at 448 nm (2.77 eV) emerges, for which the free excitons in ZnSe barriers are responsible. When the excitation density increases up to 4 mJ/cm^2, the emission line at 448 nm is dominant. Recombination luminescence of free excitons in $[(CdSe)_1(ZnSe)_3]_7$ (~2.5 eV) can also be distinguished in Figure 1(B) and Figure 1(C). One should note that the regular periodic oscillation structures in S_1 and S_2 bands is due to optical interference effects among ZnSe layers[8]. Although the periodic oscillation varies with the direction of detection, the envelopes of S_1 and S_2 bands keep unchanged. According to previous studies[5,6], these broad bands are attributed to recombination luminescence of self-trapped excitons around Te$_1$ and around Te$_n$ cluster, respectively. Moreover, the above results reveal saturation properties of self-trapped exciton states around Te centers[9].

Analysis of transient luminescence spectra reveals that after creation excitons are rapidly captured by Te isoelectronic centers within the duration of the excitation laser

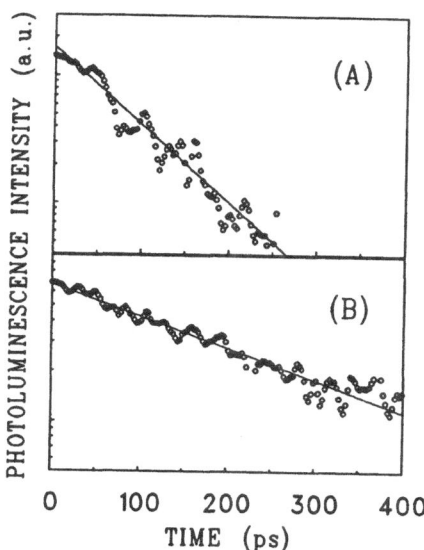

Figure 2. Photoluminescence decays of free excitons in ZnSe (A) and self-trapped excitons around Te_1 (B).

pulse. Decay of S_2 band shows a extremely long recombination luminescence lifetime of 4.2 ns for excitons self-trapped around Te_n clusters. At high enough excitation density exciton states localized around Te_n clusters are saturated because of their long-lasting decay, and most of excitation energy is transferred into exciton states localized around Te_1 centers and is emitted as S_1 band. The recombination luminescence lifetime for excitons self-trapped around Te_1 centers is determined to be 226 ps, as shown in Figure 2(B). Depicted in Figure 2(A) is the recombination luminescence decay of free excitons in ZnSe, exhibiting a lifetime of 71 ps. Recombination luminescence of free excitons in ZnSe dominates under a high excitation density (*e.g.*, 4 mJ/cm^2) due to saturation of self-trapped exciton states.

As shown in Figure 2, temporal oscillation with a period of ~22 ps can be observed in recombination luminescence decays of free excitons in ZnSe and self-trapped excitons around Te_1 centers when the QW sample is excited by 0.4 mJ/cm^2 frequency-tripled Nd:YAG picosecond pulse laser, and it does not vary with the direction of detection. The period corresponds to an energy interval of 0.18 eV, exactly equal to the energy gap between free excitons in ZnSe and self-trapped excitons around Te_1 isoelectronic centers. We therefore attribute the oscillation to quantum beats of recombination luminescence from free excitons in ZnSe and from self-trapped excitons around Te_1 centers. This interest result implies coherence between freely propagating carriers and localized carriers. A tunneling mechanism[10] can thus be proposed for excitation transfer through self-trapping barriers due to requirement of phase matching.

If coupling is supposed between transfer processes of S_1 ↔ free excitons and of S_2 ↔ free excitons, a tunneling mechanism may explain the very large spatial diffusion of the luminescence S_2 observed by Lee *et al.*[5], who interpreted it in terms of radiative transfer from S_1 to S_2. Chang *et al.*[6] suggested that the self-trapping barrier height be very low in comparison with the exciton binding energy according to the observed dramatic change of the intensities of free exciton line and S_1 band. Considering this, we suggest that the excitation transfer mechanism be temperature-dependent, and that tunneling effects dominate at sufficiently low temperature while thermal activation does at higher temperature.

In summary, picosecond photoluminescence dynamics of free excitons in ZnSe barriers and self-trapped excitons around Te_1 centers and Te_n clusters are studied, demonstrating different temporal characteristics of exciton capture and recombination. Quantum beats of recombination luminescence from free excitons in ZnSe and from self-trapped excitons around Te_1 centers is recorded, and a tunneling mechanism of excitation transfer in ZnSe/ZnSeTe superlattice QWs is proposed.

This work is supported by an NNSFC grant and a Frontier Sciences grant of Zhongshan University.

REFERENCES

1. M. A. Haase, J. Qiu, J. M. DePuydt, and H. Cheng, Appl. Phys. Lett. **59**, 1272 (1991).
2. H. Jeon, J. Ding, W. Patterson, A. V. Nurmikko, W. Xie, D. C. Grillo, M. Kobayashi, and R. L. Gunshor, Appl. Phys. Lett. **59**, 3619 (1991).
3. A. V. Nurmikko, R. L. Gunshor, and M. Kobayashi, J. Crystal Growth **117**, 432 (1992).
4. R. L. Aggarwal, J. J. Zayhowski, and B. Lax, Appl. Phys. Lett. **62**, 2899 (1993).
5. D. Lee, A. Mysyrowicz, A. V. Nurmikko, and B. J. Fitzpatrick, Phys. Rev. Lett. **58**, 1475 (1987).
6. S. K. Chang, C. D. Lee, H. L. Park, and C. H. Chung, J. Crystal Growth **117**, 793 (1992).
7. K. Dhese, J. E. Nicholls, J. Goodwin, W. E. Hagston, J. J. Davies, M. P. Halsall, B. Cockayne, and P. J. Wright, J. Crystal Growth **117**, 91 (1992).
8. J. Ren, K. A. Bowers, R. P. Vaudo, J. W. Cook, Jr. J. F. Schetzina, J. Ding, H. Jeon, and A. V. Nurmikko, J. Crystal Growth **117**, 510 (1992).
9. H. Z. Wang, X. G. Zheng, F. L. Zhao, Z. X. Yu, Z. D. Tang, Z. L. Zhang, and C. Xie, Phys. Stat. Sol. (b) **186**, K73 (1994).
10. E. I. Rashba, in *Excitons*, ed. by E. I. Rashba and M. D. Sturge (North-Holland Publishing Company, Amsterdam, 1982).

55

ULTRAFAST CARRIER TRANSFER IN QUANTUM-WELL- AND HETERO-n-i-p-i-STRUCTURES

A. Seilmeier, J. Baier, H. M. Hauenstein, N. Moritz, U. Plödereder, and I. M. Bayanov

Institute of Physics
University of Bayreuth
D-95440 Bayreuth, Germany

Carrier relaxation in 2D-semiconductor structures is of interest for applications in electrooptical devices and modulators. In many cases the relaxation process involves carrier transfer between neighboring potential wells which is studied in this paper via transient absorption changes. The mechanism for such a carrier transfer, thermionic emission, tunneling, or phonon assisted tunneling is of special interest. In a first part of this paper, carrier transfer from the quantum wells to the doping layers of type-II-hetero-n-i-p-i-structures is discussed. In the second part experimental data taken on modulation doped quantum well structures are presented. The carrier transfer from the first excited subband to potential minima in the barrier and vice versa is studied.

In a type-II-hetero-nipi-structure[1] (see Fig. 1) charge transfer and relaxation is investigated after ultrafast interband excitation of the quantum wells.[2] The structure consists of 10 $Al_{0.3}Ga_{0.7}As$ n-i-p-i-periods. The n and p regions are each 30 nm wide and are doped to a concentration of 1×10^{18} and 2×10^{18} cm^{-3} respectively. Each intrinsic region of 180 nm thickness contains a 20 nm wide GaAs quantum well in the center. Fig. 1 shows the band structure of the sample calculated self-consistently in Thomas Fermi approximation for the ground state (solid lines) and for an excited state ($n^{(2)}=1.5 \times 10^{12}$ cm^{-2} excited carriers per period, dashed line). A distinct change of the band structure with excitation density is obvious.

In time resolved experiments carriers are excited with subpicosecond pulses at frequencies slightly above the energy gap of GaAs. In this way electron-hole pairs are only injected into the GaAs quantum wells. A less intense part of the pulses is used to monitor the relaxation of the carriers and the transfer from the wells to the doped regions via transient transmission and reflection changes.

Fig. 2 shows the transient bleaching at the frequencies a) $h\nu = 1.52$ eV and b) $h\nu = 1.47$ eV and two average excitation powers of 10mW and 1.7mW corresponding to injected electron densities in the excited volume of $n_{in}^{(2)} = 1.2 \times 10^{12}$ cm^{-2} and $n_{in}^{(2)} = 0.25 \times 10^{12}$ cm^{-2} per period (T = 300K). Decay times of several 100 ps are observed representing carrier transfer from the quantum wells to the doped regions of the n-i-p-i structure.

Ultrafast Processes in Spectroscopy, edited by Svelto et al.
Plenum Press, New York, 1996

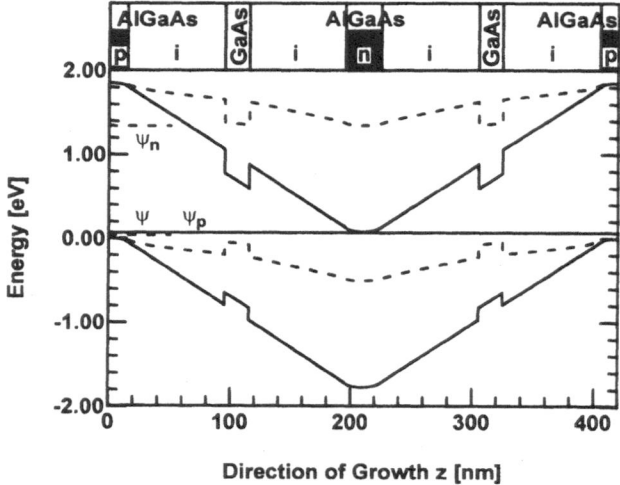

Figure 1. Composition of the sample and calculated band structures (one period) for the ground state (———; Fermi level ψ) and for a state with an excited 2D electron density of $n^{(2)}=1.5\times10^{12}$ cm^{-2} (– – –; quasi Fermi levels ψ_n and ψ_p).

There is indication that the observed transmission change represents only the dynamics of the electrons since holes are expected to contribute considerably less to the signal. The experiments reveal the following relaxation steps: i) Cooling of the charge carriers is observed within the first few picoseconds resulting in a rapidly decaying bleaching signal (first rapid decay for weak excitation in Fig. 2a). ii) A transfer of electrons from the quantum wells to the n-doped regions is found to occur on a time-scale of a

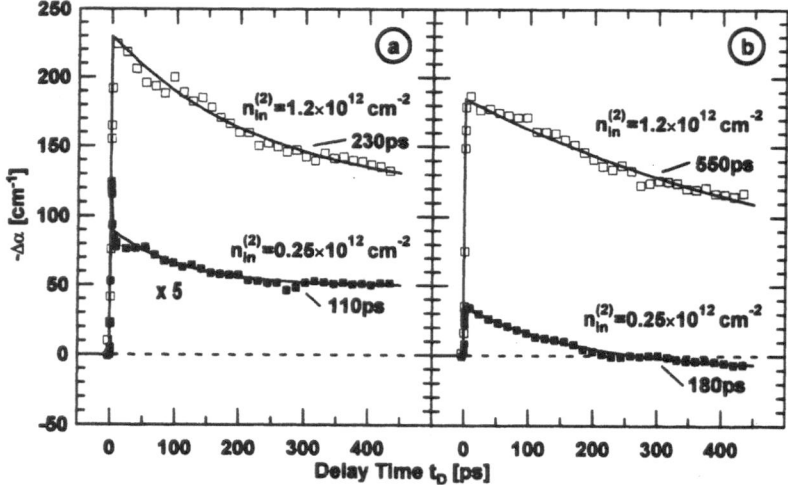

Figure 2. Transient absorption changes measured at a) hv=1.52eV (λ=815nm) and at b) hv=1.47eV (λ=845nm) for two excitation powers corresponding to injected electron densities of $n_{in}^{(2)} = 1.2\times10^{12}$cm^{-2} and $n_{in}^{(2)} = 0.25\times10^{12}$cm^{-2} per period. Thermionic emission is observed on a time scale of several 100ps.

few hundred picoseconds. A comparison with model calculations shows that thermionic emission is the most efficient transfer process. The rates for tunneling are considerably smaller on account of the relatively broad potential barriers. The signal at $t_D \geq 500ps$ in Fig. 2 is due to residual bandfilling in the quantum wells and due to partial neutralization of the space charge of ionized donors and acceptors which results in absorption changes according to the Franz-Keldysh effect. iii) Finally, the carriers return to the ground state via efficient lateral diffusion process (nanosecond time scale) and via carrier recombination with density dependent time constants between several microseconds and seconds[1,3]. The experimental findings are in agreement with detailed numerical calculations of the carrier dynamics in Thomas-Fermi approximation.[2]

A different mechanism, phonon assisted tunneling, is found to dominate the carrier transfer after intersubband excitation in a modulation-doped multiple quantum well structure.[4] The investigated $GaAs/Al_{0.31}Ga_{0.69}As$ quantum well sample consists of 50 periods. In the n-modulation doped sample 7.3 nm thick GaAs quantum wells are embedded in 38 nm wide $Al_{0.31}Ga_{0.69}As$ barriers in which the central 8 nm are silicon doped ($n=5.7\times10^{11}cm^{-2}$).

The carrier dynamics is studied by infrared bleaching experiments. Electrons are resonantly excited to the first excited subband of the quantum well by an intense infrared picosecond pulse at $\lambda \sim 11\mu m$[5]. Information on scattering processes is obtained from transient transmission changes measured by weaker pulses of the same wavelength. In this way the repopulation of the ground state is monitored.

Fig. 3 shows the conduction band structure and the subbands of the sample as obtained from selfconsistent calculations. The subbands W1 and W2 are localized in the GaAs wells (solid lines). The wavefunctions of the subbands B1 to B3 (dashed lines) exhibit a large amplitude in the AlGaAs barriers but still a small one in the wells. The hatched area C denotes one of the quasibound subbands in the continuum.

The intersubband relaxation is systematically investigated as a function of the sample temperature. Fig. 4. shows results taken with a pump pulse intensity of ~ 70 kW/cm^2. In general we observe biexponential relaxation. At $T \leq 77$ K the fast component dominates the relaxation behavior. The amplitude of the second slower component rises with temperature and is prevailing at room temperature. There is indication that this effect is due to the increasing intersubband absorption with rising temperature (see Ref. 4).

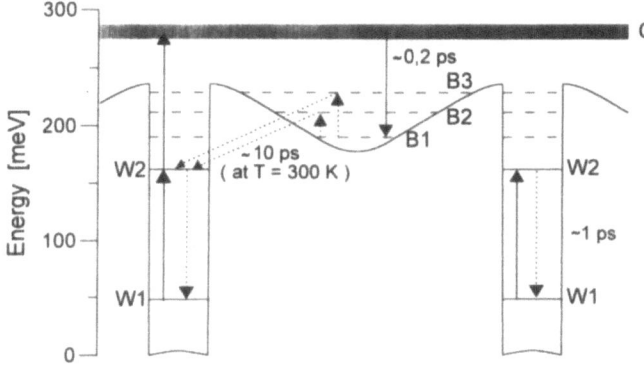

Figure 3. Band structure of the investigated modulation doped sample. The excitation processes are indicated by solid arrows, the most significant relaxation channels by dotted arrows.

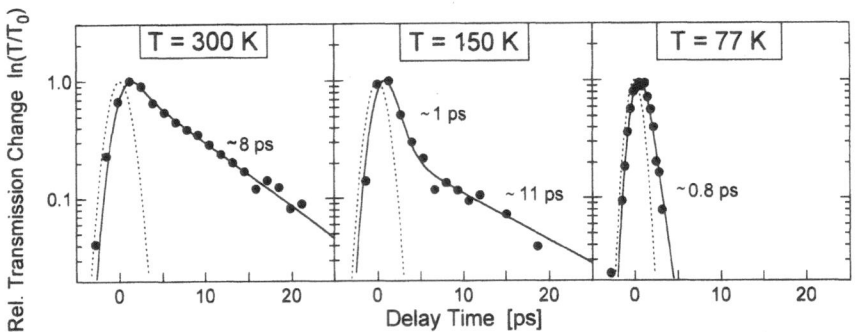

Figure 4. Normalized transmission change as a function of delay time for different temperatures at a pump intensity of ~70 kW/cm². The dotted line represents the correlation function of the pump and probe pulses.

The findings can be explained by the following model: The infrared pump pulse excites electrons both to the first excited subband of the well (W2) and to a resonant continuum band (C) via two photon excitation (see Fig. 3, solid arrows). Efficient trapping processes occupate the states B1 to B3 which are mainly localized in the barrier.

The fast relaxation component with a time constant of ~ 1 ps represents the direct intersubband relaxation W2→W1. This value is in good agreement with the calculated intersubband scattering time due to electron - LO phonon interaction.

The second slower relaxation component arises from the transfer of the trapped carriers in the barriers back into the quantum wells. Inspection of the corresponding transfer rates shows that the most efficient path is phonon assisted tunneling. Thermal excitation to the states B2 and B3 and subsequent scattering B2→W2 and B3→W2, respectively, leads to a time constant of ~10 ps at room temperature.[4,6] The temperature dependence of the activation process qualitatively explains the temperature behavior of the relaxation time of the slower component; it increases from 8 ps at T=300 K to 11 ps at T=150 K.

The experimental data discussed here show that the most efficient carrier transfer mechanism in 2D-semiconductor structures is determined by the detailed potential structure of the sample.

The authors would like to thank G. Weimann (Fraunhofer-Institut für Angewandte Festkörperphysik, Freiburg) and G. H. Döhler (University of Erlangen) for providing the samples, and G. H. Döhler and coworkers for valuable discussions.

REFERENCES

1. G. H. Döhler, CRC Crit. Rev. in Solid State and Material Sciences **13**, 97 (1987)
2. N. Moritz, H. M. Hauenstein, and A. Seilmeier, Phys. Rev. B, to be published Dec. 1995
3. H. M. Hauenstein, N. Moritz, and A. Seilmeier, Phys. Stat. Sol. (b) **188**, 581 (1995)
4. J. Baier, I. M. Bayanov, U. Plödereder, and A. Seilmeier, Superlattices and Microstructures, accepted for publication 1996
5. T. Dahinten, U. Plödereder, A. Seilmeier, K. Vodopyanov, K. Allakhverdiev, and Z. Ibragimov, IEEE J. Quant. Electron. **29**, 2245 (1993)
6. A. Seilmeier, U. Plödereder, J. Baier, and G. Weimann, in *Quantum Well Intersubband Transition Physics and Devices* ed. by H. C. Liu, B. F. Levine, and J. Y. Andersson (Kluwer, Dordrecht, 1994), p. 421

56

QUANTUM DESCRIPTION AND SIMULATION OF THE FEMTOSECOND RESPONSE OF HIGHLY PHOTOEXCITED CHARGE CARRIERS IN SEMICONDUCTORS

P. Kocevar

Inst. f. Theoretische Physik
Karl-Franzens-Universität Graz
A-8010 Graz, Austria

1. INTRODUCTION

The laser-driven two-level oscillator (LD2LO) has for many years been the standard reference for the interpretation of most coherence phenomena in the time-resolved laser spectroscopy of atomic, molecular and condensed-matter systems. Here the optical Bloch equations have been the main tool for semi-quantitative estimates[1]. Within this description the standard electronic dipole coupling of an isolated oscillator to the (classical) laser field F_L results in Rabi-type oscillations of the two state occupations and the laser-driven transition function, with a "Rabi" frequency proportional to $F_L^{1/2}$. For applications to the interband laser spectroscopy of semiconductors the most straightforward modification approximates the electronic response of the semiconductor by that of an ensemble of mutually interacting LD2LOs representing the optically coupled valence- and conduction-band states[1]. For spatial homogeneity the individual oscillators are labelled by their respective **k**-vector.

As consequence of the very high scattering rates of photoexcited free carriers in bulk semiconductors practically all experimental studies of transient coherence effects have been performed in quantum-well systems and in the excitonic regime to keep the scattering-induced dephasing as low as possible[1,2]. In the following we shall address the question whether experimentally detectable Rabi-type density oscillations could nevertheless be found in the nonexcitonic, i.e. free-carrier, regime of bulk materials.

Extending the original work of Kuhn, Rossi. et al.[3,4] we shall first report on a computationally manageable theoretical framework, in the form of generalized semiconductor Bloch equations (GSBE), for the description of highly nonequilibrium laser-driven semiconductors under conditions of dominant carrier-carrier (c-c) scattering[5,6]. Note that any such collision-dominated electronic response requires a theoretical description (such as

Ultrafast Processes in Spectroscopy, edited by Svelto et al.
Plenum Press, New York, 1996

245

the GSBE of the present approach) in which the coherent carrier dynamics and the incoherent c-c scatterings are treated on an equal footing, This requirement is in pronounced contrast to the theoretical prerequisits for the description of laser-excited atoms or molecules and of the electronic processes governing laser-driven chemical reactions and configurational changes in molecular or condensed-matter systems, where collisional effects play a secondary role and are generally taken into account by use of Bloch equations with phenomenological phase- and energy-relaxation times[7].

For the special case of the degenerate pump-and-probe absorption spectroscopy of bulk GaAs, the numerical solutions of the GSBE will enable us to specify the necessary experimental conditions for the detection of Rabi-type coherent density oscillations in the collision-dominated regime of band-to-band excitation at high excitation densities.

2. THEORY

We consider a degenerate pump-and-probe absorption experiment on intrinsic bulk GaAs with 1.7 eV / 50 fs pump pulses and excitation densities around 8_10^{18} cm^{-3} to ensure negligible excitonic and phononic contributions to the initial transient carrier response. We neglect spatial inhomogeneities, which can be experimentally minimized through appropriately structured samples. Then five basic components of the **k**-space one-particle density matrix suffice for the description of this experiment: the distribution functions $f_e(\mathbf{k})$, $f_{hh}(\mathbf{k})$ and $f_{lh}(\mathbf{k})$ of the photogenerated electrons, heavy holes and light holes and the two laser-driven interband polarizations $p_{ehh}(\mathbf{k})$ and $p_{elh}(\mathbf{k})$ (i.e. the complex expectation values of the phase-locked e-hh and e-lh pair operator). Within the GSBE approach, the time evolution of these dynamical variables is derived from the Heisenberg equations of motion (EoM) for the corresponding pair and number operators. The Hamiltonian includes the interband dipole excitation and all relevant types of c-c and carrier-phonon (c-ph) interactions[8]. All c-c and long-range c-ph interactions are taken as free-carrier screened by use of a quasidynamical extension of the usual long-wavelength limit of the RPA dielectric function[8], including the corresponding Coulomb-hole correction. As details of the bandstructure can be expected to play a minor role for the moderate excitation energies of our present concern, we treat the photoexcited carriers as quasifree particles with their standard effective masses.

The resulting BBGKY hierarchy is closed at the four-point level through appropriate truncations of six-point functions and by use of the standard adiabatic and Markov approximations[3,5,6]. Without c-c and c-ph interactions the resulting EoM just yield the the completely coherent dynamics of independent LD2LOs, and in first order one obtains the conventional mean-field renormalizations of the electric field and the particle energies[1,3]. The lowest-order scattering terms, i.e. the leading dissipative contributions, appear in 2nd order in the (screened) c-c and c-ph couplings[3]. As consequence of the above-mentioned truncation procedure, they appear in the form of Boltzmann collision integrals for both the distribution functions and the various polarizations, including some novel types of collision integrals, where polarizations p(**k**) replace the standard distribution functions ("polarization scatterings")[5,6]. The numerical solution of the resulting coupled EoM for the five basic density matrices proceeds as follows. The EoM of the polarizations are directly integrated, while the distribution functions are obtained by a parallel Ensemble-Monte-Carlo (EMC) simulation of their EoM, i.e. of their (coherent) generation/recombination and (incoherent) scattering dynamics, including the novel polarization scatterings. The parallel integration of the coupled EoM starts with the equilibrium situation just before a noticeable build up of the excitation pulse, which is assumed to be Gaussian both in time and frequency. In this way one obtains the total induced (complex) polarization $p_{tot}(\mathbf{k})$ which is directly related to the macroscopic electric polarization[1,9] and thereby to the absorption spectrum as function of time during and after the excitation pulse. To conform to the experimental detection procedure these calculated absorption spectra are finally convoluted with the temporal and spectral shape of the probe pulse.

3. RESULTS

The most remarkable outcome of these calculations is the virtual intermediate broadening of the range of photo-excited interband polarizations $p(\mathbf{k})$ beyond those band regions which would correspond to the spectral width of the external excitation pulse. These broadenings are a consequence of the increased energy uncertainty during the initial stages of the excitation pulse[3]. Note that in contrast to the semiclassical treatment of the traditional Boltzmann approach, where the energy-time uncertainty has to be included through phenomenological broadenings of the generation spectra, the fully retarded generation dynamics of the GSBE description directly yields these broadenings as functions of time, as first discussed by Kuhn and Rossi[3]. As the pump pulse develops these initially excited band regions outside the nominal excitation range of the excitation pulse will be corrected for by stimulated recombination processes which eventually restore the final excitation spectrum to coincide with the spectrum of the pump pulse. Since any scattering event of one partner in the coherently driven e-h amplitudes creates a free e-h pair, the incoherent scattering dynamics will irreversibly lead to the generation of free carries. Especially during the intermediate broadening of the combined generation/recombination rates this will lead to off-resonance creation of free particles and a correspondingly faster internal and mutual thermalization of the electron and hole distributions through ordinary c-c scatterings[4]. However, for our favourable choice of experimental conditions, the most interesting consequence of the coherent photogeneration/recombination mechanism is the fact that for two time intervals the total rates for the stimulated recombination processes exceed the total generation rates, as shown in the left part of Fig.1. This results in a non-monotonic behaviour of the carrier densities, reflecting very strongly damped Rabi oscillations of a LD2LO[1].

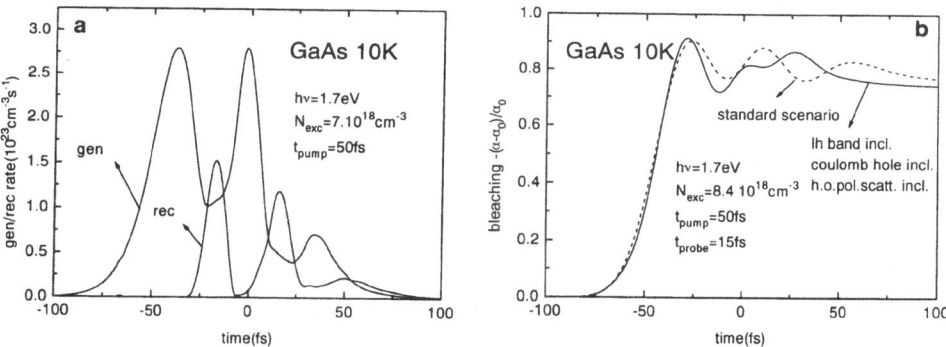

Figure 1. Left: total generation/recombination rate as function of time; right: corresponding absorption change after temporal and spectral convolution with a 15 fs probe pulse, indicating the combined effects of light holes, Coulomb hole and higher-order polarization scatterings.

The corresponding variations of the total polarization result in pronounced oscillations of the total absorption as function of time, with oscillation periods around 33 fs. To prevent a total smearing out of this oscillatory structure by the probe-pulse convolution, the duration and spectral width of the latter has to be carefully chosen. In our example, we chose 15 fs and the corresponding natural linewidth which is at the limit of present-day detection techniques. After this folding procedure the total absorption, as depicted in the right part of Fig.1, indeed shows "surviving" oscillations which should be well within the present limits of experimental resolution.

4. CONCLUSIONS

Our specific model case not only has demonstrated the possibility of experimentally detectable Rabi-type coherence effects in highly excited bulk semiconductors, but also allows us to list the decisive prerequisits for planning an experimental scenario for such a demonstration:

1. The oscillations are driven by the pump pulse; so their period, which can be estimated by their initial Rabi-oscillation period, should be smaller than half the duration of the pump pulse;
2. the laser intensity should be high enough to drive the distribution functions to values near 1 to ensure detectable oscillation amplitudes (1 being the limit of an ideal undamped LD2LO);
3. on the other hand, the laser intensity should not be too high, to prevent too short Rabi-oscillation periods (below the time resolution of the experiment) and also to avoid too high free-carrier densities and collision damping of the oscillations;
4. to prevent a temporal smearing out of the intrinsic sample response by the probe pulse, the probe-pulse duration must be smaller than one half or one third of the pump pulse duration; however, this advantage in the time regime will be inevitably counteracted by a more effective spectral smearing.

REFERENCES

1. H. Haug and S. W. Koch, *Quantum Theory of the Optical and Electronic Properties of Semiconductors* (World Scientific, 1993).
2. See e.g. the remaining contributions to the semiconductor sessions at this symposium.
3. T. Kuhn and F. Rossi, Phys. Rev. B **46**, 7496 (1992).
4. A. Leitenstorfer, A. Lohner, K. Rick, P. Leisching, T. Elsaesser, T. Kuhn, F. Rossi, W. Stolz, and K. Ploog, Phys. Rev. B **49**, 16372 (1994).
5. F. J. Adler, P. Kocevar, J. Schilp, T. Kuhn, and F. Rossi, Semicond. Sci. Technol. **9**, 446 (1994); F. Adler, G. F. Kuras, and P. Kocevar, SPIE Proc. 2142, 206 (1994).
6. G. F. Kuras, Diploma Thesis, Karl-Franzens-Universität Graz (1995).
7. B. Garraway and K.-A. Suominen, Rep. Prog. Physics **58**, 365 (1995).
8. U. Hohenester, P. Supancic, P. Kocevar, X. Q. Zhou, W. Kütt, and H. Kurz, Phys. Rev. B **47**, 13233 (1992).
9. M. Lindberg and S. W. Koch, Phys. Rev. B **38**, 3342 (1988); A. V. Kuznetsov, Phys. Rev. B **44**, 8721 (1991).

57

FEMTOSECOND SPECTROSCOPY OF HOT CARRIERS IN LOW-DIMENSIONAL SEMICONDUCTORS

M. Woerner,[1] S. Lutgen,[1] R. Kaindl,[1] T. Elsaesser,[1] A. Hase,[2] and H. Künzel[2]

[1] Max-Born-Institut für Nichtlineare Optik und Kurzzeitspektroskopie
D-12489, Berlin, Germany
[2] Heinrich-Hertz-Institut für Nachrichtentechnik Berlin GmbH
D-10587, Berlin, Germany

The rapid development of femtosecond light sources tunable in a wide spectral range allows the direct observation of elementary scattering processes in bulk or low-dimensional semiconductor structures. The non-equilibrium dynamics of carriers are commonly studied by femtosecond absorption and luminescence techniques via valence to conduction band transitions [1,2]. Such experiments give information on the combined relaxation of electrons and holes. The interaction among them leads to excitonic effects and a complex scattering scenario, making the interpretation of the data with respect to the microscopic dynamics quite difficult. Experiments with doped semiconductors allow the separate investigation of either electrons or holes [3,4]. Application of mid-infrared (MIR) excitation pulses in the range below the fundamental bandgap makes experiments possible in which the excitation process leads exclusively to a redistribution of carriers at constant total density. In this paper, we study the inter- and intra-subband relaxation of electrons in quasi-two-dimensional semiconductors structures by such a technique. The dynamics of electron redistribution after femtosecond excitation to a higher subband is monitored in spectrally and temporally resolved experiments, providing new information on the lifetimes of higher subbands and on electron thermalization. The paper is organized as follows. In the first part, we describe a novel laser system exclusively based on solid state components which allows the generation of femtosecond pulses in the MIR and is used in our experiments [5]. In the second part, we report recent results on the ultrafast relaxation of electrons in n-type modulation-doped GaInAs/AlInAs quantum wells.

A schematic of the experimental setup for the generation and characterization of our MIR pulses is shown in Fig. 1. The system is based on a 1 kHz Ti:Sapphire laser/regenerative amplifier system. A recently presented optical parametric generator (OPG)/ amplifier (OPA) system [6] consisting of a type II noncritically temperature tuned LBO seeder and a type I BBO amplifier crystal provides 30–50 fs signal and idler pulses with energies of several microjoules in the respective spectral ranges of 1200 to 1500 nm and 1700 to 2500

Ultrafast Processes in Spectroscopy, edited by Svelto et al.
Plenum Press, New York, 1996

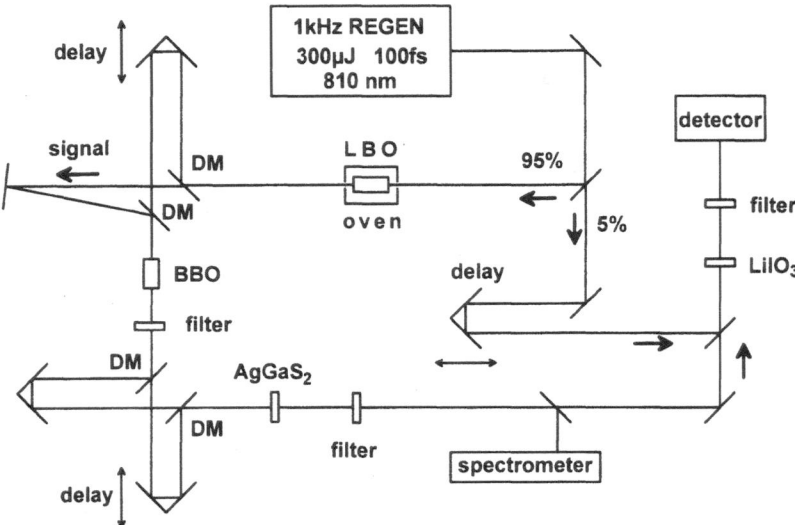

Figure 1. Experimental setup for the generation and characterization of femtosecond pulses in the mid-infrared spectral range (3 - 10 mm).

nm. The MIR radiation is generated by difference frequency mixing between the signal and idler pulses in a 2 mm long nonlinear AgGaS$_2$ crystal. A setup of two independent delay lines for the signal and idler pulses allows the correct arrangement of the input polarizations and the optimization of the temporal and spatial overlap in the AgGaS$_2$ crystal. With 1.5 µJ signal pulses we obtain MIR radiation up to 50 nJ with energy fluctuations of less than 10 %. The MIR pulses are tunable over a wide spectral range from 3.3 to 10 µm. A detailed characterization of the MIR pulses is presented in Fig. 2. The spectrum of pulses centered at 4.5 µm has a FWHM of 100 cm^{-1}. The MIR pulse duration was directly measured by upconversion with the 800 nm pump pulses in a thin LiIO$_3$ crystal. The corresponding cross-correlation trace is shown on linear and logarithmic scale s in the lower part of Fig. 2. Deconvolution yields a pulse duration of the nearly transform limited pulses as short as 130 fs.

Using the femtosecond MIR pulses discussed in the previous section, we investigated the ultrafast inter- and intra-subband dynamics of electrons in quasi-two-dimensional semiconductor structures. In our experiments, n-type modulation-doped Ga$_{0.47}$In$_{0.53}$As/Al$_{0.48}$In$_{0.52}$As multi-quantum well (QW) structures of various well widths and doping concentrations were studied. In this paper, we present data for a sample consisting of 50 GaInAs QWs of 6 nm well width separated by 14 nm thick AlInAs barriers the center of which is delta-doped with Si atoms of a concentration of 1.5×10^{12} cm^{-2}. In addition, the sample has a buffer layer of 20 intrinsic MQWs with the same structure as the modulation-doped layer packet. The stationary MIR and NIR absorption spectra of the sample are shown in Fig. 3. At low lattice temperatures (T_L = 8 K) the electrons present by doping are transferred to the n=1 subband and, consequently, the sample exhibits inter-subband absorption due to transitions to the n=2 subband. The corresponding absorption band measured with Brewster geometry is depicted in Fig. 3 (a). The electron distributions in the various subbands can be monitored via the interband absorption from the n=1 and/or n=2 valence to the conduction band the spectrum of which is plotted in Fig. 3 (b).

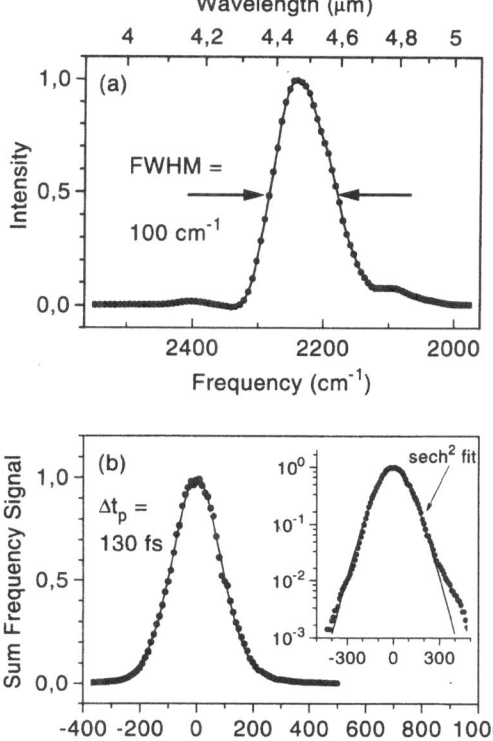

Figure 2. Characteristics of the femtosecond MIR pulses. (a) Spectrum of pulses at 4.5 μm with a bandwidth of 100 cm^{-1} (FWHM). Cross-correlation trace between the MIR pulses and pulses at 800 nm plotted on linear and logarithmic scales. Deconvolution yields a MIR pulse duration of 130 fs.

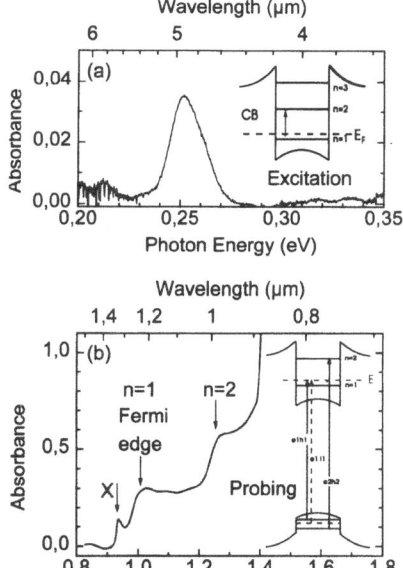

Figure 3. Stationary absorption spectra at TL = 8 K of the n-type modulation-doped GaInAs/AlInAs multi-quatum wells with a well width of 6 nm and a doping concentration of 1.5 × 10^{12} cm^{-2}. (a) MIR absorption due to transitions from the the n=1 to the n=2 subband measured with Brewster-geometry. (b) NIR absorption due to n=1 and n=2 valence to conduction band transitions. Insert: Femtosecond pump-probe scheme with n=1 to n=2 inter-subband excitation and subsequent probing of the n=1 or n=2 interband absorption.

The intrinsic layer packet shows an excitonic absorption peak at 0.935 eV. The interband absorption edge of the 50 modulation-doped QWs is shifted towards higher energies due to bandfilling in the n=1 conduction subband. The n=2 interband absorption arises at photon energies around 1.27 eV. The pump-probe scheme of our time-resolved experiments is explained in the insert of Fig. 3. Mid-infrared pulses resonant to the inter-subband transition (c.f. Fig. 3 a) promote electrons from the n=1 to the n=2 subband. The transient electron distributions in the n=1 and n=2 subbands are monitored via changes of the corresponding valence to conduction band transitions. Synchronized probe pulses of 100 fs duration which are tunable over a wide range in the near infrared are gained from a femtosecond white light continuum.

In a first time-resolved experiment the inter-subband scattering of n=2 electrons was monitored via the transient bleaching of the n=2 interband absorption. From the transient data to be discussed elsewere we derive a lifetime of 1 ps for the electrons in the n=2 subband. This value is in good agreement with theoretical work which considers intersubband scattering by emission of LO phonons [7].

The transient distribution of n=1 electrons occuring during and after back-scattering from the n=2 subband was investigated in time-resolved experiments at various probe frequencies around the Fermi edge. In Fig. 4 the transient change of transmission $\Delta T/T_0$ after MIR excitation is plotted vs the delay time between pump and probe pulses. For probe frequencies below the Fermi edge (0.93 to 0.985 eV) we observe first an instantaneous absorption increase which is caused by the depletion of n=1 states during the excitation process. Subsequently, the induced absorption rises further due to the strong heating of the initially cold n=1 distribution by carrier-carrier scattering with the back-scattered high energy electrons. The signal reaches its maximum at delay times around 1.2 ps and eventually decays on a time scale of 50 ps due to cooling of the hot quasi-equilibrium distribution by emission of optical phonons. The high energy tail of the n=1 electron distribution (probe frequencies 0.985 to 1.04 eV) shows a similar temporal behaviour but with the opposite sign of the signal, i.e. bleaching of the interband absorption. A closer inspection of the transient data shows, however, that the rise time of the bleaching signal is distinctly longer than its counterpart below the Fermi edge. The signal reaches its maximum at delay times around 2.1 ps and eventually decays in a similar fashion as below the

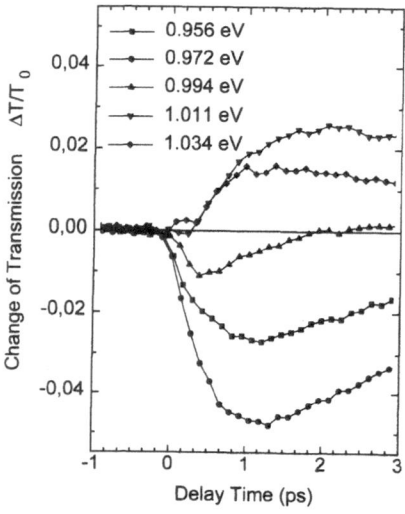

Figure 4. Transient change of transmission $\Delta T/T_0$ of the n=1 interband absorption after excitation of electrons from the n=1 to the n=2 subband with femtosecond MIR pulses. The signal at various probe frequencies (0.93 - 1.04 eV) directly reflects the transient electron distribution in the n=1 subband. The distinct kinetics above and below the initial Fermi level at 0.985 eV points to non-equilibrium distributions for early delay times $t_D < 2$ ps.

Fermi edge. This difference in the kinetics at early delay times of the transient signal below and above the initial Fermi level (i.e. for T = 8 K) directly points to non-equilibrium electron distributions in the n=1 subband during the heating period (t_D < 2 ps). In comparison to hot Fermi-like distributions, the transient n=1 distribution has a smaller population in the high energy tail and a larger below the initial Fermi level during the thermalization process. This relatively slow thermalization is in contrast to femtosecond hole burning experiments in n-type GaAs/AlGaAs in which a thermalization time of less than 60 fs after inter-band excitation was estimated [8] . The slow equilibration found in the present experiments is due to strong Pauli blocking and screening in the cold sea of n=1 electrons.

REFERENCES

1. S. Hunsche, K. Leo, H. Kurz, Phys. Rev. B 50 (1994) 5791
2. T. Elsaesser, J. Shah, L. Rota, P. Lugli, Phys. Rev. Lett. 66 (1991) 1757
3. M. Woerner, W. Frey, M. T Portella, C. Ludwig, T. Elsaesser, W. Kaiser, Phys. Rev. B 49 (1994) 17007
4. A. Lohner, M. Woerner, T. Elsaesser, W. Kaiser, Phys. Rev. Lett. 68 (1992) 3920
5. F. Seifert, V. Petrov, M. Woerner, Opt. Lett. 19 (1994) 2009
6. V. Petrov, F. Seifert, F. Noack, Appl. Phys. Lett. 65 (1994) 268
7. B. K. Ridley, Phys. Rev. B 39 (1989) 5282
8. W.H. Knox, D.S. Chemla, G. Livescu, J.E. Cunningham, and J.E. Henry, Phys. Rev. Lett. 61, 1290 (1988)

58

ULTRAFAST RECOVERY OF TRANSIENT CHANGES IN THE EXCITONIC OPTICAL PROPERTIES OF InGaAs/InP QUANTUM WELLS

D. Campi,[1] C. Coriasso,[1] M. Nisoli,[2] and S. De Silvestri[2]

[1] CSELT – Centro Studi e Laboratori Telecomunicazioni
via Reiss Romoli 274
I-10148 Torino, Italy
[2] Politecnico di Milano
piazza Leonardo da Vinci 32
I-20133 Milano, Italy

InGaAs/InP heterostructures posses strong technological relevance, particularly in view of integration of waveguide devices for interconnections and telecommunication applications. This contribution concerns the optical properties of quantum-well (QW) structures in the 1.5 μm range and the modification of these properties that can be induced by photogeneration of excitons. Beside some work already done, knowledge of those properties is far from being complete; particularly, the relative strength of optical nonlinearities induced by excitons or carriers is not properly understood. To clarify this subject we address the dynamical and spectral aspects of nonlinear behavior upon ultra-fast excitation of InGaAs/InP QWs, devoting special attention to the role of the so-called collisional broadening of the exciton peak. Experiments are done both at the exciton resonance lowest in energy and at below-resonant energies, that is in the region of strongest relevance to waveguide devices.

In general, the optical properties of semiconductors, particularly QWs, depend on the carrier density inside the semiconductor itself: this causes the optical properties of QWs to depend crucially on the intensity of an incoming light beam that photogenerates carriers and, thus, to behave non linearly.

In particular, 1), the excitonic absorption is *quenched* in the region of the peak under pumping, while, 2) the absorption coefficient is *raised* in the spectral region below the resonance. The former is due to occupation of states by the photogenerated pairs, while the latter is due chiefly to wavefunction renormalization effects which enlarge the cross-section offered by the excitons to lifetime-limiting events, thereby increasing the broadening of the peak[1]. Much of the current literature overlooks the occurrence of this collisional broadening, which is in fact the most effective one in making the exciton peak disappear under photogeneration.

Ultrafast Processes in Spectroscopy, edited by Svelto et al.
Plenum Press, New York, 1996

255

Using a widely-tunable laser source based on an optical parametric amplifier pumped by a femtosecond Ti:Sapphire laser[2,3], we performed pump-probe measurements of the dynamics of the nonlinear transmittance[4] on an InGaAs/InP multiple-QW sample grown by chemical beam epitaxy. The sample consisted of 60 period of InGaAs/InP 7.7 / 7.0 nm in width, the photon energy range addressed was around 0.8 eV and the experiments were done at room temperature. In particular, we investigated the transient transmittance variations with both pump and probe tuned at the n=1 heavy-hole exciton peak, or at energies slightly below. The observed dynamics corresponds to instantaneous generation of n=1 heavy-hole excitons, which transform, with a 150 fs time constant, into free n=1 e / h (unbound) pairs, decaying on much longer time scale. Consistently with this view, we observe a strong bleaching produced by (cold) transient excitons generated by the pump pulse inhibiting formation of further excitons by the probe pulse, which decays in time towards a plateau, and settles finally to the same value as it would in a stationary pump-probe experiment[3,4].

In Fig 2a) we also show the result of a stationary pump-probe experiment, in which carriers are directly photogenerated.

The ratio of the amplitude of the short-lived response to the long-lived plateau was found to increase with decreasing excitation density, as shown in Fig. 1. In particular, at moderate absorbed photon densities (about 4×10^{10} cm^{-2}) and at photon energies 8 meV below the exciton peak the long-lived plateau can be considered almost negligible. In addition, we generally observed that the relative strength of exciton versus carrier populations to the nonlinear transmittance is less sensitive to pump intensity variations for photon energies approaching the exciton peak.

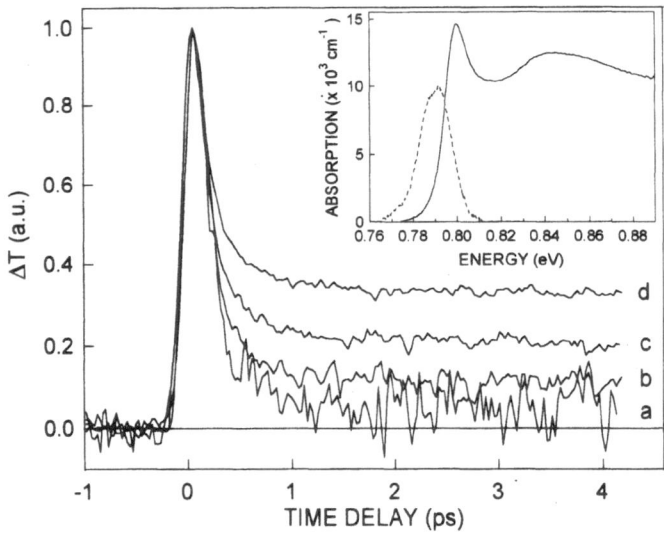

Figure 1. Transient differential transmittance ΔT measured at 8 meV pump detuning from the exciton peak as a function of probe time-delay, at varying absorbed photon density (per layer); a: 4×10^{10} cm^{-2}, b: 6.6×10^{10} cm^{-2}, c: 1.9×10^{11} cm^{-2}, d: 2.7×10^{11} cm^{-2}. Inset: room-temperature linear absorption spectrum (solid line) and spectral shape of the laser pulses. In evaluating the absorption coefficient, the total thickness of the well material (0.46 μm) was used.

We argue that the exciton versus carriers relative efficiency cannot be justified basing solely on the reduction of the oscillator strength as done in previous works; in particular we argue that collision-broadening effects, sampled by a probe of finite spectral width, become crucial in determining the relative effectiveness on the quenching of exciton absorption, as it is discussed below. In order to clarify the observed behavior, we calculated the absorption spectra for our multiple-QW system in the presence of either a exciton gas or a carrier plasma, and at varying population density. These two cases represent our limits for pump-probe delays tending towards zero and towards very long delays, respectively. Thus, we performed a many-body calculation for the optical response of an e-h plasma interacting through a screened Coulomb potential[5], where we treated the screening in a random-phase, single-plasmon pole approximation. We accounted for screened-exchange and Coulomb-hole proper self-energy contributions to the energy-gap, and distributed the carrier according to quasi-Fermi functions. This calculation yields a description of the very-long delay limit of our time evolution. In parallel, we performed a calculation for a gas of cold excitons (or, rather, exciton *constituents*) that is bound e-h pairs interacting through an unscreened Coulomb potential, actually a Hartree-Fock limit. The energy gap was renormalized through the exchange proper self energy, and the exciton constituents were distributed according to ground-state exciton wavefunctions. This yields a description of the vanishing-delay limit of our time evolution.

To obtain the optical response in the presence of Coulomb interaction, we solved the Bethe-Salpeter equation for the two-particle Green's functions in both cases. This allowed us to calculate the transmittance of our probe pulse in both cases, averaging over the finite spectral width of the pulse which we measured to be 17 meV (FWHM), see inset to Fig. 1. We note that the reduction of the oscillator strength of the n=1 exciton was found of comparable efficiency in the two limiting cases, although somewhat larger for the case of a cold exciton population (see Fig.2). The reduction can be attributed to phase-space filling, present in both cases[6]. However, we notice that the carrier-induced differential absorption changes its sign in the low energy region , as it is shown in Fig. 2a). This may be easily traced back to the density-dependent broadening of the exciton absorption peak[7] occuring under pumping. Finally, in Fig. 3), we take the ratio of the transmittance variations in-

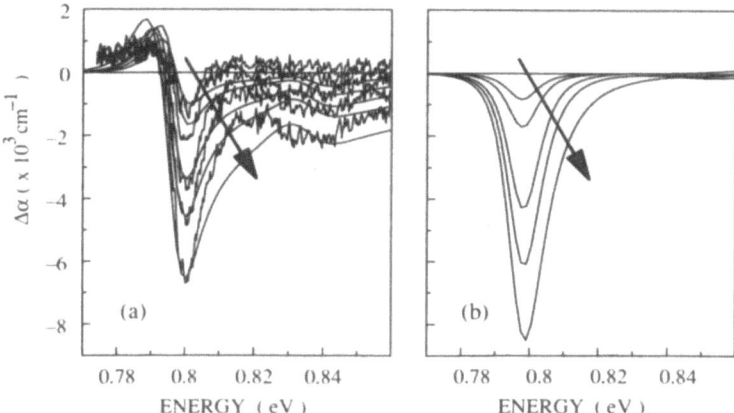

Figure 2. Calculated spectral changes induced on the absorption coefficient α of our structure at varying e-h, (a), or exciton, (b), densities. Density grows in the direction indicated by the arrows, taking the values: 7.3×10^9 cm^{-2}, 1.5×10^{10} cm^{-2}, 4.4×10^{10} cm^{-2}, 7.3×10^{10} cm^{-2}, 1.5×10^{11} cm^{-2}.

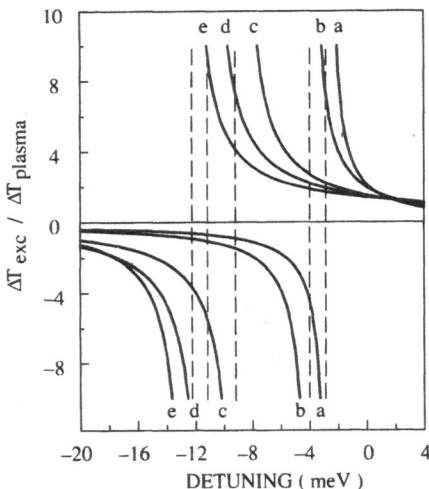

Figure 3. Calculated relative strengths of transmittance changes induced by a exciton gas or by a e-h plasma with the same density, versus the detuning on the energy axis with respect to the exciton peak. The curves are given at fixed carrier or exciton density: a) 7.3×10^9 cm^{-2}, b) 1.5×10^{10} cm-2, c) 4.4×10^{10} cm^{-2}, d) 7.3×10^{10} cm^{-2}, e) 1.5×10^{11} cm^{-2}.

duced by the exciton-gas or carrier-plasma effects calculated as in the above. We observed that the ratio can even diverge, signifying complete recovery of the nonlinear transmittance variations on ultrafast time scale, with no residual, long-lived plasma plateau. The divergence comes closer and closer to the exciton peak as the dilute limit is attained. This is because conditions there exist in which the variations induced by the free e-h pairs average to zero over the probe pulse, while the excitons always provide variations implying a reduction of the excitonic absorption coefficient.

Further detail about both experiment and theory are offered in Ref. 8.

REFERENCES

1. D.R. Wake, H.W. Yoon, J.P. Wolfe, H. Morkoc, Phys. Rev. **B46**, 13452 (1992).
2. M.Nisoli, S.De Silvestri, V.Magni, O.Svelto, R.Danielius, A.Piskarskas, G.Valiulis, A.Varanavicius Opt. Lett. **19**, 1973 (1994).
3. M. Nisoli, S. De Silvestri, O. Svelto, C.Campi, C.Coriasso, A. Varanavicius, *This volume.*
4. M.Nisoli, V.Magni, S. De Silvestri, O.Svelto, D.Campi, C.Coriasso, Appl. Phys. Lett. **66**, 227 (1994).
5. S.Schmtt-Rink, C.Ell, H.Haug Phys. Rev. **B33**, 1183 (1986).
6. S.Schmitt-Rink, D.S.Chemla, D.A.B.Miller, Phys. Rev. **B32**, 6601 (1985).
7. S.Hunsche, K.Leo, H.Kurz, K.Koler, Phys. Rev. **B49**, 16565 (1994).
8. D.Campi, C.Coriasso, M.Nisoli, S.De Silvestri, Appl. Phys. Lett. **67**, 953 (1995).

59

RECOMBINATION DYNAMICS OF NEAR-BANDEDGE EMISSION IN CUBIC GaN

R. Klann, O. Brandt, H. Yang, H. T. Grahn, and K. H. Ploog

Paul-Drude-Institut für Festkörperelektronik
Hausvogteiplatz 5–7, 10117 Berlin, Germany

The recent realization of GaN light-emitting diodes has motivated the quest for a GaN-based laser[1,2]. Compared to the commonly investigated hexagonal structure, the cubic phase of GaN offers several significant advantages, in particular when grown on GaAs: For example cleavage is readily obtained, which is important for the fabrication of a laser resonator. An essential prerequisite for determining the actual potential of this material for laser devices is the quantitative understanding of the processes dominating carrier relaxation and recombination.

We investigated the time-resolved photoluminescence of high-quality cubic GaN on (001) GaAs. The sample exhibits intense and narrow photoluminescence (PL) lines spectrally located between 3.0 and 3.27 eV. In this paper we will focus on the highest energy line at 3.27 eV. At 5 K, the decay of this line is bi-exponential with an ultrafast initial time constant of 20 to 50 ps and a second, slower component with a time constant between 100 and 400 ps.

The samples under investigation are grown by solid-source molecular beam epitaxy, employing a DC plasma discharge source for dissociating molecular N_2 into activated nitrogen species. GaAs(001) is used as substrate. The growth conditions and the structural configuration of the sample investigated are reported in detail in Ref. 3. Time-resolved PL measurements are performed using a femtosecond Ti:sapphire laser and a syncroscan streak-camera system. The 150 fs pulses with a repetition rate of 76 MHz are frequency-doubled allowing to tune the excitation from 3.10 eV to 3.40 eV (full width at half maximum 20 meV). If not stated otherwise, the excitation power is set to 60 µW, which corresponds to a peak power density of 5 kW/cm^2 (fluence 7.5 nJ/cm^2). The luminescence is dispersed by a 22 cm monochromator using a 1200 lines/mm grating and focused onto the photocathode of the streak tube. The spectral resolution amounts to 0.4 nm, the temporal resolution is 2 ps. The samples are mounted on the cold-finger of a He flow cryostat and kept at 5 K.

Fig. 1 shows the cw PL spectrum of the sample under consideration. The spectrum is dominated by an asymmetric sharp line at 3.27 eV.

Ultrafast Processes in Spectroscopy, edited by Svelto et al.
Plenum Press, New York, 1996

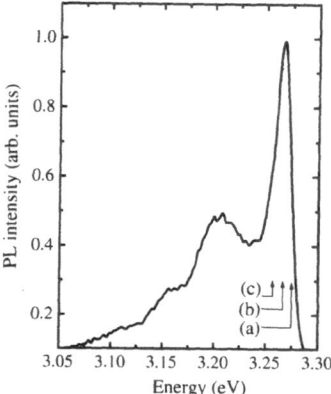

Figure 1. cw PL spectrum at 5 K with E_{ex} = 3.36 eV. The arrows labeled (a), (b), and (c) indicate the energy positions of the PL transients in Fig. 2.

The weaker lines on the low energy side are apparently impurity-related transitions. They exhibit a delayed rise of several hundreds of picoseconds and decay on a timescale of several tens of nanoseconds. In the following we will focus on the peak highest in energy. The picosecond dynamics of this peak is visualized in Fig. 2.

The energy positions of the three transients for each sample are marked by the arrows (a), (b), and (c) in Fig. 1. For all spectral positions the PL rise time is below 5 ps, i.e., of the order of or less than our time resolution. The PL decay behavior is bi-exponential with a first, fast time constant τ_F, which varies between 20 and 50 ps depending on the spectral position. This decay component is faster on the high-energy side [(transient (a)] compared to transients detected at lower-energies [(transient (c)]. The transients lower in energy (b) and (c) exhibit a delayed onset of the decay compared to transient (a) which is detected at higher energies. The second, slower time constant τ_S ranges from 100 to 400 ps. The spectral evolution during the first 30 ps at 5 K is plotted in Fig. 3. At very early times the line is broad, particularly on the high-energy side.

With increasing time delay, the spectrum becomes steeper on the high-energy side and the PL maximum shifts about 5 meV to lower energies.

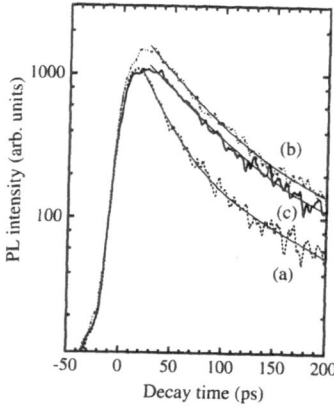

Figure 2. Transient PL decay at 5 K during the first 200 ps after excitation with E_{ex} = 3.36 eV. The three energy positions (a) (dotted line), (b) (dashed-dotted line), and (c) (full line) are indicated in Fig. 1 (spectrally averaged over 5 meV). The solid lines are least-square fits to the data using a bi-exponential.

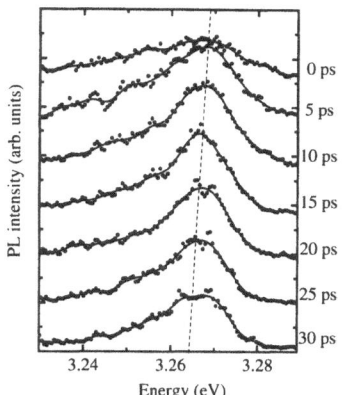

Figure 3. Transient PL spectra at 5 K for different decay times as indicated (spectral average over 2 ps). The spectra have been shifted vertically. The lines are a guide to the eye.

The transient PL spectrum at 5 K averaged over the first 100 ps is shown in Fig. 4 for an excitation energy of 3.36 eV (solid line). The fast time constant τ_F versus PL energy is denoted by solid circles in Fig. 4. The time constant, being as short as 20 ps at the higher energy side, increases up to 50 ps at the lower energy side. It is important to note that the *spectrally integrated intensity* of the PL peak in Fig. 4 increases strictly linearly with excitation power up to excitation densities of 10^{19} cm^{-3}.[3] This linear dependence demonstrate that, at 5 K no nonradiative decay channels participate in the decay process in this power range[4].

The temperature dependence of the PL transients at 3.27 eV is shown in Fig. 5.

The fast initial PL decay becomes even faster with increasing temperatures and, simultaneously, the PL intensity drops. Above 100 K the decay time is below our time-resolution. This behavior indicates that nonradiative decay channels are rapidly activated with increasing temperature.

The linear dependence of the PL intensity on the excitation density up to carrier concentrations of 10^{19} cm^{-3} is consistent with both, band-to-band and excitonic transitions,

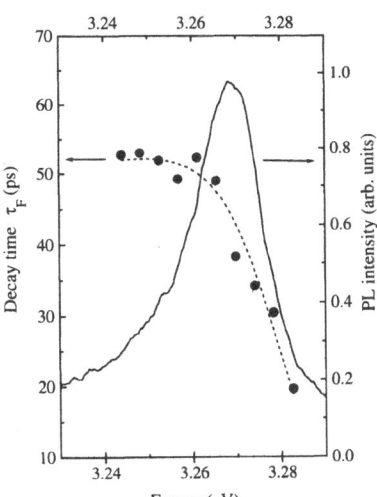

Figure 4. Time-resolved PL spectrum at 5 K with $E_{ex} = 3.36$ eV averaged over 100 ps after excitation (solid line). The fast decay time t_F at different energy positions of the PL line is indicated by solid circles. The dashed line is a guide to the eye.

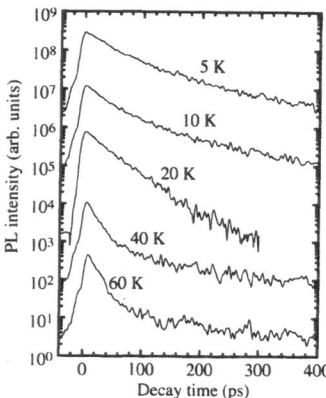

Figure 5. PL decay of the PL peak at 3.27 eV for different temperatures. The transients are integrated over 20 meV and shifted vertically.

but not with impurity-related transitions. The observed ultrafast bi-exponential decay does not depend on the excitation density and thus cannot arise from band-to-band transitions, but must be excitonic (note the low excitation density of 10^{17} cm^{-3} used for the experiments shown above). This conclusion is furthermore supported by the temperature dependence of the spectral position of the PL line, which follows the temperature behavior for the band gap of cubic GaN as reported by Ramírez-Flores et al.[5] with a constant offset of 25 meV up to room temperature. Because of its spectral width, the PL line presumably consists of a spectrally unresolved superposition of the radiative transitions of both, free and bound excitons. Only the pronounced broadening of the spectra at t = 0 ps (Fig. 3) may arise from the onset of electron-hole plasma emission (the excess energy per excitation photon is about 60 meV). We attribute the initial fast decay component (τ_F) to the free exciton decay rate, which is composed of both, the radiative decay and the relaxation of the excitons towards localized states. The second, slower decay component (τ_S) is attributed to the radiative decay of these localized exciton states.

At present, we cannot decide whether the localizing states are provided by native defects, impurities or crystal disorder. However, similar results observed in various II-VI and III-V semiconductors have been consistently interpreted by disorder-induced localization of excitons[6]. In fact, the spectral red-shift within the first 30 ps (see Fig. 3), the dependence of the PL transients on the temperature and on the energy position within the PL line as well as the delayed onset of the PL decay for transient (c) in Fig. 2 are all consistent with this picture.[7,8] Furthermore, the radiative lifetime for both, bound and localized excitons in GaN is estimated to be on the order of 400 ps[9], which is in close agreement with the slow decay time τ_S. Finally, an Arrhenius-plot of the fast decay time τ_F yields a localization energy of 3 meV for the excitons.

In conclusion, we have investigated the picosecond dynamics of the PL line highest in energy in cubic GaN. The PL signal, which rises within our time resolution, shows a bi-exponential decay behavior. The fast PL decay component of 20 to 50 ps is attributed to the radiative decay and relaxation of free excitons. The slower PL decay component of 100 to 400 ps is attributed to the recombination of localized excitons. Significant nonradiative contributions are activated already at quite low temperatures ($\approx$ 30 K).

ACKNOWLEDGMENT

We acknowledge discussions with U. Jahn, J. Menniger, and J. Müllhäuser. This work was supported in part by the Bundesministerium für Bildung und Forschung of the Federal Republic of Germany.

REFERENCES

1. S. Nakamura, T. Mukai, and M. Senoh, Appl. Phys. Lett. **64**, 1687 (1994).
2. S. Nakamura, T. Mukai, and M. Senoh, J. Appl. Phys. **76**, 8189 (1994); S.D. Lester, F.A. Ponce, M.G. Craford, and D.A. Steigerwald, Appl. Phys. Lett. **66**, 1249 (1995).
3. R. Klann, O. Brandt, H. Yang, H.T. Grahn, and K. Ploog, Phys. Rev. B **52**, R11615 (1995).
4. T. Schmidt, K. Lischka, and W. Zulehner. Phys. Rev. B **45**, 8989 (1992).
5. G. Ramírez-Flores, H. Navarro-Contreras, A. Lastras-Martinez, R.C. Powell, and J.E. Greene, Phys. Rev. B. **50**, 8433 (1994).
6. J.A. Kash, A. Ron, and E. Cohen, Phys. Rev. B **28**, 6147 (1983); J.A. Kash, Phys. Rev. B **29**, 7069 (1984).
7. E.F. Schubert and W.T. Tsang, Phys. Rev. B **34**, 2991 (1986).
8. S. Shevel, R. Fischer, E.O. Göbel, G. Noll, P. Thomas, and C. Klingshirn, J. Lumin. **37**, 45 (1987).
9. É.I. Rashba, Fiz. Tekh. Poluprovodn. **8**, 1241 (1974) [Sov. Phys. Semicond. 8, 807 (1975)]; C.J. Hwang, and L.R. Dawson, Sol. Stat. Commun. **10**, 443 (1972).

60

PHONON OSCILLATIONS IN A SPECTRUM OF REVERSIBLE BLEACHING OF GALLIUM ARSENIDE UNDER INTERBAND ABSORPTION OF A HIGH-POWER PICOSECOND LIGHT PULSE

I. L. Bronevoi,[1] A. N. Krivonosov,[1] and V. I. Perel'[2]

[1] Institute of Radioengineering and Electronics
Russian Academy of Sciences
103907, Moscow, GSP-3, Mokhovaya St., 11, Russia
[2] A.F.Ioffe Physicotechnical Institute
Russian Academy of Sciences
194021, St.Petersburg, Politekhnicheskaya St., 26, Russia

In this paper we describe the results of studies with picosecond resolution of the bleaching (enhancement of the optical transparency T) spectra of a thin layer of a high-purity GaAs exposed to a high-power light pulse having the photon energy $\hbar\omega_{ex}$, which is somewhat larger than the forbidden gap width E_g. The bleaching spectra represents the energy distribution of photoexcited charge carriers. Experiments were carried out at room temperature. The duration of both the exciting and probing pulses was about 14 ps. According to our studies[1,2] the alteration of the bleaching with the variation of the excitation intensity acquire essentially the reversible character in the spectral range $\hbar\omega > E_g^0$ (E_g^0 being the width of the band gap of a nonexcited sample), Fig.1(a). Results of the sample bleaching measurements as a function of the energy of a photon of the probing pulse $\hbar\omega_p$ nearly at the moment of the maximum of the exciting pulse intensity are shown in Fig.1(b). In this spectrum, which has been measured within the $\hbar\omega_p < \hbar\omega_{ex}$ range, we found local minima.[3] An oscillating character of the bleaching has been observed for a variety of values of $\hbar\omega_{ex}$ and the exciting pulse integral energy W_{ex}, the spectral localization of minima being retained. However at the drop of excitation and after its cessation the oscillations smoothened and then disappeared (Fig.2(a)) des-pite the bleaching degree being nearly equal to that shown in Fig.1(b) (Fig.2(a) also shows the experimental dependence $A = \lg[T^0(\hbar\omega_p = 1.379 \text{ eV}) / T^0(\hbar\omega_p)]$ which describes the spectrum of optical density of non-excited sample $A_0 \approx A \ln 10$). Oscillations have also been far less strong at $\hbar\omega_p > \hbar\omega_{ex}$, Fig.2(b). This behaviour of oscillations, as well as the fact that there is an antireflective coating on the sample surface, do not permit to attribute the oscillations to the interference in the sample.

Ultrafast Processes in Spectroscopy, edited by Svelto et al.
Plenum Press, New York, 1996

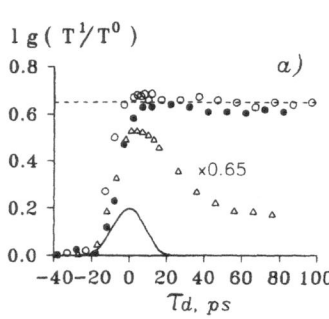
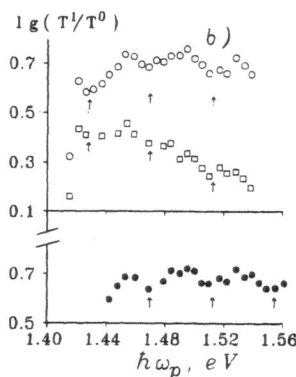

Figure 1. Transparency T variation (indices 1 and 0 show the presence and absence of the excitation, respectively) of GaAs as a function of the delay time τ_d of the probing (p) pulse with respect to the excitation one at $\hbar\omega_{ex}$=1.558 eV: $\hbar\omega_p$=1.42 eV, o -W_{ex} =1.0 a.u., -W_{ex} = 0.19 a.u.; Δ-$\hbar\omega_p$ =1.512 eV, W_{ex}=1.0 a.u, the dotted line shows the level of total bleaching at $\hbar\omega_p$ =1.42 eV, the solid line shows a cross-correlation function of exciting and probing pul-ses (a); as a function of the energy of a photon of the probing pulse at τ_d = -3 ps: $\hbar\omega_{ex}$ = 1.558 eV, o-W_{ex} =1.0 a.u., $\square$-W_{ex} = 0.19 a.u.; $\bullet$ -$\hbar\omega_{ex}$ =1.579 eV, W_{ex} =1.0 a.u. (b).

It should be noted that during the excitation pulse there appears a recombinative edge superluminescence[4,5] from the sample, Fig.3. Recombination should result in the appearance of a carrier flow in the energy space (partially due to emitting by electrons optical phonons having the energy $\hbar\Omega_0$) towards the level at which the carrier recombination occurs. The recombination must lead to a dip on the curve of electron distribution at a certain energy E_0 near the conduction band bottom. This dip is inevitably followed by another dip when $E_1=E_0+ \hbar\Omega_0$ since the energy transitions from E_1 to E_0 under phonon emission are more frequent than those in the opposite direction occurring with phonon absorption (in the absence of the dip at E_0 the transition rates should have been equal) and so on. This must lead to minima in the bleaching spectrum that are spaced by energy inter-

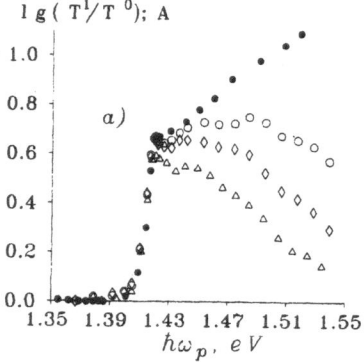
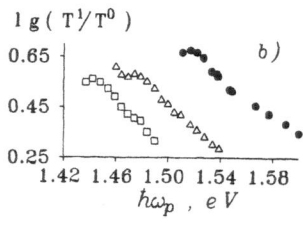

Figure 2. GaAs transparency variation as a function of the energy of a photon of the probing pulse at W_{ex}=1.0 a.u.: a) $\hbar\omega_{ex}$ =1.558 eV, o -τ_d = 13 ps, $\lozenge$ -τ_d = 31 ps, Δ -τ_d = 80 ps, the de-pendence A = f($\hbar\omega_p$) - $\bullet$ (see explanations in the text); b) $\square$ - $\hbar\omega_{ex}$ = 1.415 eV, τ_d = 5 ps; Δ - $\hbar\omega_{ex}$ =1.439 eV, τ_d = 5 ps; $\bullet$ - $\hbar\omega_{ex}$ =1.492 eV, τ_d = 0.

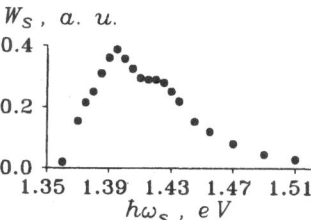

Figure 3. Integral spectrum of radiation propagating at the angle Θ = 75^{0} to the normal to sample surface at $\hbar\omega_{ex}$ =1.558 eV, W_{ex}=1.0 a.u. (radiation absorption within the non-excited range does not make it possible, regrettably, to observe radiation that propagates along the GaAs layer).

vals $\hbar\Omega_{0}$ $(1+m_{e}/m_{h})$, where m_{e} and m_{h} are the masses of electrons and holes, respectively. In Fig.1(b), the arrows indicate the positions of the supposed minima spaced by 42 meV. It is seen that they all agree well with minima observed experimentally thus proving the model suggested.

In the range of energies in which recombination takes place the absorption spectra are presented more clear in Fig.4. These spectra have been obtained from comparison of bleaching spectra with the experimental dependence A = $f(\hbar\omega_{p})$ given in Fig.2(a). It is seen that as long as the residual bleaching has not been reached, in the spectral range $\hbar\omega <$ E_{g}^{0} a small amplification appears. This amplification should result in arising the recombination superluminescence, that develops, preferentially, along the GaAs layer.[4,5] The recombination superlu-minescence must lead to a deviation of the carrier distribution from the quasiequilibrium one thus causing the distribution depletion in the spectral range of the amplification. This should to distort the absorption spectrum shape as compared to the "smooth" spectrum peculiar to quasiequilibrium distribution, which, apparently, explains in this way the appearance of a local absorption maximum at $\hbar\omega \approx$ 1.428 eV that corresponds to the above mentioned electron distribution dip reflecting in phonon oscillations.

ACKNOWLEDGMENTS

The research described in this publication was made possible in part by Grant No. 95–02–05871 from the Russian Foundation of Fundamental Research and Grant No. M3S300 from the International Science Foundation.

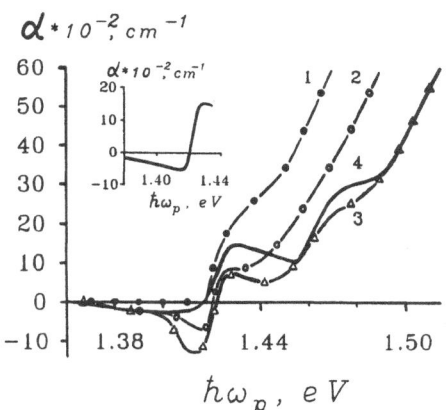

Figure 4. The dependence of the absorption coefficient α of the sample on $\hbar\omega_{p}$ at $\hbar\omega_{ex}$ = 1.558 eV, W_{ex} =1.0 a.u.: 1 -τ_{d} = 80 ps; 2 -τ_{d} = 31 ps; 3 -τ_{d} = 13 ps; 4 - τ_{d} = -3 ps (the insert gives the same dependence 4 but with regard of a correction to a finite duration of the probing pulse).

REFERENCES

1. I.L.Bronevoi et al., JETP Lett. **42**, 395 (1985); **43**, 473 (1986).
2. I.L.Bronevoi , A.N.Krivonosov, and T.A.Nalet, Solid State Commun. to be published.
3. I.L.Bronevoi , A.N.Krivonosov, and V.I.Perel`, Solid State Commun. **94**, 805 (1995).
4. N.N.Ageeva, I.L.Bronevoi et al., JETP Lett. **48**, 276 (1988); Solid State Commun. **72**, 625 (1989).
5. J.Dubard, J.L.Oudar, F.Alexandre, D.Hulin, A.Orzag, Appl. Phys. Lett. **50**, 821 (1987).

61

FEMTOSECOND REFLECTION OF BULK GaAs AND GaAs/AlGaAs MULTIPLE QUANTUM WELLS

W. Li, W. Peng, Z. Qiu, and Z. Yu

Institute for Laser and Spectroscopy
Zhongshan University
Guangzhou 510275, P. R. China

We report room temperature femtosecond time-resolved reflection studies on the carrier dynamics of a bulk GaAs and a GaAs/AlGaAs MQW samples. The experiment was based on a home-made SML Ti:Sapphire laser which delivers 82fs pulses at a repetition rate of 94MHz. Conventional two-beam pump-probe configuration with orthogonal pump-probe polarization geometry was used in the measurements. The reflection change induced on the surface of the sample by the pump pulses was monitored by the time-delayed probe pulses whose reflected intensity was detected by a lock-in amplifier working at differential mode. The samples measured were a bulk GaAs and a GaAs/AlGaAs MQW which consists of five periods of $Al_{0.3}Ga_{0.7}As(1000\text{Å})/GaAs(100\text{Å})$ sandwiched between a transparent AlGaAs cladding layer and transparent AlGaAs buffer layer grown on a GaAs substrate.

Normalized reflection changes of the bulk GaAs and GaAs/AlGaAs MQW are shown in Fig. 1 and Fig. 2 rspectively as functions of the time delay between the pump and probe pulses which were both of 1.52eV photon energy. Excitation by the 1.52eV photons avoids the contributions from intervalley scatterings and the relaxation of carriers in the barriers of the MQW sample.

The positive reflection change observed are mainly due to the increase of refractive index induced by the pump pulses[1]. This increase of refractive index can be understood by applying the concept of saturation of anomalous dispersion, which is related to the saturation of absorption by the so-called k-k relation.

The time-dependent reflection change of bulk GaAs consists at least two components, i.e., a long component of hundreds of picoseconds and a subpicosecond component (Fig.1(a)). The long picosecond component has been known as the effect of the filling of the states near the bottom of the conduction band. One can see that the relaxation of the subpicosecond component depends strongly on the carrier density. The relaxation time is shortened by approximately a factor of 10 when the carrier density is decreased from $9.6\times10^{17}cm^{-3}$ to $1.4\times10^{17}cm^{-3}$. This effect has been known for some years and has been

Ultrafast Processes in Spectroscopy, edited by Svelto et al.
Plenum Press, New York, 1996

269

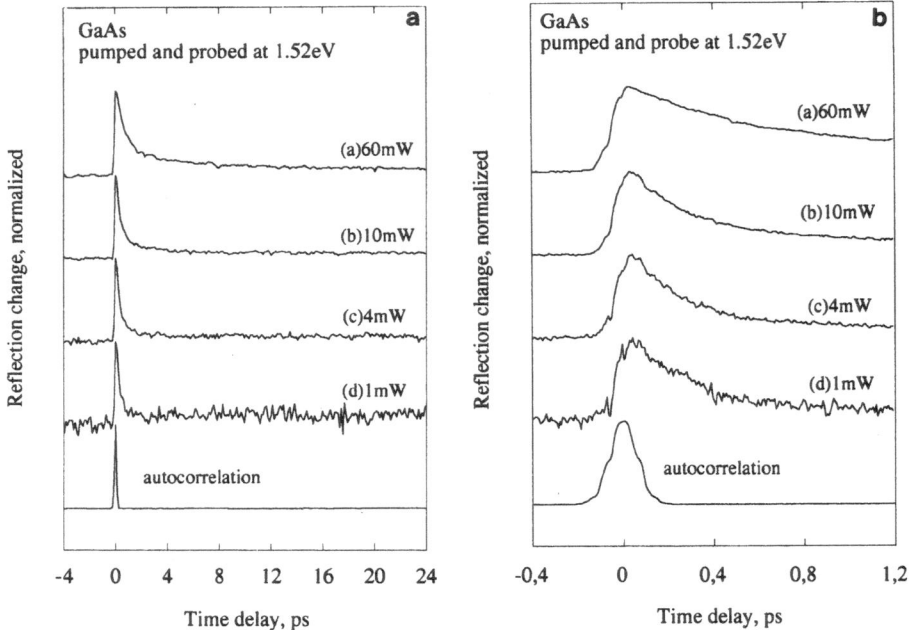

Figure 1. Normalized reflection change of bulk GaAs as a function of the time delay between the pump and probe pulses, pumped and probed at 1.52eV. (a)long time scans, (b)short time scans. The numbers attached to the curves are the average pump power used in the measurements. The corresponding carrier densities are $9.6 \times 10^{17} cm^{-3}$ (60mW), $8.0 \times 10^{17} cm^{-3}$ (10mW), $5.7 \times 10^{17} cm^{-3}$ (4mW) and $1.4 \times 10^{17} cm^{-3}$ (1mW).

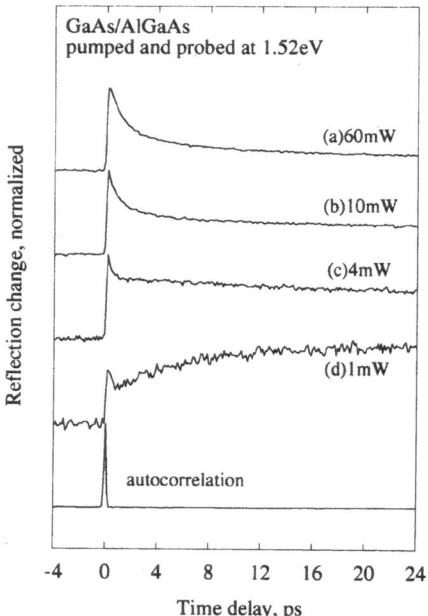

Figure 2. Normalized reflection change of GaAs/AlGaAs MQW as a function of the time delay between the pump and probe pulses, pumped and probed at 1.52eV. The numbers attached to the curves are the average pump power used in the measurements. The corresponding two dimensional carrier densities are $9.6 \times 10^{11} cm^{-2}$ (60mW), $8.0 \times 10^{11} cm^{-2}$ (10mW), $5.7 \times 10^{11} cm^{-2}$ (4mW) and $1.4 \times 10^{11} cm^{-2}$ (1mW).

Table 1. Summary of the average pump power P, number of carriers per well N, and the corresponding fractional reflection change per carrier * of the GaAs/AlGaAs multiple quantum wells

P (mW)	$N (10^{11} \text{cm}^{-2})$	Fractional reflection change per carrier (10^{-16}cm^2)*
60	9.6	3.2
10	8.0	2.9
4	5.7	3.0
1	1.4	7.3

$$* \quad \frac{1}{N}\left(\frac{\Delta R}{R}\right)\bigg|_{t=24\,ps}$$

explained as the screening of carrier-phonon interaction at high carrier density[2-4]. Carrier cooling via carrier-phonon interaction slows down due to the screening effect. Our results manifest this phenomenon in a more direct way.

The subpicosecond component of the MQW sample exhibits similar behaviour to that of the bulk sample (Fig. 2). The amplitude of the long picosecond component, however, increases as the carrier density is decreased and a rising wing begins to appear at carrier densities lower than $5\times10^{17} \text{cm}^{-3}$ ($5\times10^{11} \text{cm}^{-2}$ per well). In Table 1 we summarize the average pump power P, the number of carriers per well N and the long time (24ps) fractional reflection changes per carrier

$$\frac{1}{N}\left(\frac{\Delta R}{R}\right)\bigg|_{t=24\,ps}$$

For $N > 5\times10^{11} \text{cm}^{-2}$,

$$\frac{1}{N}\left(\frac{\Delta R}{R}\right)\bigg|_{t=24\,ps}$$

is almost constant. But it increases by approximately a factor of 2 at $N=1.4\times10^{11} \text{cm}^{-2}$, at which the rising wing is distinctively observed. This increase can not be accounted for by considering only the occupation of the states at the bottom of the conduction band by cooled free carriers, because this effect gives only slight increase of refractive index per carrier when N is decreased from $5\times10^{11} \text{cm}^{-2}$ to $1.4\times10^{11} \text{cm}^{-2}$ [5].

To explain this dependence on carrier density and the disappearance of the rising wing for higher carrier densities and in bulk GaAs, we consider the occupation of excitonic states, i.e., the formation of excitons from free $e-h$ pairs for low carrier densities. In MQW systems, the confinement of $e-h$ pairs in quasi-two-dimension layers enhances excitonic effects and makes them observable at room temperature.

The room temperature excitonic optical nonlinearity of the MQW systems of GaAs has been studied both experimentally[5,6] and theoretically[7,8]. Physically, the nonlinearity originates from the bleaching of the excitonic resonance by state-filling effect when excitons generate from free $e-h$ pairs. On the other hand, the presence of free $e-h$ pairs weakens the exciton binding energy and broadens the excitonic transition through the screening of $e - h$ Coulombic interaction. According to the calculation of Rink et al.[7] the critical carrier density N_c at which exciton vanishes in the continuum and free carrier effect becomes dominant is determined by

$$N_c a_{2D} \approx 0.2 \tag{1}$$

when

$$TE_0 \approx 2.5 \tag{2}$$

where a_{2D} is the two-dimensional exciton radius, E_0 is the exciton binding energy and T is the temperature. Using $a_{2D} = 63\text{Å}$ and $E_0 = 9meV$ for the 100Å quantum wells, we estimate $N_c = 5 \times 10^{11} cm^{-2}$ for room temperature. Experimentally, we find that when $N > 5 \times 10^{11} cm^{-2}$, the rising wing vanishes and

$$\frac{1}{N}\left(\frac{\Delta R}{R}\right)\Bigg|_{t=24ps}$$

becomes nearly constant. Further more, at the limit of low carrier densities where screening effect becomes negligible, theory predicts that the nonlinearity from excitonic state-filling effect is as two times as that from free carrier state-filling effect. This finding also agrees well with the increase of

$$N\frac{1}{N}\left(\frac{\Delta R}{R}\right)\Bigg|_{t=24ps}$$

by two times at $N = 1.4 \times 10^{11} cm^{-2}$ observed by present experiments. At this carrier density, the excitonic state-filling effect dominates.

Basing on the above dicussion we attribute the rising wing observed at low carrier density to the process of exciton formation from free $e - h$ pairs. The corresponding increased

$$\frac{1}{N}\left(\frac{\Delta R}{R}\right)\Bigg|_{t=24ps}$$

results from the increased refractive index due to the saturation of excitonic states. The rising wing of our experiment can be well fit to a single exponent, yielding a time constant of 5ps for exciton formation at room temperature.

Depicted in Fig. 3 is the time-resolved reflection change of GaAs pumped by the second harmonic and probed by the fundamental of the pulses from the Ti:Sapphire laser. Strong dependence on the pumping photon energy is observed. To understand this dependence, we have to consider the carriers excited in the satellite valley L along [1/2,1/2,1/2] direction in the k-space, where the density of states is much larger than that in the central valley Γ. Because pulses of about 1.5 eV photon energy probe only the carriers in valley Γ, the carriers in the L valley contribute to the reflection change only via L to Γ intervalley

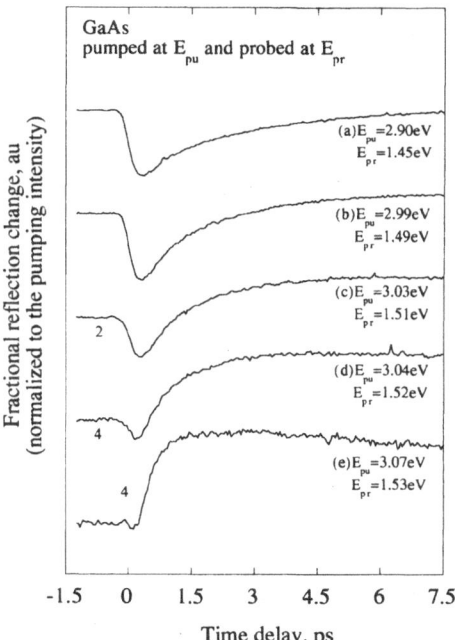

Figure 3. Reflection change of GaAs as a function of the time delay between the pump and probe pulses. The photon energies of the pump and probe pulses are indicated as E_{pu} and E_{pr} respectively.

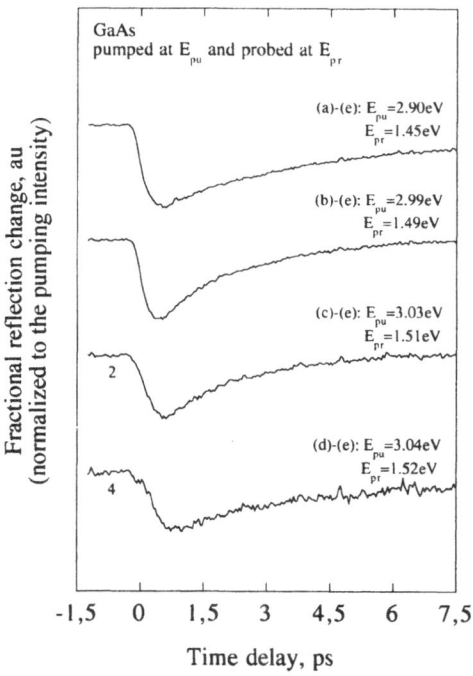

Figure 4. Time-resolved reflection change of GaAs due to the L to Γ intervalley scattering.

scattering. In fact, the highest photon energy which can excite the carriers in the L valley is 2.97eV[9]. Thus curve (e) in Fig. 3 can be considered as the reflection change without contribution from the carriers in the L valley. Subtracting curve (e) from the other curves, we obtain the net contribution of the carriers in the L valley as shown in Fig. 4. Now each curve displays similar behaviour. The dynamics of L to Γ intervalley scattering can be studied by taking into account the detailed information about the band struture as well as tne density of states in the L valley.

REFERENCES

1. A. Davidson, R. C. Compton, F. Wise, D. Mars and J. Miller, J. Appl. Phys. **76**, 2255(1994).
2. R. F. Leheny, J. Shah, R. L. Fork, C. V. Shank and A. Migus, Solid State Commun. **31**, 809(1979).
3. C. L. Collins and P. Y. Yu, Solid State Commun. **51**, 123(1984).
4. E. J. Yoffa, Phys. Rev. **B23**, 1909(1980).
5. S. H. Park, J. F. Morhange, A. D. Jeffery, R.A. Morgan, A. Chavez-Pirson, H. M. Gibbs, S. W. Koch, and N. Peyghambarian, Appl. Phys. Lett. **52**, 1202(1988).
6. W. H. Knox, R. L. Fork, M. C. Downer, D. A. B. Miller, D. S. Chemla, and C. V. Shank, Phys. Rev. Lett. **54**, 1306(1995).
7. S. Schmitt-Rink and C. Ell, J. Lumin. **30**, 585(1985).
8. S. Schmitt-Rink, D. S. Chemla and D. A. B. Miller, Phys. Rev. **B32**, 6601(1985).
9. P. Lautenschlager, M. Garriga, S. Logothetidis, and M. Cardona, Phys. Rev. **B35**, 9174(1987).

62

EVIDENCE OF INTERFACE ROUGHNESS CORRELATION IN CdTe/(Cd,Zn)Te QUANTUM WELLS

E. J. Mayer,[1] N. T. Pelekanos,[1,2] J. Kuhl,[1] N. Magnea,[2] and H. Mariette[2]

[1] Max-Planck-Institut für Festkörperforschung
Heisenbergstr. 1, 70569 Stuttgart, Germany
[2] CEA-CNRS Joint Group "Microstructures de Semiconducteurs II-VI"
17 rue des Martyrs, 38054 Grenoble, Cedex 9, France

The homogeneous and inhomogeneous linewidth of excitonic transitions characterize the interfacial quality of QW structures.[1] We measured the inhomogeneous linewidth Γ_{inh} by cw photoluminescence (PL) and reflectivity (REF) spectra on a series of undoped single CdTe/Cd$_{1-x}$Zn$_x$Te QW's with x = 0.16 - 0.18 and thickness L between 18 Å and 112 Å. We determined the zero-exciton-density- and zero-temperature-limit of the homogeneous linewidth Γ_{hom} by density and temperature dependent degenerate-four-wave-mixing (DFWM) experiments. In these experiments, a mode locked Ti:Sapphire laser operating at 76 MHz was used. For a selective, resonant excitation of the exciton transition the bandwidth of 120-fs pulses was reduced to a pulsewidth of 4 meV, corresponding to 800 fs.

For all L, the quantities for Γ_{inh} ($1 meV \leq \Gamma_{inh} \leq meV$) and Stokes shifts S ($0.1\ meV \leq S \leq 1\ meV$) remain remarkably small (Fig. 1). The striking information however is that in contrast to measurements on GaAs/(Al,Ga)As QW systems, which show an increase of Γ_{inh} with decreasing QW width, we observe smaller Γ_{inh} for smaller L. The solid line in Fig. 1 depicts the calculated values for Γ_{inh} as a function of L. The calculated Γ_{inh} contains two contributions: i) formation at the heterointerfaces (assumed here uncorrelated) of monolayer (ML) islands of lateral size comparable to the exciton Bohr radius a_B and ii) a 1 meV contribution independent of L resulting from remaining strain fluctuations in the well after relaxation of the 2μm-thick Cd$_{1-x}$Zn$_x$Te buffer layer grown on the Cd$_{0.96}$Zn$_{0.04}$Te substrates. The value of 1 meV is consistent with analysis of the x-ray diffraction linewidths of relaxed epilayers on the same system.[2] A comparison of the experimental and theoretical Γ_{inh} shows the obvious failure of the model in the present case, e.g. for the 18-Å QW a four times smaller value Γ_{inh} = 1.3 meV is observed than the theoretically predicted of 5 meV. The small Γ_{inh} and S values could be possibly attributed to the formation of pseudosmooth interfaces, i.e. ML roughness on a length scale much smaller than a_B. However, the formation of pseudosmooth interfaces alone cannot explain the decrease of Γ_{inh} and S with decreasing L.

Ultrafast Processes in Spectroscopy, edited by Svelto et al.
Plenum Press, New York, 1996

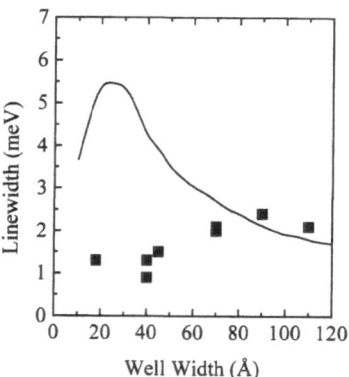

Figure 1. Measured PL (closed squares) and calculated exciton (FWHM) linewidths (solid line) in CdTe/(Cd,Zn)Te single QWs at T= 2 K as a function of QW thickness.

We present now our results from DFWM measurements. To use the homogeneous linewidth Γ_{hom} as a measure of interface roughness, one has to eliminate other contributions to Γ_{hom}, such as for example exciton-acoustic phonon-, exciton-exciton-scattering and radiative recombination. In the inset of Fig. 2, the DFWM signals of the 18-Å QW at lattice temperatures of T=9 K (open triangles) and T=25 K (open circles) are plotted. In the 9-K curve a biexponential decay is observed. The fast decay is attributed to the free excitons and the slow to a small population of localized excitons. This assumption is confirmed by the vanishing of the slow decay at T=25 K. In the rest of the article only the free excitons will be considered. The contribution of the exciton-acoustic-phonon interaction to Γ_{hom} was determined by measuring, for all samples, Γ_{hom} as a function of temperature T in the range 10–40 K for a constant exciton density ($n_x = 5 \cdot 10^9$ cm^{-2}). Γ_{hom} increases linearly up to 30 K, following Eq. 1:

$$\Gamma_{hom}(n_X, T) = \Gamma_{hom}(n_X, T = 0) + \gamma_{XP} \cdot T. \tag{1}$$

From our measurements, we are able to extract $\Gamma_{hom}(n_X, T = 0)$ and γ_{XP}. The exciton-acoustic phonon coupling parameter γ_{XP} shows no strong dependence on the QW thickness in the range between 70 Å and 112 Å with an average value of $\gamma_{XP} = 2.5$ µeV / K. Below 70 Å γ_{XP} increases, but weaker than the theoretically predicted 1/L dependence.[3] On the other hand, in analogy to the decrease of the Γ_{inh}, the $\Gamma_{hom}(n_X, T = 0)$ declines for smaller L.

The exciton-exciton scattering contribution can be evaluated by exciton-density dependent DFWM. At constant lattice temperature T = 9 K, we observe a linear increase of Γ_{hom} for excitation densities up to $n_X = 10^{10}$ cm^{-2}:

$$\Gamma_{hom}(n_{XX}, T) = \Gamma_{hom}(n_{XX} = 0, T) + \gamma_{XX} \cdot a_B^2 \cdot E_B \cdot n_{XX}. \tag{2}$$

In order to calculate the exciton-exciton coupling strength γ_{XX}, we take that, as the QW width deceases from 112 Å to 18 Å, the Bohr radius α_B decreases from 65 to 58 Å, the exciton binding energy E_B increases from 16 to 18 meV, and the exciton oscillator strength increases by 30 % over a value $f_X = 1.7 \cdot 10^{-3} \, Å^{-2}$, measured in a 100 Å QW[4]. The

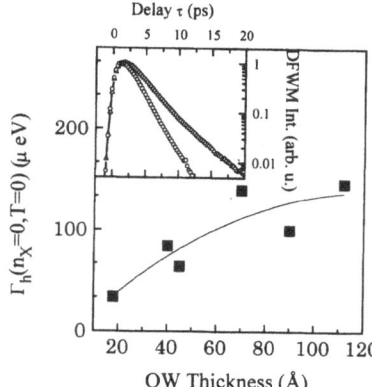

Figure 2. Zero-temperature and zero-density limit of the homogeneous linewidth versus QW thickness. The dashed line is a guide to the eye. Inset: DFWM signal for a 18-Å CdTe/(Cd,Zn)Te QW at T=9 K and 25 K.

values of γ_{XX} decrease from 3 to 1 as L decreases form 112 Å to 18 Å. Attributing this decrease to increasing localization in the narrower QW is not plausible, because of the simultaneous decrease of Γ_{inh} and S as well as the high scattering efficiency of exciton-exciton interaction at large interparticle distances (≥ 1000 Å). Considering the smaller dielectric constant of CdTe/(Cd,Zn)Te, the quantities of γ_{XX} are comparable to those of GaAs QW's.[5]

From the temperature and density dependent measurements we are now able to plot Γ_{hom} ($n_{XX} = 0$, $T = 0$) as a function of L in Fig.2. We observe a clear tendency of smaller Γ_{hom} for smaller L, indicating a strong reduction of scattering events by a reduction of the effective interface roughness. We note that other mechanisms such as radiative recombination, exciton-impurity scattering or alloy fluctuations can not explain the decrease of Γ_{hom} with smaller L: The increase of the oscillator strength with decreasing L is contrary to the observed smaller Γ_{hom}; the samples are nominally undoped; for narrower QW's the excitons increasingly penetrate into the barriers, therefore the decrease of Γ_{hom} implies negligible influence of barrier alloy fluctuations.

As an explanation for the simultaneous decrease of Γ_{inh} and Γ_{hom} for narrower QW's, we suggest a reduction of the effective interface roughness originating from a correlated growth process that preserves the roughness topography of the first heterointerface and results in a reduced average thickness fluctuation along the QW plane. Correlated interfaces have been recently detected in structural studies of SiGe/Si superlattices[6] and InAs/GaAs quantum dots[7].

REFERENCES

1. K. Ploog, A. Fischer, L. Tapfer, B. F. Feuerbacher, J. Crystal Growth **111**, 344 (1991).

2. C. Bodin et al., Phys. Rev. B **51**, 13181 (1995).

3. J. Lee, E.S. Koteles, and M.O. Vassell, Phys. Rev. B **33**, 5512 (1986).

4. Y. Merle-d'Aubigné, private communication.

5. J. Kuhl et al, Festkörperprobleme / Advances in Solid State Physics, Vol. 29, p.157 (1989).

6. Y. H. Phang et al, Phys. Rev. B **50**, 14435 (1994).

7. Q. Xie, et al.. Phys. Rev. Lett. **75**, 2542 (1995).

63

THE INFLUENCE OF A PLASMON ASSISTED STIMULATED RAMAN SCATTERING ON THE REVERSIBLE BLEACHING OF GALLIUM ARSENIDE BY A HIGH-POWER PICOSECOND LIGHT PULSE

I. L. Bronevoi and A. N. Krivonosov

Institute of Radioengineering and Electronics
Russian Academy of Sciences
103907, Moscow, GSP-3, Mokhovaya St.,11, Russia

This text describes the results of the study of the bleaching (enhancement of the optical transparency T) in the thin layer of a high-purity GaAs and the edge recombinative emission of it as a function of the energy $\hbar\omega_{ex}$ of a photon of a high-power light pulse that irradiates the sample. The length of time of both the exciting (ex) and probing (p) pulses was about 14 ps, while $\hbar\omega_p > \hbar\omega_{ex} > E_g$, E_g being the forbidden gap width. Experiments were conducted at room temperature and in an "overthreshold" range of the photoexcitation. In this range due to an abnormally fast recombination superluminescence[1,2] (that propagates predominantly along the GaAs layer), the state of the charge plasma is maintained to be like a quasi-equilibrium threshold state (when $\mu_e - \mu_h = E_g$)[1,3] and the bleaching is nearly reversibly changing with the intensity of the excitation light[4], here μ_e and μ_h are Fermy quasilevels of electrons and holes, respectively.

Step-like peculiarities are found on the $\hbar\omega_{ex}$ dependencies of bleaching, Fig.1.[5] Similar peculiarities have also been observed for the edge emission energy W_s^M in the maximum of the integral spectrum, Fig.2. The value of the photon energy $\hbar\omega_{ex}^b$, at which there appears the "step", shifts with the increase in energy W_{ex} of the exciting pulse (and, accordingly, in the charge carrier density[4] n) to the short-wave side of the spectrum. Incidentally, in the case of "steps" shown in Fig.2, the difference $\hbar\omega_{ex}^b - \hbar\omega_s^m$ (where $\hbar\omega_s^m$ is the photon energy in the maximum of the emission spectrum of the sample) when variating with n, obeys approximately the law of the plasmon energy change $\hbar\omega_{pl}$ (n), Fig.3. The value of n was found from the displacement of maximum in the emission spectrum, Fig.4. The displacement is caused by the narrowing of the band gap due to Coulomb interaction of photoexcited carriers and can be estimated as in[6]:

$$\Delta E_g \approx 0.01 \cdot [n / 10^{17} \text{ cm}^{-3}]^{1/3}, \text{ eV} \qquad (1)$$

Ultrafast Processes in Spectroscopy, edited by Svelto et al.
Plenum Press, New York, 1996

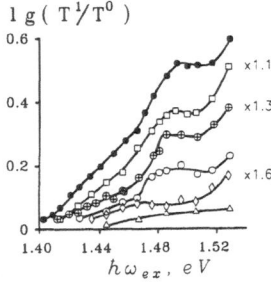

Figure 1. Transparency T variation (indices 1 and 0 show the presence and absence of excitation, respectively) of GaAs as a function of $\hbar\omega_{ex}$, the probing photon energy being $\hbar\omega_p = 1.557$ eV. Sample 1. The excitation beam diameter $\varnothing_{ex} = 0.5$ mm.• - $W_{ex} = 1.0$ a.u., $\tau_d = -1$ ps; □ - $W_{ex} = 1.0$ a.u., $\tau_d = -7$ ps; ⊕ - $W_{ex} = 0.5$ a.u., $\tau_d = -1$ ps; o - $W_{ex} = 0.135$ a.u., $\tau_d = -1$ ps; ◊ - $W_{ex} = 0.044$ a.u., $\tau_d = -1$ ps; Δ - $W_{ex} = 0.02$ a.u., $\tau_d = 18$ ps (τ_d is the delay time of the probing pulse with respect to the exciting one).

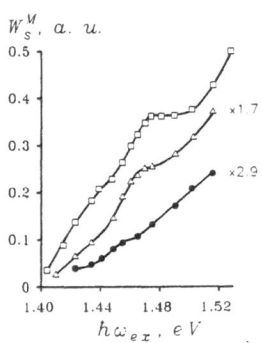

Figure 2. Dependence of the energy W_s^M in the maximum of the edge emission spectrum on $\hbar\omega_{ex}$: • - $W_{ex} = 0.1$ a.u., Δ - $W_{ex} = 0.3$ a.u., □ - $W_{ex} = 1.0$ a.u. Sample 2. The excitation beam diameter is $\varnothing_{ex} = 0.35$ mm.

In the case of "steps" shown in Fig.1, the magnitude $(\hbar\omega_{ex}^b - \hbar\omega_s^m)^2$ is a linear function of bleaching (Fig.5), the latter was estimated as proportional to n. This is also in agreement with an assumption that $\hbar\omega_{ex}^b - \hbar\omega_s^m = \hbar\omega_{pl}$.

The above said enables us to draw a conclusion that superluminescence stimulates the plasmon assisted Raman scattering of the excitation light. This, first, intensifies the edge recombination emission, which causes the "stepwise" character of $(W_s^M - \hbar\omega_{ex})$ curves. Secondly, due to plasmon generation there is a rise in the carrier plasma temperature, which (in order that $\mu_e - \mu_h = E_g$ state should be maintained) must lead to an additional photogeneration of carriers. The latter manifests itself in the enhancement of bleaching and we have "a step" when plotting the bleaching as a function of $\hbar\omega_{ex}$.

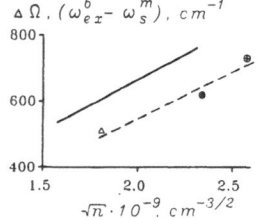

Figure 3. The difference $(\omega_{ex}^b - \omega_s^m)$ as a function of the electron density n calculated by Eq. 1.: Δ - $W_{ex} = 0.1$ a.u., • - $W_{ex} = 0.3$ a.u., ⊕ - $W_{ex} = 1.0$ a.u. Dash line is the guide for eye. Sample 2. Solid line - the dependence of the Raman shift $\Delta\Omega$ on n.[7]

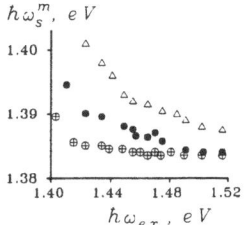

Figure 4. The dependence of the photon energy $\hbar\omega_s{}^m$ in the maximum of the emission spectrum of the sample on $\hbar\omega_{ex}$. Δ -W_{ex} = 0.1 a.u., $\bullet$ - W_{ex} = 0.3 a.u., $\oplus$ - W_{ex} = 1.0 a.u Sample 2.

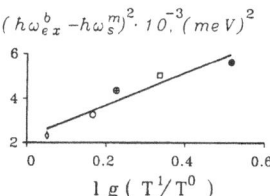

Figure 5. The magnitude of $(\hbar\omega_{ex}{}^b - \hbar\omega_s{}^m)^2$ as a function of the bleaching $\lg(T^1/T^0)$ that was measured at $\hbar\omega_{ex} = \hbar\omega_{ex}{}^b$, $\hbar\omega_p$ = 1.557 eV. Sample 1.

ACKNOWLEDGMENTS

The research described in this publication was made possible in part by Grant No. 95–02–05871 from the Russian Foundation of Fundamental Research and Grant No. M3S300 from the International Science Foundation.

REFERENCES

1. N.N.Ageeva, I.L.Bronevoi et al., JETP Lett.**48**, 276 (1988); Solid State Commun.**72**, 625 (1989).
2. J.Dubard, J.L.Oudar, F.Alexandre, D.Hulin, A.Orzag, Appl. Phys. Lett. **50**, 821 (1987).
3. I.L.Bronevoi, A.N.Krivonosov, and T.A.Nalet, Solid State Commun. to be published.
4. I.L.Bronevoi et al., JETP Lett. **42**, 395 (1985); **43**, 473 (1986).
5. I.L.Bronevoi, A.N.Krivonosov, and V.I.Perel', Solid State Commun. **94**, 363 (1995).
6. J.Shah, R.F.Leheny, and C.Lin, Solid State Commun. **18**, 1035 (1976).
7. A.Mooradian, in Laser Handbook, ed. by F.T.Arecchi and E.O.Schulz-Dubois, North-Holland Publishing Company, Amsterdam, **Vol.2,** pp. 1409–1456 (1972).

64

HIGH-BRIGHTNESS EXCIMER LASERS

S. Szatmári,[1][*] G. Almási,[2] M. Feuerhake,[3] and P. Simon[3]

[1] Department of Experimental Physics
JATE University
H-6720 Szeged, Dóm tér 9., Hungary
[2] Department of Physics
JPTE University
H-7624 Pécs, Ifjúság u. 6., Hungary
[3] Laser-Laboratorium Göttingen
D-37077 Göttingen
Hans-Adolf-Krebs-Weg 1., Germany

In most of the high-intensity experiments, it is the focused intensity that has to be considered as the major figure of merit for the performance of the pump-laser system. This is the point where the better focusability of short-wavelength gas lasers becomes a dominant advantage over long wavelength solid-state systems. The minimum focal spot area is 10–100 times smaller for excimer systems, provided by the shorter wavelength and by less optical distortion in the gaseous active medium. Therefore short-pulse excimer lasers are the best candidates to produce the highest focused intensities, especially when the problem of their limited extraction efficiency is solved, and the present performance of short-wavelength focusing optics is improved.

Recently we have made great effort how to match commercially available gain modules to the needs of efficient and distortion-free short-pulse amplification. The dynamics of optical distortions in the gaseous active medium has been studied using an interferometric method. A special "beam-relaying" technique has been developed and the operational conditions of the amplifier has been optimized to improve the beam homogeneity and minimize the effect of self-focusing. As a result of these development a brightness approaching 10^{20} Wcm^{-2}st^{-1} has been reached for the output beam of a table-top laser-system, which uses a single commercial gain module.

An attempt to focus this beam by different focusing optics has been made to produce high focussed intensities. The results were very dependent on the actual performance of the optics. The best results were obtained by one of the f/1.4 off-axis paraboloids (Optical Surfaces Ltd.), where more than 60% of the output energy is concentrated into the first

* Tel./Fax: 36/62/311–154.

Ultrafast Processes in Spectroscopy, edited by Svelto et al.
Plenum Press, New York, 1996

283

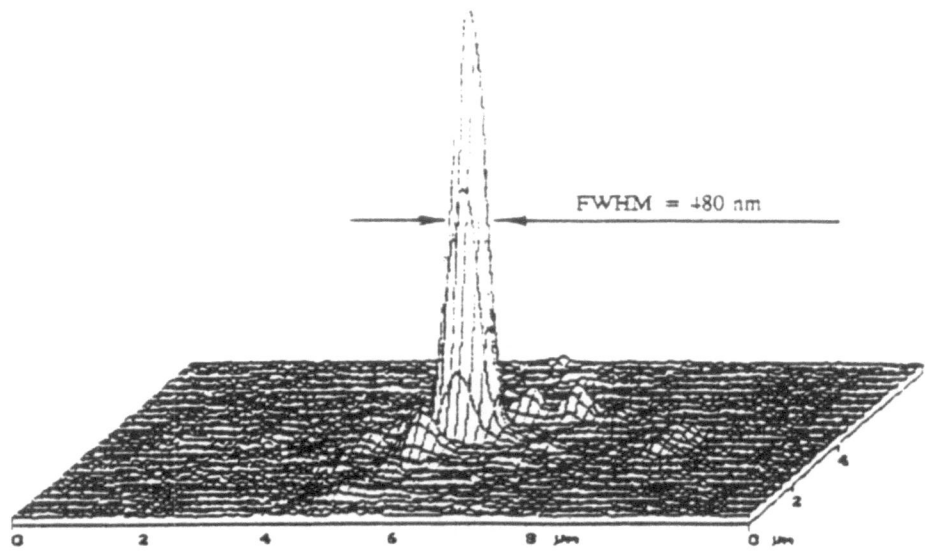

Figure 1. Intensity distibution of the focal spot of the KrF beam using a f/1.4 off-axis paraboloid mirror for focussing.

diffraction lobe, having a full width at half maximum (FWHM) of ~480 nm (fig.1). This corresponds to a maximum intensity approaching 10^{19} Wcm^{-2}.

Note, that these results are obtained in a practically single-beam amplification experiment, using not more than 10% of the available optical energy of the amplifier. It is known that the main drawback of short-pulse eximer systems is associated with the short energy-storage time of the active medium, necessitating successive replenishment of the gain. At present, this can not be realized by multiple-pass amplification or optical multiplexing of femtosecond pulses. If a generally applicable interferometric beam recombination method were to be found for optical multiplexing, it might solve the inherent energy extraction problem of excimers. This could make the excimer-based short-pulse laser systems competitive with solid-state systems even in maximum peak power. Moreover the brightness of excimer systems is expected to be significantly higher than that of long wavelength systems, which could lead to the highest focused intensities (in excess of 10^{20} Wcm^{-2}) achieved by table-top laser systems.

The details of recent measurements and some considerations for future developments are presented.

65

EFFICIENT FREQUENCY DOUBLING OF THE FEMTOSECOND TERAWATT POWER Ti: SAPPHIRE LASER PULSES

V. Sirutkaitis,[1] A. Persson,[2] E. Gaizauskas,[1] A. Piskarskas,[1] and S. Svanberg[2]

[1] Laser Research Centre
Vilnius University, Vilnius, Lithuania
[2] Division of Atomic Physics
Lund Institute of Technology, Lund, Sweden

Femtosecond optical pulses with terawatt (or subterawatt) power in the UV spectral range are of great interest for high intensity physics, but no primary laser sources for the generation of powerful femtosecond pulses in this spectral range are currently available. One widespread way to provide such pulses is the harmonic conversion of terawatt Ti:sapphire lasers. Phase-matched harmonic generation of femtosecond Ti:sapphire lasers have demonstrated their usefulness in producing femtosecond sources of light in UV and near VUV wavelength range[1,2]. The contrast ratio R of the initial pulse can be additionally increased to approximately R^2 by frequency upconversion of the fundamental pulse to its second harmonic (SH). Thus efficient SH generation with small lengthening of the femtosecond pulses of Ti:sapphire laser allows the producing of femtosecond pulses with terawatt (or subterawatt) power in UV with a sufficient contrast. Earlier experiments on frequency doubling were performed with ~ 150 fs pulses at powers lower than 0.3 TW and showed that conversion efficiency up to 33 % can be achieved[3].

We report the frequency doubling of the femtosecond terawatt power Ti:sapphire laser pulses in a large - aperture (60 x 60 mm) Type I KDP crystals with efficiency up to 49 %. SH pulses with energies more than 53 mJ and durations less than 185 fs were produced at 397.5 nm.

The pump laser used was a chirped-pulse amplification table-top terawatt (T^3) Ti:sapphire laser system. The system was operated at 795 nm with a repetition rate of 10 Hz. The amplified pulse duration after a grating compressor was about 150 fs, assuming $sech^2$ pulse shape. The energy of recompressed pulses can be as high as 220 mJ. The final beam diameter at $1/e^2$ level was 50 mm. Further details on the laser system can be found in paper[4].

KDP crystals were chosen for the current experiments because today only KDP and its isomorphs can be used for manufacturing single-crystal frequency doublers with a diameter of 60–70 mm. The calculated group-velocity mismatch (GVM) between funda-

Ultrafast Processes in Spectroscopy, edited by Svelto et al.
Plenum Press, New York, 1996

mental (ω_1) and SH (ω_2) pulses, which is the main factor influencing the doubling effi-ciency and femtosecond pulses broadening in the harmonic generation process, for KDP crystal is 79 fs/mm when the fundamental wavelength is 795 nm. For the optimal crystal length estimation numerical calculations of the coupled propagation[5] of the fundamental and second-harmonic pulses in a quadratically nonlinear medium including GVM and phase modulation effects due to the third order nonlinearity were performed. Pulse dura-tion was taken 150 fs for the calculations. The second-order nonlinear parameter σ was calculated according to the nonlinear susceptibility tensor coefficients for the appropriate phase-matching angle and the third-order nonlinearity parameter χ was taken as 3×10^{-13} esu. An analysis of the SH generation in the KDP crystal assuming exact phase-matching was made. We neglected group-velocity dispersion effects for the short KDP crystals (0.1 - 0.5 cm) to be considered as well as such transversal effects as walk-off of the extraordi-nary wave and diffraction, despite that these effects limits efficiency of SH generation. The effect of transversal intensity distribution on the SH generation efficiency was evalu-ated by simple averaging over the transversal pump profile.

By performing numerical experiments according to the model defined above, we in-vestigated how the dynamics of SH generation depends on the peak intensity I_0 of the pump pulse. The results of the numerical modelling for I_0 equal to 50 GW/cm^2 are shown in Fig. 1. As can be seen the generated SH pulse became longer when propagating at dis-tances z > 2 mm due to the GVM of the fundamental and SH pulses. Because the group velocity of the fundamental pulse in the KDP crystal is larger than the SH pulse the longer interaction is between the leading edge of the SH pulse and the fundamental pulse. At longer interaction lengths and higher intensities it causes the asymmetry of the SH pulse; the leading edge of the pulse is longer than the trailing part of the pulse. Estimation of the pulse duration at the 0.5 I_{max} level shows no large difference between SH pulses generated in crystals with 1.5 and 2.5 mm length, but the SH pulse duration increases for longer crystals if the duration is estimated at intensity level ~ 0.1 I_{max}. Thus the maximal funda-mental pulse peak intensity used is important for the selected crystal length because at the higher pump intensities the generated SH pulse has one or few prepulses at the leading edge which can be a noticable part of the pulse energy.

Frequency doubling of the 795 nm terawatt power Ti:sapphire laser pulses was investi-gated in large aperture (60x60 mm) KDP crystals. The KDP crystals used for the frequency doubling were cut at 45° to the optical axis for Type I phase - matching. Three crystals with

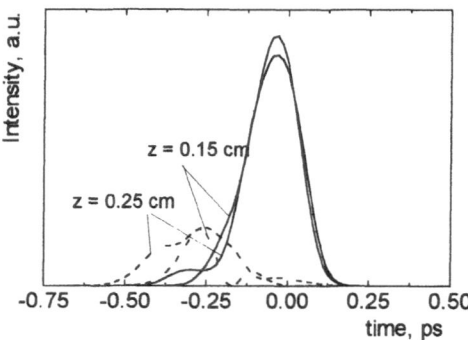

Figure 1. Numerically calculated pulses of the fundamental (dashed line) and SH (solid line) at the output of the doubler crystal with different length at the input pulse intensity 50 GW/cm^2.

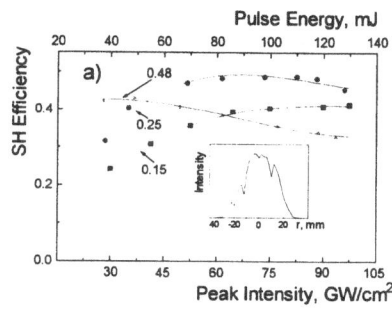

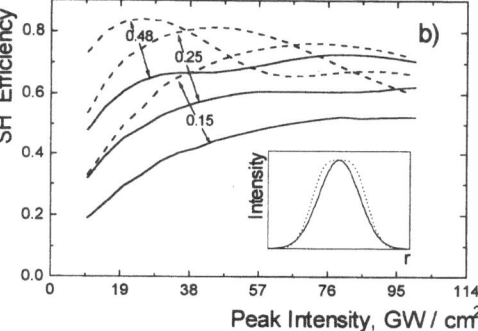

Figure 2. Experimentally measured (a) and calculated (b) second-harmonic energy conversion efficiency versus input pulse energy for 1.5 mm; 2.5 mm and 4.8 mm length crystals. SH crystal was set at optimum phase-matching. In (b) figure curves 1, 2, 3 corresponds to Gaussian and curves 1', 2', 3' to a truncated Gaussian profile in space while temporal profile is $sech^2$ in both cases.

thickness 1.5 mm, 2.5 mm and 4.6 mm were used in the experiments. For separation of the second harmonic beam from the fundamental one, two dichroic mirrors which had a high reflectivity for the second and high transmission for the fundamental were used.

Harmonic conversion measurements were performed with all three crystals of different lengths and the results are plotted in Fig. 2a. The largest second-harmonic pulse energies were obtained with the 2.5 mm long crystal. With 128 mJ in the incident pulse, a 58 mJ pulse was produced at 397.5 nm. The conversion efficiency with respect to the incident pulse energy at this incident pulse energy was 45.3 % while the largest conversion efficiency for this crystal was 49 %. SH conversion efficiency depletion as observed at incident pulse energies above 80 mJ shows that at incident pulse energies higher than 90 mJ in the 2.5 mm length crystal back conversion of SH pulse to the fundamental in the spatial and temporal center of the pulse starts. The conversion efficiency decrease in the 4.6 mm length crystal at incident energies higher than 50 mJ shows that this crystal is too long for the incident pulse intensities used. A largest conversion efficiency of 43 % in this crystal was achieved at an incident pulse energy of about 48 mJ. At higher pulse energies the process of back conversion of the SH pulse increases. For the 1.5 mm length crystal the SH conversion efficiency increased across the whole incident pulse energy range used. A largest conversion efficiency of 41.3 % was observed at a maximal incident pulse energy of 130 mJ. It means that this crystal can be used at even higher incident pulse energies without exhibiting the manifestation of SH pulse back conversion. Neglecting SH pulse back conversion shows that 1.5 mm length crystal is useful for increasing the pulse contrast in all range of the incident pulse energies used, while 2.5 mm and 4.6 mm length crystals are only useful at incident pulse energies lower than ~90 mJ and ~40 mJ, respectively. The maximum pulse intensity which is more important for conversion efficiency in our measurements varied in range 28 - 100 GW/cm^2.

The dependence of the calculated SH generation efficiency on the peak pump intensity for different crystal lengths is shown in Fig. 2b. Here Gaussian (solid line) and truncated Gaussian (dotted line) transversal distributions of the pump pulse were considered. The numerically calculated conversion efficiency has the same dependence on the pulse peak power as in the experiment but the values were higher. Really, the pulse spatial profile was not Gaussian, but more similar to a hyper-Gaussian with few deep modulation rings (see the insert in Fig. 2a). This beam modulation additionally decreased the SH conversion efficiency and could be one of the factors causing discrepancy between experimental and numerical results.

For the second-harmonic pulse-width measurements cross correlation of the fundamental beam with the second harmonic beam by using phase-matched sum-frequency generation was employed. A 0.6 mm thick crystal of KDP cut at 45° to the optical axis for Type I third harmonic generation was used in the cross-correlator. The FWHM of the cross-correlation trace between 795 and 397.5 nm pulses for a 1.5 mm doubling crystal was 320 fs. From this correlation function we derived a pulse duration of about 185 fs, taking into account the group-velocity mismatch of the interacting pulses in the correlation crystal and assuming sech^2 shaped pulse. Note that the group delay between fundamental and SH pulses in the correlation crystal used is 50 fs and a shorter crystal is necessary for the pulse duration to be measured more accurately. That measured pulse duration is limited by the correlator resolution and confirms the same (~320 fs) FWHM of the cross-correlation trace for 2.5 and 4.8 mm doubling crystals at low incident pulse energy. Numerically predicted SH pulse durations for 1.5 and 2.5 mm KDP doubling crystals in the pulse intensity range used must be shorter than 165 fs.

The spectral bandwidth for the 397.5 nm pulse depended on the doubling crystal length. At low peak intensity of the incident pulse it narrowed from 1.8 nm for 1.5 mm doubling crystal to 1.5 nm and 0.9 nm when the doubling crystal length was 2.5 mm and 4.8 mm respectively. It shows that the bandwidth tolerance of doubling crystals with a thickness of more than 1.5 mm is lower than the spectral bandwidth of the fs Ti:sapphire laser pulses. It is an additional limiting factor of the SH energy conversion efficiency. At pulse energies higher than 80 mJ self-phase modulation in the amplifier elements and air took place and caused pulse spectrum modulation. At the incident pulse energy > 90 mJ after the pass of the 4.8 mm length KDP crystal this spectrum modulation increased significantly, it was reflected on the 397.5 nm pulse spectrum as seen from Fig. 3. 1.5 and 2.5 mm length doubling crystals did not create observable spectrum modulation of the passed fundamental pulse and generated SH pulses. Some central frequency shifts in longer crystals are related to the self-phase modulation of the pump pulses and cross-phase modulation of the SH pulses in the doubling crystal. The time-bandwidth product for the 397.5 nm pulse assuming sech^2 shaped pulse with a duration 185 fs was 0.65 and 0.5 for pulses doubled in crystal with length 1.5 and 2.5 mm respectively.

ACKNOWLEDGMENT

This research was supported by the Swedish Natural Science Research Council and the Lund University.

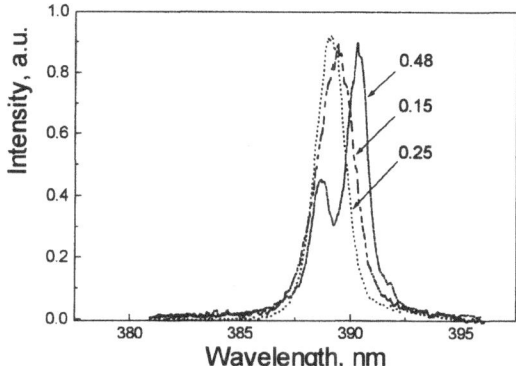

Figure 3. The second harmonic spectrum for different length doubling crystals at an incidence pulse energy of 120 mJ.

REFERENCES

1. Y. Nabekawa, K. Kondo, N. Sarukura, S. Sajiki and S. Watanabe, Opt. Lett. **18**, 1922 (1993).
2. J. Ringling, O. Kittelmann, F. Noack, G. Korn and J. Squier, Opt. Lett. **18**, 2034 (1993).
3. G. Rodriguez, J.P. Roberts and A.J. Taylor, Opt. Lett. **19**, 1145 (1994).
4. S. Svanberg, J. Larsson, A. Persson and C.-G. Wahlström, Physica Scripta **49**, 187 (1994).
5. Y.R. Shen, *The Principles of Nonlinear Optics* (Wiley, New York, 1984).

66

TIME AND SPATIAL SOLITONS WITH fs UV PULSES

Xin Miao Zhao and Jean-Claude Diels

Department of Physics and Astronomy
University of New Mexico
Albuquerque, New Mexico

The energy distribution of an intense femtosecond pulse, as it exits an amplifier chain, is shaped as a pancake: a diameter of a few cm, and a thickness of the order of 30 micron.

As it self-focuses under the influence of the nonlinear index of air, the light energy gets redistributed inside a spaghetti shaped waveguide. The transfer of energy from a pancake into a spaghetti is a problem that is not easy to formulate, and the measurements are not so unambiguous as one would like to believe. One point of debate is whether the pancake shrinks in diameter, becoming like an optical bullet launched through the spaghetti, or whether the optical energy gets redistributed uniformly inside the spaghetti? The first statement cannot be correct for the first half of the pulse. On the leading edge of the pulse, as the power increases beyond the critical power, the beam focuses at a distance decreasing as time progresses: the shortest focal distance corresponding to the peak power of the pulse. Therefore, by causality, there can be no self-trapping of the energy in the first half of the pulse into a "light bullet," but instead the focal point moves in opposite direction to that of the light. The energy density in the leading edge of the pulse, initially contained in a pancake shaped volume, gets distributed along a very long line of foci. It can easily be estimated that the energy density in the moving focus corresponding to the leading edge of the pulse is not sufficient to induce the intense nonlinear effects that have been observed.[1] Indeed, the volume of a typical "filament" 100 μm diameter $\times$ 10 m long is even larger than that of the original "pancake" from which it is issued.

The strong conical emission that has been observed[1] is evidence however that some power from the trailing edge of the pulse remains trapped in a self-induced tubular waveguide after self-focusing. Self-trapping in a stable filament requires a contribution to the nonlinear index of higher order and opposite sign to the $n_2 I$ term. Such a negative contribution to the nonlinear index can be attributed to the electrons generated by avalanche ionization.[2,1] The hollow waveguide theory can be applied to this problem to get an estimate of the characteristic length of these lossy waveguides. The result is two orders of magnitudes shorter than the distance over which the filaments have been observed. Indeed:

the loss factor for a hollow waveguide[3] (Lowest order mode EH_{11}; with fields components in polar coordinates $E_r = J_0(u_{11}\frac{r}{a})\cos\theta$; $E_\theta = J_0(u_{11}\frac{r}{a})\sin\theta$) is given by:

$$\gamma \approx \left(\frac{u_{11}}{2\pi}\right)^2 \frac{\lambda^2}{a^3} \frac{1}{\sqrt{2\Delta n}} \tag{1}$$

where u_{11} is the first solution of the Bessel function J_0, $J_0(u) = 0$, λ is the wavelength, a is the radius of the waveguide with an index discontinuity Δn. Typical numbers at 800 nm are $7 \cdot 10^{10}$ W for the power trapped (1 mJ in 150 fs), or an intensity or $2 \cdot 10^{14}$ W/cm^2, given a filament radius of 100 μm.[1] At that intensity, the index change $n_2 I = 3 \cdot 10^{-19}$ cm^2/W $\times 2 \cdot 10^{14}$ W/cm^2 $= 6 \cdot 10^{-5}$. Expressing that this positive nonlinearity has to be balanced by the negative contribution due to the field-ionized electrons leads to an estimate of $N_e \geq 10^{16}$ cm^{-3} for the electron density.[1] Using these numbers to compute the loss factor, we find that the characteristic length γ^{-1} for the filaments should not exceed 10 to 20 cm. We note here two major discrepancies: (a) filaments appear to exist over tens of meters, or much longer distances than the characteristic loss distance of 20 cm; and (b) with an electron density of 10^{16} cm^{-3}, one would expect that a discharge could be easily induced between charged electrodes — this is not the case.

A possible explanation for the contradiction between experimental observations and simple order of magnitude evaluations is that filaments are created at successive depths. The presence of non contiguous, centimeter long filaments, instead of a single continuous filament over tens of meters would explain the impossibility to induce and guide a discharge through these filaments at 800 nm.[2,4]

There are significantly more data available on the propagation of short and intense femtosecond pulses in the IR than in the UV. The physics appears at first much simpler at the longer wavelength, where dispersion, scattering and multiphoton ionization are negligible. The advantages of working at a shorter wavelength are a more gradual ionization (second to fourth order in intensity) and reduced diffraction, resulting in better waveguides.

We have made preliminary observations of filamentation in a beam of 200 fs pulses at 248 nm.[5]

With a nonlinear index $n_2 = 2 \cdot 10^{-19}$ cm^2/W at 248 nm, and using for the multiphoton ionization $N_e = 1.5 \cdot 10^{-11} I^2$ cm^{-3} (Ref. 6) for a pulse duration of 100 fs, one finds that the self-focusing and de-focusing balance each other at the intensity of $I_0 = 5 \cdot 10^{12}$ W/cm^2. Since the critical power corresponding to that nonlinear index is $2.5 \cdot 10^9$ W, one can estimate the cross section of the filament to be $A = P_{cr}/I_0 = (2.5 \cdot 10^9)/(5 \cdot 10^{12}) = 0.0005$ cm^2, or a diameter of approximately 250 μm. This value is in close agreement with our measurements.[5] Of particular interest also is the hollow waveguide loss equation (1) applied to this UV situation: we find with these numbers a characteristic distance of 5 m, which is approximately the distance over which filaments were observed.

At the wavelength of 248 nm, a pulse of 50 fs duration will double its duration after a characteristic distance of $L_D = \tau^2/(2k'') \approx 5m$, where $k'' = d^2k/d\Omega^2$ is the group velocity dispersion coefficient.[7] The phase modulation due to the Kerr effect ($n_2 I$) would result in an accelerated pulse broadening with distance. The phase modulation due to the ionization on axis should cause a downchirp, resulting in a stabilization of the pulse duration as occurs in the case of solitons in optical fibers.

To which extent can the formation or non-formation of filaments be controlled? In our previous tests,[5] the formation of filaments may have been related to the very irregular intensity profile of the beam. In the case of a Gaussian profile and for UV pulses, one expect the multiphoton absorption to flatten the intensity profile of the beam, thereby possibly preventing self-focusing.

To resolve the question of existence and characteristic lengths of *continuous* filaments, our next tests will be to send single filaments between electrodes to which a pulsed electric field of 1 μs is applied. As we have learned from our previous studies, a weak photoionization of a needle shaped volume can be made visible in an electric field, because of the multiplication of electrons and ions in applied electric field, and excitation of nitrogen which produced pink fluorescence. If the electric field is interrupted before a discharge occurs, the pink fluorescence of nitrogen makes visible the photoionized regions (we have observed previously the glow of nitrogen along a preionized path).

REFERENCES

1. E. T. J. Nibbering, P. F. Curley, G. Grillon, B. S. Prade, M. A. Franco, F. Salin, and A. Mysyrowicz. *Opt. Lett.*, 21:62–64, 1996.
2. Xin Miao Zhao, Jean-Claude Diels, A. Braun, X. Liu, D. Du, G. Korn, G. Mourou, and Juan Elizondo. In P. F. Barbara, W. H. Knox, G. A. Mourou, and A. H. Zewail, editors, *Ultrafast Phenomena IX*, pages 233–235, Dana Point, CA, 1994. Springer Verlag, Berlin.
3. E. A. J. Marcatili and R. A. Schmeltzer. *Bell Syst. Tech. J.*, 43:1783, 1964.
4. M. Miki and T. Shindo. Private communication, 1995.
5. Xin Miao Zhao, Patrick Rambo, and Jean-Claude Diels. In *CLEO, 1995*, page QThD2, Baltimore, MA, 1995. Optical Society of America.
6. L. L. Losev and V. I. Soskov. *Optica atmosfery*, 3:842–846, 1990.
7. J.-C. Diels and W. Rudolph. *Ultrashort Laser Pulse Phenomena* Academic Press, Boston, 1995.

67

HIGH HARMONIC GENERATION EFFICIENCY

A. Sanpera,[1] M. Protopapas,[2] J. B. Watson,[1] K. Burnett,[1] and P. L. Knight[2]

[1]Blackett Laboratory, Optics Section
Imperial College
Prince Consort Road, London, UK
[2]Clarendon Laboratory
Department of Physics
University of Oxford
Parks Road, Oxford, UK

1. INTRODUCTION

In the last few years the availability of very intense ultrashort laser pulses has created a new experimental domain for the study of laser atom interactions. Among many new phenomena, the generation of high order harmonics has opened up the possibility of producing coherent radiation in the extreme ultraviolet and (soft) X-ray regions. By harmonic generation we mean the conversion of the fundamental laser wavelength to high odd harmonic wavelengths when an atom is irradiated by a strong laser field. A complete description of the process should include both the response of the single atom to the laser radiation as well as the collective response of an ensemble of atoms. The harmonic conversion efficiencies obtained in experiments are dependent, to a large extent, on collective effects. Since there will be significant ionization of the medium, the refractive index will be modified and this in turn changes the phase matching between the harmonics and the fundamental. Nevertheless it is the strongly nonlinear response of the single atom that produces the large conversion efficiency from the fundamental to shorter wavelengths.

In both, experimental and theoretical work, the emitted spectrum is found to contain a comb of harmonics with approximately constant amplitude followed by an abrupt cut-off. This is commonly referred to as the harmonic generation plateau. The upper limit of this plateau is determined by the maximum kinetic energy that an electron (that has been previously promoted to the continuum) could have if it is to be driven back to the core by the action of the field. This maximum kinetic energy is approximately $3.2U_p$ ($U_p = E^2/4\omega^2$ in atomic units). It depends only on the intensity and frequency of the incident radiation, since at such intensities the interaction of the electron in the continuum with the external

Ultrafast Processes in Spectroscopy
Edited by Svelto et al., Plenum Press, New York, 1996

field dominates over the Coulomb interaction.[1] This is the origin of the well known cut-off rule

$$h\nu_{max} = U_i + 3.2U_p, \tag{1}$$

to the harmonic generation where U_i refers to the ionization potential. Harmonics corresponding to higher photon energies are produced with an exponentially diminishing efficiency thus becoming rapidly indistinguishable from the background.

From a theoretical point of view the interaction of a high-intensity field with an atom has to be calculated without introducing the usual perturbation theories; the study of this interaction entails numerical simulation that needs massive computations. In the strong regime, a much better understanding of the process have been achieved by using quasi classical models. Those models pioneered by Keldysh[2] and valid in the regime of tunneling ionization, have been very successfully applied to understand the cut-off rule in the harmonic generation. The quasi classical models assume two completely separate steps in the harmonic generation. In the first step the electron, initially in the ground state, is promoted to the continuum either by tunneling through the atomic barrier (tunneling ionization) or by absorbing several photons (multiphoton ionization) from the field. The second step deals with the evolution of the electron after it has been released, which as a first approximation, could be described classically. Once in the continuum, depending on the initial velocity and phase, there is a large probability that the electron will be driven back to the atomic core, emitting a photon in the transition back to the ground state. Since the atom has central symmetry, only odd harmonics will be produced. The first step determines that the efficiency on the harmonic signal is found to be directly proportional to the ionization rate. The fact that the dynamics of an atom driven by a strong laser field in the tunneling regime can be partially explained by means of classical trajectories suggests, therefore, that by properly modifying the electron trajectories we should be able to control the harmonic generation emission. This type of "coherence control[3]" of the atomic process can be done e.g by driving the atom by two different frequencies simultaneously with an adjustable phase between the two laser.

2. METHOD AND RESULTS

We calculate the harmonic generation from a Hydrogen atom interacting with a two color laser field. The two laser frequencies are chosen to be the fundamental and its third harmonic at the same intensity and with a relative phase between them. Our study is based on numerical solutions of the time dependent Schrödinger equation, which we solve using a partial wave decomposition of the electron's wave function, together with a split operator and Crank-Nicholson algorithm. We choose to work in the length gauge, where the interaction term can be written in the form

$$H_{in} = \vec{r} \cdot (\vec{E}_\omega \sin(\omega t) + \vec{E}_{3\omega} \sin(3\omega t + \phi))f(t). \tag{2}$$

where $\vec{E}_\omega$ and $\vec{E}_{3\omega}$ are the fundamental and third harmonic field amplitudes, ω is the fundamental frequency, and ϕ the relative phase between the two fields. The function $f(t)$ is the pulse shape function which is chosen to be $f(t) = \sin^2(\pi t/\tau)$, where τ is the pulse length; here we consider a pulse length of 32 optical cycles of the fundamental frequency. To produce the maximum enhancement of the harmonic conversion efficiency we choose the same field amplitude for both fields, i.e. $E_\omega = E_{3\omega}$. With these conditions, the time varying

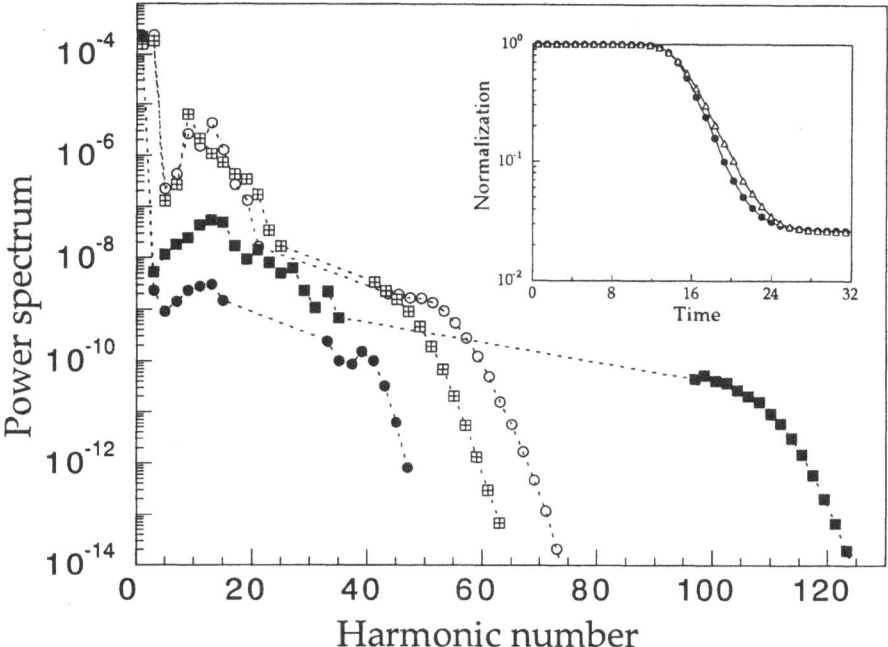

Figure 1. Comparisons between one (filled symbols) and two color (open symbols) harmonic spectra in Hydrogen in the tunneling regime. Filled circles correspond to $E_\omega = 0.05$ a.u., filled squares to $E_\omega = 0.10$ a.u. Two colors $E_\omega = E_{3\omega} = 0.05$ a.u. and $\phi = 0$ (open circles), and $\phi = \pi/2$ (open squares). The inset window shows the corresponding ionization yields.

electric field has a maximum for a relative phase difference between the two fields of $\phi = \pi$, when the maximum field strength of the combined fields reaches the value $E = 2 \times E_\omega$ twice each cycle. For a phase difference of $\phi = \pi/2$ the maximum amplitude achieved is $E = 1.8 \times E_\omega$, and finally for $\phi = 0$ the maximum field amplitude is $E = 1.6 \times E_\omega$.

The two color harmonic spectra show formally the same pattern as their one color partner: a plateau of harmonics and an abrupt cut-off where the intensity of the harmonics drops off rapidly. However, a closer inspection reveals significant differences. First of all, the plateau structure extends to higher orders than the fundamental alone at $E = E_\omega$, although the overall plateau structure appears less clearly defined. The most striking difference, however, is the different conversion efficiency in both cases. Generally speaking, the conversion efficiency is much better for the two color case regardless of the relative phase. This is especially true for low energy harmonics, those with energies below $U_i + U_p$, which are enhanced by almost four orders of magnitude compared to the one color case at $E = E_\omega$.

Furthermore, if we compare conversion efficiencies between two colors ($E_\omega = E_{3\omega}$) and the fundamental alone at $E = 2 \times E_\omega$, we still find harmonics with intensities approximately two orders of magnitude larger in the bichromatic case. Since the residual ionization for two colors is smaller than the one produced by the strong field ($E = 2 \times E_\omega$), as it is clearly shown in Fig. 1, the conversion efficiency increase seems to be partially related to the interference between the two colors. We know that harmonics are produced in transitions involving the ground state, and that implies that harmonics should be produced dominantly by the part of the wave packet closest to the nucleus. Decreasing the velocity of the particle when it "crosses" the nucleus increases the interaction time and therefore the cross section of scattering. The higher order harmonics, which correspond to electrons returning with high velocities, would be less affected by the retardation produced by the second color and the enhancement would become smaller.

The cut-off in the harmonics can be understood by applying the classical two-step model to the two color case. From this classical model the predicted cut-off energies agree reasonably well with the quantum results: $4.8U_p$ for $\phi = 0$, $5.1U_p$ for $\phi = \pi/2$ and 4.1 for $\phi = \pi$, (U_p refers to the ponderomotive potential of the fundamental). The classical model also predicts a uniform enhancement in conversion efficiency along the whole range of allowed returning kinetic energies. However, this enhancement comes from the fact that the instantaneous electric field strength of the combined fields increases. This dependence on the instantaneous electric field arises from the definition of the classical collision probability. We define the collision probability as the allowed returning trajectories weighted by the corresponding DC ionization rate, $\Gamma(t) \propto [4U_i/\xi(t)]e^{(-2/3\xi(t))}$, corresponding to the instantaneous electric field, $\xi(t)$, at which the electron escapes through or over the barrier. We have repeated our calculations for higher field intensities and the results are in agreement with those we have already presented. However, the improvement in the conversion efficiency is obviously reduced if we are above the saturation intensity or in the regime where over the barrier ionization is important. Finally, for lower ratios between the two fields $E_\omega/E_{3\omega} > 1$ the enhancement in the harmonics is also smaller.[4]

This work is financially supported by the U.K. Engineering and Physical Science Research Council, as well as the HCM program of the European Community. A.S also acknowledges Fleming/MEC program.

REFERENCES

1. P. B. Corkum, Phys. Rev. Lett. **71**,1994 (1993); K. J. Schafer, and K. C. Kulander, Phys. Rev. A **45**, 8026 (1992).
2. L. V. Keldysh, Zh. Ekps. Teor. Fiz. **47**,1945 (1964).
3. E. Charron, A. Giusty-Suzor, and F. H. Mies, Phys. Rev. Lett. **71**, 692 (1993); S. Watanabe, K. Kondo, Y. Abekawa, A. Sagisaka, and Y. Kobayashi, Phys. Rev. Lett. **73**, 2692 (1994).
4. M. Protopapas, A. Sanpera, P. L. Knight, and K. Burnett, Phys. Rev. A **52**, R2527 (1995).

68

PICOSECOND INTERFEROMETRY OF PLASMAS BEFORE AND AFTER SHORT LASER PULSE PROPAGATION

A. Giulietti,[1] M. Borghesi,[2] C. Danson,[3] D. Giulietti,[4] L. A. Gizzi,[1] A. Macchi,[1] and D. Neely[3]

[1] Istituto di Fisica Atomica e Molecolare del CNR
Pisa, Italy
[2] Imperial College of Science, Technology, and Medicine
London, United Kingdom
[3] Central Laser Facility
Rutherford Appleton Laboratory
Chilton, United Kingdom
[4] Dipartimento di Fisica
Università di Pisa, Italy

The propagation of an intense short EM wave packet through a plasma is of great relevance for both basic physics and applications, the latter including new schemes of fuel ignition for Inertial Confinement Fusion (ICF)[1], X-ray lasers[2], plasma based electron accelerators[3]. Furthermore an unexplored class of light induced effects in hot ionised matter can be studied, since the oscillating electric field on the electrons can overcome the atomic field and the electron quiver velocity can approach relativistic values. At the same time, some phenomena playing a crucial role with long pulse interaction, as hydrodynamic motion and collisional absorption, can be virtually suppressed. The experimental study of the propagation of short, intense laser pulses through preformed, underdense plasmas is therefore of great physical interest; this study requires the use of diagnostics which allow for a detailed knowledge of the parameters of the preformed plasma and which are sensitive to the effects of short pulse propagation in the plasma.

In this communication we present novel experimental results on the propagation in an underdense plasma of short (700 fs) laser pulses at irradiances from 10^{16} Wcm^{-2} to 10^{17} Wcm^{-2}. Though below relativistic values, this range of irradiances has not been explored so far for propagation in long-scalelength plasmas of interest for ICF. We shall focus our attention mainly on the use of picosecond interferometry to characterise the preformed plasma and to study the effect of short pulse propagation. The use of short (<1 ps) laser pulses for interferometry has lead to a very significant improvement in the 2-D density

Ultrafast Processes in Spectroscopy, edited by Svelto et al.
Plenum Press, New York, 1996

299

mapping of the plasma. Beside interferometry, other diagnostics included calorimetry and cross-section imaging of the interaction pulse transmitted through the preformed plasma.

The experiment was performed at the Central Laser Facility of the Rutherford Appleton Laboratory (UK). The set-up geometry is essentially the same that was used in a previous experiment on long (600 ps) pulse interaction[4]. The plasma was produced by four 600 ps, 1.053 μm laser "heating" beams focused on target in opposite pairs. The target consisted of Al disks coated on very thin (0.1 μm) plastic stripes. The disk thickness was 0.4 μm and the diameter was 0.8 mm, matching the size of the laser spot. The irradiance on each side of the target was varied in the range between 10^{13} Wcm^{-2} and 10^{14} Wcm^{-2}. A 700 fs laser pulse, obtained with the Chirped-Pulse-Amplification (CPA) technique, was interacted with the preformed plasma by 2 ns after the peak of the 600 ps laser beams. The short interaction pulse beam was focused with a large f-number along the plasma main axis in a spot much smaller than the plasma cross section, in order to avoid boundary refraction effects. A minor portion of the short pulse energy was split, frequency doubled, and used as an optical probe for interferometry.

The interferometry used a modified Nomarski configuration[5]. In this interferometer, a Wollaston prism is used to produce two separate ortogonally polarised images of the plasma surrounded by an unperturbed background. Interference between each of the two images and the background of the other image is achieved by putting a polariser before the film plane, oriented at $45°$ with respect to the polarisation axis. A spatial filter was inserted before the Wollaston prism in order to reduce noise from plasma self-emission and probe scattered light. The line of view of the interferometer, "y", was perpendicular to the interaction beam axis, "x".

Two-dimensional density profiles of the plasma were obtained from the analysis of the fringe pattern in the interferometer image plane. If the plasma has cylindrical symmetry, the phase shift induced by the plasma in the (x,z) plane, perpendicular to the interferometer axis, can be written in terms of the electron density $n_e(r,x)$ as

$$\Delta\phi(x,z) = \frac{2\pi}{\lambda n_c} \int_z^{r_0} \frac{n_e(r,x)}{\sqrt{r^2 - z^2}} r\,dr \tag{1}$$

where r is the radial coordinate from the interaction axis, r_0 is the plasma radius, λ is the probe beam wavelength, n_c is the critical density corresponding to λ. A detailed discussion of Eq.1 and its underlying approximations can be found elsewhere.[4] The fringe intensity $I(x,z)$ in the interferometer image plane can be written generally as[6]

$$I(x,z) = a(x,z) + [c(x,z) \exp 2\pi \, if_u x + c.c] \tag{2}$$

where $c(x,z) = 1/2 b(x,z) \exp(i\Delta\phi(x,z))$, f_u is the number of fringes per unit length, and $a(x,z)$ and $b(x,z)$ account for background nonuniformities and fringe visibility. In Eq.2 $c(x,z)$ contains the physical information on plasma density; the background contribution can be separated taking the Fourier transform of Eq.1:

$$F_I(f,z) = F_a(f,z) + F_c(f - f_u,z) + F_{c*}(f + f_u,z) \tag{3}$$

If the scalelength of background uniformities along x is much larger than the fringe separation, the background contribution in Eq.2 can be separated in order to obtain $F_c(x,z)$; $c(x,z)$ is then obtained via inverse Fourier transform; in practice this is carried out using a

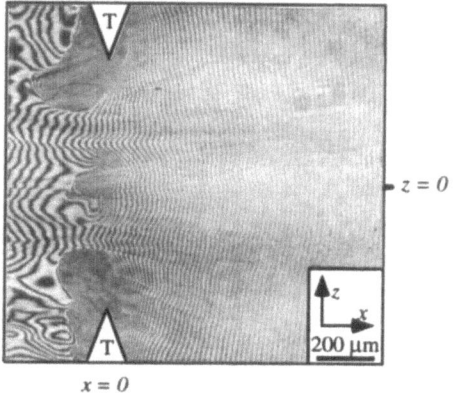

$z = 0$

$x = 0$

Figure 1. Interferogram of the preformed plasma taken 2.2 ns after the peak of the heating laser pulses, for an intensity of 5.0×10^{13} Wcm^{-2} on target.

Fast Fourier Transform code. Finally $\phi(x,z)$ is deduced from the logarithm of $c(x,z)$ and the 2-D density map $n_e(r,x)$ is obtained by Abel inversion of Eq.1. It must be noted that also small-scale perturbations of the electron density can contribute to the phase shift. The interferometer was found to be very sensitive to these density perturbations[4], making this interferometric technique very effective in detecting density inhomogeneities in the preformed plasma or perturbations induced by the interaction pulse.

A representative interferogram of the preformed plasma before short pulse interaction is shown in Fig.1. The intensity of the heating beams on each side of the Al target was 5.0×10^{13} Wcm^{-2}. The probe pulse was delayed by 2.2 ns with respect to the peak of the heating beams. The 2-D density map reconstructed from the fringe pattern in Fig.1 is shown in Fig.2.

It is very interesting to compare the interferogram in Fig.1 with the interferograms previously obtained with a very similar interferometer configuration, but using a "long" 100 ps probe pulse[4]. In those interferograms the fringe visibility vanished in the denser region of the plasma and thus it was not possible to measure the electron density in that plasma region. This effect was due to the electron density evolution which smeared out part of the pattern during the probe pulse. In fact, the density variation rate in the denser region was high enough to induce, in 100 ps, a phase change that leads to an almost complete loss of visibility. Reducing the probe duration to less than 1 ps eliminates the "smearing" effect completely, making fringes visible over the whole interferogram and thus allowing for a complete density mapping of the plasma.

The detailed density mapping of the preformed plasma at times corresponding to the interaction with the short pulse was very important in the study of the short pulse propagation. We found that the propagation is independent from the short pulse irradiance in the considered range, but strongly depends on the plasma density distribution. Fig.3 shows the fraction of the short pulse energy transmitted through the preformed plasma versus the heating beams irradiance I_H on the Al target; it is evident that short pulse transmission strongly increases over a "threshold" value of I_H. These data were compared with density maps for different values of I_H, which showed that at $I_H \approx 7 \times 10^{13}$ Wcm^{-2} the density along the plasma symmetry axis at 2 ns after plasma formation turns from a maximum to a local minimum. This effect is purely hydrodynamic and is in good agreement with 2-D simula-

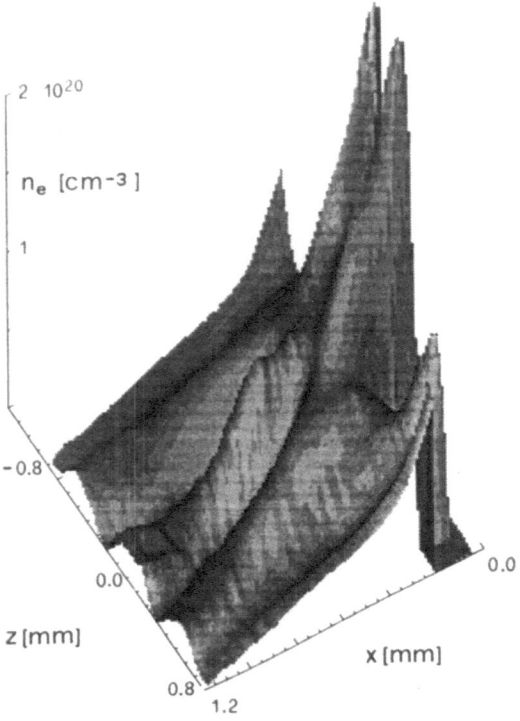

Figure 2. 2-D electron density map reconstructed from the interferogram shown in fig.1.

tions performed with the POLLUX code[7]. In this way a sort of channel is produced, allow-ing for a good propagation of the short pulse. Correspondingly, cross-beam images showed that in presence of the density minimum the short pulse beam avoids strong re-fraction from the plasma.

A number of interferograms were also taken after propagation of the short pulse through the preformed plasma. The high sensitivity of the interferometer allowed evidenti-

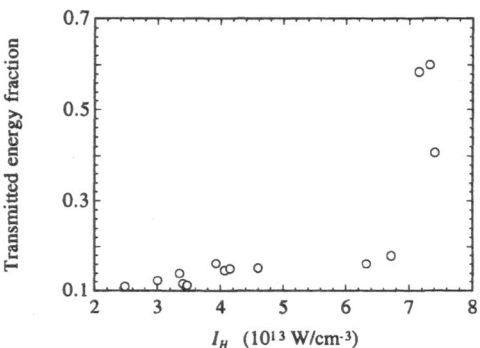

Figure 3. Fraction of the incident short pulse energy collected forward vs. heating beams irradiance I_H on each target side.

ate density perturbations of the order of 1/10 of the local electron density along the interaction axis, probably produced by ionisation. These data are presently under investigation.

In conclusion, our experimental results suggest that a short pulse of subrelativistic intensity is likely to propagate through an underdense plasma with very low losses provided a (even weak) density minimum is set along its axis. On the contrary, it was previously found that propagation of longer (600 ps) pulses through the preformed plasma was heavily absorbed, scattered and affected by self-focusing[8]. The use of picosecond interferometry was found to be very effective to characterise the preformed plasma, thus shedding light on the physics of short pulse propagation.

REFERENCES

1. M.Tabak, J.Hammer, M.Glinsky, W.Kruer, S.Wilks, J.Woodworth, E.Campbell, M.Perry, Phys.Plasmas 1, 1626 (1994)
2. C.Durfee, H.Milchberg, Phys.Rev.Lett. 71, 2409 (1993)
3. P.Sprangle & E.Esarey, Phys. Fluids B 4, 2241 (1992)
4. L.A.Gizzi, D.Giulietti, A.Giulietti, T.Afshar-Rad, V.Biancalana, P.Chessa, C.Danson, E.Schifano, S.M.Viana, and O.Willi, Phys.Rev.E 49, 5628 (1994); Phys. Rev.E 50, 4266 (1994)
5. M.Nomarski, Phys.Rad. 16, 95 (1955); R.Benattar, C.Popovics, and R.Sigel, Rev.Sci.Instruments 50, 1583 (1979)
6. M.Takeda, H.Ia, and S.Kobayashi, J.Opt.Soc.Am. 72, 156 (1982)
7. G.J.Pert, J.Comput.Phys. 43, 111 (1981)
8. A.Giulietti, E.Schifano, V.Biancalalana, C.Danson, D.Giulietti, L.A.Gizzi, and O.Willi, submitted to Europhys.Lett. (1995) for publication.

69

HIGH HARMONIC GENERATION AS A RECOLLISION PROCESS

P. L. Knight,[1] C. H. Keitel,[1] K. Burnett,[2] and J. B. Watson[2]

[1]Blackett Laboratory, Optics Section
Imperial College, Prince Consort Road
London, UK
[2]Clarendon Laboratory
Department of Physics
University of Oxford
Parks Road, Oxford, UK

ABSTRACT

High harmonics are generated when wavepackets created by tunneling recollide with the nucleus after ponderomotive acceleration. We distinguish quantum and classical contributions and identify novel relativistic features.

1. INTRODUCTION

The interaction of atoms with super intense electromagnetic fields has shown to give rise to the emergence of high harmonics as a result of tunneling and recollision with the nucleus after ponderomotive acceleration.[1,2] Many aspect of this interaction have been explained satisfactorily with purely classical models while others have required a quantum mechanical treatment of the atom. We investigate the dynamics of these strong field processes more quantitatively by comparing the time dependent phase space quantum Wigner functions with classical Monte Carlo distributions. In particular, negativities in the Wigner function will give us a clear indication of the existence of quantum mechanical behavior.[3]

2. THE MODEL

In the quantum case we solve the time-dependent Schrödinger equation in one dimension in the Kramers-Henneberger frame using the soft-core potential to obtain the wavefunction at later times

$$i\hbar\partial_t\psi_{KH}(q,t) = [-\frac{\hbar^2\partial_q^2}{2m} + V(q + \alpha(t))]\psi_{KH}(q,t). \tag{1}$$

where $V(q)$ is the potential due to the nucleus and $\alpha(t)$ the ponderomotive quiver. In this calculation we use the soft-core (or Rochester) potential $V(q) = -e^2/\sqrt{2 + q^2}$ scaled to have its lowest eigenstate at the same energy as the ground state of hydrogen. The wavefunction gives us complete information about the state of the system at any time. We use it to calculate the corresponding Wigner function, given by:

$$W(q,p,t) = \frac{1}{\pi\hbar} \int_{-\infty}^{+\infty} dq'\psi(q - q',t)\psi^*(q + q',t)\exp[2iq'p/\hbar]. \tag{2}$$

From this definition we can see that the Wigner function depends on time, and also on the spatial and momentum phase space variables q and p.

In the classical case we start with a distribution of points in phase-space corresponding to the classical orbit (1D) and then solve the classical equations of motion for each set of initial conditions. In most cases we concentrate on the motion of the quivering electron and its "recollision" with the atomic core to generate high harmonics.

In the relativistic classical calculations we have to solve the 3D trajectories of the electron via the Lorentz equation

$$m\frac{d}{dt}\frac{\dot{\vec{r}}}{\sqrt{1 - (\dot{\vec{r}}/c)^2}} = e\vec{E}(t)\cos(\omega t - \vec{k}\vec{r}) + \frac{e}{c}\dot{\vec{r}} \times \vec{H}(t)\cos(\omega t - \vec{k}\vec{r}) - \frac{e\vec{r}}{|\vec{r}|^3}, \tag{3}$$

where $\dot{\vec{r}}$ denotes the velocity of the electron, $\vec{k}$ the propagation direction of the laser field, $\vec{E}(t)$ and $\vec{H}(t)$ the electric and magnetic field vector, respectively.

3. QUANTUM SIGNATURES

In Fig. 1a and b we show the contours of the Wigner function in the tunneling and in the stabilization regime. In both cases the existence of negativities is apparent from the white areas indicating the presence of quantum features. In the tunneling regime however there are only small negativities within a fringe structure reflecting the competition of classical aspects of the recollision process versus the quantum mechanical aspects of tunneling and wave spreading. In the stabilization domain we have strong quantum signatures in the form of two symmetrically orientated strongly negative areas. The classical distributions are clearly different from the contours of the Wigner functions in both domains due to intrinsically quantum mechanical behavior; however we find that there is more resemblance between classical and quantum results in the tunneling domain. We explain the strong differences in the stabilization domain by considering the Wigner function of the corresponding dressed eigenstates, i.e., the eigenstates of the Kramers Henneberger potential. Pulsed excitation results in a superposition of KH eigenstates which interfere; the interferences generate substantial regions in which the Wigner functions are negative, something which classical dynamics cannot replicate. A superposition of a few eigenstates can not be modelled simply with a classical distribution and as a consequence we have seen total disagreement between the quantum and classical models in this domain. We find that the classical distribution matches remarkably well the Wigner function towards the end of the laser pulse in the stabilization domain as can be seen from Fig. 2.

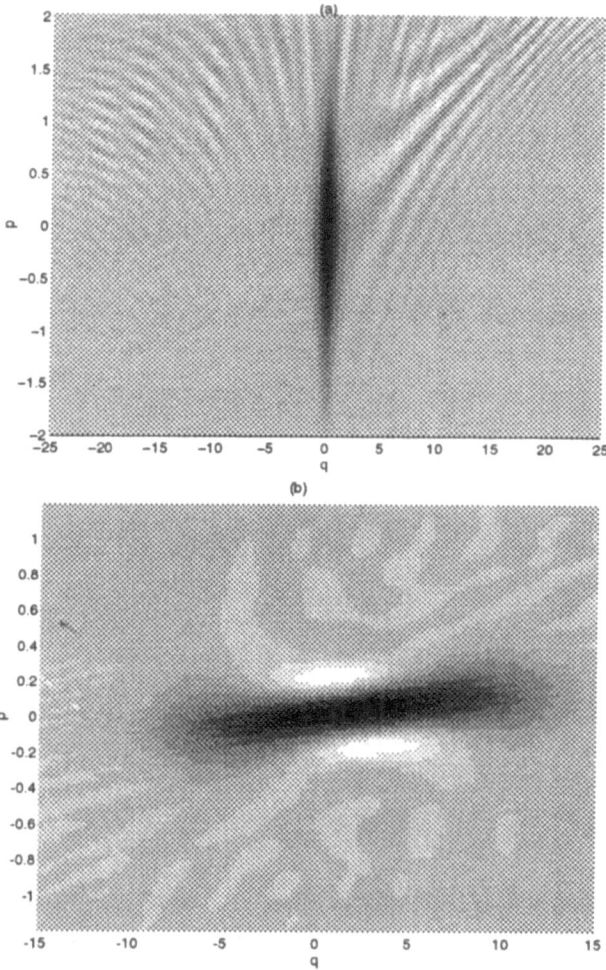

Figure 1. The contours of the Wigner function of the hydrogenic electron *after 12 cycles* of the application of a pulsed laser field with 12 turn-on cycles, 12 turn-off cycles, for (a) maximal electric field strength $E_0 = 0.6$ a.u. and angular frequency $\omega = 0.04$ a.u. (b) maximal electric field strength $E_0 = 10$ a.u. and angular frequency $\omega = 1$ a.u.

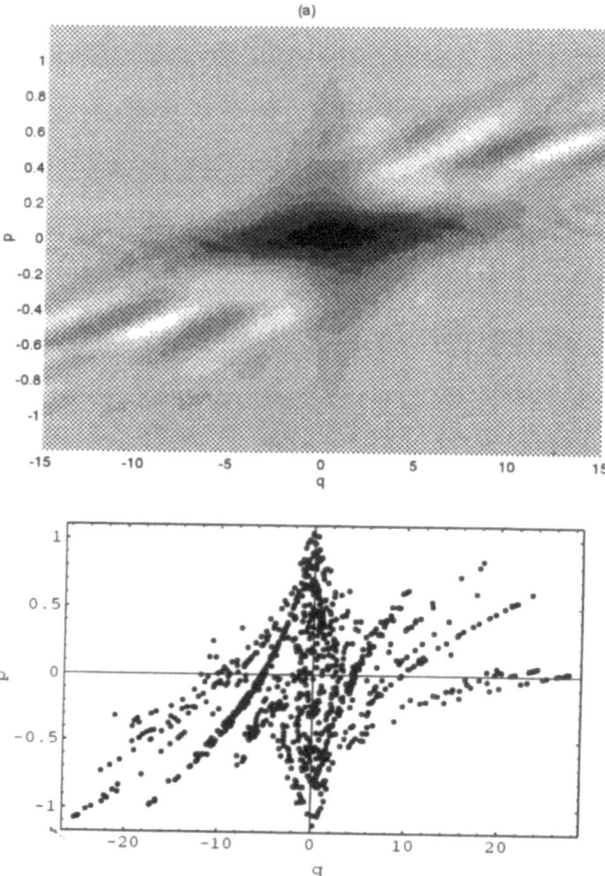

Figure 2. (a) The contour of the Wigner function and (b) classical phase space distribution of the hydrogenic electron *immediately after* the application of a pulsed laser field with 12 turn-on cycles, 12 turn-off cycles, maximal electric field strength $E_0 = 10$ a.u. and angular frequency $\omega = 1$ a.u.

4. RELATIVISTIC SIGNATURES

When we increase the intensity of the laser field or reduce its frequency, the velocity of the electron may approach that of light such that a relativistic treatment becomes necessary. The magnetic component of the laser field will then become important (see (3)) resulting in an increase of the ionization process in the forward direction of the laser field.[4] For low frequencies the ionization rate increases continuously while for high frequencies we pass first the stabilization domain before its break-down for relativistic laser field intensities (see Fig. 3). In general, few harmonics are observed in our single atom treatment because of the magnetically induced three dimensional motion of the electron for intensities approaching the relativistic regime, and because of high ionization probabilities, i.e. also the break down of stabilization for strongly relativistic laser intensities.[4]

The financial support of the European Community and of the U.K. Engineering and Physical Sciences Research Council is gratefully acknowledged.

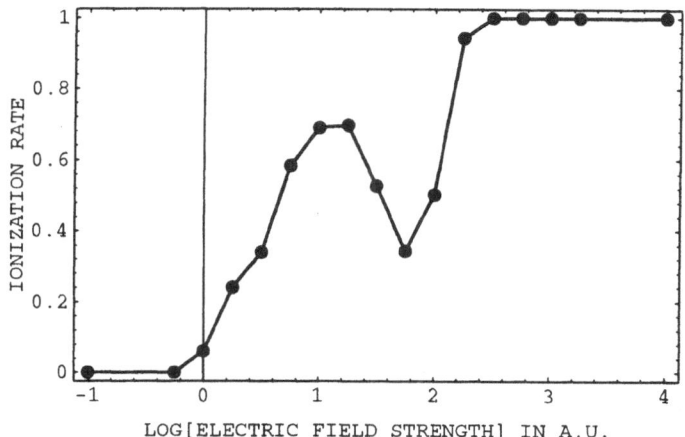

Figure 3. Ionization probability as a function of the electric field strength of the incoming laser light in a logarithmic plot for $w = 5$ a.u. for a pulse of 10 cycles.

REFERENCES

1. J. L. Krause, K. J. Schafer, and K. C. Kulander, Phys. Rev. Lett. **68**, 3535 (1992); P. B. Corkum, Phys. Rev. Lett. **71**, 1994 (1993).
2. M. Gavrila (Ed), "Atoms in Intense Laser Fields," (Academic Press, Boston, 1992); K. Burnett, V. C. Reed, P. L. Knight, J. Phys. B **26**, 561 (1993).
3. J. B. Watson, C. H. Keitel, P. L. Knight and K. Burnett, "Quantum signatures in the stabilization dynamics," Phys. Rev. **A52**, in press (1995).
4. T. Katsouleas and W. B. Mori, Phys. Rev. Lett. **70**, 1561(C) (1993); C. H. Keitel and P. L. Knight, Phys. Rev. **A51**, 1420 (1995).

70

ELECTRON IMAGING IN SHORT-PULSE STRONG FIELD MULTIPHOTON IONIZATION

V. Schyja, T. Lang, and H. Helm

Universitaet Freiburg
Freiburg i. Br., Germany

Spatial distributions of photoelectrons produced by multiphoton ionization are recorded by projecting the expanding photoelectron cloud onto a two-dimensional position-sensitive detector[1], as shown in Figure 1. The projected images provide a direct view of the squared angular part of wave function of the free electrons as well as their velocity.

We have investigated the multiphoton ionization of xenon and argon atoms at a wavelength of 800 nm and at intensities between $1*10^{13}$ and $2*10^{14}$ W/cm^2. Recent work on multiphoton ionization of atoms and molecules in intense laser fields has shown that the dynamics of ionization are governed by the modification of the electronic structure of the target by the radiation field[2-6]. The effective energies of the excited states are ac Stark shifted so strongly, that the laser, which at zero intensity is not resonant with any particular multiphoton transition, becomes resonant with individual intermediate states at specific critical intensities. An atom in an intense radiation field experiences an increase of ionization potential of

$$U_p = e^2 F_0^2 / 4\, m\, \omega^2$$

where F_0 is the peak electric and ω the frequency field of the optical wave at the position of the atom.

For short laser pulses this ponderomotive energy does not appear in the drift energy of the photoelectrons thus permitting us to analyse the energy of the intermediate excited state.

Eight photons of 1.55eV are needed to ionize xenon at zero intensity. At an intensity of $1.8*10^{13}$ W/cm^2 the 8 photon energy is shifted below the ionization potential, such that eight photon resonances can contribute to resonant multiphoton ionization. At the lowest intensities studied we find that the dominant intemediate state is the 10p state producing photoelectrons with ~1.22 eV. At higher intensities the 4f state becomes more important with a signature of ~684 meV electrons. At intensities higher than $2.6*10^{13}$ W/cm^2 the contribution of the 4f state decreases and an additional contribution appears near 400 meV, possible due to the participation of the 7p state.

Ultrafast Processes in Spectroscopy, edited by Svelto et al.
Plenum Press, New York, 1996

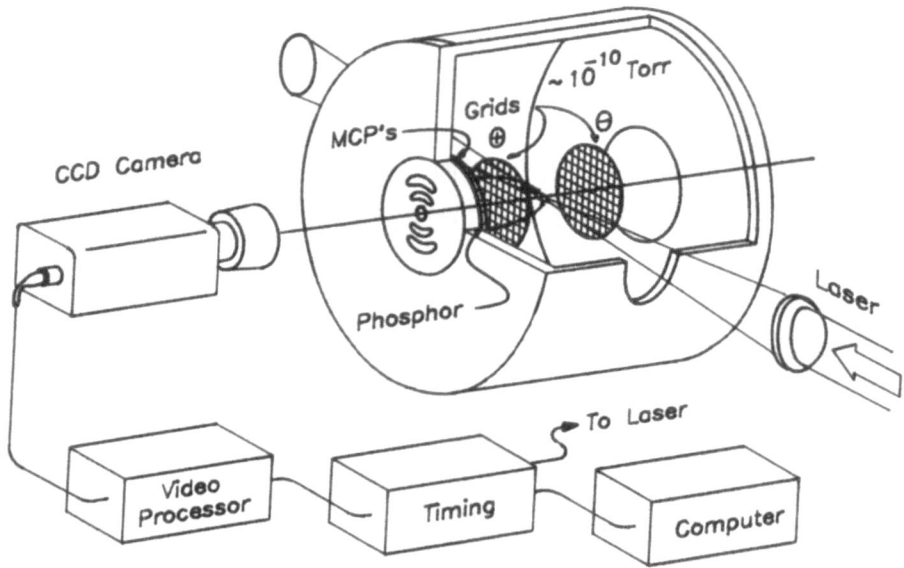

Figure 1. Schematic of the photoelectron imaging spectrometer.

In argon ten-photon resonances dominate at lower intensities and again decreasing electron energies are observed as the intensity increases. We observe electron energies of ~1 eV indicating contributions by the 5f state and contributions by the 4f state at ~700 meV. A drastic change is observed at an intensity of $4*10^{13}$ W/cm^2. At this intensity 11 photons become resonant with the 5d state of argon, producing electrons at energies of ~1 eV.

This experiment was made possible by the Deutsche Forschungsgemeinschaft

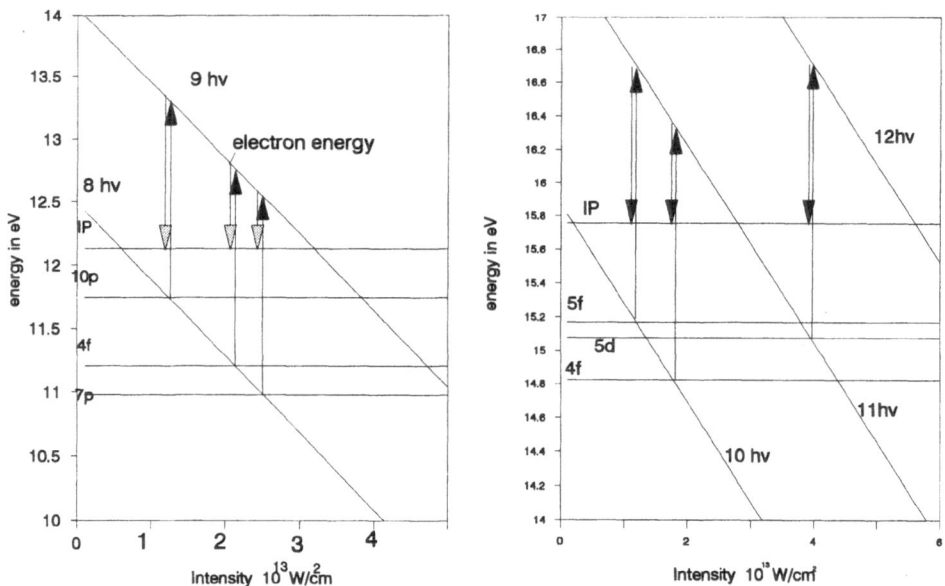

Figure 2. A simple desciption of the shift in xenon (left) and argon (right) with the most important states in this region.

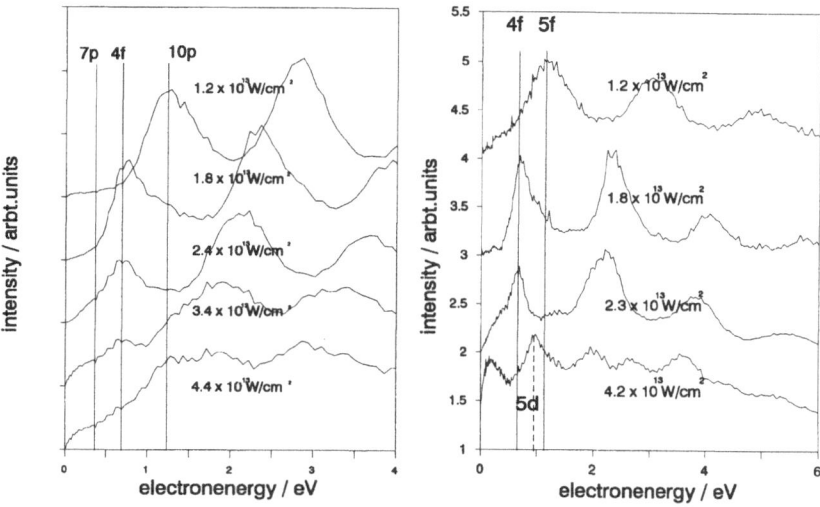

Figure 3. Kinetic energy spectra of the photoelectrons at increasing intensities downwards. Xenon left and argon right.

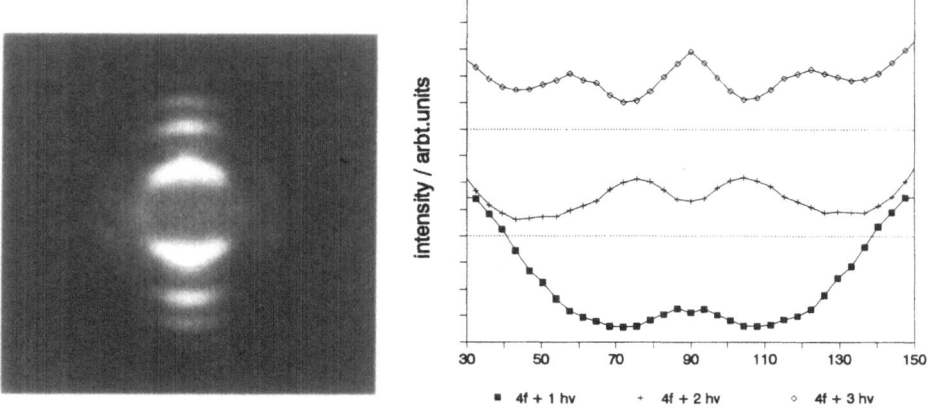

Figure 4. A raw image of xenon at an intensity of $1.5*10^{13}$W/cm^2. Right the angular distribution belonging to the 4f state and two ATI peaks is shown.

REFERENCES

1. H. Helm, N. Bjerre, M.J. Dyer, D.L. Huestis and M. Saeed, Phys. Rev. .Lett.**70**, 3221 (1993)
2. R.R. Freeman, P.H. Bucksbaum, H. Milchberg, S. Darac, D. Schumacher and M.E. Geusic, Phys. Rev. Lett. **59**, 1092 (1987)
3. T..J. McIlrath, R.R. Freeman, W.e. Cooke and L.D. van Woerkom, Phys. Rev. A **40**, 270 (1989)
4. S. Allendorf and A.Szoke, Phys. Rev. Lett **64**, 1234 (1991)
5. H. Rottke, B. Wolff, M. Trapernon, K.G. Welge and D. Feldman, Y. Phys.D **15**, 133 (1990)
6. H. Helm, M.J. Dyer and H. Bissantz, Phys. Rev. Lett. **67**, 1234 (1991)

71

GENERATION OF SHORT-PULSE TUNABLE VUV AND XUV RADIATION

B. Wellegehausen, K. Mossavi, and H. Eichmann

Institut für Quantenoptik
Universität Hannover
Hannover, Germany

For the generation of coherent radiation in the VUV and XUV spectral range low-order (perturbative) and high-order (non-perturbative) nonlinear optical processes are considered. As pump laser systems a short-pulse KrF-excimer laser, delivering 15 mJ in about 400 fs and a 150 fs Ti:sapphire laser with up to 100 mJ have been used.

With the Ti:sapphire laser, operating at a wavelength of 770 nm, high-order nonlinear processes are necessary to penetrate into the VUV and XUV. Such processes are possible and efficient at intensities above 10^{13} Wcm^{-2}, and harmonic generation up to orders of 143 and shortest wavelengths down to 7.4 nm has been demonstrated[1]. In order to achieve tunability with these processes, either the high-power pump laser system has to be tuned itself or high-order frequency mixing processes have to be applied. Because high intensity Ti:sapphire laser systems based on the chirped pulse amplification technique[2] can not be tuned easily, tuning in the XUV by frequency mixing with a more easily tunable laser system seems to be advantageous, and this possibility has been demonstrated recently[3]. Moreover, high-order frequency mixing (two-color) experiments have more degrees of freedom to manipulate the atomic response, as ω-2ω-mixing experiments with relative phase[4] or polarization control[5] have shown. In addition, frequency mixing allows a better adjustment of the collective response of the material (phase matching), which may be used for an "engineering" of the harmonic spectrum.

We have started first experiments in this direction, by considering difference-frequency mixing of fundamental (ω) and frequency doubled (2ω) Ti:sapphire laser radiation in plasmas of low charged ions[6]. For phase matching of high-order harmonics or sum-frequency mixing signals, a negative dispersive nonlinear material is needed. At the intensities typically used in harmonic generation experiments, atomic resonances play no significant role, and the dispersive characteristics of the material are mainly determined by the degree of ionization, the electron density N_e. For the process $\omega_q = q \cdot \omega$ (q=3, 5, 7,...), the phase mismatch consists of a geometrical contribution Δk_q^f, depending on the focussing parameter and a dispersive contribution Δk_q^d, which is proportional to the electron density N_e. Both contributions are positive, so that no phase matching is possible.

Ultrafast Processes in Spectroscopy, edited by Svelto et al.
Plenum Press, New York, 1996

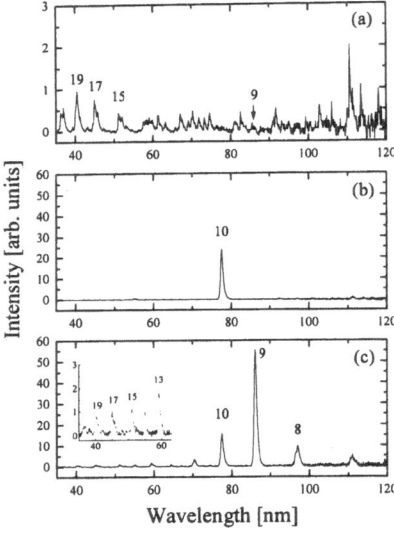

Figure 1. Harmonic spectrum of an O_2 gas jet (a) Ti:sapphire fundamental (w) pump radiation (770 nm, 25 mJ, intensity ~ 10^{16} Wcm^{-2}) (b) 2ω pump (5.5 mJ) (c) ω, 2ω pump. Length of the gas jet is about 1.3 mm.

However, difference-freqency mixing in plasmas is much more favorable and flexible in terms of phase matching, because a compensation of the geometrical and dispersive contribution is possible by an adjustment of the electron density. Experiments performed with O_2 gas jet at intensities up to 10^{16} Wcm^{-2} show evidence for phase matching of an individual harmonic. As an example, Fig 1 shows harmonic spectra of an oxygen plasma with the fundamental (ω), the frequency doubled (2ω) and the two-color (ω, 2ω) pump field. For the (ω, 2ω) pump field the 9th harmonic can be dominantly generated according to the phase matched difference-frequency mixing process $\omega_q = 6\cdot(2\omega) - 3\cdot\omega$. Further investigations on this process and recent progress in harmonic generation will be reviewed.

In order to generate VUV or XUV radiation by low-order (3rd order) processes, already short wavelength pump lasers have to be used. Here, the short pulse KrF laser system offers interesting perspectives. First, powerful VUV radiation is generated by an efficient four-wave difference-frequency mixing process in xenon, according to $\omega_{VUV} = 2\cdot\omega_{248} - \omega_{vis}$, with ω_{vis} being radiation from a short-pulse tunable laser system operating in the visible to near-infrared. Performed experiments have shown that with this process subpicosecond tunable VUV radiation in the range of 200 nm to 130 nm with energies in the µJ range can easily be generated[7]. Very recently, the output energy at 155 nm has been increased to almost 300 µJ[8] and a scaling to more than 1 mJ appears feasible. This powerful and tunable VUV radiation will now be used to generate tunable XUV radiation in the range around 50 nm, by frequency tripling or sum-frequency mixing. In first experiments, frequency tripling of the fundamental 248 nm KrF radiation has been optimized, yielding about 14 µJ at 83 nm. Details of these investigations and general perspectives for nonlinear optical generation of XUV radiation will be discussed.

REFERENCES

1. M. D. Perry and G. Mourou, Science **264**, 917 (1994)
2. C. Sauteret, G. Mainfray, and G. Mourou, Laser Focus World, October (1990)

3. H. Eichmann, S. Meyer, K. Riepl, C. Momma, and B. Wellegehausen, Phys. Rev. A **50**, R2834 (1994); M. B. Gaarde, A. Persson, C.-G. Wahlström, A. L'Huillier, P. Antoine, B. Carré, and S. Svanberg, presented at the conference "Generation and application of ultrashort x-ray pulses" in Pisa, Italy (September 1995)

4. S. Watanabe, K. Konde, Y. Nabekawa, A. Sagisaka, and Y. Koboyashi, Phys. Rev. Lett. **73**, 2692 (1994)

5. H. Eichmann, A. Egbert, S. Nolte, C. Momma, B. Wellegehausen, W. Becker, S. Long, and J.K. McIver, Phys. Rev. A **51**, R3414 (1995)

6. S. Meyer, H. Eichmann, T. Menzel, S. Nolte, B. Wellegehausen, B. N. Chichkov, and C. Momma, submitted to Phys. Rev. Lett.

7. A. Tünnermann, K. Mossavi, and B. Wellegehausen, Phys. Rev. A **46**, 2707 (1992); A. Tünnermann, C. Momma, K. Mossavi, C. Windolph, and B. Wellegehausen, IEEE J. Quantum Electron. **QE-29**, 1233 (1993)

8. K. Mossavi, L. Fricke, P. Liu, and B. Wellegehausen, Opt. Lett. **20**, 1403 (1995)

72

GENERATION OF HIGH ORDER HARMONICS FROM SOLID SURFACES BY INTENSE FEMTOSECOND LASER PULSES

D. von der Linde,[1] T. Engers,[1] G. Jenke,[1] P. Agostini,[2] G. Grillon,[3]
E. Nibbering,[3] J.-P. Chambaret,[3] P. F. Curley,[3] A. Mysyrowicz,[3] and
A. Antonetti[3]

[1] Institut für Laser- und Plasmaphysik
Universität Essen, Germany
[2] Commissariat à l'Energie Atomique
Saclay, France
[3] Laboratoire d'Optique Appliquée
Ecole Polytechnique-ENSTA, Palaiseau, France

Generation of optical harmonics of high order using intense ultrashort laser pulses is a topic of great current interest. Extensive recent work has dealt with rare gases[1], where high order harmonic generation can be attributed to the nonlinearity of the interaction of intense electromagnetic fields with the gas atoms. On the other hand, a different type of high order harmonic generation was observed more than a decade ago during *nanosecond* laser-plasma interaction[2,3]. In this case harmonic generation was explained[4,5] as being due to the anharmonicity of the electron motion across the steep plasma density gradient which develops during the expansion as a result of counteracting ponderomotive forces.

When a plasma is generated by letting an intense *femtosecond* laser pulse interact with a solid target the plasma expansion within the pulse duration is very small. During the interaction a thin surface layer of plasma is produced in which the density drops from approximately solid density to vacuum in a distance much shorter than the wavelength of light. Thus plasmas generated by femtosecond laser pulses provide an ideal situation for studying harmonic generation caused by the electronic anharmonicities in a steep plasma density gradient.

This report describes the observation of harmonics up to the eighteenth order from glass targets and aluminum films exposed to intense femtosecond laser pulses[6]. In related recent work observation of the seventh harmonics from aluminum targets has been reported[7]. Very recently, harmonic orders up to seventy have been observed[8] when picosecond pulses were focused on solid material at peak intensities up to 5×10^{18} W/cm^2.

A schematic of our experiment is shown in Fig. 1. Laser pulses from a titanium-sapphire laser system (chirped pulse amplification) were used ($\lambda = 800$ nm). The laser pulse

Ultrafast Processes in Spectroscopy, edited by
Plenum Press, New York, 1996

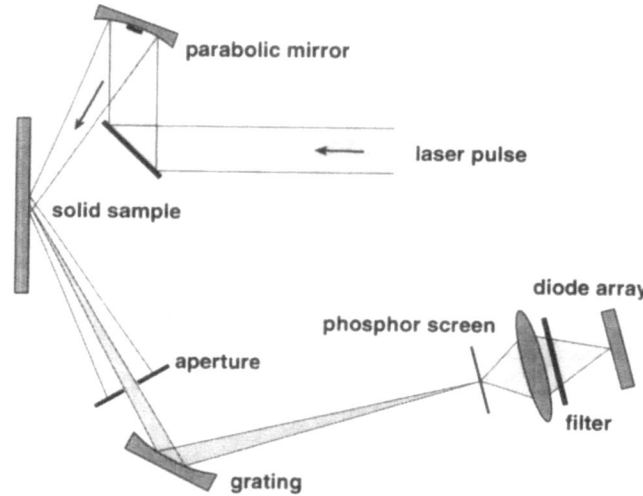

Figure 1. Schematic of the experimental arrangement for measuring harmonic generation from solid surfaces.

duration was 120 fs. Some experiments were also done at a pulse duration of about 60 fs. The laser beam was focused on target at an angle of incidence of 68 degrees using an off-axis parabolic mirror or an ordinary fused silica lens. The peak intensity on target could be varied between 10^{17} to 10^{18} W/cm^2. Large angle of incidence and p-polarization (electric vector in the plane of incidence) were chosen to ensure a strong normal component of the optical electric field at the plasma vacuum boundary. Harmonic radiation is expected in the specular direction, because of the conservation of the parallel component of the wave vector in coherent nonlinear optical surface processes. The reflected beam was made to strike an imaging toroidal grating, which produced a spectrum of the plasma emission on a phosphorescent screen. For protection of the grating and restriction of incoherent plasma emission a suitable combination of an opaque disk in the incident beam and a narrow aperture in the reflected beam was used to block the reflected fundamental light and to transmit the generated harmonics. UV and XUV spectra were recorded by imaging the

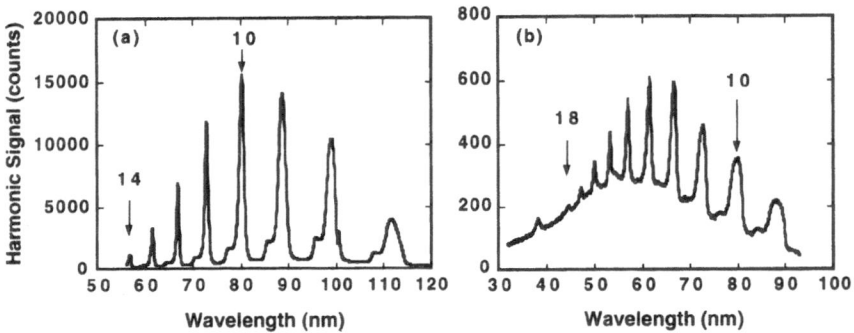

Figure 2. Examples of harmonic spectra from glass samples with weak background (a), and stronger background (b).

phosphorescent light from the screen onto a linear diode array. We used substrates of optical glass coated with a 200 nm aluminum film and bare glass substrates as targets. The samples were raster-scanned during the experiments to provide a fresh sample surface for each laser pulse.

Two examples of typical harmonic spectra from *glass* are depicted in Fig. 2 a and Fig. 2 b. In both cases 200 pulses were accumulated to obtain the spectrum. The intensity on target was estimated to be approximately 10^{17} W/cm^2. It can be seen that odd and even harmonics were observed up to a maximum order of 14 and 18, respectively. Very similar spectra were observed with *aluminum* samples.

Typically, the harmonics were superimposed on some background (see Fig. 2 b). The ratio of harmonics to background varied in different measurements. It was observed that the amount of background radiation and also the appearance of some plasma line emission depended on the temporal profile of the laser pulses. For instance, clean harmonic spectra with little background were observed with laser pulses of 120 fs duration and a temporal profile with an intensity drop of 10^{-5} in 1 ps. On the other hand, the spectra were dominated by background radiation and plasma line emission with only very weak harmonic lines when laser pulses of nominally 60 fs duration with an intensity drop of only about 3×10^{-3} in 1 ps were used. The intensity profiles for this comparison were estimated from autocorrelation measurements extended over a large intensity range. The origin and detailed nature of the background effects are yet unknown, although optical emission of the plasma in the visible and ultraviolet undoubtedly contributed.

The relative strength of the harmonic orders of the spectra as shown in Fig. 2 a and 2 b is mainly determined by the apparent spectral response of our apparatus. In particular, the spatial inhomogeneity of the phosphor screen is quite significant. Taking into account the spectral characteristics of the detection system we have obtained the variation of the harmonic intensity with the harmonic order. Results from four independent measurements are shown in Fig. 3. It should be noticed that there is no plateau and no high frequency cutoff, which are typical features of harmonic spectra produced in rare gases. Instead, a relatively smooth roll-off is observed which can be represented by an exponential function.

It is interesting to compare the high frequency cutoff of the measured harmonic spectra with the earlier theoretical predictions of Bezzerides et al.[4] and Grebogi et al.[5] An important feature of the harmonic spectra according to both models is the following. The

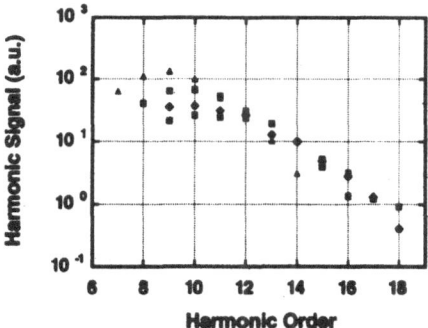

Figure 3. Variation of the harmonic signal with the order of the harmonics. Data from four different spectra are shown.

maximum harmonic order should be given by the plasma frequency corresponding to the maximum electron density at the top of the density profile. The wavelength of the highest harmonic order observed in our experiments was λ = 44 nm (18th order), which corresponds to an electron density of $N_e = m_0\varepsilon_0(2\pi c/\lambda e)^2 = 5.4\times10^{23}/cm^3$. This figure is very close to the electron density of a solid density plasma with an electron temperature of a few hundred eV. However, we believe that this agreement with the predictions in Ref.4 and 5 maybe fortuitous and that more detailed studies of the high frequency limit are necessary. In fact, recent particle-in-cell (PIC) simulations by Gibbon[9] and by Lichters et al.[10] suggested that the number of harmonic orders increases well beyond the plasma frequency limit when the intensity of the fundamental laser pulse exceeds 10^{18} W/cm^2. The key point is that at these high intensities relativistic effects play an important role in the dynamics of the electronic motion. Therefore, extension of the study of harmonic generation from solid target to relativistic laser intensities should be very interesting.

When the focused intensity of the fundamental pulses on the target surface was varied from values around 10^{17} W/cm^2 to about 10^{18} W/cm^2 the following observations were made. At low intensities a clean, undisturbed reflected beam from the target surface was obtained. The stable beam geometry readily permitted the separation of the harmonic light by an aperture as described above. The harmonic spectra shown in Fig. 2 a and 2 b were obtained under these conditions. With increasing fundamental intensity an increase of the plasma emission and background radiation resulting in a decrease of the ratio of harmonics to background was observed. At still higher intensities the reflected beam became unstable and substantial beam distortions began to develop. For intensities around 10^{18} W/cm^2 the reflected light was no longer collimated but spread out over some large solid angle around the specular direction. In addition, conspicuous generation of diffuse green 3/2 ω light was observed. Under these conditions the generation of coherent high order harmonics from the surface in the specular direction disappeared. Because only a narrow solid angle around the specular direction could be observed in our detection scheme, it cannot be excluded that some harmonic radiation was present in non-specular directions and larger solid angles as in the experiments of Moustaizis et al.[8].

We have calculated the conversion efficiency of high order harmonic generation in our experiments. For the strongest harmonic spectra the number of photons per pulse is estimated to be about 10^{11} and 5×10^9 at the tenth and the fifteenth harmonic, respectively, which corresponds to a photon conversion efficiency of approximately 10^{-6} and 5×10^{-8}. It should be kept in mind, however, that these numbers are subject to relatively large uncertainty, mainly due to the possible error in estimating the properties of the phosphor screen.

Measurements have been performed to study the variation of the efficiency of harmonic generation with the polarization and angle of incidence of the laser pulses. In particular, it was found that harmonic generation decreased strongly when s-polarization was used. The residual harmonic emission observed with a nominally s-polarized laser beam was probably due to incomplete suppression of the p-component of the laser pulses. Measurements of the harmonics after changing the angle of incidence from 68 degrees to 45 degrees did not reveal significant changes of the spectra and the efficiency.

In conclusion, we have shown that the interaction of intense femtosecond laser pulses with solid surfaces gives rise to the generation of high order optical harmonics extending well into the XUV region. Although no measurements of the pulse durations have been performed as yet, it follows from the nonlinear interaction mechanism that the durations of the generated XUV pulses should be shorter than the width of the fundamental laser pulses. The basic characteristics of the observed harmonic generation appear to be consistent with a mechanism in which the electrons perform some coherent anharmonic motion across the plasma-vacuum interface in response to the normal component of the driving optical electric field. The plasma frequency corresponding to the maximum harmonic order observed so far is in agreement with the range of electron densities of ionized solid material.

REFERENCES

1. *Atoms in Intense Fields*, ed. by M. Gavrila (Academic Press, Boston, 1992).
2. R. L. Carman, D. W. Forslund, and J. M. Kindel, Phys. Rev. Lett. 46, 29 (1981).
3. R. L. Carman, C. K. Rhodes, and R. F. Benjamin, Phys. Rev. A 24, 2649 (1981).
4. B. Bezzerides, R. D. Jones, and D. W. Forslund, Phys. Rev. Lett. 49, 202 (1982).
5. C. Grebogi, V. K. Teripathi, and H. H. Chen, Phys. Fluids 26, 1904 (1983).
6. D. von der Linde, T. Engers, G. Jenke, P. Agostini, G. Grillon, E. Nibbering, A. Mysyrowicz, and A. Antonetti, Phys. Rev. A 52, R25 (1995).
7. S. Kohlweyer, G. D. Tsakiris, C. G. Wahlström, C. Tillman, and I. Mercer, Optics Comm. 117, 431 (1995).
8. S. D. Moustaizis, M. Bakarezos, C. N. Danson, A. E. Angor, A. Dyson, P. Fews, P. Gibbon, P. Lee, P. Loukakos, D. Deely, P. Norreys, B. Walsh, J. S. Wark, M. Zepf, and J. Zhang, Euroconference on "Generation and application of ultrashort x-ray pulses, 20 - 23 September 1995, Pisa (Italy).
9. P. Gibbon, private communication and to be published.
10. R. Lichters, J. Meyer-ter-Vehn, and A. Pukhov, Euroconference on "Generation and application of ultrashort x-ray pulses", 20 - 23 September 1995, Pisa (Italy).

73

OPTICAL-HARMONIC GENERATION AND FREQUENCY MIXING IN THE FIELD OF PICOSECOND LASER PULSES IN THE PLASMA OF OPTICAL BREAKDOWN

A. B. Fedotov and A. M. Zheltikov

Physics Department
International Laser Center
Moscow State University
Moscow, Russia

1. LASERS OF MODERATE POWERS

A number of recent experiments demonstrated a high efficiency of harmonic generation in strong laser fields in excited gases[1-4] and plasmas.[5-8] Optical-harmonic generation in high-power laser fields is of considerable interest in the context of producing coherent short-wavelength radiation. Furthermore, experiments on harmonic generation give a deeper insight into processes occurring in atomic systems subjected to the influence of strong light fields. In our previous works,[5-8] we proposed a new experimental approach to optical-harmonic generation. The developed technique implies that a plasma of optical breakdown produced in gases under pressures no higher than the atmospheric pressure is used as a nonlinear medium for frequency multiplication. Among the main advantages of the proposed experimental scheme for harmonic generation, we should mention a relative simplicity of the laser setup and the possibility of achieving sufficiently high harmonic-generation efficiencies using lasers of moderate powers.

In this paper, we discuss the results of experiments on the generation of optical harmonics and sum-frequency radiation in a low-temperature laser-produced plasma. A plasma of optical breakdown is shown to be a promising nonlinear medium for producing coherent short-wavelength radiation via harmonic generation and frequency mixing. We discuss experimental data on the generation of the third, fifth, seventh, and ninth harmonics of Nd:YAG-laser radiation and describe experiments on nonlinear optical frequency mixing in a laser-produced plasma. It is demonstrated that the efficiency of optical frequency conversion in a laser-produced plasma considerably depends on phase-matching conditions.

Ultrafast Processes in Spectroscopy, edited by Svelto et al.
Plenum Press. New York, 1996

2. EXPERIMENTAL TECHNIQUE

The experimental setup for studying optical-harmonic generation in a low-temperature laser-produced plasma is described in detail in our previous papers.[6,8] To produce and to probe the plasma, we used independent synchronized laser systems. The probing picosecond laser system included a passively mode-locked Nd:YAG master oscillator with negative feedback. The master oscillator produced light pulses with the duration $\tau_p = 35$ ps. These pulses passed through three amplification cascades, which allowed us to achieve the radiation energy $E = 25$ mJ. For probing the laser-produced plasma, we used radiation at the fundamental frequency of the Nd:YAG laser with the wavelength $\lambda = 1.064$ μm and radiation of the second harmonic of the Nd : YAG laser with the wavelength of 0.532 μm. We implemented frequency doubling using a nonlinear-optical KDP crystal.

Radiation of the independent laser system for plasma excitation had the wavelength $\lambda = 1.064$ μm, the pulse duration $\tau_p = 15$ ns, and the energy E from 120 to 180 mJ. To focus laser radiation on a target, we used a cylindrical lens with the focal length $f = 10$ cm. We were able to vary the distance between the surface of the target and the region of probing with the help of a mechanical feed. To measure the radiation intensity of optical harmonics, we used the detection system, which included a vacuum monochromator and a photomultiplier.

3. EXPERIMENTAL RESULTS

We observed efficient generation of optical radiation at the frequencies of the third, fifth, seventh, and ninth harmonics of the probing radiation. The wavelengths of the corresponding signals were 355, 213, 152, and 118 nm, respectively. To detect the seventh and ninth harmonics, we performed experiments in a vacuum chamber with a residual pressure $P < 10^{-4}$ mm Hg. Along with the generation of harmonics of 1.06-μm radiation, $\omega+\omega+\omega=3\omega$, we detected the third harmonic of 0.53-μm probing light, $2\omega+2\omega+2\omega=6\omega$, and observed frequency-mixing processes occurring through the third-order optical nonlinearity of the medium according to the schemes $2\omega+\omega+\omega=4\omega$, $2\omega+2\omega+\omega=5\omega$, and $2\omega+2\omega-\omega=3\omega$. We experimentally studied the properties of the detected optical harmonics and analyzed specific features of frequency-mixing processes in the plasma of optical

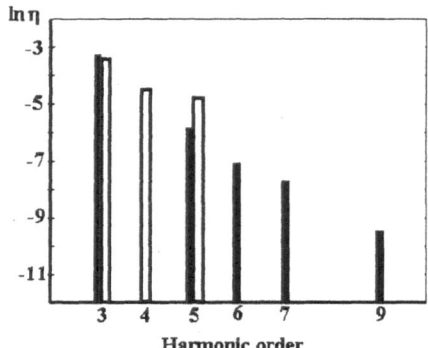

Figure 1. Diagram of the efficiencies of optical-harmonic generation (shaded bars) and frequency mixing (unshaded bars).

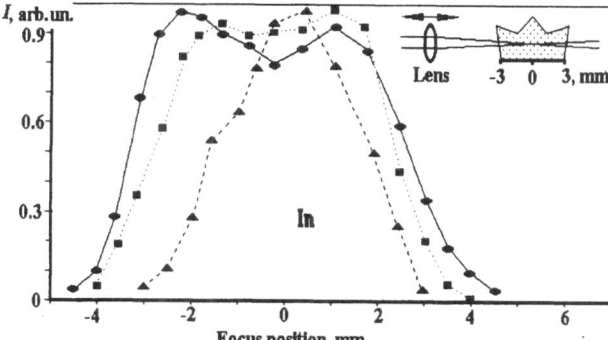

Figure 2. The normalized power of the third harmonic as a function of the focus position of the probing light beam for τd = 250 ns (solid curve), 650 ns (dotted curve), and 3.5 µs (dashed curve).

breakdown. These experiments allowed us to optimize the conditions for each of the considered nonlinear optical processes.

Figure 1 displays conversion efficiencies for harmonics (shaded bars) and signals corresponding to frequency mixing (unshaded bars) observed in the spectrum of Nd:YAG-laser radiation (with the energy E = 25 mJ and the intensity in the focus I = 5*1012 W/cm2) transmitted through the plasma of optical breakdown produced on the surface of an indium target.

To investigate phase-matching conditions,[9-12] we varied the geometry of focusing of laser radiation into the plasma by shifting the focus of the spherical lens along the plasma axis. For studying third-harmonic generation, we produced plasma on the surface of an indium target at the atmospheric pressure. Figure 2 shows the normalized power of the third harmonic as a function of the focus position of the probing light beam for various delay times. The solid curve in Fig. 2 displays the dependence of the power of the third harmonic on the position of the focus (f = 6 cm) with respect to the center of the plasma for the delay time τ = 250 ns. As can be seen from this plot, the studied dependence is nonmonotonic near the center of the nonlinear medium. In this region, the power of the third harmonic increases as we displace the focus of the probing beam toward the boundaries of the medium. When we increased the delay time, which corresponded to lowering the density and the temperature of the plasma, the dip between these maxima became less pronounced and, finally, vanished.

Figure 3 shows the dependences of the radiation intensity at the frequency of the fifth harmonic on the position of the focus (f = 6 cm) for direct fifth-harmonic generation and for frequency mixing according to the scheme $5\omega = 2\omega + 2\omega + \omega$. The delay time τd between producing and probing the plasma was equal to 400 ns. As can be seen from curves displayed in Fig. 3, the dependence of the radiation intensity at the frequency of the fifth harmonic on the position of the focus considerably differs from the corresponding dependence for third-harmonic generation. As the focus of the probing beam is displaced to the output boundary of the plasma plume, the intensity of the fifth harmonic features a considerable increase.

Qualitatively, all the above-described effects can be satisfactorily described within the framework of the theoretical model developed in our previous studies.[13,14]

4. CONCLUSION

Thus, we demonstrated that phase-matching effects may have a considerable influence on the efficiency of generating the third and fifth harmonics of Nd:YAG-laser radia-

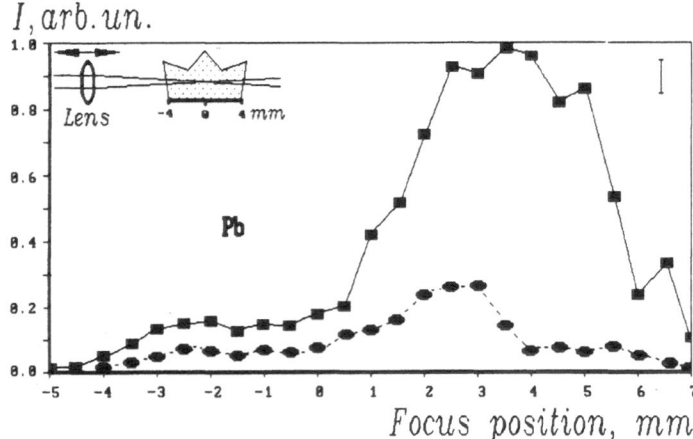

Figure 3. Intensity of radiation at the frequency of the fifth harmonic as a function of the position of the focus in a plasma produced on the surface of a lead target at the atmospheric pressure: (solid curve) two-frequency pumping and (dashed curve) single-frequency pumping.

tion in the plasma of optical breakdown produced on surfaces of metal targets. We performed elementary calculations of phase-matching integrals for optical-harmonic generation using model approximations for simulating the temporal dependence of the electron concentration and spatial distribution of electrons in the plasma. These calculations qualitatively reproduce the main experimental results. We showed that a detailed investigation of phase-matching effects and correct analysis of optimal conditions for optical-harmonic generation in the laser-produced plasma require consideration of absorption at the frequency of probing radiation and at frequencies of harmonics.

ACKNOWLEDGMENTS

We are grateful to Prof. N.I.Koroteev for valuable discussions and to D.A. Sidorov-Biryukov and A.N.Naumov for assistance in experiments. A part of this work was carried out within the framework of the Agreement between the International Laser Center of Moscow State University and Lawrence Livermore National Laboratory (USA).

REFERENCES

1. A. McPherson, G. Gibson, H. Jara, et al., *J. Opt. Soc. Am. B*, **4**, 595 (1987).
2. P. Balcou, C. Coruaggia, A.S. Gomes, et al., *J. Phys. B*, **25**, 4467 (1992).
3. J.J. Macklin, J.D. Kmetec, C.L. Gordon III, *Phys. Rev. Lett.*, **70**, 766 (1993).
4. N. Sarakura, K. Hata, T. Adachi, et al., *Phys. Rev. A*, **43**, 1669 (1991).
5. S.M. Gladkov, A.M. Zheltikov, N.I. Koroteev, et al., *Pisma Zh. Tekh. Fiz.*, **14**, 1399 (1988).
6. A.B. Fedotov, S.M. Gladkov, N.I. Koroteev, et al., *J. Opt. Soc. Am. B*, **8**, 373 (1991).
7. A.M. Zheltikov, N.I. Koroteev, and A.B. Fedotov, *Bull. Russ. Acad. Sci., Phys.*, **58**, no. 2, 276 (1994).
8. A.M. Zheltikov, N.I. Koroteev, and A.B. Fedotov, *Laser Phys.*, **4**, 569 (1994).
9. G.C. Bjorklund, *IEEE J. Quantum Electron.*, **QE-11**, 287 (1975).
10. J.F. Ward and G.H.C. New, *Phys. Rev.*, **185**, 57 (1969).
11. P. Balcou and A. L'Huillier, *Phys. Rev. A*, **47**, 1447 (1993).

12. A. L'Huillier, K.J. Schafer, and K.C. Kulander, *J. Phys. B*, **24**, 3315 (!991).

13. A.B. Fedotov, N.I. Koroteev, and A.M. Zheltikov, *Laser Phys.*, **5**, no. 4 (1995).

14. N.I. Koroteev, A.N. Naumov, and A.M. Zheltikov, *Laser Phys.*, **4**, 1062 (1994).

74

REFLECTION AND TRANSMISSION OF HIGH INTENSITY FEMTOSECOND LASER PULSE FOCUSED ON VERY THIN PLASTIC FOILS

D. Giulietti,[*,1,2] L. A. Gizzi,[1] V. Biancalana,[1] T. Ceccotti,[1] P. Audebert,[4] J. P. Geindre,[4] and A. Mysyrowicz[3]

[1] Istituto di Fisica Atomica e Molecolare del CNR
via del Giardino,7, 56100, Pisa, Italy
[2] Dipartimento di Fisica
Università di Pisa
Piazza Torricelli, 2, Pisa, Italy
[3] Laboratoire d'Optique Appliquée
ENSTA, Ecole Polytechnique
91120, Palaiseau, Cedex, France
[4] Laboratoire puor l'Utilisation des Lasers Intenses
Ecole Polytechnique
91120, Cedex, Palaiseau, France

ABSTRACT

The 150 fs pulse of the LOA Ti:Sapphire laser has been focused at an intensity of 5×10^{17} W/cm^2 on a 800Å plastic foil. The reflected and transmitted laser radiation resulted strongly affected by Self Phase Modulation effects. Even if the measurements were time integrated, the analysis of the spectra give information on laser plasma interaction at different times.

A peculiar characteristic of laser produced plasmas with femtosecond pulses is the negligible hydrodynamic expansion during the pulse. In fact, the scale length of the plasma density perpendicular to the target surface, given by $L = c_s \Delta t$, is much shorter than the vacuum wavelength λ of the impinging laser radiation, that is : $\lambda \gg L$, where c_s is the sound velocity, and Δt the pulse duration. In these experimental conditions the interaction geometry is particularly well defined. Thus the laser radiation impinges at a given angle θ on a fairly flat plasma surface having a sharp density profile.

* Tel: 050 911 237; fax: 050 48277.

Ultrafast Processes in Spectroscopy, edited by Svelto et al.
Plenum Press, New York, 1996

331

The majority of the experiments reported so far have been performed by focusing the laser pulse on thick targets, or films coated on transparent massive supports. The plasma produced in this conditions can be divided in three regions. The first region consists of the plasma expanding in the vacuum, whose typical extension is few hundred of Ångstrom. The second, whose electron density is that of the solid target times the average ionization degree, extends in the overdense plasma over a length of the order of the skin depth, typically $l_s \approx 200$Å. The third region, not directly reached by the laser e.m. field, is determined by heat diffusion processes and extends over several thousands of Ångstroms.

However, this structure can be strongly modified if target pre-heating occurs, due to the presence of laser pre-pulse. In the case of laser systems based on the chirped pulse amplification technique[1], laser pre-pulse can arise from imperfections in the compression stage of the laser. In addition, pre-pulse can also arise from amplified spontaneous emission[2]. This pre-pulse can produce a tenuous plasma in front of the target, before the arrival of the main pulse, deeply changing the interaction conditions.

In order to minimize this effect we used targets consisting of very thin (d< 1000Å) plastic (FORMVAR) foils. Due to the high transparency of these targets, the threshold intensity for damage and consequent plasma formation is expected to be relatively high compared to the estimated pre-pulse intensity level in our experimental conditions. Another characteristic of this type of target for the laser pulse regime considered here is the higher temperature achievable, during the interaction, as a consequence of minor energy losses due to heat conduction. In fact, the thermal conduction length is much larger than the target thickness.

The experiment we present here was performed at the Laboratoire d'Optique Appliquée by using a femtosecond Ti:Sapphire laser system[3]. We studied the interaction processes of an intense femtosecond pulse in a very short density scale length plasma. The experimental results, concerning the correlation of X-ray and second harmonic emissions with the laser polarization, have been presented elsewhere[4]. In this paper we report on reflection and transmission of laser radiation impinging on the very thin foil. The data add a piece of new information on laser plasma interaction in experimental conditions not yet investigated enough.

The Ti:Sapphire laser pulse ($\lambda \approx$ 8040Å, $\Delta t \approx$ 150fsec, E $\approx$ 10 mJ on target) was focused ($\phi \approx 5\mu$m, spot diameter) on 800Å FORMVAR foil at intensities up to 5×10^{17}W/cm^2. The angle of incidence was 20 deg. The beam was focused by means of an f/4 reflective optics in an off-axis configuration. The specularly reflected and transmitted laser radiation was collected and sent part to photo diodes, part to CCD cameras for spectral analysis or imaging. All measurements were time integrated.

The measured transmittivity resulted $\approx$5%, that is at least 100 times higher than the expected theoretical value, even taking into account the pre-pulse energy passing through the target, before plasma formation. This suggests different possibilities : a) the transmitted radiation is partially refracted, passing through the target externally to the critical region; b) the leading front of the laser pulse passes through the target, before the plasma formation. On the other side, if we exclude effects of overdense penetration, or relativistic self-focusing[5] since the laser intensity is below the relativistic threshold, it is even more difficult to consider that radiation could pass the target through channels due to ordinary self-focusing/filamentation. In fact the filament formation times range in the hydrodynamic time scale, much larger than the 150fsec laser pulse duration.

We observed that the reflectivity increases with the laser intensity. According to the Drude model applied to the produced plasma, this corresponds to an increase of the plasma temperature with the laser intensity.

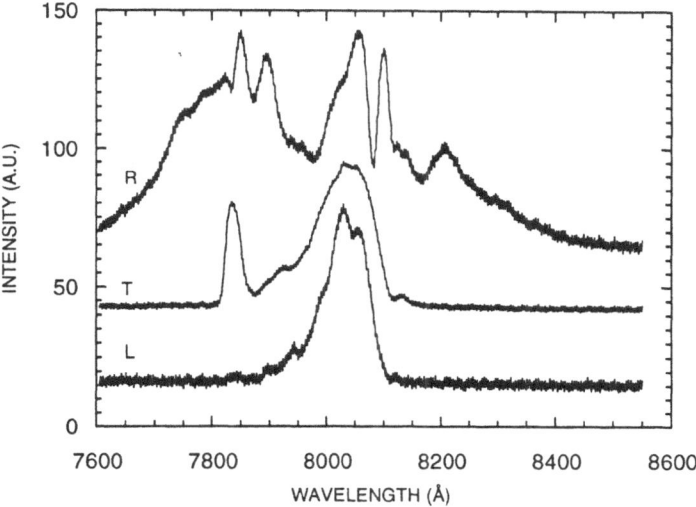

Figure 1. The spectra of the impinging (L), reflected (R), and transmitted (T) laser radiation. The intensity of the impinging laser radiation is $5 \times 10^{17} W/cm^2$.

The laser radiation transmitted through the target has basically the same spectrum as the one of the impinging radiation (see Fig.1). However, an intense up-shifted component, as previously observed in similar experiments[5], and a down-shifted component much weaker are also visible in the spectrum.

Such spectral components can be explained in terms of Self Phase Modulation (SPM) of the laser radiation passing through a plasma whose electron density increases in time during the leading front of the laser pulse (up-shifted component) and decreases during the tailing front (down-shifted component). A qualitative but quite exhaustive explanation of the observed phenomenon can be obtained by plotting, versus the time, the laser pulse, the electron density n_e and the per cent shift of laser frequency, due to SPM: $\Delta\omega/\omega = (Z/2c(1-N)^{1/2}) \, dN/dt$. In the previous formula Z represents the longitudinal plasma extension, c the speed of light in vacuum, and $N = n_e/n_c$ (n_c being the critical density for the impinging laser radiation). In agreement with simulation results, the electron density increases with a characteristic time of the same order of that of the leading front of the laser pulse, and decreases with a much longer time scale (≈ 10 times longer). The laser radiation can be transmitted until the electron density is below the critical density ($n_e \, ^2 \, n_c$), resulting progressively more up-shifted. At later times the plasma is completely opaque, until the density decreases below the critical value. This happens on the tail of the laser pulse, due to the longer time of the density decrease. The resulting down-shift is smaller due to the minor value of dN/dt. Also the intensity of such component is lower, because it is produced in the tail of the laser pulse.

Therefore the spectral analysis of transmitted radiation actually provides information on the interaction with the target at different times: the unperturbed component corresponds to the laser radiation passing through the target before the plasma formation; the up-shifted component is transmitted during the leading front of the pulse; the down-shifted during the tailing front.

We notice that the laser spectrum was as wide as two times the Fourier transform limit. On the other hand the blue and red components result definitely narrower than such a limit. Therefore they have to be generated in times longer than the main laser pulse, namely also during the arrival of its pedestal on the target.

Finally we observe that, in the conditions of the present experiment the radiation transmitted through the target interacts with the plasma less strongly than the reflected one. In the spectrum of the reflected radiation the unperturbed component as well as the up and down component are still recognisable, but more structured and immersed in a wider background spectrum. The wide spectrum can be attributed to SPM, but in this case, in contrast with the transmission spectrum, the cut-off due to the critical density is absent.

In conclusion, the interaction of high intensity femtosecond laser pulses with very thin plastic targets has been investigated experimentally. The spectral properties of the transmitted light provide valuable information on the plasma formation and evolution. The spectral broadening of the reflected laser light reveals the high degree of coupling of such a radiation with the sharp boundary of the plasma.

REFERENCES

1. M. Pessot, P. Maine, G. Mourou, Opt. Comm., **62**, 419, (1987).
2. H.M. Milchberg, R.R. Freeman, Phys. Rev. A, **41**,2211, (1990).
3. C. Le Blanc, G. Grillon et al, Optics Letters, **18**, 140, (1993).
4. L.A. Gizzi, D. Giulietti, et al, preprint.
5. B. Corkum, Phys.Today, **48**, 36, (1995).

75

LASER PLASMA INTERACTION AT RELATIVISTIC INTENSITIES

Meenu Asthana

Laser Programme
Centre for Advanced Technology
Indore - 452013, India

There has been considerable recent technological advances of high intensities (of the order of 10^{18} W/cm^2), short pulsed lasers (1psec or less). The major theme in ultra-fast and ultra-intense laser plasma includes the basic laser plasma phenomena that arise when laser fields are so strong that relativistic electron effects become important. Relativistic self-focusing of radiation in a plasma is applicable to pulse length τ_R in the range $1/\omega < \tau_R < r/C_s$, where "r" is the effective beam width and C_s is the ion acoustic speed[1], and can arise owing to the creation of density depression in the plasma as well as to an increase in the electron mass due to relativistic effects. The time scale for self-focusing due to density depression is longer than that the self-focusing due to relativistic mass increase. The latter time scale is $\tau_R \approx \omega^{-1}$ where ω is the radiation frequency and is assumed to be much greater than the electron plasma frequency. Here we are concerned with relativistic self-focusing on time scale short enough such that the plasma density profile does not evolve significantly under the influence of radiation beam. This implies that the pulse length τ_R of the radiation beam must be short compared with $\tau_S = r/C_s$ (the time scale for the density depression to occur) and, of course, long compared with a radiation period τ_R . Based on steady state paraxial theory Relativistic self-focusing of rippled laser beams in a time harmonic plane wave at arbitrary intensity is studied. Nonlinear dielectric constant due to relativistic variation of mass, self-focusing equation relating the variation of beam width parameter with distance of propagation, self-trapping condition and critical power are evaluated followed by numerical calculations[2]. Because of saturating nature of non linear di-electric constant the analysis leads to two values for critical beam power to self focus P_{cr1} and P_{cr2} . When power $P < P_{cr1} < P_{cr2}$, the beam diverges. When $P_{cr1} < P < P_{cr2}$, it first coverges, then diverges, and so on. When $P > P_{cr2}$, it first diverges, then converges and so on. Using WKB approximation and paraxial ray theory the non linear dielectric constant for a rippled laser beam can be expressed as

Ultrafast Processes in Spectroscopy, edited by Svelto et al.
Plenum Press, New York, 1996

$$\Phi(\langle EE \rangle) \approx \left[\left(\frac{k(0)E_{oo}^2}{k(f)2f^2} \right) \left(1 + \left(\frac{E_{1o}r_o}{E_{oo}r_{1o}} \right) \cos\Phi_p \right) \right] -$$
$$\left[\left(\frac{r^2 E_{oo}^2 k(o)}{2r_o^2 f^4 k(f)} \right) \left\{ 1 - \left(\frac{E_{1o}r_o}{E_{oo}r_{1o}} \right) \cos\Phi_p \left(1 - \frac{1}{2} \left(\frac{r_o^2}{r_{1o}^2 + 1} \right) \right) - \left(\frac{E_{1o}r_o}{E_{oo}r_{1o}} \right)^2 \right\} \right] \Phi'$$

$$(1)$$

where Fp is the phase difference between the main beam and the ripple, and r1o indicates the position and width of the ring ripple; the maximum of the ring is at r = r1o . The variation of beam width parameter "f" with distance of propagation "z" (self-focusing equation) comes out as

$$d^2f / dz^2 = \left(1/k^2(f)r^4 f^2 \right) - \left(\omega_p r_o / c \right) \left[\left(k(o)E_{oo}^2 / k(f)4r^4 f^2 \right) \left(1 + k(o)\alpha E_{oo}^2 X / 8\lambda^2 k(f)f^2 \right) XY \right]$$
$$/ \left[1 + \left(k(o)\alpha E_{oo}^2 / k(f)2f^2 \right) X \left(1 + \left(k(o)E_{oo}^2 X / k(f)32\lambda^2 f^2 \right) \right) \right]^{3/2}$$

$$(2)$$

where

$$X = \left[1 + \left(\frac{E_{1o}r_o}{E_{oo}r_{1o}} \right) \cos\Phi_p \right]$$

$$Y = \left[1 - \left(\frac{E_{1o}r_o}{E_{oo}r_{1o}} \right) \cos\Phi_p \left(1 - \frac{1}{2} \left(\frac{r_{oo}^2}{r_{1o} + 1} \right) \right) - \left(\frac{E_{1o}r_o}{E_{oo}r_{1o}} \right)^2 \right]$$

The condition for self-trapping and critical power of beam obtained are respectively

$$\left(\frac{\omega_p r_o}{c} \right)^2 = 2 \left[1 + \left(\frac{\alpha E_{oocr}^2}{2} \right) X \left(\frac{1 + \alpha E_{oocr}^2}{32\lambda^2} \right) X \right]^{\frac{3}{2}}$$
$$\left[1 + \left(\frac{\alpha E_{oocr}^2}{8\lambda^2} \right) X \right] \left(\frac{\alpha E_{oocr}^2}{2} \right) XY$$

$$(3)$$

$$P_{cr} = \left(\frac{cr_o^2 E_{oocr}^2}{8} \right).$$
$$\frac{1}{2} \left[1 - \left(\frac{\omega_p^2}{\omega^2} \right) \left\{ 1 + \left(\frac{\alpha E_{oocr}^2}{2} \right) X \left(1 + \left(\frac{\alpha E_{oocr}^2}{32\lambda^2} \right) X \right) \right\}^{-\frac{1}{2}} \right]^{\frac{1}{2}}$$

$$(4)$$

For computation a typical set of parameters $\omega_{po} = 0.5 \, \omega$, r = 1–3μm and $\omega = 1.7$ x 10^{14} rad/sec (Co_2 laser) $E_{1o} / E_{oo} = 0.2$; $\Phi_p = 0$ and $r_o / r_{1o} = 1$ has been chosen. Figure 1 and

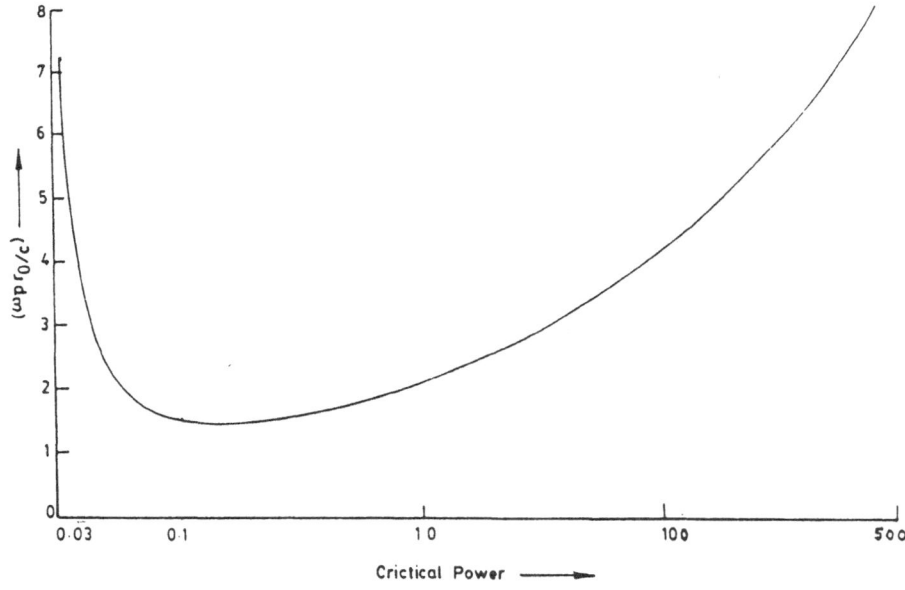

Figure 1.

2 display respectively the variation of critical power P_{cr} with dimensionless radius $(\omega_{po}r_o/c)$ and beam width parameter "f" with normalised distance of propagation. The medium behaves as an oscillatory waveguide in the regime $P_{cr1} < P < P_{cr2}$ and $P < P_{cr2}$. Thus, by controlling the power of the main beam and phase angle $_p$, self-focusing of ripple gaussian beam changes significantly. Substantially, we say that self-focussing of rippled gaussian laser beam can be analysed like self-focusing of gaussian beams in plasmas. The

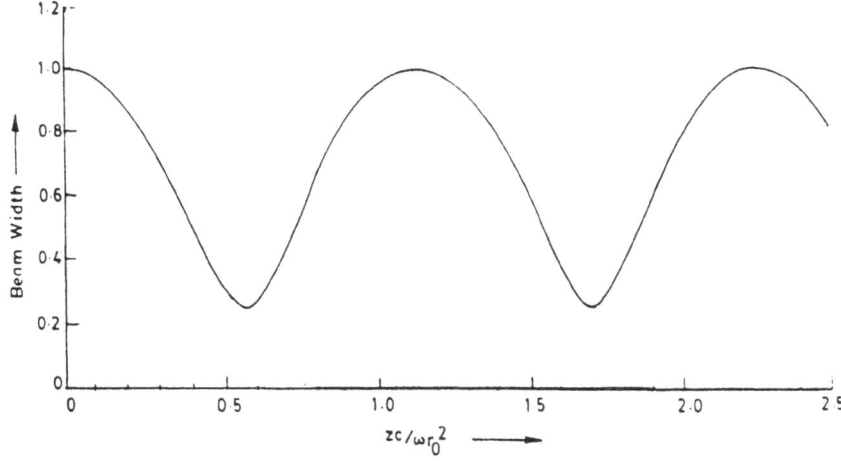

Figure 2.

power of the beam and phase difference between electric vector of the main beam and the ripple is found to change significantly the nature of self-focusing of the ripple. Due to higher intensities necessary for relativistic self-focusing to occur the saturation effects of nonlinearity become predominant, making the nonlinear refractive index in the paraxial region have slower radial dependence, and thus the ripple extract relatively less energy from its neighbourhood.

REFERENCES

1. X.Liu and D.Umstadter, Phys. Rev.Lett. 69, 1935 (1992) ; P.Maine IEEE J.Quantum Electronics QE -24, 398 (1988); E.Esarey, A.Ting and P.Sprangle, Appl. Phys.Lett. 152, 1266 (1988).
2. M.Asthana, Indian Journal of Pure and Appl. Phys. 31, 564 (1993); M.Asthana, M.S.Sodha and K.P.Maheswari, Journal of Plasma Phys. 51, 155 (1994); . M.Asthana,M.S.Sodha and K.P.Maheswari, Laser and Part. Beams 12, 623(1994).

76

ELECTRON GAS KINETICS IN A STRONG ELECTROMAGNETIC FIELD

A. V. Zinoviev, A. V. Lugovski, and T. Usmanov

NPO "Akadempribor" 7001433
Tashkent, Uzbekistan

Any investigation of solid state response to laser pulse irradiation demands some information about the electronic distributions function (EDF). The EDF and dynamics of its changing can strongly influence the characteristics of observed processes such as charge particle emission, "hot" luminescence, linear reflection and non-linear optical phenomena[1-4]. Simple estimations show that, at pulse magnitude $10^{10} - 10^{14}$ W/m^2 the average energy absorbed by one electron during the electron-electron relaxation time may be equal to 1 eV, which is comparable to its kinetic energy. So in our opinion the statement that EDF is an equilibrium function even with temperature different from that of the lattice[5] needs more serious argumentation than reference to the smallness of the electron-electron relaxation time τ_{ee}.

Consider the system of the homogeneous and isotropic electron gas in the homogeneous potential of positive lattice charge (the jelly model).

To analyze the effect of a strong high-frequency electromagnetic field on the electron distribution in the metal, we employ an equation for the one-particle density matrix;

$$\frac{\partial f(p,t)}{\partial t} = S_{ee} + S_{ep} \tag{1}$$

where $f(p,t)$ is the one-particle density matrix (which, in what follows, we refer to as the electron distribution function), p is the canonical momentum, and S_{ee} and S_{ep} are the electron-electron and electron-phonon collision integrals, respectively.

Equation (1) was derived in the standard manner from the Heisenberg equation under the assumption that the distribution function is spatially uniform. This assumption is justified in the case of the normal skin effect. In the case at hand it is applicable if

$$v / \omega << \delta$$

where v is the characteristic electron velocity, δ is the depth of the skin layer, and ω is the frequency of the field. At optical frequencies ($\omega \propto 10^{14} - 10^{15}$ sec^{-1}) for metals $\delta \propto 10^{-6}$cm

Ultrafast Processes in Spectroscopy, edited by Svelto et al.
Plenum Press, New York, 1996

and $v/\omega \propto 10^{-7}$. Therefore, the dependence of the distribution function on the spatial coordinates and the transport processes associated with this dependence can be neglected.

We obtain an equation for the slowly varying component of the distribution function by averaging. Eq. 1 over one period of the field:

$$\frac{\partial f(p)}{\partial t} = -\frac{1}{\tau_{ee}}\left[f(p)-f_0(p)\right]+\frac{2\pi}{\hbar}\sum_{n=-\infty}^{\infty}\sum_{k}|C|^2$$

$$\left[2N(k,t)+1\right]J_n^2(a,k)\times\left[f(p+k,t)-f(p,t)\right]\delta\left(\varepsilon_{p+k}-\varepsilon_p-n\hbar\omega\right) \qquad (2)$$

where $f_0(p)=\left[\exp\left(\varepsilon_p-\varepsilon_F/kT\right)+1\right]^{-1}$ is the equilibrium EDF; N(k,t) is the distribution function of phonons in a state with quasimomentum k. Since the electron-phonon coupling constants are small, we assume below that this distribution is the equilibrium distribution with time-dependent temperature T(r,t); Jn are Bessel functions; $a = eE_0/m\hbar\omega^2$ is the amplitude of the electron oscillations in high-frequency field $E = E_0 \sin(\omega t); \varepsilon_p = p^2/2m$ is the electron energy. The quantity Ck is related to the matrix element for scattering of an electron by acoustic phonon from the state p into the state p";

$$\left|M_{p,p'}^k\right|^2 = \delta_{p,p-k}\frac{1}{L^3}|C_k|^2$$

The first term on the right-hand side of Eq. 1 is the electron-electron collision integral in the τ approximation. We note that the electron-electron collision integral does not contain any terms which correspond to induced absorption (emission) of photons, since the field may be regarded as uniform over distance of the order of screening length.

The Eq. 2 shows that the induced absorption (emission) of photons is possible in electron-phonon collisions (the second term on the right-hand side of Eq. 2).

In studying the kinetic processes in the electron gas of metal, the fact that, as a rule, the distribution function over the direction at the momentum relaxes much more quickly (t $\sim 10^{-14}$ sec) must be taken into account. Averaging Eq. 2 over the directions of the momentum and neglecting the contribution of the electron-phonon scattering (the term with n = 0 in the electron-phonon collision integral) to energy relaxation, we obtain

$$\frac{\partial f(\varepsilon)}{\partial t} = -\nu_{ee}\left(f(\varepsilon)-f_0(\varepsilon)\right)+\nu_{10}^{(+)}$$

$$\times\left(f(\varepsilon+n\hbar\omega)-f(\varepsilon)\right)+\nu_{10}^{(-)}\left(f(\varepsilon-n\hbar\omega)-f(\varepsilon)\right) \qquad (3)$$

where $\nu_n^{(\pm)}$ are the electron-phonon collision frequencies. The argument of the Bessel function X is a parameter which determines the multiphoton nature of the process. Since in our problem the field strength is limited by the plasma formation threshold, X always holds, and the contribution of multiphoton processes can be neglected compared to one-photon processes.

After further simplifications we can write the kinetics equation in final form

$$\frac{\partial f(\varepsilon, x)}{\partial x} = -f(\varepsilon, x) + f_F(\varepsilon, T_e) + \kappa(x)\{f(\varepsilon + \hbar\omega, x)$$
$$-f(\varepsilon, x) + U(\varepsilon - \hbar\omega)(f(\varepsilon - \hbar\omega, x) - f(\varepsilon, x))\} \tag{4}$$

where $U(x)$ is the Heaviside step function.

Its solution can be obtained in the form of series in small parameter $\kappa = \nu\tau_{ee}$

$$f(\varepsilon, t) = \sum_{n=0}^{\infty} \kappa_0^n f_n(\varepsilon, t) \tag{5}$$

where κ is a small dimensionless parameter equal to the number of incident field photons per electron over the time τ_{ee}:$\kappa_0 = W_0 \tau_{ee} / (n_1\hbar\omega\delta)$ where W_0 is the peak power density of the absorbed radiation and $f_n(et)$ has the form.

$$f_n(\varepsilon, t) = \int_{-x}^{x} dy_1\, \beta(y_1) \int_{-x}^{y_1} dy_2\, \beta(y_2)... \int_{-x}^{y_n} dy_n\, \beta(y_n) \times \Lambda^n \big[f_0(\varepsilon, T(x))\big] \exp(y_n - x) \tag{6}$$

where $x = t/t$; Λ^n is the result of displacement operator Λ operating n times on the function f.

$$\Lambda\big[f(\varepsilon)\big] = f(\varepsilon - \hbar\omega) - \big[f(\varepsilon) + U(\varepsilon - \hbar\omega) - f(\varepsilon)\big] \tag{7}$$

and $\beta = W(t)/W_0$; $T_e = T_e(r, t)$ is the temperature of the electron gas. The graph of the distribution function is show in the Fig. 1.

Then we analyzed how the nonequilibrium state of the electron distribution influence the electron emission current j_e. Using Eq. 5 we found the following expression for j_e:

$$j_e = ADT_e^2 \left\{ \sum_{n=0}^{K-1} \kappa_0^n \theta_n(t) F\left(\frac{\phi - n\hbar\omega}{T_e}\right) + \kappa_0^K \theta_K(t) \right.$$
$$\left. \times \left[\frac{\pi^2}{6} + \frac{(\phi - K\hbar\omega)^2}{2T_e^2} - 2KF\left(\frac{\phi - (K-1)\hbar\omega^2}{T_e}\right) - F\left(\frac{K\hbar\omega - \phi}{T_e}\right) \right] \right\} \tag{8}$$

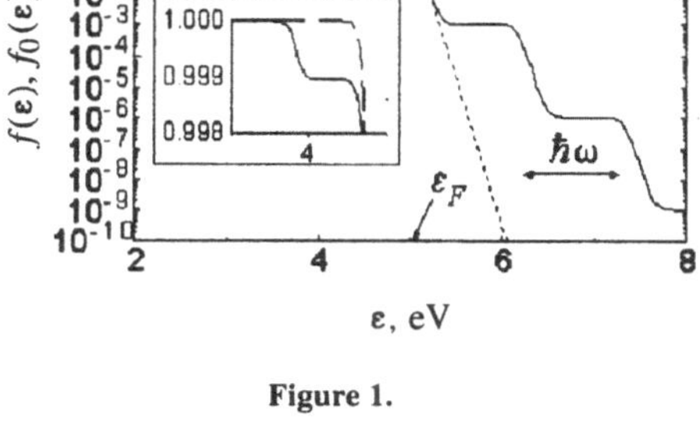

Figure 1.

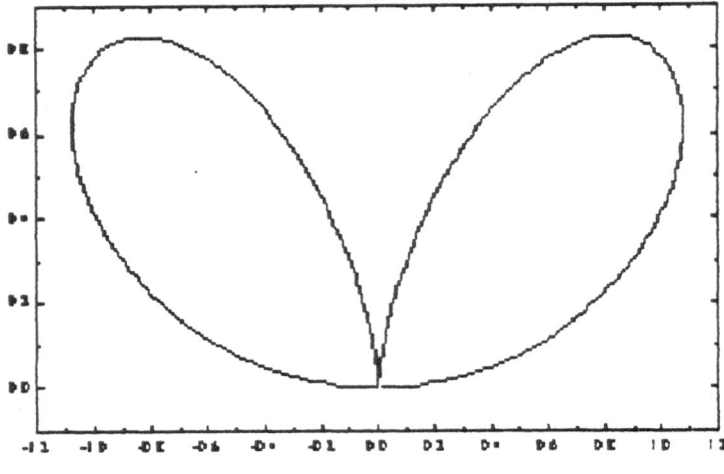

Figure 2.

where K corresponds to the photoemission multiplicity, $F(x)$ is the Fowler function, ϕ is the work function of the metal, A is the Zomerfield constant. Factually, this formula coincides with the expression found within the phenomenological Fowler-DuBridge approach.

At first glance the isotropy of the electron distribution is contradict to the special anisotropy of the photoelectron current observed in the experiment. Meanwhile this contradiction can easily be understand if we notice that the frequency of the triple collision - electron, photon and phonon - is anisotropy (see Eq. 2). Considering that only electrons in the layer of metal with thickens equal to the mean free path length can participate in the emission process and taking into account the mentioned anisotropy we numerically calculated the spatial distribution of the photocurrent (see Fig. 2).

REFERENCES

1. W. Knecht Appl. Phys., 6,99, (1965).
2. U.A. Arifov, V.B. Lugovskoi, V.A. Makarenko. Fiz. Tverd. Tela, 20, 1505, (1978)
3. M.B. Agranat, S.I. Anisimov and B.I. Makshantsev. Appl. Phys. B47, 209, (1988).
4. G.L. Eesley Phys. Rev. Lett., 51, 2140 (1983).
5. S.I. Anisimov et al. Usp. Fiz. Nauk 122, 185 (1977) (Sov. Phys. Usp. 20, 467, (1077)).

77

OPTICAL HARMONICS GENERATION IN LOW-TEMPERATURE LASER-PRODUCED PLASMA

R. A. Ganeev, V. I. Redkorechev, and T. Usmanov

NPO "Akadempribor"
Tashkent 700143, Uzbekistan

Nonlinear optical effects (harmonics generation, CARS, etc.) in low-excited laser-produced plasma have a number of distinctive features from those in a dense laser plasma produced by high-intensity 10^{16}–10^{18} W/cm^2 light fields[1,2]. In particular optical harmonics generation (HG) in visible and near ultraviolet (HV) ranges was investigated in laser plasma generated after an optical breakdown in gases and at metal surfaces with the intensity of 10^{10}–10^{12} W/cm^2 [3,4]. One of the mechanisms increasing significantly an efficiency of multiphoton processes (for example, harmonics generation) under optical breakdown (or other types of excitation) is an enhancement of high-order nonlinear susceptibilities[5,6]. The latter caused by occupancy of the excited atomic and ionic levels in low-temperature plasma.

In this paper an investigation of HG in low-temperature plasma in UV and vacuum ultraviolet (VUV) ranges is described. Here we consider high-order HG and optimization of the frequency conversion depending on parameters of radiation and nonlinear media. We have observed HG in (i) an atmospheric plasma, when the target plasma serves as an operating mechanism for atmospheric plasma formation and (ii) in target plasma placed in vacuum chamber.

We have used the radiation of Nd:glass laser both for heating and probing of low-temperature plasma. The cavity of the master oscillator operated in a negative feedback regime utilizes neodymium phosphate glass (GLS-22P) is formed by spherical (r = 1.7m, R = 100%) and plane (R = 50%) mirrors. For passive mode-locking we used a solution of dye 3274-u in an ethanol with an initial transmission of 80%. The dye cell was in contact with the curved mirror. Negative feedback loop consists of an electrooptical modulator (ML-102A model) biased by an electrical signal from a high-current fast photomultiplier (14ELU-FK model). Long train (4 µs) of picosecond pulses (τ = 1.5 ps) is generated[7]. After 2 µs from the beginning of oscillation a short train of 15 pulses was sliced from the prolonged one by Pockels cell. The short train of transform limited pulses is directed to preamplifier represented by a Q-switched Nd: glass laser[8]. Amplified pulse train is split into two parts. First beam ("heating" pulse train) was focused by cylindrical lens (f =

Ultrafast Processes in Spectroscopy, edited by Svelto et al.
Plenum Press, New York, 1996

100 mm) onto the target (Al, W, Ta), located in front of the input slit of WMR-2 type vacuum monochromator. Plasma with dimensions 3 0.15 mm^2 was generated at the target surface. Second beam ("probing" radiation) was propagated transversely to the direction of plasma development and focused by spherical lens. Intensity of probing radiation (10–12 pulses) in an interaction zone was varied from 3 10^{11} to 5 10^{13} W/cm^2 (pulse duration of 3 ps). Probing and harmonic radiations were directed in monochromator and registered by photomultiplier.

In the case of experiments with the target plasma in a vacuum, the target was placed in a chamber mounted at the input slit of vacuum monochromator and registered by photomultiplier.

In the case of experiments with the target plasma in a vacuum, the target was placed in a chamber mounted at the input slit of vacuum monochromator. Here we suppressed an influence of atmospheric gases on plasma development. Thus we were able to detect a high-order harmonics (higher than fifth) generation, which wavelengths are shorter than 200 nm. The experiments were conducted at pressure in a chamber of 10^{-5} torr.

A study of a probing (λ = 1.054 μm) and generated (λ = 351.3 nm) radiation oscillograms showed that an efficient third harmonic generation has a certain delay with respect to the beginning of propagation of probing radiation through plasma. Time delay between the two pulse trains is caused by the laser plasma evolution processes. Each of the successive pulses of a probing radiation comes to the target simultaneously with pulses of the heating radiation. For the first pulses of a probing radiation an optimal conditions of frequency conversion in laser-generated plasma are not attained (a "nonlinear" plasma is not yet produced). But with each of the later following pulses the best conditions for HG are produced due to plasma absorption and it's excitation. Time delay between the heating and generated radiation varied from shot to shot in the range of 75–125 ns. We have investigated a dependence of the Δ parameter on conversion efficiency where Δ is a distance between the target and focusing point of probing radiation. The $\eta(\Delta)$ dependence showed a strong effect of parameter Δ on conversion processes connected with optimization of plasma's nonlinear characteristics such as concentration and degree of excitation of higher laying levels of molecules. Note that optimization of parameter Δ may be connected with the phase-matching for four-photon processes. Calculation of phase matched integral for four-photon processes[5] requires definition of electron density in the interaction zone. At the same time it is difficult to calculate a conversion efficiency because of uncertainty in contribution of the each of heating pulses in the process of plasma excitation. Time delay optimization between the probing and generated pulse trains can also be connected with the provided phase matching conditions.

During the experiments we observed a phenomena when an enhancement of the conversion efficiency was followed by the appearance of spectral lines in the spectrum of excited gases in plasma. Being an isotropic medium atmospheric plasma yields in generation of only odd laser radiation harmonics[1]. In our experimental conditions an efficiency of second harmonics generation is much smaller compared to the third harmonic efficiency. Second harmonic generation was probably connected with a gradient of optical characteristics of plasma[2].

When the wavelength of converted radiation is close to VUV range it is necessary to change the experimental scheme, namely, place the target into the vacuum chamber. The general relationships which characterized the experiments described above, re-observed in this case too. Thus the probing radiation conversion efficiency dependence on parameter Δ in the case of the ninth harmonic generation (λ = 117 nm) was the same as for the above described experiments with an atmospheric plasma. The summary diagram of conversion efficiencies is shown in Table 1.

Table 1. Efficiencies of harmonics generation in atmospheric
and target plasma

Wavelength of harmonics, nm	Number of harmonics	Efficiencies harmonic generation in	
		Atmospheric plasma	Target plasma
527	2	$2\ 10^{-8}$	10^{-7}
353	3	$6\ 10^{-4}$	10^{-3}
211	5	$6\ 10^{-5}$	$2\ 10^{-5}$
151	7	–	$8\ 10^{-5}$
117	9	–	$2\ 10^{-6}$
96	11	–	$8\ 10^{-9}$

Maximum efficiency was achieved for third harmonic generation ($\eta = 10^{-3}$). Conversion efficiencies in similar processes ($3\ 10^{-2}$, see Ref. 5) has been achieved to date with the use of nanosecond pulses as a heating radiation synchronized accurately with respect to picosecond pulses of probing radiation. In our case a delay between the probing and heating radiation has been automatically realized. This leads the same time both to simplification of experimental scheme and reduction of efficiency of nonlinear processes.

Odd harmonics up to the eleventh ($\lambda = 96$ nm) were registered. Here we note two peculiarities. First, an essential influence of low-temperature plasma excitation on the nonlinear optical processes was seen especially in the case of high harmonics (9,11) generation.

The intensity of the eleventh harmonic is compared with the plasma spectral intensity in the same spectral range. In this case we have observed both the linear spectra and plasma continuum generation. Second, a modification of target surface after interaction with the intense heating radiation plays an important role in these experiments. In order to remove this modification a shift of a target from shot to shot was made.

REFERENCES

1. E.A. McLean, J.A. Stampel, B.H. Ripin, H.R. Griem, J.M. McMahon, S.E. Bodner. Appl. Phys. Lett., **31**, 825 (1977).

2. R.L. Carman, C.K. Rhodes, R.F. Benjamin, Phys. Rev. A., **24**, 2694 (1981).

3. A.M. Brodnokovsky, S.M. Gladkov, V.N. Zadkov, M.G. Karimov, N.I. Koroteev, Pis'ma v JTP, **8**, 497 (1982).

4. A.M. Zheltikov, N.I. Koroteev, A.B. Fedotov, Laser Physics, **4**, 569 (1994).

5. A.B. Fedotov, S.M. Gladkov, N.I. Koroteev, A.M. Zheltikov, JOSA B, **8**, 373 (1991).

6. R.A. Ganeev, V.V. Gorbushin, I.A. Kulagin, T. Usmanov, Sov. J. Quantum Electron, **16**, 115 (1986).

7. R.A. Ganeev, F.S. Ganikhanov, I.G. Gorelik, A.B. Dakhin, D.G. Kunin, T.Usmanov, A.V. Zinoviev, Optics Commun., **114**, 432 (1995).

8. S.A. Ametov, R.A. Ganeev, F.S. Ganikhanov, D.G. Kunin, V.I. Rekoretchev, T. Usmanov, Optics Commun, **96**, 75 (1993).

9. R.B. Miles, S.E. Harris, IEEE, J. Quantum Electron, **QE-9**, 470 (1973).

78

SPATIAL SOLITONS IN A BULK KERR-MEDIUM UNDER THE INFLUENCE OF SELF-INDUCED PHOTOIONIZATION

S. Henz and J. Herrmann

Max-Born-Institut
Berlin, Germany

In a pure Kerr-medium spatial solitons can only exist in two-dimensional geometries as for example in planar waveguides, limiting the diffraction to one spatial direction. Recently in a three-dimensional geometry the self-channeling of a small-scale filament formed from a femtosecond laser pulse of about $20 GW$ over a distance of 30 meters in air has been observed.[1] This effect was attributed to the combined effects of Kerr-nonlinearity and nonlinear index change by plasma creation.

In this communication we consider the evolution of an optical high power beam in a Kerr-medium under the influence of self-induced multiphoton-ionization by using a variational approach. The refraction index change by the Kerr-effect and the self-defocusing can be described by the relation $\Delta n = n_2 |A|^2 - N_e/(2 n_0 N_{cr})$ where A is the slowly varying amplitude of the electric field strength, n_0 the linear refractive index change, n_2, the nonlinear coefficient due to the Kerr-effect, $N_{cr} = m\omega_0^2 \varepsilon_0/e^2$ the critical plasma density, and N_e the number density of free electrons. In the following we restrict ourself on the region of input intensity where multiphoton-ionization dominates compared with tunneling ionization. The spatial and temporal dynamics of a pulse in a bulk Kerr-medium can therefore described by the reduced Maxwell equation

$$ \tag{1} $$

where r is the radial coordinate, z the longitudinal coordinate, k the wave number, and $\eta = t - z/v$ the time of the moving frame of the pulse maximum, further $\kappa = k^2 n_2/n_0$, n is the number of quanta necessary to ionize the molecules and $\rho = k^2 N n^{1.5} \varepsilon_0^n/(8^n n_0^2 N_{cr}^{n+1} I_i^n)$ with the number density of molecules N and the ionization energy I_i. We solve approximately equation (1) by means of the variational approach based on the construction of the Lagrangian of Eq. (1) and a suitable choice of a trial function $A(z,r,\eta) = T(\eta)B(z,r)$ with a Gaussian profile $B(z, r)$ having an amplitude $C(z)$, a beam radius $w(z)$, and a curvature pa-

Ultrafast Processes in Spectroscopy, edited by Svelto et al.
Plenum Press, New York, 1996

rameter $b(z)$ and where $T(\eta) = \sec h(1.76\eta/\tau_0)$ describes the input pulse shape. Introducing the normalized variable $\zeta = z/(kw_0^2)$ and the relative beam radius $y = w(z)/w_0$ the equation

$$(2)$$

can be derived with the input parameters b_0, w_0 at $z = 0$, the power P_0, the critical power of self-focusing $P_{cr} = \pi n_0^2 \varepsilon_0 c/(n_2 k^2)$, and the parameter

$$\gamma = 0.5^n n^{1.5} N P_{cr}^n g(\eta) / \left([n+1]^2\right) * * 1 / \left(n_0^{n-3} k^3 N_{cr}^{n+1} c^{n-1} I_i^n\right)$$

where $g(\eta) = \int_{-\infty}^{\eta} T^{2n}(t)dt$.

In Fig. 1 the potential $U(y)$ is depicted for air parameters under normal conditions with $n_0 = 1.0$, $n_2 = 7.4*10^{-26}\ m^2/V^2$, $P_{cr} = 1.82 GW$, $I_i = 14.54\ eV$, $n = 10$ and the laser parameters $w_0 = 52.6$ mm, $k = 7.85*10^6\ m^{-1}$, and $b_0 = 0$ for the maximum of the input pulse ($\eta = 0$), with the width $\tau_0 = 200 fs$, and three different input powers $p_0 = P_0/P_{cr}$ and the parameter $\gamma = 2.725*10^{-4}$. In curve 1 the power lies with $p_0 = 0.8$ below the critical one and the potential decreases monotonously since the diffraction dominates over the self-focusing. In curve 3 for a power larger than the critical one ($p_0 = 1.5$) the combined effects of self-focusing and self-defocusing due plasma creation lead to a minimum at $y = y_{min}$ where both effects balance each other. A minimum in the potential at $y = 1$ is a necessary condition for the existence of a spatial soliton or a self-trapped beam. Therefore we find from Eq. (2) the relation for a spatial soliton $p_0 = p_s$:

$$p_s - 1 = \frac{1}{2}\gamma m p_s^n.$$

$$(3)$$

For fixed material parameters and a fixed time, e.g., $\eta = 0$, Eq. (3) represents a relation between the power and the beam radius w_0 (via the coefficient γ) analogous to the re-

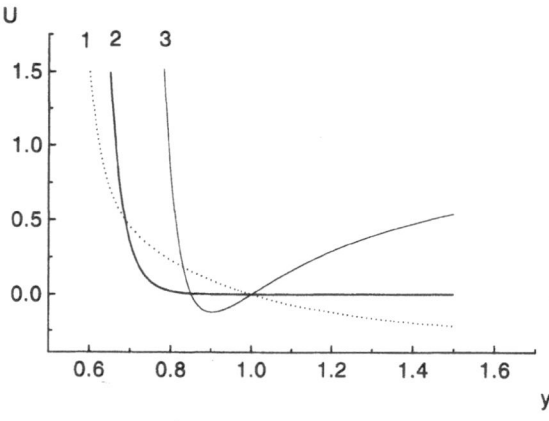

Figure 1. Potential $U(y)$ for the input parameter $\gamma = 2.725*10^{-4}$.

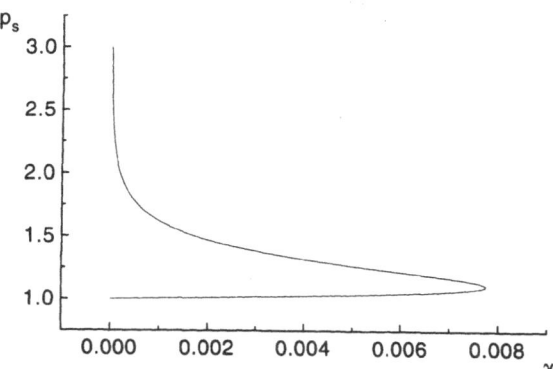

Figure 2. Soliton power in dependence on the parameter γ.

lation for time-like solitons in fibers between the soliton peak power and the pulse duration.

In curve 2 of Fig. 1 for $p_s = 1.0014$ the soliton condition Eq. (3) for $\eta = 0$ is fulfilled and the potential has its minimum at $y = 1$ and therefore the "particle" rests in this point. For a given pulse duration and a fixed time (e.g $\eta = 0$) the normalized soliton power p_s, is just a function of the parameters γ depending on the beam radius w_0 and the number of ionization-quanta n.

In Fig. 2 for $n = 10$ the power p_s is plotted. As seen for small values of γ one gets a soliton power p_s only slightly larger than unity, and with increasing γ the soliton power becomes a two-valued function of γ. But for a γ greater than a critical one $\left(\gamma_{cr} = 2[n-1]^{n-1} / n^{n+1}\right)$ no soliton solution exists. A necessary and sufficient criterion for the stability of the soliton-states of Eq. (1) is the condition $dp_s/d\beta > 0$ where β is the wave number shift $d\Phi/d\zeta = <196>\beta$. We calculated the function $p(\beta)$ and found a monotonously increasing $p(\beta)$ in dependence on increasing β (i.e. $dp/d\beta > 0$) in the whole parameter region. Consequently both soliton solution branches in Fig. 2 are stable.

In the curves of Fig. 3 we finally show still the behavior of the beam radius for the same parameters as in the curves of Fig. 1. For $p_0 = 0.8$ the beam radius increases monoto-

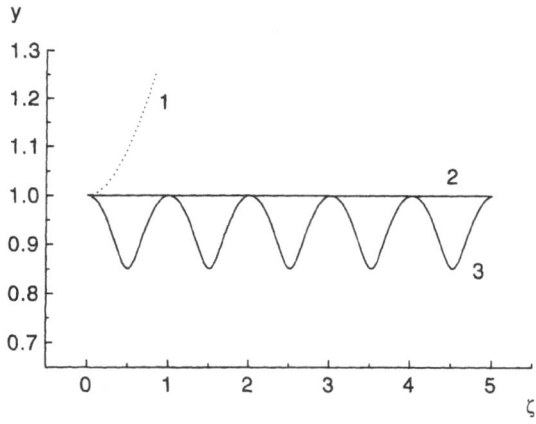

Figure 3. Beam radius in dependence on the propagation distance.

nously during the propagation (curve 1), for $p_0 = 1.5$ the beam radius reveals an oscillating behavior (curve 3), and if the soliton condition is fulfilled the beam radius stays constant during the propagation (curve 2).

We must bear in mind that in our discussion we have mainly referred to a fixed time as e.g. the time of the pulse maximum $\eta = 0$, but the general time-dependence of all relations is included in the input parameter $p_0(\eta)$ and $\gamma(\eta)$. Using the explicit time-dependence of the relation (3) we calculated the pulse shape $I(\eta)$ and found a nearly rectangular shape with a normalized intensity near unity. The sharp cut-off of this shape is caused by the decrease of the input function $T(\eta) = \sec h(1.76 \, \eta/\tau_0)$ which yields an increase of the parameter $\gamma(\eta)$ on the pulse wings because the function $g(\eta)$ changes only slowly with time. Therefore at the pulse wings the parameter γ reaches its critical value $y_{cr} = 7.748*10^{-3}$. For the subcritical parts of the pulse the soliton relation (2) cannot be fulfilled and the pulse wings are suppressed by diffraction-induced erosion.

REFERENCES

1. A. Braun et al., Opt. Lett. **20**, 73 (1995).

79

COMPENSATION OF THE GROUP-VELOCITY MISMATCH IN A BLUE-PUMPED NON-COLLINEAR PARAMETRIC GENERATOR OF FEMTOSECOND PULSES

P. Di Trapani,[1] A. Andreoni,[1] C. Solcia,[2] G. P. Banfi,[2] P. Foggi,[3]
R. Danielius,[4] and A. Piskarskas[4]

[1] Istituto di Scienze Matematiche, Fisiche e Chimiche
   Universita' di Milano a Como, Como, Italy
[2] Dipartimento di Elettronica
   Universita' di Pavia, Pavia, Italy
[3] European Laboratory for Nonlinear Spectroscopy
   Firenze, Italy
[4] Laser Research Center
   Vilnius University, Vilnius, Lithuania

The recent development and spread of high-power femtosecond amplified solid-state laser systems has greatly stimulated the interest in Travelling-Wave Optical Parametric Generators[1] as the most suitable means for achieving a very broad wavelength-tunability from such sources[2,3]. In particular, the extension of the accessible spectrum to the visible-near UV range for 100 fs pulses seems very promising for spectroscopic applications[4,5].

In the present experiment we combine a non collinear seed generator[6] and a collinear parametric amplifier in a blue pumped device that delivers 78 fs signal pulses at 1 kHz repetition rate, tunable between 460 and 700 nm (930nm-3μm for the idler), with single pulse energy as high as 17 μJ, the highest values so far reported from parametric conversion of sub-100 fs pulses in the visible range.

The device lay-out is shown in fig. 1; the pump beam is provided by a frequency doubled, amplified $Ti:Al_2O_3$ system (BMI Alpha 1000), delivering 400 nm pulses at 1 kHz repetition rate, with an estimated full-width at half-maximum (FWHM) duration of 100 fs.

The parametric seed is generated in two passes in a 4 mm long BBO crystal, cut for type I non-collinear phase matching at 33°; the first pump (P1, ~30 μJ) is focused by the telescope T1 down to an estimated averaged intensity on the crystal, assuming a gaussian profile in time, of 85 GW/cm^2. Parametric superfluorescence is generated off-axis, along a cone shaped surface, in those directions and at those wavelengths that simultaneously ful-

Ultrafast Processes in Spectroscopy, edited by Svelto et al.
Plenum Press, New York, 1996

pump beam

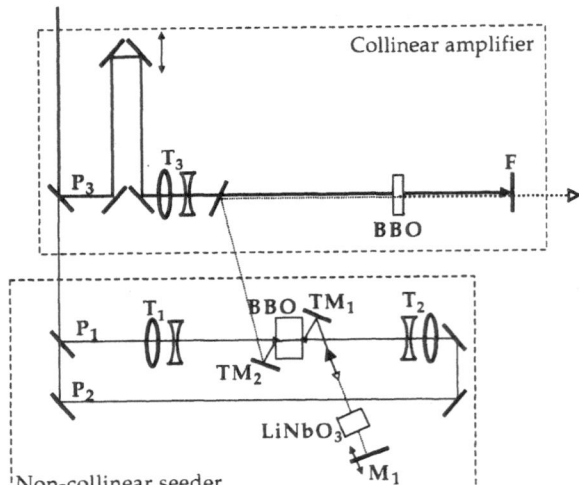

Figure 1. Experimental set-up. The mirrors TM1 and TM2, mounted on rotating holders, are used to compensate for the variable direction of signal emission in the non-collinear seeder.

fil the phase matching condition and minimise the GVM[7]. The half aperture of the signal cone, as measured in the horizontal plane containing the crystal, varies from 3° to 9°, while the generated signal wavelength (λ_s) sweeps the 460–700 nm range, when the (internal) tuning angle between the crystal optical axis and the pump beam is changed from 29.5° to 41°. The Al-coated mirror TM1 is used to direct a portion of the signal superfluorescence on the back reflector M1 (Al coated, flat), which is mounted on a linear translation micrometric stage for fine adjustment of the pulse delay. The second-pass pump (P2, 15 μJ), focused by the telescope T2 to an averaged intensity of 40 GW/cm^2, amplifies a small portion of the superfluorescence, leading to a collimated beam of fairly good spatial quality which feeds the amplifier; a signal energy of approximately 1 μJ is measured throughout the whole tuning range after the second step. We remark that the chosen parameters for the seeder allow a fair compromise between the best beam quality and the

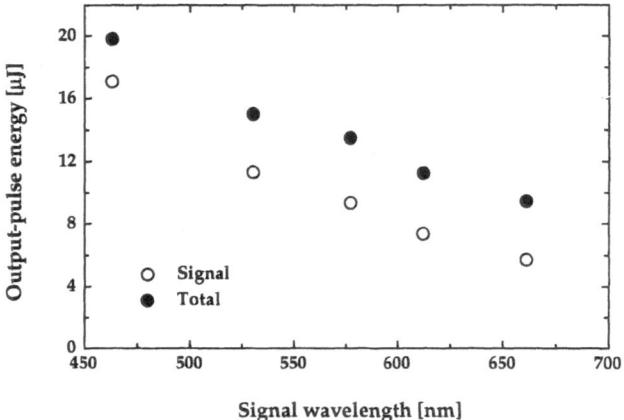

Figure 2. Measured output energy per pulse of the TOPG. Open dots: signal energy; full dots: total energy.

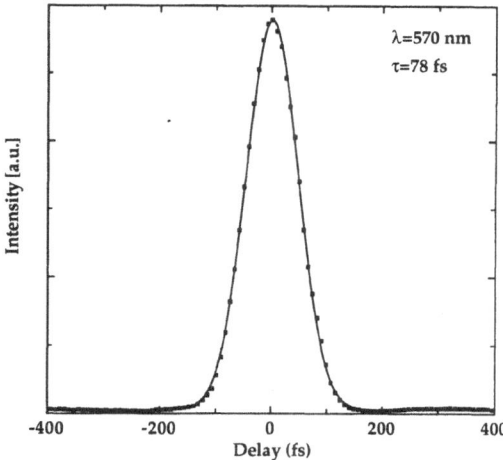

Figure 3. Signal autocorrelation traces measured with a 600 μm BBO I crystal at 570 nm. The dots are measured values, the line is a gaussian fits.

maximum generated energy. Replacement of the mirror M1 with a diffraction grating would allow to control the bandwidth to values close to the Fourier Transform limit[6], but would complicate the tuning procedure. We placed a 15 mm long $LiNbO_3$ crystal in the signal path, acting as a dispersive medium, to chirp the superfluorescence seed, so that the pump P2 "picks" a narrower spectral portion of the re-injected light.

The signal is then amplified in a 1 mm long BBO crystal, cut for type I collinear PM at 26°, with a greater portion of the pump (P3, 120 μJ, focused by the telescope T3 to an averaged pump intensity of about 150 GW/cm^2). This pump level is close to the limit of observable beam distortion induced by pump self phase-modulation.

The output signal energy (fig. 2) is highest close to the blue edge, 17 μJ at 460 nm with an overall conversion efficiency of the pump into signal+idler of 17% in the amplifier stage, and smoothly drops towards the red-end part, with about 6 μJ still produced at 660 nm. We remark that the tuning range is limited on both sides by the seeder performance: for signal wavelengths close to 700 nm, the condition for optimum GVM compensation is met at very wide non collinear angles, so that the seeder efficiency is gradually reduced: the amplifier output energy falls below 1 μJ above 700 nm. On the other side, towards the tuning edge, the signal is emitted almost collinearly, a condition which is not supported by the present seeder layout.

A signal autocorrelation trace is presented in fig. 3. The autocorrelations have been recorded with a single-shot autocorrelator equipped with a CCD detector. The typical FWHM pulse duration was around 90 fs, with a shortest value of 78 fs at 570 nm. Bandwidths of 8.3 nm (387 cm^{-1}) at 465 nm up to 10.7 nm (245 cm^{-1}) at 660 nm have been measured, for typical time-bandwidth products, smaller than that of the pump, of about 0.7. The output beam profile appears to be similar to that of the pump in the far field. In conclusion we have shown how efficient parametric generation of highly energetic, ultrashort pulses, broadly tunable in the visible, can be accomplished by combining a GVM-compensated seeder and a strongly-pumped collinear amplifier.

REFERENCES

1. R. Danielius, A. Piskarskas, A. Stabinis, G.P. Banfi, P. Di Trapani and R. Righini, J.Opt. Soc. Am. B **10**, 2222 (1993).
2. F. Seifert, V. Petrov and M. Woerner, Opt. Lett. **19**, 837 (1994).
3. M. Nisoli, S. De Silvestri, V. Magni, O. Svelto, R. Danielius, A. Piskarskas, G. Valulis and A. Varanavicius, Opt. Lett. **19**, 1973 (1994).
4. V. Petrov, F. Seifert, and F. Noak, Appl. Optics **33**, 6988 (1994).
5. M. K. Reed, M. K. Steiner-Shepard and D K. Negus, Opt. Lett **19**, 1855 (1994).
6. P. Di Trapani, A. Andreoni, P. Foggi, C. Solcia, R. Danielius and A. Piskarskas, Opt. Commun. **119**, 327 (1995).
7. P. Di Trapani, A. Andreoni, G.P. Banfi, C. Solcia, R. Danielius, A. Piskarskas, P. Foggi, M. Monguzzi, and C. Sozzi, Phys. Rev. A **51**, 3164 (1995).

80

FEMTOSECOND PULSE COMPRESSION BY SUM FREQUENCY GENERATION IN BBO

M. Nisoli,[1] S. DeSilvestri,[1] G. Valiulis,[2] and A. Varanavicius[2]

[1] Centro di Elettronica Quantistica-CNR
Dipartimento di Fisica Politecnico di Milano, Italy
[2] Laser Research Centre
Vilnius University, Vilnius, Lithuania

In the past few years it was recognised that frequency upconversion in presence of group velocity mismatch (GVM) and strong energy exchange between interacting pulses is an effective pulse compression method. Pulse compression by SH generation in KDP crystal for Nd laser pulses was studied in a number of papers. Compression factors of 4–5 were achieved starting with 0.5–1.2 ps pulses of Nd:glass lasers[1–3]. Recently 11 ps pulses of Nd:YAG laser were compressed down to 500 fs using this technique[4]. In a theoretical work of Stabinis et al. the conditions of pulse compression in general case of sum frequency (SF) generation were found[5].

In this paper we present the results on femtosecond pulse compression by SF generation in beta-barium borate (BBO) crystal. The results of computer simulation of frequency mixing of femtosecond pulses taking into account GVM and group velocity dispersion are also presented. The generation of pulses as short as 35 fs by SF generation are demonstrated experimentally.

SF generation by mixing of femtosecond light pulses in nonlinear crystal can be described by set of equations with GVM and group velocity dispersion taken into account:

$$\frac{\partial A_1}{\partial z} + \frac{1}{u_1}\frac{\partial A_1}{\partial t} - \frac{i}{2}g_1\frac{\partial^2 A_1}{\partial t^2} = -i\sigma_1 A_2^* A_3,$$

$$\frac{\partial A_2}{\partial z} + \frac{1}{u_2}\frac{\partial A_2}{\partial t} - \frac{i}{2}g_2\frac{\partial^2 A_2}{\partial t^2} = -i\sigma_2 A_1^* A_3,$$

$$\frac{\partial A_3}{\partial z} + \frac{1}{u_3}\frac{\partial A_3}{\partial t} - \frac{i}{2}g_3\frac{\partial^2 A_3}{\partial t^2} = -i\sigma_3 A_1 A_2,$$

$$(1)$$

Ultrafast Processes in Spectroscopy, edited by Svelto et al.
Plenum Press, New York, 1996

where $A_j=a_j \exp(i\varphi_j)$ is the pulse complex amplitude (j=1,2,3), z and t are the longitudinal and time coordinates, u_j is the group velocity, g_j is the group velocity dispersion coefficient, σ_j is the nonlinear coupling coefficient.

In femtosecond time scale GVM between interacting pulses affects essentially the process of SF generation. It can be described by the parameter $m=\tau_{20}v_{13}/\tau_{10}v_{23}$, where τ_{j0} are the initial duration of pump pulses, $v_{j3}=u_j^{-1}-u_3^{-1}$, (j=1,2). If m is negative efficient SF pulse compression can be obtained in presence of strong energy exchange between the interacting pulses. The optimum compression occurs when value of m is close to -1.

Considering dispersion characteristics of BBO crystal we have found that pulse compression could take place by mixing infrared pulses of 1.2–2 μm wavelength with orthogonal polarizations. Parametric light generator based on type II phase matching BBO crystal and pumped by the fundamental wavelength of Ti:sapphire laser emits pulses with wavelengths which fits well into above mentioned region. Accomplishing frequency mixing of signal and idler pulses from the output of the parametric generator one can expect efficient SF pulse compression at $\lambda \sim 800$ nm. The parameter m assuming equal pulsewidths of incident pulses differs from the optimum one and is in range of $-(1.1–2)$ tuning signal from 1.2 μm to degeneracy. Nevertheless, efficient SF pulse compression could be obtained adjusting intensities of pump pulses[5]. Dispersion spreading length for all interacting pulses are above 6 m and influence of this parameters was estimated to be negligible for pulses longer than 20 fs.

In Fig. 1 the results of computer simulation of SF generation crystal accomplished solving numerically the set of equations (Eq.1) are presented. They were obtained with intensities of 23 GW/cm^2 and 15 GW/cm^2 at 1.3μm and 1.95μm respectively.

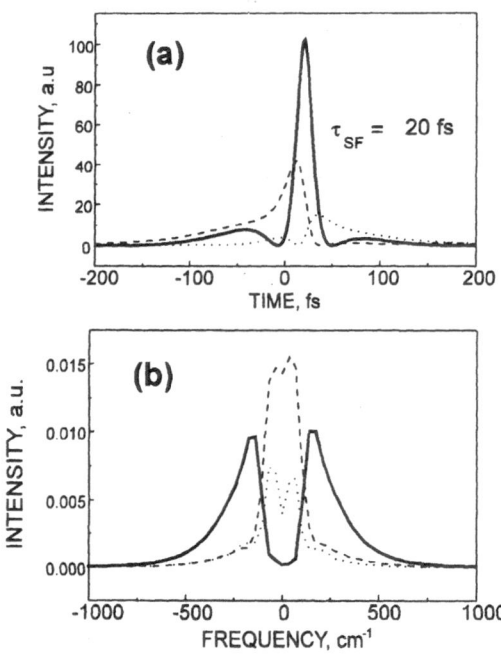

Figure 1. Calculated pulse envelope (a), spectrum (b) of pump and SF pulses at the output of 3 mm BBO crystal. Solid line - SF pulse (λ=0.78μm), dashed line - first pump pulse (λ=1.3μm), dotted line - second pump pulse (λ=1.95μm).

Initial temporal separation of 220 fs between 160 fs pump pulses was assumed. The SF pulse duration at the output of 3 mm long BBO crystal was found to be ~ 20 fs yielding 8 times pulse compression. Calculation of the pulse phase derivative shows that pulse is slightly phase modulated. This modulation is caused by dispersion of group velocities. However, in the case considered here, it does not influence final pulse duration significantly. When performing calculation with omitted equation terms which describe dispersion spreading of pulses SF pulses becomes shorter by ~10% only.

The experimental set up consists of a femtosecond Ti:sapphire laser with chirped pulse amplification (Clark-MXR, mod.CPA-1) providing 150 fs pulses at 780 nm and energy up to 750 μJ at 1 kHz repetition rate used as a pump source for parametric light generator. Triple pass travelling wave parametric generator (Light Conversion Ltd., mod. TOPAS) was tuned to generate ~160 fs pulses at 1.3 μm (signal wave , vertical polarization) and 1.95μm (idler wave, horizontal polarization). Parametric energy conversion was ~16% yielding 70μJ and 50μJ energy in signal and idler pulse respectively. From the output of parametric generator pulses of different wavelength were separated by dichroic mirror and after delay lines were recombined using a second dichroic mirror. Predelay between pulses was changed by means of variable delay line. Sum frequency of 1.3μm and 1.95μm was generated in 3 mm length BBO crystal cut for type II phase matching. Pulse intensity autocorrelation functions were measured by background free noncollinear second harmonic generation in a 1 mm long KDP crystal. SF pulse spectrum was monitored by a multichannel spectrum analyser. In order to achieve conditions of efficient pulse compression the intensity of pump pulses was raised up by means of telescope with reducing factor of 1:3.2. The intensity of 23 GW/cm^2 for 1.3μm and 15 GW/cm^2 for 1.95μm pulses at the entrance of BBO crystal were evaluated.

In Fig. 2 the SF pulse duration, spectrum and energy are presented as a function of predelay between pump pulses. Maximum conversion efficiency of ~20% was achieved at initial pump pulse separation in time scale of 90 fs. In order to obtain shortest SF pulse it was needed to increase predelay value up to 200 - 240 fs. In this range the spectrum of SF pulse exhibits specific two-peak structure. We should note, that the spectrum shape was the main reference point when aligning frequency mixer for the generation of shortest pulses. The autocorrelation of shortest measured SF is presented in Fig. 3.

Autocorrelation fit assuming sech^2 pulse shape a 35 fs FWHM pulsewidth of SF pulse. The pulse duration of the incoming pump pulse at 1.3μm was measured to be of 162 fs. The pulsewidth of other pump pulse (λ = 1.95μm) relaying on the data of computer simulation of parametric generation should be the same as of signal pulse with accuracy of 15%. Thus, we pointed out the SF pulse compression factor of 5 in this experiment.

Computer simulation of SFG generation process in type II BBO using the parameters from the experiment gave SF pulse shortening down to ~20 fs with conversion efficiency of 45%. We believe that the main reason of difference from results obtained experimentally is that model does not account the spatial modulation of interacting waves. The dynamic of SF generation is quite sensitive to initial intensity of pump pulses. Therefore, in real beams the optimum length of nonlinear medium is dependent on spatial coordinate in beam cross section. Obviously, in the experiment when pulse parameters are measured integrating signals over whole beam the resulting pulse duration should be larger as compared with that one in case of plane waves interaction.

Data of numerical modelling show that more higher compression ratio can be obtained increasing intensity of pump pulses. In experiment we were limited by energy of pulses produced by parametric generator. Pulse intensity increasing by means of telescope is not acceptable because in type II nonlinear crystal beam spatial walk-off effect de-

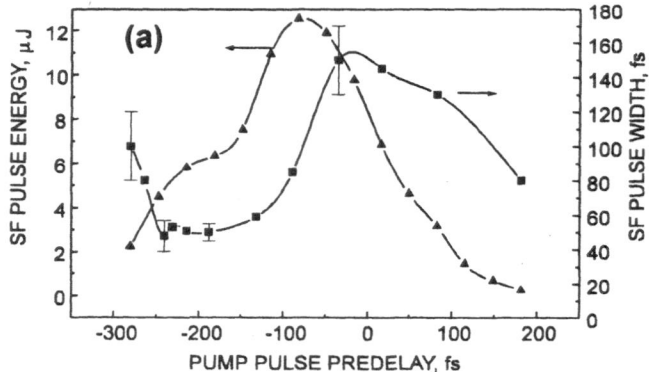

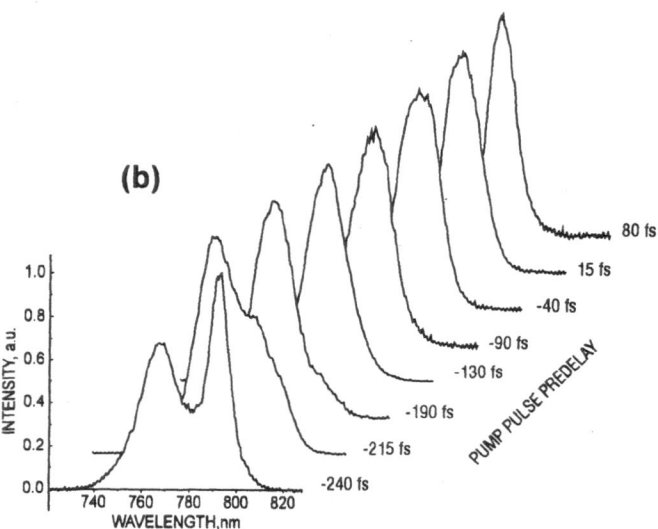

Figure 2. Experimental results: (a) SF pulse energy and pulsewidth versus initial pump pulse predelay, (b) SF pulse spectra versus pump pulse predelay.

presses parametric interaction. Relaying on computer simulation data pulse compression down to 10 - 15 fs can be obtained at higher intensities. For SF pulses shorter than 20 fs dispersion spreading of pulse becomes important factor limiting compression efficiency. One more parameter affecting SF pulse compression is phase modulation of pump pulses. In case when pump pulse chirp are of the same sign the generated SF pulse usually are longer as compared to transform limited pump case. However, when the signs pump chirp are opposite the compression rate is close to that when pump pulses are unchirped. Finally, we should note that this method allows to obtain tunable sub-50 fs pulses in case when the wavelength of one pulse is constant and other pump pulse is frequency tunable but the variation of it's group velocity is small in tuning range.

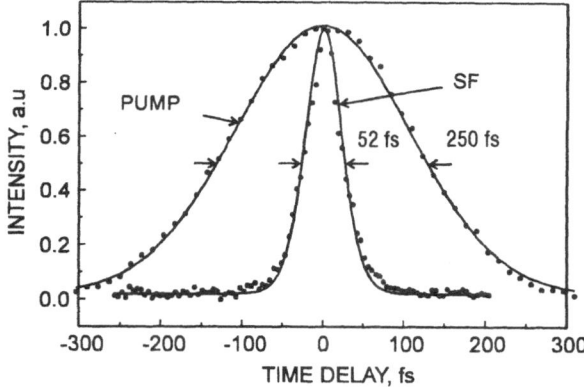

Figure 3. Autocorrelation traces of pump and SF pulse. Circles - experimental data, solid line - fit curve assuming sech^2 pulse shape.

ACKNOWLEDGMENT

A. Varanavicius wants to acknowledge financial support from the NATO Collaborative Research Grant n. 941165.

REFERENCES

1. Y.Wang, R.Dragila, Phys.Rev.A **41**, 5645 (1990); Y.Wang, B.Luther-Davies, Opt.Lett. **17**, 1459 (1992).
2. P.Heinz, A.Laubereu, A.Dubietis, A.Piskarskas, Lithuanian Journal of Physics, **33**, 314 (1994).
3. C.Y.Chien, G.Korn, J.S.Coe, J.Squeir, G.Mourou, R.S.Craxon, Opt.Lett. **20**, 353(1995).
4. A.Umbrasas, J.-C.Diels, G.Valiulis, A.Piskarskas, J.Jacob, in Conference on Lasers and Electro-Optics, Vol.15, 1995 OSA Technical Digest Series (Optical Society of America, Washington, D.C., 1995), p.130.
5. A.Stabinis, G.Valiulis, E.A.Ibragimov, Opt.Commun. **86**, 301 (1991).

81

FEMTOSECOND HIGH CONTRAST TUNABLE PULSES FROM BBO TOPG PUMPED BY SELF-COMPRESSED SECOND HARMONIC OF Nd:GLASS LASER

R. Danielius, A. Dubietis, G. Valiulis, and A. Piskarskas

Laser Research Center
Vilnius University, Lithuania

Second harmonic (SH) pulse shortening is observed in type II phase matching crystals with appropriate group delay relations between two orthogonally polarized pump pulses and a SH pulse[1,2]. So far, the best conditions for SH pulse compression have been found in a KDP crystal with picosecond Nd:glass 1055 nm pulses at input [3,4]. However, the self-compressed pulses exhibit a pedestal that contains a significant amount of the pulse energy and has the width comparable with that of the input pulse. Recently it has been shown that the contrast may be improved by introducing a parametric amplifier (PA) stage[1,5].

In this communication we report on application of self-compressed SH pulses of a picosecond Nd:glass laser to generation of femtosecond tunable high contrast pulses by a traveling-wave optical parametric generator (TOPG). The tuning range of 635–3000 nm was covered by pulses as short as ~200 fs with the time-bandwidth product of <0.6. The pulse shape was measured to be nearly Gaussian over the dynamic range of autocorrelator (10^3). Numerical simulation indicated that the amount of energy contained in the TOPG output pulse wings was reduced by a factor of ~$4\cdot10^4$ as compared with that in the pump pulse satellites.

As the laser source, we used a fiberless CPA-based Nd:glass system, delivering up to ~4 mJ, 1.3 ps pulses at 1055 nm. The SH pulse compressor was made of two 2 cm long KDP crystals cut for type II second-harmonic phase matching ($\theta=58^{\circ}$). The first crystal was used to introduce an appropriate delay between the o and e components of the pump pulse, the second one was a frequency doubler. The TOPG we used was a triple-pass device with independent pump of all three stages (TOPAS 501, Light Conversion Ltd.). The nonlinear crystal employed was 4 mm long type II phase matching BBO cut at $\theta=30^{\circ}$.

We have performed numerical simulations of SH compression and TOPG operation upon the conditions of the experiment. Assuming 1.3 ps input pulses and intensity 12 GW/cm^2, the calculation yielded 200 fs wide central peak and two satellite pulses shifted

Ultrafast Processes in Spectroscopy, edited by Svelto et al.
Plenum Press. New York. 1996

361

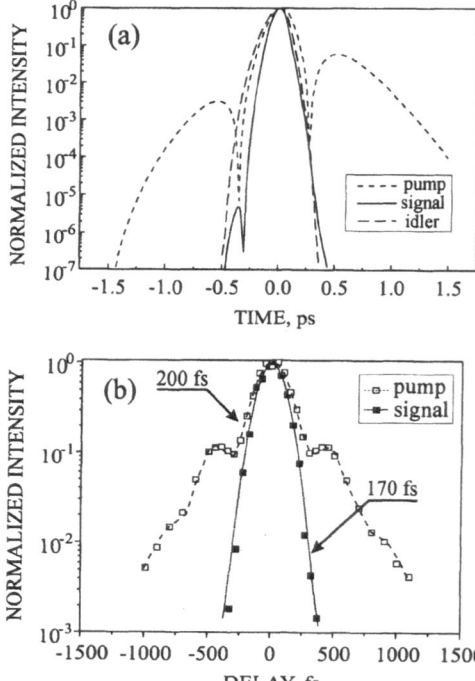

Figure 1. (a) Numerically simulated pulse shapes: the SH pulse at compressor output, the TOPG signal at 1040 nm and idler at 1071 nm pulses. (b) Autocorrelation measurements of the SH pulse and TOPG signal pulse. Dashed and solid curves show Gaussian fits for data points. Numbers indicate pulse duration at FWHM.

by ~500 fs from the main pulse [Fig.1(a)]. The satellite pulses contained about 11% of the total energy. The presence of two shoulders on each side of the autocorrelation trace indicated that the SH pulse consisted of three sub-pulses [Fig.1(b)]. Decomposition of the SH autocorrelation trace into three Gaussian curves yields 280 fs wide central peak, the corresponding pulse width is then 200 fs. Energy conversion as measured in the whole beam was ~45% at the frequency doubler input pulse energy of 2.5 mJ.

Numerical simulation of the parametric amplification was performed in the plane-wave approximation with the group velocity mismatch and pulse spreading effects taken into account. At pump wavelength of 527 nm a rather special situation with the group velocity mismatch takes place in the BBO type II phase matching crystal, when both signal and idler pulses move faster than the pump pulse (Fig.2). The walk-off length of the idler pulse is rather small, so it escapes the pump pulse at the output of the crystal [Fig.2(b)]. Under these conditions one may expect somewhat longer idler pulses, as compared with the signal pulse duration. We assumed pump intensities of three subsequent TOPG stages as 115, 50, and 25 GW/cm^2 for seeder, preamplifier, and power amplifier stages, respectively. In the 4 mm long BBO crystal, this provided corresponding gain factors of 10^7, 10^4, and 200. At the input of the seeder, instead of quantum noise, we assumed a 1.5 ps long Gaussian-shaped pulse. Energy conversion was kept about 10% in first two passes and ~15% in the last one. The optimal delay between the signal and pump pulses for the highest energy conversion was found to be 270 fs. The profile of the pulse exiting the TOPG is depicted in Fig.1(a). It is interesting that only a weak prepulse in the signal with magnitude of ~10^{-5} is generated on the leading wing. This is because the idler pulse escapes the main pump pulse, thereby producing radiation at the signal wavelength. The

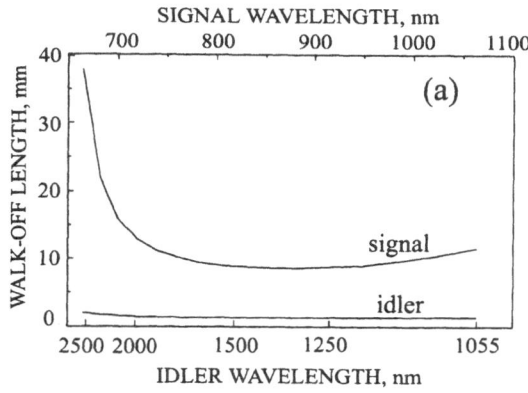

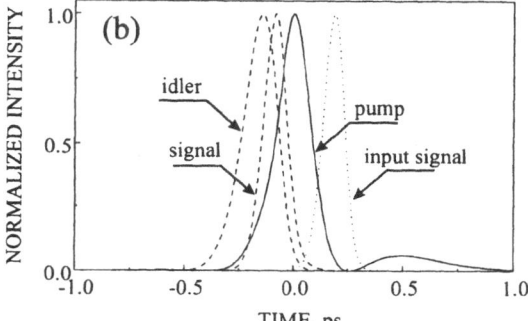

Figure 2. (a) Walk-off lengths of signal and idler waves, calculated for 200 fs pulse duration. (b) Numerically simulated third pass in the TOPG at signal wavelenght of 1040 nm. In order of figurativeness all amplitudes are normalized.

idler pulse then is somewhat longer (~200 fs), but also of high contrast. The experimentally measured autocorrelation trace [Fig.1(b)] followed closely the Gaussian shape over 3 orders of magnitude, which was the dynamic range of the autocorrelator.

We have measured the TOPG output energy ($E_s + E_i$) to be close to 100 µJ for entire tuning range [Fig.3 (a)]. Within measurement range of the autocorrelator (635–1300 nm) we did not observed strong deviation from the value of 200 fs, even for idler pulses at 1071–1226 nm [Fig.3 (b)].

In conclusion, we have demonstrated femtosecond BBO TOPG pumped by self-compressed second harmonic of a picosecond Nd:glass laser. Dramatic suppression of the pump pulse wings during the parametric process was observed. An important result is that we can operate the TOPG close to the saturation, thereby keeping the energy conversion and fluctuations at reasonable level. We believe that high contrast and spatial quality (<1.1 diffraction limit), sub-gigawatt peak power of the tunable pulses make the TOPG combined with the SH compressor an attractive source for pump and probe experiments as well as for studies in nonlinear optics.

ACKNOWLEDGMENT

Authors acknowledge the support by the International Science Foundation (Grants LA4000 and LED000, and generous support by Dutch Foundation for Fundamental Research of Matter (FOM).

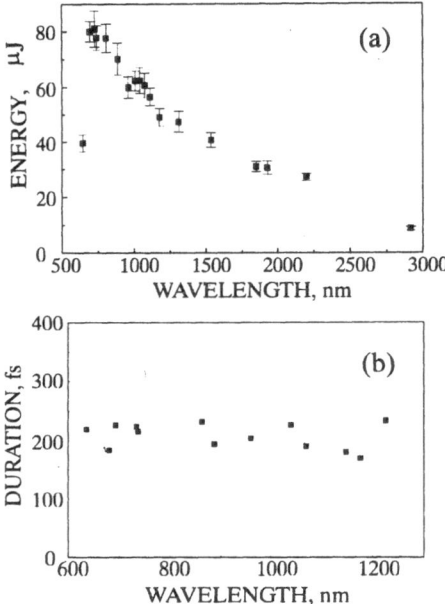

Figure 3. (a) TOPG output energy vs wavelength. (b) TOPG pulse duration vs wavelength.

REFERENCES

1. P. Heinz, A. Laubereau, A. Dubietis, A. Piskarskas, Lithuanian J. Phys. **33**, 314 (1993).
2. Y. Wang and B. Luther-Davies, Opt. Lett. **17**, 1459 (1992).
3. Y. Wang and R. Dragila, Phys. Rev. A **41**, 5645 (1990).
4. A. Stabinis, G. Valiulis, E. A. Ibragimov, Opt. Commun. **86**, 301 (1991).
5. Y. Wang and B. Luther-Davies, J. Soc. Am. B **11**, 1531 (1994).

82

GREEN-LIGHT GENERATION FROM PICOSECOND PULSES VIA FIRST-ORDER QUASI-PHASE-MATCHED LITHIUM NIOBATE

V. Pruneri, S. D. Butterworth, J. Webjörn,[*] P. St. J. Russell, and D. C. Hanna

Optoelectronics Research Centre
University of Southampton
Southampton United Kingdom

Recently periodically poled lithium niobate (PPLN) has become a very attractive material for efficient second order nonlinear interactions such as second-harmonic generation (SHG) [1] and optical parametric amplification/oscillation [2] both in CW and Q-switched regimes. Compared to birefringence phase-matching, quasi-phase-matching (QPM) can give access to higher nonlinear coefficients and new wavelengths, in particular where the birefringence cannot compensate the dispersion, by simply choosing the appropriate period of the domain grating. It also avoids spatial walk-off problems since fundamental and generated waves have the same polarisation in order to exploit the highest component of the second-order nonlinear tensor. Unfortunately for such a polarisation geometry, lithium niobate is highly dispersive. In the case of pulsed applications, this limits the effective interaction length (l_{eff}) in PPLN. On the other hand PPLN can offer a very high effective nonlinear coefficient, d_{eff} up to ~21 pm/V, resulting in a large product of $d_{eff}*l_{eff}$ despite the small l_{eff}. This material can therefore be very interesting for processes involving short pulses. Recent papers have reported on blue-light generation via SHG of pulses of duration >20 ps in PPLN [3] and in periodically inverted lithium tantalate [4]. The average harmonic powers generated were <10 mW. Here we report on efficient SHG of green light of shorter pulses (2–3 ps) in PPLN at a higher level of harmonic power (average SH power >300 mW).

A sample of lithium niobate with 0.3 mm thickness was periodically inverted using two high-voltage (~21 kV/mm) pulses of ~300 ms duration applied via liquid electrodes [1,2]. The sample had a period of domain reversal of 6.35 μm and a length of 3.2 mm, suitable for frequency doubling at 1 μm of ~2.6 ps pulses (l_{eff}~2.1 mm).

Initially, CW second harmonic measurements were carried out using a Nd:YLF laser at 1.047 μm. The measured nonlinear coefficient of ~17 pm/V is close to the theoretical limit of ~21 pm/V.

---

* Now at, Spectra Diode Labs, San Jose, California.

Ultrafast Processes in Spectroscopy, edited by Svelto et al.
Plenum Press. New York. 1996

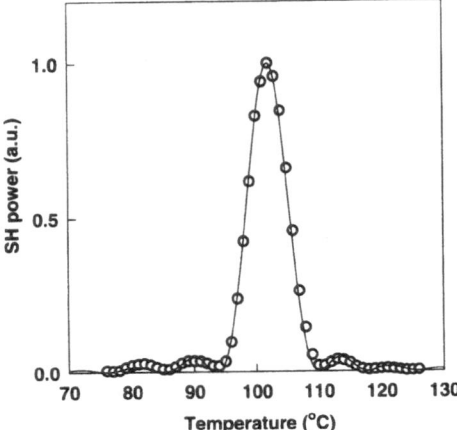

Figure 1. SH power as a function of the temperature of the crystal. The continuous line corresponds to a grating of length equal to the crystal length, 3.2 mm.

Fig.1 shows the SH intensity as a function of the temperature of the crystal. The experimental curve agrees well with the expected sinc^2 shape for a perfect grating 3.2 mm long, indicating that the whole length of the sample is involved in the interaction and that there was no evidence of any significant randomness in the position of the domain walls.

The same Nd:YLF laser when additive pulse mode-locked gave transform-limited pulses of ~2 ps duration, at a repetition rate of 105 MHz. These pulses were amplified in a diode-pumped bulk Nd:YLF amplifier used in a double pass configuration. These amplified pulses of ~2.6 ps duration were focussed in the PPLN sample to a spot size of 36 μm ($1/e^2$ intensity radius).

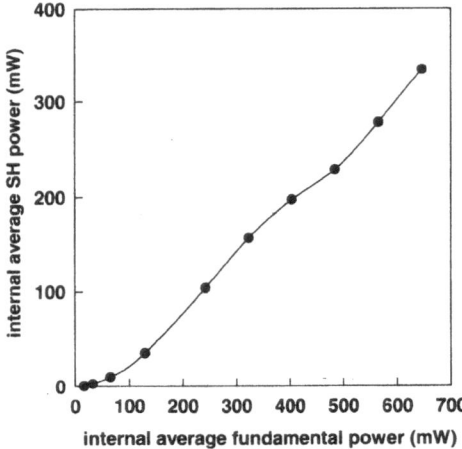

Figure 2. Dependence of the average SH power on the average fundamental power. These values are internal in the uncoated crystal.

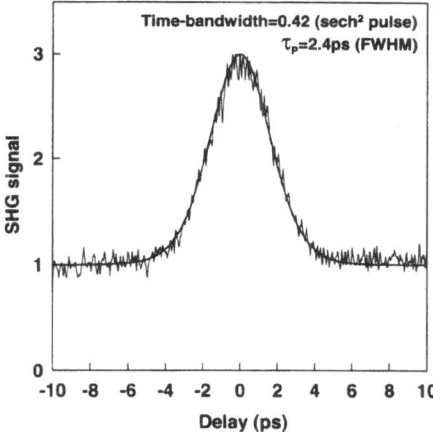

Figure 3. Autocorrelation of the SH pulses. The fit has been made assuming a sech^2 shape for the pulses.

Fig.2 shows the dependence of the SH power on the fundamental power at the QPM temperature of ~102 °C. For an average fundamental power of ~640 mW (peak intensity of ~110 MW/cm^2), the average SH power was ~330 mW, giving an average conversion efficiency of ~52%. Despite the high average and peak powers of the generated green-light, there was no sign of photorefractive damage as confirmed by measured M^2 beam quality factor of ~1.1 for both output fundamental and SH beams. The absence of photorefractive damage is consistent with our previous observation that the photorefractive effect was drastically reduced by operating above room temperature [5].

Fig.3 shows the intensity autocorrelation of the SH pulses. Assuming a sech^2 shape, the pulse duration was estimated to be ~2.4 ps while the bandwidth of the corresponding frequency spectrum was ~175 GHz. The time-bandwidth product was thus ~0.42, indicating the presence of only a weak chirp.

In conclusion we have confirmed that PPLN can be a very effective nonlinear medium for picosecond pulses, with further prospects for power scaling from the values we have reported here.

REFERENCES

1. V. Pruneri, R. Koch, P. G. Kazansky, W. A. Clarkson, P. St. J. Russell and D. C. Hanna, to be published in Optics Lett (1 Dec. '95).
2. V. Pruneri, J. Webjörn, P. St. J. Russell and D. C. Hanna, Appl. Phys. Lett. **67**, 2126 (1995).
3. L. Goldberg, R. W. McElhanon, W. K. Burns, Elec. Lett. **31**, 1576 (1995).
4. K. Yamamoto, K. Mizuuchi, Y. Kitaoka and M. Kato, Opt. Lett. **20**, 273 (1995).
5. V. Pruneri, P. G. Kazansky, J. Webjörn, P. St. J. Russell and D. C. Hanna, Appl. Phys. Lett. **67**, 1957 (1995).

83

GENERATION OF HIGHLY COHERENT TUNABLE FEMTOSECOND PULSES AT 82 MHz IN THE VISIBLE AND MID-INFRARED USING A BLUE-PUMPED OPTICAL PARAMETRIC OSCILLATOR

F. Hache and G. M. Gale

Laboratoire d'Optique Quantique du CNRS
Ecole Polytechnique, route de Saclay
91128 Palaiseau, France

The recent development of reliable Ti:Sapphire-driven femtosecond optical parametric oscillators (OPO) and optical parametric amplifiers (OPA) has opened up new domains of application in various fields. Most of these developments employ a fundamental Ti:Sapphire pump beam in the near infrared. We describe here the advantages, potential and realized, of using the *second-harmonic* of the Ti:Sapphire laser to pump a visible-range OPO. This technique has allowed the production of what are to our knowledge the shortest tunable visible pulses obtained directly from an oscillator and may herald sub-10-fs OPO operation in the visible. In a first application, the tunable visible femtosecond OPO has been parametrically frequency-mixed with the fixed-wavelength Ti:Sapphire fundamental pump beam to produce highly-coherent femtosecond pulses tunable in the mid-infrared. A theoretical model for the parametric amplification and frequency-mixing process has been developed. This model describes accurately the frequency-mixing results and is currently employed in the design of a high-energy mid-infrared OPA.

The optimization of ultrashort pulse OPO's involves a large number of parameters from which we single out here two major bandwidth limiting factors; phase-matching and group-delay dispersion (GDD) compensation. For example, in non-critically-phase-matched KTP, phase-matching occurs at a single wavelength, independent of crystal angle, and this type of OPO can only increase bandwidth, over and above the pump bandwidth, through self-phase-modulation. On the contrary, for 400-nm pumped BBO in a non-collinear type I pumping geometry, a single crystal orientation can give phase matching over as much as 170 nm around 620 nm. This phase-matching is highly critical and oscillation bandwidth is limited only by mirror reflectivity and GDD compensation. Fortunately, the high-bandwidth non-collinear pumping angle and the spatial walk-off compensation angle are almost identical in BBO allowing large bandwidth and high parametric gain to be achieved simultaneously.

Ultrafast Processes in Spectroscopy, edited by Svelto et al.
Plenum Press, New York, 1996

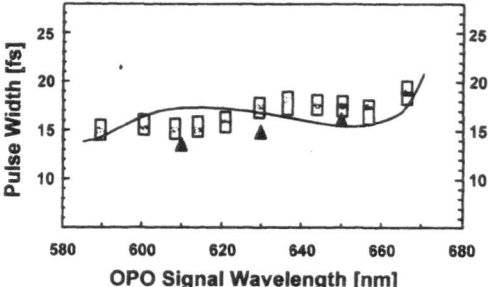

Figure 1. Measured pulse durations of the BBO OPO output in fs as a function of OPO wavelength in nm. Open rectangles: from intensity autocorrelation of the cavity with 570 mm intra-prism spacing. Full circles: Fourier transform of the corresponding spectra. Full triangles: from fringe-resolved autocorrelation of the cavity with 380 mm intra-prism spacing. Full line: Calculated duration incorporating the effects of total cavity group-delay-dispersion and mirror reflectivity for an intra-prism spacing of 570 mm.

The OPO is synchronously pumped by 1 W of radiation at 400 nm, generated by efficient frequency-doubling of a 2 W, 90 fs, 82 Mhz Ti:Sapphire oscillator output. The angle between the extraordinary pump beam and the ordinary resonated signal beam is 3.7° internal to the BBO crystal. Fig. 1 shows the results obtained using fused silica prisms for GDD compensation[1]. Tunable pulse widths averaging about 16 fs are observed in the 590 to 666 nm range and the shortest pulse width (created by minimizing intra-cavity material and prism-spacing) is 13 fs. Calculation (full line in figure) indicates that GDD is limiting pulse duration in this OPO and that better GDD compensation, which can in principle be obtained using chirped mirrors, should allow the production of pulses significantly shorter than 10 fs in the visible from this type of OPO.

We note that an interesting phenomenon of self-compression occurs in the BBO OPO *without* prisms[2,3]. Normally, in this configuration, the OPO produces highly chirped pulses due to the normal dispersion of the BBO crystal. These pulses have a duration close to that of the blue pump pulse (≈ 200 fs). However, as the signal wavelength is tuned towards 640 nm, *signal* second-harmonic-generation phase-matching conditions are approached and an induced non-linear refractive index is produced by a cascade process. At certain signal wavelengths this induced index is negative, leading to negative self-phase-modulation and negative chirp sufficient to compensate the positive chirp produced by normal dispersion. In this case short (≈ 30 fs), near-transform-limited OPO output pulses are observed. The underlying hypothesis of self-compression in the OPO, a strong non-linear phase-shift caused by cascaded $\chi^{(2)}$ effects, has been independently verified in the femtosecond regime by Z-scan experiments and detailed calculation for BBO[3].

Operating the OPO well into the negative GDD region allows the production of longer transform-limited pulses suitable for generation of low-power mid-infrared radiation by frequency-mixing. Non-collinear parametric frequency-mixing of a 70 fs tunable OPO output and the 800 nm Ti:Sapphire beam remaining after frequency-doubling was performed in two crystals of KTP of thickness 0.5 and 1.5 mm respectively, in an **e - o** ➔ **o** geometry[4]. The mid-IR beam, in the 100 μW power range, was detected by a PbSe room-temperature detector, behind an IR monochromator. IR duration was measured by *in situ* sum-frequency generation of the IR and a visible delayed OPO probe beam. In the shorter crystal, pulse durations are 100 fs in the 2.8 to 3.6 μm range, and the time-bandwidth product is always < 0.4, indicating that the mid-IR radiation is highly coherent. On the contrary, in the longer (1.5 mm) crystal time-bandwidth products always exceed 0.5. Fig. 2 shows the wavelength variation of the measured pulse durations for the two crystals. Also shown (full curves) is the duration calculated using the non-linear coupled parametric propagation equations,

$$\frac{\partial E_s}{\partial u} = \frac{i\omega_s \chi_{eff}^{(2)}}{2n_s c} E_p E_{ir}^*$$

$$\frac{\partial E_{ir}}{\partial u} + (v_{ir}^{-1} - v_s^{-1})\frac{\partial E_{ir}}{\partial t} = \frac{i\omega_{ir} \chi_{eff}^{(2)}}{2n_{ir} c} E_p E_s^*$$

$$\frac{\partial E_p}{\partial u} + (v_p^{-1} - v_s^{-1})\frac{\partial E_p}{\partial t} = \frac{i\omega_p \chi_{eff}^{(2)}}{2n_p c} E_s E_{ir}$$

$$(1)$$

which are solved using a fourth-order Runge-Kutte technique. The v^{-1} coefficients represent the inverse group velocities of **p**ump, **s**ignal and mid-**IR**. The effects of intra-pulse group dispersion and phase-modulation are also introduced into the calculation between spatial (u) steps in the solution. Agreement between calculation and experiment is excellent, and a closer inspection of the results indicates that pulse-broadening in the longer KTP crystal is mainly due to group velocity walk-off (GVW) between the short-wavelength pump pulses and the long-wavelength IR pulse. This observation suggests that GVW effects may be alleviated by displacing the pump and signal wavelengths into the near-infrared.

The high-repetition rate frequency-mixing process allows the production of tunable femtosecond mid-infrared pulses with $\approx 10^7$ - 10^8 photons per pulse. Certain experiments, however, require a much higher pulse energy, albeit at a lower repetition rate. We are currently constructing a high energy mid-infrared femtosecond pulse generator. The principle of this new device involves the generation and amplification of a *near*-infrared pulse, tunable around 1.1 μm, using an amplified Ti:Sapphire pump pulse at 0.8 μm, and a subsequent frequency-mixing of the near-infrared pulse with the Ti:Sapphire pump pulse to produce tunable femtosecond mid-infrared pulses as in the above experiment. In order to produce high-energy and coherent mid-infrared pulses the frequency-mixing process is required to be very efficient and to conserve the high pulse quality observed in the low energy frequency-mixing experiment.

Fig. 3 shows the result of a calculation, using eq. 1, of mid-IR energy and pulse duration over a wide range of pump powers in KTP (injected pump and signal energies are held constant and the surface area is varied). This calculation shows that mid-IR pulse duration remains constant, even in the presence of pump saturation and a more than 50% signal to mid-IR energy conversion ratio. Furthermore, the calculated time-bandwidth

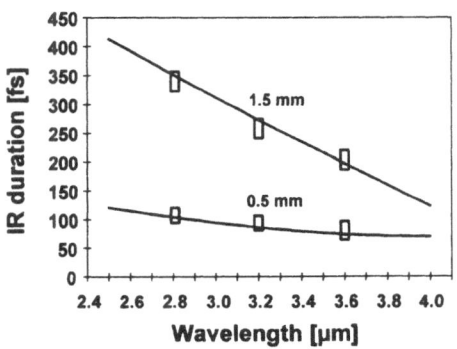

Figure 2. Frequency-mixing mid-infrared pulse duration in fs as a function of infrared wavelength in μm. Open rectangles: Measured pulse duration in (upper) 1.5 mm of KTP and (lower) 0.5 mm of KTP. The full curves are calculated using eq. 1 without adjustable parameters. The effects of group-velocity-walk-off can be clearly seen on the upper curve.

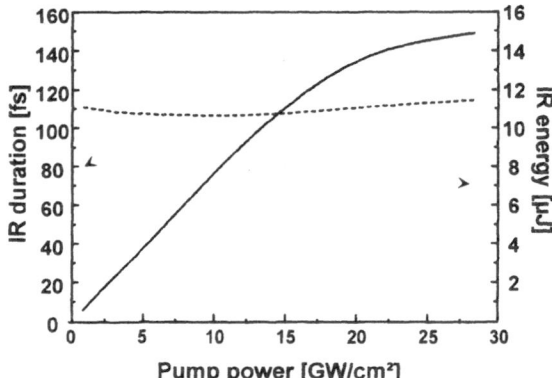

Figure 3. Calculated mid-infrared energy in μJ at 3 μm (right scale-full line) and duration in fs (left scale-dashed line) obtained by frequency mixing a 60 μJ, 120 fs pump pulse at 0.8 μm and a 30 μJ, 200 fs signal pulse at 1.1 μm in a 1 mm thick crystal of KTP. The pump/signal power density is varied by changing the surface area of the beams. The lower left-hand corner corresponds to the *power* conditions of the high repetition-rate experiment. Although pump beam saturation and significant signal gain occur at high pump power, the mid-infrared pulse duration remains close to 110 fs at all power levels. The time-bandwidth product at 3 μm is 0.44 0.01 across the figure (gaussian pulses were employed in this calculation).

product of the mid-IR pulse remains close to 0.44 over the whole range of pump powers, indicating that the mid-IR pulse remains highly coherent even at high conversion efficiencies.

In conclusion, a novel β-barium borate optical parametric oscillator pumped by the second harmonic of a femtosecond Ti:Sapphire oscillator has produced broadly tunable pulses in the visible with a duration down to 13 fs. This OPO also exhibits pulse self-compression in the presence of normal dispersion, due to cascaded $\chi^{(2)}$ effects. Frequency-mixing of the OPO output with the near-infrared synchronous Ti:Sapphire pump produces highly coherent tunable mid-infrared pulses with a duration inferior to 100 fs. Calculation shows that the frequency-mixing process can be extended to high energy without loss of coherence. This result is significant for the design of optical parametric amplifier systems.

REFERENCES

1. G. M. Gale, M. Cavallari, T. J. Driscoll and F. Hache, Opt. Lett. **20**, 1562 (1995)
2. T. J. Driscoll, G. M. Gale and F. Hache, Optics Comm. **110**, 638 (1994)
3. F. Hache, A. Zéboulon, G. Gallot and G. M. Gale, Opt. Lett. **20**, 1556 (1995)
4. G. M. Gale, M. Cavallari, T. J. Driscoll and F. Hache, Optics Comm. **119**, 159 (1995)

84

OPTICAL PARAMETRIC OSCILLATOR WITH CHIRPED MIRRORS FOR DISPERSION COMPENSATION

J. Hebling,[1,3] E. J. Mayer,[1] J. Kuhl,[1] and R. Szipöcs[2]

[1] Max-Planck-Institut für Festkörperforschung
Stuttgart, Germany
[2] Optical Coating Laboratory
Budapest, Hungary
[3] JATE University
Szeged, Hungary

Ti:sapphire laser-pumped optical parametric oscillators (OPO's) are attractive sources for tunable femtosecond light pulses in the visible[1,2] and near-infrared[3-5] spectral ranges. Generation of femtosecond transform-limited pulses shorter than the pump pulse necessitates compensation of the group-delay dispersion (GDD) that originates from both material dispersion and self-phase modulation in the nonlinear crystal. Previously this GDD compensation was accomplished by insertion of prism pairs into the resonator[1-4].

In this paper, we report on the generation of transform-limited pulses as short as 73 fs from an OPO that uses chirped mirrors (CM's)[6]. The main advantages of CM's compared with prism pairs are the small insertion losses that lead to high conversion efficiency of the OPO. The independence of the GDD from cavity alignment and the attainable reduction in the cavity length are attractive from practical points of view. Finally, CM's may become important for the generation of extremely short pulses (<20 fs) that require compensation of higher-order dispersion.

The cavity configuration of our noncritically phase-matched KTP OPO[4,7] depicted in Fig. 1 has two-times-smaller material dispersion per round trip than a linear cavity. The focal length of folding mirrors M2 and M3 is relatively long (10 cm) in order to reduce the sensitivity to resonator misalignment. The transmission of the output coupler (M1) is 5%, and the back high reflector (M4) is mounted upon a piezoelectric translator (PZT) for fine adjustment of the cavity length. The pump beam is focused by an achromatic lens (L; f=10 cm) into the 2-mm-thick KTP crystal. Tuning of the pump laser from 820 nm to 920 nm shifts the signal pulse wavelength from 1.18 to 1.32 μm. The spectra of the OPO output pulses and the pump pulses are characterised by a HP 70950A optical spectrum analyser and the pulse widths by autocorrelation measurements in a 1mm long BBO crystal.

Ultrafast Processes in Spectroscopy, edited by Svelto et al.
Plenum Press, New York, 1996

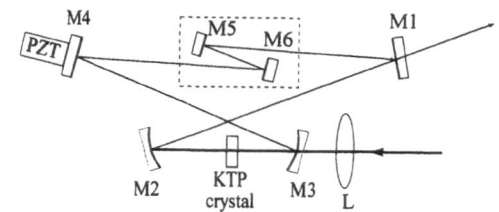

Figure 1. Set-up for the chirped mirror GDD-compensated OPO.

To study the influence of the CM's on the pulse duration and spectrum, we have varied the negative intracavity GDD in steps by increasing the number of reflections from such mirrors. For comparison, data were taken first for a cavity with standard single-stack mirrors. Mirrors M2 and M4 were then replaced by CM's introducing a negative GDD. We obtained an increase in the negative GDD by inserting two additional CM's (M5 and M6) into the cavity. Using multiple reflections from these two mirrors, we can provide as much as eight times the GDD of one CM in the cavity. Until now we have performed experiments using two types of CM's. These mirrors will be referred as #1 and #2, and the measured GDD for these CM's are depicted on Fig. 2. The GDD of the CM's was determined by the method described in Ref. 8. For both types of CM's, the reflectivity reaches at least 99.5% over the spectral range from 1.1 to 1.4 μm. In the experiments using #1 type CM's, the OPO was pumped by 120–150 fs pulses from a Ti:sapphire laser (Coherent Mira 900) configured for 12 W argon laser pumping. In the experiments using #2 type CM's, the Ti:sapphire laser was configured for 8 W pumping and it produced 150–170 fs pump pulses for the OPO.

In the experiments using #1 CM's[9], the pump power was 760 mW. For this case, the spectral tuning range of the OPO from 1.18 to 1.32 μm was independent of the number of CM's and was limited by the sharp reflectivity drop of the standard single-stack mirrors at

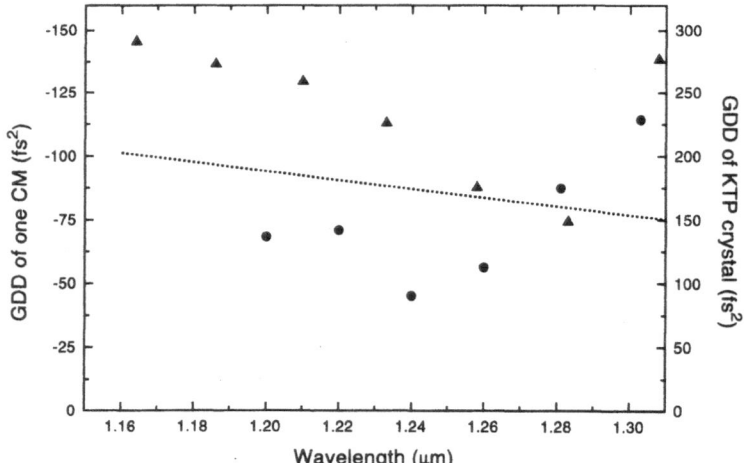

Figure 2. Measured GDD for one #1 (square), and #2 (triangle) type CM. The dashed curve shows the material dispersion of the 2 mm long KTP crystal.

short wavelengths and the tunability of the pump laser at long wavelengths. The losses associated with the exchange of the single-stack mirrors for CM's or the additional insertion of CM's are negligible for less than four reflections from CM's. For six and eight reflections from CM's, we obtained a 20% and 50% drop in the OPO power. This drop is due to the increase in the reflection loss and (most significantly) in the diffraction loss on M5 and M6, which had only half an inch diameter. For the configuration that uses four reflections from CM's, the signal power is >100 mW over the entire tuning range and reaches its maximum of 180 mW at 1.2 μm. The corresponding energy conversion efficiency at 1.2 μm is η=24% for the signal beam alone. Such a high conversion efficiency has been reported for prism-pair-compensated OPO's only for significantly higher pump power[2].

As a result of the GDD compensation, the OPO pulse duration decreased to about half of its value without the GDD compensation, and the spectrum was symmetric without any strong structures contrary to the asymmetric and structured spectrum of the femtosecond OPO's without GDD compensation[4]. At 1.315 μm when four CM's were used the OPO pulse duration was 73 fs and the spectrum was 22 nm broad. These two values give a time-bandwidth product of 0.28, which is slightly smaller than the value for a transform-limited pulse.

A thorough investigation of the time-bandwidth product over the whole tuning range indicated that this product is close to the transform limit at long wavelengths, but around 1.24 μm, (at the position of the dent in the GDD of the #1 type CM's) it is significantly larger. To push the product bellow 0.5 over the whole tuning range it was necessary to use eight reflections[9] from CM's.

Very recently some preliminary experiments were carried out using #2 type CM's. The GDD of these mirrors (see Fig. 2.) does not show any significant drop at 1.24 μm and its value is higher than that of the #1 over almost the whole tuning range. Because of these improved properties the time-bandwidth product of the pulses generated by the OPO consisting of four #2 type CM's is below 0.5 over almost the whole tuning range (see Fig. 3.). One reason for the sharp increase of the time-bandwidth product at the short- wavelength end of the tuning range is the rapidly increasing positive GDD of the single-stack mirrors (M1, M3).

For wavelength longer than 1.20 μm, the OPO pulse duration is shorter than the pump pulse. In the present preliminary experiment the time-bandwidth product of the pump laser pulses was 0.41, and the pulse duration was about 50% longer than the optimum value provided by the MIRA. Further improvement in both the duration and the time-bandwidth product of the OPO pulses is thus expected after optimisation of the pump laser.

The average pump power of the OPO in the latter experiments was only 400–450 mW. In spite of this relatively low pump power, the energy conversion efficiency of the OPO exceeded 10 % across the signal wavelength range 1.18–1.28 μm and its maximum value at 1.19 μm was 27%. The smallest threshold was as low as 280 mW. This small value demonstrate the advantage of using low loss CM's for GDD compensation in OPO's. To the best of our knowledge, such a low threshold has been achieved only for OPO without any GDD compensation[10]. For prism-pair compensated OPO's the threshold is typically two times higher. The low threshold of the CM compensated OPO would make it possible to use Ti:sapphire lasers, which can generate significantly shorter pulses on the expense of a lower pulse energy, as the pump laser. Such systems probably allow generation of shorter OPO pulses.

Another strategy to create shorter OPO pulses is to use of high pump power in order to increase self-phase-modulation in the nonlinear crystal, and compensation of the in-

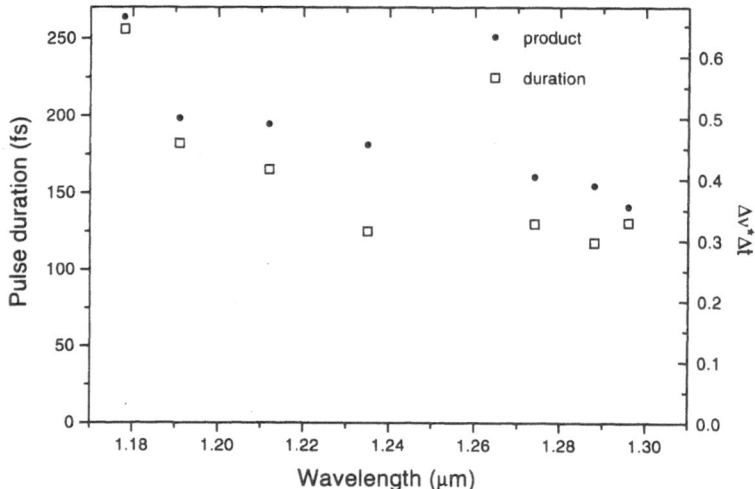

Figure 3. The pulse duration (square) and the time-bandwidth product (circle) of the pulses generated by the OPO consisting four #2 type CM's for GDD compensation.

creased chirp by a hybrid GDD compensator consisting of a prism-pair and CM's. The prism-pair can compensate the large GDD, while the third order dispersion compensation can be accomplished by specially designed CM's.

This work was partly supported by the Deutsche Forschungsgemeinschaft (DFG). J. Hebling thanks the DFG for financial funding of his stay at the Max-Planck-Institut in Stuttgart.

REFERENCES

1. G. M. Gale, M. Cavallari, T. J. Driscoll, and F. Hache, Opt. Lett. **14**, 1562 (1995).
2. P. E. Powers, R. J. Ellingson, W. S. Pelouch, and C. L. Tang, J. Opt, Soc. Am. B **10**, 2162 (1993).
3. W. S. Pelouch, P. E. Powers, and C. L. Tang, Opt. Lett.**17** , 1070 (1992).
4. J. M. Dudley, D. T. Reid, M. Ebrahimzadeh, and W. Sibbett, Opt. Commun. **104**, 419 (1994).
5. A. Nebel, C. Fallnich, R. Beigang, and R. Wallenstein, J. Opt. Soc. Am. B **10**, 2195 (1993).
6. R. Szipöcs, K. Ferencz, C. Spielmann, F. Krausz, Opt. Lett. **19**, 201 (1994).
7. A. Stingl, C. Spielmann, F. Krausz, and R. Szipöcs, Opt. Lett. **19**, 204 (1994).
8. K. Osvay, G. Kurdi, J. Hebling, R. Szipöcs, A. P. Kovács, and Z. Bor, Opt. Lett. **20** (1995 november 15.).
9. J. Hebling, E. J. Mayer, J. Kuhl, and R. Szipöcs, Opt. Lett. **20**, 919 (1995).
10. G. Mak, Q. Fu, and H. M. van Driel, Appl. Phys. Lett. **60**, 542 (1992)

85

KERR LENS MODE-LOCKING OF A SUB-PICOSECOND OPTICAL PARAMETRIC OSCILLATOR

R. Laenen, C. Rauscher, and A. Laubereau

Physik Department E11
Technische Universitaet Muenchen
James-Franck-Str., 85748 Garching, Germany

Synchronous pumping of singly resonant, optical parametric oscillators (OPO) yields intense, tunable radiation from the VIS to the near IR.[1,2] Especially OPO's working in the pulsed regime generate light pulses in extended tuning ranges with significant shortening compared to the pump duration. A high spatial and spectral quality is achieved, close to the diffraction and Fourier transform limitations. In this paper we want to present a new approach for the generation of sub-ps parametric pulses starting with pump pulses of a few ps and introducing Kerr lens mode locking[3] as a additional shortening mechanism besides temporal gain narrowing. Furthermore we will demonstrate results on the OPO working in the saturation regime and offering pulses down to 260 fs.

Our experimental approach starts with pulse trains lasting for 5 microseconds generated by a feedback-controlled (FCM) and additive-pulse modelocked (APM) Nd:YLF laser with high amplitude stability and repetition rate of 50 Hz. The pulse duration is 2.7 ±0.2 ps for a single pulse energy of 1 microjoule at 1047 nm. The laser emission is subsequently amplified in a single stage - double pass scheme and frequency-doubled in a 5 mm BBO crystal. At the second harmonic of the laser we get single pulse energies up to 7 microjoule with pump duration of 3.5 ± 0.3 ps. The linear resonator of the OPO represents a two focus configuration. In one focus, accomplished by a highly reflecting mirror and a lens, the 6 mm KTP crystal is placed, cut for type II phasematching in the xz-plane (o-oe) with phase-matching angle of 62 degree and equipped with anti-reflection coating. This crystal is employed for parametric amplification inside the OPO. A noncollinear pumping scheme ensures singly resonant operation of the OPO with fedback of the idler. The second focus is realized by the help of a telescope and incorporates the Kerr medium. An aperture between the output mirror and the telescope discriminates the wings of the parametric emission in the case of Kerr lensing. The output mirror is mounted on a precision translation stage to adjust the round trip time of the parametric component to the temporal spacing of the pump pulses (10 ns) by the help of a stepping motor. We obtain tunable pulses between 1200 and 1700 nm by angle tuning of the KTP-crystal, readjusting

Ultrafast Processes in Spectroscopy, edited by Svelto et al.
Plenum Press, New York, 1996

the OPO length and tilding the highly reflecting mirror. The parametric emission is monitored by the help of a 0.5 m monochromator and a standard autocorrelation setup, both integrating over the OPO pulse train.

We want to present data on the OPO operated in two different configurations: OPO with and without a Kerr medium inside the telescope.

With a KTP crystal of length 10 mm as Kerr medium inside the telescope we have found strong evidence for Kerr lens mode-locking of the OPO-pulses[4]: For an aperture diameter of 2.8 mm in front of the output mirror we measured the idler pulse duration in dependence of the position of the KTP crystal inside the telescope. The wavelength of the OPO was adjusted to 1300 nm while pumping the OPO with 2.2 microjoule of single pulse energy. Moving the crystal in direction of the focus one can find a shortening of the OPO pulses from 0.8 ps to 0.5 ps at optimum position. Our experimental data give rise for a double minimum of the idler pulse duration in dependence on the position of the Kerr medium position. Each minimum is located at a distance of about 3 mm from the focus of the telescope.

For the shortest pulses the measured spectral bandwidth amounts to 29 ± 3 cm^{-1} corresponding to a bandwidth - pulse duration product of 0.40 ± 0.06, close to the limit for Gaussian pulses. The measured dependence of the pulse duration on the position of the nonlinear medium is a strong hint for Kerr lens mode-locking of the OPO, as the position of the induced nonlinear lens is a critical parameter of the technique.[5]

A second proof for Kerr lensing is shown in Fig. 1. In this case we determine the duration of the idler component versus aperture radius with optimum position of the Kerr medium for pulse shortening. One can see clearly a minimum of the duration at a diameter of 2.8 mm, while at substantially higher or lower values a duration of 0.8 ps is achieved, representing the case of vanishing Kerr lensing.

The shortest pulse duration obtained fine adjustment of the aperture amounts to 460 ± 50 fs and the measured autocorrelation trace is shown in Fig. 2. We have fitted the data with a hyper-gaussian profile $\exp(t/tp)^{2.3}$ indicating steep slopes of the parametric emission with a decay time of approximately 100 fs. From dispersion data the amplification bandwidth of the crystal is calculated to be 25 cm^{-1} in the plane wave approximation, in

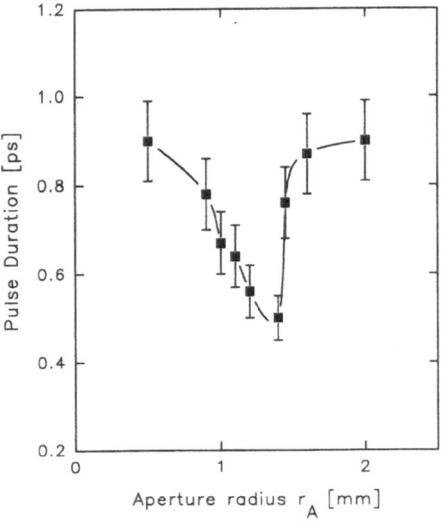

Figure 1. Measured duration of the OPO emission versus aperture radius of the Kerr lens setup. A minimum idler duration of 0.46 ps is obtained, as compared to 0.8 ps for a opened aperture; experimental points, solid line is a guide for the eye.

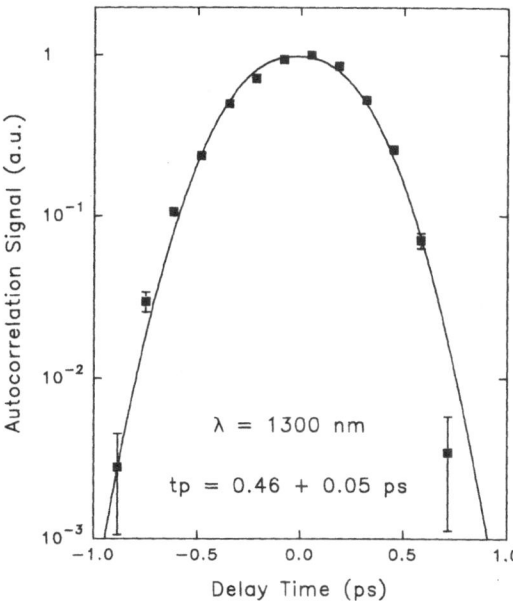

Figure 2. Autocorrelation data for idler pulses at 1300 nm; a pulse duration of 460 ± 50 fs is measured, assuming a hyper-gaussian shape (solid curve); experimental points.

fair agreement to the measured bandwidth of the pulses. The actual amplification bandwidth in the OPO may be higher because of the strong focusing of pump and idler into the nonlinear crystal. It is suggested that the duration of the shortest pulses is limited by the amplification bandwidth of the nonlinear crystal, since a shorter crystal provides smaller pulse duration (see below).

We measure 55 nJ single pulse output energy in the case of the shortest pulses from the OPO. This value corresponds to an internal energy conversion of 7% from the pump to the idler. In the second version we insert a shorter KTP-crystal with length of 3 mm in the OPO while the nonlinear Kerr medium inside the telescope is omitted (other parameters are equal to the former situation). Pumping the OPO a factor of 2 to 4 above threshold significant shortening can be found for the circulating idler component with a measured duration of as short as 260 fs, equivalent to a shortening factor up to 15 relative to the pump.

Fig. 3 depicts the measured idler duration versus applied pump energy with respectively optimized resonator lengths (filled squares). An almost linear decrease of the pulse width with pump intensity is observed which is accompanied with an increasing spectral bandwidth (open circles) of the parametric emission. The minimum idler duration of 260 fs seems to be restricted again by the amplification bandwidth. The latter is estimated to approximately 50 cm^{-1} for the KTP crytal of 3 mm. The optimum length of the OPO is found to increase with the applied pump intensity because of the constant cavity roundtrip time for synchronous pumping, the observation corresponds to a faster propagation through the nonlinear crystal.

An explanation of the dominant shortening mechanism was proposed by Akhmanov et al.[6] for parametric amplification in the saturation regime using the constant pump field approximation: The leading edge of the parametric component rises rapidly due to strong

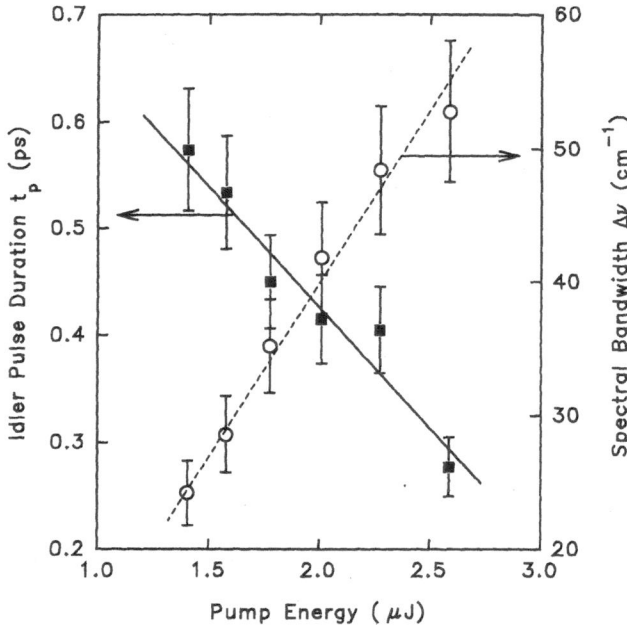

Figure 3. Measured idler duration (filled squares, left hand ordinate scale) and spectral bandwidth (open circles, right hand ordinate scale) versus applied pump pulse energy for a wavelength setting of 1300 nm. A significant shortening of the idler with rising pump intensity is observed (lines are guides for the eye).

parametric interaction while the trailing edge of the propagating pulse finds no amplification due to a strongly depleted pump field. The mechanism also leads to an increased propagation velocity of the amplified pulse too, depending on the parametric gain. The process appears to be dominant for the measured pulse shortening in the saturation regime because of the strong dependence of the parametric pulse duration and optimum OPO length on the applied pump intensity. A minor Kerr lens efect of the parametric crystal is not excluded by our data.

We have measured the idler duration versus the adjusted wavelength position and found a lengthening from 260 fs at 1300 nm to 370 fs at 1600 nm.[7] This feature may be explained by the wavelength dependence of the parametric gain and cavity losses. All idler components generated in this strong saturation regime exhibit extended wings appearing a factor of 25 below the maximum of the autocorrelation curve and are indicative for the imperfect pump depletion during the leading edge of the idler. This behaviour is illustrated by Fig. 4 depicting the measured autocorrelation trace of the idler pulse of 260 fs in a semilogarithmic plot versus delay time. A Gaussian shape can be deduced for the central peak of the autocorrelation, only. The contrast factor of 25 (intensity ratio of peak to pedestal) exceeds the numbers reported for previous investigations.[8] All measured idler pulses exhibit transform limited bandwidths (bandwidth - pulse duration product: 0.46 ± 0.04) for optimum cavity length setting. The high quality of the diffraction limited beams and the internal photon conversion efficiency of 11% inside the OPO should be noted.

In summary we have shown for the first time Kerr lens mode-locking of a synchronously pumped OPO resulting in pulses of 460 fs with steep wings extending over several orders of magnitude using pump pulses of 3.5 ps. We show further pulse shortening down

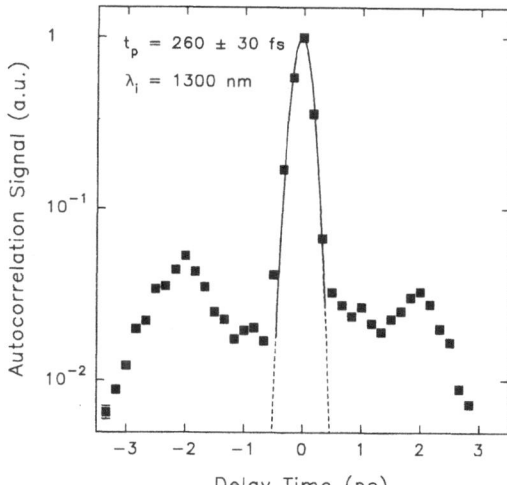

Figure 4. Autocorrelation trace of the idler component at 1300 nm of a modified OPO setup with duration of 260 ± 30 fs, experimental points, solid line calculated for a Gaussian shape of the central peak.

to 260 fs working with a shorter parametric crystal in the strong saturation regime. The almost diffraction and fourier transform limited idler pulses exhibit pedestals appearing a factor of 25 below maximum.

REFERENCES

1. T.J. Driscoll, G.M. Gale, and F. Hache, Optics Commun. 110, 638 (1994).
2. R. Laenen, K. Wolfrum, A. Seilmeier, and A. Laubereau, J. Opt. Soc. Am. B10, 2151 (1993).
3. D.E. Spence, P.N. Kean, and W. Sibett, Opt. Lett. 16, 42 (1991).
4. R. Laenen, C. Rauscher, and A. Laubereau, Optics Commun. 115, 533 (1995).
5. V. Magni, G. Cerullo, and S. De Silvestri, Opt. Commun. 101, 365 (1993).
6. S.A. Akhmanov, A.S. Chirkin, K.N. Drabovich, A.I. Kovrigin, R.V. Khokhlov, and A.P. Sukhorukov, IEEE J. of Quantum Electr. QE-4, 598 (1968).
7. C. Rauscher, T. Roth, R. Laenen, and A. Laubereau, Opt. Lett. 20, 2003 (1995).
8. J.D.V. Khaydarov, J.H. Andrews, and K.D. Singer, Opt. Lett. 19, 831 (1994).

86

ESSENTIALLY TRANSIENT RAMAN AMPLIFICATION IN EXCITED MEDIUM

I. A. Kulagin and T. Usmanov

NPO "Akadempribor" of Acad. Sci.
Tashkent, Uzbekistan

The self-induced gain modulation is inherent in the process of the transient stimulated Raman scattering (e.g.[1]). This modulation is connected with the finite relaxation time and is caused by an interference between the pump radiation depletion and the medium response. As a result the interacting pulses assume the structured form, the connection between Stokes and fundamental pulses is reduced, the energy conversion efficiency into Stokes pulse decreases. The analysis of those effects bases on numerical calculation of temporal and spatial self-congruent problem[2-4].

In our work on a base both analytical solutions and numerical calculations the analysis of self-induced modulation influence on the energy exchange in essentially transient Raman amplification is carried out for excited medium.

The variation of interacting pulses in the process of propagation into Raman media is described on a base of coupled equations account for fundamental E_p and Stokes E_s fields and motion equation for matrix density. From the last the equations for material excitation σ and population changes η are evident. So, in the resonant approximation the set of equations has well known view:

$$\frac{\partial E_p^*}{\partial z} + v_p^* \frac{\partial E_p^*}{\partial t} = -i\beta_p E_s^* \sigma^* \ ,$$

$$\frac{\partial E_s}{\partial z} + v_s^{-1} \frac{\partial E_s}{\partial t} = i\beta_s E_p \sigma^* \ ,$$

$$\frac{\partial \sigma}{\partial t} + T_2^{-1} \sigma = \frac{i}{\hbar} M E_p \, E_s^* \eta \ ,$$

$$\frac{\partial \eta}{\partial t} + T_1^{-1}(\eta - \eta_0) = -\frac{4}{\hbar} \text{Im}\left(M E_p \, E_s^* \sigma^* \right),$$

$$(1)$$

Ultrafast Processes in Spectroscopy, edited by Svelto et al.
Plenum Press, New York, 1996

where $\beta_i = 4\pi\omega_i \, NM / cn_i; v_i$ and ω_i group velocities and frequencies of interacting waves; $T_{1,2}$ longitudinal and transverse relaxation times; M scattering matrix element; n_i and N index of refraction and medium density.

In the case of small motion of populations and assuming $E_p >> E_s$ and an equality of group velocities the homogenous hyperbolic equation is followed from Eq. 1^{5-7}. Similarly the attenuation of high frequency component (inverse Raman scattering) is described. The solutions for the envelopes of the low-intense component E_i and of the material excitation σ can be obtained by the Riemann method and have the following form in the case when pulse duration are much less than relaxation times $\tau_i << fs24 \, T_{1,2}$ (cf.[5,6]).

$$E_t(z',t') = E_i(0,t') \pm \beta_i \gamma z'^{1/2} \, E_j(t') \int\limits_{-\infty}^{t'} Z_1\left(2\left[\{z'\}\tau(t') - \tau(t'')^{1/2}\right]\right)$$

$$\times [\tau(t') - \tau(t'')]^{-1/2} \, E_i(0,t'') \, E_j^*(t'')dt''$$

$$+ i\beta_i E_j(t') \int\limits_0^{z'} Z_0\left(2\left[\{\tau(t')[z'-z'']\}^{1/2}\right]\right)\sigma^*(z'',-\infty)dz'',$$

$$\sigma(z',t') = \sigma(z',-\infty) \pm \tau^{1/2} \int\limits_0^{z'} Z_0\left(2\left[\{\tau(t')[z'-z'']\}^{1/2}\right]\right)$$

$$\times [\sigma^*(z'',-\infty)(z'-z'')]^{1/2} \, dz'' + i\gamma \int\limits_{-\infty}^{r} Z_1\left(2\left[\{z'[\tau(t')-\tau(t'')]\}^{1/2}\right]\right) E_t^*(0,t') E_j(t')dt', \tag{2}$$

where

$$\tau(t) = \beta_i\gamma \int\limits_{-\infty}^{t} |E_j(t')|^2 \, dt' ; \gamma = M\hbar^{-1}\eta_0 ; z' = z; t' = t - z / v_s.;$$

E_j high-intense field. For the case $E_p >> E_s$, the upper sign is used and $Z_i(x)$ is a Bessel function with an imaginary argument; if $E_p << E_s$, the lower sign is used and $Z_i(x)$ is Bessel function with a real argument.

If initial coherent excitation of the Raman media is linear $\sigma(z) = \sigma_0 + \sigma_1 z$ and the interacting pulses have similar temporal profiles at the input one can find the amplitudes of the low-intense component and the excitation wave from Eq. 2:

$$E_i(z',t') = E_i(0,t')Z_0(G) + iE_j(t')\beta_i[z'/\tau(t')]^{1/2}$$

$$\times \left\{\sigma_0^* Z_i(G) + \sigma_1^*[z'/\tau(t')]^{-1/2} Z_2(G)\right\},$$

$$\sigma(z',t') = i\beta_i^{-1}[z'/\tau(t')]^{-1/2} Z_i(G) + \sigma_0 Z_0(G) + 2\sigma_1 z' G^{-1}Z_1(G), \tag{3}$$

where $G = 2[z' \tau(t')]^{1/2}$.

From Eq. 3 it follows that in the case of attenuation of high frequency component the variations of its envelope and the excitation become oscillatory, which relates to behavior of a Bessel function with a real argument and is typical of self-induced modulation[2]. In the cases of various initial excitation types for energy density of the low-intensity component

$$P_i(z) = \kappa \int_{-\infty}^{\infty} |E_i(z,t)|^2 \, dt, (\kappa = n_i c / 8\pi)$$

it follows from Eq. 3:

$$P_i(z) = \begin{cases} P_i(0)\left[Z_0^2(G_1) \mp Z_1^2(G_1)\right], \\ \kappa \dfrac{\beta_t z}{\gamma}|\sigma_0|^2\left[-Z_1^2(G_1) \pm Z_0^2(G_1) \mp 1\right], \\ \kappa \dfrac{\beta_t z^3}{3\gamma}|\sigma_1|^2\left\{-4G_1^{-2}\left[Z_2^2(G_1) \mp Z_1^2(G_1) \mp 1\right]\right\}, \\ 2\dfrac{n_t}{n_j}\beta_t^2 z P_j(0)F^2\left[Z_0^2(G_1) \mp Z_1^2(G_1) \pm G_1^{-1}Z_0(G_1)Z_1(G_1)\right], \end{cases}$$

(4)

where $G_1 = 2[z\tau(\infty)]^{1/2}$.

The first term in the right-hand side of Eq. 4 describes amplification of low-intense component, the second and the third terms represent generation due to constant and linear excitations of the nonlinear medium, respectively. The last term describes generation as a result of δ-correlated initial excitation: $\langle \sigma(z)\sigma(z') \rangle = F^2 \delta(z - z')$[7].

It is evident from Eq. 4 that different mechanisms of amplification relate with different rates of change of energy density versus the length of the nonlinear medium. If the medium is initially excited the growth of Stokes component can be faster. A similar mechanism is accounted for the presumed increase in the intensity of Stokes component with an additional fundamental pulse[8]. It is evident from Eq. 4 that inverse Raman scattering is characterized by step-like change in the energy density. The initial excitation of medium can alter the direction of energy conversion in this processes.

From Eq. 3 it follows that in the presence of several mechanisms of generation and amplification (attenuation) the efficiency of the process is determined by the initial phases of the interacting waves. The ratio of the intensities of the initial interacting waves and degree of excitation governs the deviation of the gain from equilibrium one, whereas the initial phase relation between the interacting waves determines the nature of variation of Stokes pulse gain.

In the case of comparable intensities of interacting optical waves and strong change of populations going to the real amplitude $E_i = \rho_i \exp(i\varphi_i)$ and $\sigma = \sigma_p \exp(i\phi)$ from Eq. 1 one can derive well known set of six equations (e.g.[9]). But if phase mismatch between three interacting waves is absent and $\tau_i < \text{fs}24 \, T_{1,2}$ Eq. 1 may be represent in handy form:

$$\frac{\partial f}{\partial z'} = \alpha \sin g,$$

$$\frac{\partial g}{\partial t'} = S \sin f,$$

(5)

where

$$\alpha = \left(\beta_p \beta_s\right)^{1/2} \mu; \; f = 2\sin^{-1}\left(\beta_p^{1/2} \rho_s / R\right); g = \sin^{-1}\left(2\sigma_p / \mu\right); S = R^2 M / \alpha \hbar;$$

$$\mu^2 = \eta_0^2 + 4\sigma_{p0}^2; R^2 = \beta_s \rho_p^2 + \beta_p \rho_p^2..$$

The numerical calculation of Eq. 5 gives following. Evolution of Stokes and fundamental pulse energy has also step-like character. So, the frequency conversion process repetitively reaches the saturation. The energy conversion of the Stokes radiation into fundamental wave in various region of the pulse is occurred. Just the same, the energy decreasing of the Stokes pulse is not takes place. With the increasing of the gain of the Stokes pulse the amplitude and width of the steps are reduced. The level of saturation achievement is determinate by the relation between the initial intensities of interacting waves and practically no dependencies from initial values of the intensities. The required lengths of the nonlinear medium are varied in the last case only.

The gain also increases in excited medium. For example, in 25 cm-length gas cell with hydrogen under 15 atm pressure the energy efficiency of amplification of injected Stokes pulse from 10^{-4} level of 4 GW/cm^2 power density of 50 ps fundamental pulse of second harmonic of a neodymium laser can rise from 9% with equilibrium medium up to 55% with $\sigma_{p0} = -0.25$.

The influence of a temporal delay between the pump and amplified injected Stokes pulses is considered. It is obtained that it is worthwhile to delay somewhat the fundamental pulse to obtain a more efficient energy conversion into the Stokes pulse and the injected pulse for more compression of this pulse.

REFERENCES

1. A. Laubereau, in *Iltrashort Laser Pulses and Applications*, ed. by W. Kaizer (Springer-Verlag, Berlin, 1988).
2. G.I. Kachen and W.H. Lowdermilk. Phys. Rev. **A14**, 1472 (1976).
3. M.D. Duncan, R. Mahon, L.L. Tankersley et al., J. Opt. Soc. Amer. **B7**, 202 (1990).
4. G. Hilfer nd C.R. Menyuk, J. Opt. Soc. Amer., **B7** 739 (1990).
5. R.L. Carman, F. Shimizu, C.S. Wang and N. Bloembergen, Phys. Rev. **A2**, 60 (1990).
6. Y.R. Shen, *Principles of Nonlinear Optics* (Wiley, New York , 1984).
7. A.P. Sukhorukov, *Nelinejnye Volnovye Vzaimodejstvya v Optike i Radiofizike (Nonlinear Wave Interactions in Optics and Radiophysics)* (Nauka, Moscow, 1988).
8. T.P. Mirtchev, N.I. Minkovski and I.V. Tomov, Opt. Commun. **80**, 143 (1990).
9. V.A. Gorbunov, Sov. J. Quant. Electron., **12**, 98 (1982).

87

EFFECTIVE COMPRESSION OF ULTRASHORT PULSES DURING SUM-FREQUENCY GENERATION IN BBO CRYSTAL

R. Danielius, A. Dubietis, G. Valiulis, and A. Piskarskas

Laser Research Center
Vilnius University, Lithuania

Transient sum-frequency (SF) generation in presence of the special conditions of the group velocity mismatch (GVM) and strong energy exchange can lead to efficient SF pulse compression[1]. Second harmonic generation in type II phase matching KDP crystal, as a partial case of the SF generation, recently has been used as an effective method of the pulse shortening[2,3]. The potential of this technique can be rather extended by employment of frequency-tunable sources such as a traveling-wave optical parametric generators (TOPG)[4,5]. This should yield a number of the new features of compressed pulses: tunability, new wavelength region, and other nonlinear crystals to be used.

In this communication we have calculated that the GVM conditions in BBO crystal of type II phase matching favor the SF compression ranging in the interval of $600 \div 720$ nm, when mixing a fundamental Nd:glass wave with TOPG output at 1400–2200 nm. Experimentally we have obtained SF pulses, generated in the 8 mm long BBO crystal, as short as 160 fs at 645 nm and 185 fs at 665 nm, starting with 1.3 ps pulses of the Nd:glass laser.

Compression of a SF pulse may take place when one of the pump pulses moves faster and another slower with reference to the generated SF pulse. The GVM can be described by a parameter $m = Lv_2/Lv_1$, where $Lv_j = \tau_{j0}/v_{j3}$ is the walk-off length, assuming $\tau_{j0} = \tau_3$; $v_{j3} = u_j^{-1} - u_3^{-1}$, $(j=1,2)$, u_1, u_2 are the group velocities of the pump pulses, u_3 is the group velocity of the SF pulse, τ_{j0} are the initial durations of the pump pulses. The optimum SF compression occurs when the value of m is close to -1. Nevertheless, a considerable compression factor can be achieved also with m values, different from the optimum one. In this case the ratio of intensities of the pump pulses should be optimized [1]. Obviously, this condition can be satisfied in the nonlinear crystals for a certain region of wavelengths. Thus, an implementation of a frequency-tunable pulse source as a TOPG, for the subsequent SF generation, allows one to fulfill the conditions for the pulse compression while maintaining a tunability of the compressed SF pulses.

We have performed the computer simulation of the SF pulse compression using the software for modeling of the three-wave interaction. The set of equations for the plane waves with the GVM and group velocity dispersion effects taken into account:

Ultrafast Processes in Spectroscopy, edited by Svelto et al.
Plenum Press, New York, 1996

$$\frac{\partial A_j}{\partial z} + \frac{1}{u_j}\frac{\partial A_j}{\partial t} - \frac{i}{2}g_j\frac{\partial^2 A_j}{\partial t^2} = -i\sigma_j\frac{\partial W}{\partial A_j^*}, \quad (W = A_1^* A_2^* A_3 + A_1 A_2 A_3^*) \tag{1}$$

where $A_j = a_j \exp(i\varphi_j)$ is the pulse complex amplitude ($j=1,2,3$), z and t are the longitudinal and time coordinates, u_j is the group velocity, g_j is the group velocity dispersion coefficient, σ_j is the coupling coefficient, were solved numerically with the boundary conditions: $A_{1,2}(t)|_{z=0} = A_{10,20}(t+T_{1,2})$, $A_3(t)|_{z=0} = 0$, where $T_{1,2}$ is the predelay of the pump pulses. The initial envelopes of the pump pulses with the carrier frequencies ω_1 and ω_2 ($\omega_3 = \omega_1 + \omega_2$) were assumed to be Gaussian $A_j = a_{j0}\exp(-2\ln 2 t^2/\tau_{j0}^2)$.

In Fig.1 the group velocity walk-off lengths in BBO crystal for the interaction of type eo-e are presented. From the values of m one can expect a compression of the SF pulses in the wavelength range 600÷720 nm. The main factor that limits the tuning range of the SF pulses from the short-wave side is a break-up point of o-polarized pulse at ~1200 nm (namely TOPG idler pulse). From the long-wave side the compression efficiency vanishes due to the decrease of the effective nonlinearity and drop of TOPG idler pulse energy, which is determined by the Manley-Rowe equation.

As it was shown in Ref.1, maximum compression of the SF pulse can be achieved in the case of symmetric interaction, i.e. in the case of equal intensities, durations and absolute values of GVM of the pump pulses. The treatment of the Eq.1 showed that symmetry requirements for the SF generation in the case of strong energy exchange can be approximately fulfilled choosing an appropriate intensity ratio of the pump pulses: $I_{idl}/I_F = |m|$ $\kappa/(\kappa-1)$, where κ is the degeneracy parameter defined as ω_{idl}/ω_{SF}. The computer simulation for the BBO crystal length of 8 mm yielded SF pulses as short as 145 fs at 645 nm (Fig.2) with an energy conversion of ~30% when the initial delay between pump pulses was fixed to be 0.9 ps, the TOPG idler pulse intensity was 2 GW/cm², and $I_{idl}/I_F = 0.8$.

We have examined the numerical predictions using a fiberless CPA Nd:glass system, that delivered up to 4 mJ, 1.3 ps pulses at 1055 nm. A small portion of the laser radiation (150 µJ) has been taken as the first beam for the frequency mixing. The main part of pulse was frequency doubled in a 1.2 cm long KDP crystal, and served for the TOPG pump. In the TOPG used in our experiments (TOPAS 501, Light Conversion Ltd.) three amplification stages with an independent pump were arranged in a single 8 mm long BBO type II phase matching crystal. As the second pump of the SF generator we used the idler pulse of TOPG, that has tunablity within 1055 ÷3000 nm, an energy of 100 ÷30 µJ (the dependence on wavelength is determined by Manley-Rowe equation) and duration of ~0.95

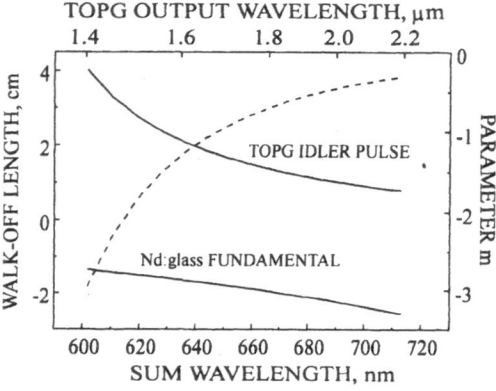

Figure 1. Group velocity walk-off lengths of the Nd:glass fundamental pulse (1.3 ps, e-polarization) and the TOPG idler pulse (0.95 ps, o-polarization) in the BBO crystal. Dashed curve shows values of the parameter m.

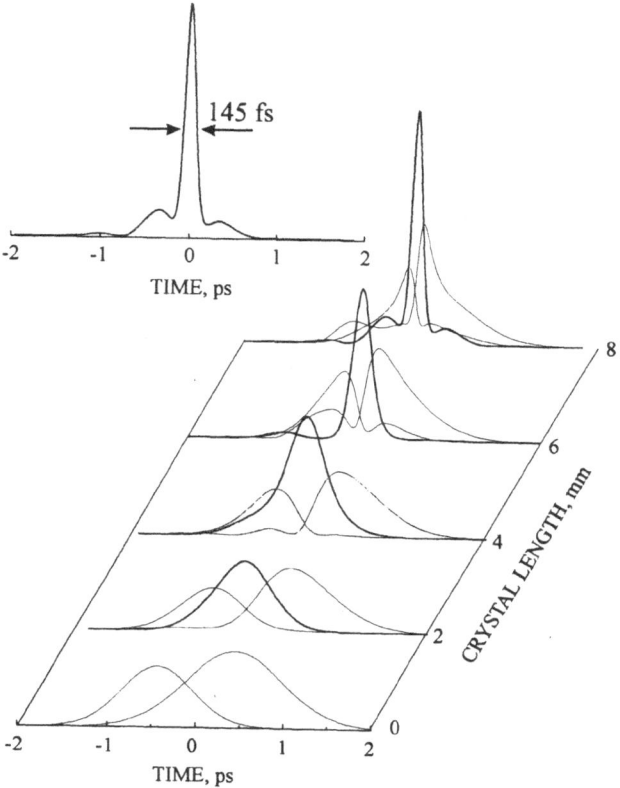

Figure 2. Computer simulated dynamics of the SF pulse at 645 nm (1055+1660 nm) compression.

ps. The SF pulse has been generated in a 8 mm long BBO crystal cut for the type II phase matching. We have measured the SF pulse at 645 nm (Fig.3); the decomposition SF auto-correlation trace into three Gaussian curves yielded 224 wide central peak that corresponded to the pulse duration of 160 fs. The energy of the SF pulse up to ~50 μJ was measured (energy conversion of ~30%), it is in a good agreement with the results of the computer simulation.

In conclusion, we have demonstrated for the first time to our knowledge the compression of the SF pulses, generated by mixing the pulses of different wavelengths. We have found that BBO crystal is highly beneficial for the effective compression of tunable SF pulses in visible. Experimentally we have achieved an 8-fold compression factor, obtaining tunable pulses as short as 160 fs, during mixing picosecond pulses from Nd:glass laser and TOPG.

ACKNOWLEDGMENT

Authors acknowledge the support by the International Science Foundation (Grants LA4000 and LED000, and generous support by Dutch Foundation for Fundamental Research of Matter (FOM).

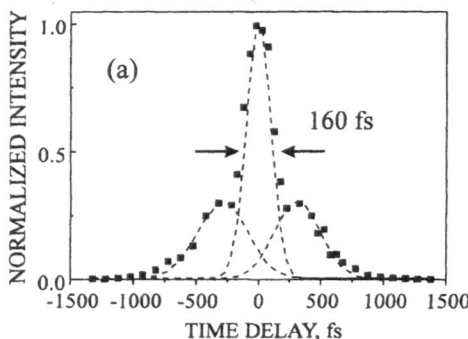

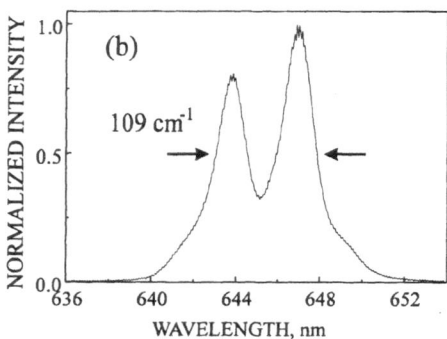

Figure 3. Autocorrelation trace and spectrum of the compressed SF pulse at 645 nm. Dashed curves show Gaussian fit for the measured data.

REFERENCES

1. A. Stabinis, G. Valiulis, and E. A. Ibragimov, Opt. Commun. **86**, 301 (1991).
2. Y. Wang and B. Luther-Davies, Opt. Lett. **17**, 1459 (1992).
3. P. Heinz, A. Laubereau, A. Dubietis, A. Piskarskas, Lithuanian J. Phys. **33**, 314 (1993).
4. I. M. Bayanov, R. Danielius, P. Heinz, A. Seilmeier, Opt. Commun. **113**, 99 (1994).
5. M. Nisoli, S. De Silvestri, V. Magni, O. Svelto, R. Danielius, A. Piskarskas, G. Valiulis, and A. Varanavicius, Opt. Lett. **19**, 1973 (1994).

88

OPTICAL HARMONICS GENERATION FROM CHIRAL SURFACES WITH FEMTOSECOND LASER PULSES

A. A. Angelutc, A. V. Balakin, N. I. Koroteev, I. A. Ozheredov, A. V. Pakulev, A. Yu. Resniansky, and A. P. Shkurinov

Physics Department and International Laser Center
Lomonosov Moscow State University
Moscow, Russia

It is well recognized that surface second-harmonic generation is a method that gives detailed information about the properties of thin films and surfaces[1]. This method has gained particular attention because a second harmonic signal may originate from a surface of a material, whereas little or no signal comes from the bulk of the sample. Because of macroscopic inversion asymmetry the second-order optical susceptibility tensor of chiral media has nonvanishing components for sum- and difference frequency generation nonlinear process[2]. Due to the symmetry reasons, the second harmonic generation process is forbidden in the bulk of noncentrosymmetric sample. However, on a surface the symmetry is broken and the second harmonic radiation may occur both for chiral and achiral media. Chirality of the sample under study leads to the appearance of new additional nonvanishing susceptibility tensor components.

Recently nonlinear analogs of optical rotation[3] and circular dichroism[4] techniques have been proposed. In this paper we present experimental results of determination of relative values of various second-order electric-dipole susceptibility tensor components for the surface of chiral molecules solved in centrosymmetric solvent. The technique is based on polarization dependence measurement of second harmonic intensity from the surface. In the case of linear polarized incident light, the polarization vector of generated wave is tilted in respect to incident one. The tilting angle may be determined by relative correlation between "chiral" and "regular" components and orientation of the incident wave polarization.

The nonlinear surface polarization $\vec{P}(2\omega) = \chi_{ijk}^{(2),S} E(\omega) E(\omega)$ is responsible for the SHG. The macroscopic tensor elements $\chi_{ijk}^{(2),S}$ relate to the microscopic description of the interfaces, and thus to the information such as the chirality, orientation and the structure of individual molecules. If the surface is isotropic but lacks inversion symmetry[5], the susceptibility tensor can be written in the systems of coordinates (x,y,z) of the sample and of the laboratory (p,s,k) as:

Ultrafast Processes in Spectroscopy, edited by Svelto et al.
Plenum Press, New York, 1996

$$\chi^{(2)} = \begin{bmatrix} 0 & 0 & 0 & \chi_{xyz} & \chi_{zzx} & 0 \\ 0 & 0 & 0 & \chi_{zzx} & -\chi_{xyz} & 0 \\ \chi_{zxx} & \chi_{zxx} & \chi_{zzz} & 0 & 0 & 0 \end{bmatrix} \qquad a_{ijk} = \begin{bmatrix} a_{ppp} & a_{pss} & a_{psp} \\ a_{spp} & a_{sss} & a_{sps} \end{bmatrix}$$

The nonlinear susceptibility tensor $\chi_{ijk}^{(2),S}$ can be transformed completely from the sample system into the laboratory frame where the tensor a_{ijk} has only 6 components resulting in 5 nonvanishing elements for the case of free surface with C_∞ symmetry:

$$\begin{cases} a_{ppp} = 8\pi ik\left[2F_x f_x f_z \chi_{zzx} + F_z\left(f_x^2 \chi_{zxx} + f_z^2 \chi_{zzz}\right)\right]; \\ a_{pss} = 8\pi ikF_z f_y^2 \chi_{zxx}; \\ a_{psp} = 16\pi ikF_x f_y f_z \chi_{xyz}. \end{cases} \qquad \begin{cases} a_{sps} = 16\pi ikF_y f_z f_y \chi_{zzx}; \\ a_{spp} = -16\pi ikF_y f_x f_z \chi_{xyz}. \\ a_{sss} = 0; \end{cases}$$

where F and f denote the respective Fresnel-like coefficients for the harmonic and the fundamental waves[5]. The **k**-direction is assumed to be parallel to the propagation direction of the fundamental and the harmonic beam while the indices **s** and **p** denote field components normal and parallel to the plane of incidence. In the case of centrosymmetric sample ($C_{\infty,v}$ type of symmetry), one have to suppose χ_{xyz}, a_{psp}, and a_{spp} to be equal to zero, i.e. this components are responsible for the macroscopic chirality.

The outcoming electric field at the harmonic frequency 2ω can be decomposed as follows:

$$E_p(2\omega) = a_{ppp}\left(E_p^0\right)^2 + a_{pss}\left(E_s^0\right)^2 + a_{pps}E_p^0 E_s^0; \qquad E_s(2\omega) = a_{spp}\left(E_p^0\right)^2 + a_{sss}\left(E_s^0\right)^2 + a_{sps}E_p^0 E_s^0;$$

In our experiments the incident beam was linearly polarized, with the polarization angle which could be determined as $E_p^0 = E^0 \cos(\varphi)$; $E_s^0 = E^0 \sin(\varphi)$ (the angle is defined relative to the p-direction). Then the intensities $I_p(2\omega), I_s(2\omega)$ can be written as:

$$I_p(2\omega) \propto \left|E_p(2\omega)\right|^2 = \frac{\left|E^0\right|^4}{4}\left|a_{ppp}\left[1+\cos(2\varphi)\right] + a_{pss}\left[1-\cos(2\varphi)\right] + \frac{1}{2}a_{pps}\sin(2\varphi)\right|^2; \tag{1}$$

$$I_s(2\omega) \propto \left|E_s(2\omega)\right|^2 = \frac{\left|E^0\right|^4}{4}\left|a_{spp}\left[1+\cos(2\varphi)\right] + a_{sps}\sin(2\varphi)\right|^2; \tag{2}$$

The second harmonic light generation measurements were performed with the set-up based on Ti:Sapphire femtosecond laser working at the wavelength of 800 nm with repetition rate of 100 MHz, pulse duration of 100 fs and overage power of 200 mW. The fundamental beam was aligned relative the p-direction with a polarizer and a Fresnel Rotator. The laser light was focused to the free surface of the sample by the 5 cm lens. The second harmonic signal from the sample is further collected, separated from the fundamental by appropriate filters and, finally imaged via a single stage monochromator on the cooled CCD camera and then analyzed on its frequency content. Using a polarizer in the SH beam, it was possible to detect only the p- or s-polarized component of the harmonic radiation.

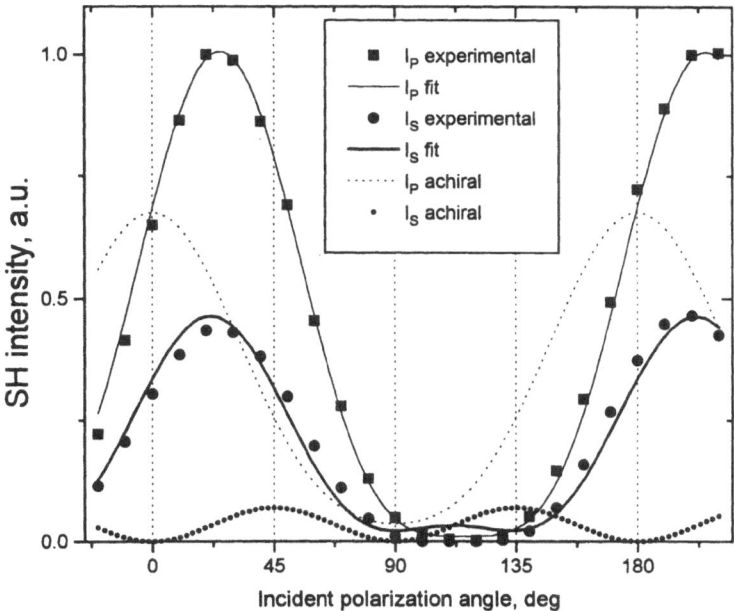

Figure 1. Measured and fitted dependence of the SHG intensity for described sample as a function of incident polarization angle for various sets of output analyzer.

The SH signal was observed from the noncentrosymmetric saturated native sugar solution. A single measurement series consisted of incident beam polarization rotation by an angle φ at a constant sample inclination angle. A typical result of such an experiment for chiral sugar solution and a computer simulation for achiral sample are shown in Fig.1. The experimental data were fitted according to Eq.1, Eq.2 and the relative values of the second order nonlinear susceptibility of sugar sample was determined as the following:

$$a_{ppp}=1, \ a_{pss}= 0.25 \pm 0.02, \ a_{pps}= 0.45 \pm 0.01, \ a_{spp}= - 0.45 \pm 0.01, \ a_{sps}= - 0.20 \pm 0.01$$

Assuming $\chi_{ijk}^{(2),S}$ - components to be real, their relative values may be obtained from our measurements and found to be:

$$\chi_{zzz}/\chi_{zxx} = 5.5 \pm 0.8; \ \chi_{xzx}/\chi_{zxx} = -0.41 \pm 0.07; \ \chi_{xyz}/\chi_{zxx} = -1.2 \pm 0.1$$

The summary of this work confirms the relatively large hyperpolarizability tensor components of chiral media which could be measured by the SHG technique. It opens new perspectives for nonlinear spectroscopy in noncentrosymmetric media, because of surface SH signal delivers such unique information as the chirality, orientation and the structure of individual molecules.

This work has been partially supported by the RFBR. (The Russian Foundation for Basic Research, grant 93–02–15026) and by ISF (The International Science Foundation, grant MCG000). We wish to acknowledge the support provided to one of us (A.V.Balakin) by the Samsung Electronics Co.

REFERENCES

1. Y.R. Shen, *The Principles of Nonlinear Optics* (Wiley, New York, 1984).
2. N.I. Koroteev, in *Frontiers in Nonlinear Optics. The Sergei Akhmanov memorial volume*, ed. by H. Walther, N. Koroteev, and M. Scully (I.P.P., Bristol 1993), p. 228.
3. J.D. Byers, H.I. Lee and J.M. Hicks, J. Chem. Phys. **101**, 6233 (1994).
4. J.D. Byers, H.I. Lee, T. Petralli-Mallow and J.M. Hicks, J. Phys. Chem. **97**, 1383 (1993).
5. B.U. Felderhof, A. Bratz, G. Marowsky, O. Roders, F. Sieverdes. J.Opt.Soc.Am.B, **10**, 1824 (1993)

89

SRS OF CHIRPED PULSES AS A METHOD OF WAVELENGTH CONVERSION OF FEMTOSECOND PULSES

V. G. Bespalov

Research Center "S.I.Vavilov State Optical Institute"
Sankt-Petersburg, Russia

The stimulated Raman scattering (SRS) of femtosecond pulses is strongly transient with high threshold intensity which is comparable to threshold of optical element damage. As a rule the SRS of femtosecond pulses accompanies many other nonlinear effects like a self-phase modulation [1] and a white light continuum generation [2]. To prevent these undesirable effects we propose to use femtosecond pulses stretched by diffraction gratings up to duration of 0.5–1 ns for SRS conversion. In this case SRS process is quasi stationary in the field of broadband pump and has some peculiarities [3]. Thus for original pulse duration of 100 fs spectrum of stretched emission is $f = 150$ cm^{-1} and because of that SRS will be generated effectively only in forward direction as the coherent length of backward interaction is order $1/f = 6$ μm. The SRS threshold is also increased in the comparison with that of monochromatic pump. For gain ($M = I g z$) calculation of forward SRS in the field of broadband pump we can use the expression [3]:

$$M = Mp - \ln\{[(fp + fc)df](Mp)1/2\} \tag{1}$$

where Mp - is a gain in the field of monochromatic pump, fp, fc - are pump and Stokes emission bandwidths respectively, df - is a linewidth of spontaneous Raman scattering. From equation (1) we see that the gain of stretched pump decreases with the increase of fp and fc. It is important that with SRS of broadband pump, pump spectrum is reproduced in Stokes spectrum [3] i.e. fc = fp and the chirped structure of stretched pulse is preserved and it is possible to compress the SRS pulse up to the original (100 fs) duration.

The threshold power for gaseous media most suitable for SRS conversion of stretched pulses is calculated for diffraction limited focused pump of Ti:sapphire laser emission (780 nm) with the help of (1) and are shown in Table 1. These powers correspond to pump energy of millijoules level for 1 ns pulse and are simply reached in regenerative amplification.

Later we plan to study this conception in detail including numerical simulation of SRS processes with broadband chirped pump and amplification in alternative Nd:glasses. The result of investigation will be in determining the optimal conditions of these processes.

Ultrafast Processes in Spectroscopy, edited by Svelto et al.
Plenum Press, New York, 1996

Table 1. Characteristics of active media and SRS conversion of stretched femtosecond
pulses of Ti:sapphire laser (780 nm) , g - is a steady-state gain coefficient
for monochromatic pump.

Active media	Stokes shift cm^{-1}	Linewidth cm^{-1}	Gain coefficient	Threshold power MW
H_2	4155	0.09	1.7 cm/GW	0.24
HD	3631	2	0.1 cm/GW	4.46
D_2	2997	0.11	0.27 cm/GW	1.67
CH_4	2917	0.38	0.75 cm/GW	0.58
N_2	2330	1	0.01 cm/GW	44.6
SiH_4	2186	0.5	1.5 cm/GW	0.31
GeH_4	2111	0.5	1.2 cm/GW	0.39

REFERENCES

1. J.-K.Wang, Y.Siegal et al., J.Opt.Soc.Am.B, v.11, 1031(1994)
2. SPIE Proceeding vol.2041, "Mode-locked and solid state lasers, amplifiers and applications," Session 2, 1993, Quebec, Canada.
3. M.G.Raymer, J.Mostowski, J.L.Carlsten, Phys.Rev.A, v.19, 2304(1979).

90

RESONANCE RAMAN SCATTERING OF LASER DYES IN A TRAVELLING WAVE AMPLIFIER

J. Klebniczki,[1] J. Seres,[2] and J. Hebling[1]

[1] Department of Optics and Quantum Electronics
JATE University, Szeged, Hungary
[2] Department of Physics
Juhász Gy. College, Szeged, Hungary

Investigation of the Raman scattering (RS) of fluorescent molecules is generally difficult because the fluorescent light on the Stokes side is much stronger than the scattered light. In order to have an easier detection of RS there are different methods of increasing it e.g. using (a) the enhanced ratio of RS to fluorescence on colloidal silver surface[1], (b) an active material with high gain[2,3], (c) microdroplets with high-Q microcavities[4]. We investigate here the resonance RS of three highly fluorescent laser dyes when they are pumped with picosecond pulses. A travelling wave (TW) amplifier is used, where the short optical pump pulse and the amplified signal propagate together inside the active medium resulting an extremely high ($\sim10^9$) gain[5,6]. Another advantage of our experimental arrangement is that the interaction length is long even when the absorption of the RS medium is high. Since in these experiments the dyes are the active material of an amplifier, the investigated dye molecules are simultaneously the source of scattering and the gain medium. In an earlier experiment with TW amplifier we have observed RS line for the case of Rhodamine 6G dye[7].

In this paper we report on strong RS lines in the amplified spontaneous emission (ASE) spectra of three laser dyes using TW amplifier.

In our experiment the TW amplifier is practically the same that is detailed in [8,9]. It consists of two dye cells as it is shown in the insert of Fig. 2. A 1 cm long dye cell containing the investigated dye solution is placed into an other one, which contains Glycerol. Sulforhodamine 101, Cresyl Violet and DODCI dyes are dissolved in Methanol in the concentration of 10 g/l, 5 g/l and 5 g/l, respectively. Changing the angle of incidence of the pump beam (α) the sweep velocity of the pump pulse across the active volume changes, and the synchronisation between the pump and the ASE pulse can be obtained[8].

The pump pulse is generated in a XeCl excimer laser pumped dye laser-amplifier system, which is very similar to the one described in [10]. The pump pulse duration is 6 ps and the pump pulse energy is 30 µJ at a wavelength of 608 nm. The pump beam is horizontally polarised and it is focused to a line into the dye solution with a cylindrical lens

Ultrafast Processes in Spectroscopy, edited by Svelto et al.
Plenum Press. New York, 1996

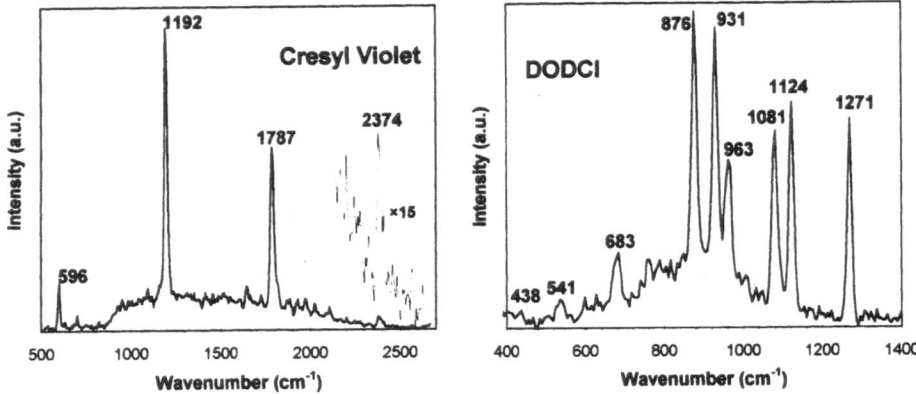

Figure 1. Spectra of TW amplifier for Cresyl Violet and DODCI dyes.

producing 10^9 W/cm^2 pump intensity in the dye solution. The length of the active volume is approximately 7 mm. The low repetition rate (0.8 Hz) makes possible to detect single pulses.

The ASE beam generated in the TW amplifier is spectrally dispersed in a STE-1 spectrograph and the spectra are detected by a CCD camera. The spectral resolution of the system is 5 cm^{-1}, while the RS lines have a typical bandwidth of 10–20 cm^{-1}.

In order to have TW synchronisation we adjusted the incidence angle of the pump beam (α) according to the equation in [8]. The ASE spectra are rather broad with a bandwidth of 30–40 nm, depending on the dyes and have a structure containing 20–30 spectral lines, which changed randomly shot to shot. However there are a few spectral lines that appear in every pulses. After averaging 10–15 pulses these lines rise out from the ASE background as it is displayed in Fig. 1, where the horizontal axis shows the wavenumbers related to that of the pump. Changing the pump wavelength these lines change as well remaining the frequency difference between the pump and these lines constant. For Cresyl Violet 4 lines, for DODCI 9 lines, and for Sulforhodamine 101 6 lines have been found, respectively (see Fig. 1 and Table 1). Raman lines at 596 and 1192 cm^{-1} have been identified in the literature[3] for Cresyl Violet.

It was found that decreasing the pump intensity the lines became weaker and finally disappeared, typically first on the longer side of the spectrum.

Table 1. Measured Raman lines together with some earlier observed lines

Cresyl Violet		596	1192				1787	2374	
			2·596				3·596	4·596	
Cresyl Violet [3]	530	590	1180	1220	1240	1350			
Sulforhodamine 101	422	472	774	1353	1512	1659			
Rhodamine 6G [1]			776	1365	1509	1652			
DODCI	438	541	683	876	931	963	1081	1124	1271
				2·438			2·541	438+683	

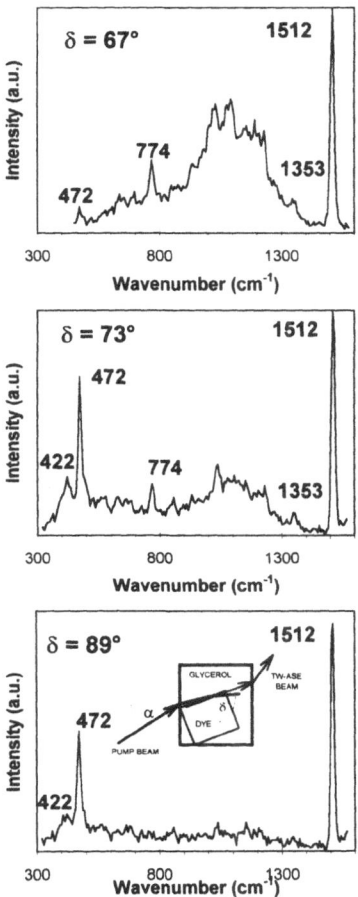

Figure 2. Spectra for three different values of the angle δ. Please notice the dependence of the RS/ASE background ratio on the angle δ.

In Fig. 2 we show how the spectrum changes by changing the angle of incidence (α). This causes a change both in the TW synchronisation and in the pump beam direction inside the dye solution. The latter is measured by the angle of refraction (δ) of the pump beam in the active material. Increasing angle of δ, one can observe a decrease in the ASE intensity while the RS lines become more pronounced. At δ = 89 degrees - i.e. when the pump beam is almost parallel to the ASE beam direction - the lines 1512 cm^{-1} and 472 cm^{-1} saturate the TW amplifier. We repeated this measurement by replacing the Glycerol with CS$_2$, when the TW synchronisation is not fulfilled perfectly. For this case too, the RS/ASE ratio was largest for the largest value of δ. The intensity of the RS, however, was significantly smaller then for the TW condition.

Evaluating the averaged spectra some well defined RS lines have been observed, which are listed in Table 1. For Cresyl Violet two RS lines are very near to the ones published in [3], but two new lines were identified at longer wavelength (1787 and 2374 cm^{-1}). In [3] the pump wavelength was the same as in our experiment, but the concentration of the solution was more than 100 times lower. Moreover Ethanol was used as solvent instead of Methanol. The line of 590 cm^{-1} in [3] had medium intensity and this line may be amplified in our TW amplifier. The weak 530 cm^{-1} line is out of the gain region of the TW amplifier, so its detection can not be expected. Around the 1180 cm^{-1} line the TW amplifier has a maximal gain, that is why we detected this weak 1192 cm^{-1} line as the strongest one. It is important to note that all four lines are very near to the multiples of the 596 cm^{-1} line.

We can compare our results for Sulforhodamine 101 only with another dye, Rhodamine 6G. However, the two dyes have basically the same structure thus we should find spectral lines at the same places. In Table 1 we listed some RS lines[1] for Rhodamine 6G, which are near to our measuring results.

Most RS lines have been found for DODCI (see Table 1), some of them are combinations of others. We could not find data in the literature for the RS of this dye.

It is noteworthy that we were not able to detect any Raman lines of the solvent, although some of them (1171, 1456 cm^{-1}) fall into the gain region of the dye solutions. This shows the great enhancement of the scattering cross section because of the resonance. A model calculation for RS in the TW amplifier is under investigation.

We can summarise our results as follows: we have studied resonance RS lines of three laser dyes. Two new lines were found for Cresyl Violet. RS lines of Sulforhodamine 101 and DODCI were measured for the first time. Moreover it was demonstrated that RS lines can be very strong in TW amplifier.

REFERENCES

1. P. Hildebrandt, M. Stockburger, J. Phys. Chem. **88**, 5935 (1984)
2. J. Herrmann, J. Wienecke, Opt. Commun. **11**, 261 (1974)
3. W. Werncke, A. Lau, M. Pfeiffer, H.-J. Weigmann, G. Hunsalz, K. Lenz, Opt. Commun. **16**, 128 (1976)
4. A. S. Kwork, R. K. Chang, Opt. Lett. **18**, 1703 (1993)
5. Z. Bor, S. Szatmari, A. Müller, Appl. Phys. B **32**, 101 (1983)
6. J. Hebling, J. Kuhl, J. Klebniczki, J. Opt. Soc. Am. B **8**, 1089 (1991)
7. J. Hebling, J. Klebniczki, P. Heszler, Z. Bor, B. Racz, Apl. Phys. B **48**, 401 (1989)
8. Z. Bor, B. Racz, Opt. Commun. **54**, 165 (1985)
9. J. Hebling, J. Kuhl, Opt. Lett. **14**, 278 (1989)
10. J. Seres, J. Klebniczki, J. Hebling, Optical and Quantum Electr. **26**, 933 (1994)

91

EFFICIENT BROADBAND SUM FREQUENCY GENERATION BY CHIRPED PULSES

K. Osvay[1,2] and I. N. Ross[1]

[1] Central Laser Facility
Rutherford Appleton Laboratory, Chilton
Didcot, Oxon OX11 0QX, United Kingdom
[2] Department of Optics and Quantum Electronics
JATE University
Dóm tér 9, H-6720 Szeged, Hungary

Much recent interest in the development of lasers has centred around the generation of very broad spectral bandwidth pulses, either because this is required for ultrashort pulses or because it is demanded by the application physics. The main limitation to efficient broad bandwidth frequency conversion originates from the dispersion of the non-linear crystal[1]. To overcome these restrictions, various techniques have been developed (see e.g. Ref. 2–5).

In this paper an alternative technique is proposed in which the generating beams to be mixed in the nonlinear material are chirped at different chirp rates such that the phase matching condition can be met at all times.

The principal of the scheme is demonstrated in Figure 1. The phase-matching condition is satisfied, at some angle in the nonlinear medium, for the central frequencies of the generating beams (ω_1^0 and ω_2^0) and sum frequency beam (ω_s^0). Then for a frequency offset at the sum frequency there can be found a pair of frequencies $\omega_{1,2}^-$ and $\omega_{1,2}^+$ in the generating beams which are also phase matched at the same crystal angle. If the sum frequency has its frequency varying linearly with time (chirped) there are corresponding chirps on both generating beams for which the phase matching condition is met over the full bandwidth.

For the analytical description we consider a collinear Type I mixing process, where the well collimated generating ordinary $\omega_1 < \omega_2$ waves propagating in z direction produce the generated extraordinary ω_s wave. Following from the general solution of the coupled wave equations in a nonlinear crystal, the mismatches affecting the efficiency, and hence the bandwidth of the frequency conversion, can be described by the total phase mismatch of the process

$$\Delta\psi(t,z) = \psi_s(t,z) - \psi_1(t,z) - \psi_2(t,z), \quad where \quad \psi_i(t,z) = \omega_i t + k_i z . \tag{1}$$

Ultrafast Processes in Spectroscopy, edited by Svelto et al.
Plenum Press, New York, 1996

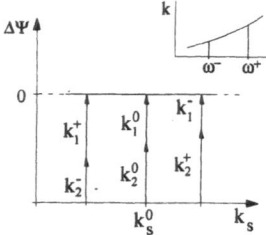

Figure 1. The phase matching condition for the proposed new frequency conversion scheme.

For perfect generation $\Delta\psi=0$ should be satisfied in time and in space simultaneously, that is

$$\omega_s(t) = \omega_1(t) + \omega_2(t), \quad k_s(z) = k_1(z) + k_2(z).$$

(2)

To consider the more general case of broad bandwidth pulses, the frequencies are assumed varying with time while the wave vectors change in space. Expanding them into series, the equations following from the time and space dependence of $\Delta\psi$ in eq.(1) are

$$\omega_{s0} = \omega_{10} + \omega_{20}, \quad \beta_s = \beta_1 + \beta_2$$

(3)

and

$$k_{s0} = k_{10} + k_{20}, \quad k'_{s0}\beta_s = k'_{10}\beta_1 + k'_{20}\beta_2$$

(4)

respectivelly, where β_i is the linear chirp parameter defined by $\omega_i = \omega_{i0} + \beta_i t$, $\omega_{i0} = \omega_i(t=0)$ and k'_{i0} is the reciprocal group velocity at ω_{i0}. Note, that eq.(3.a) and (4.a) give the ordinary phase matching condition and determine the phase matching angle. The solution of eq.(3.b) and (4.b) at the phase matching angle gives the first order compensation, that is

$$\beta_1 = \left(1 - \frac{\Delta k'_{s1}}{\Delta k'_{21}}\right)\beta_s = A_1\beta_s, \quad \beta_2 = \frac{\Delta k'_{s1}}{\Delta k'_{21}}\beta_s = A_2\beta_s$$

(5)

where $\Delta k'_{21} = k'_{20} - k'_{10}$ and $\Delta k'_{s1} = k'_{s0} - k'_{10}$ are the reciprocal group velocity mismatches between the generating and generated waves. This solution is in full agreement with the qualitative physical picture described above. Eq.(5) means that (i) the requirement of first order compensation can be fulfilled by introducing linear chirp (b1, b2) into the initial fields, so it can be interpreted as chirp-assisted group velocity matching (CGV); (ii) the ratio of these necessary chirps is determined by the group velocities of the generating and generated waves.

Assuming perfect CGV matching, the residual total phase mismatch due to second order effects (add second order terms to eq.(1)-(2)) at the end of a crystal of length L is

$$\Delta\psi^{CGV} = \frac{\beta_s^2 L^3}{2c^2}\Delta k''$$

(6)

where $\Delta k'' = k_{s0}'' - A_1^2 k_{10}'' - A_2^2 k_{20}''$ and k_{i0}'' is the group velocity disperion at ω_{i0}. Equation (6) is exactly what could be expected, namely that upon propagation through the crystal there is a residual phase mismatch due to uncompensated group velocity dispersion.

The chirp of the generated field seems to be arbitrary. It can be shown, however, that β_s has an optimum value determined by group velocity dispersion[6] through eq.(6) as well as by the need for high conversion efficiency[7].

The calculations presented are made for a frequency tripling process via sum frequency generation ($\omega_1 + 2\omega_1 \rightarrow 3\omega_1$) by Gaussian pulses in BBO with length L=1000 μm. The input transform limited pulse duration is τ_{10}=50 fs. If the acceptable total phase mismatch is $\pi/_2$, Fig.2.a shows the generated bandwidth $\Delta\bar{v}_s^{CGV}$ ($= \Delta\omega_s^{CGV}/2\pi c$) vs. wavelength of λ_1. For comparison, the bandwidth achievable by the traditional method $\Delta\bar{v}_s^0$ ($= \Delta\omega_s^0/2\pi c$) under the same conditions are also displayed. Since the dispersion increases with frequency, the gain in generated bandwidth compared to the traditional method is more significant at shorter wavelengths.

The bandwidth of the generated pulse is dependent on the crystal length through β_s and the duration of the generated pulse in a quite complicated way[7]. In Fig.2.b the bandwidths $\Delta\bar{v}_s^{CGV}$ and $\Delta\bar{v}_s^0$ are plotted as a function of crystal length for BBO. The limitation of increasing the crystal length to an arbitrary high value, at which broadband conversion is achieved, is set by chirping the pulses by the crystal itself (see (ii) of Section 2).

The analysis can be extended to include higher order dispersion[7]. As can be seen from equation (1)-(2), if higher order terms are included, there is still a solution to the phase matching condition which yields nonlinear equations relating the required chirps on each beam and hence to a requirement for slightly nonlinear chirps on the generating beams. Other than for extreme cases of high dispersion and crystal length linear chirps can be used efficiently without exceeding the 'normal' bandwidth limit of the process.

The main conclusions established are: (i) the bandwidth generated has an asymptotic value as a function of crystal length; (ii) the CGV technique is even more superior to the traditional one in the cases where the dispersion is higher, and (iii) the phase matching condition can theoretically be satisfied for arbitrarily wide bandwidth by designing the higher order chirp coefficients appropriately.

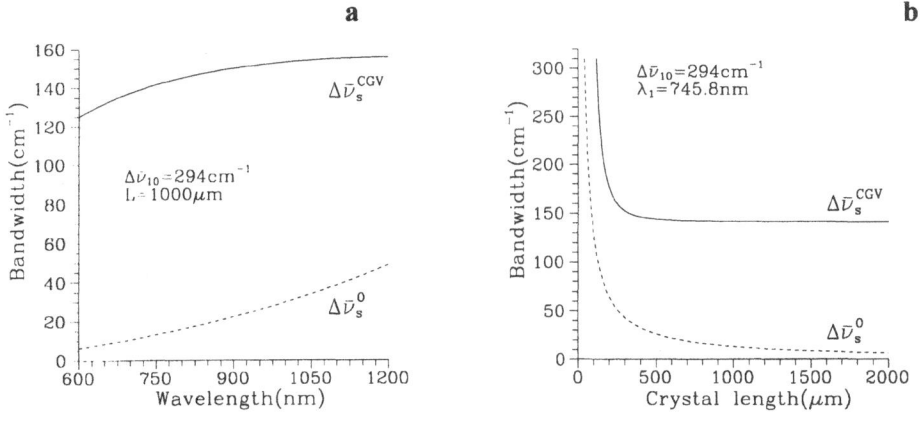

Figure 2. The generated bandwidth in BBO with and without CGV compensation in function of wavelength of λ_1 (a) and of crystal length (b).

ACKNOWLEDGMENT

One of the authors (K.O.) acknowledges the support of the OTKA Foundation of Hungary under grant #F014271.

REFERENCES

1. A. Penzkofer, F. Ossig, P. Qiu, Appl.Phys. B **47,** 71 (1988)
2. O. E. Martinez, IEEE J.Quant.Electron. **QE-25,** 2464 (1989)
3. M. J. T. Milton, T. J. McIlveen, D. C. Hanna, P. T. Woods, Optics Comm. **87,** 273 (1992)
4. G. Szabó, Z. Bor, Appl.Phys. B **58,** 237 (1994) (review)
5. T. R. Zhang, H. R. Choo, M. C. Downer, Appl.Optics **29,** 3927 (1990)
6. I. N. Ross, K. Osvay, Opt.Quant.Electr. accepted
7. K. Osvay, I. N. Ross, submitted to JOSA B

92

A SUB-PICOSECOND SOURCE OF COHERENT, PARTIALLY TUNABLE, VUV RADIATION OPERATING AT HIGH REPETITION RATE

C. de Lisio,[1] C. Altucci,[1] R. Bruzzese,[1] S. Solimeno,[1] F. Vigilante,[1] M. Bellini,[2] and P. Foggi[2]

[1] Dipartimento di Scienze Fisiche
Università di Napoli "Federico II" and Istituto Nazionale di Fisica della Materia (INFM)
Sezione di Napoli, Napoli, Italy
[2] European Laboratory for Nonlinear Spectroscopy (LENS)
Firenze, Italy

We report the realization of a VUV radiation source based on high order harmonic generation in gases. High pulse repetition rate, tunability, spatial and temporal coherence, short pulse duration and, in turn, extremely high brightness and considerable average photon fluxes are some of its characteristics which make it very feasible for spectroscopy, ultrafast processes and surface physics experiments. We observed up to the 13th harmonic ($\lambda \approx 61$ nm) of a Ti:Sa laser ($\lambda \approx 800$ nm), operating at 1 kHz repetition rate, after its interaction with a Xe gas target. In spite of the low pulse energy (≈ 1 mJ) of the Ti:Sa source, intensity in excess of 10^{14} Wcm^{-2} can be easily obtained due to the short pulse duration (≈ 100 fs) and the good optical quality of the laser beam, which allows its focusing near the diffraction limit. At such intensity level, odd harmonics of the laser fundamental frequency are produced up to very high orders[1]. We report here the spectrum of VUV radiation generated during the laser-gas interaction and a study of the spectral profile of the 9th harmonic.

The laser source of our experiment consists of a Ti:Sa oscillator delivering (after chirped-pulse-amplification) 900 μJ-80 fs pulses at 1 kHz repetition rate. The nonlinear medium is realized by injecting the Xe gas sample through a piezoelectric valve operating at 500 Hz. The local pressure in the interaction region is about 30 Torr and the gas jet length approximately 1 mm. A complete characterization of piezo-valve performances and the main features of gas jets is described elsewhere[2]. The laser beam is focused onto the gas target by a 200 mm focal length lens and reaches a radius of about 20 μm in the beam waist. After the BK7 focusing lens and interaction chamber window, the actual pulse duration turns out to be 110 fs. Thus the intensity in the focal region can be as high as 6×10^{14} Wcm^{-2}. The VUV radiation emerging from the laser-gas interacion region is analyzed by a nearly normal incidence grat-

Ultrafast Processes in Spectroscopy, edited by Svelto et al.
Plenum Press, New York, 1996

405

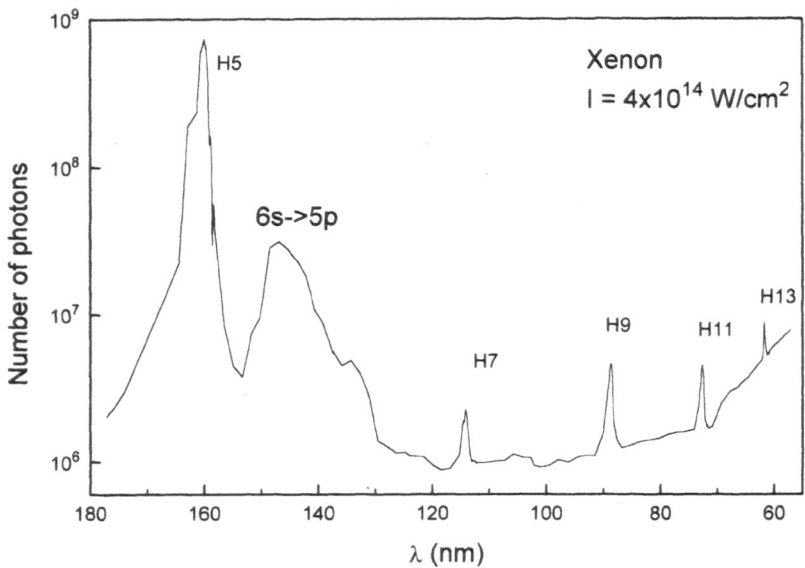

Figure 1. Radiation spectrum in the region 60–180 nm following the interaction of a Ti:Sa laser pulse with Xe gas sample at about 30 Torr. The laser intensity is 4×10^{14} Wcm^{-2}. The vertical axis represents the absolute number of photons generated per laser shot.

ing monochromator with 1 m arms; the spherical holographic grating has 2160 grooves/mm and a diffraction efficiency grater than 10% up to about 60 nm. Its spectral resolution is of the order of 0.1 nm. VUV radiation from the monochromator exit slit is down-converted into visible light by a TPB phosphor deposited on a 200 μm thik glass substrate. More details on the TPB spectral and angular response and its efficiency are deduced from literature[3]. Fluorescence light emitted by the phosphor is detected by a photomultiplier tube faced with the quartz window which closes the exit arm of the monochromator.

In Fig.1 a spectrum of the radiation emerging from the interaction region is reported. Odd harmonics from the 5th to the 13th are clearly evident. The broad peak at 140 nm corresponds to the transition 6s→5p of neutral Xenon.

The number of photons per laser shot is estimated by taking into account the overall collection efficiency of our apparatus. The average photon flux per second is easily obtained by multiplying by a factor of 500 the number of photons reported in Fig.1. The saturation intensity for multiphoton ionization has been estimated to be about 8×10^{13} Wcm^{-2}, in good agreement with published values obtained in similar experimental conditions4. Thus, an intensity of 4×10^{14} Wcm^{-2} corresponds to a quite deep saturation. At such intensity, harmonics from the 7th to 13th lie in the plateau region. In Fig.1 harmonic peaks are overlapped to a pedestal due to recombination radiation from 120 nm to 90 nm and to fundamental laser light at wavelength below 90 nm.

In view of the application of harmonic generation in gases as a VUV light source, the spectral charactrization of the generated radiation is of particular interest. As an example, we show in Fig.2 the spectral profile of the 9th harmonic ($\lambda \approx 89$ nm) at three different laser intensities.

The center of the spectral profile is shifted towards shorter wavelengths by increasing the laser intensity. Moreover, the spectral width broadens almost by a factor of 2. Both

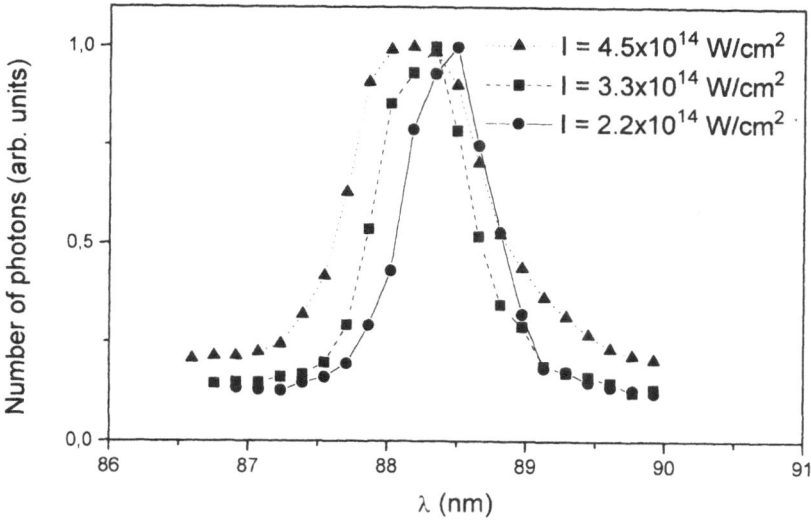

Figure 2. Spectral profile of 9th harmonic ($\lambda \approx 89$ nm) at a laser intensity of 2.2×10^{14} Wcm^{-2} (—●—), 3.3×10^{14} Wcm^{-2} (---■---) and 4.5×10^{14} Wcm^{-2} (···▲···), respectively.

blue-shifting and broadening are due to the propagation of the laser fundamental as well as harmonic beams through an ionized medium: time dependent photoelectron density and, in turn, refractive index are responsible for such phenomena[4].

In conclusion, we have realized a VUV radiation source based on high order harmonic generation in noble gases. Short pulses (below 100 fs) at the nanojoule level are efficiently produced at 500 Hz repetition rate with VUV peak power ling in the kW range. The short pulse duration together with the high repetition rate make the source described here very suitable for atomic and molecular physics experiments in the VUV region.

REFERENCES

1. C.G. Wahlström, Physica Scripta, **49**, 201 (1994).
2. C. Altucci, C. Beneduce, R. Bruzzese, C. de Lisio, G.S. Sorrentino, T. Starczewski, and F. Vigilante, J.Phys.D, in press (1995).
3. G. Naletto, E. Pace, L. Placentino, and G. Tondello, "Fluorescence of Metachrome in the far and vacuum ultraviolet spectral region", SPIE Proc. 2519, 31–38, (1995).
4. C.-G. Wahlström, J. Larsson, A. Persson, T. Starczewski, S. Svanberg, P. Salières, Ph. Balcou, and A. L'Huillier, Phys. Rev. A, **48**, 4709 (1993).

93

NARROW-BAND SEEDING OF A FEMTOSECOND OPTICAL PARAMETRIC AMPLIFIER

V. Petrov and F. Noack

Max Born Institute for Nonlinear Optics and Ultrafast Spectroscopy
D-12474 Berlin, Germany

In the last few years considerable progress in the field of infrared femtosecond pulse generation through optical parametric generators (OPG's) and optical parametric amplifiers (OPA's) has been reported[1,2]. These extremely simple techniques possess broad tunability, but the generation of bandwidth limited pulses turns out to be a serious problem in both OPG's and OPA's on one hand because of the possibility of off-axis parametric generation in OPG's and on the other hand because of the continuum seeding used in OPA's.

Narrow-band seeding avoids the problems associated with the bandwidth limitation[3] and has the potential of preserving the good temporal and spectral quality of the pump pulses at reduced pump intensities as compared to the case of OPG's. In the present work we apply for the first time to our knowledge quasi-cw narrow-band seed to an OPA instead of continuum in the femtosecond regime. We employ a commercial tunable femtosecond Ti:sapphire regenerative amplifier system as a pump source near 800 nm whereas the seed signal is derived from the fundamental of the Q-switched Nd:YLF laser whose second harmonic pumps the regenerative amplifier.

Our 1-kHz repetition rate regenerative amplifier provides 0.3 mJ pulses at 150...175 fs pulse duration between 740 and 840 nm. It is seeded by an Ar^+-laser pumped Ti-sapphire laser oscillator which produces sub-100 fs pulses operating at 82 MHz repetition rate. By substituting one of the total reflectors of the Nd:YLF laser by a 2.5% (1.053 μm) transmitting mirror which is totally reflecting at the second harmonic, nearly 1 W of average power at 1.053 μm is available as a seed signal. The seed energy in the time window defined by the pump pulse can be adjusted up to 80 pJ. After telescoping of the seed beam it is recombined with ≈200 mW of the pump beam near 800 nm using a dichroic mirror. The narrow-band seeded OPA (Fig.1) consists of only a single 2-mm long BBO crystal (θ=21°, type I phase-matching) placed between two lenses and before the waist formed by the first focusing lens. The pump intensity in the OPA could be easily varied by translating the nonlinear crystal.

At the maximum seed level the BBO-OPA produced ≈0.5 μJ idler pulses with average pump intensity of 65 GW/cm^2 and ≈100 nJ with 50 GW/cm^2. Tuning is achieved by

Ultrafast Processes in Spectroscopy, edited by Svelto et al.
Plenum Press, New York, 1996

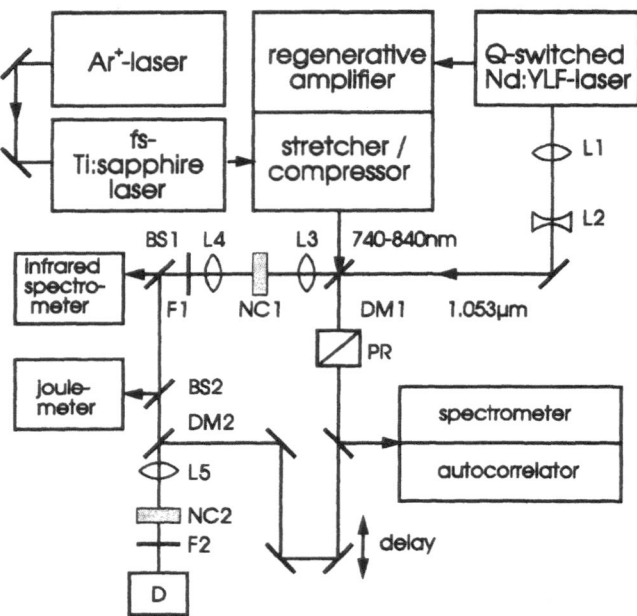

Figure 1. Experimental setup, L1-L5, lenses; NC1, BBO-OPA; NC2, sum-frequency generation crystal; F1-F2, filters; DM1-DM2, dichroic mirrors; BS1-BS2, removable silver mirrors; PR, polarization rotator; D, Si photodiode.

changing the pump wavelength and slight readjustment of the phase-matching angle. The measured cross-correlation widths (Fig. 2) are shorter than the autocorrelation width of the pump pulses and indicate a compression factor of >1.5 in the whole tuning range achieved (2.56–3.16 μm). Estimation of the idler pulse duration yields about 110 fs resulting in a pulselength/bandwidth product of 0.5 which reproduces the value for the pump pulses. As can be seen from the cross-correlation trace the pump pulse shape can be resolved and a satellite behind the main peak of the pump pulse can be identified.

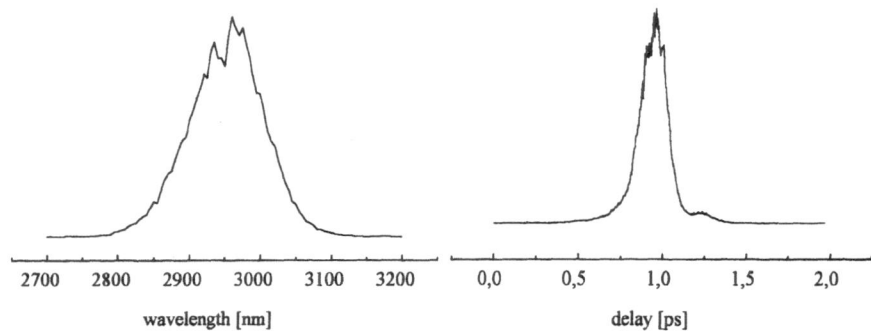

Figure 2. Spectrum and cross-correlation trace of the generated infrared pulses with the pump pulses for an idler wavelength of 2.95 μm. The pump intensity amounts in this case to 50 GW/cm^2.

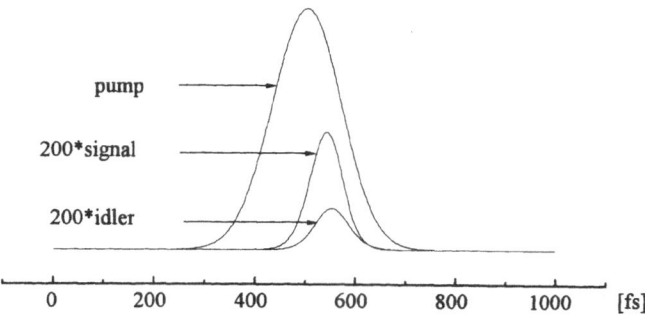

Figure 3. Calculated pump, signal and idler temporal profiles behind the nonlinear crystal. The pump wavelength is 776 nm and the pump intensity amounts to 50 GW/cm^2.

Neither parametric generation from noise nor saturation effects could be observed at the pump levels applied. The estimated gain achieved at the signal wave exceeds 3000 at 50 GW/cm^2 pump intensity. A numerical simulation of the amplification process in the small signal case asssuming plane waves and taking into account the experimentally estimated seed and pump levels (Fig. 3) yielded gain factors and output energy levels very close to the experimental ones, whereas the calculated idler pulse duration was of the order of 80 fs. The generation of idler pulses shorter than the pump pulses is attributed to temporal gain narrowing at crystal lengths comparable to the walk-off lengths[2,4].

The output energy could be increased using a longer (4 mm) BBO crystal. Such a crystal yielded infrared energy as high as 2 µJ at 2.7 µm when pumped at 50 GW/cm^2. One sacrifices, however, in this case the short duration of the idler pulses which amounts to ≈250 fs and exceeds the pump pulse duration.

In conclusion we generated tunable femtosecond pulses with well defined narrow spectral profiles by a single stage OPA due to the difference-frequency generation which initiates the optical parametric amplification. An important advantage of our system is the ultimate simplicity and ease of alignment. Further energetic scaling without pulse broadening is possible using additional amplification stages whereas the wavelength of the generated idler pulses could be shifted further into the mid infrared using nonlinear crystals with higher transparency above 3 µm. The 1 kHz repetition rate and the simultaneous presence of synchronized femtosecond pulses at the pump wavelength facilitate pump and probe spectroscopy in conjunction with lock-in modulation and signal averaging techniques.

REFERENCES

1. R. Danielius, A. Piskarskas, A. Stabinis, G. P. Banfi, P. Di Trapani, and R. Righini, J. Opt. Soc. Am. B **10**, 2222 (1993).
2. V. Petrov, F. Seifert, O. Kittelmann, J. Ringling, and F. Noack, J. Appl. Phys. **76**, 7704 (1994).
3. G. A. Massey, J. C. Johnson, and R. A. Elliot, IEEE J. Quantum Electron. **QE-12**, 143 (1976).
4. S. A. Akhmanov, A. S. Chirkin, K. N. Drabovich, A. I. Kovrigin, R. V. Khokhlov, and A. P. Sukhorukov, IEEE J. Quantum Electron. **QE-4**, 598 (1968).

94

TEMPERATURE TUNING OF A FEMTOSECOND TRAVELING WAVE OPTICAL PARAMETRIC GENERATOR BASED ON UNCRITICALLY PHASE-MATCHING IN A TYPE-II LITHIUM TRIBORATE CRYSTAL

V. Petrov,[1] F. Seifert,[2] and F. Noack[1]

[1] Max Born Institute for Nonlinear Optics and Ultrafast Spectroscopy
D-12474 Berlin, Germany
[2] Federal Institute of Physics and Technology
D-10587 Berlin, Germany

Traveling wave optical parametric generators (OPG's) represent a powerful tool for generating tunable ultrashort light pulses in the near-infrared spectral region. The reproducibility and the spectral width which often extends beyond the Fourier limit are the main problems associated with femtosecond traveling wave OPG's[1]. We demonstrated recently that type-II phase matching in a lithium triborate (LBO) seeder could provide narrow spectral widths at room temperature[2] even in the vicinity of the degeneracy point. Such a scheme has the potential of intrinsic spectral limitation. Besides the advantage of possible temperature tuning without spatial walk-off, noncritical phase-matching in LBO does not result in off-axis parametric generation as observed in critical phase-matching. The latter effect can be enhanced by group velocty walk-off as well as by the birefringence and lead to pump depletion. As compared to type-II interaction in β-barium borate (BBO) the chosen LBO type-II seeder exhibits much larger acceptance angles which is an important advantage when tight focusing is applied for scaling to lower pump energies.

Here we report on the temperature tuning of this femtosecond OPG which is based on such a LBO-seeder and a type-I BBO parametric amplifier. Operation of the amplifier stage at high pump levels provides an effective temporal gain narrowing mechanism at reduced group velocity mismatch which results in parametric pulses shorter than the pump pulses in the small signal limit[3].

The pump source is a commercial femtosecond 1 kHz Ti:sapphire laser/regenerative amplifier system (Spectra Physics Tsunami/Quantronix Model 4810/20 RGA) exhibiting less than 1% peak to peak noise (Fig.1). The pulse duration/bandwidth product of the pump pulses amounts to ≈0.4 and pump pulse durations of 150...200 fs yield peak powers ≥1.5 GW. The positions of both LBO and BBO crystals can be varied in order to control

Ultrafast Processes in Spectroscopy, edited by Svelto et al.
Plenum Press, New York, 1996

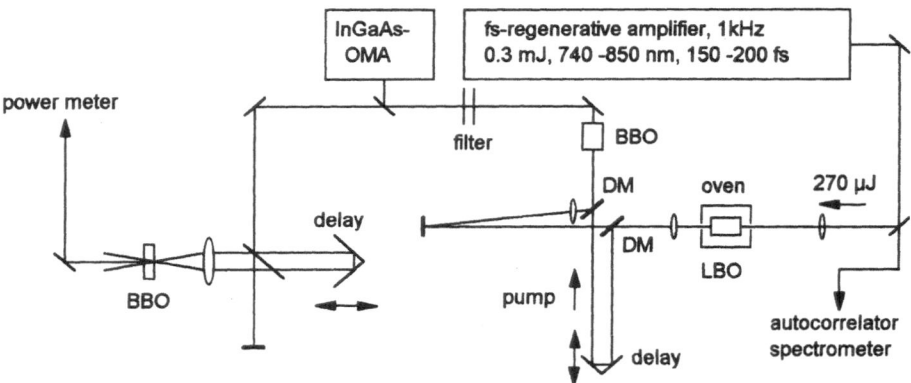

Figure 1. Experimental setup. DM - dichroic mirrors. The telescope consists of a 50-cm focusing lens and an 11-cm collimating lens mounted on a translation stage.

the pump intensity and typical values used were 130 GW/cm^2 in the LBO seeder and 50 GW/cm^2 in the BBO amplifier. The 5-mm-long LBO crystal is cut along the z axis for type-II noncritical phase-matching (pump and idler polarized along the x axis and signal polarized along the y axis). The 4-mm-long BBO crystal is cut for type-I phase matching at $\theta=20°$ and seeded only by the signal pulse generated in LBO. A delay line adjusts the signal/pump temporal separation between the two stages. The threshold for parametric fluorescence (on the nJ level) was about 100GW/cm^2 in the LBO-seeder and such levels were sufficient to seed the BBO amplifier stage.

Tunability is accomplished by changing the wavelength of the pump system and/or heating the LBO crystal and adjusting the angle of the BBO amplifier. Fig. 2 shows the achieved temperature tuning range at a fixed pump wavelength. It can be seen that temperatures as high as 300°C are sufficient to approach the upper wavelength limit for the idler which is imposed by the transparency edge of LBO. Shorter pump wavelengths permit operation closer to the degeneracy point at room temperature whereas longer pump wavelengths result in longer idler wavelenths at a fixed temperature.

With our background-free autocorrelator (based on BBO) it was possible to measure the autocorrelations of the signal and idler pulses as well as their cross-correlation. Both the amplified signal and the idler that is generated in the second stage are nearly bandwidth limited (pulselength/bandwidth product of the order of 0.5) at sub-100 fs pulse durations and > 1 µJ energies.

The pulse shortening obtained with the present arrangement is in good agreement with numerical simulations in the plane-wave approximation in which group velocity mismatch effects are taken into account.

Fig. 3a shows typical pulse profiles calculated behind the LBO-II seeder for our experimental pump intensities. As can be expected from group velocity data the signal pulse is delayed with respect to the pump pulse. Because of the complex three-pulse interaction, the idler pulse is also delayed with respect to the pump pulse, although its group velocity should be larger than the group velocity of the pump pulse. The signal pulses generated by seeder are somewhat shorter (by ≈25%) than the pump pulses and this has been taken into account in Fig. 3b. In Fig. 3b we fitted the intensity level of the seed signal pulse in such a way in order to reproduce the experimentally achieved energy conversion efficiency in the

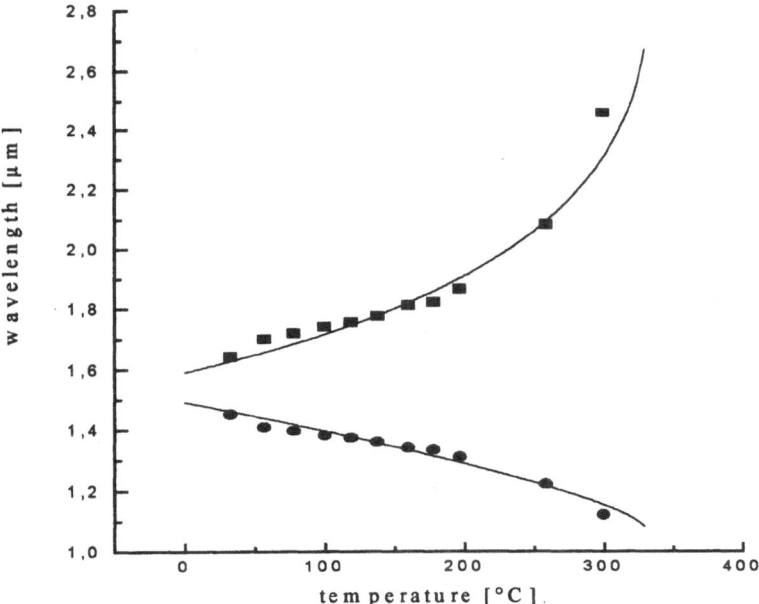

Figure 2. Idler (upper curve) and signal (lower curve) wavelength versus temperature of the LBO seeder for a pump wavelength of 770 nm. Calculation (solid line) and experimental data (squares and circles).

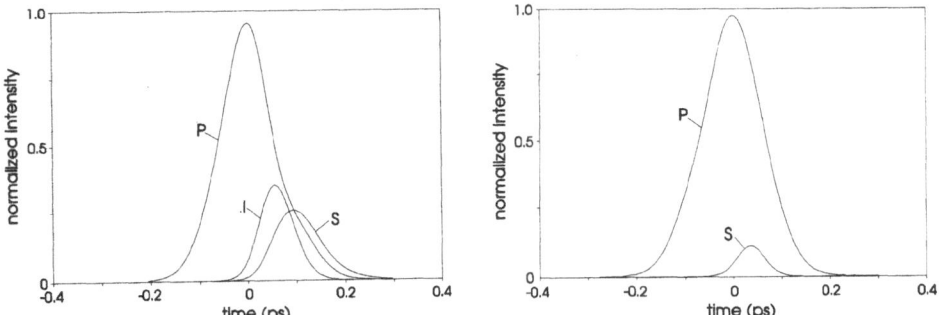

Figure 3. a: Numerical simulation of the LBO seeder. The pump wavelength is 805 nm and the signal wavelength is 1460 nm. The initial pump pulse duration is 150 fs. Behind the seeder we calculated a pump pulse (P) duration of 150 fs, a signal pulse (S) duration of 113 fs and an idler pulse (I) duration of 83 fs. In order to visualize the signal and idler pulses the parametric fluorescence level has been exaggerated in this case; very similar pulse durations and maximum positions are obtained, however, also at negligible pump depletion. b: Numerical simulation of the BBO amplifier. The pump pulse duration is 150 fs and the initial signal pulse duration is shorter (110 fs) in accordance with Fig. 3a. The idler generated in the amplifier nearly concides temporally with the signal.

amplifier stage ($\approx$6% for signal + idler). The calculated amplified signal pulse duration (55 fs) results in a compression factor of the order of 3 which is very close to the experimentally obtained value at this wavelength. It should be noted that the main contribution to the compression is due to the parametric amplifier: even if seeded with a cw signal the simulation of the amplifier yields compression factors of the order of 2.5.

In conclusion we applied noncritical type-II phase-matching in a LBO-based OPG with temperature tuning for the generation of bandwidth limited sub-100 fs optical pulses in the near infrared. The simple combination with a broadband BBO amplifier provides pulse durations shorter than the pump pulse duration.

REFERENCES

1. R. Danielius, A. Piskarskas, A. Stabinis, G. P. Banfi, P. Di Trapini, and R. Righini, J. Opt. Soc. Am. **B 10**, 2222 (1993)
2. V. Petrov, F. Seifert, and F. Noack, Appl. Phys. Lett. **65**, 268 (1994)
3. S. A. Akhmanov, A. S. Chirkin, K. N. Drabovich, A. I. Kovigin, R. V. Khokhlov, and A. P. Sukhorukov, IEEE J. Quant. Electron. **QE-4**, 598 (1968)

95

FEMTOSECOND DYNAMICS IN CONJUGATED POLYMERS

T. Kobayashi

University of Tokyo
Tokyo, Japan

1. INTRODUCTION

The optical nonlinearities of the one-dimensional conjugated polymers [1, 2] are closely related to localized excitations such as a pair of solitons, polarons, and a self-trapped exciton (STE) formed via a strong coupling between electronic excitations and lattice vibrations [3]. The localized excitations modulate dielectric properties of the conjugated polymers, inducing the nonlinear changes of absorptive and refractive indices. The temporal response of the macroscopic nonlinear properties is quite essential and the most fundamental subject to be investigated since it directly reflects the dynamical behavior of the localized excitations. The soliton [3, 4] is a topological domain-wall between two energetically-equivalent dimerization-phases symbolically represented as $\cdots$ - A=B-A=B-A= $\cdots$ and $\cdots$= A-B=A-B=A- $\cdots$. It was experimentally confirmed to be excited in *trans*-polyacetylene, where A=B=CH, by the appearance of mid-gap absorption due to the two-fold degeneracy in the ground state associated with decreasing ground-state absorbance as a chemical charge-transfer doping proceeded [5].The mid-gap absorption was well explained by the theoretical work by Su, Schrieffer and Heeger [6], who employed a tight-binding model Hamiltonian modified with an electron-phonon coupling to investigate a spatial shape and an excitation energy of the soliton. Nonlinear spectroscopies were applied to this material [7–9], revealing that the third-order nonlinear susceptibility amounts to as large as $\sim 10^{-7}$ esu. The dynamical properties of the soliton excitations were also studied by means of quasi-static modulation spectroscopies [10–12] and time-resolved absorption and/or reflection spectroscopies with nanosecond [13], picosecond [14–17], and femtosecond [18–20] time-resolutions. It has been revealed that a photoexcited electron-hole pair in a chain created by a π-π^* interband transition is unstable to relax to an intrachain, charged soliton-antisoliton pair within an order of 100 fs [21], which recombines geminately after random walk showing a power-law behavior ($\sim t^{-1/2}$) of the photoinduced absorption (PA) signal [18]. A polaron pair is also created from an electron and a hole in different chains via interchain photoexcitation, and it decays in a nanosecond to microsecond time scale by an intersite hopping of polarons [13, 22].

Ultrafast Processes in Spectroscopy, edited by Svelto et al.
Plenum Press, New York, 1996

In nondegenerate conjugated polymers such as polydiacetylenes and polythiophenes, on the other hand, the soliton excitation does not occur because of an inherent strong confinement of photoexcitations in the forms of polarons, bipolarons, and STE's[23]. Particularly in typical polydiacetylenes the excitonic transition rather than the interband transition dominates the ground-state absorption, and a relaxation of the photoexcited one-dimensional excitons has been discussed in terms of the self-trapping process of the excitons [24–26]. The idea was applicable to photoexcited species in some derivatives of polythiophenes as well [25, 27]. The form of photoexcitations and their relaxation processes are strongly dependent upon both the main-chain structure and the conjugation length of the polymer. Therefore, a systematic study on polymers with different main-chain and/or side-group structures is, therefore, indispensable to understand a microscopic mechanism underlying a various nonlinear optical properties of this group of one-dimensional semiconductors.

In this paper we have further applied the scheme to a group of the substituted polyacetylenes, i.e. poly[(4-tert-butyl-2,6-dimethylphenyl)acetylene](hereafter abbreviated as PMBPA) and poly[(o-isopropylphenyl)acetylene](PPPA), and measured the transient spectra over a wide spectral region covering from near-ultraviolet to near-infrared.

2. TRANSIENT SPECTRA OF PMBPA AND PPPA THIN FILMS

Figures 1 and 2 show difference absorption spectra of thin films of PMBPA and PPPA, respectively, measured at room temperature for several delay-times between the pump and probe pulses. The pump- and probe-polarizations were parallel to each other and the excitation density was ca. 2×10^{16} photons/cm^2 at 3.1 eV ($\lambda = 0.40$ μm) for both samples.

In both samples a broad PA is observed below the stationary absorption band. A strong photoinduced bleaching of the ground-state absorption is also observed above 2.1 eV for PMBPA and above 2.3 eV for PPPA, though it is not spectrally resolved due to

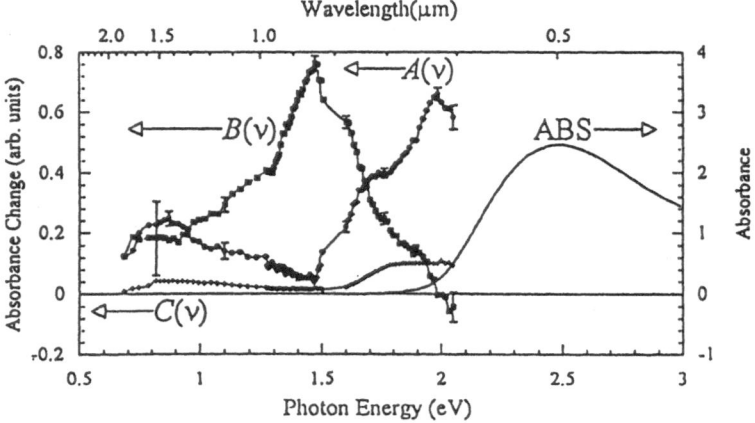

Figure 1. Difference absorption spectra of a thin film of PMBPA from visible to NIR region plotted at several delay times. Measurements were done at room temperature, and the excitation density was ca. 2×10^{16} photons/cm^2 at 3.1 eV ($\lambda = 0.40$ μm). Stationary absorption spectrum (ABS) of the PMBPA film is also shown.

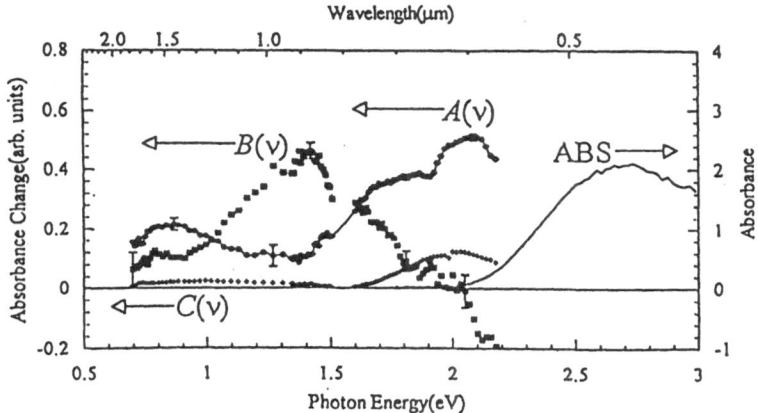

Figure 2. Difference absorption spectra of a thin film of PPPA. The experimental conditions were the same as those in Fig. 1. ABS denotes the stationary absorption spectrum of the PPPA film.

strong absorption of the samples. At early delay-times three PA peaks in regions of 0.8 - 0.9 eV, 1.4 - 1.5 eV, and 1.9 - 2.0 eV can be found for PMBPA. The middle PA peak at 1.4 - 1.5 eV rapidly decays, while the other two peaks rather grow after photoexcitation and then decay more slowly as compared to the middle peak, resulting in the time-dependent spectral shape clearly seen at delay times shorter than 1.0 ps. The transient spectra at longer delay-times are characterized by two broad peaks around 0.8 - 0.9 eV and 1.9 - 2.0 eV. A quite similar behavior of the transient spectra is observed for PPPA with some shift in energy. That is, a broad PA peaking at 0.8 - 0.9 eV, 1.4 - 1.5 eV, and 2.0 - 2.1 eV is observed at zero delay-time. The middle peak rapidly disappears within a picosecond, while the lower-and higher-energy peaks grow within ~ 0.3 ps after photoexcitation and decay more slowly, dominating the transient spectra at longer delay-times.

Figures 3 and 4 show time-dependence of the absorbance change for PMSPA and PPPA, respectively, at several probe photon energies. The temporal behavior depends strongly on the probe photon energies in both samples. In the region of 1.4 - 1.5 eV the PA instantaneously rises and then rapidly decays to a long-lived residual absorption within ~ 1.0 ps. The PA most likely decays exponentially in this region. As is seen in Figs. 1 and 2, the exponential and power-law components spectrally overlaps particularly at early delay-times. The energy-dependent temporal behavior imply a conversion of photoexcited states occurring within a sub-picosecond time scale. The dynamics can be drawn out by clarifying a decomposed spectrum proper to each decay component. The decomposed spectra corresponding to the soliton-antisoliton pair and hot STE were derived by fitting the time-dependence of the absorbance change (ΔA) to the following function at each probe photon energy v:

$$\Delta A(t,v) = A(v) \cdot erf\left[\left\{\sigma(t - t_d)\right\}^{-n}\right] + b(v) \cdot \exp(-t/\tau) + C(v) \tag{1}$$

Here erf [ ] is the error function, which term can be approximated with a power function of $(t - t_d)$ for $t - t_d > 1/\sigma$, and is associated with the spatially-confined soliton-antisoliton pair. Taking into account a finite time-resolution of the measurements, i.e. by con-

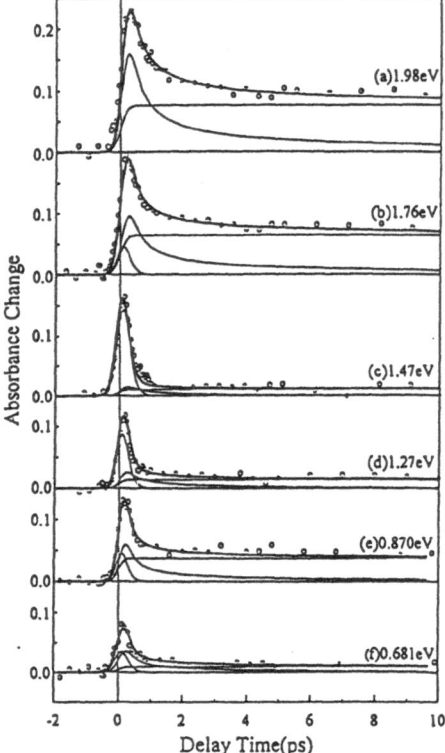

Figure 3. Time-dependence of the absorbance change in the PMBPA film shown at several probe photon energies. The circles are experimental data, and the fitted curves to Eq. (1) are shown with bold curves. The magnitudes of the three components in Eq. (1) are also shown with thin curves.

voluting the cross-correlation trace, a time-constant (τ) of the exponential decay was estimated. It was essentially the same within the region, and $\tau = 130 \pm 20$ fs for PMBPA and $\tau = 115 \pm 35$ fs for PPPA were evaluated. The exponential component is responsible for the middle PA peak around 1.4 - 1.5 eV observed at early delay-times in Figs. 1 and 2. Its instantaneous rise implies that the component is due to excited species which are directly created by photoexcitation. Furthermore it is commonly observed in various conjugated polymers even with different back-bone structures such as polydiacetylenes [24–26], polythiophenes [25–27], polyphenylacetylene [28], polythienylenevinylenes [29], and so on. It is, therefore, not relevant to the ground-state degeneracy just as the case for a soliton formation. These experimental facts enable us to attribute it to a hot STE, i.e. a nonthermal bound electron-hole pair captured in a lattice distortion, formed via the intrachain photoexcitation. The decay dynamics of the one-dimensional excitons in nondegenerate systems has been widely discussed so far in terms of the self-trapping process of excitons [26]. It is supposed that a free exciton is quickly coupled to the lattice vibrations (especially to carbon-carbon stretching modes of the main-chain) within a few phonon periods of order 10 - 20 fs (self-trapping). The STE is still vibrationally excited (hot STE) and decays to a thermal STE. The successive monomolecular decay processes of the STE's are observed as a sum of a few exponential functions, though the initial self-trapping process is too fast to be time-resolved. In conjugated polymers with a ground-state degeneracy, however, the hot STE is quite unstable toward an energetically favored soliton-antisoliton pair before the phonon emission takes place. The spatial separation of the electron and hole on the same chain vanishes the hot STE and then forms the soliton-antisoliton pair

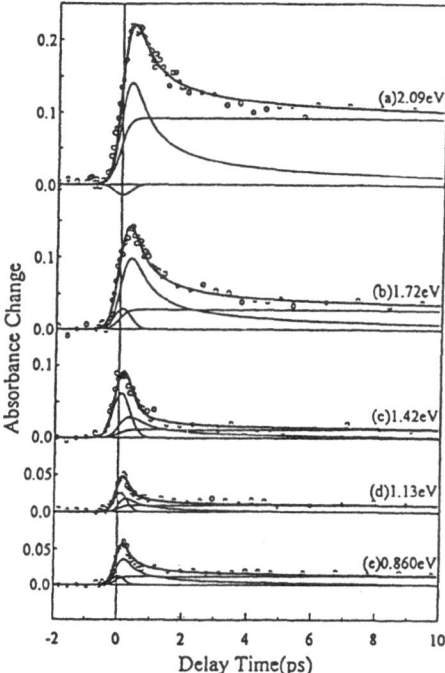

Figure 4. Time-dependence of the absorbance change in the PPPA film shown at several probe photon energies. The circles are experimental data, and the fitted curves to Eq. (1) are shown with bold curves. The magnitudes of the three components in Eq. (1) are also shown with thin curves.

discussed below. The exponential decay observed above results from a monomolecular disappearance of the hot STE.

In contrast, the lower- and higher-energy PA peaks respectively in the region of 0.8 - 0.9 eV and 1.9 - 2.0 eV for PMBPA and of 0.8 - 0.9 eV and 2.0 - 2.1 eV for PPPA slowly decay with a long trail lasting up to 10 ps or longer. The temporal behavior well follows a power-law decay ($\sim t^{-n}$) rather than the exponential decay, as seen by logarithmic plots up to 100 ps in Fig. 5 where the PA at 1.92 eV for PMBPA and that at 2.12 eV for PPPA are shown as examples. The power n was estimated from the slope in the figure as $n = 0.78 \pm 0.07$ for PMBPA and $n = 0.86 \pm 0.07$ for PPPA, and they were almost the same within the above region. The power-law behavior is known to be characteristic of a geminate recombination process [30] between two excited species which randomly walk in a quasi-one-dimensional chain. It was observed in the PA signal of a *trans*-polyacetylene thin film [18] and also reported for PMSPA in our previous work [31]. The power-law behavior observed in a thin film of PMSPA has been attributed to an intrachain, oppositely-charge, overall-neutral, spatially-confined soliton-antisoliton pair which is formed after a photogenerated electron-hole pair spatially separates from each other. The nearly degenerate ground-state conformation in PMBPA and PPPA makes the soliton formation possible as in another polyphenylacetylene previously studied [28], though the soliton and antisoliton might be confined within a segmented conjugation chain or bound to each other by the Coulomb attractive force between them [32].

In conclusion we could explain the experimental results of two phenylpolyacetylene PMBPA and PPPA together with a previously studied PMSPA in terms of the STE and confined soliton-antisoliton pair in nearly degenerate systems. These systems are considered to form an intermediate case with weak nondegeneracy between the degenerate system of *trans*-polyacetylene and strongly degenerate polydiacetylenes.

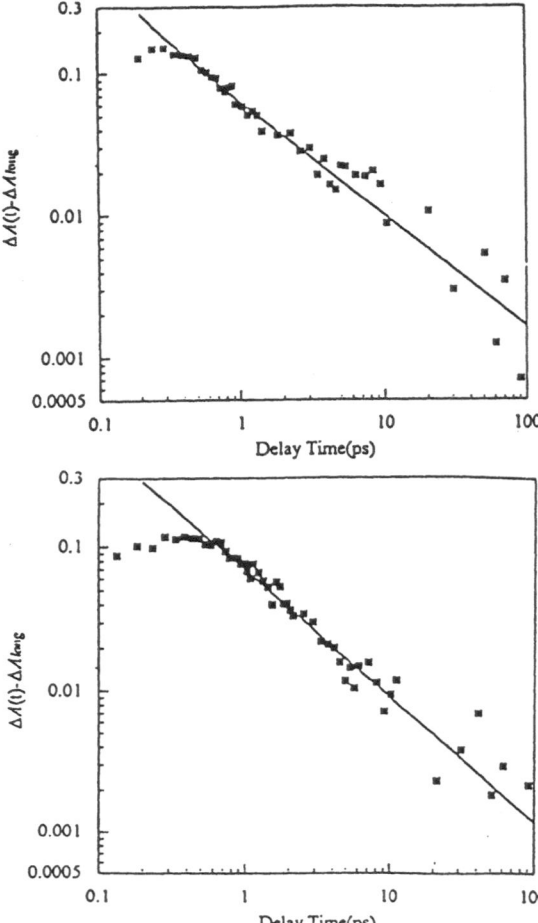

Figure 5. Logarithmic plots of the time-dependence of the absorbance change at 1.92 eV for PMBPA (top) and at 2.12 eV for PPPA (bottom). The power-law behavior is found up to a delay-time of 100 ps. The power constant n estimated from the slope is 0.78 ± 0.07 and 0.86 ± 0.07 for PMBPA and PPPA, respectively.

ACKNOWLEDGMENTS

The author is grateful to Dr. S. Takeuchi for his essential contribution in both experiment and discussion and to Prof. T. Masuda for sample preparation. This work is supported in part by the Grant-in-Aid for Specially Promoted Research No. 05102002.

REFERENCES

1. T. Kobayashi ed.: *Nonlinear Optics of Organics and Semiconductors*, Springer Proceeding Physics, Vol. **36** (Springer, Berlin, Heidelberg 1989).
2. T. Kobayashi ed., *Relaxation in Polymers* (World Scientific, Singapore, 1993).
3. A. J. Heeger, S. Kivelson, J. R. Schrieffer, and W. P. Su, Rev. Mod. Phys. **60**, 781 (1988).
4. W. P. Su, J. R. Schrieffer, and A. J. Heeger, Phys. Rev. Lett. **42**, 1698 (1979).
5. A. Feldblum, J. H. Kaufman, S. Etemad, A. J. Heeger, T. C. Chung, and A. G. MacDiarmid, Phys. Rev. **B26**, 815 (1982).
6. W. P. Su, J. R. Schrieffer, and A. J. Heeger, Phys. Rev. **B22**, 2099 (1980).

7. C. Halvorson and A. J. Heeger, Chem. Phys. Lett. **216**, 488 (1993).

8. W. S. Fann, S. Benson, J. M. J. Madey, S. Etemad, G. L. Baker, and F. Kajzar, Phys. Rev. Lett. **62**, 1492 (1989),

9. S. D. Phillips, R. Worland, G. Yu, T. Hagler, R. Freedman, Y. Cao, V. Yoon, J. Chiang, W. C. Walker, and A. J. Heeger, Phys. Rev. **B40**, 9751 (1989).

10. J. Orenstein and G. L. Baker, Phys. Rev. Lett. **49**, 1043 (1982).

11. N. F. Colaneri, R. H. Friend, H. E. Schaffer, and A. J. Heeger, Phys. Rev. **B38**, 3970 (1998).

12. P. D. Townsend and R. H. Friend, Phys. Rev. **B40**, 3112 (1989).

13. M. Yoshizawa, T. Kobayashi, H. Fujimoto, and J. Tanaka, J. Phys. Soc. Japan **56**, 768 (1982).

14. Z. Vardeny, J. Strait, D. Moses, T. C. Chung, and A. J. Heeger, Phys. Rev. Lett. **49**, 1657 (1982).

15. Z. Vardeny, Physica, **127B**, 338 (1984).

16. L. Rothberg, T. M. Jedju, S. Etemad, and G. L. Baker, Phys. Rev. Lett. **57**, 3229 (1985).

17. L. Rothberg, T. M. Jedju, S. Etemad, and G. L. Baker, Phys. Rev. **B36**, 7529 (1987).

18. C. V. Shank, R. Yen, R. L. Fork, J. Orenstein and G. L. Baker, Phys Rev. Lett. **49**, 1660 (1982).

19. C. V. Shank, R. Yen, J. Orenstein, and G. L. Baker, Phys. Rev. **B28**, 6059 (1983).

20. L. Rothberg, T. M. Jedju, P. D. Townsend, S. Etemad, and G. L. Baker, Mol. Cryst. Liq. Cryst. **194**, 1 (1991).

21. W. P. Su and J. R. Schrieffer, Proc. Natl. Acad. Sci. USA, **77**, 5626 (1980).

22. H. Y. Choi and E. M. Conwell, Mol. Cryst. Liq. Cryst. **194**, 23 (1991).

23. S. A. Brazovskii and N. N. Kirova, JETP Lett. **33**, 4 (1981).

24. M. Yoshizawa, M. Taiji, and T. Kobayashi, IEEE J. Quantum Electron. **QE-25**, 2532 (1989).

25. T. Kobayashi, M. Yoshizawa, U. Stamm, M. Taiji, and M. Hasegawa, J. Opt. Soc. Am. **B7**, 1559 91990).

26. M. Yoshizawa, A. Yasuda, and t. Kobayashi, Appl. Phys. **B53**, 296 (1991).

27. U. Stamm, M. Taiji, M. Yoshizawa, K. Yoshino, and T. Kobayashi, Mol. Cryst. Liq. Cryst. **182A**. 147 (1990).

28. S. Takeuchi, T. Masuda, and T. Kobayashi, to be published in Phys. Rev. **B52**, 7166 (1995).

29. S. D. Halle, M. Yoshizawa, H. Murata, T. Tsutusi, S. Saito, and T. Kobayashi, Synth. Met. **49–50**, 429 (1992).

30. I. V. Zozulenko, Solid State Comm. **76**, 1035 (1990).

31. S. Takeuchi, M. Yoshizawa, T. Masuda, T. Higashimura, and T. Kobayashi, IEEE J. Quantum electron. **QE-28**, 2508 (1992).

32. M. Grabowski, D. Hone, and J. R. Schrieffer, Phys. Rev. **B31**, 7850 (1985).

96

FEMTOSECOND RELAXATION DYNAMICS IN THIOPHENE OLIGOMERS

G. Lanzani,[1] M. Nisoli,[2] and S. De Silvestri[2]

[1] Institute of Mathematics. and Physics
University of Sassari, Sassari, Italy
[2] Quantum Electronics Center
Physics Department
Polytechnic, Milan, Italy

Using methyl substituted thiophene rings as starting monomers for controlled polymerization interesting materials are obtained whose optical properties depend on the length and the regiospecificity of substitution. The ground state conformation is determined by the opposite roles exerted by π-electron conjugation (favoring coplanar arrangements) and the side groups non bonded interactions (favoring twisted arrangements). Solubility is quite good for these systems even for fairly large conjugation (6 units). In this paper we present a complete characterization of the ultrafast optical dynamics of α-conjugated regiospecifically substituted $3,3',4'',3''',4'''',3'''''$ hexamethyl $2,2':5',2'':5'',2''':5''',2'''':5'''',2'''''$ sexithiophene(HMT$_6$)[1]. Femtosecond pump-probe experiments were carried out on HMT$_6$ in chloroform solution (10^{-3} mol/l) in a cuvette of 1 mm thickness. The sample was excited by 390 nm femtosecond pump pulses with 10 nJ energy. The pump beam was focused at the sample to a spot of 250 μm diameter. The probe wavelength was varied in a wide spectral region ranging from 490 nm to 1500 nm. We used a Ti:Sapphire oscillator in Kerr-lens modelocking followed by chirped pulse amplification. The pump pulses were obtained by frequency doubling in a LiB$_3$O$_5$ crystal (1 mm length) a fraction of the fundamental beam at 780 nm. Probe tunability from 490 nm to 1000 nm was obtained by white light supercontinuum generation. Probe pulses tunable in the near infrared region up to 1500 nm were obtained by using a three-pass optical parametric generation and amplification system (OPA), which consists of two angle tuned β-BaB$_2$O$_4$ (BBO) crystals pumped by the laser beam at 780 nm[2].

The cw absorption spectrum of HMT6 in chloroform solution, shown in Fig. 1 is structureless and peaks at 370 nm, blue shifted with respect to unsubstituted sexithiophene peaking at 432 nm[3]. This is consistent with a reduction of $\pi-\pi$ conjugation due to a non planar conformation of the molecular skeleton. The cw photoluminescence (PL) spectrum, excited at 390 nm, peaks at 498 nm and it has an asymmetric shape probably originating from the envelope of the vibronic structure. The large apparent Stokes shift between the

Ultrafast Processes in Spectroscopy, edited by Svelto et al.
Plenum Press, New York, 1996

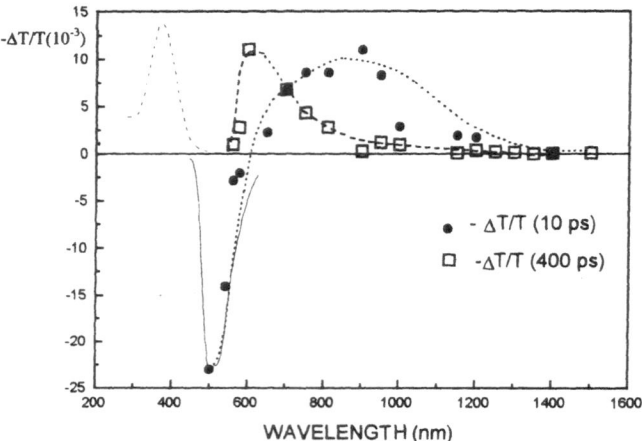

Figure 1. Differential transmission spectra (-ΔT/T) of HMT$_6$ at pump-probe delays of 10 ps (solid circle) and 400 ps (open square). Dotted lines through the data points are guides to the eye. The cw absorption spectrum (dashed line) and the stimulated emission band as obtained from cw PL (solid line) are also shown. PA1 and PA2 are assigned in the text.

absorption and emission maxima (twice that of unsubstituted oligomers) indicates that substantial energy relaxation is taking place before emission. The PL quantum yield is 0.1, evaluated by comparison with a quater-thiophene sample.[4]

From a series of pump-probe data measured at different probe wavelengths, we have obtained a time dependent differential transmission spectrum -ΔT/T. In Fig.1 we show -ΔT/T spectra for two significant probe time delays namely 10 ps and 400 ps. Let us first discuss the data reported at 10 ps probe delay. We observe a transmission increase from 500 nm to 560 nm and a photoinduced absorption (PA1) which starts to develop at wavelength longer than 560 nm and apparently peaks at 900 nm. Since cw absorption is negligible below 490 nm we assign the transmission increase to stimulated emission (SE) from transient population inversion between the vibrational manifold of the radiative state and that of the ground state. The SE spectrum matches reasonably well that of the cw PL. This means that SE and cw PL come from the same state which we assign to the lowest (dipole allowed) excited singlet state S$_1$. The -ΔT/T spectrum for large probe delay (400 ps) shows (see Fig.1) a rather narrow PA band peaking at 650 nm (PA2) while both the SE and the broad absorption band around 900 nm, observed at small probe delay, are absent. No significant absorption features are present at wavelength up to 1500 nm.

The observed behavior of -ΔT/T(λ,t) can explained taking into account three simultaneous phenomena: (i) SE from the hot and thermalized optically excited S$_1$ state; (ii) PA from S$_1$ to higher lying singlet states; (iii) PA from a new state which is associated to PA2. A high lying vibronic state of S$_1$ is initially populated by the pump photons at 390 nm. Subsequent energy relaxation down to the bottom of S$_1$ occurs on a rather long time scale as can be monitored by the SE dynamics. The SE risetime observed at 540 nm is reported in Fig.2 together with the pump-probe cross-correlation.

The transient data taken throughout the SE band (not shown) are consistent with a red shift of the whole gain band as occurring due to a spectral relaxation of the emissive state[5]. In Ref.6 the thermalization kinetics has been modeled by a two step process which

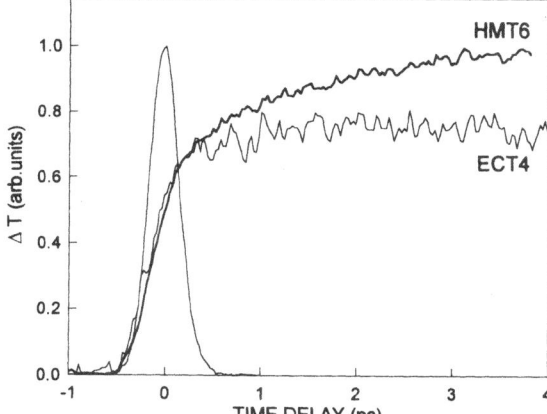

Figure 2. Stimulated emission build up at $\lambda=540$ nm (thick line) for HMT_6 and at $\lambda=550$ nm (thinner line) for ECT_4. The cross-correlation function between pump and probe (at 540 nm) is also shown.

leads to a more planar excited state geometry. In the first step, dissipation of the vibrational quanta takes place with a characteristic time constant of 850 fs. This results into a new equilibrium position associated to a larger contribution of the quinoid form to the dimerization pattern. In the second step, driven by the new quinoid structure, where double bonds link adjacent rings, the molecule reaches the minimum of the potential surface by dissipating quanta of inter-ring motion. This last process is described by a time constant of 4 ps. Because of the large difference between the characteristic frequencies, temporal separation between vibrational and torsional relaxation phenomena is justified. We note that this rather fast spectral change might have some applications in optical switching devices. In order to further support our relaxation model in Fig.2 we also show the SE dynamics of end-capped quater-thiophene[7] (ECT_4) which is almost *planar* in chloroform solution, as demonstrated by the absorption band maximum[8]. In this case *we do not see, as expected, any slow formation time*. A slighlt delay in the SE formation, of about 600 fs, maybe assigned to vibrational redistribution. Once the intraband relaxation process is accomplished, SE at 540 nm decays monoexponentially with a time constant of 140 ps. Alternative informations on S_1 lifetime τ_s can be obtained by studying the kinetics of PA1 which we assign to $S_1 \rightarrow S_n$ transitions. A value of $\tau_s=160$ ps can be obtained from the fitting of the decay curves in the whole band. From the measured PL quantum yield and by using the value for τ_s, estimated above, we obtain a radiative lifetime of 1.6 ns.

At 400 ps probe delay the $-\Delta T/T$ spectrum shows only one PA feature centered at 650 nm (PA2). This band is not singled out at shorter delays, being buried in the broad initial PA structure and possibly overwhelmed by SE at shorter wavelength. Consequently its dynamics is not directly measurable. However the transient data shows that 700 nm is an isosbestic point in the PA spectrum, indicating that: i)PA2 is formed at the expenses of PA1; ii) PA2 formation time is 160 ps (i.e. the decay of PA1). Since the lack of any PA band of comparable intensity with PA2 in the $-\Delta T/T$ spectrum rules out the formation of charged states[9] PA2 can be assigned to triplet state absorption $T_1 \rightarrow T_n$. The high inter system crossing efficiency can be accounted for by the heavy atom effect due to sulfur heteroatom, which is common to the whole thiophene series, and the high torsional mobility peculiar of the substituted systems. Moreover the planarization of the excited state implies a large displacement of the equilibrium position with respect to the ground state increasing the Frank-Condon overlaps.

REFERENCES

1. G. Barbarella, A. Bongini, and M. Zambianchi, Tetrahedron **48**, 6701 (1992).
2. M. Nisoli, S. De Silvestri, V. Magni, O. Svelto, R. Danielus, A. Piskarskas, G. Valiulis, and A. Varanavicius, Opt. Lett. **19**, 1973 (1994).
3. H. Chosrovian, S. Rentsch, D. Grebner, D. U. Dahm, E. Birckner, and H. Naarmann, Synth. Met. **60**, 23 (1993).
4. Y. Kanemitsu, K. Suzuki, Y. Masumoto, Y. Tomiuchi, Y. Shiraishi and M. Kuroda, Phys. Rev. **B50**, 2301 (1994).
5. G. Lanzani, M. Nisoli, S. De Silvestri, and R.Tubino, Synth. Met. in press.
6. G. Lanzani, M. Nisoli, S. De Silvestri, R. Tubino, G. Barbarella, and M. Zambianchi, Phys. Rev **B51**, 13770 (1995).
7. G. Barbarella, M. Zambianchi, A. Bongini and L. Antolini, Adv. Mater. **4**, 282 (1992).
8. G. Lanzani, M. Nisoli, S. De Silvestri, R. Tubino, Chem. Phys. Lett., in press.
9. D. Fichou, G. Horowitz, and F. Garnier, Synth. Met. **39**, 125 (1990).

97

SUB-PICOSECOND OPTICAL NON-LINEARITIES IN EXCITED STATES OF DIPHENYL-POLYENES AND "PUSH-PULL" POLYENES

J. Oberlé, G. Jonušauskas, E. Abraham, and C. Rullière

CPMOH, Université Bordeaux I
U.A. CNRS N°283
351 Cours de la Libération
33405 Talence Cedex, France

Diphenyl-polyenes are known for their large third order optical non-linearity $\chi^{(3)}$ in the ground state [1]. This property is attributed to delocalisation of π-electron system along the polyenic chain. Recently the possibility to obtain larger optical non-linearities on populating excited states of organic molecules was demonstrated for diphenyl-hexatriene [2]. The $\chi^{(3)}$ enhancement is due to a more important delocalisation of π-electron system in excited states. Possible changes of excited state symmetry as compared with ground state can occur and influence value of $\chi^{(3)}$. Substituted polyenes (such as "push-pull" stilbenes for example) may show important enhancement of optical non-linearities in excited states as well. Charge transfer processes may indeed play an important role in strong delocalisation of π-electron system and, as a consequence, in the enhancement of optical non-linearities of the excited states.

We have studied third order non-linear optical properties in excited states of a series of diphenyl polyenes with increasing chain lengths (*trans*-stilbene (tS), 1,4 diphenyl-butadiene (DPB), 1,6 diphenyl-hexatriene (DPH) and 1,8 diphenyl-octatetraene (DPO)) and two "push-pull" polyenes (4-dimethylamino-4'-cyano-t-stilbene (DCS) and 4-dimethylamino-4'-cyano-1,4-diphenyl -butadiene (DPB)).

The technics were Degenerate Four Wave Mixing (DFWM) [3] and Optical Kerr Ellipsometry [4]. In both experiments, the second harmonics (300 nm) of an amplified hybridly mode-locked dye laser pulse was used to create excited state population.

In DFWM experiment, two pump beams orthogonally polarised (600 nm, pulse energy $\approx$ 3–4 μJ and pulse duration $\approx$ 500 fs) are used to form a transient polarisation grating in the sample. To test the grating, a third probe beam (600 nm, pulse energy $\approx$ 1 μJ) is sent to the sample in non-coplanar configuration and diffracted by the grating in a direction which corresponds to momentum conservation. Changing the optical delay of the probe

Ultrafast Processes in Spectroscopy, edited by Svelto et al.
Plenum Press, New York, 1996

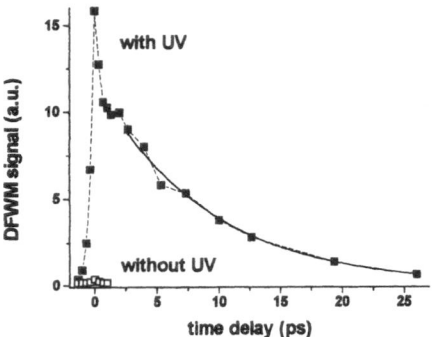

Figure 1. Time evolution of the diffracted signal as a function of the probe delay for excited (the UV pulse delay Δt_{UV} was 10 ps) and unexcited (without UV pulse) solution of *trans*-stilbene (10^{-3}M in heptane).

pulse relative to the 600 nm pump pulses allows measurement, as a function of time, of the intensity-grating signal which is proportional to $|c^{(3)}(-w;w,-w,w)|^2$.

Figure 1 shows that strong enhancement can really be observed in excited states in our experimental conditions. Our evaluations of an effective $\chi^{(3)}$ of excited diphenyl-polyenes (concentration 10^{-3} M/l) compared to CS_2 are presented in table 1 for different compounds [5]. We have compared the enhancements at constant excited state population and corrected the values given in table 1 for the reabsorption process at 600 nm [5]. It is necessary to point out that in all studied materials DFWM was in resonance conditions for excited states, and results were influenced by electronic transitions [5]. The maximum enhancement is obtained for DPB, this enhancement decreasing slightly for t-St and more strongly for DPH and DPO. In our experimental conditions, the signal does not increase with the chain length as in the ground state. Resonance conditions might be taken into account to explain this behaviour. But the resonance conditions are not drastically different for t-St, DPB and DPH. Only DPO is out of resonance [5]. For t-St, DPB and DPH resonance cannot explain such large differences. Depending on the symmetry of the lowest excited state, more or less strong enhancement can be expected in diphenyl-polyenes derivatives, B_u symmetry being in favour of a stronger enhancement as theoretically shown in trans-octatetraene and diphenyl-hexatriene [6,7].

Optical Kerr Ellipsometry allows to extract the spectral dependence of the real (bire-fringence) and imaginary (dichroism) part of the optical non-linearity [4]. In this experiment the pump light (UV pulse) is linearly polarized and the sample is placed between two polarisers. The entrance polariser polarises the probe light (a continuum of light ($\approx$1ps) extending from 450 to 650 nm,) at angle ψ with respect to the pump polarisation. The analyser behind the sample is oriented at angle θ with respect to the pump polarisation. For a classical Kerr gate configuration, the pump polarisation makes angles of +45° and -45° with entrance and analyser polarisers respectively (ψ=-45°, θ=45°). Now, if the analyser is twisted by a small angle α, the transmitted probe intensity through the analyser at frequency ω_c and time τ after pumping is given by [4]:

$$T(\omega_c,\tau) = E^2 \exp\left[-2\phi''(\omega_c,\tau)\right]\left\{\left[\Delta\phi''(\omega_c,\tau)-\alpha\right]^2 + \left[\Delta\phi'(\omega_c,\tau)\right]^2\right\} \quad (1)$$

Table 1. $\chi_{exc}^{(3)}/\chi_{CS_2}^{(3)}$ is the ratio between the DFWM signal observed at 600 nm in the studied solution and pure CS_2 sample and normalised to an equivalent concentration of 10^{-3} M

Polyenic chain length	Molecule	$\left\| \chi^{(3)}(10^{-3}M)/\chi_{CS_2}^{(3)} (pur) \right\|$	Symmetry of the lowest excited state
1	tS	0.3	Bu
2	DPB	0.5	Bu
3	DPH	0.04	Ag
4	DPO	0.005	Ag

where $\Delta\phi''(w_c,t)$ and $\Delta\phi'(w_c,t)$ are the imaginary and real parts of phase delay, and E^2 is the intensity of the probe. The analysed signal in Eq.1 is a quadratic function on α whose minimum abscissa is the pump induced dichroism $\Delta\phi''(w_c,t)$ and whose minimum ordinate is the pump induced birefringence magnitude $\left|\Delta\phi'(w_c,t)\right|$. Measurement of T for at least three different angles α provides angles $\Delta\phi''$ and $\left|\Delta\phi'\right|$ as solutions of eq. 1 without knowing the incident probe intensity E^2, detector sensitivity and sample transmission $\exp\left[-2\phi''(w_c,t)\right]$ in either the ground nor the excited state.

Our Kerr Ellipsometry measurements confirm that the strongest non-linearities concerning the diphenyl-polyenes series were obtained for DPB and the same dependence on chain length was observed as in the DFWM experiments [5]. Nevertheless higher values of induced birefringence were measuerd for the excited push-pull molecules as shown in figure 2. In these molecules the strong enhancement of the non-linear response amplitude can be correlated to the charge transfer which occurs between donor and acceptor electron substituents in a picosecond time-scale.

DFWM and Kerr Ellipsometry, with subpicosecond time resolution, appears to be powerful methods for photophysical studies. They allow not only to measure amplitude and kinetic of effective $\chi^{(3)}$ in excited states, but also give information about temporal behaviour of excited states related to molecular transformations.

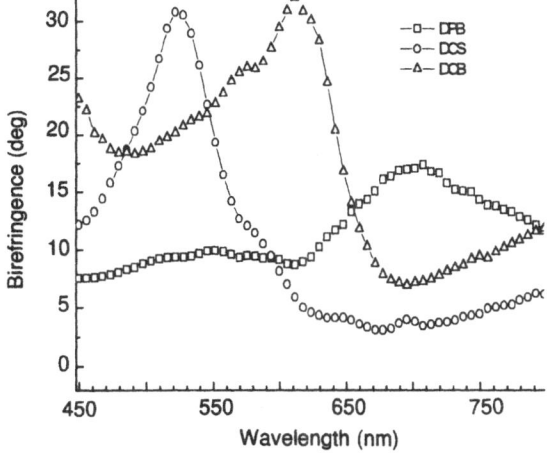

Figure 2. Spectral dependence of the induced birefringence (real part) for the different molecules, using OKE .The UV pulse delay Δt_{UV} is 20 ps.

REFERENCES

1. C. Maloney, H. Byrne, W. M. Dennis, W. Blau, J. M. Kelly, Chem.phys., **121** (1988) 21.
2. Q. L. Zhou, J. R. Heflin; K. Y. Wong, O. Zamani-Khamiri, A. F. Garito, Phys.Rev. A, **43** (1991) 1673.
3. J. Oberlé, G. Jonusauskas, E. Abraham, C. Rullière, Chem. Phys. Lett., **241**, (1995) 281.
4. N. Pfeffer, F. Charra, J. M. Nunzi, Opt.Lett., **16** (1991) 1987.
5. J. Oberlé, G. Jonusauskas, E. Abraham, C. Rullière, "Third-Order Optical Non-Linearities Of Excited States In Diphenyl-Polyene Derivatives: A Sub-Picosecond Study", Opt. Comm., in press.
6. Q. L. Zhou, J. R. Heflin, K. Y. Wong, O. Zamani-Khamari, A. F. Garito, Phys. Rev. A **43** (1991) 1673
7. J. L. Brédas, C. Adant, P. Tackx, A. Persoons, B. M. Pierce, Chem. Rev. **94** (1994) 243.

98

ULTRAFAST OPTICAL PROBES OF EXCITED STATES IN CONDUCTING POLYMERS

Z. V. Vardeny and S. Frolov

Department of Physics
University of Utah
Salt Lake City, Utah 84112

Within the theoretical picture of excitonic energy levels in π-conjugated polymers[1-3], the photo-luminescence (PL) efficiency γ and the third-order nonlinear optical properties are determined by the energy locations and nature of a subset of the electronic excited states, including a series of singlet excitons with odd (B_u) and even (A_g) parity symmetries lying below the continuum threshold band. In particular, the relative energies of the dipole forbidden character of the lowest B_u ($1B_u$) and A_g ($2A_g$) excitons determine γ: if $E(2A_g) < E(1B_u)$, then γ is small because of the dipole forbidden character of the lowest singlet. Conversely, for $E(2A_g) > E(1B_u)$, γ is large.

In this paper we show how various picosecond optical probe techniques can elucidate the nature of the electronic excited states and primary photoexcitations in luminescent and non-luminescent π-conjugated polymers, in relation to their PL efficiency.

The ultrafast techniques reviewed here are the transient photomodulation (PM), PL, and the transient strain spectroscopy. Most of the picosecond (ps) transient responses have been measured by the pump and probe correlation technique using two dye lasers synchronously pumped by a frequency-doubled modelocked Nd:YAG laser at a repetition rate of 76 MHZ with 5 ps resolution. The PL transient was measured by a streak camera with 10 ps resolution[4]. For the PM measurements, transient spectra of photoinduced changes ΔT in transmission T were obtained by fixing the pump wavelength at 570 nm (2.17 eV) and varying the probe wavelength between 1.2 and 2.2 eV.

We will discuss separately the two main classes of the π-conjugated polymers: luminescent and non-luminescent. As a representative of the luminescent class we have chosen the poly(2,5-di hexyl paraphenylene-acetylene) (PPA), for which PL and electroluminescence with relatively high efficiencies have been recently demonstrated[5]. The non-luminescent class is represented by poly (2,5-thienylene-vinylene) [PTV][6]. Both polymers were in the form of thin films deposited on glass substrates.

Ultrafast Processes in Spectroscopy, edited by Svelto et al.
Plenum Press, New York, 1996

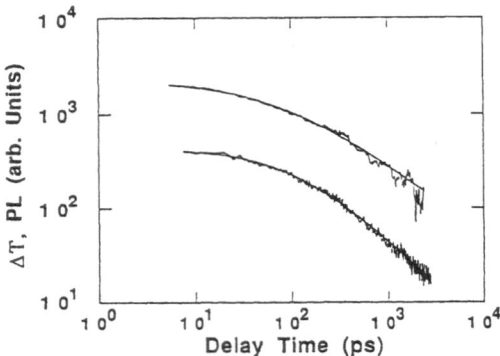

Figure 1. Ps transient decays of PA (at 1.8 eV) and PL (at 2.1 eV), respectively, in PPA at 300 K.

PS TRANSIENTS IN PPA

The PA and PL transient decays in PPA at 300K are shown in Fig. 1; the two transients decay together up to 3 ns. We can fit the decays with a functional form: $[1+(t/\tau)^p]^{-1}$, with $\tau \simeq 80\pm5$ ps and $p \simeq 0.8\pm0.1$. This decay form is borrowed from that observed in disordered materials[7]. It is typical of dispersion recombination which is probably associated with the exciton inhomogeneity in the PPA film. The similarity of the PA and PL decays in Fig. 1 leaves little doubts that in PPA they share a common origin, namely they both originate from the photogenerated excitons. We believe that the ps response of all polymers with large γ is dominated by photogenerated excitons. The different ps PL and PA decays found in PPV[8] can be simply caused by different radiative recombination pathways. These may be associated with impurities and defects known to strongly influence the PL transient response but less so the PA transient.

PS TRANSIENTS IN PTV

The transient PM signal in PTV films is composed of three ΔT components: a very fast component ΔT_a, decaying within several ps, a slow component, ΔT_b, with a time constant of order 10 ns and an oscillatory component, ΔT_c, with a decay time of several hundreds ps. Typical decay of the fast component ΔT_a at 625 nm measured with 100 fs time resolution is shown in Fig. 2, $\Delta T_a>0$ and it decays within 4 ps into a persistent plateau. However, within about 150 ps ΔT changes sign into PA and shows an oscillatory response (ΔT_c) riding of top of a slow PA signal (ΔT_b), as shown in Fig. 3. We can observe in Fig. 3 about 12 periods of oscillation in ΔT_c, with period τ_{os} of about 200 ps. We found that τ_{os} depends linearly on the sample thickness d, and this indicates that it is caused by the strain waves launched in the film following the laser pulse absorption[9]; the ps oscillation and its associated spectrum have been therefore dubbed transient strain spectroscopy (TSS)[9].

The PM spectra associated with ΔT_a (t=o) and ΔT_b (t=1ns) are shown respectively in Fig. 4(a). We believe that the PA band at 1.4 eV is an electronic response, most probably due to photogenerated $2A_g$ excitons. The PA feature at 1.8 eV, on the contrary, is due to photoinduced static strain in the film[9], because it does not carry any polarization memory.

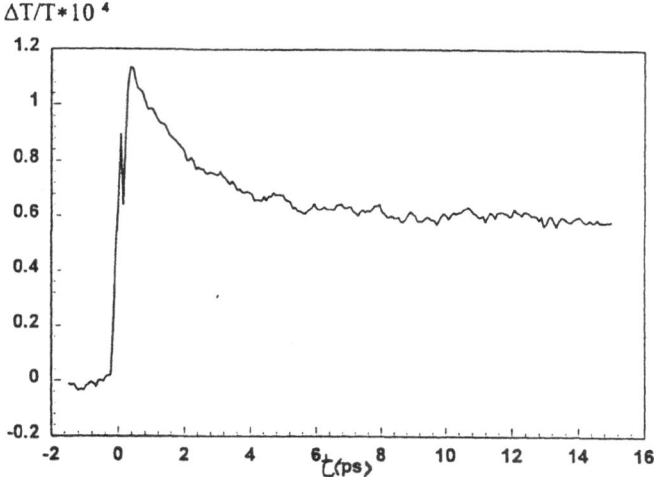

Figure 2. $\Delta T(t)$ in a PTV film at 625 nm up to 4 ps.

Its spectrum however, does not follow the first derivative of α, $\partial\alpha/\partial\omega$, but is similar to the electro-absorption spectrum of PTV[10]; the reason for this apparent similarity is not known at the present time.

The PM spectrum of ΔT_c component (TSS) is shown in Fig. 4(b). It follows almost exactly the spectrum of $\partial\alpha/\partial\omega$, as is theoretically predicted[9]. From the maximum in the TSS spectrum we can therefore extract the energy level of $1B_u$ in PTV; we find $E(1B_u) = 1.8$ eV.

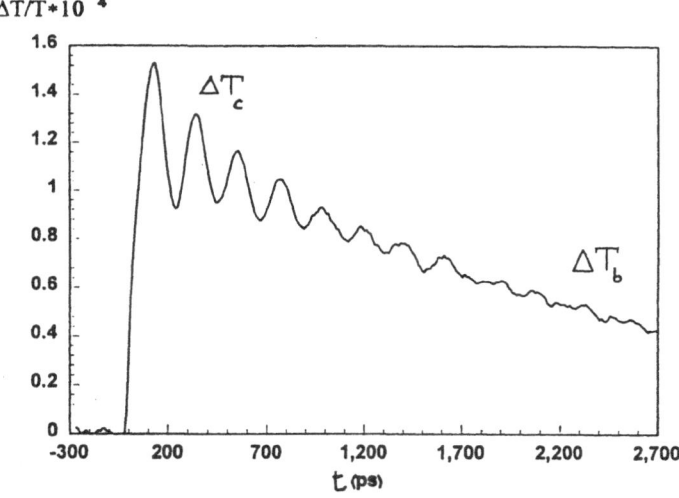

Figure 3. $\Delta T(t)$ in PTV at 625 nm up to 2700 ps.

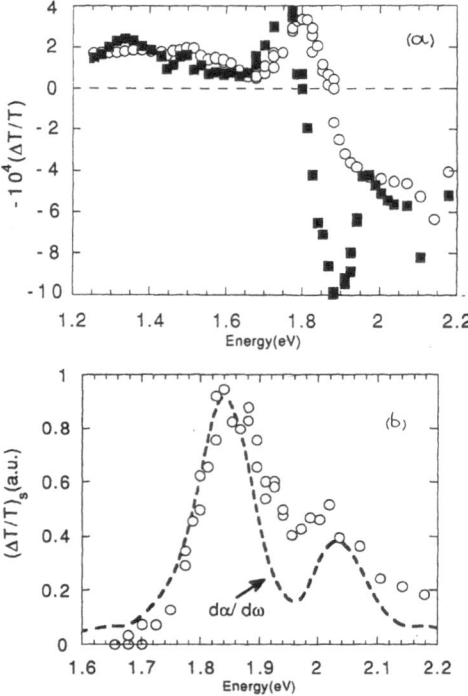

Figure 4. The ps spectra of the various ΔT components in PTV. (a) ΔT_a (t=o) and ΔT_b (t=1ns). (b) ΔT_c at 170 ps.

DISCUSSION AND SUMMARY

We showed here that the primary photoexcitations in both luminescent and non-luminescent π-conjugated polymers are *excitons*, with a strong PA band in the visible spectral range regardless of their respective γ. This band has been recently attributed[4] to an optical transition from the photogenerated excitons into a biexcitonic level. In luminescent polymers the PA occurs instantaneously and is from photoexcited $1B_u$ exciton into a biexcitonic state with A_g symmetry. In non-luminescent polymers the PA is formed following internal conversion of order 200 fs and is from photoexcited $2A_g$ excitons into a biexcitonic state with B_u symmetry.

We also showed that ps TSS is a powerful technique in detecting electronic states of both B_u and A_g symmetries[9]. In non-luminescent polymers it may be used to show that the small γ is associated with the reverse order of the electron excited states ($E(2A_g) < E(1B_u)$), rather than with defects in the material.

ACKNOWLEDGMENTS

We thank Y. Ding and T. Barton for the PPA and PTV samples, and G. Kanner and J.M. Leng for useful discussions. The Work supported by DOE, Grant No. FG-03–93ER45490 and ONR, 94–1–0853.

REFERENCES

1. S.N. Dixit, D. Guo, and S. Mazumdar, Phys. Rev. B 43, (1991) 6781.
2. A.G. Soos, S. Etemad, D.S. Galvo, and S. Ramasesha, Chem. Phys. Lett. 184, (1992) 341.
3. D. Guo et al., Phys. Rev. B 48, (1993) 1433.
4. J.M. Leng et al., Phys. Rev. Lett. 72, (1994) 156.
5. L.S. Swanson, J. Shinar, Y.W. Ding, and T.J. Barton, SPIE Conf. Proceedings 1910 (1993) 101.
6. A.J. Brassett et al., Phys. Rev. B 41 (1990) 10586.
7. G.S. Kanner, Ph.D thesis, University of Utah, 1991 (unpublished).
8. J.W. Hsu et al., Phys. Rev. B. 49 (1994) 712.
9. G.S. Kanner, S. Frolov, and Z.V. Vardeny, Phys. Rev. Lett. 74, 1685 (1995).
10. D.D.C. Bradley and O.M. Gelsen, Phys. Rev. Lett. 67, 2589 (1991).

99

FEMTOSECOND PUMP AND PROBE EXPERIMENTS ON FILMS AND SOLUTIONS OF CONJUGATED POLY(*PARA*-PHENYLENE)-TYPE LADDERPOLYMERS

W. Graupner,[1] G. Leising,[1] G. Lanzani,[2] M. Nisoli,[3] S. De Silvestri,[3] and U. Scherf[4]

[1] Institut für Festkörperphysik
Technische Universität Graz, Graz, Austria
[2] Istituto di Matematica e Fisica
Universita di Sassari, Sassari, Italy
[3] Quantum Electronics Center
Physics Department, Polytecnico, Milano,Italy
[4] Max-Planck-Institut für Polymerforschung
Mainz, Germany

The wide-bandgap semiconductor poly(*para*-phenylene) (PPP) was applied successfully as the active layer in blue light emitting devices[1]. The synthetic routes to PPP result in a material with a broad distribution of effective conjugation lengths due to a wealth of chemical defects created during synthesis. By incorporating a PPP-backbone into a soluble ladderpolymer-structure (LPPP)[2] a compound of high intrachain order[3], small Stokes shift[4], high photoluminescence quantum yield[5] and excellent film forming properties was synthesized (see Fig. 1), which was used in blue and yellow light emitting devices[6].

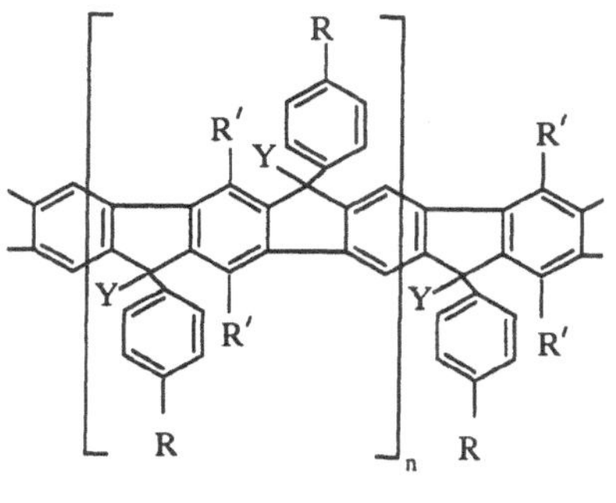

Figure 1. Chemical structure of m-LPPP; $Y=CH_3$, $R=C_{10}H_{21}$, $R'=C_6H_{13}$, $n_{average}=26$.

Ultrafast Processes in Spectroscopy, edited by Svelto et al.
Plenum Press, New York, 1996

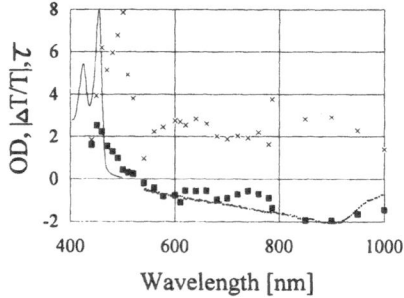

Figure 2. Maximum of 10ΔT/T (), τ (, in ps, see text), cw-PA spectrum (dashed, from ref. 6) and optical density (solid, OD) of a m-LPPP film.

The focus of this work is the nature of the primary photoexcitations in the well defined conjugated LPPPs. To study the influence of chain-chain interactions, time-resolved pump and probe experiments were conducted on films and solution of the LPPPs by the usual interferometric geometry. For the experiments we chose m-LPPP (methyl group at the Y-position of the methin bridge, Fig. 1) since it shows the highest degree of intrachain order[7] and the highest PL quantum yield[5] of the various chemical derivatives available from the LPPP family. The solution experiments were done in toluene while films were produced by casting the solution onto sapphire. The sample was excited by pulses of 120 fs duration at 390 nm with pulse energies of 10 nJ, which were derived by frequency doubling the output of an amplified Ti:Sapphire laser oscillator. Probe tunability was obtained using a white light continuum generated by focusing the fundamental beam into a flowing water cell. Desired wavelengths are obtained using 10 nm bandwidth interference filters or a monochromator. A polarization memory artifact is avoided by using an unpolarized probe beam. In order to obtain additional information about the photogenerated states polarization decay measurments are performed by placing a polarizer in the incident probe beam and an analyzer in the transmitted probe beam.

The spectral dependence of the relative change of transmission (ΔT/T) together with the decay time τ - obtained via measuring the slope of the initial decay of ΔT/T - for the film is shown in Figure 2. For probe wavelengths larger than 530 nm we observe photoinduced absorption (PA) in accordance with cw-photomodulation experiments[8], the result of which is also shown in Fig. 2. For probe wavelengths below 530 nm, ΔT/T is positive. It is seen that τ

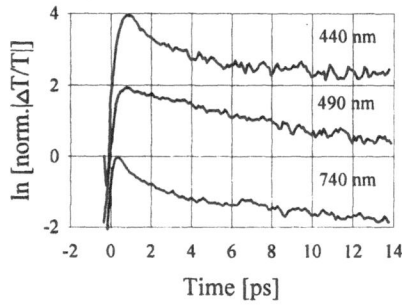

Figure 3. 15 ps traces of ΔT/T in a m-LPPP-film at selected probe wavelengths.

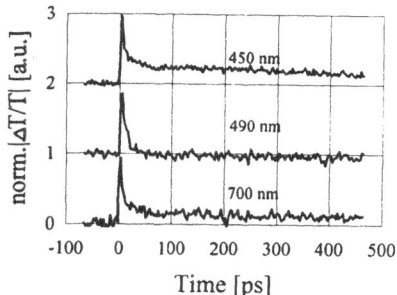

Figure 4. 500 ps traces of ΔT/T in a m-LPPP-film at selected probe wavelengths.

reaches a maximum of 8 ps in the region of around 490 nm, while it decreases to 2 ps at 440 nm (absorbing region of the sample) and 530 nm (region of PA). From the traces of ΔT/T at selected probe wavelengths it can be seen, that (i) the initial decay follows a single exponential behaviour only around 490 nm (Fig. 3) and that (ii) only in this region no long living component (Fig. 4) is seen. Since the interband PL occurs in this spectral region and since no ΔT/T signal (i.e. no spontaneous PL emission) is observed for the blocked probe beam, we assign the positive ΔT/T around 490 nm to stimulated emission (SE). The different ΔT/T-kinetics in the absorbing region of the sample (λ<460 nm) identifies the positive ΔT/T-signal between 440 and 460 nm to be due to photobleaching (PB).

In solutions we find the same qualitative behaviour as in films (Fig. 5). The decay time τ, describing the single exponential decay at the probe wavelength of 490 nm (SE), rises up to 190 ps in the solution (Fig. 6), indicating a 24-fold increase of the quantum yield for SE in solution. This increase in quantum yield results in a very strong SE-signal relative to the PB signal. Therefore one can see a fairly good spectral match of the fs-SE-signal and the cw-PL spectrum in m-LPPP solutions (Fig. 5).

Both in solution and in the film we find *no spectral overlap* of the stimulated emission with photoinduced absorption in the timescale of 0 to 500 ps. This is very important since such an overlap was observed in poly(*para*-phenylene vinylene) (PPV)[9] and would preclude the fabrication of solid state PPV-lasers[10] even though photopumped lasers have been demonstrated with soluble members of the phenylenevinylene family[11]. Another important difference between PPV and m-LPPP is the relative magnitude of the observed stimulated emission. We find stimulated emission and photoinduced absorption to be of the same size in m-LPPP films, whereas in PPV-films the stimulated emission is found to be only one tenth of the maximum photoinduced absorption[9].

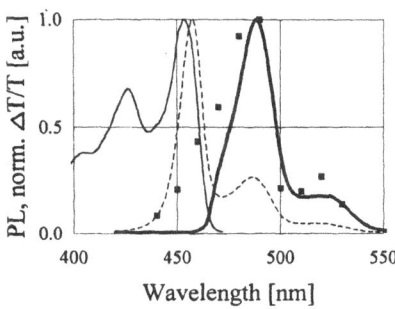

Figure 5. Maximum of ΔT/T () and cw-PL spectrum (bold) of a concentrated m-LPPP solution in toluene; cw-PL excitation (solid) and emission (dashed) for a dilute m-LPPP solution in toluene.

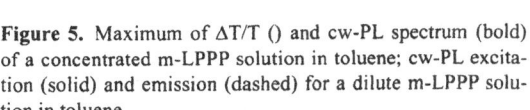

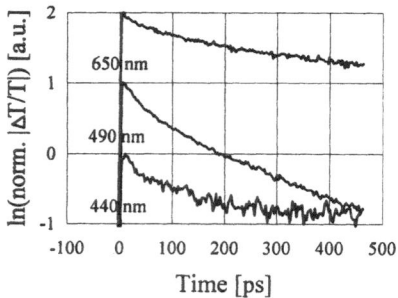

Figure 6. 500 ps traces of $\Delta T/T$ of a m-LPPP-solution in toluene at selected probe wavelengths.

The different kinetics of PA and SE show that the emitting species are not the absorbing species. This conclusion is supported by the observed polarization memory behaviour in the m-LPPP films probed by polarized detection of $\Delta T/T$ parallel and perpendicular to the pump beam. While we observe no polarization memory for the SE, the PA signal at 680 nm (Fig. 7) shows an anisotropy, which decays within a few ps. We conjecture the absorbing species to be hot intrachain charged polarons $P^{\pm}$, which can separate due to the high degree of intrachain order characteristic for the m-LPPP[7]. $P^{\pm}$s diffuse on neighbouring chains at a slower hopping rate with respect to singlet excitons, losing the initial orientation in about 1 ps, and get ultimately localized into long living interchain species[12,13,14]. Since the interaction of charged species is amongst the possible mechanisms for aggregate formation in organic crystals[15], we assign the long living species to aggregates, whose fluorescence is indeed observed in m-LPPP[5]. Within this model the initial fast decay ($\tau = 2$ps) of the PA can be ascribed to the thermalization of $P^{\pm}$, which stabilize into an aggregate. The subsequent non-exponential decay is attributed to relaxation of the aggregate population, with a distribution of possible interchain coupling geometries. The PB dynamics is also consistent with this picture.

To conclude, we report a strong stimulated emission from m-LPPP-films and solution ($\tau = 8$ and 190 ps respectively). This is the first observation of stimulated emission for conjugated polymers in the solid state without competition with PA. We assume, this to be an intrinsic property of conjugated polymers with high fluorescent quantum yield and well defined conjugation length. PA has been observed below the spectral region of SE and as-

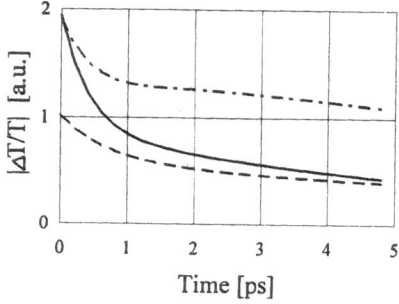

Figure 7. Polarization memory loss of the PA at 680 nm in a m-LPPP film. $\Delta T/T$ parallel (solid) and perpendicular (dashed) and ratio parallel/perpendicular (dash-dotted).

signed to intrachain charged polarons. Our observations indicate that it should be possible to develop a laser based on conjugated ladder-type polymers.

ACKNOWLEDGMENT

This work was supported by the Austrian *Bundesministerium für Wissenschaft und Forschung* under contract 30.507/2-IV/8/94. We acknowledge the contribution of the cw-PA-spectrum by Klaus Petritsch[8] and the Human Capital and Mobility Programme of the European Community under the contract n. ERB-CHRXCT-940561.

REFERENCES

1. G. Grem, G. Leditzky, B. Ullrich, G. Leising, Adv. Mat. **4**, 36 (1992).
2. U. Scherf, K. Müllen, Makromol. Chem., Rapid Commun. **12**, 489 (1991).
3. W. Graupner, M. Mauri, J. Stampfl, O. Unterweger, G. Leising, U. Scherf, K. Müllen, Mol. Cryst. Liq. Cryst. **256**, 431 (1994).
4. J. Stampfl, W. Graupner, G. Leising, U. Scherf, Journal of Luminescence **63**, 117 (1995).
5. J. Stampfl, S. Tasch, G. Leising, U. Scherf, Synth. Met. **71**, 2125 (1995).
6. G. Grem, G. Leising, Synth. Met. **57**, 4105 (1993).
7. W. Graupner, S. Eder, M. Mauri, G. Leising, U. Scherf, Synth. Met. **69**, 419 (1995).
8. K. Petritsch, Diploma Thesis, Technical University of Graz, 1995.
9. M. Yan, L. J. Rothberg, F. Papadimitrakópoulos, M. E. Galvin, T. M. Miller, Phys. Rev. Lett. **72**, 1104 (1994).
10. M. Yan, L. Rothberg, B. R. Hsieh, R. R. Alfano, Phys. Rev. B **49**, 9419 (1994).
11. D. Moses, A. J. Heeger, Appl. Phys. Lett. **60**, 3215 (1992).
12. J.W.P. Hsu, M. Yan, T.M. Jedju, L.J. Rothberg, B. R. Hsieh, Phys. Rev. B **49**, 712 (1994).
13. E. L. Frankevich, A. A. Lymarev, I. Sokolik, F. E. Karasz, S. Blumstengel, R. H. Baughman, H. H. Hörhold, Phys. Rev. B **46**, 9320 (1992).
14. M. Gailberger, H. Bässler, Phys. Rev. B **44**, 8643 (1991).
15. J.B. Birks, *Organic Molecular Photophysics*, p. 409, John Wiley & Sons, New York 1975.

100

LASER INDUCED PROCESSES IN OLIGOTHIOPHENES STUDIED BY PS- AND FS-TIME RESOLVED SPECTROSCOPY

S. Rentsch, D. V. Lap, D. Grebner, and M. Helbig

Institute of Optics and Quantum Electronics
Friedrich-Schiller-University Jena
Max-Wien-Platz 1, 07743 Jena

Oligothiophenes are model substances for polythiophene which is a promising material for opto-electronic devices. Because of the expected fast response compared with semiconductor materials a reliable study of light induced fast processes at thiophenes is necessary. To this aim we studied a series of thiophene oligomers nT with n = 2 to 6 monomer units in solution by means of ps- and fs-spectroscopy [1,2].

We used a Nd-YLF laser system with third harmonic (349 nm, 10 ps) for excitation and white light continuum for probing the optical density alteration after excitation with ps pulses. For spectroscopy with fs time resolution we used a colliding pulse mode locked dye laser amplifier system at 616 nm and the second harmonic 308 nm with pulse duration of 80 fs (at 616 nm) and 350 fs (at 308 nm), respectively for excitation and white light continuum for test. The transient spectra were recorded from about 400 to 1000 nm with an optical multichannel analyzer in the time range between the excitation and about 2 ns.

At first each oligomer (n = 2 - 6) exhibit a transient absorption A1 which arises during excitation with ps and fs pulses as well (Fig. 1.). A second absorption A2 increases at the same time when A1 decreases for all studied nT and then remains unchanged up to the terminal measuring time of 2 ns (Fig. 2).

A1 decays single-exponentially with a lifetime, which increases with size n: 2T: 50 ps, 3T: 135 ps, 4T: 530 ps, 5T: 880 ps, 6T: 1018 ps (with 10 % accuracy). The A1 lifetimes correspond to the fluorescence lifetimes independent measured elsewhere[3] and in conclusion they are to assign to excited state absorption (ESA) from the S1 state (see Fig. 3).

The A2 bands have been proved to be identical to triplet triplet absorption bands (TTA) observed by flash photolyse studies with μs-time-resolution[4]. The T1-state occupation is caused by an intersystem-crossing-process (ISC) followed by internal conversion IC within the triplet manifold as shown in Fig. 3.

We were able to demonstrate the emergence of triplets for n = 3 - 6 on the picosecond time scale for the first time. A1 is located on the red side of A2 and both transient absorptions, A1 and A2 shift to longer wavelengths with size.

Ultrafast Processes in Spectroscopy, edited by Svelto et al.
Plenum Press, New York, 1996

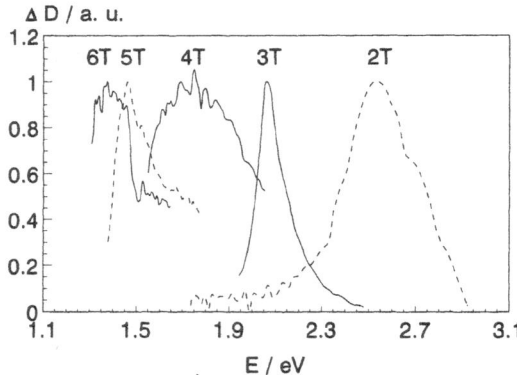

Figure 1. Excited state absorption (A1) of nT recorded 40 ps after excitation.

By fs-pump-probe-technique the first ps was studied in detail. The first observed spectrum is the transient absorption A0 located at the same spectral range as the induced fluorescence. Later this A0 band is superposed by increasing fluorescence. The nature of this band is not clear up to now. We assume it stems from a higher electronic state which should be occupied after excitation with hν=4.02 eV (308 nm). Between this energy level and the lowest excited (fluorescing) level not only vibrational but electronic states are available which can play the role of absorber states [6].

As an example of fs spectroscopy the transient behaviour of 5T after excitation with UV-pulses is demonstrated in Fig. 4.

Further we were able to investigate 6T with a time resolution below 100 fs by a two photon excitation process[2]. Here the transient absorption A0 appears more strongly than after one photon absorption (308 nm). Moreover, we found a further increase of induced fluorescence in the time region up to about 15 ps. We explain this behaviour by higher [1]A-states which exhibit a finite lifetime and lead to a storage effect. The existence of such states of 6T was received within quantum chemical CNDO/S-calculations [6].

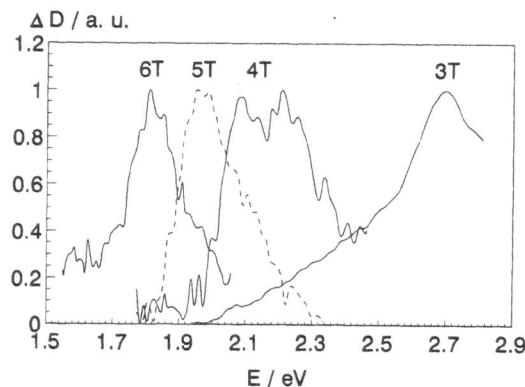

Figure 2. Triplet-triplet absorption (A2) of nT recorded 2 ns after excitation.

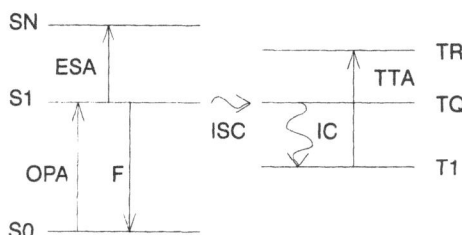

Figure 3. Singlet and triplet level scheme of nT including excitation by one photon absorption (OPA) and detected transient absorptions between singlet states (ESA) and triplet states (TTA) and fluorescence (F).

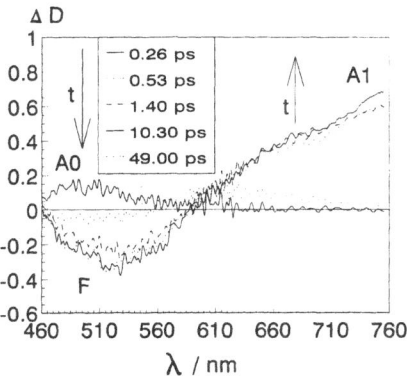

Figure 4. Change of optical density ΔD of 5T between 460 nm and 760 nm after excitation with UV-pulses (λ = 308 nm, τ = 350 fs).

The size dependence of the transient absorption bands A1 and A2 can be described with the same equation which was designed by Kuhn[5] for HOMO-LUMO-transitions of polyenes called the extended FEMO-model. Fig. 5 shows the dependence of transition energy ΔE versus the number of double bonds for the fluorescence (F), excited state absorption (A1) and triplet-triplet absorption (A2) of nT and the fitted curves using Kuhn's formula of the extended FEMO-model.

The extension of the function $\Delta E(DB)$ allows a comparison with the well known fluorescence and absorption bands of polythiophene. These values match the curve at a number of 20 double bonds which corresponds to 10 thiophene rings. That means that oligomers with more than 10 thiophene rings show no more optical relevance.

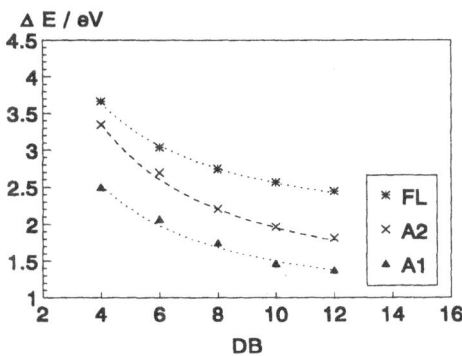

Figure 5. Transition energy ΔE in dependence of number of double bonds for the fluorescence (F), excited state absorption (A1) and triplet-triplet absorption (A2) of nT, fitted with extended FEMO-model.

REFERENCES

1. H. Chosrovian, D. Grebner, S. Rentsch, H. Naarmann, Synth. Met. 52 (1992) 213.
2. D. Grebner, D. V. Lap, S. Rentsch, H. Naarmann, Chem. Phys. Lett. 228 (1994) 651.
3. H. Chosrovian, S. Rentsch, D. Grebner, D. U. Dahm, E. Birckner, Synthetic Metals 60 (1993) 23.
4. V. Wintgens, P. Valat, F. Garnier, J. Phys. Chem. 98 (1994) 228.
5. H. Kuhn, J. Chem. Phys. 17 (1949) 1198.
6. R. Colditz, D. Grebner, S. Rentsch, Chem. Phys., in press.

101

RELAXATION TIMES OF NEW PASSIVE POLYMER SWITCHES FOR 1.06μm

R. Grigonis,[1] M. Eidenas,[1] V. Sirutkaitis,[1] V. Bezrodnyi,[2] A. Ishchenko,[3] and Yu. Slominskii[3]

[1] Laser Research Centre
Vilnius University
Vilnius, Lithuania
[2] Institute of Physics
Kiev, Ukraine
[3] Institute of Organic Chemistry
Kiev, Ukraine

The method of using a saturable absorber to mode-lock the solid state laser is now a well established and simple technique[1,2]. A number of polymethine dyes are presently known to be useful for passive mode locking of the solid state lasers in the 1.06 μm range[3,4]. So far, reports on application of these dyes for mode-locked operation consider mostly dyes in solution rather than in plastic, although the polymethine dyes in plastic have some advantages. First, they can simplify the design of a mode-locked laser and, second, increase reproducibility of temporal and energetic parameters of the radiation. But this can be achieved only in the case if spectral, luminescence, and nonlinear optical characteristics of the dyes in the polymer matrix are not significantly different from the characteristics in liquid solutions. It was a difficult problem for a long time, and only now it has been solved by appropriate selection of the dye structure and the nature of the matrix containing it[5].

In this contribution, we report measurements of the relaxation times and some modelocking characteristics of new passive switches for 1.06 μm on the basis of polymethine dyes in polymer matrices.

The passive switches under study were made on the basis of known and new cyanine dyes hosted in polyurethane. Their absorption maximum was in the range 1.05 − 1.09 μm. The polymer used was a multicomponent urethane composition with hyper elastic state in the temperature range from - 40 to 160^0 C. This is the main reason for the significant increase of the damage threshold in the polymer used in this work in comparison with the ones used earlier. The components of the polymer were chosen such that prevented the dye molecules from destruction during the polymerization process, and the dyes used had a low aggregation level in the polymer matrix. The polymer matrix with the dye was

Ultrafast Processes in Spectroscopy, edited by Svelto et al.
Plenum Press, New York, 1996

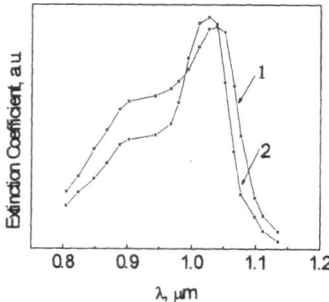

Figure 1. Absorption spectra of the dye No. 4965-u in the polyurethane matrix (1) and ethanol (2).

placed between glass plates, and its thickness was 30–50 μm. As can be seen from Fig. 1, the spectral characteristics of the dyes under study in the polymer matrix are not significantly different from the characteristics in ethanol.

We determined the relaxation times of the bleached states of these passive polymer switches by a direct excite- and - probe method. Picosecond OPO tunable in the range of 1.05 – 1.09 μm was used for the probe and excitation pulse wavelength adjustment with the absorption maximum of the passive switches. The synchronously pumped singly resonant picosecond OPO was used similar to that described previously[6] but without resonant pump reflection. The master oscillator was a Nd:glass laser with passive mode locking (PML) and active negative feedback (NFB). By applying active NFB, stable and flat trains consisting of approximately 800 pulses as short as 1.2 ps were generated. About 180 transform-limited pulses were cut from the middle of the pulse train, further amplified to an energy up to 40 mJ by two double-pass amplifiers and one single-pass amplifier and then frequency-doubled with conversion efficiency of 40% in a KDP crystal. The OPO consisted of a flat mirror with metal coating, dielectric mirror with reflection coefficient of 80% in the spectral region 750–1060 nm and the KDP crystal (l = 4 cm, type II phase matching). To achieve precise matching of the OPO cavity to the pump laser cavity, the output mirror was placed on a micropositioner. Non-collinear pumping of the OPO ensured that only one of the two generated OPO pulses was coupled back into the resonator, i.e. the OPO was singly resonant. The OPO was tunable over the range of 755 – 1700 nm. The OPO pulse duration depended on mismatch between lengths of the OPO and pump laser cavities and could be varied from 0.65 ps to 2 ps. Pulse shortening was related with the energy decrease, so for further experiments the pulse duration about 1.5 ps was chosen. The single pulse energy at 1050 nm was 18 μJ. The spectral bandwidth of the OPO pulses in this case was 16 cm^{-1}. Exciting and probing pulses for the shame of the dye relaxation time measurements were formed by splitting the OPO pulse train with mirrors. The homogeneity of the excitation of the probe volume was ensured by selecting the diameter of the probe beam to be half the diameter of the excitation beam. A filter placed in the path of the probe beam was used to ensure the ratio of the intensities of the excitation and probe beams to be at least 500. In our experiment, we used electromagnetic shutter to measure transmission of the sample in the excited and unexcited states. This made it possible to calculate in real time the change of optical density $\Delta A = \log[T_0/T(t)]$, where T(t) is the transmission of the solution at a moment t in the presence of excitation and T_0 is the transmission of the solution in the absence of excitation[4]. The computer-controlled delay line performed the probing pulse temporal delay within 1 ns.

Table 1. The relaxation time τ of dyes hosted in polyurethane

Dye, No.	3274-u	4805-u	4871-u	4965-u	4973-u	5053-u	5490-u	5577-u	5596-u
τ, ps	11 ± 2	16 ± 2	10.5 ± 2	20 ± 2	12 ± 2	37.5 ± 2	18 ± 2	31 ± 3	12 ± 2

The main results of our measurements are presented in Table 1. As seen, the relaxation times of the dyes under study hosted in the polyurethane matrix are in the range of 6 – 40 ps. The shortest relaxation time among the investigated dyes had dye 4915-u (an analog of dye 3321-u, which has relaxation time in solutions in the range of 1–6 ps). The other dyes had the relaxation times longer than 10 ps. It is worth mentioning that relaxation times of dyes 3274-u, 4871-u, and 5596-u do not change significantly while introduced into solid state matrices. Comparison of relaxation time τ for the polymer sample measured in our work with τ for a solution of the same dye measured in the work[7] shows that the former value is intermediate and sometimes even close to the minimum obtained for different solvents. For passive mode locking of Nd:YAG and Nd:YAL lasers, suitable are polymer switches of all dyes investigated. For creating the polymer switches for passive modelocking of Nd:glass lasers, among the investigated dyes, the most suitable are dyes No. 4915-u, 3274-u, 4871-u, 4805-u, and 5596-u, because they have the shortest relaxation times.

The passive polymer switches were tested in Nd:glass and Nd:YAG lasers with PML and NFB. The experimental set-up of our laser with PML and NFB is depicted schematically in Fig. 2. A 1.5 m long laser resonator was formed by an output plane mirror (2) deposited on a wedge-shaped substrate and a nontransmitting 3.5 m radius spherical mirror (1). A mirror with reflectivity of 55 % at 1.06 μm was used as the output coupler. The passive switch (6) was placed at the Brewster angle near the output mirror. Laser rods (5) of Nd:phosphate glass (GLS-22, 100×4 mm) and Nd:YAG (65x4 mm) were used as the lasing materials. The electro-optical amplitude stabilization was provided by a NFB circuit consisting of a Pockels cell (3), dielectric polarizer (4), fast photodiode (7), and a fast high-voltage electronic amplifier (8). In order to prevent spectral selection, the faces of the rods and the Pockels cell were cut at an angle of 87 deg. A portion of the intracavity radiation, which had been selected by the permanent voltage applied to the Pockels cell was directed onto the photodiode (7) from which the signal for the feedback control was derived. After a delay time of 15 ns, the control signal amplified up to 500 V was applied to the Pockels cell for electro-optical amplitude stabilization.

Some parameters of the Nd:glass laser with PML and NFB using the polymer passive switches under operation conditions for the best combination of temporal and energetical parameters are listed in a Table 2. T is the pulse train duration, T_0 is transmission

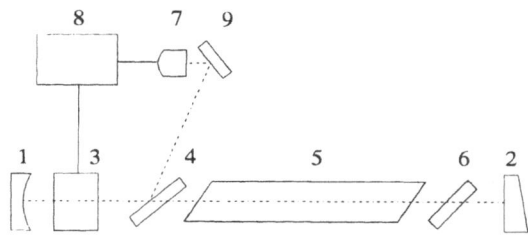

Figure 2. Experimental setup of solid-state laser with PML and NFB.

Table 2. Modelocking characteristics of the Nd:glass laser with PML and NFB
using polymer passive switches with different saturable absorbers

Saturable absorber	T_0 %	T µs	E_t mJ	τ ps	E_a µJ
3274-u	60	16	2.3	0.9	1.3
4915-u	67	8	1	1.1 - 1.4	1.1
4965-u	70	10	1.8	1.4	1.6
4973-u	74	17	0.8	1.6	0.9

of the switch for a small signal, E_t is the pulse train energy, E_a is the single pulse energy, τ is the pulse duration measured by single shot autocorrelator. The used passive polymer switches had a low light transmission of 60 − 75 % at 1055 nm for such experiments.

The minimal duration of the pulses in the Nd:glass laser with PML and NFB was 0.9 ps by using the passive polymer switch with dye 3274-u. Stable pulses with duration of 1.4 − 1.6 ps were generated also using passive polymer switches with dye 4965-u or dye 4973-u. The single pulse energy and duration could be changed by choosing the value of initial transmission and the value of NFB. Increase of the initial transmission of the saturable absorber No. 4973-u from 79 % to 74 % at the same pulse stabilization level resulted in the shortening of the pulse train from 23 µs to 17 µs and in the decrease of the pulse duration from 1.9 − 2 ps to 1.6 − 1.7 ps. The similar changes were also observed with the dye No 3274 when the sample with $T_0 = 79$ % was replaced by an other one with $T_0 = 60$ %. For this dye, the decrease of the initial transmission allowed us to short the pulse train from 16 µs to 9 µs and to shorten the pulse duration from 1.5 ps to 0.9 ps. Note, however, that in the latter case, the pulse stabilization level was about twice lower. Although the polymer passive switch with dye 4915-u had the shortest relaxation time, the mode locking process with this switch was complicated. In most cases the process of mode locking was not stabile and took longer than 10 µs for short single pulse generation. This is, probably related with the short relaxation time and low absorption cross-section of this dye that requires for the saturation of such a passive switch a special cavity with increased intensity at the place where the switch[2] is located. Comparison of operation of the Nd:glass laser with PML and NFB using polyurethane and solution-based switches shows similarity of the laser parameters. Up to 10^6 laser firings, no noticeable degradation of the sample was observed.

In the case of the YAG:Nd laser with PML and NFB, the train of approximately 3000 pulses was investigated. For PML we used polymer switch with dye 3274-u. The initial transmission of the passive switch was 64%. The single pulse duration was ~30 ps, the energy ~1 µJ. Using an output mirror with the reflection coefficient of 18%, we achieved the pulse train energy more than 6 mJ.

The modelocking characteristics of the passive switch with dye 4965-u were also tested in the YAG:Nd laser with NFB and additional Q-switching. In this case the train of pulses was 300 ns long. The pulse energy in this case was ~ 100 µJ. In both cases no noticeable degradation in the sample has been seen up to 10^5 laser firings.

According to our investigations, the polyurethane passive switches present a competitive alternative to the saturable absorbers in solution, and allow one to design neodymium-doped solid state lasers with PML and NFB using solid-state merit of such switches.

The research described in this work was supported in part by Grants No. LED000 and LHD100 from the International Science Foundation.

REFERENCES

1. G.H.C. New, Rep. Prog. Phys. 46, 877 (1983).
2. A. Piskarskas and V. Sirutkaitis, Bulletin of the Academy of Sciences of the USSR. Physical Series, **54**, 9 (1990).
3. B.Kopainsky, W. Kaiser and K.H. Drexhage, Opt. Commun. **32**, 451 (1980)
4. B.F. Bareika, R.V. Danielius, G.A. Dikchyus, G.G. Dyadyuscha, A.A. Ishchenko, M.A. Kudinova, A. P. S. Piskarskas, V.A. Sirutkaitis, and A.I. Tolmachev, Sov. J. Quantum Electron. **12**, 1485 (1982).
5. A.A. Ishchenko, Kvantovaya Elektron. (Moscow) **21**, 513 (1994).
6. K. Gardziulis, R. Grigonis, J. Jaseviciute, G. Sinkevicius, and V. Sirutkaitis, Lithuanian J. Phys. **33**, 296 (1993).
7. R. Grigonis, A. Ischenko, G. Sinkevicius, V. Sirutkaitis, and Yu. Slominskij, in *Lasers and Ultrafast Processes*, **4**, 197 (Vilnius University Press, 1991).

102

PHOTODISSOCIATION OF POLYMETHINE DYES IN LANGMUIR-BLODGETT FILMS

A. Baltuska,[1] R. Gadonas,[1] A. Pugzlys,[1] and K. -H. Feller[2]

[1] Laser Research Center
Vilnius University, Sauletekio 10
2054 Vilnius, Lithuania
[2] Faculty of Physics and Medical Engineering
Technical University Jena
Tatzendpromenade 1b, D-07745 Jena, Germany

One of the main factors preventing up to now the wide application of molecular LB films in nonlinear optics is limited photostability of the layers [1]. Therefore, study of the photodegradation processes of LB films is of great practical importance.

Up to now there has been no data on the role of the light intensity in photodecomposition of LB films. This is of particular importance for fast data processing since intense pulsed light signals are required to ensure high enough efficiency of the nonlinear optical process.

We investigated the photodissociation of different multilayer structures of monomeric and aggregated dyes. Two different irradiation regimes were used in order to clarify the role of light intensity: illumination by megawatt peak power picosecond pulses as well as by miliwatt cw argon ion laser .

Langmuir-Blodgett films of different polymethine dyes have been prepared according to standard technology on a glass substrate. The films of following dye molecules were investigated: D282 – 1,1'-dioctadecyl-3,3,3',3'-tetramethylindocarbocyanine perchlorate, D275 – 3,3' dioctadecyloxacarbocyanine perchlorate, D1125 – 3,3' dihexadecyloxacarbocyanine perchlorate, 1,1'-dioctadecyl-2,2'-cyanine iodide (called pseudoisocyanine and refered below as PIC), and one of xanthene dyes D455 – acridine orange-10-dodecyl bromide.

From the absorption and fluorescence spectra it can be assumed that D455 film consists of monomeric molecules, films of the dyes D282, D275 and D1125 contain monomers and dimers. The LB film of PIC and D282 consist of J–aggregated dye molecules as indicate the narrow and strong absorption bands red-shifted with respect to monomer absorption as well as fluorescence without Stokes shift.

In pulsed regime the samples under study were irradiated by the second harmonic (SH) of the output radiation of optical parametric oscillator OPO pumped by SH mode-locked Nd:YAG laser [2]. The energy of single 10 ps pulse was 20 μJ. The maximum in-

Ultrafast Processes in Spectroscopy, edited by Svelto et al.
Plenum Press, New York, 1996

455

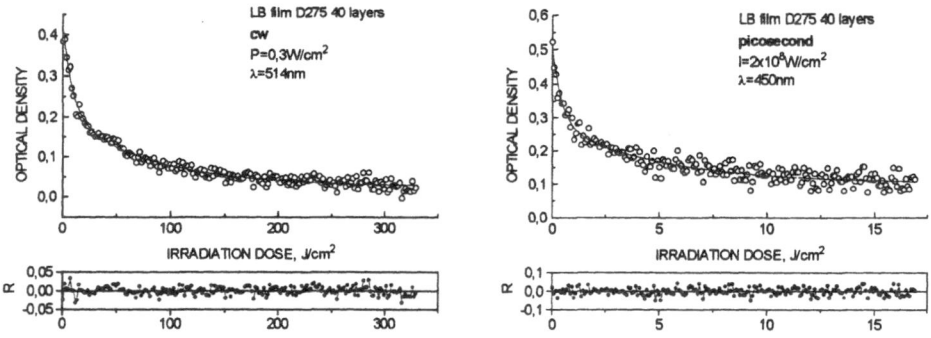

Figure 1. Typical decay of optical density of the LB films under illumination by cw light and picosecond pulses. Experimental data (open circles), two exponential fits (solid line) and residuals (R) are shown.

tensity used in the experiment was 2×10^8 W/cm^2. A weak delayed probe pulse was used to measure the sample transmission during the illumination to exclude the contribution of absorption changes originated from excited state population. In the cw regime an argon ion laser radiation at the wavelength 514 nm with 200 mW power was used.

The changes in the absorption spectra of the LB film of PIC J-aggregates within the irradiated area were measured using a spectrophotometric consisting of light source, two multimode fibers for measurements of the absorption in the central part of the photobleached area, polychromator and OMA-3 for detection. To calculate the anisotropy absorption spectra were measured with linearly polarized light of two perpendicular polarizations.

Under both pulsed and cw irradiation regimes all the investigated LB films showed the decrease in optical density with complex dependence on the irradiation dose applied. All decay curves were satisfactorily fitted by double exponential function of the dose applied (Fig. 1).

The latter means the integral light energy (or integral number of photons) applied per unit area at any time after beginning of light exposition. The rates of two decay components for optical density of every film are by order of magnitude different, furthermore the general character of the curves do not show any obvious dependence on the nature of the film (monomeric, dimeric or aggregated) and on the wavelength of irradiation. Under pulsed irradiation the photobleaching of all studied films is by order of magnitude faster compared to the cw case although the character of the curves remains the same.

The complexity of the decay of optical density is supposed to reflect the continuous decrease in photodissociation yield of the molecules with increasing degree of decomposition of the film. It is supposed to that the appearance of more torsional freedom and increase of vibrational energy exchange rate with the bath for individual monomers makes them more photostable.

The intensity dependence of decay rates can be qualitatively explained in the following way. At high intensities (in our experiments one photon absorbed per ten molecules) of the pulses the light energy is absorbed and subsequently released as vibrational excitation during the excited state decay. The vibrational excitation is locally accumulated in the "active" planes of chromophore molecules due to relatively slow exchange with the surrounding. It rises the local vibrational temperature and the probability of appearance of structural defects in the film as well as of molecule dissociation becomes high.

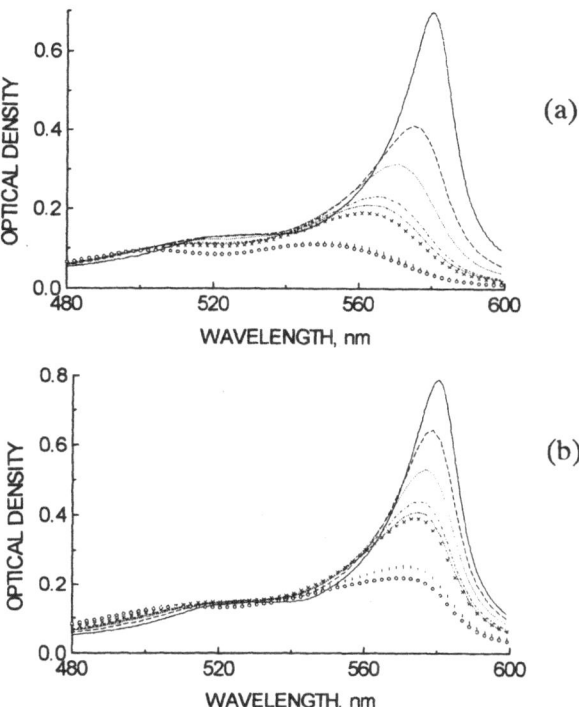

Figure 2. Dependence of absorption spectrum on irradiation dose for 13 layers of PIC J-aggregates for parallel (a) and perpendicular (b) polarizations of probe light.

In pulsed irradiation regime we observed significantly higher decay rates for PIC J-aggregates compared to other dyes. It can be supposed from these data that in aggregate films the exciton annihilation plays an important role in the process of photodecomposition being dominant mechanism of exciton decay in solutions at the intensities used in present experiments [3].

The slower component in the case of cw irradiation can be attributed to the photodecomposition of the molecules which have some vibrational bath produced by the fragments of destroyed molecules and, as a consequence, less photodissociation yield. This component should exist in the case of pulsed irradiation too, however the estimation of quantitative characteristics in this case is connected with high error due to small irradiation doses applied.

In order to provide with more data on the photodissociation of PIC J-aggregates, absorption spectra of the LB film containing 13 aggregate monolayers under irradiation by linearly polarized pulsed light were measured using fiber spectrometer. The evolution of the absorption spectra is presented in Fig. 2.

As seen from the figure, the of J-band is continuously bleached during the irradiation. Besides, the maximum of the band shifts to the blue and the J-band broadens. The impact is more pronounced for the polarization parallel to that of excitation. The increase in optical density observed in the short wavelength tail of J-band as well as at the wavelengths $\lambda < 520$ nm obviously does not compensate the bleaching in integral.

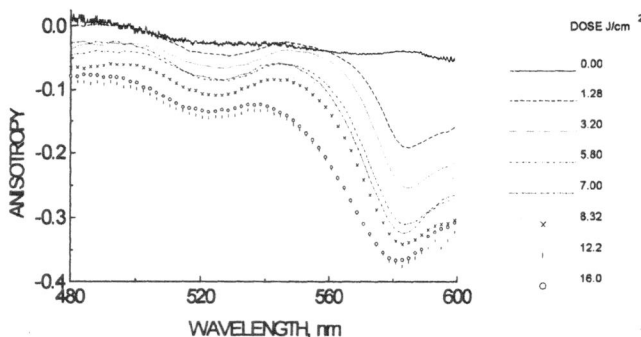

Figure 3. Spectra of the induced steady state anisotropy of the LB film built from J-aggregates of PIC-C$_{18}$ (13 monolayers) for different doses of irradiation by picosecond pulses at wavelength 579 nm.

From the obtained data we calculated the anisotropy of the sample:

$$r = \frac{\alpha_{\parallel} - \alpha_{\perp}}{\alpha_{\parallel} + 2 \cdot \alpha_{\perp}}$$

where $\alpha_{\parallel}$ and $\alpha_{\perp}$ are absorption coefficients for the light with polarization parallel and perpendicular to that the excitation respectively. The anisotropy (Fig. 3.) has two maxima the positions and values of which depend on the irradiation dose. In both "red" and "blue" maxima the anisotropy grows with the dose applied. Besides, the positions of both maxima shift to the blue.

As can be seen from Fig. 4. the value of anisotropy approaches its theoretical limit on the red side with the highest irradiation dose of 16 J/cm^2. The magnitude of the shift of the both maxima is approximately 8 nm. Note that in the beginning of irradiation the "red" peak of the anisotropy is observed not in the maximum of J-band. It is shifted by 8 nm to the longer wavelength side. During the illumination it asymptotically approaches the J-band maximum (580.5 nm).

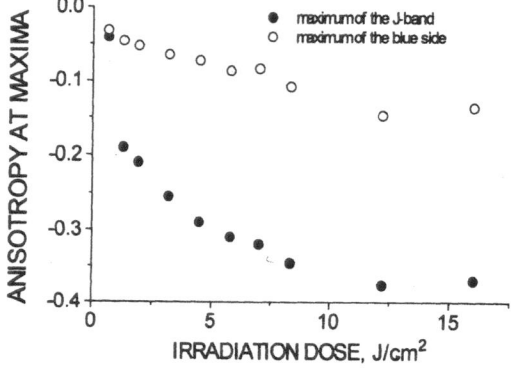

Figure 4. Dependence of absorption anisotropy at "**red**" and "**blue**" maxima on irradiation dose.

The results obtained indicate, that under irradiation of the J-aggregate film by picosecond pulses the regular linear molecular structures responsible for the coherent exciton transitions are irreversibly destroyed. The bleaching of the J-band which begins in the long wavelength tail as well as the initial position of the maximum of the anisotropy indicates that in the beginning of irradiation mainly the longest optical aggregates which have absorption in the red wing of the J-band are decomposed. As a result of dissociation shorter aggregates, dimers and monomers are produced with blue shifted absorption transitions and less anisotropic orientational distribution.

The aggregates with the transition moments parallel to the light electric field vector are destroyed preferentially what results in considerable anisotropy within J-band. The fact that the steady state anisotropy within J-band is close to theoretical limit indicates that the primary processes resulting in photodecomposition of the aggregates occur before the orientational distribution of the optical aggregates excited is randomized due to energy transfer within a domain. The other possible explanation of this is high orientational order in the domain, i.e. the optical aggregates in a domain are aligned parallel each other so that the energy transfer does not change considerably the orientational distribution of the excited aggregates.

REFERENCES

1. D. S. Chemla and J. Zyss, *Nonlinear Optical Properties of Organic Molecules and Crystals*, (Academic Press, Orlando, 1987).
2. A. Baltuska, R. Gadonas, A. Pugzlys, A. Piskarskas, *Exper. Techn. Physics* (in press).
3. V. Sundstrom, T. Gillbro, R. Gadonas, A. Piskarskas, J. Chem. Phys. 89 (5) (1988) 2754.

103

DEGENERATE FOUR WAVE MIXING IN J-AGGREGATES OF PSEUDOISOCYANINE IN SOLUTION AND LANGMUIR-BLODGETT FILM

A. Baltuska,[1] R. Gadonas,[1] A. Pugzlys,[1] K. -H. Feller,[2] G. Jonusauskas,[3] J. Oberle,[3] and C. Rulliere[3]

[1] Laser Research Center
Vilnius University, Sauletekio 10
2054 Vilnius, Lithuania
[2] Faculty of Physics and Medical Engineering
Technical University Jena
Tatzendpromenade 1b, D-07745 Jena, Germany
[3] Centre de Physique Moleculaire Optique et Hertzienne
Universite de Bordeaux I
351 Course de la Liberation, 33405 Talence Cedex, France

J-aggregates of polymethine dyes have become focus of interest in the interdisciplinary field of molecular nonlinear optics [1–3] and Langmuir-Blodgett (LB) engineering [4]. Polymethine dye J-aggregates were formed in LB layers, storage of optical information in monolayers and mixed multilayers has been recently demonstrated [4]

The complexity of time- and intensity dependence for absorption changes of PIC aggregates are known to be caused by exciton annihilation and subsequent changes in aggregate spectrum due to vibrational excitation [5]. Therefore, annihilation as well as secondary vibrational energy dynamics in aggregate influence the nonlinearity and its time response in degenerate four wave mixing (DFWM) [1].

Compared to other thin layer techniques Langmuir-Blodgett technique offers the unique possibility to align the structure on a molecular level. Special devices can be relatively easily designed by introducing sensitive to a certain property molecules into the layer structure, changing the molecular substituents controlling the intermolecular distances, the angular distribution, the symmetry and other parameters important for applications [6]. Study of primary processes of photoexcited LB films of dye aggregates as well as investigation of nonlinear properties of LB mono- and multilayers is of great importance aiming at the understanding the relationship between the optical nonlinearity, relaxation rates and mechanisms, and structure, dimensionality of molecular assembly [6].

Two different concentration samples of J-aggregates of PIC chloride in water solution and J-aggregates of PIC iodide in LB film of 40 layers were used in our experiments.

Ultrafast Processes in Spectroscopy, edited by Svelto et al.
Plenum Press, New York, 1996

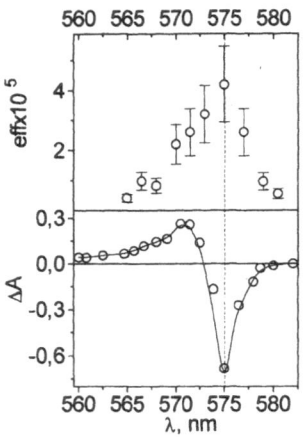

Figure 1. Wavelength dependence of conversion efficiency *eff* in PIC J aggregates at excitation intensity I_{exc} = 100 MW/cm^2 and correlation with the spectrum of optical density changes.

Resonance experiments within the J-band were performed with 5 10^{-4} M PIC aqueous solutions. For measurements within the wavelength range 580–605 nm we used the samples of concentration 10^{-2} M.

DFWM experiments were carried out using both forward (phase conjugation) and backward (BOXCARS) geometries. Optical parametric oscillator (OPO) pumped by second harmonics of modelocked Nd:YAG laser [7] and hybridly mode locked dye laser (Coherent Dye Laser 702, Rhodamine 590 as amplifier and Pinacyanol chloride as saturable absorber pumped by second harmonic of a CW active mode locked Nd:YAG laser ("Coherent Antares") were used for the measurements..

DFWM in J-aggregates is an efficient process. The maximum efficiency observed at 575 nm is slightly red shifted compared to the absorption maximum (573 nm). It is in rather good correlation with the electronic part of light induced absorption changes at low excitation intensity (Fig.1).

From the measured efficiency we calculated for PIC J-aggregate in a solution at concentration 5×10^{-4} M the third order susceptibility at 579 nm to a value of 1.2·10^{-11} esu.

The size of the PIC molecule in the solution was measured by means of taking the isotherms of the monomolecular layer produced on the water surface of a Langmuir-Blodgett through. For the area of the PIC molecule we calculated 0.48 nm^2. The height of the PIC molecule was estimated to 0.15 nm taking into account the length of the C -H bond as well as the torsion angle of the PIC molecule (50.6° in ethanol). The volume per molecule at closest packing amounts to 0.072 nm^3. The water molecules which are not active in mixing process take an essential part of the volume. Therefore the active PIC molecules take only 2.2′10^{-5} volume part of solution. Taking this into account we estimate (5±2)×10^{-7} esu for third order nonlinear susceptibility of molecular LB solid containing J-aggregate at 579 nm. Although the above estimation is evidently very doubtful, this high values can be considered as an estimation from the experiments of the nonlinear susceptibilities of bulk material under the situation of high dilution in solution.

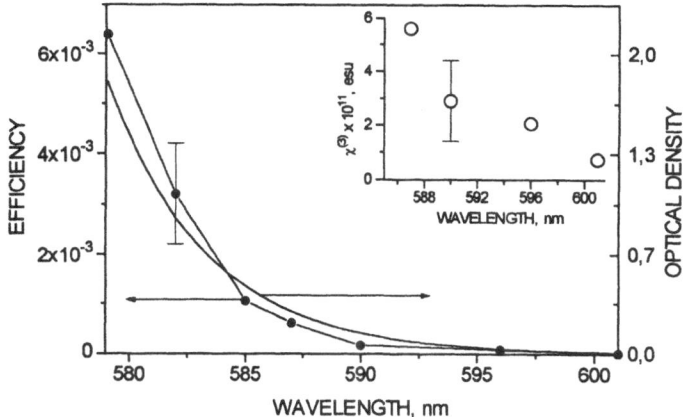

Figure 2. Spectra of efficiency of DFWM and the dispersion of $\chi^{(3)}$ of PIC J-aggregates at concentration 10^{-2}M in 0.2 mm cell.

Going out of resonance DFWM efficiency well correlates with the absorption spectrum (Fig.2) what shows that the nonlinearity of PIC J-aggregates is mainly determined by real excitations.

From the obtained $\chi^{(3)}$ value at 590 nm $(3\pm2)\cdot10$–8 we estimate $(7\pm2)\cdot10$–8 esu for third order nonlinear susceptibility at 590 nm of molecular LB solid containing J-aggregate.

The values of the DFWM efficiency of J-aggregates in LB films are comparable with those in solution. Nevertheless the spectrum of the DFWM efficiency is well correlated with the absorption spectrum (Fig.3) and has red shifted maximum. The similar red shift of efficiency spectrum has been observed in J-aggregates in solution. This let us suppose the similar ultrafast processes contributing to the nonlinearity of J-aggregates in solution and LB film.

From the measured DFWM efficiency the $\left|\chi^{(3)}\right|$ value of 1.5×10^{-7} esu was estimated for the LB film of PIC-C$_{18}$ at the wavelength 590 nm .The LB film contains not

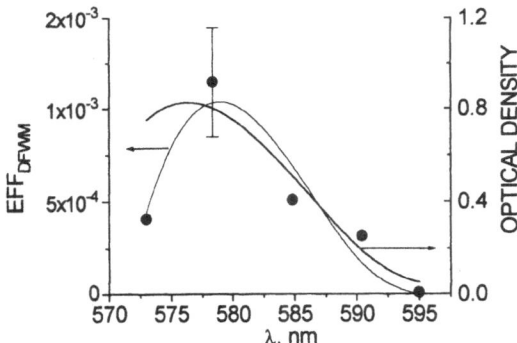

Figure 3. Spectra of DFWM efficiency (solid circles and dotted line) and absorption (solid line) of PIC J-aggregates in LB film of 40 layers.

only the "active" PIC molecules but also long aliphatic chains $C_{18}H_{37}$ between the layers which do not contribute to the nonlinearity. The thickness of the interlayer can be estimated as the length of aliphatic chain of ~25 Å. Taking into account that the height of molecule of 1'1-diethyl-2'2-cyanine chloride is about 1.5 Å it can be deduced that "active" molecules take only ~6% of the volume of LB film. This allows to estimate the $|\chi^{(3)}|$ value for highest packing density of "active" molecules (nonlinear susceptibility of bulk material). It is $(9\pm5)\times10^{-7}$ esu.

The above estimated values of $|\chi^{(3)}|$ give the possibility to compare the results obtained for PIC in solution and in LB film. The higher value of $|\chi^{(3)}|$ at highest packing density in the case of LB film compared to that in solution $((7\pm2)\times10^{-8}$ esu) can be explained by the fact that the wavelength 590 nm is closer to the exact resonance of the LB film (nonlinearity of PIC decreases very strong going out of resonance).

The dynamics of the $\chi^{(3)}$ was studied in BOXCARS geometry with 0.5 ps pulses at wavelength 595 nm in case of J-aggregates in solution at consentration 10^{-2} M and with 5 ps pulses in case of J-aggregates in LB film. The interpretation of the kinetics of DFWM is easier in the case of population grating furthermore the pump and probe experiments deals with relaxation of real excitations. The dynamics of exciton population n_e can be described using Paillotin's approach:

$$\frac{dn_e}{dt} = \sigma \cdot I(t) \cdot (1-n_e) - k \cdot n_e - \gamma \cdot n_e^2 \,,$$

where σ is absorption cross section, I is excitation intensity, k is annihilation free decay constant and γ is annihilation constant. The term $\sigma \cdot I(t) \cdot (1-n_e)$ is responsible for the excitation as well as terms $k \cdot n_e$ and $\gamma \cdot n_e^2$ describe annihilation free relaxation and relaxation caused by annihilation, respectively.

The population of vibrational modes n_v which appears in the consequence of annihilation can be described by the following differential equation:

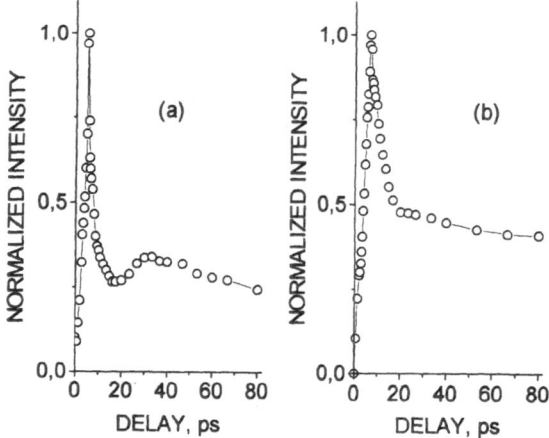

Figure 4. Relaxation kinetics of component χ_{1122} at pump intensities I = 44 GW/cm^2 (a) and I = 30 GW/cm^2 (b). Excitation wavelength - 595 nm.

$$\frac{dn_v}{dt} = \gamma \cdot n_e + \gamma^2 \cdot n_e^2 - k_v \cdot n_v,$$

where k_v is equilibration rate of vibrational modes.

For numeric simulations we used: annihilation free rate - $k = 2.5 \times 10^9$ s^{-1}, annihilation constant - $\gamma = 3 \times 10^{-3}$ cm^3 s^{-1}, pulse duration - $\tau = 1$ ps, equilibration rate of destructive modes - $k_v = 2.5 \times 10^{11}$ s^{-1}, excitation intensity - $\gamma N(0)/k = 1...7500$, (N(0) is the integral absorbed photon density in 1 ps pulse).

Kinetics of relaxation of population in DFWM experiments wit J-aggregates in solution and LB film as well as results of theoretical modelling are presented in Fig. 4 and 5.

As seen from the figure the relaxation has complex character and is intensity dependent. The hollow at delay 17 ps in the case of higher intensity can be explained by destructive interference of the population grating and vibrational modes arising in the consequence of annihilation.

Taking in to account that interference of the exciton population and population of vibrational modes is destructive, we calculated the difference n_e-n_v for different excitation intensities and for different relative weights of vibrational modes (relative efficiency of creating the vibrational modes in the consequence of the annihilation). As seen from the Fig. 5b with increasing the excitation intensity some hollow in relaxation kinetics appears. By varying the excitation intensity and relative weight of vibrational modes the relaxation kinetics similar to the experimental curve (Fig.4) can be obtained.

In case of LB films the transfer of vibrational excitation to the surrounding is slower than in solution. On the other hand the limited photostability of the J-aggregates in LB films can cause the photodegradation acts instead of reversible disordering. In this case delayed transient absorption characteristic for J-aggregates in solution at high excitation intensities can be not observable. Due to these reasons direct comparison of the relaxation kinetics of J-aggregates in solution and LB film is less ore more complicated.

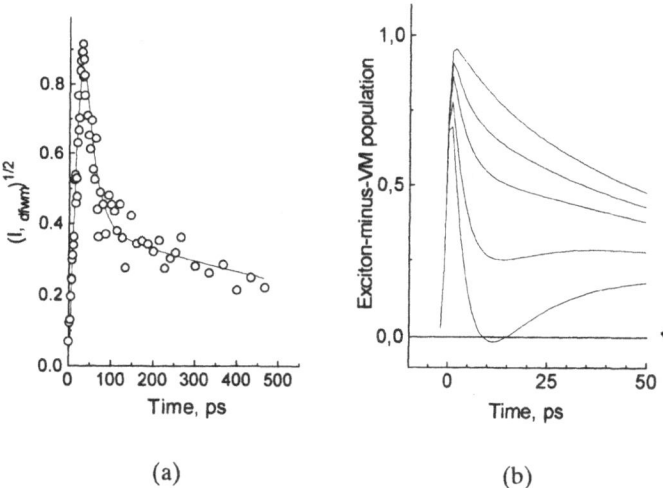

(a)   (b)

Figure 5. Kinetics of the relaxation of population in PIC J-aggregates in LB films (a) and results of the modelling of destructive interference of exciton population and population of vibrational modes (VM) (above to below with increasing of relative weight of VM) (b).

From the obtained data we conclude that J-aggregates of PIC have high cubic non-linear optical susceptibility comparable to that of polydiacetylenes and other self organized multimolecular structures with conjugated π-electron systems. Efficient degenerate four wave mixing in J-aggregates is the resonance phenomenon where real exciton population, decay of biexciton states, reversible aggregate disordering contribute to nonlinear interaction resulting in complex spectral, temporal, and intensity scale behaviour. Exciton annihilation is the mechanism responsible for intensity dependent decay rate of exciton population and for decrease in power of intensity dependence of wave mixing efficiency. Single-exciton-associated transient changes and those of reversible vibrational aggregate disordering result in destructively interfering amplitude and phase gratings. This interference results in further decrease in power of intensity dependence of diffracted signal and appearance of oscillatory features in time response. Estimated value of $|\chi^{(3)}|$ for J-aggregates in LB film is similar to that calculated for highest packing density from $|\chi^{(3)}|$ value obtained in for J-aggregates in solution. The temporal behaviour of DFWM process of J-aggregates in LB layers is similar to that in solution.

REFERENCES

1. R. Gadonas, K.-H. Feller and A. Pugzlys, Opt. Commun. 112 (1994) 157.
2. Ying Wang, J. Opt. Soc. Am. B 8 (1991) 981.
3. F. C. Spano, S. Mukamel, Phys. Rev. A 40 (1989) 5783.
4. T. Kawaguchi, K. Iwata, Thin Solid Films 165 (1988) 323.
5. V. Sundström, T. Gillbro, R. Gadonas, A. Piskarskas, J. Chem. Phys., 89 (1988) 2754.
6. D. S. Chemla and J. Zyss, *Nonlinear Optical Properties of Organic Molecules and Crystals*, (Academic Press, Orlando, 1987)
7. Gadonas, A. Pugzlys, A. Piskarskas, Exp. Techn. Phys. (in press).

104

ANNIHILATION ENHANCED FOUR-WAVE MIXING IN J-AGGREGATES

E. Gaizauskas[1] and K. -H. Feller[2]

[1] Vilnius University
Laser Research Center
Vilnius, Lithuania
[2] Fachhochschule Jena
Fachbereich Medizintechnik/Physikalische Technik
Jena, Germany

Exciton-exciton annihilation processes are known as influencing the nonlinear response of molecular aggregates in a great deal. Recently, by using simplified model of a collection of two-level quantum systems, significant enhancement of the four wave mixing efficiency due to the annihilation processes was demonstrated[1]. The comparison with experimental findings[2] of DFWM at the excitonic resonance in J-aggregates of pseudoisocyanine chloride (PIC), showed good qualitative agreement for all data, but the wavelength dependence of the nonlinearity power index $p = \partial(\log_{10} S)/\partial(\log_{10} I)$ of the DFWM signal S on the pump intensity I. In the two-state (single-exciton) approximation p is reduced from the expected value 3 even far in the wings of the J-band due to the power-broadening of the spectrum.

Because for four-wave mixing process at high field intensities both single-exciton and two-exciton subspace need to be considered, we modeled the J-aggregates of PIC in a standard way[3] as a linear chain of N = 20 two-level molecules including the near-neighbor dipole-dipole coupling J = -633 cm^{-1}. The description of light and matter interaction was performed in the semiclassical way to account for the interplay of nonlinear effects for high pump intensities. Let us notice at this point, that the contributions of the single-exciton aggregated states to both four-wave mixing (one- and two-photon resonant) processes under consideration are well separated in the energetic space. As it is shown in Fig. 1, the lowest excitonic state of the first excited band may to be considered as only one contributing to one-photon resonant DFWM, whereas two-photon resonance enhanced DFWM appears mainly through the higher aggregated states of the first excitation band. This fact is of great relevance for the strategy of the theoretical consideration of the problem, and allows in principle to perform nonperturbative treatment of both single as well two-exciton resonances.

Ultrafast Processes in Spectroscopy, edited by Svelto et al.
Plenum Press, New York, 1996

467

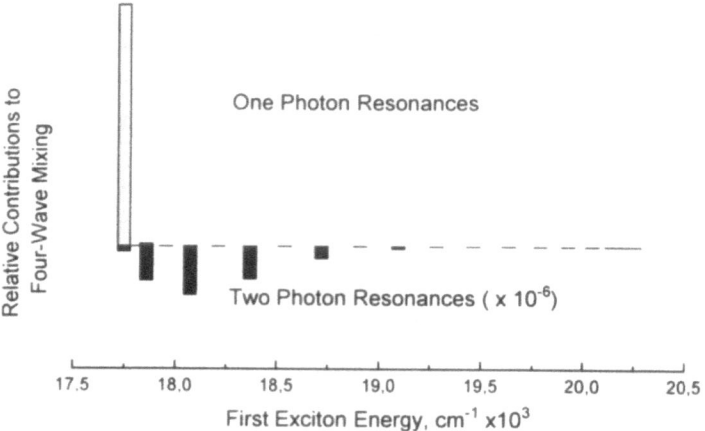

Figure 1. Relative contributions of the lowest excitonic states to one-photon (light gray columns) and two-photon (dark gray columns) resonance enhanced DFWM in J-Aggregate.

The modified optical Bloch equations for the description of the excitations in two excitonic bands may be written as follows:

$$\frac{\partial \rho_{12}}{\partial t} = i\Delta_1 \rho_{12} + i\Lambda n_{21} - \frac{\rho_{12}}{T_2},$$ (1A)

$$\frac{\partial n_{21}}{\partial t} = -4\,\text{Im}[\Lambda \rho_{12}] - 2W_{21}\rho_{22} + W_{32}\rho_{33} - 3\gamma\rho_{22}^2,$$ (1B)

$$\frac{\partial n_{32}}{\partial t} = W_{21}\rho_{22} - 2W_{32}\rho_{33} + 3\gamma\rho_{22}^2,$$ (1C)

$$\frac{\partial \rho_{13}^{(k)}}{\partial t} = i\Delta_2^{(k)}\rho_{13}^{(k)} - 2\widetilde{\mu}^{(k)}\Lambda^2\rho_{11},$$ (1D)

where:

$$\Lambda = \frac{\mu}{\hbar}E(t),\ n_{ij} = \rho_{ii} - \rho_{jj},\ \ \rho_{11} + \rho_{22} + \rho_{33} = 1,\ \Delta_1 = \omega_1^{(1)} - \omega\ ,\ \Delta_2^{(k)} = \omega_2^{(k)} - 2\omega\ ,$$

and the dipole matrix element μ responds for the single-exciton transition to lowest excitonic state (having energy $\hbar\omega_1^{(1)}$), whereas $\widetilde{\mu}^{(k)}$ describes transition to the k-th two-exciton state in the second excited band (having energy $\hbar\omega_2^{(k)}$) and contain summation over all virtual excitations in the lowest one. By deriving Eqs.1 the slowly varying amplitudes of the non-diagonal matrix elements were introduced in a usual way and rotating wave approximation was made. Additionally, the terms describing annihilation and relaxation processes with characteristic constants γ , W_{ij} and T_2 were added into Eqs. 1 phenomenologically.

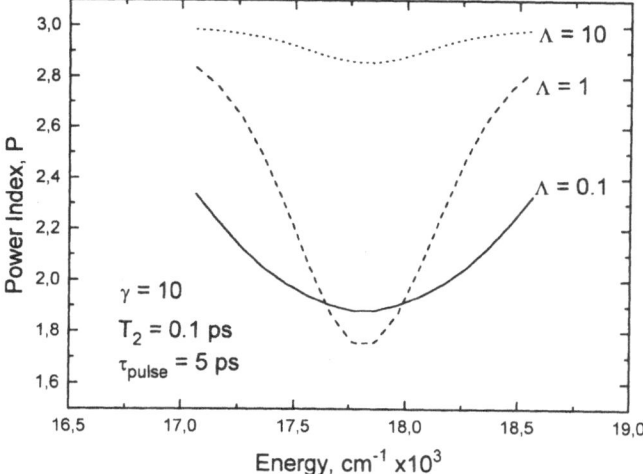

Figure 2. Calculated wavelength dependencies of nonlinearity power index p by considering transitions to two-exciton states of PIC aggregates.

We analyze numerically the DFWM efficiency when light pulses as long as 5 ps are used in the experimental geometry with two pump beams forming small angle and the test beam of the same intensity propagating in the direction opposite to one of the pump. Following procedure described in [1] we expanded material variable describing population of the single-exciton state into a spatial Fourier series to account for the periodicity of the media created by the pump pulses. The measured DFWM signal can be evaluated then as proportional to the polarization waves induced in the direction from both single-exciton and two-exciton states.

The influence of annihilation processes can be clearly demonstrated by considering nonlinearity power index p dependencies on the pump intensity I. As a result of numerical modeling we have found that main features of the DFWM in the simplified two-level approximation[1] holds for the more complicated model under consideration. Exactly at the resonance, as it was found in two-level approximation too, nonlinearity power index becomes "expected 3" only for low level of excitation. As a result of saturation the nonlinearity index p reduces here to 0 in the absence of annihilation, whereas for high annihilation level ($\gamma > 3$) it keeps independent of the wavelength at some initial level (p<3) and does not change at all. This feature (saturation for the high pump and low annihilation) remains to some extend far away from the J-band provided transitions to two-exciton states are not considered. Otherwise, dramatic changes of the nonlinearity power index p dependencies on the pump intensity I appears in the in comparison to our previous[1] results. Two-exciton states were found as contributing significantly to the DFWM in the wings of the J-band. Inclusion of the two-photon resonant transitions to these states through the one-exciton intermediate states improves as it is shown in Fig.2. in a great extent the agreement between the experimentally received nonlinear power dependence of the DFWM and the theoretically estimated one.

REFERENCES

1. E. Gaizauskas, K.-H. Feller, and R. Gadonas, Optics Comm. **118**, 360 (1995).
2. R. Gadonas, K.-H. Feller, and A. Pugzlys, Optics Comm. **112**, 157 (1994).
3. F.C. Spano, Phys. Rev. **B46**, 13017 (1992).

105

ULTRAFAST WINDOW OF $\chi^{(3)}$-EFFECT IN CONJUGATED POLYMER

Ji-Ping Yang[1,2]

[1] FB Physik, Freie Universität Berlin
Arnimallee 14, 14195 Berlin, Germany
[2] Department of Applied Physics and Institut for Laser-Technology
Hefei University of Technology
Hefei/Anhui, Peoples Republic of China

Currently there is a strong interest in exploring the third-order nonlinear processes of conjugated polymers since they provide mechanisms for all-optical switching of light. Some groups have investigated two-photon absorption (TPA) of the conjugated polymer[1,2] because TPA is the dominant mechanism for the imaginary part of the third-order non-linearity. But the ultrafast feature of TPA is accompanied by a weaker long lived excited state absorption[1,2], which is not appropriate for ultrafast optical applications. We report the measurements of the TPA-spectrum of poly(p-phenylenevinylene) (PPV) by the pi-cosecond pump-probe technique, our experimental results clearly show an 'ultrafast window' (UW) for TPA without slow absorption tail.

The high molecular weight PPV sample was prepared via the sulphonium polyelec-trolyte preccursor route with a film thickness of 4.1 μm or 0.7 μm. For the transient TPA experiments a mode-locked Nd:YAG laser was used that produced pulses with 35 ps pulse width (FWHM). The fundamental wavelength of the laser (1.17 eV) was used as the pump beam. This photon energy (1.17 eV) is out of the residual absorption of PPV. Part of the light of the fundamental wavelength was focused into a 10 cm long D_2O flow cell to gen-erate a white light continuum. In our experiment the TPA around zero time delay is due to a TPA process where one photon is provided by the pump beam. Thus, by means of the white probe beam from 1.38 eV to 2.76 eV we can directly obtain the spectral dependence of the transient TPA from 2.55 eV (1.38 eV(probe) + 1.17 eV (pump)) to 3.93 eV (2.76 eV(probe) + 1.17 eV(pump)).

Fig. 1 shows the result of transient TPA-spectrum at zero time delay (open circles), plotted against the total (pump + probe) enegy. The linear absorption spectrum (solid line) is also shown in the figure for easy comparison of the energy position of the one- and two-photon excited states. The transient TPA-spectrum rises sharply above about 3.3 eV to a peak at 3.58 eV (two-photon energy). The TPA is further confirmed by the very clear lin-ear intensity dependence of the transient absorption signals on the pump beam at five

Ultrafast Processes in Spectroscopy, edited by Svelto et al.
Plenum Press, New York, 1996

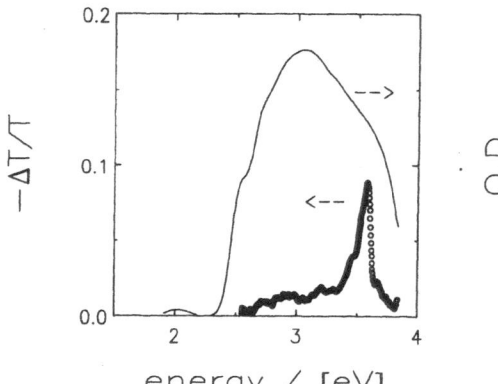

Figure 1. Linear absorption spectrum (solid line) and transient TPAspectrum (open circles) of PPV (sample thickness 0.7 μm) at zero time delay. The transient TPA spectrum is plotted versus two-photon energy (pump + probe).

probe wavelengths (2.32 eV, 2.28 eV, 2.14 eV, 1.94 eV and 1.70 eV) from the resonance-to non-resonance region at zero time delay for λ_{pump} = 1.17 eV. Consequently, the energy of the lowest even parity excited singlet can be deduced to be 1.22 eV (3.58 eV - 2.36 eV) above the lowest odd parity excited singlet state (2.36 eV)[3].

As can be seen from Fig. 2, the spectral waveforms of the photoinduced absorption change around the absorption peak is almost same for both shape and amplitude at the same positive- and negative delay times relative to zero time delay. This means that between about 2.14 eV (580 nm) - about 2.35 eV (527 nm) the TPA is not accompanied by a long lived component (note the 35 ps pulse width of the laser). This conclusion can further be verified by a Gauss-shaped fitting procedure for temporal waveform of the TPA signal at 2.34 eV. The later (solid line in Fig. 3) shows that the Gauss-width at FWHM is 48 ps, this 48 ps width correspondens a 34 ps (0.7 * 48 ps) Gauss-pulse width. This is identical with 35 ps laser pulse width at FWHM.

Therefore, the temporal shape at 2.34 eV shows that we just observe the two pulse correlation function. This indicates that no long lived intermediate state exists at that wavelength. In contrast to that, a long lived component can be clearly seen in Fig. 2 if λ > ~ 2.14 eV (580 nm), and thus the rise time of the signal as depicted in Fig. 3 is diffrent

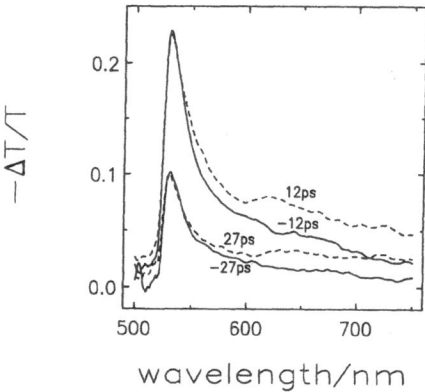

Figure 2. Transient TPA spectrum of PPV (sample thickness 4.1 μm) at various rise- and decay time after excitation.

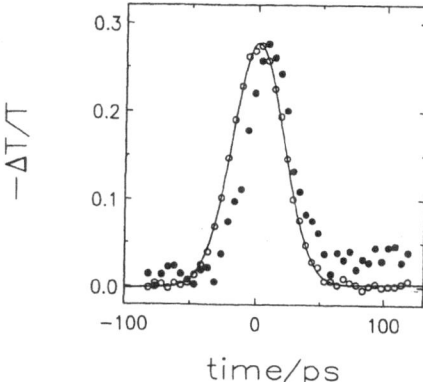

Figure 3. Time dependence of TPA at 2.34 eV (open circles) and 1.70 eV (solid circles). Solid line is Gauss-fitting, Gauss-width at FWHM is 48 ps.

from the slow decay time at long wavelengths. Because the signal at the low energy tail (> ~ 2.14 eV) at longer time delays comes from a real excitations due to TPA of the intense pump pulse, we think that there is an UW between about 2.14 eV and about 2.35 eV (note that for λ_{pump} = 1.17 eV) in which the ultrafast $\chi^{(3)}$-effect due to TPA is not accompanied by a one-photon absorption with a comparatively long lived excited odd parity state.

In summary, using a picosecond pump-probe technique we measured the energy position of the lowest even parity excited singlet (two-photon state) of PPV film to be 3.58 eV which is 1.22 eV above the lowest odd parity excited state (2.36 eV). Also, there is an UW for the ultrafast $\chi^{(3)}$-effect due to TPA in the ps-range between ~ 2.14 eV - ~ 2.35 eV for λ_{pump} = 1.17 eV. This experiment work was performed by Prof. Haarer's group of university Bayreuth, Germany.

REFERENCES

1. Y. Pang, M. Samoc and P.N. Prasad, J. Chem. Phys. **94**, 5282 (1991).
2. U. Lemmer et al, Chem. Phys. Lett. **203**, 28 (1993).
3. U. Rauscher, H. Bässler, D.D.C. Bradley and M. Hennecke, Phys. Rev. B **42**, 9830 (1990).

106

IMAGING THROUGH DIFFUSING MEDIA WITH TIME-RESOLVED TRANSMITTANCE

R. Cubeddu, A. Pifferi, P. Taroni, A. Torricelli, and G. Valentini

CEQSE-CNR and Dipartimento di Fisica
Politecnico, Milan, Italy

In the last years, in the field of medicine great attention has been paid to the possible development of noninvasive diagnostic techniques, such as the optical ones. Of special interest is the time-domain transillumination of the female breast for the early diagnosis of cancer. In particular, illuminating with a short light pulse and collecting only the first-arriving transmitted photons by means of time-gating, promising results have already been obtained by several research groups.[1-7]

Time-resolved measurements interpreted with the diffusion theory of photon transport allow a complete optical characterization of turbid media, through the simultaneous evaluation of the absorption and transport scattering coefficients (μ_a and μ'_s, respectively). Therefore this technique can be used, at least in principle, to detect differences in the optical properties, such as the ones observed between neoplastic areas and healthy surrounding tissue. A matrix of point measurements may allow one to obtain bidimensional information, and eventually an image in which pathological areas could be localized, due to their different scattering or absorption properties.

A cavity dumped dye laser, synchronously pumped by a mode-locked Argon laser, was used as the illumination source (<10 ps, $\cong$10 MHz, 650 nm). Time-resolved transmittance curves were collected by means of a system for time-correlated single photon counting with an impulse response function of less than 50 ps FWHM.[8] One-millimeter plastic-glass fiber optics allowed both illumination and light collection. With an incident power <1 mW, the typical acquisition time was $\cong$1 sec/curve, enough to collect 10^5 counts/curve. Measurements were performed every 2 mm over an area of 4 cm x 4 cm. The solution of the diffusion equation for a homogeneous slab[9] was best-fitted to the experimental curves. Since the samples were not homogeneous, it is not straightforward to interpret the fitted values of μ'_s and μ_a. Nevertheless they are related to the optical properties of the traversed medium and, at least in principle, sensitive to changes in the actual μ'_s and μ_a. Images were constructed by plotting either of the fitted optical coefficients as a function of position. All the images are displayed expanding the range of measured values to occupy a full 64-gray scale. A piecewise bilinear interpolation is performed to smooth the discontinuities between adjacent pixels. For easier comparisons, when necessary the

Ultrafast Processes in Spectroscopy, edited by Svelto et al.
Plenum Press, New York, 1996

images are inverted to represent the inclusion always darker than the background. To render quantitative information, ticks are reported every 4 mm and a contour line, corresponding to 50% of full scale, is overlapped.

Time-gated images were constructed from the same experimental data, plotting the gated intensity within a time-gate Δt, starting the instant when the first-arriving unscattered photons are expected. In our experimental conditions, the best results were obtained with Δt = 200 ps, corresponding to the best compromise between spatial resolution and signal to noise ratio.

The performances of the proposed method based on the fitting were compared to the ones achievable through time-gating.

Measurements were performed on realistic tissue phantoms. The selected optical properties were typical for tissues at near infrared wavelengths (700–900 nm)[10] and the thickness (4.3 cm) was comparable with the average value for compressed tissue in x-ray mammography, which is presently the most widespread technique for the clinical diagnosis of breast cancer. The phantoms were solid slabs made of 1% agarose in distilled water, containing appropriate amounts of ink (absorber) and Intralipid® (diffuser). The presence of a tumor mass was simulated through embedded agarose cylinders of various optical properties and dimensions, from 11-mm diameter and 7-cm height down to 6-mm diameter and 8-mm height. The optical coefficients of the background, simulating the healthy tissue, were kept constant at μ'_{sBKG} = 8 cm^{-1} and μ_{aBKG} = 0.1 cm^{-1}, while the parameters of the inclusion were varied separately. The transport scattering coefficient of the inclusion (μ'_{sINC}) was either increased or decreased by a factor of two or less with respect to the background value (μ'_{sINC} = 4, 12, and 16 cm^{-1}) and the absorption coefficient μ_{aINC} by a factor of three or less (μ_{aINC} = 0.03, 0.05, and 0.2 cm^{-1}). The optical difference between inclusion and background was kept small enough to be realistic.

Aqueous solutions of Intralipid and ink as well would have provided realistic optical properties. However the liquid models require a solid wall, such as a plastic or glass tube, to isolate the two regions with different optical coefficients, used to simulate healthy and neoplastic tissues, respectively. As reported by other authors[11–12] and confirmed also by our group, the detection system is sensitive to the presence of such a wall in the solution, and this causes unavoidable artifacts. On the contrary, the solid agarose model allows us to avoid any problems related to the interfaces between adjacent regions of different optical properties.

The method based on the fitting of the transmittance curves proved sensitive to changes in the scattering properties (both increase and decrease in μ'_s) and allowed us to localize the inclusion with good contrast in all the experimental conditions considered. In general, the noise level is lower and image quality better than with time-gating, especially for small scattering differences (i.e. 50% difference between inclusion and background). A typical example is reported in Fig .1, referring to the inner end of a cylindrical inclusion (6 mm in diameter), with μ'_s = 16 cm^{-1} (i.e. twice the background value). The fitted μ'_s is displayed on the left and the gated intensity in a 200-ps gate on the right.

The 50% of full scale contour line was usually considered as indicative of the shadow boundary. In the case shown in Fig. 1, in the plot of the fitted μ'_s the dimensions of the shadow are rather larger than the actual dimensions of the inclusion. This inaccuracy can be at least in part explained by taking into account that measurements were performed every 2 mm and the inclusion diameter is only 6 mm. A higher spatial resolution would be expected if a smaller step-size were considered. This hypothesis is supported by the experimental finding that, when images of 11-mm diameter inclusions are acquired with 2-mm step, the fit allows an acceptable estimate of the inclusion size.

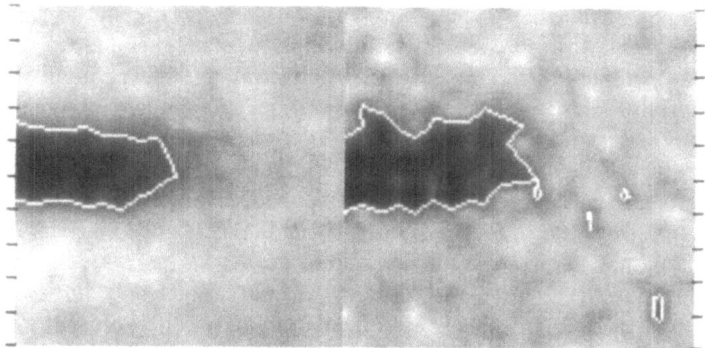

Figure 1. Fit (left) and gated (right) images of a scattering inhomogeneity. Inclusion diameter = 6 mm. A contour line at 50% of full scale is overlapped.

When absorption inhomogeneities are considered, with μ_{aINC} either higher or lower than μ_{aBKG}, a very broad shadow, much larger than the actual size of the inclusion, is obtained with both the fit and the gate. Moreover, if the absorption difference between inclusion and background is small, the noise level may become critical. Even when the absorption coefficient of the inclusion differs from the absorption of the background by a factor of 3, it is not possible to define the exact shape and size of the inhomogeneity. An example is given in Fig. 2, showing the images of the inner end of a cylindrical inclusion (11-mm diameter) with $\mu_{aINC} = 0.03$ cm$^{-1} = 0.3\mu_{aBKG}$. The fitted μ_a is displayed on the left and the gated intensity on the right.

In general, with absorption inhomogeneities, the image quality is comparable for the two methods or slightly better with the gate. Both methods allow the detection of an absorption inhomogeneity, but its accurate description (shape, size, and position) through the acquisition of good quality images is not possible.

Finally it should be noted that the plot of the absorption coefficient for a scattering inclusion is not perfectly uniform (data not shown). This behavior is not completely unexpected, since we have used a theoretical model valid for a homogeneous medium to interpret experimental data collected from heterogeneous samples.. However, the absorption plot is definitely more noisy that the image displaying the scattering coefficient. Similar

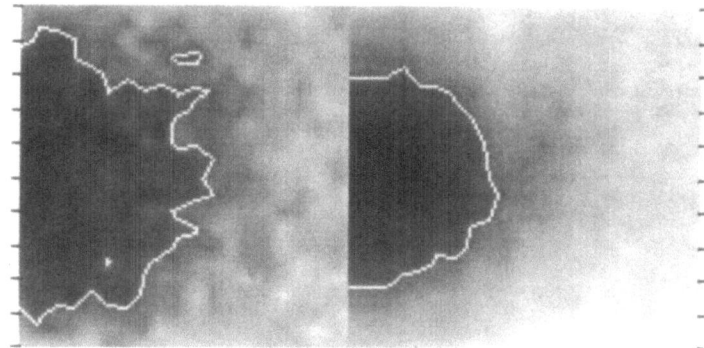

Figure 2. Fit (left) and gated (right) images of an absorption inhomogeneity. Inclusion diameter = 11 mm. The contour line at 50% of full scale is overlapped.

observations can be made when dealing with absorption inclusions: in this case, the plot of the fitted μ'_s is characterized by a significantly higher noise level and the localization of the inclusion is even more difficult than with the absorption plot. The higher quality of the scattering (absorption) image for scattering (absorption) inhomogeneities can be explained by considering the absolute values of the fitted coefficients. The contrast in the plot of the optical parameters was determined through the following procedure. The average value of the fitted coefficients was evaluated along the inclusion axis ($<\mu_{aINC}>$ and $<\mu'_{sINC}>$) and in the background ($<\mu_{aBKG}>$ and $<\mu'_{sBKG}>$), far from the inclusion. The contrast C was defined as the relative variation of the fitted coefficient: $C = (<\mu_{INC}> - <\mu_{BKG}>)/<\mu_{BKG}>$. For scattering inclusions, the contrast C_s for the transport scattering coefficient is always higher than the contrast C_a for the absorption coefficient, usually at least by a factor of 2. Similarly, for absorption inhomogeneities, $C_a > 2C_s$.

In conclusion, the proposed method, based on time-resolved transmittance data interpreted with a diffusion model, allows one to acquire good quality images of scattering inhomogeneities with realistic optical parameters. Absorption inclusions can be detected, even though images cannot be obtained. With respect to time-gated imaging, the fit method has the interesting potential to discriminate scattering from absorption inhomogeneities. This feature can be particularly useful in clinical applications, to discriminate malignant from non-malignant lesions. With the present experimental set-up, the acquisition time would already be acceptably low for a clinical examination. Moreover, the illumination power could be increased without exceeding the safety rules, and preliminary measurements seem to suggest that the number of counts/curve could be decreased without sensible worsening of the fit outcomes. Therefore promise is shown for a possible future application in clinical diagnosis.

REFERENCES

1. B. C. Wilson, E. M. Sevick, M. S. Patterson, and B. Chance, Proc. IEEE **80**, 918 (1992).
2. G. Mitic, J. Kölzer, J. Otto, E. Piles, G. Sölkner, and W. Zinth, Appl. Opt. **33**, 6699 (1994).
3. J. C. Hebden and D. T. Delpy, Opt. Lett. **19**, 311 (1994).
4. S. Andersson-Engels, R. Berg, S. Svanberg, and O. Jarlman, Opt. Lett. **15**, (1990).
5. M. R. Hee, J. A. Izatt, E. A. Swanson, and J. G. Fujimoto, Opt. Lett. **18**, 1107 (1993).
6. M. D. Duncan, R. Mahon, L. L. Tankersley, and J. Reintjes, Opt. Lett. **16**, 1868 (1991).
7. B. B. Das, K. M. Yoo, and R. R. Alfano, Opt. Lett. **18**, 1092 (1993).
8. R. Cubeddu, M. Musolino, A. Pifferi, P. Taroni, and G. Valentini, IEEE J. Quantum Electron. **30**, 2421–2430 (1994).
9. M. S. Patterson, B. Chance, and B. C. Wilson, Appl. Opt. **28**, 2331 (1989).
10. V. G. Peters, D. R. Wyman, M. S. Patterson, and G. L. Frank, Phys. Med. Biol. **35**, 1317 (1990).
11. J. C. Hebden and K. S. Wong, Appl. Opt. **32**, 372 (1993).
12. S. Andersson-Engels, R. Berg, and S. Svanberg, J. Photochem. Photobiol. B **16**, 155 (1992).

107

TIME AND SPACE RESOLVED REFLECTANCE FOR THE MEASUREMENT OF OPTICAL COEFFICIENTS OF TISSUES

J. -M. Tualle , B. Gélébart , and S. Avrillier

Laboratoire de physique des lasers
Université Paris XIII
93430 Villetaneuse, France

Average coefficients measured in the near infrared spectral range on relatively uniform volumes of biological tissues can give information about their structure and their functions [1]. This spectral range is particularly convenient for this purpose since biological tissues present a transmission window between 700 and 1500nm. Our purpose was to show that a global analysis of the spatial and temporal map of the reflectance registered with a streak camera can be used to obtain quantitative values of the optical coefficients in a very short time [2]. To enlarge the number of possible medical applications, we focalise on diffuse light reflectance measurements, in the semi-infinite geometry, which can be applied as a non-contact technique to thick tissue.

Short light pulses of about 100 fs (FWHM) were produced by a self mode-locked Ti-Sa laser, at λ=800 nm, with a 100 MHz repetition rate. The light reached the medium through a short optical multimode fiber (about 15 cm long) applied at the surface of the sample (see figure 1). The average delivered power was 100 mW which is compatible with in vivo applications.

The time and space-resolved reflectance was recorded with a Hamamatsu streak camera, model C4334. This single shot model was synchronised with the laser source through a fast photodiode. Time resolution was guaranteed by Hamamatsu to be 15 ps, but we measured a 35 ps resolution with our set-up, which cannot be interpreted only by the fiber dispersion. The fact that we were working at 100 MHz, a little more than the 80 MHz prescripted by Hamamatsu, could be responsible for this lower resolution.

The streak camera directly measured the time and space resolved diffuse reflectance $R(\rho,t)$ along a segment of the sample's surface imaged on the input slit of the camera. In all experiments we explored the medium for a distance ρ from the source ranging from 2 to 10 mm, that is in areas corresponding to the validity range of the diffusion approximation.

The diffusion equation depends only on the absorption coefficient μ_a, ranging from 0.01 to 1 cm^{-1}, and on the reduced scattering coefficient μ_s' ranging from about 10 to 100

Ultrafast Processes in Spectroscopy, edited by Svelto et al.
Plenum Press, New York, 1996

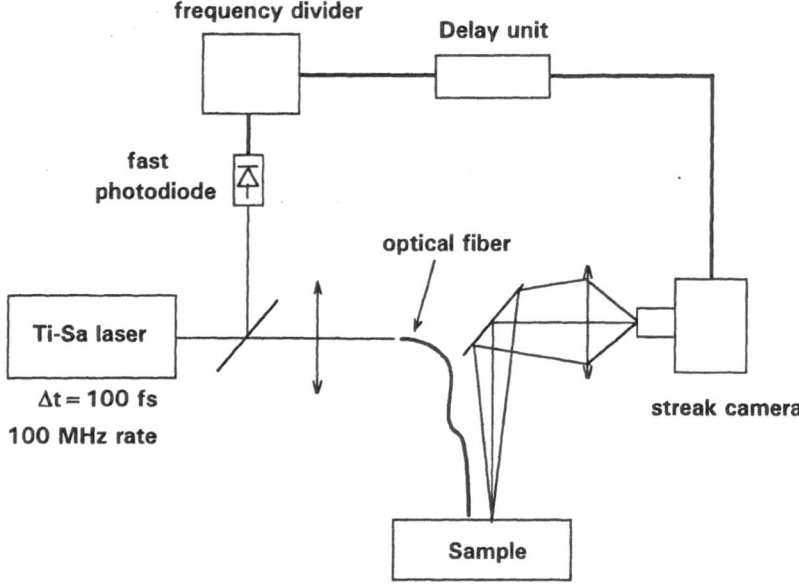

Figure 1. Experimental setup.

cm^{-1}. As it has been suggested by Patterson et al [3], we deduced these coefficients from our measurements by using equations

$$\frac{\partial}{\partial t}\ln R(\rho,t) \approx -\frac{5}{2t} - \mu_a c + \frac{\rho^2}{4Dct^2} \qquad (1)$$

and

$$\rho^2 \approx 4D\mu_a c^2 t_{max}^2 + 10Dct_{max} \qquad (2)$$

where $t_{max}(\rho)$ is the time position of the maximum of $R(\rho,t)$ at fixed ρ, and $D = [3(\mu_a + \mu_s')]^{-1}$ is the diffusion coefficient. According to equation (2) a parabolic fit of function $\rho^2 = f(t_{max})$ could have led *a priori* to the determination of μ_a, D and time zero, but the dispersion of experimental results forbade an accurate estimation of the second order term and only allowed a reasonable linear fit. Assuming that $\mu_a \approx 0$, we obtained a first evaluation of D, but this estimation had to be improved. Using this rough estimation of D, equation (1) allowed an evaluation of μ_a, which was then reported to equation (2) giving another estimation of D and time zero. This process was automatically reiterated showing a fast convergence (less than 1 second computation time).

The accuracy of this method was tested with calibrated suspensions of latex microspheres in water. The mean diameter of the spheres was 435nm with a standard deviation of 7%. The concentration of these suspensions was adjusted to obtain μ_s' values ranging from 10 to 100 cm^{-1}. Results for μ_s' and μ_a obtained with these suspensions are shown on fig 2.

μ_s' follows a linear law versus concentration, as predicted from theory at low concentrations. The experimental value of the corresponding linear fit is 30.6±0.5 cm^{-1}/%,

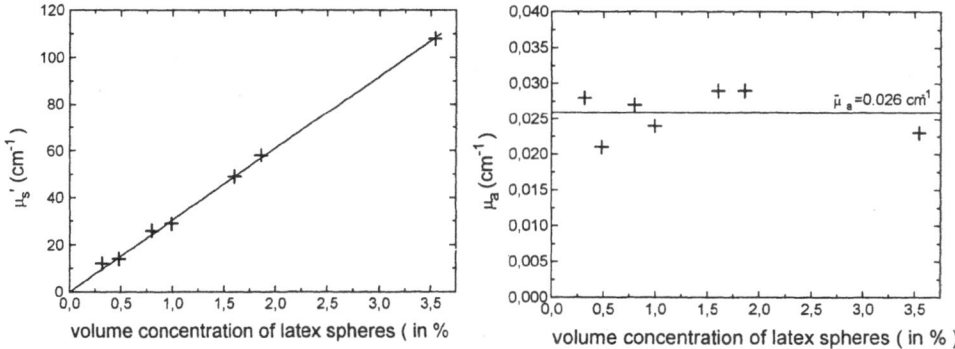

Figure 2. Experimental values of μ_s' and μ_a versus latex concentration.

which compares well with the 30 ± 1 cm^{-1}/% calculated from the Mie Theory. Since latex's absorption is very low [4], the main absorption came from water and μ_a should be expected to be constant. The measured mean value of μ_a was $\bar{\mu}_a \approx 0.026\,cm^{-1}$ (figure 2) which is in good agreement with the value of $\mu_a = 0.023$ cm^{-1} reported by Hale et al [5] at 780 nm.

The uncertainty on the μ_s measurements was estimated to be about 1 to 2.5 cm^{-1}. The precision of the measurements of μ_a was estimated to be 15~20%.

In vitro optical coefficients of different freshly excised tissues were measured with this technique.

Due to the small value of μ_s the uncertainty is of course quite high in the case of bovine liver. These results are consistent with the little data previously published for tissues in the near infrared spectral range [see for exemple 3,6–9].

One of the main advantages of the method presented here is the rapidity of measurements and the possibility of its automation. We have shown that by using this non invasive technique, almost real time and accurate measurements were possible. The next step should be in vivo experimentation.

ACKNOWLEDGMENT

The authors wish to thank Hamamatsu for the supply of the streak camera and the French Army Health Authority (Service de Santé des Armées) for financial support.

Table 1.

Tissue	μ_s' (cm^{-1})	μ_a (cm^{-1})
Porcine brain (grey matter)	31±2	0.07±0.01
Bovine liver	8±2	0.26±0.05
Whole milk	39±2	0.032±0.005

REFERENCES

1. B. Chance, J.S. Leigh, M. Miyake, D.S. Smith, S. Nioka, R. Greenfeld, M. Finander, K. Kaufman, W. Levy, M. Young, P. Cohen, H. Yoshioka and R. Boretsky, Proc. Natl. Acad. Sci. USA **85** (1988) 4971.
2. J.-M. Tualle, B. Gélébart, E. Tinet, S. Avrillier and J-P. Ollivier, Optic Communications, in press.
3. M.S. Patterson, B. Chance and B.C. Wilson, Applied optics **28** (1989) 2331.
4. G. Zaccanti and P. Donelli, Applied optics **33** (1994) 7023.
5. G.M. Hale and M.R. Querry, Applied optics **12** (1973) 555.
6. F. Bevilacqua, P. Marquet and C. Depeursinge, SPIE **2326** (1994) 173.
7. F. Bevilacqua, P. Marquet and C. Depeursinge, SPIE **2326** (1994) 173.M.H. Eddowes, T.N. Mills and D.T. Delpy, SPIE **2326** (1994) 117.
8. G. Zaccanti, A. Taddeucci, P. Bruscaglioni, F. Martelli, A. Sassaroli, SPIE **2389** (1995) in press.
9. A. Roggan, H. Albrecht, K. Dörschel, O. Minet, G. Müller, SPIE **2323** (1994) 16.

108

APPLICATION OF ULTRASHORT X-RAY PULSES TO BIOLOGICAL AND MEDICAL IMAGING

S. Svanberg,[1] M. Grätz,[1] K. Herrlin,[2] I. Mercer,[1] A. Persson,[1] C. Tillman,[1] and C. -G. Wahlström[2]

[1] Department of Physics
Lund Institute of Technology
P.O. Box 118, S-221 00 Lund, Sweden
[2] Department of Diagnostic Radiology
Lund University Hospital
S-221 85 Lund, Sweden

Medical imaging is developing very swiftly with x-ray radiography, magnetic resonance imaging, ultrasound and scintigraphy as well established techniques widely used in the clinical praxis [1]. During recent years a number of laser-based techniques have been put forward, some of which are under clinical testing. Thus, laser-induced fluorescence imaging using emission from tissue endogenous chromophores or tumour-seeking administered agents is emerging as a tool, e.g. for early cancer detection. (See, e.g. [2,3]). Quasi-elastic light scattering is used for blood perfusion imaging utilising minute Doppler shifts due to blood cell movements [4]. Both these techniques interrogate only superficial layers because of the limited penetration of light into tissue. For deeper sounding, x-rays are normally employed. During the last few years this very powerful and safe technique has come under some discussion, because of hypersensitivity to ionising radiation in patients carrying the AT gene [5]. Partly because of that there is a considerable interest for optical transillumination of tissue, particularly for mammography. Conventional transillumination is hampered by image blurring due to heavy multiple scattering in tissue. This has spurred the development of several techniques attempting to reduce the influence of scattering (for an overview, see e.g. [6]). One of the methods, that appear promising is the gated-viewing technique using single-photon counting [7,8]. At this point it is not clear how far this optical transillumination technique will be clinically applicable. X-rays of medical interest can be generated by focusing high-peak-power, short-pulselength radiation onto a high-Z target material [9]. This is a new method, that might complement the traditional way of accelerating electrons into a high-Z anode. We have used such laser-produced x-rays for imaging of biological samples onto standard image plates, which are employed in radiography. Initial experience [10,11] shows, that very sharp images can be obtained and that

Ultrafast Processes in Spectroscopy, edited by Svelto et al.
Plenum Press, New York, 1996

483

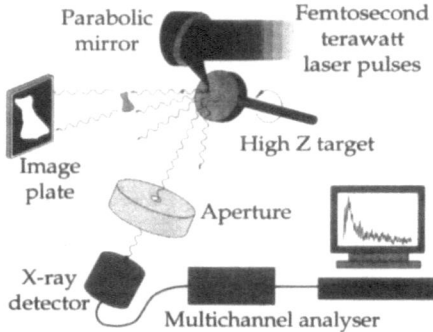

Figure 1. Experimental set-up for experiments with laser-produced x-rays. (From [14].)

high-magnification radiography, and even single-pulse imaging, can be performed. In contrast to the optical region, a non-negligible component of non-scattered radiation gives rise to clear shadows. However, also here scattering occurs, that yields a strong background, especially for whole-body transillumination. Gated viewing might then again be interesting, since a reduced radiation dose would be needed when detrimental broadband background can be eliminated, thus reducing the noise in the images [12]. Another aspect,

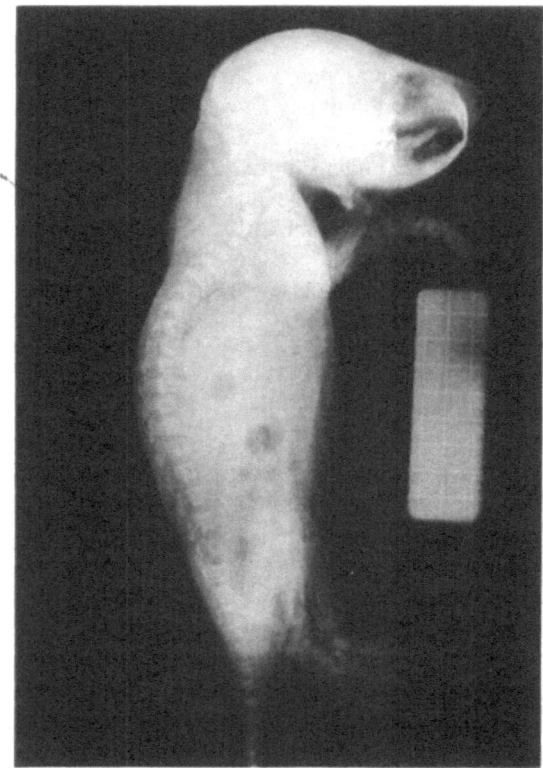

Figure 2. X-ray image of a 4 day old Syrian hamster. (From [12]).

which is also related to dose reduction, is differential imaging in connection with the use of contrast agents, such as iodine or gadolinium. It is then advantageous to use monochromatic radiation bridging the K absorption edge of the agent. Synchrotron radiation has proved to be very useful in this respect [13]. However, since such installations are very costly and scarce we have investigated the possibilities for differential imaging using laser-produced plasmas [14]. It is then important to optimise the emission of characteristic x-ray lines, and hard x-ray spectrometer techniques must be employed. Since such methods are not readily available, we have also engaged in such development using differential absorption [15] and crystal spectrometry [16].

The experimental set-up used in our x-ray experiments is shown in Fig. 1. Laser pulses from the Lund terawatt titanium-sapphire laser system [17], having typically 150 fs duration and an energy of 200 mJ/pulse, were focused with an f/1 off-axis parabola at a rotating target, normally consisting of a tantalum or a gadolinium foil. The target is kept at the laser focus during rotation using a piezo-electric servo system, and the optics is protected from plasma debris by a thin microscope slide, being automatically replaced when required.

An example of an x-ray image of a 4-day-old Syrian hamster is shown in Fig. 2, featuring much structure. The exposure time needed for this image, 5 minutes at 10 Hz, shows that a practical X-ray radiography system must be much improved, which seems feasible.

The principle of differential-absorption imaging is shown in Fig. 3, where schematic x-ray emission spectra for a tantalum and a gadolinium target are shown together with a

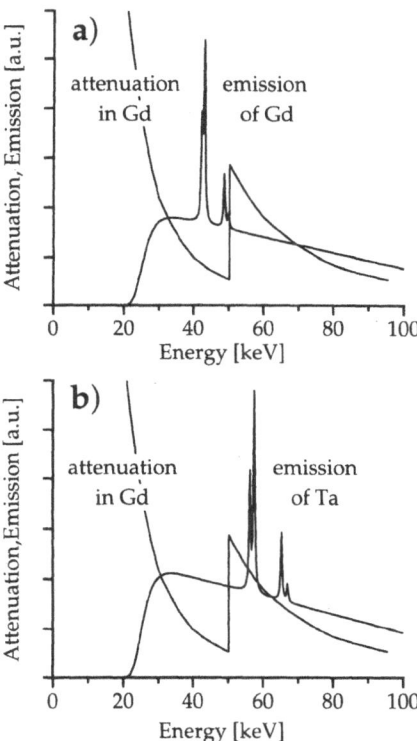

Figure 3. Schematic emission spectra for gadolinium (upper) and tantalum (lower) targets together with absorption spectrum for gadolinium. (From [14]).

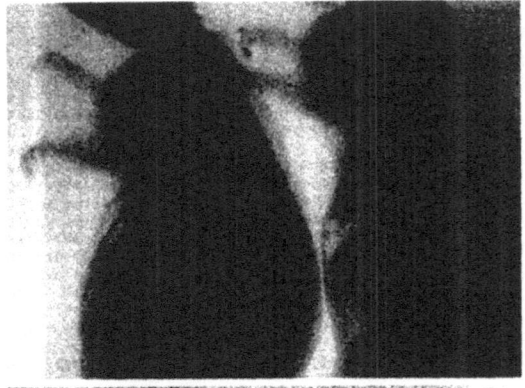
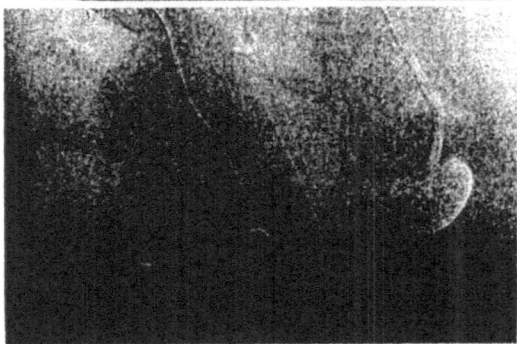

Figure 4. Image of sacrificed rats, with gadolinium and cerium contrast agents. After division of images taken with a gadolinium and a tantalum target, the stomach containing gadolinium (to the right) appears brighter. (From [14]).

gadolinium absorption profile. If the strength of the characteristic emission lines is substantial, images of a biological specimen containing a gadolinium contrast agent will differ at locations where the agent is localised. A demonstration of this effect is shown in Fig. 4, where rats, featuring gadolinium and cerium contrast agent in their stomachs, show up differently when images for the two target materials are divided by each other. In the right hand case the emission lines are bridging the K edge, in the left hand case they both fall on the lower side of the K-edge. Clearly, the effect would appear much more clearly for monochromatic radiation. Conditions for optimum generation of characteristic lines will be explored when proper hard x-ray spectrometry has been established.

The ultra-sharp imaging, related to the small extent of the source, and differential-imaging features might for the future be combined with suppression of Compton-scattered background photons using gated viewing with picosecond x-ray bursts [12].

ACKNOWLEDGMENT

This work was supported by NFR, MFR and the EC Programme (CHRX-CT93–0346).

REFERENCES

1. R.S. McKay, *Medical Images and Displays: Comparisons of Nuclear Magnetic Resonance, Ultra-sound, X-Rays and Other Modalities* (Wiley, New York 1984).

2. S. Svanberg, in *Lasers in Medicine*, ed. by R. Waynant and G. Pettit (Plenum, New York, to appear).
3. S. Andersson-Engels and B.C. Wilson, J. Cell Pharmacol. **3**, 48 (1992).
4. K. Wårdell, A. Jacobsson and G.E. Nilsson, IEEE Trans. Biomed. Eng. **40**, 309 (1993).
5. M. Swift, D. Morrell, R.B. Massey, and C.L. Chase, New Engl. J. Med. **325**, 1831 (1991)
6. G. Müller, B. Chance, R. Alfano, S. Arridge, J. Beuthan, E. Gratton, M. Kaschke, B. Masters, S. Svanberg and P. van der Zee, eds., *Medical Optical Tomography: Functional Imaging and Monitoring* (SPIE, Bellingham, VA 1993), Vol. IS11.
7. S. Andersson-Engels, R. Berg, O. Jarlman, and S. Svanberg, Opt. Lett. **15**, 1179 (1990).
8. R. Berg, O. Jarlman, and S. Svanberg, Appl. Opt. **32**, 574 (1993).
9. J.D. Kmetec, C.L. Gordon III, J.J. Macklin, B.E. Lemoff, G.S. Brown and S.E. Harris, Phys. Rev. Lett. **68**, 1527 (1992).
10. K. Herrlin, G. Svahn, C. Olsson, H. Pettersson, C. Tillman, A. Persson, C.-G. Wahlström and S. Svanberg, Radiology **189**, 65 (1993).
11. C. Tillman, A. Persson, C.-G. Wahlström, S. Svanberg, and K. Herrlin, Appl. Phys. **B61**, 333 (1995).
12. C.L. Gordon III, G.Y. Yin, B.E. Lemoff, P.M. Bell, and C.P.J. Barty, Opt. Lett. **20**, 1056 (1995).
13. W. Thomlinson, N. Gmür, D. Chapman, R. Garret, L. Lazarz, H. Moulin, A.C. Thompson, H.D. Zeman, G.S. Brown, J. Morrison, P. Reiser, V. Padmanabahn, L. Ong, S. Green, J. Giacomini, H. Gordon, and E. Rubinstein, Rev. Sci. Instruments **63**, 625 (1992).
14. C. Tillman, I. Mercer, S. Svanberg, and K. Herrlin, J. Opt. Soc. Amer. **13** (1996).
15. C. Tillman, A. Persson, C.-G. Wahlström, S. Svanberg and K. Herrlin, *Proc. High Field and Short Wavelength Generation Conference*, St. Malo, August 21–25, 1994; and I. Mercer et al., to appear.
16. M. Grätz, C. Tillman, C.-G. Wahlström, S. Svanberg, G. Höltzer and E. Förster, in *Generation and Application of Ultrashort X-Ray Pulses II*, Pisa, September 20–23, 1995.
17. S. Svanberg, J. Larsson, A. Persson and C.-G. Wahlström, Phys. Scripta **49**, 187 (1994).

109

FEMTOSECOND INFRARED AND VISIBLE SPECTROSCOPY OF BIOMOLECULES

P. Hamm, M. Zurek, T. Röschinger, and W. Zinth

Intstitut für Medizinische Optik
Ludwig Maximilians Universität München, Germany

1. INTRODUCTION

The technical progress in the generation of ultrashort infrared pulses in a wide spectral range up to 11 μm allows now vibrational spectroscopy with femtosecond time resolution [1,2]. At present, visible-pump and infrared-probe experiments are performed which give new insights into the structural rearrangement in molecules during photoreactions. In this paper we present two examples where femtosecond IR spectroscopy is used to obtain new information on fast processes in molecules of biological relevance. (i) New data on the photo isomerisation of retinal will be presented showing that different reaction channels exist. A preliminary assignments of these channels will be given. (ii) For the primary charge separation in photosynthetic reaction centers femtosecond IR spectroscopy has revealed an additional fast reaction process - unresolved in visible and near IR spectroscopy- which could be of relevance for some of the still open questions of photosynthetic electron transfer.

2. FEMTOSECOND SPECTROSCOPY OF THE PHOTOISOMERISATION OF RETINAL

The light induced isomerisation of the retinal molecule (more exactly of a protonated Schiff base of retinal, PSBR) plays an essential role in several biological systems like rhodopsin (responsible for vision), bacteriorhodopsin (photosynthetic light induced proton pump) and halorhodopsin (light induced chloride pump). These systems were investigated by femtosecond spectroscopy in the visible range showing fast reactions on the S_1 potential surface prior to the transition to the ground state. Isomerisation is completed extremely fast in 200 fs in rhodopsin [3], in 500 fs in bacteriorhodopsin [4] and 1.5 ps in halorhodopsin [5]. The reaction speed and the selectivity of the isomerisation motion depends strongly on the protein surrounding of the retinal chromophore. We chose isolated retinal (PSBR) to study the photoisomerisation in solution (ethanol) by femtosecond IR and visible spectroscopy.

Ultrafast Processes in Spectroscopy, edited by Svelto et al.
Plenum Press, New York, 1996

2.1. VIS Experiments

In a first step, VIS pump/probe experiments (pump pulse at 407 nm: SHG of a Ti:sapphire amplifier, probing pulse: white light continuum, 450–750 nm, cross correlation width: 150–200 fs) were used to resolve the basic reaction scheme of retinal: Immediately after electronic excitation of all-trans retinal a reaction takes place which leads the system on the 100 fs time scale into a region on the S_1 potential surface from where stimulated emission (with large Stokes shift of 8000 cm^{-1}) takes place. Afterwards, the molecules decay biexponentially with 2 ps and 7.5 ps into the ground state S_0. This can be deduced from both the decay of the excited state absorption and the stimulated emission. The transient absorbance changes at late delay times (> 50 ps) suggest that a fraction of the retinal molecules has isomerized from the all-trans to the 11-cis configuration. Since the wavelength dependencies of the 2 ps and the 7.5 ps kinetics behave differently we conclude that the decay of the excited state occurs from two points on the S_1 potential surface and that this splitting decides whether the molecule relaxes back to the all-trans state or to the isomerised 11-cis state (see below).

2.2. Mid IR-Experiments

In a second experiment, time resolved vibrational spectra (probing pulse generated by difference frequency mixing in AgGaS$_2$ crystal, cross correlation width 300 fs) are used to follow the pathway from the excited S_1 state back to the all-trans configuration or forward to the 11-cis configuration. Both kinetics (the 2 ps and the 7.5 ps transients) coupled to the decay of the S_1 state are clearly detected in the mid IR range between 1150–1300 cm^{-1} (C-C range, see Fig. 1) and between 1480 -1720 cm^{-1} (C=C and C=N range). It is interesting to note that the amplitude spectra related to these two kinetics are completely different. The 7.5 ps transient dominates near to absorption lines of all-trans retinal while the 2 ps transient is observed at frequency positions which are assigned to 11-cis retinal absorption lines. A first analysis of the data leads to the conclusion that the 7.5 ps process is related to the reaction back to the all-trans ground state while the 2 ps decay is connected to the reactive channel with the isomerisation into the 11-cis state.

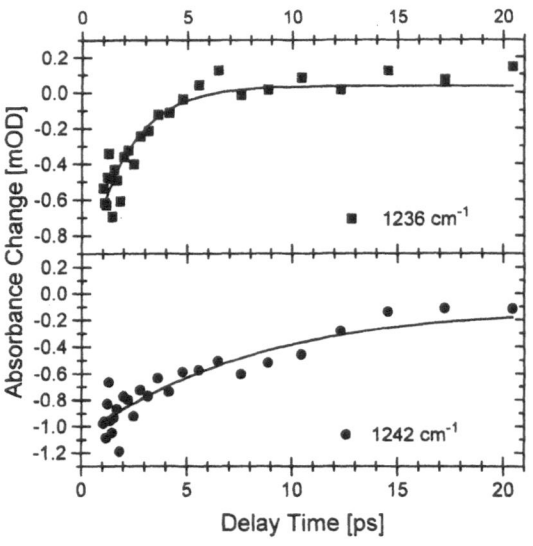

Figure 1. Transient absorbance changes at 1236 cm^{-1} (near to a absorption line of 11-cis retinal) and at 1242 cm^{-1} (near to a absorption line of all trans retinal). A fit of the first experiment yields a time constant of 2 ps, a fit of the second experiment a time constant of 7.5 ps.

3. THE INITIAL PHOTOREACTION IN BACTERIAL REACTION CENTERS

The reaction center (RC) from purple bacteria is the key unit to perform the primary charge separation and energy storage in bacterial photosynthesis. Two basic experimental techniques have been applied to investigate the function of this unit: (i) The structure of bacterial RC's, a large chromophore-protein complex, was determined by x-ray scattering with atomic resolution. (ii) Time resolved spectroscopy in the visible or the near IR was applied to characterize the kinetics of the photoreaction as well as the electronic states of the chromophores. After light absorption, an electron is transferred from a dimer of two bacteriochlorophyll molecules (special pair P) to a bacteriochlorophyll monomer B in $\approx$3–4 ps, to a bacteriopheophytin H in $\approx$1 ps and finally to a quinone Q in $\approx$200 ps. However, only one of the two, almost symmetric electron transfer (ET) branches seems to be essential for this reaction. This so-called 'unidirectionality of ET' is one of the remaining open questions of bacterial photosynthesis which are not answered by the known visible experiments. Femtosecond spectroscopy in the mid-IR is used to give supplementary information on the reaction mechanism, on the dynamics of the protein residues surrounding the chromophores and on the electronic structure of the excited state P* and of the oxidized state P^+.

In this paper, we will focus on an initial photo reaction of the primary electron donor P. Recent femtosecond IR-experiments gave first evidences for a process which occurs on a time scale of 200 fs and which was not detected in previous visible experiments [6]. Consequently, an initial state P** is populated prior to the well known excited state P*. Fig. 2 shows the difference spectrum of this state P** in comparison with the difference spectra of the 'relaxed' excited state P*. Especially in the low frequency range between 1000 and 1500 cm^{-1}, the overall absorption cross section increases considerably during the 200 fs reaction from P** to P*. We assign this absorption increase to a change of the configuration of the two bacteriochlorophyll molecules of the dimer P along a low frequency molecular mode causing a considerable change of its electronic structure (see ref. [6] for details). This motion is able to disturb the symmetry of the special pair leading to an increased charge transfer character of P*. In the same time it may facilitate energy trapping from the antenna system into P. In addition, the charge transfer character of P* may be one reason for the unidirectionality of the further charge transfer along the active chromophore branch.

Figure 2. Difference spectra of the electronically excited states P** (t_D=0 ps) and P* (t_D=0.5 ps). Pronounced absorption changes are observed during the reaction from P** to P* within 200 fs which are explained by an intramolecular charge separation within the special pair.

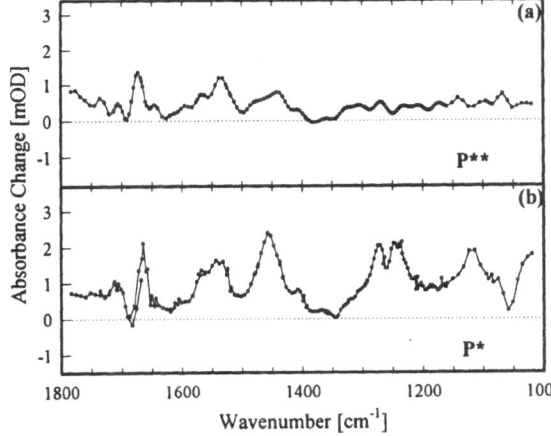

REFERENCES

1. P. Hamm, C. Lauterwasser, W. Zinth Opt. Lett. **18,** 1943, (1993)
2. P. Hamm, S. Wiemannm, M. Zurek, W. Zinth, Opt.Lett. **19,** 1642, (1994)
3. Q. Wang, R. W. Schoenlein, L. A. Peteanu, S. J. Rosenthal, R. A. Mathies and C. V. Shank in ´Ultrafast Phenomena IX´ eds. P.F. Barbara, W. H. Knox, G. A. Mourou A. H Zewail, Springer 1994, pp. 425
4. J. Dobler, W. Zinth, W. Kaiser, Chem. Phys. Lett., **144,** 215, (1988)
5. T. Arlt, S. Schmidt, W. Zinth, U. Haupts, D. Oesterhelt, Chem. Phys. Lett., **241,** 559, (1995)
6. P. Hamm, W.Zinth, J. Phys. Chem. 99, 13537, (1995)

110

ULTRAFAST RELAXATION OF EXCITONS IN THE BACTERIOCHLOROPHYLL ANTENNA PROTEINS FROM GREEN PHOTOSYNTHETIC BACTERIA

A. Freiberg,[1,2] S. Lin,[2] W. Zhou,[2] and R. E. Blankenship[2]

[1] Institute of Physics
Tartu, Estonia
[2] Center for the Study of Early Events in Photosynthesis
Arizona State University
Tempe, Arizona

Exciton dynamics in molecular nanoaggregates is currently an active research area. Photosynthetic antenna proteins present an example of native nanoaggregates of (bacterio)chlorophyll molecules in specific protein surrounding. Besides this cognitive interest, investigations of energy relaxation rates and pathways in antenna proteins are needed for better understanding of highly efficient light harvesting, excitation transfer, and energy trapping processes in photosynthesis[1]. Unfortunately, the optical spectroscopy of chromoproteins is tangled due to poor resolution of the spectra of most pigment-proteins. Little is usually known also about three-dimensional structure of these proteins.

In this work, we report on ultrafast (subpicosecond and picosecond) dynamics of excitons in a relatively simple bacteriochlorophyll -protein aggregate, known as FMO protein. The FMO proteins exhibit clear-cut absorption and fluorescence emission spectra at low temperatures (≤ 200 K)[2]. This allows selective excitation and probing of spectrally defined exciton states. The spatial structure of the FMO protein is known as well[3]. It consists of three identical 47 kDa protein subunits arranged around C_3 symmetry axis. Each protein subunit hosts 7 bacteriochlorophyll a (Bchl) molecules, the whole protein, correspondingly, 21 Bchl molecules. Tight packing of molecules results in strong excitonic coupling between them. As calculated in[4], the pair-wise dipolar coupling energy reaches 200 cm^{-1} when the Bchl molecules are in the same protein subunit. For the Bchl molecules from different subunits, the coupling is substantially weaker, ≤ 20 cm^{-1}. In green sulphur bacteria FMO proteins play a double role: as an antenna and as an efficient mediator of light energy from the extra-membrane chlorosome antenna to the cell membrane where long-term energy storage is carried out by the photochemical reaction center.

Ultrafast Processes in Spectroscopy, edited by Svelto et al.
Plenum Press, New York, 1996

A two-colour transient differential absorption spectrometer[5] with 150 fs long pump and probe pulses and operating at 540 Hz has been utilized. The pump and probe beam polarizations were at the "magic angle", 54.7 degrees. Care was taken to avoid the multiple excitation effects. At pulse intensities 30 $\mu J/cm^2$, statistically less than one photon is absorbed by a trimeric pigment aggregate. Another concern was a high optical (low scattering) quality of the sample at cryogenic temperatures. Sticking of proteins against the quvette walls and severe cracking of the sample upon cooling below the glass transition temperature of the protein has long been a trouble. We have found that coating by Repel-Silane (a solution of dimethyldichlorosilane in trichloroethane) of inner surfaces of the sample holder almost removes this problem. Only a few cracks were usually observed over the 4 cm^2 sample surface area. The FMO proteins were isolated from *Chlorobium tepidum*, as described[5]. The protein solution in the 20 mM Tris-HCl buffer (pH 8.0) was diluted with glycerol in the 1:2 volume ratio and was fixed between two fused quartz plates. Optical density of the 2-mm thick sample was ≤1 at the low-temperature maximum 805 nm. The spectral width of the transform-limited excitation pulses was 5 nm that matches the linewidths in the FMO steady-state absorption spectrum[2]. A white light continuum generated in a water cell served for probe and reference pulses. Although measurements have been carried out at several temperatures between 15 K and 160 K, here we will focus only on the 15 K results.

Four different pump wavelengths have been used. Three excitation wavelengths (801 nm, 815 nm, and 832 nm) were chosen in close resonance with distinct absorption features in the Bchl Q_y transition region (805 nm, 814 nm, 825 nm). In addition, a non-selective excitation at 590 nm to the less structured Bchl Q_x absorption band was used. Figure 1 shows femtosecond transient absorption spectra of FMO proteins at 801 nm excitation. The spectra present a clear evidence of relaxation of initially excited exciton state(s). Only a time-dependent intensity redistribution between the bands and not the shift or broadening of the bands (except the 805 nm one that shifts and narrows) is observed. This implies a small reorganization of the protein matrix in response to optical excitation on the Bchl aggregate. The 805 nm band shift may be due to its composite origin and initial predominant excitation of the blue component of the band. The relaxation dynamics is complex and can not be represented by a single-exponential process. There is a very fast relaxation phase proceeding already during the excitation pulse. This is evidenced by the fact that bleaching at the 825 band appears almost simultaneously with the bleaching at

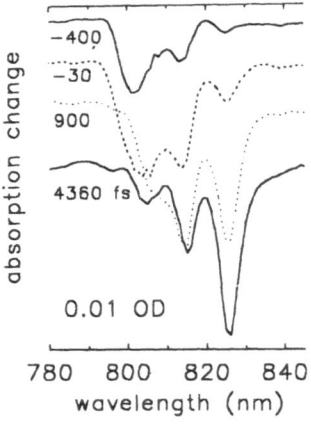

Figure 1. Transient absorption difference spectra of FMO proteins at several probe pulse time delays indicated. Spectra were acquired with 1 nm interval between data points and by averaging over 2160 laser shots. Excitation at 801 nm, temperature 15 K.

the excitation wavelength. Also, a rather slow relaxation phase, which has not completed yet on the time-scale of the experiment is observed.

The transient spectra have remarkable clear-cut structure. Bandwidths seem to be excitation pulse-limited. The bleaching maxima at longer delays appear at resonance with the peaks in the steady-state absorption spectrum. The stimulated emission, that at low temperatures contributes only to the 825 nm band, shifts its maximum by mere 1 nm. This is well in accordance with the small Stokes shift of the fluorescence band (maximum at 827 nm at 5 K[2]). A weak sideband at 840 nm is entirely due to stimulated emission. The structure of the differential absorption spectrum becomes much less pronounced at the 590 nm excitation when the 805 nm and 815 nm bands merge to a single band at all delay times (not shown). The pump pulse-induced absorption from the one-exciton states to two-exciton states is extremely weak at the pulse intensities used in Fig. 1. It can distinctly be observed only at higher pump intensities and at short delay times in spectral regions where bleaching is weak: between 815 nm and 825 nm and to the blue from the 805 nm band.

A global analysis of the data has been performed. Lifetimes were calculated using a least-squares fitting algorithm assuming multiexponential kinetics and taking into account the finite instrument response function. Transient absorption difference spectra at different time delays, δt, were built as follows: $\Delta OD(\lambda, \delta t)=\Sigma A_i(\lambda)\exp(-\delta t/\tau_i)$, where the sum is over the number of kinetic components used in the fitting. A plot of A_i versus wavelength is called a decay associated spectrum (DAS). The negative amplitude DAS corresponds to the signal decay while the positive one- to the rise. As seen in Fig. 2, at least three kinetic components are needed to fit the time-resolved spectra. The initially excited state lifetime is 0.3±0.1 ps. Absorption at 815 nm appears with the same time constant, but decays mostly within 1.0±0.2 ps. Both the 0.3 ps and 1.0 ps time constants are distinct in the lowest exciton state population kinetics. The apparent lifetime of this energetically well separated state is 150 ps, much shorter than the 1.1 ns spontaneous fluorescence emission lifetime[2]. If to exclude annihilation of excitons or some trivial experimental errors at long delay times, this discrepancy is difficult to understand. However, an observation that a substantial part of higher energy excitons relaxes very slowly is in qualitative agreement with previous low-temperature studies,[2] (fluorescence) and[6] (photon echo).

The data obtained at 815 nm and at 832 nm pump wavelengths are in agreement with the results at 801 nm excitation. The only difference is that in the former case, not all but certain lower-energy exciton states are involved. The 590 nm pump revealed 0.3 ps, 1.4 ps, and 32 ps kinetic constants. DAS of the 0.3 ps component has only positive ampli-

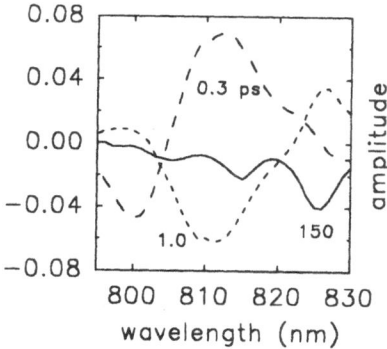

Figure 2. DAS of FMO proteins at 803 nm excitation and at 15 K.

tude and spreads over the whole measured spectrum. This allows its attribution to the $Q_x \Rightarrow Q_y$ internal conversion process. At room temperature a longer, 600 ps internal conversion time, was measured[7].

In conclusion, a complex non-exponential phonon-induced exciton relaxation process was uncovered in FMO proteins. The states at the top of the Q_y exciton band relax almost completely within few hundred femtoseconds. A small part of high-energy states and a major part of intermediate-energy states, however, reveal unexpectedly slow relaxation phases. The characteristic time constants of these phases span about a picosecond to few hundred of picoseconds. Low density of the Bchl nanoaggregate exciton states and their weak coupling to vibrational-librational modes of the aggregate and the host protein are proposed as a probable cause of the slow relaxation. Although generalization of the present results to other pigment-proteins is not clear, a hypotesis that certain long-lived librational modes exist in the circular antenna aggregates of Bchl molecules, has been put forward recently. To the best of our knowledge, this is the first time, femtosecond dynamics of spectrally well defined exciton states in any pigment-protein complex have been studied in the time domain.

REFERENCES

1. A. Freiberg, in *Anoxygenic Photosynthetic Bacteria*, ed. by R. E. Blankenship, M. T. Madigan, and C. E. Bauer (Kluwer Academic Publishers, Dordrecht, 1995).
2. A. Freiberg, P. Kukk, M. Tars, and M. Miller Chem. Phys. Lett., submitted.
3. D. E. Tronrud, M. F. Schmidt, and B. W. Matthews, J. Mol. Biol. **188**, 443 (1986).
4. X. Lu and R. M. Pearlstein, Photochem. Photobiol. **57**, 86 (1993).
5. S. Savikhin, W. Zhou, R. E. Blankenship, and W. S. Struve, Biophys. J. **66**, 110 (1994).
6. R. J. W. Louwe and T. J. Aartsma, J. Lumin. **58**, 154 (1994).
7. S. Savikhin and W. S. Struve, Biochemistry **33**, 11200 (1994).
8. R. Danielius, V. Novoderezhkin, and A. Razjivin, FEBS Lett. **345**, 203 (1994).

111

HEMOGLOBIN OXYGENATION DYNAMICS ON PICOSECOND TIME SCALE

B. M. Dzhagarov, N. N. Kruk, S. A. Tikhomirov, and V. A. Galievsky

Institute of Molecular and Atomic Physics
Belarus Academy of Sciences
220072, 70 F. Skoryna Ave., Minsk, Belarus

The binding of ligands by proteins is a fundamental biological process. The sequential binding (oxygenation) of four oxygen molecules to the tetrameric protein hemoglobin is a basic reaction for the study of a cooperative ligand binding and allosteric interactions in protein and enzymes[1-4]. The affinity of Hb to O_2 rises with the ligand saturation and tuned by different heterotropic effectors. In particular the proton concentration determines the hemoglobin affinity to O_2 (Bohr effect) and regulates the CO_2/O_2 exchange by Hb in the tissues[1-4]. Every subunit of tetrameric hemoglobin molecule has a Fe(II)-protoporphyrin-IX (heme) as an active site, which is situated inside the folded polypeptide chain. Motion of the unbound oxygen molecule in the interior of the globin matrix is mediated by the protein moiety. The laser time-resolved spectroscopy provides a powerful method for the study of the kinetics parameters of the oxygenation because the photoexcitation leads to the very rapid dissociation of O_2[5]. The oxy-Hb photodissociation triggers events which can be followed by time-resolved spectroscopy monitoring of the transient optical spectrum[5-19]. The principal point of such experiments is a study of the direct oxygen rebinding with the paternal heme from the bulk protein. This phenomenon is known as a geminate rebinding (GR) and provides the important information about the protein function and structure. This paper is devoted to the investigation of the protein control of the ligand geminate rebinding under the different pH values. The main goal of these studies was to measure the primary quantum yield of the oxy-Hb photodissociation and to elucidate the dynamics of the oxygen motion in the interior of protein. Earlier we presented the results of the measurements of the efficiences and rates of the geminate and nongeminate rebinding processes in the large range of the pH values[19,20]. Now we present the refined kinetic data obtained for the geminate stages of the ligand rebinding and discuss the protein control of the GR.

The measurements were carried out with the help of three experimental sets-up: (i) The absorption picosecond double-beam spectrometer based on the feed-back controlled passive mode-locked Nd-phosphate glass laser (the pulse halfwidth is 3.5 ps)[18,21]. The second harmonic (λ=528 nm) was used for the excitation. The picosecond continuum gener-

Ultrafast Processes in Spectroscopy, edited by Svelto et al.
Plenum Press, New York, 1996

497

ated in the cell with heavy water was used as a probe light. (ii) The tunable pump-probe absorption spectrometer having 10-ps time resolution[22]. Pumping of the objects was provided by the second harmonic of the YAlO$_3$:Nd solid-state master laser (λ_{pump}=540 nm), while output of an optical parametric oscillator was used as a probing beam tunable over the spectral region 360–500 nm. Intracavity negative-feedback electrooptical system in combination with the intracavity pulse selection have provided sufficienly high laser intensity stability (intensity scattering < 2%) and resulted in sufficiently high sensivity and accuracy of the spectrometer (3·10^{-4} OD units). (iii) The nanosecond laser photolysis set-up having the second harmonic of Q-switched ruby laser as a photolytic sourse[19,20]. The pulse halfwidth was 30 ns. In our experiments the photoexcitation energy was adjusted to dissociate no more than 10% of the oxygen molecules. It is the case that when every Hb(O$_2$)$_4$ loses only one oxygen molecule and we study the oxygen GR with triliganded hemoglobin Hb(O$_2$)$_3$ with the tetrameric protein being in the high-affinity R-state. The measurements of the primary quantum yield of the photodissociation γ_o were performed by the method described earlier[23,24]. We have used the stationary extinction coefficients for both oxy- and deoxy-Hb for the γ_o calculation. For the calculation of pK values from pH-dependences obtained the authors wrote the PASCAL procedure based on the Henderson-Hasselbalch equation. The procedure allowed to calculate three pK and their weights simultaneously. The stripped human hemoglobin HbA in oxy-form were prepared from the fresh donor blood by the method[25]. Concentrations of the Hb samples were varied in the range of 1·10^{-5} 3·10^{-4} M. All measurements were carried out at the temperature 21±1°C and under the controlled atmospheric pressure.

The correct interpretation of the 2–4 ps photoinduced transient absorption is very essential for the unambigous understanding of the ligand dissociation and intramolecular relaxation processes on HbO$_2$[5,12,15]. Earlier, we have associated this short-lived intermediate with the absorption of HbO$_2$ in the lowest excited ^{1,3}CT-state[15,18,19,26]. The population of this state is a result of the radiationless transition $^3T_{\pi\pi} \rightarrow {}^{1,3}$CT. The orbital origin of this state was discussed elsewere[27]. Earlier it was also shown that the direct optical excitation to this state didn't lead to the photodissociation[23,24]. But now we are inclined to believe that this transient absorption is likely to be that of HbO$_2$ absorption in the vibrationally excited ground state, caused by the heating of heme by the absorbed pump pulse[28]. This question will be discussed in details in the accompaning paper[29]. In any case, the triplet $\pi\pi^*$-state of HbO$_2$ is a dissociative one and the transitions $^3T_{\pi\pi} \rightarrow {}^{1,3}$CT or $^3T_{\pi\pi} \rightarrow S_o$ compete with the dissociation process and decrease the primary quantum yield γ_o. Therefore its value should be principally less than 1.0. The direct measurements of γ_o proved this conclusion.

The picosecond photolytic measurements were carried out in the whole alkaline Bohr effect region. We have obtained the value γ_o=0.23±0.03 which does not depend on the pH value. Figure 1 presents the example of the original kinetic data. The measurements have confirmed our previous results that the GR of O$_2$ is more efficient for the higher pH values and the rebinding kinetics can not be fitted by a single or double exponentials. But it was found that some of the kinetic parameters do not coincide with those previously published.

The processing of the obtained experimental data set for GR shows that the kinetics can be statisfactory described by the following equation:

$$A(t)=A_1 \exp(-t/\tau_1)+A_2 \exp(-t/\tau_2)+A_3,$$

where A(t) is a normalized number of the deoxy-Hb molecules. Correspondingly, $A(0)=A_1+A_2+A_3=1$. The data were fitted to this functional form by a least-squares proce-

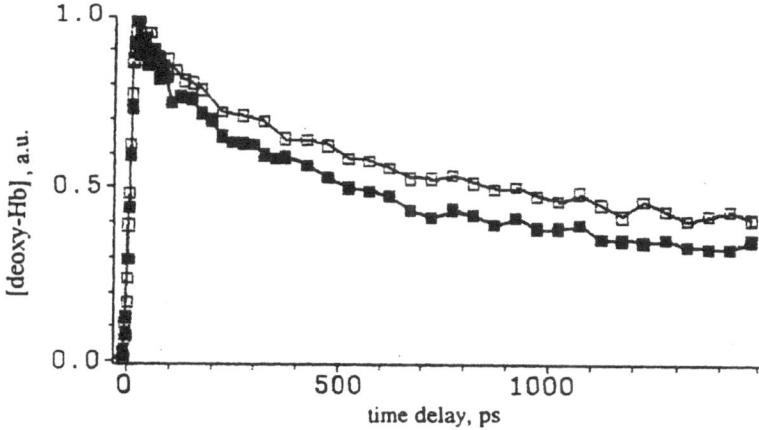

Figure 1. The experimental GR kinetics: $\square$ - pH=6.0, $\blacksquare$ - pH=8.5. Protein concentration $2.4 \cdot 10^{-4}$ M calculated at the heme basis. λ_{exc}=540 nm, λ_{obs}=435 nm.

dure. The two long rebinding processes: non-geminate association (k'_4-rate constant) and the protracted part of the geminate rebinding with ~100 ns duration (see below) form the non-zero base line A_3. The value of A_3 corresponds to two exponentials, but their durations too long for the 1650 ps temporal measurements. It is neccesary to note that this largest delay time in the ps-measurements and 100-ns time resolution of the nanosecond absorption spectrometer used for the measurements of the bimolecular nongeminate association have limited kinetical measurements. As a result of that we had the closed window for the measurements of the slowest stages of GR with the decay time ca. several tens ns (so-called 100-ns stage). For the fitting procedure and the calculations of the A_3 value we used our values of γ listed also in Table 1. The A_3 value was taken as 1.43$^{\cdot}$ in order to take into account the efficiency of the 100 ns stage. The value of 1.43 is based on the literature data[8]. We used this value for all the pH-values because we did not find the data on the pH-effect on this stage. Thus given the data on primary quantum yield γ_o, picosecond time scale GR kinetics, apparent quantum yield γ, nongeminate association rate constant k'_4 at different pH, we can reconsruct the whole rebinding prosess. The decay times τ_1 and τ_2 the weights of the different components, the γ values and parameter α which determins the efficiency of the O_2 escape from the protein matrix and corresponds to the ratio γ/γ_o for a set of pH are summarized in Table 1.

Table 1.

pH	A_1	A_2	A_3	τ_1, ps	τ_2, ps	α	γ, %
6.0	0.280	0.505	0.215	190	1600	0.150	3.45
6.8	0.300	0.500	0.200	190	1600	0.139	3.20
7.0	0.310	0.500	0.190	190	1600	0.133	3.05
7.2	0.320	0.500	0.180	190	1600	0.126	2.90
7.7	0.355	0.490	0.155	190	1600	0.109	2.50
8.0	0.385	0.470	0.145	190	1600	0.102	2.35
8.5	0.395	0.470	0.135	190	1600	0.096	2.20
9.4	0.360	0.490	0.150	190	1600	0.104	2.40

Figure 2 displays the rate constants of the bimolecular nongeminate stage of the oxygen rebinding with the triliganded hemoglobin k'_4 and the apparent quantum yields γ corresponding to the portion of the oxygen molecules left the bulk globin plotted versus pH. The calculations with the Henderson-Hasselbalch equation gave the $pK_{app}=5.5\pm0.1$, $pK_{app}=7.3\pm0.1$ and $pK_{app}=9.1\pm0.1$ for both curves. This means that the same amino acid residue which controls exit/entrance from/into protein responsible for its pH dependence. Figure 2 displays also the pH-dependence of the $(1-A_1)$ value corresponding to the fraction of the unbound O_2 at time when the first stage of GR was finished. It is obvious that the same amino acid residues which control exit/enrance from/into the protein is responsible for the pH-effect on this stage of GR.

From the analysis of the obtained parameters of the Eq.1 at the different pH the following major conclusions can be made. There exists a fast component of the geminate re-

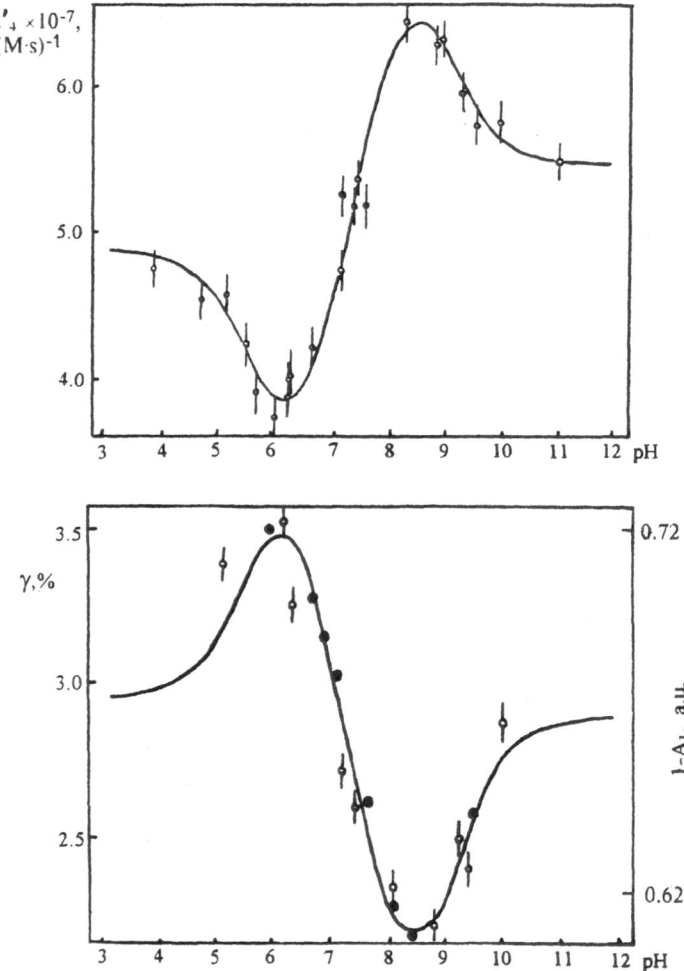

Figure 2. The pH-dependence of the nongeminate bimolecular rebinding rate constant k'_4 (top) and those for the apparent quantum yield γ (O) and for the value $1-A_1$ (●) (bottom). The solid curves represent the best fit with the Henderson-Hasselbalch equation as described above.

binding ($\tau_1=190\pm10$ ps), the weigth of which A_1 is increasing with the increase of the pH value, but the rate constant of the stage does not change with pH change. We attribute this stage to the oxygen rebinding from the region adjacent to heme, called heme pocket. A more protracted stage of the geminate rebinding (1600 ± 50 ps) exists too and is likely to correspond to the oxygen rebinding from the protein matrix. Both the rate constant of this stage ($1/\tau_2$) and the weight A_2 do not depend on the pH-value. In the large extent these two fast stages determine the fraction of O_2 molecules which can escape from the globin to the solvent. Since the primary quantum yield γ_0 does not depend on pH value the Bohr effect observed has dynamic origin. It is customary to begin an analysis of the kinetic observation within the framework of a simple phenomenological sequential model with a number of geminate intermediates $A\Leftrightarrow B\Leftrightarrow C\Leftrightarrow D\Leftrightarrow S$ with the single path for oxygen molecule leaving and entering the bulk protein[8–10,13,16,30–32]. In this model oxygen must overcome several barriers. In the photodissociation experiments the system starts from A state which corresponds to oxygen bound to the heme. The ligand is in the solvent in the state S. There are also several geminate states B, C, D. State B corresponds to the binding to heme from the heme pocket. States C and D represent two additional geminate intermediates formed by the protein matrix. Every individual observed rebinding kinetics can be explain within the framework of this model but, we believe, all the data pH effect related cannot be. Actually, the weight of the first component of GR increases with the rise of pH and control the GR and the oxygen exit. But at the same time the weight and the rate constant of the second that does not practically change. It is possible to suggest that the internal globin space can be divided into two parts. All the oxygen molecules which can find themselves in the first part should be rebound in 1600 ps and have no chance to escape from the protein matrix but the another part has the opportunity for oxygen to exit. But the fact that τ_1 is constant produces the new difficulties for the interpretation. Generally, for the diatomic ligands O_2, NO and CO the geminate rebinding data are not fit well by two or even three exponentials. It is suggested and widely discussed that such nonexponential behaviour can result on the different physical phenomena and the other functional forms and the various kinetics schemes with the different pathways for the oxygen should be considered. The discussion will be given elsewhere.

The results presented are sufficient to allow definite conclusion that the oxygen motion in the bulk protein in both directions is controlled by the same amino acid residues with the earliest stage of GR determining the pH-dependence of the oxygen escape efficiency.

ACKNOWLEDGMENT

This work is supported by the grant 'B5–338 from Foundation of Fundamental Research of the Republic of Belarus and partly by the International Soros Science Education Program.

REFERENCES

1. E. Antonini , M. Brunori, *Hemoglobin and Myoglobin in Their Reactions with Ligands* (North-Holland Publ.Comp., Amsterdam, 1971).
2. M. F. Petutz, G. Fermi, B. Luisi, B. Shaanan, R. C. Liddington, *Cold Sprihg Harbor symposia on quantitative biology*, **52**, 555 (1987).

3. Q. H. Gibson, *The porphyrins, v.VII, Biochemistry, Part B*, ed. by D. Dolphin (Academic Press, N.Y.-San-Franc.-London, 1979).
4. L.J. Parkhurst, Ann.Rev.Phys.Chem. **30**, 503 (1979).
5. R. M. Hochstrasser, C. K. Johnson, in *Ultrashort laser pulses and applications*, ed. by W. Kaiser (Springer-Verlag, Berlin, 1988).
6. B. Alpert, R. Banerjee, L. Lindqvist, Proc. Natl. Acad. Sci. USA, **71**, 558 (1974).
7. D. A. Duddell, R. J. Morris, N. J. Muttucumaru, J. T. Richards, Photochem. Photobiol., **31**, 479 (1980).
8. J. S. Olson, R. J. Rohlfs, Q. H. Gibson, J. Biol. Chem., **262**, 12930 (1987).
9. E.R. Henry, J.H. Sommer, J. Hofrichter, W.A.Eaton, J.Mol.Biol. **166**, 443 (1983).
10. L.P. Murray, J. Hofrichter, E.R.Henry et al., Proc.Natl.Acad.Sci.USA, **85**, 2151 (1988).
11. D. A. Chernoff, R. M. Hochstrasser, A. W. Steele, Proc. Natl. Acad. Sci. USA, **77**, 5606 (1980).
12. J. W. Petrich and J. L. Martin, Chem.Physics, **131**, 31 (1989).
13. J. W. Petrich, J.-C. Lambry, K. Kuczera, M. Karplus, C. Poyart, J.-L. Martin, Biochemistry, **30**, 3975 (1991).
14. J. M. Friedman, T. W. Skott, G. J. Fisanik et al., Science, **229**, 187.
15. B. M. Dzhagarov, V. Gulbinas, V. Kabelka, Zh. Savickiene, Izv. Akad. Nauk USSR , ser. fiz., **53**, 1504 (1989).
16. K. N. Walda, X. Y. Liu, V. S. Sharma, D. Magde, Biochemistry, **33**, 2198 (1994).
17. M. R. Chance, S. H. Courtney, M. D. Chavez et. al., Biochemistry, **29**, 5537 (1990).
18. B.M. Dzhagarov, N.N. Kruk, V. Gulbinas et al., Lithuanian J.of Physics. **34**, 108 (1994).
19. B.M. Dzhagarov, N.N. Kruk, S.A. Tikhomirov, I.I. Stepuro, in *5th International Conference on Laser Applications in Life Sciences*, ed. by P.A. Apanasevich, N.I. Koroteev, Yu.V. Zadkov, S.G. Kruglik, (Proc. SPIE, **2370**, 1995).
20. B.M. Dzhagarov, N.N. Kruk, Biophysica, **41**, (1996), in press.
21. N. A. Lysak, S.V. Mel'nichuk, S. A. Tikhomirov, G. B. Tolstorozhev, Zh. Prikl. Spectroscopii (Russian). **47**, 267 (1987).
22. S.G. Kruglik, V.A. Galievsky, V.S. Chirvony et al., J.Phys.Chem., **99**, 5732 (1995).
23. B. M. Dzhagarov, P. N. Dyl'ko, G. P. Gurinovich, Dokl. Akad. Nauk USSR, **275**, 765 (1984).
24. B. M. Dzhagarov, V. S. Chirvony, G. P. Gurinovich, in *Laser picosecond spectroscopy and photochemistry of biomolecules*, ed. by V.S. Letokhov, (Adam Hilger, Bristol, 1987).
25. A.A. Khachaturyan, E.P. Vyazova, G.M. Morozova, G.Ya. Rozenberg, Problems of Hematology (Russian). **1**, 58 (1979).
26. V. Gulbinas, B. M. Dzhagarov, V. Kabelka, Zh. Savickiene, Dokl. Akad. Nauk USSR, **293**, 987 (1987).
27. W. A. Eaton, L. K. Hanson, P. J. Stephens, et al., J. Amer. Chem Soc., **100**, 4991 (1978).
28. E.R. Henry, W.A. Eaton, R.M. Hochstrasser, Proc. Natl. Acad. Sci. USA, **83**, 8982 (1986).
29. N.N. Kruk, S.A. Tikhomirov, G.M. Andreyuk, B.M. Dzhagarov, this volume.
30. R.H. Austin, K.W. Beeson, L.Eisenstein et al., Biochemistry, **14**, 5355 (1975).
31. A. Ansari, C. M. Jones, E. R. Henry, J. Hofrichter, W. A. Eaton, Biochemistry, **33**, 5128 (1994).
32. D.G. Lambright, S. Balasubramanian, S.M. Decatur, S.G. Boxer, Biochemistry, **33**, 5518 (1994).

112

FEMTOSECOND TRANSIENT ABSORPTION SPECTROSCOPY ON THE ISOLATED REACTION CENTER FROM Rb. *sphaeroides* R-26

Marc G. Müller and Alfred R. Holzwarth

Max-Planck-Institut für Strahlenchemie
Stiftstr. 34–36
45470 Mülheim/Ruhr, Germany

Femtosecond transient absorption spectroscopy has been applied to study the primary electron transfer processes in the isolated reaction center of the purple bacterium *Rhodobacter sphaeroides* strain R-26.1. In particular the wavelength ranges of the bacteriopheophytin (BPh) Q_x absorption at 545 nm and of the bacteriochlorophyll (BChl) Q_y absorption between 795 and 915 nm were investigated in order to clarify the multi-exponential behavior of the charge separation [1,2].

A high speed camera which is capable of measuring 3000 spectra per second was developed for MHz-real-time data acquisition. Each spectrum contains 2×256 pixels corresponding to a wavelength range of approximately 125 nm. The system allows measurements with very high signal to noise ratio and high spectral resolution of 0.5 nm simultaneously. The rms noise of the decay trace of one individual pixel was as low as 5×10^{-5} OD units. Measurements were performed with excitation pulses at 860 nm (FWHM 13 nm) from a Ti:S-regenerative amplifier/optical parametric generator with a width of 70 fs (FWHM) and an energy of 3 nJ. Sodium ascorbate and PMS were added to the sample to keep the reaction centers open (PHQ_A). The three-dimensional plot of the data surface for the 500–620-nm range is shown in Fig. 1.

The global analysis of the data resolves three components with lifetimes of ≈ 1 ps, 4.2 ps and 140–260 ps. The longest one is not determined precisely on the time-range of 20 ps. The shortest lifetime is markedly dependent upon the detection wavelength range analyzed simultaneously in the global analysis. Analyzing only the 520 to 560 nm range leads to a lifetime of 900 fs whereas an analysis which includes also the near-infrared range results in a lifetime of 1.7 ps. Thus the kinetics appears to be more complex in the first few picoseconds than reported so far. The decay associated absorption spectra (DAAS) reveal in the BPh range at 545 nm a clearly shifted shape of the ≈ 1-ps component with respect to the 4-ps and even stronger to the 200-ps component. Its minimum occurs at 538 nm as compared to 544 nm for the 200-ps component (Fig. 2A). In the near infrared range the DAAS of the two shortest-lived components are quite similar except in the

Ultrafast Processes in Spectroscopy, edited by Svelto et al.
Plenum Press, New York, 1996

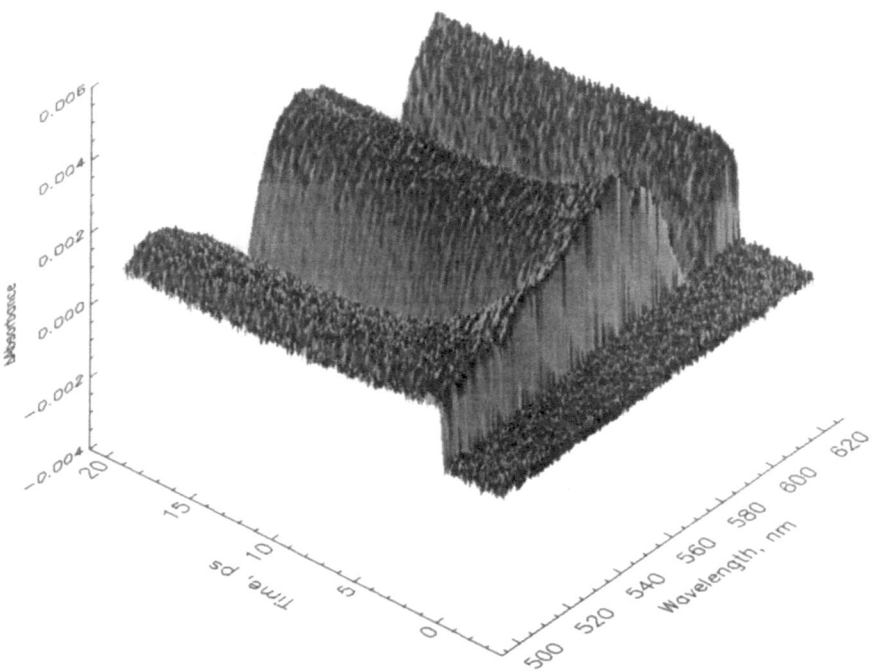

Figure 1. Hyper surface of the femtosecond transient absorption raw data for isolated RCs from *Rb. sphaeroides* strain R-26. The excitation wavelength was 860 nm. Note the delayed rise of the BPh bleaching at 538 nm.

stimulated emission region above 880 nm and in the 800-nm region where the DAAS of the 4-ps component is strongly influenced by the electrochromic shift of the BChl band upon formation of P^+ (Fig. 2B).

The spectral difference of the two shortest-lived components in the 545-nm range appears to be a key feature for understanding the nature of the 1-ps component with respect to the highly controversial discussion of the sequential model[2] employing an intermediate B- on the one hand and the superexchange model [3] on the other hand. To clarify which model is consistent with our data we calculated the rate constants and species associated absorption spectra (SAAS) (Fig. 3). The sequential model is in very good agreement with our data. The most reasonable SAAS are obtained if back transfer of the

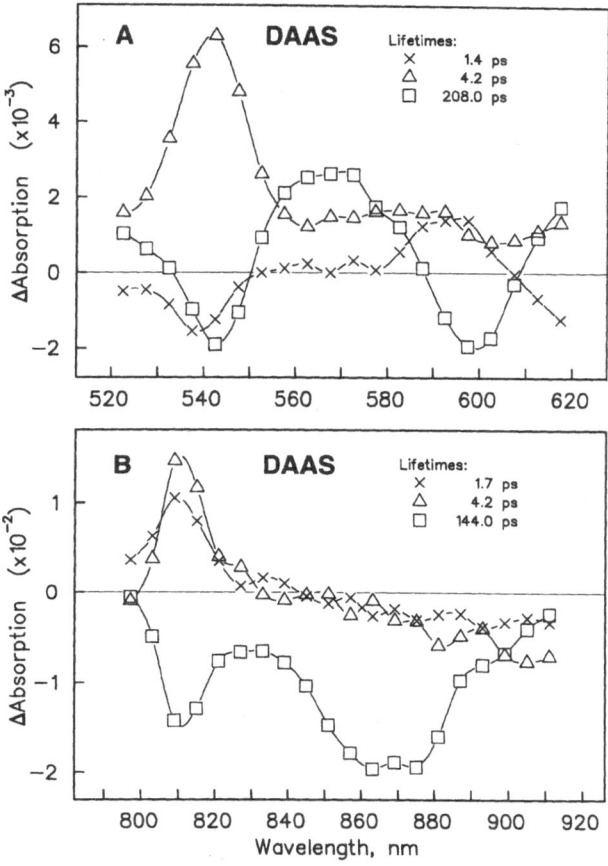

Figure 2. Decay associated absorption spectra for isolated RCs from *Rb. sphaeroides* R-26. The two wavelength ranges were analyzed separately.

electron is allowed, in contrast to the original model presented in Ref.[2]. The analysis leads to a determination of the energies DG of the P⁺B⁻ and P⁺H⁻ states. Relatively small driving forces for the first two electron transfer steps are obtained. Interestingly the ΔG values resemble quite well those determined by fluorescence decay measurements of P* for the first steps[4]. The exact numbers should be considered preliminary in view of the wavelength dependence of the two fast lifetimes. The SAAS in Fig. 3 provide strong support for the sequential model. For example the SAAS of state 2 (P⁺B⁻) bears some sine-like behavior with low amplitude on the red side of the BPh Q_x band and high amplitude on the blue side, containing an inflection point right at the maximum of the BPh band at 544 nm. This strongly suggests an electrochromic blue-shift of the BPh band upon formation of the intermediate state indicating a high electric dipole moment for the intermediate. This could in fact be a P⁺B⁻ state since the SAAS of the intermediate (Fig. 3B) shows a strong bleaching at 810 nm i.e. on the red side of the Qy band of the accessory BChl. The determination of the energies of the intermediates answers a fundamental question in the discussion of sequential transfer vs. superexchange mechanism.

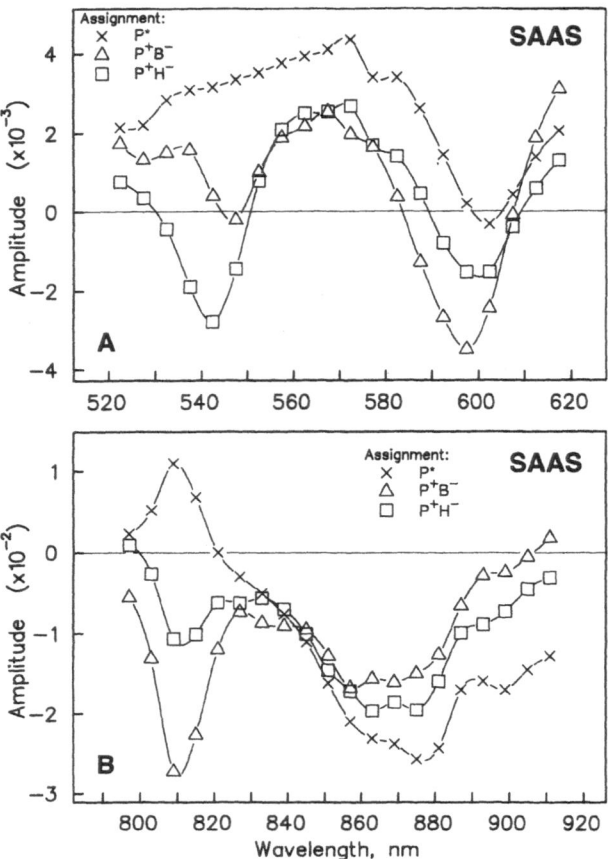

Figure 3. Species associated absorption spectra for isolated RCs from Rb. sphaeroides R-26.

ACKNOWLEDGMENTS

Partial financial support by Deutsche Forschungsgemeinschaft (Sonderforschungsbereich 168, Univ. Düsseldorf) is acknowledged.

REFERENCES

1. M.G. Müller, K. Griebenow, and A.R. Holzwarth, Chem. Phys. Lett. 199, 465–469 (1992).
2. W. Holzapfel, U. Finkele, W. Kaiser, D. Oesterhelt, H. Scheer, H.U. Stilz, and W. Zinth, Proc. Natl. Acad. Sci. USA 87, 5168–5172 (1990).
3. M. Bixon, J. Jortner, and M.E. Michel-Beyerle, Chem. Phys. 197, 389–404 (1995).
4. M.G. Müller, D. Dorra, and A.R. Holzwarth, in Proceedings of the Xth international Photosynthesis Congress, ed. by P. Mathis, in press (1995).

113

PRIMARY PHOTOPROCESSES IN BIOLOGICALLY ACTIVE PIGMENTS RELATED TO PHOTOSENSITIZED TUMOUR THERAPY

R. Rotomskis

Laser Research Center
Vilnius University
Vilnius, Lithuania

The primary photoprocesses, occurring after the light absorption by biologically active molecules, play an important role in the elucidation of some processes in photobiology and photomedicine. Interest in hematoporphyrin (Hp) and its derivatives (HpD) is caused by their successful application in tumour treatment. In aqueous solution porphyrins have tendency to associate and create the equilibrium aggregates which together with covalently linked structures in HpD play an important role in photosensitized tumour therapy (PTT). The differentiation between these components by time resolved absorption spectroscopy was performed. Two-step photoprocesses realized under powerful ultrashort pulse laser irradiation of sensitizers used in PTT are discussed and studies of Hp and its derivatives: HpD and photosan (PS) by time resolved spectroscopy in ethanol and aqueous solution at concentrations close to these used in practical treatment of tumours are presented.

The spectral and temporal characteristics of the absorption changes (ΔA) of Hp, HpD, and PS in aqueous and ethanol solutions were measured by picosecond pump-and-probe spectrophotometer based on mode-locked Nd-YAG laser and optical parametric oscillators[1]. The use of parametric oscillators enables continuous tunability of pump and probe pulses in the region of 400–1500 nm. The system ensured measurements of ΔA with an accuracy higher than 10^{-3} optical units and time resolution of about 10 ps.

On the basis of the results of the time-resolved and steady state absorption spectroscopy it was suggested that different structures of the aggregates of HpD aggregates could be clarified from specific life times of excited state. In addition to the exponential excitation energy relaxation time constant for monomeric Hp (12–15 ns), the time constants 1.5–4 ns and 100–200 ps for complex mixture of HpD and PS were detected. It was suggested that time constant 1.5 - 4 ns reflects the excitation energy relaxation in covalently linked linear dimers and aggregates and time constant 100–200 ps is related to relaxation processes in "sandwich" type aggregates. The clarification of the nature of the shortest component (< 50 ps) in the excitation energy relaxation kinetics of HpD in aqueous solution is still under the question.

Ultrafast Processes in Spectroscopy, edited by Svelto et al.
Plenum Press. New York. 1996

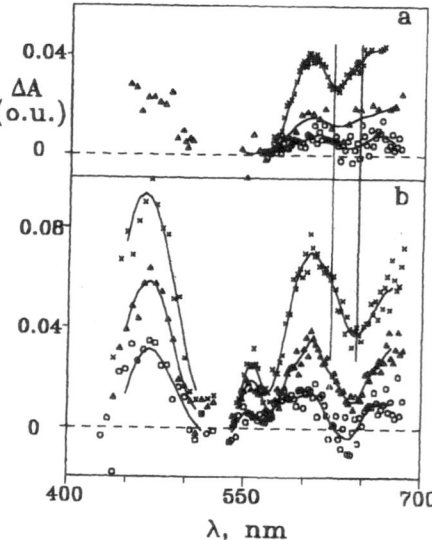

Figure 1. The difference absorption spectra of: a) HpD in phosphate buffer solution, pH-6.0, b) irradiated HpD in phosphate buffer solution, pH-8.5. Concentration 10^{-4} M, λ_{ex}-532 nm, delay 0 (x), 100 (Δ) and 1230 (O) ps.

As well known, an irradiation of porphyrins with visible light sources causes the formation of chlorin type stable photoproducts[2] which shows photodynamic activity[3]. Spectral and kinetic characteristics of non-irradiated and irradiated HpD (where the photoproduct formation was detected) were measured (Fig.1).

The analysis of ΔA spectra of HpD revealed that in the absorption changes measured with delay between pump and probe pulses (excitation at 532 nm dominantly absorbed by HpD molecules) the depletion of S_0 manifests as the bleaching bands (on the transient absorption background) with a maximum at the same position as that of the stationary absorption of HpD molecules. The maximum of the bleaching of the band, which is found at 623 nm (Fig. a) in the absorption transient spectra, is close to the maximum of the first absorption band of HpD. In the absorption difference spectrum of irradiated aqueous solution of HpD (Fig. b) without delay between pump and probe pulses (the excitation at 532 nm is dominantly absorbed by the molecules of HpD because the photoproduct has negligible absorption at this wavelength) the depletion of the S_0 state yields a bleaching band with a maximum at around 640 nm, as that of the corresponding stationary absorption band of photoproduct molecules. Thus, the time of excitation energy transfer from HpD to photoproduct (indicated by the appearance of the photoproduct absorption bleaching during the action of about 30 ps excitation pulse) is comparable or even shorter than the duration of the excitation pulse. Such fast excitation energy transfer to photoproduct shows that the photoproduct molecules are close to the HpD molecules. It should be noted that the photoproduct molecules are tightly packed in aggregates of HpD or exist as covalently linked Hp-photoproduct structures. So the photoproduct may act as an intermediate of energy transfer from porphyrin molecules to cells substrate (Fig.2) because in the time of treatment of tumour cells incubated with porphyrins the photoproduct is produced by light.

In PTT the most common is to use a constant irradiation intensity from continuous wave lasers. The primary process in the usual photosensitized tumour therapy is sensitization from the sensitizers lowest triplet state. Alternatively, the laser light can be pulsed so that excitation occurring over short interval at high peak power can induce two-photon ex-

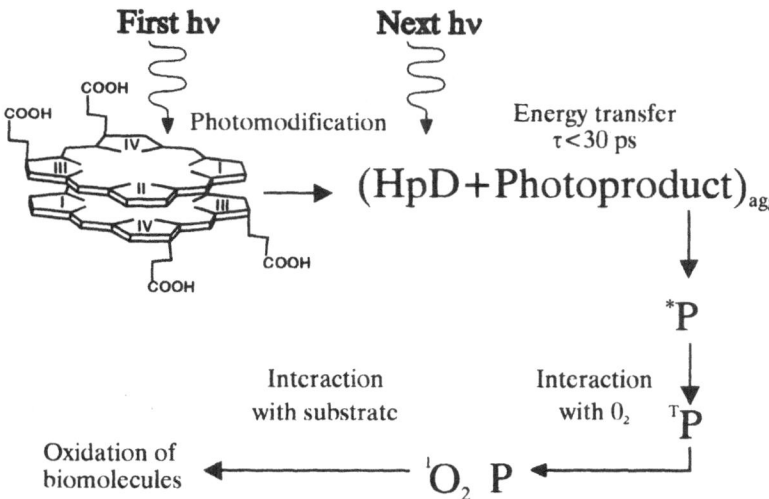

Figure 2. Proposed scheme of photosensitization in PTT.

citation of sensitizers via non-resonant absorption and initiate new photochemical or photobiological processes. Two-step excitation of the high-lying electronic states of porphyrins can be realized via intermediate state S_1 and via intermediate level T_1. Thus it is important to know the absorption cross-section σ_2 and σ_5 from the levels S_1 and T_1. The absorption cross-sections were measured by time resolved absorption spectrometer when absorption changes were probed by 30 ps and 34 ns delayed probe pulse excitation respectively[5]. The singlet excited state cross-section and triplet state cross-section were determined assuming that all excited molecules are in the S_1 and T_1 states (we neglect the population of highly excited singlet and triplet states, assuming that their lifetimes are less than 1 ps as in the case for many other organic molecules). On the basis of our model[4], the parameters of the laser pulse and the measured characteristics of the Hp the population density of the excited singlet and triplet states dependence on excitation pulse duration and energy for the hematoporphyrin like molecules are calculated (Fig.3).

The solid lines show the irradiation pulse energy and duration parameters at which the population densities of S_0, S_1 and T_1, S_0 are equal. The irradiation pulse energy and duration parameters ensure the domination of two-step excitation through the singlet manifold in region (a), because the S_1 state population density is higher than T_1 and S_0. In the region (b) the two-step excitation through the triplet manifold is dominant. The dotted line in the inset of Fig.3 presents the irradiation pulse parameters that ensure the equal population density of the excited singlet and triplet state. The theoretical model has been developed and laser pulse parameters that ensure efficient two-step photochemistry have been calculated[4]. It seems, that it is possible to initiate the two-step process in PTT by today available pulsed lasers without destroying the biological tissue.

This work partially has been supported by the Lithuanian State Foundation of Science and Studies. The author is thankful to prof.A.Piskarskas and Dr.J.Didziapetriene (Lithuanian Oncological Center) for fruitful discussions, Dr.'s R.Gadonas, V.Vaicaitis, and R.Kapociute for their assistance in making experiments on picosecond laser spectroscopy. Much appreciated also is valuable assistance from V.Mickunaitis for this good help with computer simulation of two step processes.

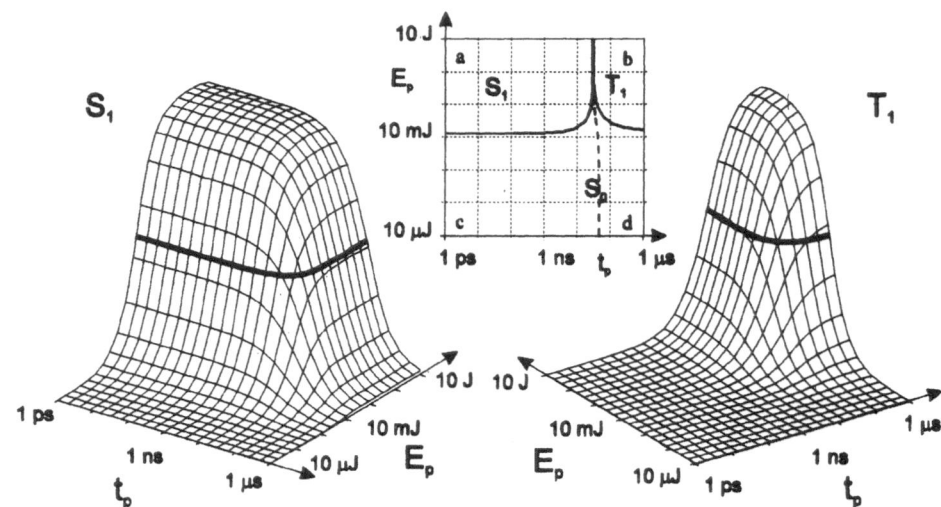

Figure 3. The excited singlet and triplet state population density as function of the incident pulse energy E_p and duration t_p. Inset. The dotted line presents the incident pulse energy and duration when population densities of the excited singlet and triplet states are equal. a) $S_1 > T_1$, $S_1 > S_0$; b) $T_1 > S_1$, $T_1 > S_0$; c,d) $S_0 > S_1$, $S_0 > T_1$. {c) $S_1 > T_1$, d)$T_1 > S_1$}. The solid line shows 0.5 level of populations.

REFERENCES

1. R.Gadonas, A.Piskarskas and R.Rotomskis, in *Laser Scattering Spectroscopy of Biological Objects,* ed. by J.Stepanek, P.Anzenbacher, and B.Sedlacek (Elsevier, Amsterdam, , 1987).

2. R.Rotomskis, J.Rotomskiene, T.J.Aartsma, A.J.Hoff and G.Varanaviciene, Lithuanian Journal of Physics **35**, 35 (1995).

3. L.Giniunas, R.Rotomskis, V.Smilgevicius, A.Piskarskas, J.Didziapetriene, L.Bloznelyte and L.Griciute Lasers in Med. Sci. **6**, 425 (1990).

4. R.Rotomskis, V.Mickunaitis, D.Juodzevicius and A.Piskarskas, in *Photodynamic Therapy of Cancer SPIE 2325,* ed. by D.Brault, G.Jori, J.Moan and B.Ehrenberg (1994).

5. R.Gadonas, R.Kapociute, V.Krasauskas, A.Piskarskas and R.Rotomskis Chem.Phys.Let. **29**, 603 (1986).

114

DETERMINATION OF SPATIAL DISTRIBUTION OF CHLOROPHYLL SPECTRAL TYPES IN PHOTOSYSTEM I ANTENNAE BASED ON PICOSECOND FLUORESCENCE KINETICS

G. Trinkunas[1] and A. R. Holzwarth[2]

[1] Institute of Physics
Vilnius, Lithuania
[2] Max-Planck-Institut für Strahlenchemie
Mülheim a.d. Ruhr, Germany

It is now generally accepted that all photosynthetic antenna systems are spectrally inhomogeneous. This leads to complex multiexponential kinetics of the excitation decay, covering time scales from subpicosecond to nanosecond. This kinetics can no longer be described by the conventional compartment kinetic models. Rather detailed description of light energy delivery to the reaction center (RC) requires knowledge about the spatial distribution of the chlorophyll (Chl) spectral types in light-harvesting antenna. As it was shown in excitation dynamics modeling[1,2] the latter is very well reflected in so-called decay associated spectra (DAS). The results of the global analysis of the time resolved fluorescence measurements at hand[3] the inverse problem solution suggest the way to find the relevant spatial spectral type arrangement.[2]

We report on the mathematical procedure and the results of the search for spatial distributions of spectral types of Chls in the cyanobacterial photosystem (PS) I core antenna/RC particle (100 Chls/P700) based on genetic algorithms[4]. A striking feature of PS I antenna is the presence of "red" Chls that absorb at longer wavelengths than the energy trap (P700 pigment). The structure of the PS I core particle was modeled by a three-dimensional regular pigment lattice. The spectral content was determined from the deconvolution of the absorption spectrum in terms of Chl absorption spectra. Seven Chl spectral forms, identical in shape but shifted in their absorption maximum, including two "red" ones, have been used to describe the inhomogeneous broadening of the PS I-100 particle absorption spectrum.[2] Kinetic modeling of the excitation migration has been performed by an exact solution of the Pauli master equation for excitation transfer to the neighbor or the next to neighbor chromophores assuming a Förster mechanism. All the spacings between the pigments were characterized by the same lattice constant which was a fit parameter. Exceptions to this regular spacing concern the nearest neighbors of the P700 and of the

Ultrafast Processes in Spectroscopy, edited by Svelto et al.
Plenum Press, New York, 1996

511

"red" pigments. Due to the lack of exact data describing the pigment transition dipole orientations and the distances between the P700 or "red" pigments and their nearest neighbors, we introduced two extra scaling parameters for the rates of excitation energy transfer involving the P700 and/or the "red" pigments. The P700 pigment is an excitation energy sink while the "red" pigments are the most likely excitation residence sites. The rate constant of charge separation in the P700 site was a variable parameter also.

The search for the optimal parameter set was organized as follows: A randomly generated population of 1000 different pigment arrangements and continuous numerical parameters was encoded in the so-called chromosomes and was altered using the genetic operators of random mating selection, crossing and random mutation[4] to produce the population for the next generation. The least deviation of the calculated DAS as well as the steady state fluorescence spectrum from the results of the global analysis of experimental data at room temperature[3] served as criterion for the selection of the relevant spatial distributions of the spectral types. The evolution of the continuous parameters for the best individual in the population is presented in Fig. 1.

The level of good agreement between the model and experimental kinetics was already reached in 50 generations. More than 10% of individuals in the final population exceeded that level. The obtained lattice spacings are in good agreement with the available structural data[5] for PS I (8Å-15Å). Due to the fact that the intrinsic excitation traping rate scales the trapping dependent part of the lowest momentum of the excitation decay function k_{P700} value provides the estimate for the rate-limiting process. The charge separation rate converged to the value ($\cong 4$ ps^{-1}) more than twice as high as the lowest limit (1.8 ps^{-1}) determined by the spectral content of the PS I particle. The models with this charge separation rate value most frequent in the final generation predict an intermediate kinetics between the migration- and the trapping-limited ones. However, few optimal individuals with low and high charge separation rates were in the final generation as well. This indicates that on the base of available experimental kinetics data[3] both close to extreme rate-

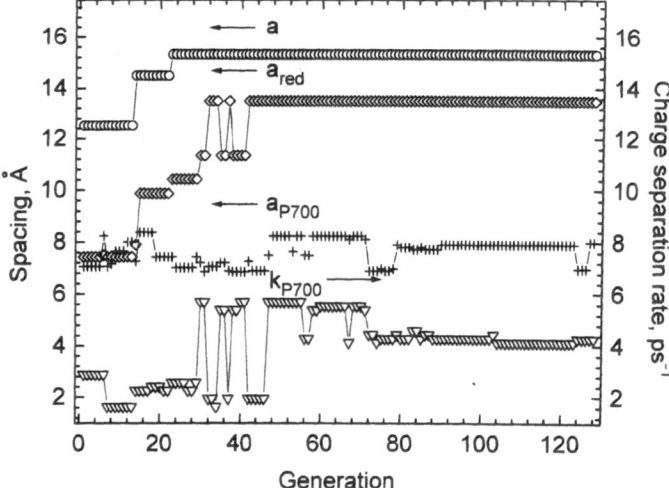

Figure 1. The evolution of the continuous parameters for the best individual in population. a stands for the lattice constant; a_{red} - for mean spacing between the "red" and surrounding pigments; a_{P700} - for mean spacing between the P700 and surrounding pigments; and k_{P700} - for the charge separation rate in P700 pigment.

Figure 2. The Chl spectral type arrangement for the best individual in the final generation. The seven spectral types[2] (5 pigments of spectral type 1 with absorption maximum at 654.1 nm, 23 - 2 - 667.3 nm, 38 - 3 - 678.3 nm, 22 - 4 - 686.7 nm, 7 (one of which is P700) - 5 - 700 nm, 4 - 6 - 712.2 nm, 1 - 7 - 724 nm) are distinguished by the grey levels indicated also by the spectral type numbers. The corresponding continuous parameters can be found in Fig.1 from the last generation point (k_{P700}= 4.28ps^{-1}; a=15.3Å ; a_{P700}=8Å ; a_{red}=13.5 Å).

limiting cases can be valid. The higher time resolution kinetic evidence as well as the temperature dependence study are required in order to find the actual rate limiting process in the PS I core antenna.

The most populated PS I spectral arrangement model of the last generation is shown in Fig.2.

In contrast to the common view about the "red" pigments located just next to the P700 pigment the obtained model predicts the "red" Chl pool to be spread on the surface of the antenna system in the vicinity of the P700 pigment, but without direct contact to it. The low a_{P700} value indicates also the higher Chl molecule density around the P700 pigment This particular spectral type distribution is quite characteristic while the models with the lower charge separation rates show less compact "red" pigment distribution.

ACKNOWLEDGMENT

The research described in this publication was made possible to GT in part by Grant No LE 6000 from the Interational Science Foundation.

REFERENCES

1. M.Beauregard, I.Martin and A.R.Holzwarth, Biochim.Biophys.Acta **1060**, 271 (1991).
2. G.Trinkunas and A.R.Holzwarth, Biophys.J. **66**, 415 (1994).
3. A.R.Holzwarth, G. Schatz, H. Brock and E. Bittersmann, Biophys.J., **64**, 1813(1993).
4. D.E.Goldberg, *Genetic Algorithms in Search, Optimization and Machine Learning* (Addison-Wesley, 1989).
5. N.Krauss, W.Hinrichs, I.Witt, P.Fromme, W.Pritzkow, Z.Dauter, C.Betzel, K.S.Wilson and W.Saenger, Nature **361**, 326 (1993).

115

TIME-RESOLVED FLUORESCENCE OF AGGREGATED CHLOROPHYLLS

H. Z. Wang,[1] X. G. Zheng,[1] S. Y. Yang,[2] Z. X. Yu,[1] and Z. L. Gao[1]

[1] Institute for Laser and Spectroscopy
Zhongshan University, Guangzhou, China
[2] Shanghai Institute of Plant Physiology
Chinese Academy of Sciences, Shanghai, China

Most of chlorophylls exist as aggregated chlorophyll-protein complexes in thylakoid membranes of plant chloroplasts, in which chlorophyll-chlorophyll interaction and processes such as charge separation and recombination contribute much to their physical, chemical and biological properties. In order to study effects of chlorophyll-chlorophyll molecular interaction and processes such as charge separation and recombination *in vivo*, an *in vitro* model system of aggregated chlorophylls is an approach, and it has therefore received much attention in the past[1-4]. In this communication, picosecond time-resolved fluorescence of aggregated chlorophylls, together with fluorescence of aggregated chlorophylls during deaggregation, is studied.

Chlorophyll *a* is isolated from fresh spinach leaves, and purified by filtration in a microcrystal cellulose column. Aggregated chlorophylls *a* are prepared in the following procedure. Chlorophyll *a* is dissolved in ethyl ether by appropriately heating. The solution is kept in darkness for several days until precipitate yields. Pure chlorophyll aggregations are then collected through ultracentrifugation.

It is found that aggregated chlorophylls posses peculiar spectroscopic properties such as delayed luminescence, which relates to long-lived vibronic states. Fluorescence with a lifetime longer than several minutes is usually too weak to detect. A luminescence measurement during deaggregation has been designed as an alternative experimental method for investigation of long-lived and weak fluorescence in aggregated chlorophylls. When aggregated chlorophylls are injected into acetone, deaggregation of aggregated chlorophylls occurs. If aggregated chlorophylls have been kept in darkness for 30 min, no fluorescence can be observed during the deaggregation process. If aggregated chlorophylls are illuminated before injected into acetone, fluorescence is recorded during deaggregation process. The relationship between fluorescence intensity and duration of darkness before injection of aggregated chlorophylls into acetone demonstrates that the absorbed light energy can be stored up in aggregated chlorophylls for about 20 min.

Ultrafast Processes in Spectroscopy, edited by Svelto et al.
Plenum Press, New York, 1996

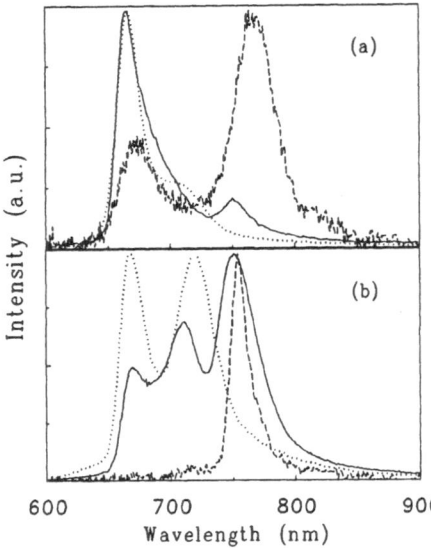

Figure 1. Fluorescence spectra of monochlorophyll and aggregated chlorophylls at room temperature (a) and 77 K (b). The dotted lines denote the time-integral fluorescence spectra of monochlorophyll, the solid lines the time-integral fluorescence spectra of aggregated chlorophylls, and the dashed lines the fluorescence spectra of aggregated chlorophylls during the first 100 ps of the decay.

Picosecond time-resolved fluorescence experiments on aggregated chlorophylls are performed with a synchronously CW mode-locked and dumped dye laser as an excitation source. The pulse duration of the dye laser is 6 ps. Time-resolved fluorescence is detected by a synchroscan streak camera (Hamamatsu Model C1587, 10 ps time resolution) connected to a polychromator, thus the spectral and temporal characteristics can be recorded simultaneously. Fluorescence is measured at a sample temperature of 77 K and at room temperature, respectively.

As shown in Figure 1(a), there are two fluorescence peaks in the time-integral fluorescence spectrum of aggregated chlorophylls at room temperature, one of which is intensive around 668 nm similarly to that of monomolecular chlorophyll, and the other is weak around 750 nm and absent in monochlorophyll fluorescence spectrum. Similarly, there are two fluorescence peaks in the time-integral fluorescence spectrum of monochlorophyll at 77 K, while in the time-integral spectrum of aggregated chlorophylls at the same temperature, three fluorescence peaks characterize, and the maximal fluorescence peak is around 750 nm, as shown in Figure 1(b). This means that fluorescence around 750 nm results from chlorophyll-chlorophyll molecular interaction in aggregated chlorophylls.

The fluorescence decays during the first 2 ns are shown in Figure 2(a) and 2(b), which are respectively recorded around 668 nm and around 750 nm at room temperature. The lifetime of the transient fluorescence at 668 nm is 4.8 ns, while the decay time of the transient fluorescence component at 750 nm is only about 280 ps. Time-resolved fluorescence of aggregated chlorophylls at 77 K during the first 300 ps of the decay is shown in Figure 3, in which a sharp band emission peaking at 750 nm emerges at the beginning and is followed by red-shifting and broadened weak emission.

The experimental results above are summarized as that there are four kinds of fluorescence decays, in which the first one with a decay time of several hundred picoseconds is located around 750 nm, the second one with a decay time of several nanoseconds at 668 nm, the third one with a decay time of dozens of microseconds[3], and the fourth one with a decay time of tens of minutes. The complexity of excited-state dynamics of aggregated chlorophylls is attributed to strong chlorophyll-chlorophyll interaction and radiationless

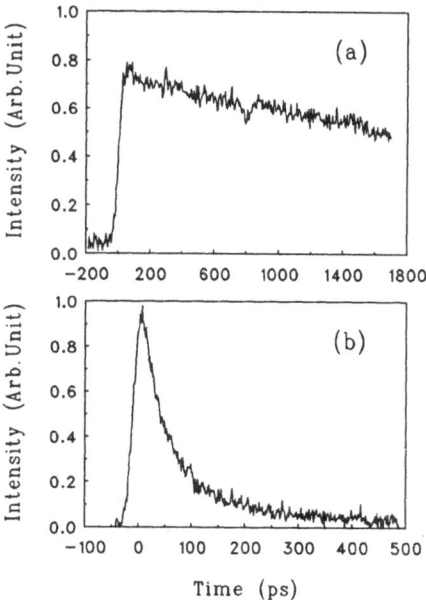

Figure 2. Fluorescence decay curves of aggregated chlorophylls at room temperature: (a) recorded at 668 nm; (b) recorded at 750 nm.

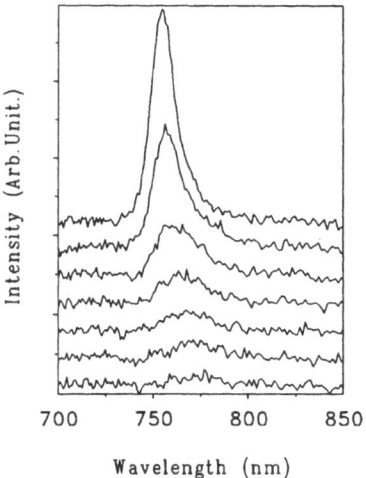

Figure 3. Time-resolved fluorescence of aggregated chlorophylls at 77 K. From top to bottom, t = 0, 48, 96, 145, 193, 242, and 290 ps.

processes such as charge seperation and recombination, intersystem crossing and hot luminescence. The experimental results above also imply that chlorophylls are important media not only for light harvesting and energy transfer but also for storage of absorbed light energy. It should be noted that the spectral and temporal behaviors of fluorescence in aggregated chlorophylls *in vitro* at 750 nm are simillar to that *in vivo*, and that the long life fluorescence in aggregated chlorophylls is similar to delayed fluorescence in photosynthetic system[5].

Electrochemical experiments have also been performed on aggregated chlorophylls upon illumination, in which when light is shedded on a platinum electrode covered with aggregated chlorophylls, a variation of electrochemical voltage is observed. In an electrophoresis experiment, aggregated chlorophylls is found to be chagred. Moreover, pH variation of aggregated chlorophylls suspension in water is also observed upon illumination. These experimental results confirms our proposal that an energy storage state with a long lifetime ($\sim 10^1$ min) relates to charge separation in aggregated chlorophylls or chlorophyll-solvent complexes. When aggregated chlorophylls are illuminated, charge separation in aggregated chlorophylls occurs. These charge separation species may undergo a series of processes, such as singlet-triplet crossing and charge transport, followed by charge recombination and luminescence.

In summary, the experimental results show that there are four kinds of fluorescence decays in aggregated chlorophylls *a*, with lifetimes ranging from $\sim 10^2$ ps to $\sim 10^1$ min. The complicated excited-state dynamics of aggregated chlorophylls *a* is related to chlorophyll-chlorophyll molecular interaction and processes such as charge separation and recombination.

ACKNOWLEDGMENT

This work is supported by an NNSFC grant and a Frontier Sciences grant of Zhongshan University.

REFERENCES

1. K. Uehara, M. Mamoru, Y. Fujita, and M. Tanaka, Photochem. Photobiol. **48**, 725 (1988).
2. K. Uehara, M. Mimuro, and M. Tanaka, Photochem. Photobiol. **53**, 371 (1991) .
3. A. Planner, and D. Frackowiak, Photochem. Photobiol. **54**, 445 (1991) .
4. J. Gottstein, A. Scherz, and H. Scheer, Biochim. Biophys. Acta **1187**, 413 (1993).
5. J. Amesz and H. J. van Gorkom, Ann. Rev. Plant Physiol. **29**, 47 (1978).

116

SCATTERING OF FRENKEL EXCITONS IN R-PHYCOERYTHRIN SINGLE CRYSTAL

X. G. Zheng,[1] H. Z. Wang,[1] J. C. Zhu,[2] Z. X. Yu,[1] Z. L. Gao,[1] and L. J. Jiang[2]

[1] Institute for Laser and Spectroscopy
Zhongshan University, Guangzhou, China
[2] Institute of Photographic Chemistry
Chinese Academy of Sciences, Beijing, China

One of the most suprising features in primary process of photosynthesis is that in the light-harvesting anntenae excitation is rapidly and efficiently transferred between chromophores and finally to chemical reaction centers with a remarkable high efficiency of 90% or higher within so short a duration that in an order of 10^{-12} sec[1]. Investigation into photophysical properties of photosynthetic materials is of significant importance to better understanding of photosynthesis nature and efficient artificial utilization of solar energy. R-phycoerythrin is one of the chromophore-protein complexes in photosynthetic system of red algae[2]. Peculiar optical behaviors such as coherent spontaneous emission of radiation (superradiance) in R-phycoerythrin single crystal have been experimentally investigated[3,4]. An excitation created in R-phycoerythrin single crystal coherently propagates among chromophores due to the translational symmetry of the crystal and interaction between transition dipole moments of the chromophores. This elementary excitation is a Frenkel exciton, and analogous to a reversed 1/2-spin wave in a Heisenberg ferromagnet. In a high excitation field, more than two excitons are created in the crystal, and they interact with each other. The scattering dynamics of Frenkel excitons has aroused interest in research[5,6]. We present in this work both theoretical and experimental studies on scattering of Frenkel excitons in R-phycoerythrin single crystal. The results demonstrate exsistence of multi-exciton bound states which relate to the observed superradiance in the crystal.

In order to describe n-exciton states of Frenkel excitons, a basis set for 1- and 2-exciton states[8] as follows is introduced in analogy to that for spin deviations in a Heisenberg ferromagnet in an external field:

$$| \mathbf{j} = S_j^+ |0\rangle \tag{1}$$

$$| \mathbf{jk} = S_j^+ S_k^+ |0 \tag{2}$$

Ultrafast Processes in Spectroscopy, edited by Svelto et al.
Plenum Press, New York, 1996

where S_j^+ is the quasi-spin operator for excitation localized at j site. After tranformation of site to wave-vector representations, the Green's function for the Schrödinger equation describing the scattering process of Frenkel excitons

$$G = (E - H_0)^{-1} \qquad (3)$$

are derived in the wave-vector representation. The corresponding Lippmann-Schwinger equation is numerically solved, yielding the exciton scattering amplitude and cross section. The above theoretical treatment is similar to that of Boyd and Callaway,[8] who studied scattering between three-dimentional spin-waves in a Heisenberg ferromagnet, but different from that of Hanamura et al.[5], who treated one-dimentional Frenkel excitons as quantum spins in a model of antiferromagnetic chain[9], which is appropriate only for complet up-conversion or a high exciton density. The theoretical results reveal that bound exciton states can exist when two or more excitons are created in the crystal. This is in good agreement with the experimental results of time-resolved excitation-density-dependent fluorescence in R-phycoerythrin single crystal as follows.

R-phycoerythrin is isolated from *Porphyra yezoensis* red alga. The procedures of isolation, purification and crystallization were previously described in details[7]. Time-resolved excitation-density-dependent fluorescence experiments on R-phycoerythrin single crystal are perfomed by means of picosecond single-pulsed laser spectroscopic technique at room temperature. A frequency-doubled passively mode-locked Nd:YAG is employed as an excitation source, and the spectral and temporal characteristics of the sample fluorescence are simultaneously recorded by a single pulse sweep streak camera (2 ps time resolution) coupled to a polychromator.

The experimental results show that the bandwidth, peak intensity and duration of the exciton emission in R-phycoerythrin crystal is dependent upon excitation density. As shown in Figure 1, the exciton emission bandwidth decreases with increasing of excitation density. Under extremely low excitation, the exciton emission band is determined to be 15 nm in bandwidth after deconvolution with other emission bands. When the single pulse excitation density is 12 GW/cm^2, the exciton emission bandwidth is 13 nm, and it decreases to 7 nm when the excitation density increases to 30 GW/cm^2 (see Figure 2). However, the emission bandwidth is not smaller than 4.4 nm in spite of higher excitation density. Moreover, the peak of the exciton emission band slightly red-shifts when the excitation density increases. The peak intensity (I_{peak}) of the emission is found to be almost proportional to the squared excitation density (N_{ex}) as $I_{peak} \propto N_{ex}^{2-\beta}$ ($\beta < 0.5$). This excludes

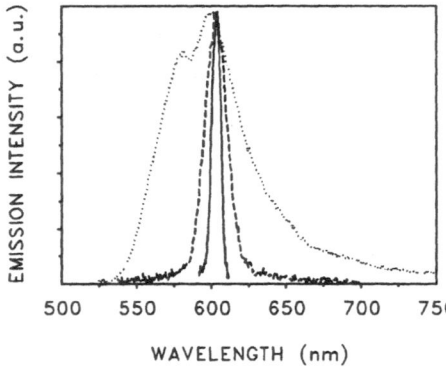

Figure 1. Exciton emission spectra of R-phycoerythrin single crystal at room temperature under various excitation densities: 1 GW/cm^2 (dotted curve), 12 GW/cm^2 (dashed curve), and 30 GW/cm^2 (solid curve).

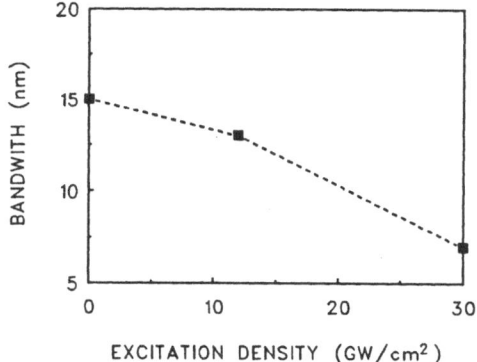

Figure 2. Dependence of R-phycoerythrin single crystal exciton emission bandwidth on excitation density.

case of exciton annihilation or amplified spontaneous emission. In addition, analysis of the emission decay shows that the emission duration of R-phycoerythrin is as long as 320 ps at low excitation density, but as short as the excitation pulse duration with an excitation density higher than 7 GW/cm^2, as shown in Figure 3. At sufficiently high excitation density, exciton superradiance occurs.

The above experimental results can be interpreted as consequences of bound excitons. A physical picture of bound states of multi-excitons is that two or more than two Frenkel excitons bind together and coherently propogate through a lattice. Behaviors of the time correlation function $\langle S^+(t)S^-(t)\rangle$ demonstrate that the states of bound n-excitons may enhance coherence of interacting excitons and therefore relate to the observed exciton superradiant emission. Our results also advocate q-deformed boson nature and condensation properties of collective excitations and interacting Frenkel excitons firstly proposed by Birman[10] and mentioned by Suzuura $et\ al.$[11]

ACKNOWLEDGMENT

This work is supported by an NNSFC grant and a Frontier Sciences grant of Zhongshan University.

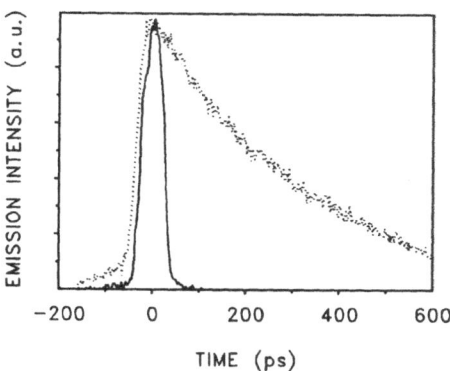

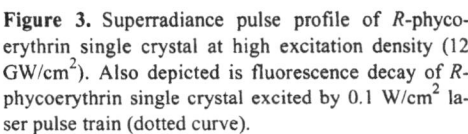

Figure 3. Superradiance pulse profile of R-phycoerythrin single crystal at high excitation density (12 GW/cm^2). Also depicted is fluorescence decay of R-phycoerythrin single crystal excited by 0.1 W/cm^2 laser pulse train (dotted curve).

REFERENCES

1. R. van Grondelle, J. P. Dekker, T. Gillbro, and V. Sundström, Biochim. Biophys. Acta **1187**, 1 (1994).
2. A. N. Glazer, Biochim. Biophys. Acta **768**, 29 (1984).
3. H. Z. Wang, X. G. Zheng, F. L. Zhao, Z. L. Gao, Z. X. Yu, J. C. Zhu, J. L. Jiang, J. P. Zhang, and D. C. Liang, Chem. Phys. Lett. **222**, 204 (1994).
4. H. Z. Wang, X. G. Zheng, F. L. Zhao, Z. L. Gao, and Z. X. Yu, Phys. Rev. Lett. **74**, 4079 (1995).
5. Y. Manabe, T. Tokihiro, and E. Hanamura, Phys. Rev. B **48**, 2773 (1992).
6. T. Tokihiro, Y. Manabe, and E. Hanamura, Phys. Rev. B **51**, 7655 (1995).
7. L. Liang and N.-J. Zhu, Scientia Sinica B **35**, 58 (1992).
8. R. G. Boyd and J. Callaway, Phys. Rev. **138**, A1621 (1965).
9. E. Lieb, T. Schultz, and D. Mattis, Ann. Phys. **16**, 407 (1961).
10. J. L. Birman, Sol. State Comm. **84**, 259 (1992).
11. H. Suzuura, T. Tokihiro, and Y. Ohta, Phys. Rev. B **49**, 4344 (1994).

117

THE STUDY OF EXCITED STATES OF HIGH- AND LOW-SPIN DERIVATIVES OF METHEMOGLOBIN BY PICOSECOND SPECTROSCOPY

N. N. Kruk,[1] S. A. Tikhomirov,[1] G. M. Andreyuk,[2] and B. M. Dzhagarov[1]

[1] Institute of Molecular and Atomic Physics, Belarus Academy of Sciences
70 F.Skoryna Ave., 220072 Minsk, Belarus
[2] Institute of Bio-Organic Chemistry, Belarus Academy of Sciences
Zhodinskaya St. 5, 220041 Minsk, Belarus

Heme proteins and related enzymes play a fundamental role in biochemistry. For example, hemoglobin is involved in uptake, transport and release of molecular oxygen. All heme proteins have an identical active site or prostetic group composed of an iron-porphyrin complex (heme). The heme iron in hemoglobin exists in two oxidation states. In the reduced ferrous state (Fe^{2+}) hemoglobin can bind small ligands such as O_2, CO, NO and some other molecules. In the oxidized ferric state (Fe^{3+}) hemoglobin cannot bind the oxygen molecule and is physiologically inactive but it can bind a water molecule and a number of different anions[1]. The oxidation and spin states of the heme iron cation in different heme proteins can also be varied and impact on the optical spectra and intramolecular energetics. In this situation the knowledge of the electronic spectra and the structure of the heme ground and low-lying excited states is a great help in the detailed understanding of the protein functions. In particular such data are principally important for the understanding of the ligand binding and autooxidation mechanisms. In this paper the transient absorption spectra and lifetimes for the high-, low- and intermediate-spin human methemoglobin derivatives are reported.

We carried out the experiments with the absorption picosecond double-beam spectrometer based on the feed-back controlled passive modelocked Nd-phosphate glass laser (the pulse halfwidth is 3.5 ps)[2-4]. The second harmonic (λ=528 nm) was used for the excitation. The picosecond continuum generated in the cuvette with heavy water was used as a probe light. The experimental accuracy was $1 \cdot 10^{-3}$ O.D. units. The transient absorption spectra were measured in the spectral range from 430 to 850 nm. We have studied the methemoglobin derivatives with the following axial ligands: H_2O, OH^-, H_2O/OH^- mixture, Im, N_3^-, F^-, CN^- and hemichrome which has the imidasole of the distal histidine E7 as a second axial ligand. The stripped human hemoglobin in oxy-form extracted from fresh donor blood[5] with impurities being removed[6] on a DEAE-Sephadex A-50 ("Pharmacia",

Ultrafast Processes in Spectroscopy, edited by Svelto et al.
Plenum Press, New York, 1996

Sweden) has been taken for the methemoglobin derivatives preparation. Aquomethemo-globin (or methemoglobin proper - metHb) was prepared by oxyhemoglobin oxidation with potassium ferricyanide. Excess oxidizing agent and ferrocyanide were removed by dialysis against 50 mM potassium phosphate buffer with pH 7.2 over night at 4°C. In the case of hemoglobin complexes with Im, N_3, F⁻, CN⁻ the ratio of ligand to heme was 10:1. To prepare OH⁻ derivative the 50 mM Tris-HCl buffer with pH 9.0 was used. Hemichrome was prepared in 50 mM Tris-HCl buffer with pH 7.2 by adding a solution of potassium oleate to methemoglobin in a 10:1 molar excess per heme[7].

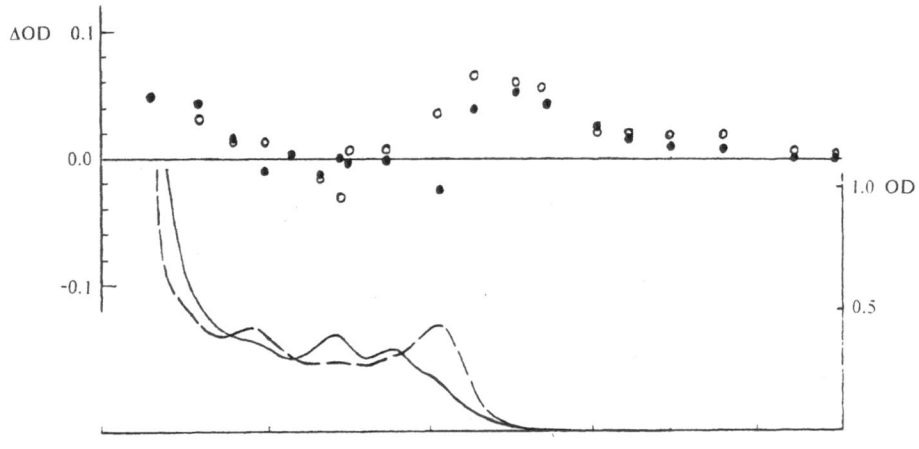

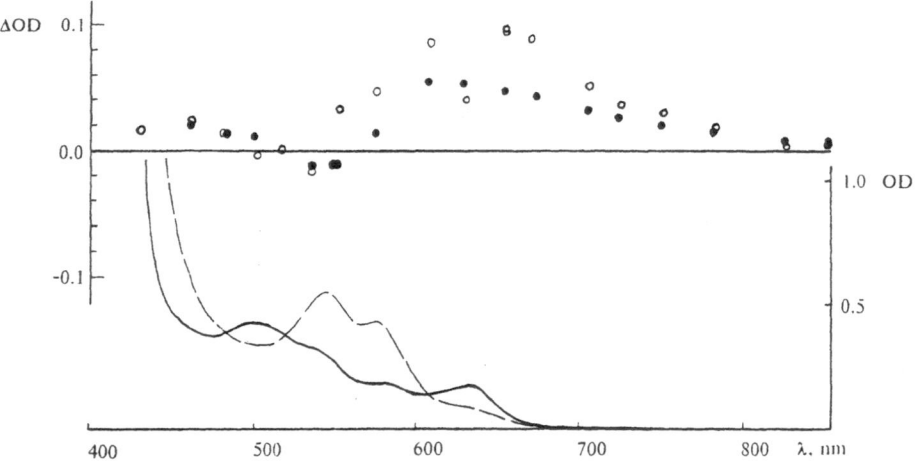

Figure 1. The ground state absorption spectra recorded in 1 mm path length cuvette and the measured difference transient that for the same samples: (top) $Hb(Fe^{3+})F^-$ (S=5/2) indicated by dotted line and closed circles, respectively, $Hb(Fe^{3+})OH^-$ (S=1/2) - solid line and open circles; (bottom) $Hb(Fe^{3+})N_3^-$ (S=1/2) - dotted line and closed circles, $Hb(Fe^{3+})H_2O$ (S=5/2) - solid line and open circles. In case $Hb(Fe^{3+})H_2O$ sample the difference spectrum was measured in 3 mm path length cuvette. Exept $Hb(Fe^{3+})OH^-$ which was prepared in 50 mM Tris-HCl buffer pH 9.0, all the samples were prepared in 50 mM potassium phosphate buffer, pH 7.2.

Fig.1. shows the absorption stationary and difference spectra for four derivatives of methemoglobin. It is evident that all the spectra are similar in the red and near-IR spectral region where the ground state absorption is absent ($\lambda > 620$ nm). In the spectral region of Soret and visible Q-bands the transient absorption is diffuse and follows the "landscape" of the ground state absorption for all the compounds studied. The general view of the transient spectra resembles the stationary ground state absorption spectra for all the studied proteins but the transient absorption bands are broaden and red-shifted. Therefore it is practically impossible to attribute this photoinduced absorption to that of low-lying excited state. The orbital origin of the low-lying excited states is the following[8]: this state arises from a configuration-interaction mixture of the porphyrin-iron charge-transfer state and the lowest singlet and triplet $\pi\pi^*$-states for the high-spin derivatives, at the same time for the low-spin derivatives this state is assigned to the porphyrin $e_g(\pi) \rightarrow$ iron d_π charge-transfer state. The energies of the $\pi\pi^*$ and excited charge-transfer states are different and depend on the ligand nature too[8]. Thus neither spin state nor energy of the lowest excited states and their orbital origin influence the transient absorption observed.

The similarity of the decay rate constants complements that of the general view of the transient spectra. All the measured decay curves are fit with a double exponential, having $\tau_1 = 3.5 \div 4.0$ ps and $\tau_2 \approx 30 \div 40$ ps in the weight ratio $\approx 10:1 \div 20:1$ (Fig.2.). Therefore it is very difficult to link the observed recovery kinetics with the electronic relaxation caused by the radiationless transition from the low-lying triplet $\pi\pi^*$ or CT-states to the ground state. The experimental values of τ_1 and τ_2 are in a good agreement with the molecular dynamics calculations[9,10] which predict the photoinduced heating of the heme in the proteins and the subsequent cooling having two stages: $1 \div 4$ ps and $20 \div 40$ ps. The cooling phenomenon and its spectral manifestation in the vibrational and absorption time-resolved spectroscopy is widely discussed[11-18]. Thus we associate the observed transient absorption with that of the protein heme in the vibrationally excited (hot) ground state.

It is obvious that for the final interpretation of the observed absorption the data of the time-resolved Stokes/anti-Stokes Raman studies and the femtosecond absorption studies are required. But the spectral-kinetic parameters of the observed transient absorption incline us to believe that not the the electronic relaxation but the heme cooling is responsible for the photoinduced spectral transformation of the all methemoglobins studied.

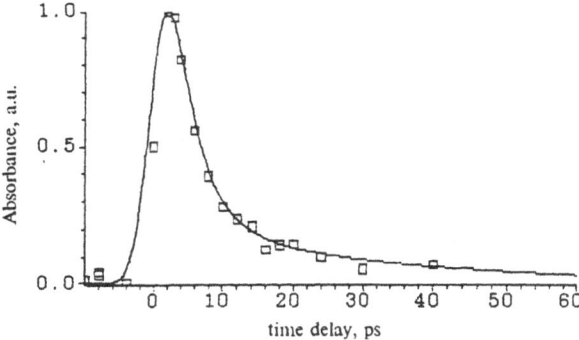

Figure 2. The observed decay kinetics at 650 nm for cyanide methemoglobin Hb(Fe^{3+})CN⁻ (S=1/2) (□). The solid line is the best fit with the following formula: A(t)=0.8*exp(-t/3.5ps)+0.2*exp(-t/30ps).

ACKNOWLEDMENT

This work is supported by the grant No. B5–338 from Foundation of Fundamental Research of the Republic of Belarus and partly by the International Soros Science Education Program.

REFERENCES

1. E. Antonini , M. Brunori, *Hemoglobin and Myoglobin in Their Reactions with Ligands* (North-Holland Publ.Comp., Amsterdam, 1971).
2. N. A. Lysak, S.V. Mel'nichuk, S. A. Tikhomirov, G. B. Tolstorozhev, Zh. Prikl. Spectroscopii (Russian). **47**, 267 (1987).
3. B.M. Dzhagarov, N.N. Kruk, V. Gulbinas et al., Lithuanian J.of Physics. **34**, 108 (1994).
4. B.M. Dzhagarov, N.N. Kruk, S.A. Tikhomirov, I.I. Stepuro, in *5th International Conference on Laser Applications in Life Sciences*, ed. by P.A. Apanasevich, N.I. Koroteev, Yu.V. Zadkov, S.G. Kruglik, (Proc. SPIE, **2370**, 1995).
5. A.A. Khachaturyan, E.P. Vyazova, G.M. Morozova, G.Ya. Rozenberg, Problems of Hematology (Russian). **1**, 58 (1979).
6. T.H.J. Ruisman, A.M. Bozy, J.Chromatography, **19**, 160 (1965).
7. A.A. Akhrem, G.M. Andreyuk, M.A. Kisel, P.A. Kiselev, Bochim. Biophys. Acta **992**, 191 (1989).
8. M.W. Martinen, A.K. Churg, in *Iron Porphyrins*, ed. by A.B.P. Lever and M.B. Gray, Part 1 (Addison-Wesley, 1983).
9. E.R. Henry, W.A. Eaton, R.M. Hochstrasser, Proc. Natl. Acad. Sci. USA, **83**, 8982 (1986).
10. E.R. Henry, W.A. Eaton, R.M. Hochstrasser, in *Ultrafast Phenomena V*, ed. by R.J. Fleming, A.E. Siegman (Springer-Verlag, Berlin, 1986).
11. J.W. Petrich, J.L. Martin, Chem.Phys., **131**, 31 (1989).
12. R. M. Hochstrasser, C. K. Johnson, in *Ultrashort laser pulses and applications*, ed. by W. Kaiser (Springer-Verlag, Berlin, 1988).
13. P. Li, J.T. Sage, P.M. Champion, J.Chem. Phys., **97**, 3214 (1992).
14. L. Zhu, P. Li, P.M. Champion, J. of Luminescence, **60–61**, 503 (1994).
15. R. Lingle Jr., X. Xu, H. Zhu et al., J. Am. Chem. Soc., **113**, 3992 (1991).
16. R.G. Alden, M.D. Chavez, M.R. Ondrias et al., J. Am. Chem. Soc., **112**, 3241 (1990).
17. P.A. Anfinrud, C. Han, R.M. Hochstrasser, Proc. Natl. Acad. Sci. USA, **86**, 8387 (1989).
18. J. Rodriguez, Ch. Kirmaier, D. Holten, J.Chem. Phys., **94**, 6020 (1991).

118

NORMALIZED FLUORESCENCE YIELD AS A FUNCTION OF LASERPULSE INTENSITY IN THE LIGHT HARVESTING COMPLEX OF PHOTOSYSTEM II OF HIGHER PLANTS AND CHLOROPHYLL A IN SOLUTION

René Schödel,[1] Klaus-D. Irrgang,[2] Joachim Voigt,[1] and Gernot Renger[2]

[1] AG Molekulare Biophysik und Spektroskopie
Institut für Physik der Humboldt Universität zu Berlin, Germany
[2] Max Volmer Institut für Biophysikalische und Physikalische Chemie
Technische Universität Berlin, Germany

The trimeric light harvesting complex II (LHC II) constitutes the major part of the Photosystem II (PS II) antenna in green plants and accounts for about 50% of the total chlorophyll content of the thylakoid membrane [1]. In this contribution we present new experimental and theoretical investigations of non-linear (intensity dependent) fluorescence properties of the nonaggregated LHC II preparation isolated from spinach. The LHC II complexes used in this study were obtained by β-DM solubilization of salt washed (NaCl, $CaCl_2$) PS II membrane fragments and separation by sucrose density gradient centrifugation in the presence of 0,05% β-DM as outlined by Irrgang et al. [2]. Commercial (Sigma Chemical Co.) Chl a from spinach was dissolved in acetone (80%). The concentration was about 0.2 mg Chl/ ml.

Monochromatic 645 nm laser pulses of 3.5 ns duration were used for sample excitation. Special efforts were made to collect data only from sample regions exposed to similar photon densities of the laser pulse. Therefore, the fluorescence from highly excited areas was spatial separated from lower excitation by a special display method. Furthermore the scattered laser light was spectral "discarded" with a double monochromator so that only the pure fluorescence was monitored as a function of the photon density of the excitation pulses refered to as I_p.

The fluorescence $F(I_p)$ emitted at 680 nm was devided by the exciting photon density (I_p/cm^{-2} pulse^{-1}) to obtain the relative fluorescence yield $\Phi_F(I_p)$. Φ_F was normalized to the value $\Phi_F(I_p \to 0)$ gained at the limit of weak intensities. In this study $\Phi_F(I_p)$ was measured over a wide range covering six orders of I_p-values (up to about 10^{19} photons per cm^2 per pulse). The magnitude of $\Phi_F(I_p)$ drops down to levels of about 10^{-4} at highest photon densities.

Ultrafast Processes in Spectroscopy, edited by Svelto et al.
Plenum Press, New York, 1996

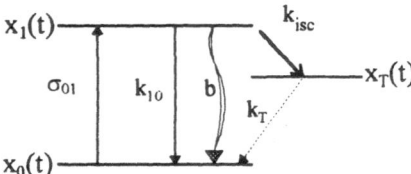

Figure 1. Picture of populations and channels.

The general feature of Φ_F decrease is well known in literature [3] but to our knowledge it has never been investigated over such a wide range. It is usual to ascribe the phenomenon of $\Phi_F(I_p)$ decline to two processes: (i) bimoleculare exciton- exciton-annihilation leading to a radiationless decay of excited states (within domains of large pigment-protein complexes and therefore absent in dilute chlorophyll solutions) and (ii) ground state depletion. Furthermore the influence of triplet states has to be considered. In order to account especially for annihilation processes, comparative measurements were performed under the same conditions in dilute Chl a solution. The experimental data obtained were numerically analyzed whitin the framework of a system comprising singlet and triplet levels as described in Fig. 1.

This scheme gives rise to the following kinetic equations for the relative occupation densities of excited singlet states (x_1), excited triplet states (x_T) and remaining ground states $x_0 = 1 - x_1 - x_T$:

$$\frac{d}{dt} x_1(t) = \sigma_{01} \cdot I(t) \cdot \left(1 - x_1(t) - x_T(t)\right) - x_1(t) \cdot \left(k_{10} + k_{isc}\right) - b \cdot x_1(t)^2 \tag{1}$$

$$\frac{d}{dt} x_T(t) = k_{isc} \cdot x_1(t) - k_T \cdot x_T(t) \tag{2}$$

The differential equations were numerically solved assuming a Gaussian shape excitation pulse with a FWHM of 3.5 ns ($\Delta = 1.1$ ns) :

$$I(t) = I_p / \left(\Delta \cdot \sqrt{2\Pi}\right) \cdot \exp\left[-(t - t_0)^2 / \left(2 \cdot \Delta^2\right)\right].$$

$\Phi_F(I_p)$ was calculated by the following integration:

$$\Phi(I_p) = \frac{\int_{-\infty}^{\infty} x_1(t) \cdot dt}{I_p} \left/ \left| \frac{\int_{-\infty}^{\infty} x_1(t) \cdot dt}{I_p} \right|_{I_p \to 0} \right. \tag{3}$$

The numerical fits for Chl a were performed with values taken from literature: $k_{10} + k_{isc} = 1/(\tau = 5$ ns$)$ and $k_{isc} = 0.66 \times (k_{10} + K_{isc})$ [see Ref. 4] and the lifetime $\tau = 4.3$ ns of excited singlet states in nonaggregated LHC II [5]. The extinction coefficient of Chl a

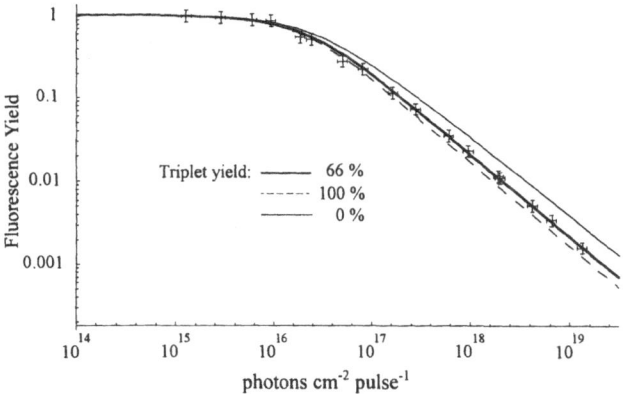

Figure 2. Measured fluorescence yield of Chl a fitted by the following parameters: $\sigma_{Chl\,a}(645\ nm) = 0.62\times10^{-16}\ cm^2$, $b = 0$, $\tau = 1/(k_{10} + k_{isc}) = 5$ ns. The triplet yield $\Phi_T = k_{isc}/(k_{10} + k_{isc})$ was varied.

in acetonic solution at 646.6 nm is 18580 l $mol^{-1}\ cm^{-1}$ [6], i,e. the absorption cross section is $\sigma(646.6\ nm) = 0.71\times10^{-16}\ cm^2$. From the ratio OD(646.6 nm)/OD(645 nm) one obtains: $\sigma_{Chl\,a}(645\ nm) = 0.63\times10^{-16}\ cm^2$.

With the above mentioned values τ and k_{isc} the best fit of fluorescence data in Chl a solution were obtained with: $\sigma_{Chl\,a} = 0.62\times10^{-16}\ cm^2$. This value is fully consistent with that determined from the extinction coefficient. In this case parameter b was set to zero because no exciton-exciton-annihilation contributes to excited singlet state decay in dilute Chl a solution. In order to illustrate the influence of the triplet formation on the $\Phi_F(I_p)$ curves, two extremes were analyzed, i.e. triplet yields of 0 % and 100 %, respectively.

The results depicted in Fig. 2. reveal that triplets do not cause drastic effects.

The same type of analysis performed for LHC II led to surprising results: The shape of the $\Phi^{LHC\ II}(I_p)$ curve can only be fitted if effects due to exciton-exciton-annihilation are marginally small, otherwise the calculated curves are too flat to be consistent with the experimental results. This conclusion, i.e. the absence of bimoleculare exciton-exciton annihilation process in nonaggregated LHC II preparations is also supported by measurements of the absorption as a function of I_p [7, 8].

An average absorption cross section of Chl a and Chl b within LHC II was determined from absorption measurements and entire chlorophyll concentration within the sample (N^{sample}):

$$\sigma_{Chl\,a\_Chl\,b}^{LHC\,II}(645\ nm) = \alpha(645\ nm)\Big/N^{sample} = 0.75\times10^{-16}\ cm^2.$$

This value, however, did not permit any data fit. In order to achive this goal by using eqns. 1-3 the value $\sigma_{Chl\,a\_Chl\,b}^{LHC\,II}(645\ nm)$ has to be scaled by a factor of 10-20, depending on the assumption about the triplet yield ($8 \times 10^{-16}\ cm^2$ for 100 % triplet yield and $1.7 \times 10^{-15}\ cm^2$ for zero triplet yield). Interestingly this factor corresponds with the number of chlorophylls bound per monomeric subunit of LHC II [1].

Based on the results of the present study it is concluded that the pigments inside an LHC II monomer are strongly coupled thus giving rise to excitonic states which can be ascribed only to the whole pigment ensemble. In this case "diffusion controlled" bimolecu-

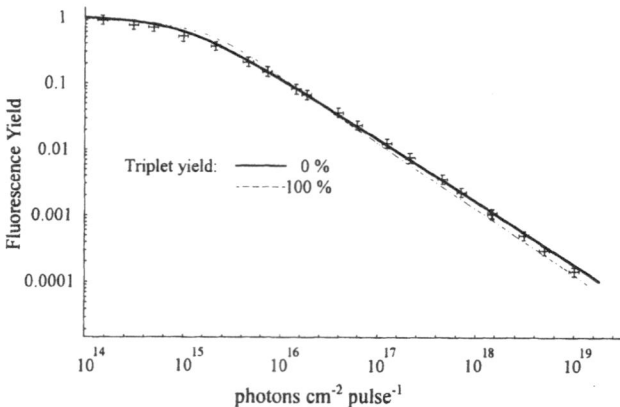

Figure 3. Measured fluorescence yield of LHC II fitted with the following parameters: $\tau = 1/(k_{10} + k_{isc}) = 4.3$ ns, $b = 0$. The two borderline cases for triplet yied are considered: $\Phi_T = 0$ leads to $\sigma_{Chl\,a}(645\ nm) = 1.7 \times 10^{-15}\ cm^2$, $\Phi_T = 100\ \%$ leads to $\sigma_{Chl\,a}(645\ nm) = 0.8 \times 10^{-15}\ cm^2$.

lar annihilation processes of excited states do not occur within the pigment ensemble of each monomeric subunit of LHC II. The implecations of this basic conclusion will be analyzed in forthcoming studies.

ACKNOWLEDGMENT

The financial support by Deutsche Forschungsgemeinschaft (Sfb 312) is grateful acknowledged.

REFERENCES

1. W.Kühlbrandt (1994) Current Biol. 4, 519-528
2. K.D. Irrgang, E.J. Boekema, J. Vater, G. Renger (1988); Eur. J. Biochem. 178, 209-217
3. J. Breton, N.E. Geacintov (1980); Biochim. Biophys. Acta 594,1-32
4. G.Renger (1992) in: J. Barber (ed.) The Photosystems: Strukture, Funktion and Moleculare Biology, pp.45-99, Elsevier, Amsterdam
5. B. Liu, A. Napiwotzki, H.-J. Eckert, Eichler, H.J. and Renger, G. (1993) Biochim. Biophys. Acta 1142, 129
6. R. J. Porra, W. A. Thompson and P. E. Kriedemann (1989), Biochim. Biophys. Acta 975, 384-394
7. G. Kehrberg, T. Schrötter, J. Voigt, and G. Renger (1995) Biochim. Biophys. Acta 1231, 147
8. D. Leupold, S. Mory, R. König, P. Hoffmann, B. Hieke (1977) Chem. Phys. Lett. 45 No.3, 567-571

119

MOLECULAR DYNAMICS OF MICROBIAL LIPASES IN DIFFERENT SOLUBILIZATION SYSTEMS AS DETERMINED BY PHASE-RESOLVED FLUOROMETRY

M. Graupner,[1] L. Haalck,[2] F. Spener,[2] F. Paltauf,[1] and A. Hermetter[1]*

[1] Institute of Biochemistry and Food Chemistry
Spezialforschungsbereich Biokatalyse
Technische Universität Graz, Graz, Austria
[2] Institut für Chemo- und Biosensorik
Universität Münster, Germany

Lipases are fat-cleaving enzymes. Recently we were able to show that lipase activity as well as their stereoselectivity strongly depend on the composition of the reaction medium, e.g. the presence of detergent or organic solvent in an aqueous system[1]. To obtain information on how lipase conformation is affected in such environments, we studied fluorescence properties of the tryptophans in the lipases from *Chromobacterium viscosum* (CVL) and *Pseudomonas species* (PSL) in aqueous buffer, detergent micelles and water-*iso*-propanol mixtures $(1/1)^2$.

Fluorescence lifetimes and time-resolved anisotropies were performed using a multifrequency phase and modulation fluorometer. From phase angles and modulations determined at 25 different modulation frequencies between 11 to 500 MHz, continuous Lorentzian lifetime distributions were recovered[3] that were bimodal for CVL and trimodal for PSL in all reaction systems under investigation. Assuming that lifetime distribution widths are representative for the heterogeneity of the tryptophan microenvironment, the presence of detergent or iso-propanol leads to protein conformational changes that result in increased fluorophore microheterogeneity in CVL, whereas the opposite effect is observed for PSL.

From phase angles and demodulations, determined at parallel and perpendicular orientations of the emission polarizer relative to the vertically adjusted excitation polarizer, time-resolved fluorescence anisotropy decays were obtained[4]. The differential polarized phase angles increase toward a maximum value (not shown), depending on motional rate and modulation frequency. A maximum associated with larger phase angles and higher

---

* Corresponding author.

Ultrafast Processes in Spectroscopy, edited by Svelto et al.
Plenum Press, New York, 1996

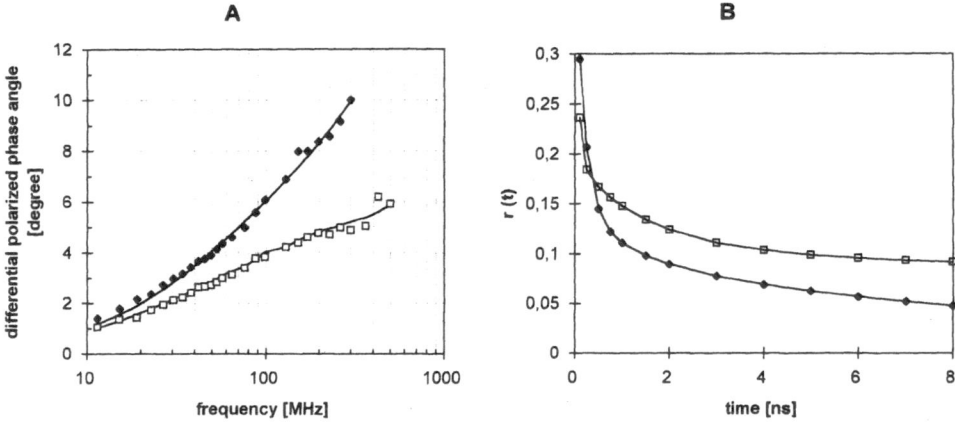

Figure 1. Frequency-domain (A) and time-domain (B) representation of anisotropy decays of CVL (♦) and PSL (□) in aqueous buffer.

frequencies is observed for CVL as a consequence of faster molecular motions. Fig. 1B shows the corresponding time-dependent anisotropy decays, indicating again faster CVL and slower PSL motions[5].

In table 1 the rotational correlation times for CVL and PSL in buffer, detergent micelles and water-iso-propanol (1/1) are listed. We found three correlation times for both lipases in all systems. The long correlation time reflects the motion of the entire particle. The short and medium correlation times are probably due to internal motions of protein segments to which the fluorophors are bound. The long correlation time reveals, that the degree of association of CVL and PSL is different depending upon "solvent" composition. It can be inferred, that CVL is monomeric in all systems. In contrast, PSL is aggregated in aqueous buffer and dissociates into smaller particles upon addition of detergent or iso-propanol. It is striking that only 25% loss of anisotropy (long correlation time) accounts for the motion of the entire particle. Thus, both lipases seem to exhibit high internal flexibility, because most of the anisotropy decays as a consequence of very fast motions in the polypeptide on a picosecond time scale. In the presence of iso-propanol, the fast internal motions become sightly more restricted, whereas detergents sightly enhance the short rotational times of the fluorophor.

Table 1. Rotational correlation times (θ) and g-values (in per cent) of CVL and PSL in buffer, detergent micelles and water-iso-propanol (1/1)

		Rotational correlation times			
		$\theta_1[v\sigma]$	$\theta_2[ns]$	$q_3[ns]$	χ^2
CVL	buffer	0.20 (64%)	1.34 (13%)	12.26 (23%)	3,12
	micelles	0.06 (46%)	0.61 (23%)	14.56 (31%)	1,79
	50% propanol	0.26 (56%)	3.31 (14%)	32.52 (30%)	3,39
PSL	buffer	0.07 (52%)	1.52 (22%)	61.60 (26%)	1,83
	micelles	0.08 (53%)	1.98 (22%)	33.24 (25%)	9,60
	50% propanol	0.10 (57%)	2.19 (20%)	18.92 (23%)	10,1

In summary, both lipases (CVL and PSL) showed significant alterations of their gross conformations and molecular dynamics under different solubilization conditions that also impart different catalytic properties on the enzymes (activity, stereoselectivity). It remains to be established as to how and what extent the observed changes in enzyme structure are associated with conformational changes in the active site that finally lead to changes in lipase function.

ACKNOWLEDGMENT

Financial support by the Fonds zur Förderung der wissenschaftlichen Forschung (project No. F0107 to A. H.) is gratefully acknowledged. We are indebted to the Center of Fluorescence, University of Maryland Medical School, Baltimore/USA, for providing support and facilities for the time-resolved fluorescence measurements.

REFERENCES

1. G. Zandonella, L. Haalck, F. Spener, K. Faber, F. Paltauf, and A. Hermetter *Eur. J. Biochem.* 231, 50–55 (1995).
2. G. Carrea, G. Ottolina, and S. Riva *TIBTECH* 13, 63–70 (1995).
3. J. R. Alcala, E. Gratton. and F. G. Prendergast *Biophys. J.* 51, 925–936 (1987).
4. B. Maliwal, and J. R. Lakowicz *Biochimica and Biophysica Acta* 873, 161–172 (1986).
5. J. R. Lakowicz, and I. Gryczynski, in *Topics in Fluorescence Spectroscopy*, Vol.1, ed. by J. R. Lakowicz (Plenum Press, New York and London, 1991)

120

EARLY LIGHT-INDUCED CHARGE DISPLACEMENT PROCESSES IN BACTERIORHODOPSIN

G. I. Groma,[1] J. Hebling,[2*] C. Ludwig,[2] and J. Kuhl[2]

[1] Institute of Biophysics
Biological Research Centre of Hungarian Academy of Sciences
Szeged, H-6726, Hungary
[2] Max-Planck-Institut für Festkörperforschung
Stuttgart, D-70506, Germany

Bacteriorhodopsin (bR) is the only protein of the purple membrane found in *Halobacterium salinarium* and related species. By means of the retinal chromophore attached to it via a protonated Schiff base, this protein utilizes the energy of light to build up a proton electrochemical potential for ATP synthesis. In the physiologically important light-adapted state, every chromophore is in the all-trans form. In this state, photon absorption initiates a photocycle of different intermediates, resulting in a pump of protons across the membrane[1,2]. The primary events of the photocycle of light-adapted bR, studied in ultrafast absorption kinetic experiments, can be summarized as follows[2]:

$$bR_{570} + h\upsilon \longrightarrow bR*(trans) \xrightarrow{100-200\,fs}$$

$$bR*(cis) \xrightarrow{300-500\,fs} J_{625} \xrightarrow{3-5\,ps} K_{590} \cdots,$$

where the indices indicate the absorption maxima of the corresponding forms, and the asterisks denote excited states. The existence of a reverse photoreaction K→bR has been demonstrated[3], but the details of that route are still unknown.

Since bR functions as a proton pump, investigation of charge movement processes in the course of the primary events is also of crucial importance. Excitation of electrically oriented purple membrane samples allows the detection of such charge movements by direct electric methods[4]. Picosecond electric studies in recent years have demonstrated the occurrence of an ultrafast phase in the light-induced protein electric response signal

---

* Permanent address: Institute of Optics and Quantumelectronics, JATE University, Szeged, H-6701, Hungary.

Ultrafast Processes in Spectroscopy, edited by Svelto et al.
Plenum Press, New York, 1996

(PERS) of bR[5–8]. Here we briefly summarize our new results on the simultaneous detection of the early charge movement processes corresponding to the forward and the reverse photoreaction, with an improved time resolution. The details of this study will be published elsewhere[9].

Electrically oriented purple membranes were deposited and dried on an indium-tin-dioxide thin layer as described previously[7]. This sample was incorporated into a modified commercial SMA panel mount jack receptacle ensuring a capacitive junction between the intra-membrane charge motion and the external electric detector.

The excitation pulses were generated by a home-made CPM laser, amplified by a Cu vapor laser. This system emitted a train of 150 fs, 620 nm pulses with 7 kHz repetition rate. The maximum average intensity at the site of the sample was 100 mW/cm^2. The light-induced electric response signal of the sample was detected by a Tektronix 11801A digital sampling oscilloscope, equipped with a 40 GHz sampling head.

The ultrafast component of the PERS generated by the 7 kHz pulse train (Fig. 1) consists of a positive and a negative peak. This shape is in sharp contrast with that of the previously published traces, which were generated at low repetition rate (typically 10 Hz) and displayed only a single negative phase[5–8]. The relative weight of the unusual positive peak:

- decreased by chopping the exciting laser beam (Fig. 1),
- decreased by reducing of the excitation intensity,
- decreased by applying a background illumination of the sample with a 647 nm Kr$^+$ laser line,
- and increased by a background illumination with a 511 nm Cu vapor laser line.

All of these effects can be explained by the presence of a relatively high concentration of the K intermediate, populated by a photoequilibrium which is controlled by the above factors. The negative peak of the PERS trace corresponds to the transition bR→K, and the unusual positive one to the reverse photoreaction K→bR. The observed shape of the signal indicates that the rate of the reverse reaction is higher than that of the forward one, in accordance with the findings of previous optical studies[10].

Statistical analysis based on least square model fitting resulted in a 5 ps upper and a 2.5 ps lower limit for the time constant of the charge displacement process, corresponding

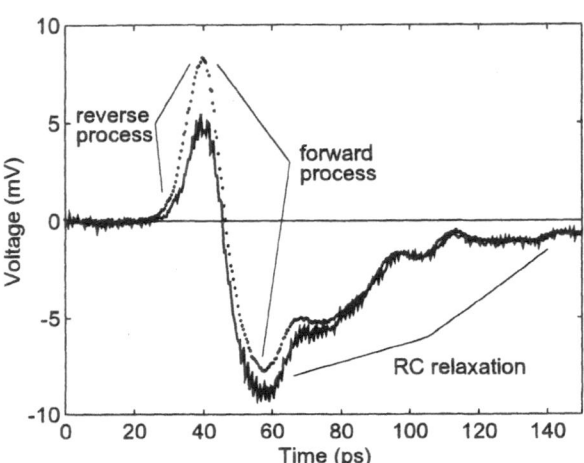

Figure 1. Light-induced protein electric response signal (PERS) of bacteriorhodopsin excited by a 620 nm amplified CPM laser of 150 fs pulsewidth. Dotted line: excitation with 7 kHz repetition rate. Solid line: excitation with chopped laser. A train of 40 pulses with the original repetition rate passed through the hole of the chopper every 100 ms.

to the forward reaction. This is in good agreement with the formation time of K measured in absorption kinetic experiments[2]. The opposite charge separation process, driven by the reverse phototransition K→bR, has an upper limit of 3.5 ps for the time constant. This limit is much lower than that determined by optical methods[10]. The difference in the rates indicates the existence of different routes for the forward and the reverse photoreactions.

ACKNOWLEDGMENT

This work was supported by the National Scientific Research Foundation of Hungary, OTKA-T6401 (G. I. G).

REFERENCES

1. W. Stoeckenius, R. H. Lozier, and R. A. Bogomolni, Biochim. Biophys. Acta, **505**, 215 (1979).
2. R. A. Mathies, S. W. Lin, J. B. Ames, and W. T. Pollard, Annu. Rev. Biophys. Biophys. Chem. **20**, 491 (1991).
3. R. R. Birge, Biochim. Biophys. Acta **1016**, 293 (1990).
4. L. Keszthelyi, and P. Ormos, J. Membrane Biol. **109**, 193 (1989).
5. G. I. Groma, G. Szabó, and G. Váró, Nature (Lond.) **308**, 557 (1984).
6. H.-W. Trissl, Biochim. Biophys. Acta **806**, 124 (1985).
7. G. I Groma, F. Ráksi, G. Szabó, and G. Váró, Biophys. J. **54**, 77 (1988).
8. R. Simmeth, and G. W. Rayfield, Biophys. J. **57**, 1099 (1990).
9. G. I. Groma, J. Hebling, C. Ludwig, and J. Kuhl, Biophys. J. **69**, *in press* (1995).
10. P. G. Kryukov, Yu. A. Lasarev, Yu. A. Matveets, E. L. Terpugov, L. N. Chekulaeva, and A. V. Sharkov, Stud. Biophys. **83**, 101 (1981).

121

TEMPERATURE DEPENDENCE OF SECOND-ORDER SUSCEPTIBILITY OF NOBLE METALS NEAR THE INTERBAND TRANSITION THRESHOLD

J. Hohlfeld,[1] U. Conrad,[1] D. Grosenick,[2] and E. Matthias[1]

[1] Fachbereich Physik
Freie Universität Berlin, Germany
[2] Max-Born-Institut
Berlin, Germany

In noble metals, the time dependence of linear and nonlinear reflectivity in the interband transition region is dominated by the electron temperature dependent broadening of the Fermi-Dirac distribution. In this contribution, we report on the influence of non-equilibrium electron dynamics in noble metals by means of reflection second-harmonic generation (SHG) with the photon energy of the fundamental or second-harmonic (SH) near resonance with the interband transition threshold. Previously,[1] we used this technique for studying the electron relaxation of polycrystalline copper in air, and demonstrated that femtosecond time-resolved SHG is far more sensitive for measuring transient electron temperatures of metals compared to linear Transient Thermo Reflection Spectroscopy (TTRS). Here, we present TTRS and pump-probe SH measurements on polycrystalline copper, silver and gold, and show that these measurements provide information about the electron temperature dependence of the second-order susceptibility $\chi^{(2)}(T_e)$.

Following a phenomenological model presented by Sipe et al.[2] the electric field of second-harmonic radiation generated in reflection at a surface can be expressed within the electric-dipole approximation as:

$$\vec{E}(2\omega) = \vec{f}(\omega) \cdot \ddot{\chi}^{(2)} \cdot \vec{F}(2\omega) \cdot \vec{E}(\omega) \cdot \vec{E}(\omega).\tag{1}$$

The Fresnel factors $\vec{f}(\omega)$ and $\vec{F}(2\omega)$ connect the external fields in vacuum to local fields in the sample. They are determined by the linear dielectric functions $e(\omega)$ and $e(2\omega)$ and sample the material within their optical penetration depth. The second-order susceptibility $\ddot{\chi}^{(2)}$, a tensor of third rank, describes the intrinsic ability of the sample to generate second-harmonic radiation. In centrosymmetric media as the noble metals, it is restricted to the near surface region where inversion symmetry is broken. For polycrystalline sam-

Ultrafast Processes in Spectroscopy, edited by Svelto et al.
Plenum Press, New York, 1996

ples, the elements of $\bar{f}(\omega)$, $\bar{F}(2\omega)$ and $\bar{\chi}^{(2)}$ that will contribute to the second-harmonic field are determined by the polarization of both, fundamental and second-harmonic.

The electron temperature dependence of $\chi^{(2)}(T_e)$ is of fundamental interest for three reasons. It may serve as a crucial test for theoretical calculations of $\bar{\chi}^{(2)}$. Secondly, it is required for proper data interpretation in case intense laser pulses, which cause significant heating of the sample, are used in SH measurements. The third reason is that $\bar{\chi}^{(2)}$ is exclusively sensitive to the electron temperature at the surface. In contrast, linear TTRS measurements sample the material within the optical penetration depth and only provide information about the relaxation of an averaged electron temperature. Hence, the temperature dependence of $\bar{\chi}^{(2)}$ opens the way to uncover possible differences between the electron dynamics at the surface and the bulk. However, since all three factors $\bar{f}(\omega)$, $\bar{F}(2\omega)$ and $\bar{\chi}^{(2)}$ will be affected by electron temperature, the determination of $\chi^{(2)}(T_e)$ from time-resolved pump-probe SHG data requires the knowledge of the time dependence of electron temperature and its influence on $\bar{f}(\omega)$ and $\bar{F}(2\omega)$. This information was obtained from analysis of additional linear TTRS measurements at fundamental and second-harmonic frequencies.

The experimental setup is reported in Ref.1. Here, only the most important parameters will be given. All measurements were carried out on polycrystalline Cu, Ag and Au samples under normal conditions, using p-polarized laser pulses of 100 fs duration at a wavelength of 630 nm (2 eV) and incident angles of 43° (pump) and 48° (probe). In all SHG measurements identical pump and probe pulse energies between 12 and 16 µJ and foci of about 0.2 mm were used. The p-polarized probe pulse SHG yield was recorded as a function of pump-probe delay. In TTRS measurements, probe pulse energies were attenuated by a factor of 5000 and the pump and probe beams were focused to about 0.5 mm and 0.2 mm, respectively. To obtain $R(2\omega)$ frequency doubled probe pulses were used. All data points shown below correspond to an average of about 500 shots.

The analysis of the TTRS data, based on theoretical models presented in Ref.1, showed that the models have to be refined in two points. Firstly, diffusion of hot electrons needs to be included and, secondly, the approximation of electron momentum independent transition matrix elements instead of oscillator strength ought to be used. The calculations refined in this way lead to good agreement with the TTRS data for all three metals. Their results were used to predict the time dependence of SHG intensity due to the Fresnel factors only. Since polarization dependent SHG measurements reveal χ_{zzz} to be the only significant tensor element, the relative change of $|\chi_{zzz}|^2$ with T_e can be extracted from the differences $\Delta SH = SH(T_e) - SH_0$ of measured (SH_m) and predicted (SH_p) second-harmonic intensities:

$$\frac{\Delta|\chi_{zzz}|^2}{|\chi_{zzz}|^2} = \frac{\Delta SH_m - \Delta SH_p}{SH_p(T_e)}.$$

(2)

Here, SH_0 is the yield for the probe pulse alone. Results of pump-probe SHG measurements for Au are shown in Fig. 1.

The initial peak around zero delay corresponds to a coherent artifact and will not be discussed here. The difference between measured and predicted relaxation behavior indicates, that the temperature dependence of SHG is dominated by χ_{ZZZ}. Evidently, SHG yields an effect which is about two orders of magnitude larger compared to changes in linear reflectivity.

The huge variation of χ_{ZZZ} originates from the fact, that the large temperature independent contribution to linear reflectivity due to the free conduction electrons is absent in case of χ_{ZZZ}.

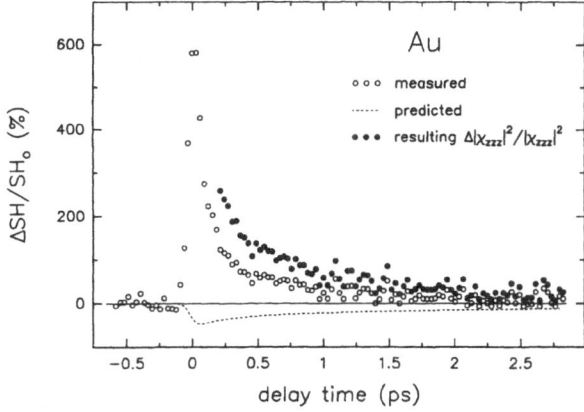

Figure 1. Normalized differences of measured and predicted SH yield and the extracted $\Delta|\chi_{zzz}|^2/|\chi_{zzz}|^2$ as a function of pump-probe delay for polycrystalline Au.

The knowledge of the time dependence of T_e can now be used to transform the determined time dependence of χ_{ZZZ} into a temperature dependence. Corresponding results are shown in Fig. 2 for all three metals.

In case of Cu, there is no change of $|\chi_{zzz}|^2$ with electron temperature for the photon energy used, and the relaxation behavior is well described by the Fresnel factors. For Ag and Au, we find a linear dependence of $|\chi_{zzz}|^2$ on T_e. The slopes have different sign which can be attributed to the different sign of the mismatch between the photon energy and the interband transition threshold.

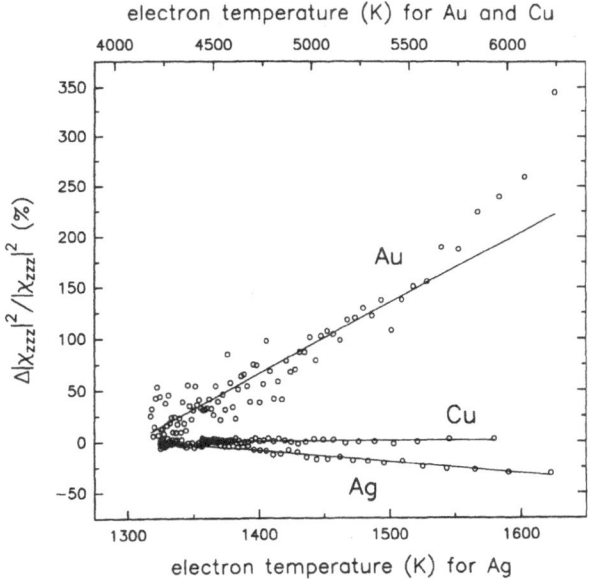

Figure 2. Relative change of $|\chi_{zzz}|^2$ with electron temperature for Cu, Ag, and Au.

The energy mismatch of -150 meV in case of Cu is small compared to the width of the Fermi distribution caused by one pulse alone (kT_{e0} = 390 meV). Thus, the density of vacant states and the accompanying resonance enhancement of χ_{ZZZ} induced by the first pulse is already saturated and therefore nearly unaffected by additional heating due to the second pulse. For Au, the mismatch is -380 meV and hence no saturation occurs due to heating of the first pulse (kT_{e0} = 310 meV). Consequently, the second causes a strong increase of χ_{ZZZ}.

For Ag, the interband transition threshold is close to $2\hbar\omega$ and therefore heating is significantly smaller. However, $|\chi_{zzz}|^2$ is reduced by even a small electron temperature increase due to occupation of the initially vacant states (energy mismatch=+20 meV) and loss of resonance enhancement of SHG.

Although no theoretical prediction about the temperature dependence of $|\chi_{zzz}|^2$ is reported in the literature, it is reasonable to assume that $|\chi_{zzz}|^2$ as well as the linear reflectivity depends linear on small temperature variations. Thus, the linear dependencies of $|\chi_{zzz}|^2$ on T_e for Ag and Au (cf. Fig. 2) indicate similar electron dynamics at the surface and in the bulk. It remains an open question, whether the deviations seen for Au at temperatures above 5500 K originate from a nonlinear temperature dependence of χ_{ZZZ} or from differences of electron dynamics at surfaces and in the bulk during the first 200 fs.

ACKNOWLEDGMENT

This work was supported by Deutsche Forschungsgemeinschaft, Sfb 290.

REFERENCES

1. J. Hohlfeld, D. Grosenick, U. Conrad, and E. Matthias, Appl. Phys. A **60**, 137 (1995)
2. J.E. Sipe, D.J. Moss, and H.M. van Driel, Phys. Rev. B **35**, 1129 (1987)

122

ELECTRONIC EXCITATION ENHANCED CRYSTALLIZATION OF AMORPHOUS FILMS?

J. Solis,[1] C. N. Afonso,[1] S. C. W. Hyde,[2] N. P. Barry[2] and P. M. W. French[2]

[1] Instituto de Optica
CSIC. C/Serrano 121
28006-Madrid, Spain
Tel:+34–1–5616800, Fax:+34–1–5645557
e-mail: IODJS37@PINAR1.csic.es
[2] Femtosecond Optics Group, Optics Department
Imperial College of Science and Technology
Prince Consort Rd, London SW7–2BZ, United Kingdom

The study of the interaction of ultrashort laser pulses with solids has raised some important questions concerning the nature of the transformations occurring at the material surface in the presence of very strong levels of electronic excitation. Even when the presence of electronic excitation induced metastable transient phases has been demonstrated in several materials upon irradiation with femtosecond laser pulses [1,2,3,4,5], the role of electronic excitation in permanent phase transitions is controverted and different non-thermal effects have been invoked to explain permanent structural changes induced in amorphous Si and Ge upon ns and ps laser pulse irradiation [6,7]. The aim of the work presented here is to elucidate whether the high electronic excitation induced by irradiation with subnanosecond laser pulses, can enhance the crystallization process of an amorphous phase.

We have chosen for that purpose high Sb-content GeSb thin films which have been shown to crystallize and amorphize upon irradiation with subnanosecond (ps and fs) laser pulses [8]. The existence of electronic excitation crystallization enhancement should lead to appreciable changes in the energy density threshold for crystallization as the laser pulse duration changes, according to the differences in the excited carrier populations induced.

We have performed what we believe to be the first experiment to determine the minimum energy density required for crystallization of an amorphous phase upon irradiation with laser pulses whose duration extend over three orders of magnitude, covering the range 400 fs-8 ns. An scheme of the experimental setup has been included in Fig.1. The irradiations were performed using a CPA system operating at 830 nm to single shot irradiate 50 nm-thick bilayered films formed by an amorphous $Ge_{0.07}Sb_{0.93}$ surface layer and an underlayer slightly richer in Sb and therefore polycrystalline. The laser beam, appropriately

Ultrafast Processes in Spectroscopy, edited by Svelto et al.
Plenum Press, New York, 1996

543

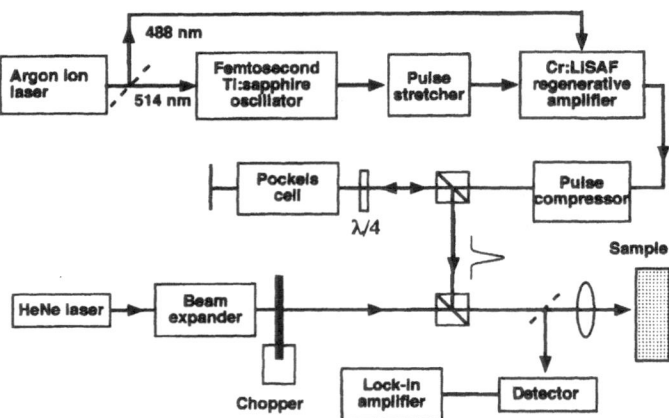

Figure 1. Experimental setup for pulsed laser irradiation. The laser pulse source operates at 830 nm and may produce pulses with durations ranging from 400 fs to 8 ns. The reflectivity of the sample at 633 nm is monitored before and after irradiation.

attenuated, was focused on the sample surface to a spot size of the order of 15 μm by means of a NA 0.4, 10x microscope objective. Since crystallization is evidenced by significant changes in the reflectivity of the surface, a CW HeNe (633 nm) laser beam arranged to be collinear with the irradiation beam, was used to monitor the reflectivity of the surface before and after irradiation. The HeNe laser beam was expanded before overlapping in order to focus it to a size of ≈ 2μm at the sample site. From the reflectivity values observed before and after irradiation it is possible to determine to minimum pulse energy required to crystallize the surface for a given pulse length. A combination of this measurement with a careful determination of the energy density distribution at the sample site allows to determine the minimum energy density required for crystallization.

The evolution of the energy density crystallization threshold as a function of the pulse duration is shown in Fig.2. It shows a significant increase for pulse durations in the ns range which is related to the existence of heat transfer to the substrate while the laser pulse is still being absorbed. This can be verified by comparing the pulse duration to the minimum time required for heat to diffuse to the substrate. We have estimated this time to

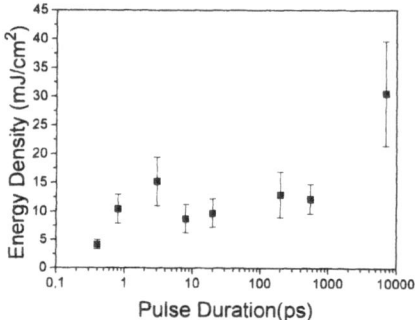

Figure 2. Energy density crystallization threshold of GeSb films versus pulse duration.

be in the order of at least 200 ps assuming that the thermal diffusion length equals the film thickness, and using the thermal properties of Sb films. Obviously, for a pulse duration of the order of the thermal diffusion time or longer, heat is transferred to the substrate while the pulse is being absorbed by the film. The energy density crystallization threshold remains constant within experimental resolution as the pulse length is decreased down to 800 fs. For the shortest pulse length used, (400 fs), a decrease in the threshold is observed, thus suggesting the existence of electronic excitation enhanced crystallization for this pulse duration. These results clearly demonstrate that electronic excitation enhanced crystallization can only occur for pulses shorter than 800 fs in this material and indicate that the time for energy transfer between the excited carriers and the lattice is in the range of a few hundreds of fs. Other possible effects that might lead to the observed behavior -like a non-linear response in the sample reflectivity for very short pulses- have to be tested. Additional measurements for pulse durations in the range from 100 fs to 1 ps are on their way in order to further confirm the existence of electronic excitation enhanced crystallization for pulses shorter than 800 fs.

REFERENCES

1. H.W.K.Tom, G.D.Aumillier, C.H.Brito-Cruz; Phys.Rev.Lett.**60**, 1438 (1988).
2. Y.Siegal, E.N.Glezer, E.Mazur; Phys.Rev.**B49**, 16403 (1994).
3. S.V.Govorkov, Th.Schröder, I.L.Shumay, P.Heist; Phys.Rev.**B46**, 6864 (1992).
4. P.Saeta, J.-K Wang, Y.Siegal, N.Bloembergen, N.Mazur; Phys.Rev.Lett.**67**, 1023 (1991).
5. D.H.Reitze, X.Wang, H.Ahn, M.C.Downer; Phys.Rev.**B40**, 11986 (1989).
6. W.Sinke, T.Warabisako, M.Miyao, T.Tokuyama, S.Roorda, F.W.Saris;J.non Crys.Sol.**99**, 308 (1988).
7. J.Marfaing, W.Marine; Phase Transitions **14**, 225 (1989).
8. C.N.Afonso, J.Solis, F.Catalina, C.Kalpouzos; Appl.Phys.Lett.**60**, 3123 (1992).

123

TIME-RESOLVED STUDIES OF FREE CARRIERS IN INSULATORS

Ph. Daguzan, S. Guizard, P. Martin, and G. Petite

Service de Recherche sur les Surfaces et l'Irradiation de la Matière
Commissariat à l'Energie Atomique, DSM/DRECAM
CEN Saclay, 91191 Gif sur Yvette, France

Electronic excitation in insulators is known to originate numerous defects in these materials. Today, ultrashort laser pulses allow the observation of the different stages of electron relaxation and can bring new informations about these processes and especially about defect formation time. In the present set of experiments, interferometry in the frequency domain[1] has been applied to three different wide band gap materials. The principle of the experiment and the laser description have been given in details elsewhere[2,3]. Let us just recall that the sample is probed by two twin laser pulses separated by a fixed time delay, one impinging the solid before and the second after the pump pulse. A phase shift in the interference pattern appears in the frequency spectrum of the sequence of the two probe pulses. This phase shift, induced by the high intensity pump pulse, is proportional to the modification of the refractive index, and its temporal evolution is obtained by changing the delay between the pump pulse and the two probe pulses. The absorption of the second probe pulse can also be deduced from the variation of the fringe contrast. The laser pulses duration and wavelength are respectively 70 fs and 310 nm (4 eV photons) for the pump beam, 60 fs and 560 nm (2.2 eV) for the probe beam.

The variations of the phase shift as a function of time in SiO_2, Al_2O_3 and MgO are shown in Fig.1. When the time delay is close to zero, we observe a positive phase shift in both SiO_2 and in Al_2O_3. This is due to the phase modulation induced by the pump pulse and can be observed as long as the pump and the probe pulses temporarily overlap[3]. The phase shift is proportional to the product $n_2 I_p$ where n_2 is the second order non-linear index and I_p the pump laser intensity. The absorption of two photons is enough to cross the band gap of MgO, while three photons are needed in SiO_2 and in Al_2O_3, since the band gap of these materials is larger than 8 eV. Therefore, the pump laser intensity used in MgO is about 10 times smaller than in the two other samples, and the initial positive phase shift is very reduced. Then, just after the pump pulse, the phase shift becomes negative in the three materials. As we will see later, this is due to the photoinjection of electrons in the conduction band (CB) from the valence band (VB) via multiphoton excitation. The phase shift goes rapidly back to zero and then remains positive in SiO_2 while

Ultrafast Processes in Spectroscopy, edited by Svelto et al.
Plenum Press, New York, 1996

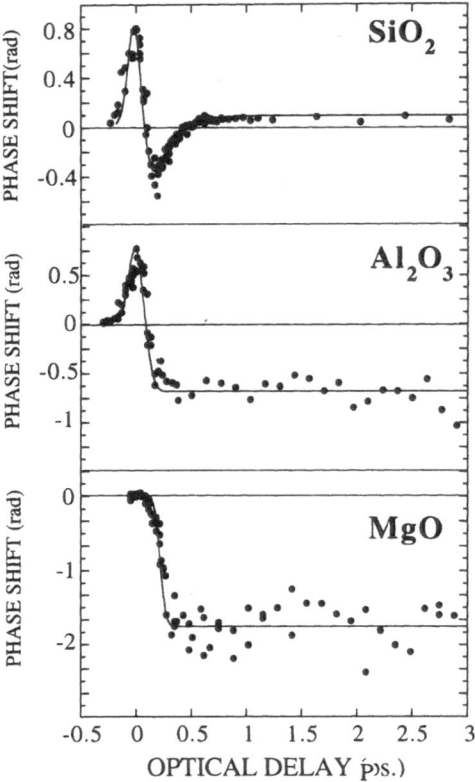

Figure 1. Phase shift in SiO$_2$, Al$_2$O$_3$ and MgO as a function of time. Dots : experimental data, full line: numerical simulation.

it remains negative and almost constant in Al$_2$O$_3$ and in MgO. The fact that electrons disappear from the CB and that the phase shift does not go back to zero suggests that electrons are trapped in the band gap. The evolution of the absorption of the second probe pulse is shown in Fig. 2. The absorption is important in the three samples (several tens of percent). In SiO$_2$, it disappears very rapidly, whereas it remains constant in Al$_2$O$_3$ and in MgO.

In order to model the results, a three level system is considered (VB, CB where the electrons are supposed to be free and a level in the band gap where the electrons can be trapped). The variations of the different electron populations are evaluated by using a set of rate equations, describing the temporal evolution of $n_{cb}(t)$ and $n_{tr}(t)$ (electron densities in the CB and in the trap level respectively). If the electron densities are small, the phase shift can be written as[4]:

$$\Delta\phi(t) = \frac{2\pi}{\lambda} L\left[n_2 I_p(t) + \left(\frac{e^2}{2n_0\varepsilon_0}\right)\left(-\frac{n_{cb}(t)}{m^*}\frac{1}{\omega^2} + \frac{n_{tr}(t)}{m_{tr}^*}\frac{1}{\omega_{tr}^2-\omega^2}\right) - n_0\right] \quad (1)$$

m* and m_{tr}^* stand respectively for the effective electronic masses in the CB and in the trap level, w is the probe laser frequency, n_0 is the usual refractive index (non perturbed me-

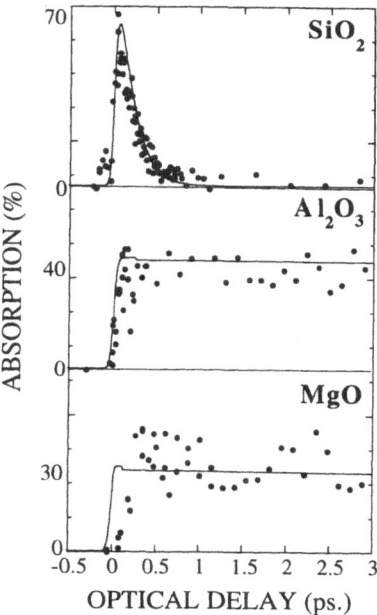

Figure 2. Absorption of the second probe pulse in SiO_2, Al_2O_3 and MgO as a function of time. Dots : experimental data, full line: numerical simulation.

dium), ε_0 is the static dielectric constant of the material and e the electronic charge. $\hbar\omega_{tr}$ is the trap energy measured from the bottom of the CB. L is the effective length over which the index is modified (70 μm), λ is the wavelength of the probe beam. We can see that free electrons in the CB induce a negative phase shift and free electron densities of about 10^{19} cm^{-3} are found in the three samples. The contribution of trapped electrons depends on the respective magnitudes of the trap and the probe photon energies (2.2 eV): trapped electrons into deep levels (>2.2 eV) will induce a positive phase shift. The result of this model is drawn in full lines in Fig. 1. In SiO_2, experimental points can be well fitted by assuming a mean trapping time of 150 fs. In Al_2O_3 and MgO, electrons remain free or are trapped in levels no deeper than the photon energy. The long time behaviour can be observed by letting the two probe pulses propagate through the sample after the pump pulse (relative mode measurement)[1]. The electron density is found to decrease with a time constant of 50 ps in MgO and 100 ps in Al_2O_3. A zero trap energy has been assumed in Eq. 1 to fit the experimental results.

In Fig. 2, we note that the absorption is high as long as electrons are present in the CB. Therefore, we observe directly Free Carrier Absorption (FCA). This process implies simultaneously the interaction of an electron with a photon and a phonon. If we use a Drude model as in the phase shift simulation, an additional term, describing the electron-phonon coupling, must be included. The imaginary part of the dielectric constant writes[5]:

$$\varepsilon_2 = \frac{n_{hc}(t)e^2}{m_*\varepsilon_0} \frac{\omega_c}{\omega(\omega^2 + \omega_c^2)}$$

(2)

where ω_c is the electron-phonon collision frequency. The result of this calculation is shown in full line in Fig. 2. Values of the order of 10^{15} s^{-1} must be used to reproduce the experimental points. These values are the typical ones for such materials[6]. Furthermore, note that this observation confirms the results of recent laser induced photoelectron spectroscopy experiments, that have allowed the observation of FCA in SiO$_2$[7]. Electron-photon-phonon collision cross sections can be computed with second order perturbation theory[8], and values about 10^{-18}–10^{-19} cm^2 are obtained. The FCA cross sections deduced from the interferometry experiment are about 6.10^{-18} cm^2. Therefore, there is a very good agreement between experimental and theoretical results. Detailed knowledge of FCA in wide band gap materials is very important because it is the key point for the study of optical breakdown[9].

The most striking result of this interferometry experiment is the ultrafast trapping in SiO$_2$, and the nature of the defect must identified. Radiation induced colour centers have been widely studied in SiO$_2$, both from the experimental and theoretical point of view. In particular, the self-trapped exciton (STE), corresponding to the localisation of an electron-hole pair associated with a lattice distortion[10–13], is known to have an absorption band around 240 nm. In order to check if the electron trap we observed is correlated with the STE, a time-resolved absorption measurement, based on the classical pump probe experiment scheme, has been performed. The transmission through the sample is obtained by measuring the ratio of the probe (240 nm) beam intensity before and after the sample as a function of the delay with the pump pulse (310 nm, 2 TW/cm^2). Fig. 3 shows the temporal evolution of the absorption. We found that the rise time of the absorption is 150 fs[14]. Therefore, we can conclude that the electron trapping in SiO$_2$ is due to the formation of STEs.

Interferometry in the frequency domain is a well suited tool to study photoexcited electron in the CB of wide band gap oxides. In MgO and Al$_2$O$_3$, the electrons remain

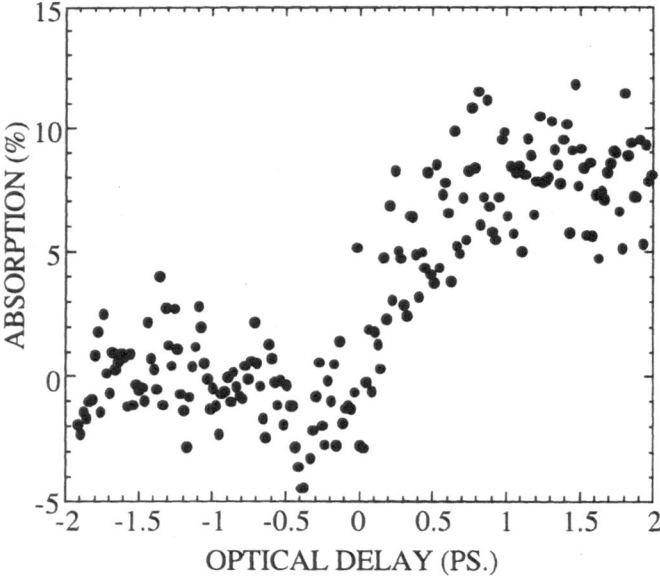

Figure 3. Absorption in SiO$_2$ at 240 nm as a function of the time delay between the pump and the probe pulses.

highly polarizable at optical frequencies for more than 50 ps, while they are trapped in a few hundreds of fs in SiO_2. Time resolved absorption measurements reveal that this trapping is due to the formation of STEs. Furthermore, in the three materials, a high free carrier absorption can be observed because of a strong electron-phonon coupling.

REFERENCES

1. J.P. Geindre, P. Audebert, A. Rousse, F. Falliès, J.C. Gauthier, A. Dos Santos, G. Hamoniaux, and A. Antonetti, Opt. Lett. **19**, 1997 (1994).
2. P. Audebert, Ph. Daguzan, A. Dos Santos, J.C. Gauthier, J.P. Geindre, S. Guizard, G. Hamoniaux, K. Krastev, P. Martin, G. Petite, and A. Antonetti, Phys. Rev. Lett. **73**, 1990 (1994).
3. S. Guizard, P. Martin, Ph. Daguzan, G. Petite, P. Audebert, J.P. Geindre, A. Dos Santos, and A. Antonetti, Europhys. Lett. **29**, 401 (1995).
4. Y. R. Shen, *The Principles of Non Linear Optics* (John Wiley and Sons, New York, 1984).
5. O. Madelung, *Introduction to Solid State Theory* (Springer-Verlag, Berlin,1989).
6. E. Cartier and F. R. McFeely, Phys. Rev. B **44**, 10689 (1991).
7. Ph. Daguzan, S. Guizard, K. Krastev, P. Martin, G. Petite, A. Dos Santos, and A. Antonetti, Phys. Rev. Lett. **73**, 2352 (1994).
8. B. K. Ridley, *Quantum Processes in Semiconductors* (Clarendon Press, Oxford, 1993).
9. S. C. Jones, P. Braunlich, R. T. Casper, X. A. Shen, and P. Kelly, Opt. Eng. **28**, 1039 (1989).
10. T.E. Tsai, D.L. Griscom and E.J. Friebele, Phys. Rev. Lett. **61**, 444 (1988).
11. T.E. Tsai and D.L. Griscom, Phys. Rev. Lett. **67**, 2517 (1991).
12. R.A.B. Devine, Phys. Rev. Lett. **62**, 340 (1989).
13. K. Tanimura, T. Tanaka and N. Itoh, Phys. Rev. Lett. **51**, 423 (1983).
14. S. Guizard *et al*, J. Phys. C (to be published).

124

SATURABLE ABSORPTION DYNAMICS OF A SEMICONDUCTOR DOPED GLASS

M. Wittmann and A. Penzkofer

Institut für Experimentelle und Angewandte Physik
Universität Regensburg, Regensburg, Germany

Semiconductor doped glasses are used as short-wavelength cut-off filters. They are applied in nonlinear optical studies[1]. Here the commercially available colour filter glass RG645 (CdS_xSe_{1-x} microcrystallites embedded in crown glass) from Schott is investigated. Fresh (not exposed to laser light) and photodarkened samples (exposed to 2000 J cm^{-2} accumulated energy density of 5 ns pulses at 532 nm) are studied. Relevant parameters (composition, x; particle diameter, d; volume fraction of microcrystallites, p; number density of microcrystallites, N_{cr}; number density of CdS_xS_{1-x} formula units in particles, N_0) have been determined and are listed in Table 1. The microcrystallite radii are larger than the effective Bohr radius, a_B ($a_{B,CdS} \approx 3$ nm, $a_{B,CdSe} \approx 4.5$ nm). Therefore the microcrystallites behave like bulk semiconductor material and quantum size effects may be neglected[1].

Nonlinear transmission and absorption recovery measurements are carried out with femtosecond light pulses generated in an argon ion laser pumped cw mode-locked dye laser oscillator[2] and in a Nd:YAG laser pumped dye laser amplifier[3] (pulse duration $\Delta t = 350$ fs, wavelength $\lambda_L = 625$ nm, single pulse energy up to 200 µJ). The number density of interacting states, N_i, the ground-state absorption cross-section, σ_L, and the excited-state absorption cross-section, σ_{ex}, are determined from saturable absorption measurements. Relaxation rates of excited states are derived from absorption recovery studies using femtosecond pulse excitation and time delayed femtosecond pulse probing.

The absorption dynamics is analysed with the aid of an energy-momentum band structure and deep trap level diagram shown in Fig. 1. This model is supported by fluorescence quantum yield and fluorescence polarisation studies[4]. The momentum selective excitation of electrons from the valence band to the conduction band is indicated by solid arrows and hatched regions. In the conduction band a fast relaxation (τ_{FC}) occurs to the centre of the Brillouin zone. The conduction band electrons relax mainly to the valence band and to deep donor traps. It follows a relaxation from deep donor to deep acceptor traps and a final return to the conduction band.

The experimental intensity dependent femtosecond pulse transmission through 1 mm thick fresh (a) and photodarkened (b) RG645 samples is shown by circles in Fig. 2. The

Ultrafast Processes in Spectroscopy, edited by Svelto et al.
Plenum Press, New York, 1996

553

Table 1. Parameters of RG645 colour filters

Parameter	Fresh	Photodarkened	Method	Reference
x	0.37	0.37	absorption edge	Ref. 5
d (nm)	13.7 ± 1.5	15.8 ± 1.5	Rayleigh scattering	Ref. 6
p	$(2.4 \pm 0.2) \times 10^{-3}$	$(2.6 \pm 0.2) \times 10^{-3}$	absorption coefficient	Ref. 7
N_{cr} (cm^{-3})	$(1.8 \pm 0.6) \times 10^{15}$	$(1.3 \pm 0.5) \times 10^{15}$	$N_{cr} = 6p/pd^3$	
N_0 (cm^{-3})	$(4.4 \pm 0.4) \times 10^{19}$	$(4.8 \pm 0.4) \times 10^{19}$	$N_0 = prN_A/M$	
N_i (cm^{-3})	2.5×10^{16}	2.3×10^{16}	saturable absorption	this work
s_L (cm^2)	$(1 \pm 0.2) \times 10^{-15}$	$(1.2 \pm 0.2) \times 10^{-15}$	saturable absorption	this work
s_{ex} (cm^2)	$(1.15 \pm 0.2) \times 10^{-16}$	$(1.9 \pm 0.3) \times 10^{-16}$	saturable absorption	this work
t_{FC} (fs)	100	100	assumed	Ref. 7
t_{ex} (fs)	55	55	assumed	Ref. 8
t_{n1} (ps)	700	40	absorption recovery	this work
t_{DT} (ps)	1500	60	absorption recovery	this work
t_{AT} (ps)	1500	1500	absorption recovery	this work
t_{n2} (ps)	≈ 1500	≈ 55	fluorescence quantumdistribution	Ref. 4

$r = 5.33$ g cm^{-3}, density of CdS$_{0.37}$Se$_{0.63}$ microcrystallites; $M = 174$ g mol^{-1}, molecular mass of CdS$_{0.37}$Se$_{0.63}$; $N_A = 6.022045 \cdot 10^{23}$ mol^{-1}, Avogadro number.

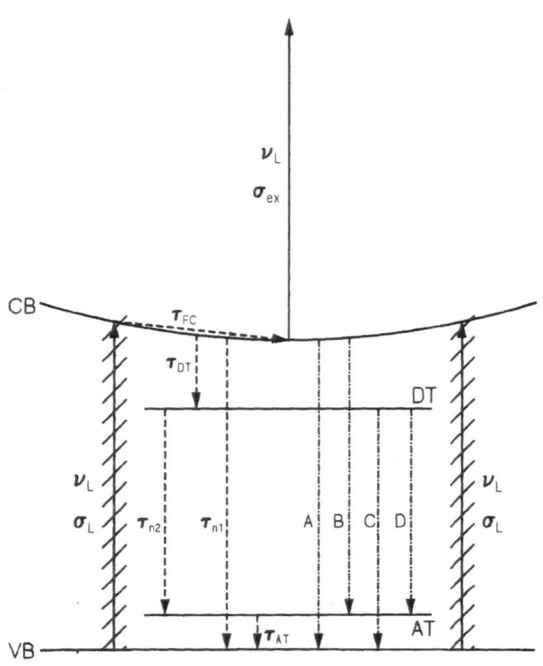

Figure 1. Schematic energy-momentum diagram. *VB*, valence band; *CB*, conduction band; *A, B, C, D* show radiative decay channels.

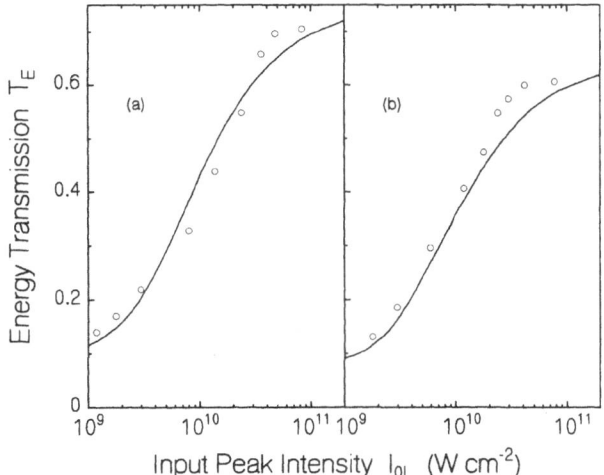

Figure 2. Saturable absorption behaviour. (a) Fresh sample. The curve is calculated using $N_i = 2.5 \times 10^{16}$ cm^{-3}, σ_L $= 10^{-15}$ cm^2, $\sigma_{ex} = 1.15 \times 10^{-16}$ cm^2. (b) Photodarkened sample. The parameters of the curve are $N_i = 2.7 \times 10^{16}$ cm^{-3}, $\sigma_L = 10^{-15}$ cm^2, $\sigma_{ex} = 1.6 \times 10^{-16}$ cm^2.

solid curves are obtained from numerical calculations[4]. The best fitting parameters, N_i, σ_L, and σ_{ex}, to the experimental data are listed in Table 1.

In Fig. 3 the experimental time-delayed probe pulse transmission through fresh (a) and photodarkened (b) samples is shown by circles. The solid curves are calculated[4]. The fit parameters are listed in Table 1. τ_{AT} is determined from the tail of the absorption recovery curve. τ_{n1} and τ_{DT} are determined by fitting the initial absorption recovery. τ_{n2} is estimated from fluorescence quantum distribution measurements[4].

Saturable absorption measurements and absorption recovery studies together with fluorescence measurements allow a good optical characterisation of semiconductor doped glasses (energy level diagram, number density of interacting states, ground-state and excited-state absorption cross-sections, relaxation dynamics).

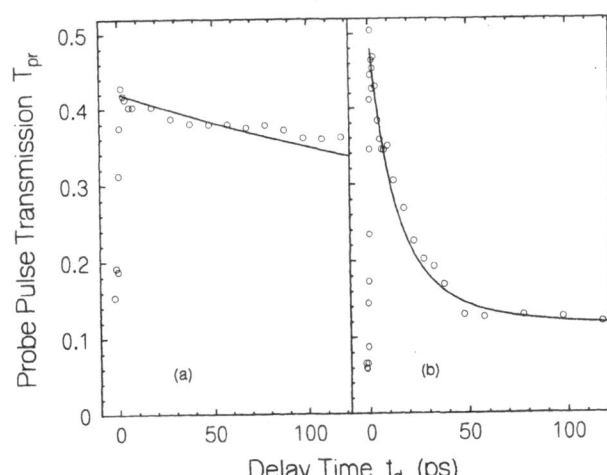

Figure 3. Ground-state absorption recovery monitored by probe pulse transmission. Circles, experimental data. Curves calculated with the parameters of Table 1. (a) Fresh sample. (b) Photodarkened sample.

REFERENCES

1. C. Flytzanis and J. Hutter, in *Contemporary Nonlinear Optics*, edited by G. P. Agrawal and R. W. Boyd (Academic Press, San Diego, 1992) p. 297.
2. W. Bäumler and A. Penzkofer, Opt. Quantum Electron. **24**, 313 (1992).
3. M. Wittmann, A. Penzkofer, and G. Gößl, Appl. Opt. **34**, 5287 (1995).
4. M. Wittmann and A. Penzkofer, Opt. Quantum Electron. **27**, 705 (1995).
5. M. P. Lisitsa, V. N. Malinko, and S. F. Terkehova, Sov. Phys. Semicond. **3**, 491 (1969).
6. H. Gratz, A. Penzkofer, and P. Weidner, J. Non-Cryst. Solids **189**, 50 (1995).
7. M. C. Nuss, W. Zinth, and W. Kaiser, Appl. Phys. Lett. **49**, 1717 (1986).
8. D. von der Linde, in *Ultrashort Laser Pulses and Applications*, 2nd edn., edited by W. Kaiser (Springer-Verlag, Berlin, 1994) p. 113.

125

GAIN DYNAMICS IN CdSe QUANTUM DOTS

U. Woggon,[1] H. Giessen,[2] B. Fluegel,[2] G. Mohs,[2] Y. Z. Hu,[2] S. W. Koch,[3] and
N. Peyghambarian[2]

[1] Institut für Angewandte Physik
Universität Karlsruhe, Germany
[2] Optical Sciences Center of the University of Arizona
Tucson, Arizona
[3] Fachbereich Physik
Philipps-Universität Marburg, Germany

Quantum dots in the strong confinement regime, i.e. where the confining radius is much smaller than the excitonic Bohr radius show discrete energy levels separated by several meV and are being considered as new gain media for laser applications. Using a very simple approach, one expects optical gain in quantum dots exactly at the energies of the transitions between the quantum confined electron and hole levels.However, the more detailed investigation of quantum dots with sizes below the Bohr radius has revealed as new intriguing peculiarities the strong influence of Coulomb interaction and the existence of stable two-electron-hole pair states (2EHP), as well as the relaxation of the selection rules due to the Coulomb effects and the mixing of the valence bands (see e.g. [1–3] and references therein). Consequently, it can be expected that all these features will also be reflected in the stimulated optical transitions.

Fig. 1 shows a principal sketch of the possible stimulated transitions contributing to the gain spectrum. The basic excitations of the system are the zero-, one-, and two-electron hole pair states. Due to the Coulomb interaction and valence band mixing, the ideal selection rules of the noninteracting electron-hole pairs in quantum dots are modified in such a way that transitions appear for the 1EHP and 2EHP states where the holes populate excited states. Calculating the energies and oscillator strength of this multitude of new optical transitions a rather broad gain spectrum is expected. The gain spectrum and its femtosecond dynamics have been investigated for CdSe quantum dots of an average radius R~2.5 nm, embedded in glass.

The dynamics of the gain spectrum have been measured in a fs-pump- and probe experiment for different delay times of the probe pulse (pulse width 115 fs, pump fluence 25 mJ/cm^2) and the result is shown in Fig. 2. The overlap of all transitions and the appearance of a broad gain spectrum can be clearly seen. At early times (320 fs) the gain develops from the low energy side of the spectrum and reaches its maximum after 2 ps. At later times the gain spectrum extends from below 660 nm to 560 nm. The magnitude of the optical gain below below the absorption edge clearly exceeds the absorption value of the lin-

Ultrafast Processes in Spectroscopy, edited by Svelto et al.
Plenum Press, New York, 1996

557

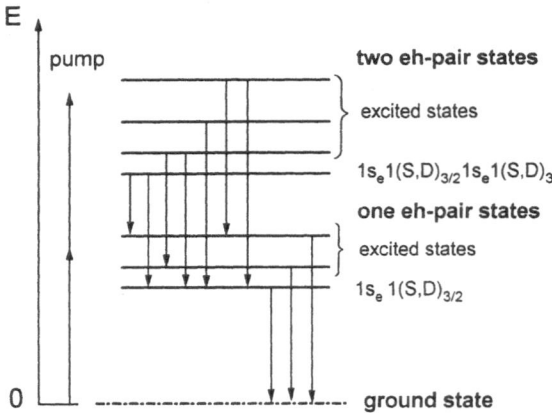

Figure 1. Scheme of the optical transitions involved in the gain model. A multitude of energy states is obtained due to the consideration of Coulomb interaction and valence band mixing in the case of strong confinement.

ear spectrum, thus excluding the possibility that the gain arises from trap states in the absorption tail. Thus the spectrally broad gain spectrum measured at CdSe quantum dots verifies the multilevel feature of the gain in zero-dimensional systems. A further result is the extremely fast gain buildup below 1 ps.

To clarify the fast population dynamics of the excited 1EHP and 2EHP states, the femtosecond differential absorption has been measured at low excitation and the results are shown in Fig. 3.

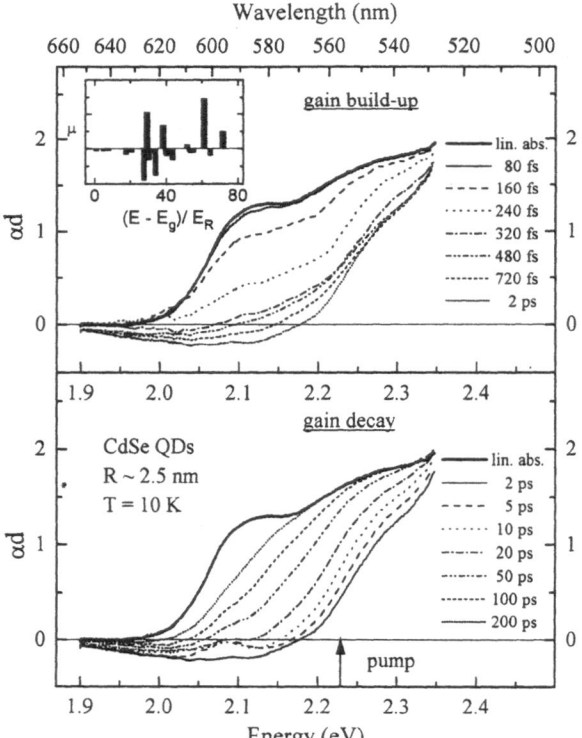

Figure 2. Gain buildup and decay measured at R~2.5 nm CdSe quantum dots. The gain is maximum after 2 ps and characterized by a broad quasicontinuous spectrum which extends around and below the absorption edge. It has a life time of about 200 ps. The inset shows the calculated dipole matrix elements for the 1EHP transitons (above the base line) and 2EHP transitons (below the baseline).

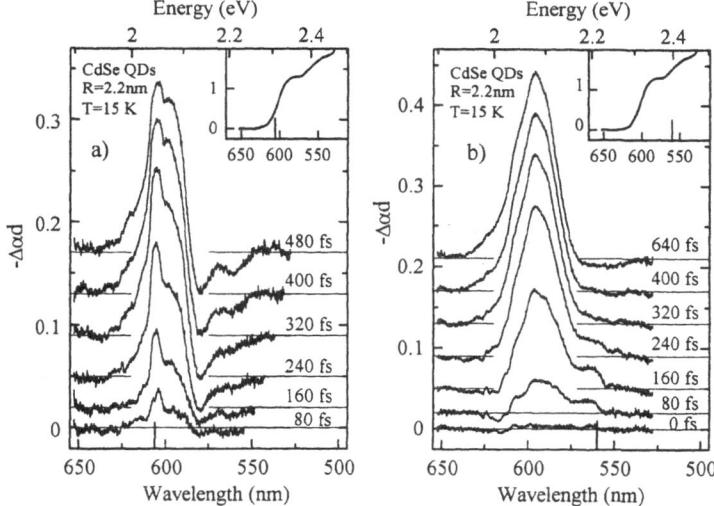

Figure 3. Femtosecond, low-density differential absorption spectra of CdSe quantum dots measured at T = 15 K and different delay times of the probe beam. The pump beam (vertical bar) is tuned a) to the energetically lowest 1EHP transition and b) high energetically to the (s,d)-type excited 1EHP transitions.

In the experiment the pump has been tuned both resonant to the ground 1EHP transition (Fig. 3a) and energetically slightly above to one of the higher excited (s,d)-type 1EHP transitions (Fig. 3b). In Fig. 3a the pump beam rxcites an electron-hole pair at the energy of the lowest one-pair transition and the strong peak around the pump corresponds to the bleaching of this transition. In differential absorption spectroscopy all one-pair transitions will be bleached simultaneously which belong to the same ladder of excited states. For example, all s-like excited hole states will contribute to the bleaching spectrum if the selectron state is optically populated etc. Therefore, the background in the bleaching high energetically to the pump shows the contribution of (for that size) near neighbouring one-pair transitions composed from the same electron state and (s,d)-type excited hole states. The induced absorption feature high energetically develops already after 240 fs and after 480 fs two pronounced maxima can be distinguished. That means, the absorption process of a probe photon and the formation of a two-pair state appears already at very early times. After 480 fs the whole differential absorption spectrum is formed by a combination of bleaching of the one-pair states and induced absorption into two-pair states. In Fig. 3b the pump is tuned to an energy of one of the weak one-pair states with the hole populating an excited state. From that experiment the relaxation time of the hole to its ground state can be derived by measuring the time until the induced absorption feature high-energetically appears in the spectrum which belongs to the probe-photon absorption process out of the one-pair ground state (see the level scheme displayed in Fig. 1). At that time, the hole which is deposited by the pump at an energy of an excited state has arrived its ground state and the probe photon will be nearly instantaneously (see Fig. 3a) absorbed causing the induced absorption. From the experiment presented in Fig. 3b the relaxation has been derived to ≈0.5 ps. No hint to the existence of the so-called phonon bottleneck has been obtained, i.e. the suppression of the relaxation rate within the ladder of the confined energy levels because of the lack of phonons matching the level separation [4,5]. Obviously the occurence of a sufficient number of new, low-energy acoustic phonons introduced by the spherical confinement removes the problem of the phonon bottleneck.

The appearance of quasicontinous gain even in systems exhibiting discrete level schemes is a completely new aspect of gain processes and of general interest for all zero-dimensional semiconductor systems. Examples of such systems are excitons localized at well width fluctuations in wide-gap II–VI quantum wells, excitons in spherical nanocrystals or excitons confined in small, three-dimensional islands, formed during MBE growth on lattice mismatched substrates. The selection rules imposed by the strong confinement are lifted, resulting in a multitude of optical transitions between the 1EHP and 2EHP states. The gain buildup time is mainly determined by the 2EHP state formation time and lies in the range around 1 ps for the CdSe dots investigated. No hints to the existence of the so-called phonon bottleneck have been obtained from the presented experiments.

REFERENCES

1. Y.Z. Hu, M. Lindberg, and S.W. Koch, Phys. Rev. B **42**, 1713 (1990).
2. A.I. Ekimov, F. Hache, M.C. Schanne-Klein, D. Ricard, C. Flytzanis, I. A. Kudryavtsev, T.V. Yazeva, A.V. Rodina, and Al.L. Efros, J. Opt. Soc. Am. B, 100 (1993).
3. S. Nomura and T. Kobayashi, Phys. Rev. B **45**, 1305 (1992).
4. H. Benisty, C.M. Sotomayor-Torres, and C. Weisbuch, Phys. Rev. B **44**, 10945 (1991).
5. U. Bockelmann and G. Bastard, Phys. Rev. B **42**, 8947 (1990); U. Bockelmann, Phys. Rev. B **48**, 17637 (1993).

126

FEMTOSECOND STUDIES OF COLLOIDAL METAL NANO-PARTICLES

Dependence of Electronic Energy Relaxation on the Liquid-Solid Interface and Particle Size

J. Z. Zhang, B. A. Smith, A. E. Faulhaber, J. K. Andersen, and T. J. Rosales

Department of Chemistry
University of California
Santa Cruz, California, 95064

Metal nano-particles have interesting electronic and optical properties relative to bulk metals due to their small size and extremely large surface-to-volume ratio. Of fundamental importance are the effects of particle size and surface on the electronic relaxation dynamics. Our recent studies of Ag and Au colloidal particles have established, at a qualitative level, that the small size renders electronic relaxation slower than in bulk while the large surface area, coupled with surface adsorbates and defects, facilitates electronic relaxation. To understand the size dependence and surface dependence of electronic relaxation at a quantitative level, we have extended the study to Au particles in a broader range of sizes and in organic solvents and to Pt particles. In this paper we report new results that provide further insight into the influence of particle size and the liquid-solid interface on electronic relaxation in metal nano-particles.

The preparation of Au particles has been reported elsewhere [1]. Particles in a broader size range (14–40 nm) have been obtained in the present study by varying the concentration ratio between the metal salt ($HAuCl_4$) and the reducing agent, sodium citrate. Phase-transfer of particles prepared initially in aqueous solution into cyclohexane was carried out using sodium oleate as an emulsifier and with $MgCl_2$ added to facilitate phase separation [2]. The Pt particles are prepared following the procedure in Ref. [3]. A solution of 1.5 mM H_2PtCl_6 (10 ml) was brought to 90°C and 4.8 ml of 1% aqueous sodium citrate solution was added. The solution was heated for 4 hours and allowed to lose half its volume. The solution turned from a clear to a dark yellow-brown solution. The UV-Visible absorption spectra of both Au and Pt particle are found to be in excellent agreement with literature data. The particle sizes were determined using TEM. Four Au colloids are discussed in this paper: (a) 14 nm particles (average size) in water, (b) 40 nm particles in water, (c) 18 nm particles in water, and (d) 18 nm particles in cyclohexane. The average particle size of the Pt colloids studied is 35 nm.

Ultrafast Processes in Spectroscopy, edited by Svelto et al.
Plenum Press, New York, 1996

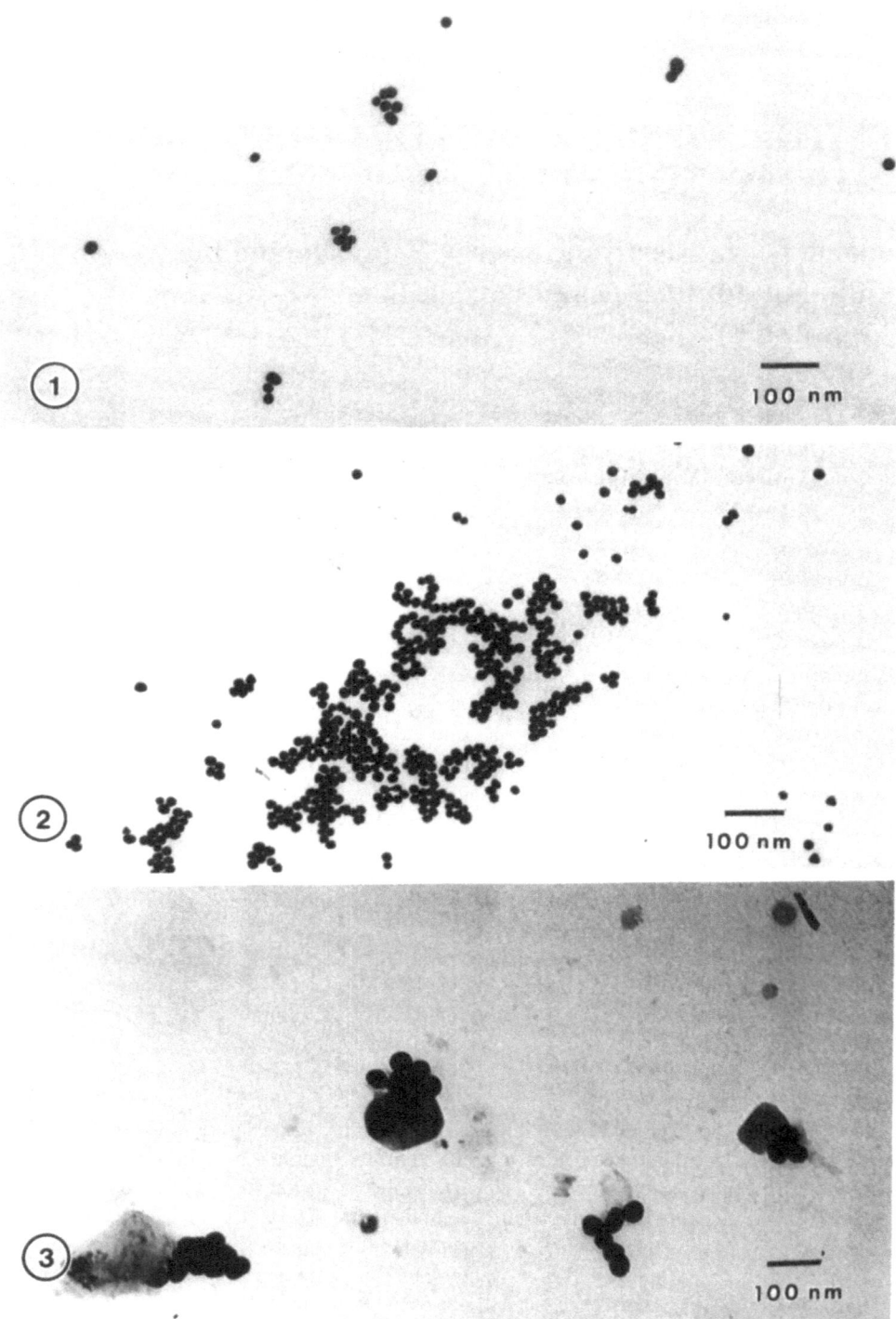

Figure 1. Transmission Electron Micrographs of three Au colloidal particles with an average size of: (1) 18 nm, (2) 14 nm, and (3) 40 nm.

The femtosecond Ti-sapphire laser system and spectrometer have been described previously [4]. The electron dynamics were measured using a pump-probe scheme. The measurement can be performed in the transmission mode because the particle size is much smaller than the wavelength of light. Fig. 2 shows the transient absorption spectra of four gold colloids with different particle sizes and in different solvents probed at 780 nm following excitation at 390 nm. All spectra feature a fast rise followed by a fast and slower exponential decay. The rise time is limited by the laser pulse represented by a Gaussian with 400 fs FWHM. For all the aqueous colloids, the decay time constants for the fast and slow components are 7 ps and 400 ps, respectively, and they are essentially independent of the probe wavelength and pump intensity. The amplitude of the fast component increases linearly with excitation intensity while the amplitude of the slow component increases with the excitation intensity to the power of 1.7. This slow decay has been attributed to electrons ejected from the gold particles into the liquid through a two-photon process, as has been described elsewhere [1]. The fast decay has been attributed to electronic relaxation due to electron-phonon coupling [5] The fast decay for gold particles is slower by a factor of 7 than the electronic relaxation time of 1 ps observed in bulk gold films [6]. This slower decay in particles is attributed primarily to weaker electron-phonon coupling. This observation is qualitatively consistent, but quantitatively much faster than, a theoretical prediction based on a much smaller electron-phonon coupling constant [7]. The discrepancy between experiment and theory is attributed to the effect of liquid-solid which was not accounted for in the theoretical model explicitly [5].

One of the most important observations is the size independence of the decays. For the size range between 14 to 40 nm, the dynamics can all be fit with the same time constants. This is in contrast with the size dependence predicted by theory that seems to suggest that the lifetime should be proportional to 1/R, where R is the radius of the particles. (The decay time constant is inversely proportional to the rate constant, which is equal to the particle volume times the electron-phonon coupling constant; the volume is propor-

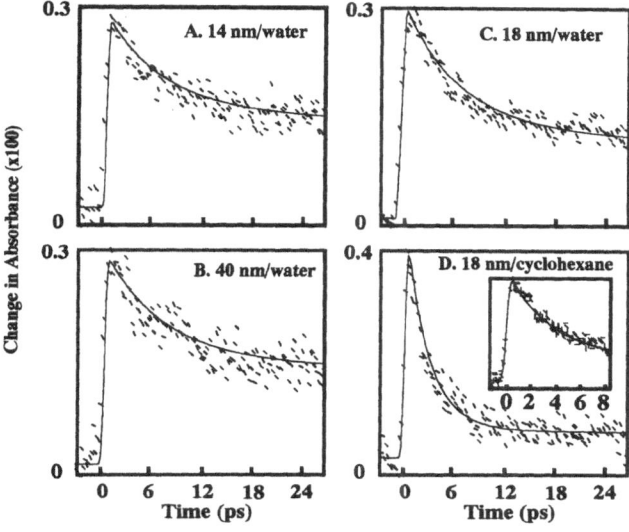

Figure 2. Photoinduced electron relaxation dynamics of Au particles for (A) 14 nm, (B) 40 nm and (C) 18 nm particles in water and (D) 18 nm particles in cyclohexane.

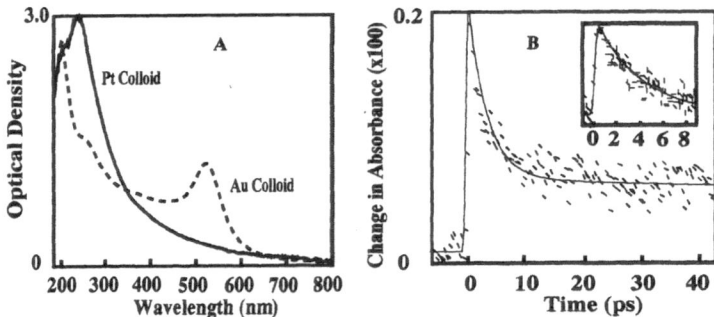

Figure 3. (A) Representative electronic absorption spectra of Au and Pt colloidal particles. (B) Photoinduced electron relaxation dynamics of Pt particles.

tional to R^3 while the electron-photon coupling constant is proportional to $1/R^2$ [7]). One possible rationalization is that the lengthening of lifetime, as predicted by theory, is canceled by the faster relaxation caused by the increasing surface-to-volume ratio in conjunction with surface adsorbates and defects. The surface-to-volume ratio increases with $1/R$ and the lifetime decrease due to the surface effect could conceivably be proportional to $1/R$. It is thus possible that these two competing factors cancel out exactly when R changes. The possibility of a significant surface effect is supported by the observed solvent dependence of the relaxation dynamics.

Fig.2D shows the photoinduced electron dynamics for Au particles in cyclohexane for 18 nm particle size. The most prominent feature is that the fast decay is much faster in cyclohexane than in water (Fig.2C). The slower decay has a smaller amplitude than in water under the same excitation intensity but can still be fit with the same 400 ps time constant, indicating less electron ejection in organic solvents. The faster decay of the fast decay component in cyclohexane than in water is significant since it indicates that the electronic relaxation is indeed sensitive to the surface environment and supports the proposal that the liquid-solid interface is responsible for the discrepancy between experimentally observed size independence and the theoretically predicted size dependence.

The slower relaxation rate in Au particles than in bulk is similar to that observed for Ag particles [8]. To explore the generality of the observation made on Au and Ag nanoparticles, we have extended the study to another important noble metal, Pt. Figure 3 shows some preliminary data on the electronic relaxation dynamics of Pt colloids probed at 780 nm and excited at 390 nm. The fast decay has a time constant of about 3.5 ps, which is much faster than in Au (7 ps) and slightly slower than in silver (2.5 ps) [8]. A slow component, fit with a 400($\pm$50) ps time constant, is also present, which is similar to that observed in Au colloid and is attributed to ejected electrons in water. A more quantitative study of the dependence of the electronic relaxation dynamics in Pt particles on pump power, particle size, and solvent is currently under investigation.

The above results indicate that the surface plays an important role in electronic relaxation in metal nano-particles and competes with the size effect. This suggests the need and the intriguing possibility of manipulating the electron behavior by controlling the surface properties. We are currently developing techniques to modify the particle surface to length the electron lifetime, which is desirable for applications in photocatalysis and photoelectrochemistry.

ACKNOWLEDGMENTS

This work is supported in part by PRF administered by ACS, the Faculty Research Fund by UCSC, a grant from the University of California Energy Institute, the Summer Undergraduate Research Fund by NSF, and the California Alliance for Minority Participation Summer Research Program of UCSC.

REFERENCES

1. B.A. Smith, D.M. Waters, A.E. Faulhaber, M.A. Kreger, and J.Z. Zhang, J. Sol-Gel Sci. Tech., submitted, 1995.
2. H. Hirai and H. Aizawa, J. Coll. Int. Sci. 161, 471 (1993).
3. R.M. Wilenzick, D.C. Russell, R.H. Morriss, and S.W. Marshall, J. Chem. Phys. 47, 533 (1967).
4. J.Z. Zhang, R.H. O'Neil, T.W. Roberti, J. Phys. Chem. 98, 3859 (19994).
5. A.E. Faulhaber, B.A. Smith, J.K. Andersen, J.Z. Zhang, Mole. Cryst. Liquid Cryst., in press, 1995.
6. S.D. Brorson, J.G. Fujimoto, E.P. Ippen, Phys. Rev. Lett. 59, 1962 (1987).
7. E.D. Belotskii and P.M. Tomchuk, Int. J. Electronics 73, 955 (1992).
8. T.W. Roberti, B.A. Smith, J.Z. Zhang, J. Chem. Phys. 102, 3860 (1995).

127

ULTRAFAST THERMALIZATION PROCESSES IN PbTe QUANTUM-SIZE FILMS

Studying by Biharmonic Pumping Technique

A. G. Kornienko,[1] V. M. Petnikova,[1] V. V. Shuvalov,[1] L. N. Vereshchagina,[2] and A. N. Zherikhin[2]

[1] International Laser Center
M.V. Lomonosov Moscow State University, Moscow, Russia
[2] Scientific Research Center for Technological Lasers
Troitsk, Moscow Region, Russia

The electronic structure and the relaxation rates for thin layers are not the same as for the bulk samples due to the quantum-size effect (QSE)[1]. The QSE results in quantitative (renormalization of bands) and qualitative (splitting of 3D bands into series of 2D subbands) changes of the electronic structure. Similar effects have been found in superlattices as well as in quantum wires and quantum dots[2]. For today, optical spectroscopy methods have been proved to be a powerful tool in experimental studying of the QSE in wide-band semiconductors. However, in narrow-band semiconductors, one needs to use lasers of the middle- or far-IR ranges. This problem can be solved by means of nonlinear spectroscopy techniques based on two-photon excitation of electronic transitions, for example, biharmonic pumping (BP)[3]. Using BP, one can obtain detailed information about spectra of states of electron and phonon subsystems, constants of electron-electron (e-e) and electron-phonon (e-p) interactions, relaxation times, etc. In BP, due to e-e and e-p interactions, two electromagnetic waves with different frequencies $\omega_{1,2}$ and wave vectors $\mathbf{k}_{1,2}$ result in electron and phonon excitations at frequency $\Delta = \omega_1 - \omega_2$ with wave vector $\Delta\mathbf{k} = \mathbf{k}_1 - \mathbf{k}_2$. Scattering of the same waves on such a transient grating leads to generation of two new waves with frequencies $\omega_4 = 2\omega_{1,2} - \omega_{2,1}$ and wave vectors $\mathbf{k}_4 = 2\mathbf{k}_{1,2} - \mathbf{k}_{2,1}$. It is this process - the selfdiffraction - that enables one to investigate low-frequency transitions using tunable lasers in the visible range.

For our study, we choose ultrathin PbTe films because laser deposition technique enables us to produce continuous monocrystalline PbTe films of high optical quality with appropriate thickness L=3–30 nm[4]. Moreover, optical phonons in PbTe are inactive in Raman scattering[5] that simplifies analysis of experimental data. The design of our setup was typical for BP measurements[3]. We used two tunable (wavelengths in the range of 590–645 nm) picosecond dye lasers (DLs) with pulse duration τ_p=20 ps. For PbTe films

Ultrafast Processes in Spectroscopy, edited by Svelto et al.
Plenum Press, New York, 1996

567

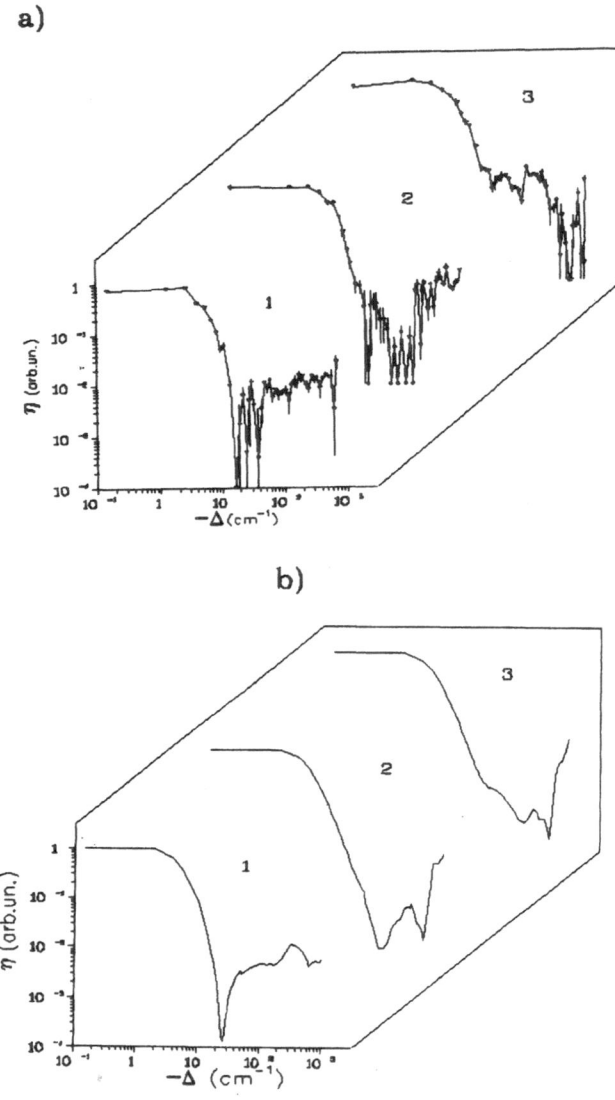

Figure 1. Experimental (a) and theoretical (b) dependencies η(Δ) for PbTe films with thicknesses L = 6 (1), 18 (2), and 30 (3) nm.

with L = 6, 18, and 30 nm, experimental dependencies of the self-diffraction efficiency η(Δ) are shown in Fig.1a and demonstrate a narrow central peak at $|\Delta| < 10$ cm^{-1} and a wide side wing for 10 cm$^{-1} < |\Delta| < 800$ cm^{-1}. The spectral width of the central peak is about twice as large as the spectral width of DL radiation. There are some specific features that change their positions along the Δ-axis with changing L and we attribute these features to two-photon excitation of the electron subsystem.

To interpret our data, we used an advanced version of the model developed for the bulk samples[7]. This version takes into account all main specific features of studied PbTe quantum-size films. The basis of our calculations of the nonlinear susceptibility $\chi^{(3)}$ was the Liouville equation for density matrix. "Writing" the interference grating by BP components with $\mathbf{k}_1 \neq \mathbf{k}_2$ is possible only due to the transfer of quasi-momentum deficiency $\Delta\mathbf{k} = \mathbf{k}_1 - \mathbf{k}_2$ from the "active" electron to electrons and phonons of the reservoir. We described these processes by the Hamiltonians

$$\hat{H}_{FE} = \int dK_{FE} \int d\Delta K_{FE} A_E(\Delta K_{FE})\hat{a}^+(K_{FE}) \exp[i\Delta K_{FE}(\hat{Q}_{FE} - \hat{Q}_{AE})]\hat{a}(K_{FE}) + h.c.,$$

$$\hat{H}_{FP} = \sum_{\lambda} \int dK_{FP} A_P^{\lambda}(K_{FP})\hat{b}_{\lambda}(K_{FP}) \exp(iK_{FP}\hat{Q}_{AE}) + h.c. \tag{1}$$

Here, $A_{E,P}$ are the constants of e-e and e-p interactions, $\hat{Q}_{RE,AE}$ are the operators of coordinates of an electron in the reservoir and in the active subsystem. Operators of creation (annihilation) $\hat{a}^+$ ($\hat{a}$) and $\hat{b}^+$ ($\hat{b}$) correspond to the electron and phonon in the reservoir with quasi-momenta $\mathbf{K}_{RE}$ and $\mathbf{K}_{RP}$, respectively. The integral is evaluated inside the first Brillouin zone.

We divided $\chi^{(3)}$ into three additive components. The first component was the resonant electron part χ_{ee} connected with $\Delta\mathbf{k}$ transfer due to one-particle e-e scattering. The second component was the resonant phonon part χ_{ep} determined by similar e-p processes. The third component was the nonresonant part χ_{nr} describing contributions of multiparticle processes. We found χ_{ee} as

$$\chi_{ee} \propto (\delta k_z)^3 P_0(K_+P_+ + K_-P_-), \tag{2}$$

where

$$P_0 = \sum_{i,s,i',s'} \int \frac{n_i(k_{x,y},s)[1 - n_{i'}(k_{x,y},s')]}{[\omega_1 - \Omega_{i,s,i',s'}(k_{x,y}) + i\Gamma_{i,s,i',s'}(k_{x,y})]^2} dk_{x,y}, \tag{3}$$

$$P_{\pm} = \sum_{i,s,i',s'} \int \frac{n_i(k_{x,y},s)[1 - n_{i'}(k_{x,y},s')]}{[\omega_1 \pm \Delta - \Omega_{i,s,i',s'}(k_{x,y}) \pm i\Gamma_{i,s,i',s'}(k_{x,y})]^2} dk_{x,y}, \tag{4}$$

$$K_{\pm} = \sum_{i,s,i',s'} \int \frac{n_i(k_{x,y},s)[1 - n_{i'}(k_{x,y},s')]}{[\pm\Delta - \Omega_{i,s,i',s'}(k_{x,y}) \pm i\Gamma_{i,s,i',s'}(k_{x,y})]^2} dk_{x,y}. \tag{5}$$

Here, s and s' represent the numbers of 2D subbands in bands i and i', $\hbar\Omega_{i,s,i',s'} = E_{i'}(k_{xy},s') - E_i(k_{xy},s)$, and $\Gamma_{i,s,i',s'}(k_{xy})$ are the relaxation rates. Factor $\delta k_z \propto L^{-1}$ results from the conservation of the number of electronic states. Further, we took into account a small value of the photon quasi-momentum in (3)-(5) that leads to the conservation of k_z and the selection rule s=s'.

To compute χ_{ee}, one must determine $\Omega_{i,s,i',s'}(k_{x,y})$, that is, simulate PbTe electronic structure. To do this, we interpolated to 3D Brillouin zone the data calculated by the pseudopotential technique[8]. Taking into account 3 upper bands of the PbTe valence band (i=1,...,3) and 2 bottom bands of the PbTe conduction band (i=4,5), we used a set of power polynomials for fitting. Then, we took into account QSE in a potential well with infinitely high walls and computed $E_i(k_{xy},s)$ for layers.

Our next step was connected with simulation of $\Gamma_{i,s,i',s'}(\mathbf{k}_{xy})$. To do this, we included in our consideration characteristic times of some relaxation processes.

1. The time T_3 described thermalization of the valence or conduction band as a whole. We assumed $T_3 = 1$ ps that is in good agreement with experimental data[9].

2. The time T_3' described thermalization within an isolated subband. We supposed T_3' changes in the range 100–1000 fs with change of the concentration of free carriers N, the temperature Θ, etc.

3. The time T_1 characterized the recombination of free carriers. Because $T_1 \gg \tau_p$, we did not take into account such processes.

4. The time T_2 characterized the decay of the interband polarization. This relaxation results from all processes listed above. It is why we supposed $T_2 \sim T_3' \sim$ 100–1000 fs.

Furthermore, we assumed that $\Gamma_{i,s,i',s'}(\mathbf{k}_{xy}) \sim 1/T_2$ consists of two additive items. We determined the first item by N and relative velocity v_{rel} of free carriers as $\Gamma^{(\alpha)}Nv_{rel}$. We considered the cross section of scattering $\Gamma^{(\alpha)} = 10^{-12}$–$10^{-13}$ cm^2 as an adjustment parameter of our model. Nonzero thickness of edges of the potential well and spatial fluctuations result in random phase distortions of the wave functions. Such an averaging results in additional broadening of electronic transitions by a factor $\Gamma^{(\beta)}\exp\{[s(L_0/L)]^2\}$. We considered $\Gamma^{(\beta)} = 10^{-12}$–$10^{-13}$ s^{-1} as one more adjustment parameter of our model and supposed $L_0 = 6$ nm. This enabled us to describe the kinetics of free carriers with taking into account the change of T_2 with change of $(i,s,\mathbf{k}_{xy})$.

All initial populations in (3)-(5) should be governed by the Fermi distribution. However, $n_i(\mathbf{k}_{xy},s)[1- n_{i'}(\mathbf{k}_{xy},s')] \cong 0$ for i, i'=1–3 or i, i'= 4,5 at $\Theta < 10^3$ K. Therefore, because of "hole burning" in the distribution functions, any corrections to $n_i(\mathbf{k}_{xy},s)$ can significantly change the total response. We found these corrections using the technique of a modified density matrix[10].

Using the model described above for ultrathin PbTe films with thicknesses 6, 18, and 30 nm, we performed computer simulation of our experiments. First, taking into account both broadening of electronic transitions and "hole burning", we computed χ_{ee} (2) using the interpolation of the real band structure. We revealed that observed specific features we can see only if $T_2 \cong T_3 \geq 300$ fs. Then, we carried out approximation of our experimental data. We attributed the central peak region where $\chi^{(3)}(\Delta)$ can be approximated by $\chi_{ac}/(\Delta-i\Gamma_{ac})^2$ to the excitation of acoustic phonons and thermal nonlinearity. For PbTe, this component completely defines χ_{ep} owing to forbidden two-photon excitation of optical phonons. Then, we associated the side wing with χ_{nr} and used the complex amplitudes of these two components χ_{ac} and χ_{nr} as the adjustment parameters. The result is represented in Fig.1b. It is easy to see good agreement between our experimental and theoretical data.

ACKNOWLEDGMENT

This work was initiated owing to the Sloan Foundation Grant awarded by the American Physical Society and supported in part by the International Science Foundation (grants #RIP000 and #RIP3000), and the Russian Foundation for Fundamental Research (grant #95–02–03639a).

REFERENCES

1. L.L. Chang, in: Highlights in Condenced Matter, Plenum, New York (1991).
2. Quantum Well and Superlattice Physics, Proc. SPIE **943** (1988).
3. B.A. Grishanin et al., Sovremennye Problemy Lazernoi Fiziki, Moscow, **2** (1990) (in Russian).
4. A.N. Zherikhin, Sovremennye Problemy Lazernoi Fiziki, Moscow, **1** (1990) (in Russian).
5. J. Raisland, Physics of Phonons, Mir, Moscow (1975) (in Russian).
6. L.N. Vereshchagina et al., Kvantovaya Electronika **21**, 855 (1994) (in Russian).
7. B.A. Grishanin et al., Laser Physics **3**, 121 (1993).
8. P.J. Lin, L. Kleinman, Phys. Rev. **142**, 478 (1966).
9. K. Tanaka, H. Ohtake, T. Suemoto, Phys. Rev. Lett. **71**, 1935 (1993).
10. P.A. Apanasevich, Resonant Interaction of Light with Matter, Nauka i Tekhnika, Minsk (1977) (in Russian).

128

PICOSECOND ALL-OPTICAL SWITCHING IN THIN-FILM FABRY–PEROT INTERFEROMETERS AND CAVITYLESS DEVICES

Olga V. Goncharova and Sergey A. Tikhomirov

Institute of Molecular and Atomic Physics
Belarus Academy of Sciences, Minsk, Belarus

Thin films activated by nanometer-sized microcrystallites (nanocrystals — NCs) are the promising materials for optical data processing due to picosecond room-temperature nonlinearities.[1–2] To investigate the possibility of using photoinduced changes in their refractive index and absorption coefficient in the region of the excitation wavelength (528 nm) for photonic switching, we examined the nonlinear response of thin-film Fabry–Perot interferometers (TFIs) as well as of mirrorless microdevices by use a control beam at 518 and 650 nm. NC-composites consisting of ZnSe-, CdS- or CdSe-NCs embedded in amorphous SiO_2 and microcrystalline CaF_2 thin-films were used as nonlinear media.

NC-composites were prepared using special multicomponent targets[1]. The NCs in test samples under investigation have the average size about 10 nm, the concentrations of NCs is around 20%. The thickness of test NC-samples is about 0.1–0.5 μm. 3D confinement effect in test samples was evidenced by observation of both the dynamic blue-shift and the bleaching in their nonlinear transmission spectra in the region near the quantum-well (QW) levels position.

The samples of the next group were designed as thin-film Fabry–Perot interferometers (TFIs) with composite QZD intermediate layers.[1] Such interference structures can perform optical switching and modulation. The principle of such operations is the sweeping of the cavity transmission peak in response to control pulses. For a high-finesse Fabry–Perot cavity, a small change in the intracavity refractive index can translate into a large change in the cavity transmission. Until now, conventional TFIs have the microsecond response times because the dominate nonlinearity mechanism is thermo-optic.[3] The thickness of the TFIs under investigation is about 0.5–1 μm. The dielectric thin-film mirrors consist of ZnS and cryolite layers.

We have used of the mode-locked Nd–glass laser to perform pump-probe measurements on test samples at room temperature. 3-ps-pulses of the second and the third harmonic (λ = 528 and 352 nm) are used as the pump beam. The pulses of the fundamental

Ultrafast Processes in Spectroscopy, edited by Svelto et al.
Plenum Press, New York, 1996

frequency are focused into a cuvette with D_2O producing a white-light continuum, which is used as a probe pulse. Experimental technique has been briefly described in Ref. 4.

We have performed a series of measurements on time-resolved absorptive and re-fractive nonlinearities induced by 3-ps-pulses in the QZD samples. Time evolution of the photoexcited absorption changes probed at λ_0 = 475, 500, 550 and 675 nm in the case of $CdS+SiO_2$-sample (λ_{exc}= 352 nm) is shown in Fig. 1. The insets of Fig. 1 show the initial relax of the optical nonlinearities. Probing of the transmission spectra in the other NC-samples (e.g., $CdSe+CaF_2$)[4] found similar dynamical features. Measurements near the QW-levels result in the observation of fast ($\tau_{relax} \leq 10$ ps) bleaching (Fig. 1,a). Studies of QW-bleaching in $CdSe+CaF_2$ and $CdS+SiO_2$ -samples reveal the importance of elec-tron–phonon and electron–electron interactions in shortening of bleaching decay up to ap-pearance of the subpicosecond components. Measurements near the band edge show that optical nonlinearities on picosecond time scale are governed by two different effects, such as band-filling (Fig. 1,b) and free-carrier absorption (Fig. 1,c). Picosecond (1–15 ps) tran-sient absorption observed near the middle of the forbidden gap of host semiconductor (Fig. 1,d) we attribute to filling of intrinsic surface states inside of the NCs[5].

The estimations using Kramers–Kronig correlation give the values of nonlinear re-fraction index change Δn of about -0.5 at pump power of 100–150 MW/cm^2 for bleach-ing near the QW-levels (so the parameter of nonlinearity n_2 is about -10^{-5} cm^2 /kW). So, the picosecond QW-bleaching under investigation connected with spatial confinement of electron–hole motion can be used to realize all-optical switching in TFIs with the same QZD media as intermediate layer.

The results of the experimental investigation of the possibilities of the all-optical pi-cosecond switching at 528 nm are shown in Fig. 2. We can see, that the use of thin films activated by one kind of NCs as nonlinear media may result in reduction of τ_{off} down to tens of picoseconds at the wavelengths near the 528 and 650 nm.

An intermediate layer, consisting of CdS-NCs and SiO_2-matrix, made it possible to record the short-wavelength dynamic shift ($\Delta\lambda \approx 3$ nm) and bleaching ($\Delta T \approx 30\%$) of the interferometer spectral profile at λ_{exc}= 528 nm (Fig. 3 a). In addition as shown in Fig. 3,b, when the power is varied from 2 MW/cm^2 to 20 MW/cm^2, the bleaching of the TFI spec-tral profile changed into its darkening while the amplitude of the dynamic shift decreases but is still visible. It is considered that this tendency as well as the presence of subpicosec-ond component in complicated relaxation kinetics of induced nonlinearity in TFI at $I_{exc} \approx$ 5 MW/cm^2 (Fig. 4 a) are the result of the competition between the saturated 100-ps band-edge bleaching and the increased absorption with growth of I_{exc}. In accordance with our experimental results, the origin of the fast short-wavelength dynamic shift of the TFI spec-tral profile may be attributed to the QW-level filling.

The results of investigation of the possibilities of picosecond TFI switch-off times are shown in Fig. 4. We can see, that the use of intermediate layers activated by one kind of NCs in TFI-switches results in reduction of τ_{off} down to tens of picoseconds (Fig. 4,a). At the same time, in the case of prototype of cavityless microdevice in form of dielectric film activated by both CdS- and CdSe-NCs (Fig. 4,b) we have been able to detect subpi-cosecond bleaching ($\tau_{off} \leq 1$ ps) as well as ultrafast transition from bleaching to short-lived transient absorption with growth of I_{exc}. In this case, the nonlinear response probed at the wavelength 650 nm is made of two components, a fast (tens of picoseconds) QW-bleaching in CdSe-NCs (see Fig. 2,d) and a fast (1–35 ps) transient absorption in CdS-NCs (Fig. 2,c).

In conclusion, using composite NC-sample, we have demonstrated an all-optical switch in a TFI configuration, which worked with a switching power of ≤ 10 MW/cm^2.

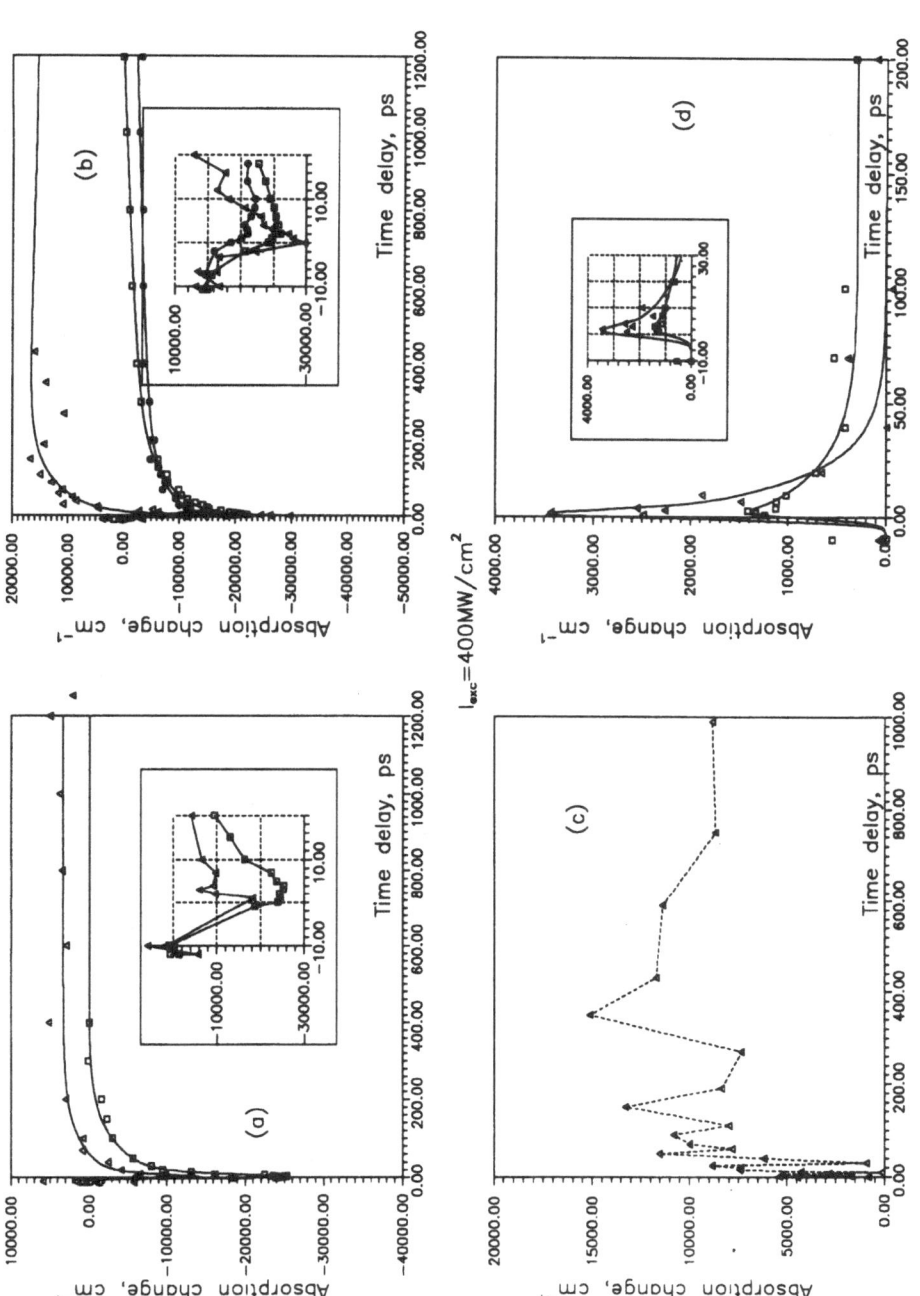

Figure 1. Absorption change decay dynamics of the CdS+SiO$_2$ sample at the probe wavelengths $\lambda_0 = 475$ (a), 500 (b), 550 (c) and 675 nm (d) as a function of the delay time. $\lambda_{exc} = 352$ nm. $I_{exc} \approx 20$ (filled circles), 100–150 (boxes) and 400 MW/cm^2 (triangles). The solid lines give the reasonable multiexponential fit. The dashed line as well as solid lines at the a and b insets connect the experimental data.

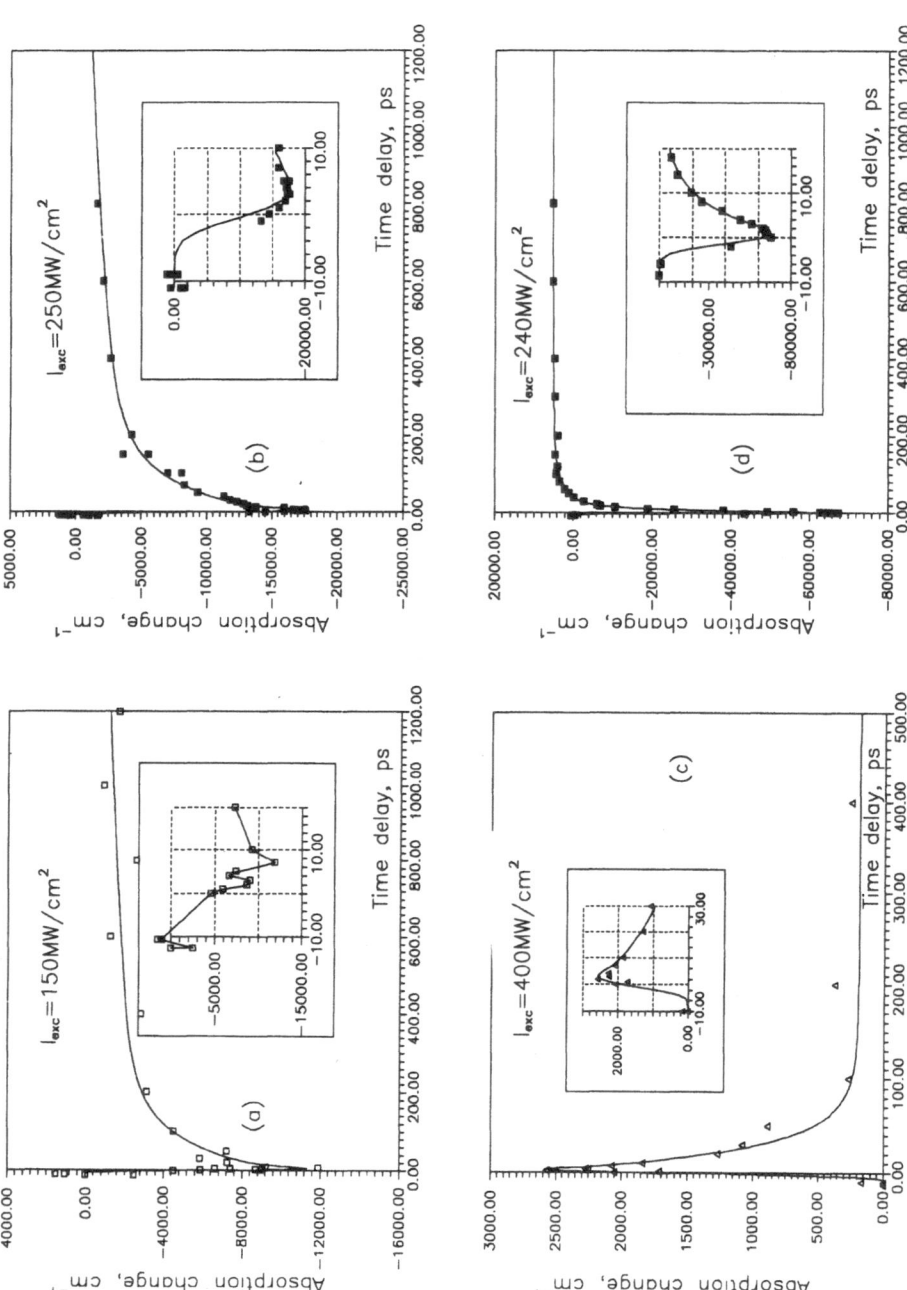

Figure 2. Absorption change decay dynamics of both the CdS+SiO$_2$-sample at the probe wavelengths $\lambda_0 = 518$ (a,b), 675 nm (c) and the CdSe+CaF$_2$-sample at the probe wavelength $\lambda_0 = 630$ nm (d) as a function of the delay time. $\lambda_{exc} = 528$ nm. $I_{exc} \approx 100–150$ (boxes), 200–250 (filled boxes) and 400 MW/cm^2 (triangles). The solid lines give the reasonable multiexponential fit with the exception of a solid line connected the experimental data at the a inset.

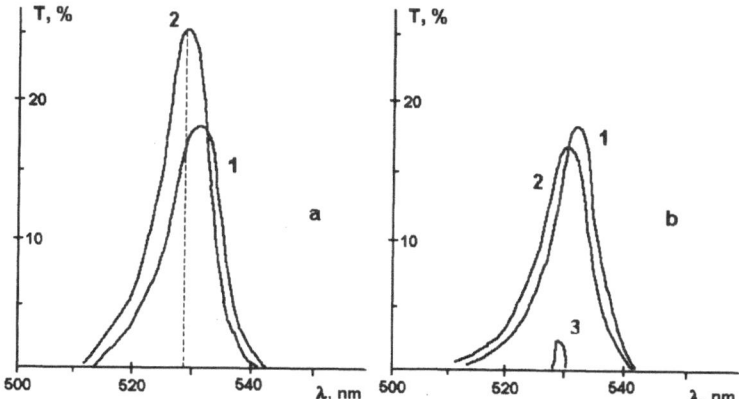

Figure 3. Amplitude characteristics of the prototype of TFI-switches with QZD nonlinear layer. (*a,b*) — The transmission spectra for an interference filter before (1) and 300-fs after (2) the 3-ps pulse excitation at $I_{exc} \approx 2$ (*a*) and 20 (*b*) MW/cm^2. Curve 3 is the spectrum of the excitation light.

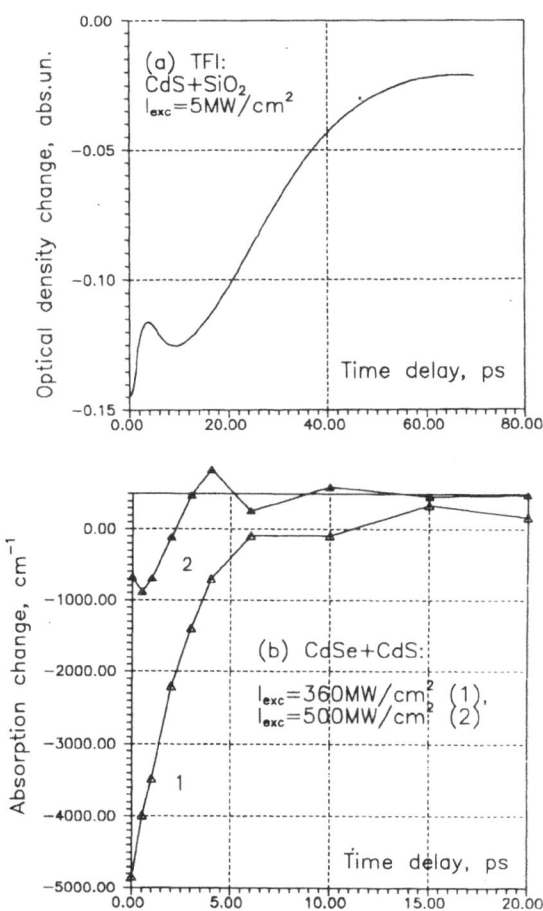

Figure 4. Amplitude and kinetic characteristics of the prototypes of TFI- and cavity-less switches with QZD nonlinear layer. *a* — The relaxation kinetics of induced nonlinearity in TFI at $I_{exc} \approx 5$ MW/cm^2. $\lambda_0 = 518$ nm. *b* — Bleaching dynamics of the dielectric film activated by both CdS- and CdSe-NCs at $I_{exc} \approx 360$ (1) and 500 (2) MW/cm^2. $\lambda_0 = 650$ nm.

The TFI-switches exhibit short-wavelength picosecond shift of spectral profile and have a transmission gain ($\Delta T \approx 30\%$). Besides that, we made a prototype of similar device, but without cavity, with the assistance of layer activated by both CdS- and CdSe-NCs. Ultra-fast switching with a switching power of $I_{exc} \geq 300$ MW/cm^2 was successfully demonstrated.

REFERENCES

1. O. V. Goncharova, in: *New Materials for Thin-Film Electronic-Technological Components*, ed. V. A. Labunov (Nauka i Tekhnika, Minsk, 1994).
2. O. V. Goncharova and S. A. Tikhomirov, Quantum Electronics, **25**, 357 (1995).
3. S. P. Apanasevich, O. V. Goncharova, F. V. Karpushko, G. V. Sinitsyn, Bull Acad. Sci. USSR, Phys. Ser., **47**, 82 (1983).
4. O.V. Goncharova and S. A. Tikhomirov, this issue.
5. O. V. Goncharova, F. V. Karpushko, and G. V. Sinitsyn, Sov. Phys.-Tech. Phys. **28**, 1142 (1983).

129

ULTRAFAST LASER SPECTROSCOPY OF SEMICONDUCTOR QUANTUM DOTS AND WIRES

V. Dneprovskii[*]

Physical Faculty
Moscow State University
119899 Moscow, Russia

Kinetics and relaxation of nonequilibrium charge carriers in semiconductor quantum wires (QWrs) of GaAs and QWrs and quantum dots (QDs) of porous silicon have been investigated and the origin of strong and fast nonlinearity in QWrs and QDs has been determined.

The picosecond laser spectroscopy pump and probe method has been utilized. The samples were excited by intense ultrashort pulse of the mode-locked laser and probed by a delayed pulse of a picosecond continuum. Discrete bleaching bands have been observed in the time-resolved nonlinear transmission spectra of GaAs wires in the crysotile asbestos nanotubes [1,2] and crystalline porous silicon [3,4]. They are attributed to a saturation of optical transitions between quantum-size energy levels in a system of spatially localized carriers in QWrs and QDs. In bulk semiconductors different nonlinear processes compete and coexist: the Burstein-Moss saturation effect, renormalization of the energy gap at a high density of carriers, bleaching or broadening of the exciton absorption line, etc. The increase or decrease of absorption dominates in the vicinity of the energy gap at different temperatures of the sample. In semiconductor QDs and QWrs the only surviving mechanism for nonlinear absorption near the lowest resonance is state filling [5].

In some cases the saturation of discrete optical transitions could be observed while they were not pronounced in the linear absorption spectra. The latter is obviously caused by inhomogeneous broadening of the quantum-size energy levels due to the size dispersion and variation in the shape of nanostructures. This broadening could be substantially suppressed in the nonlinear transmission spectra by selective (resonant) excitation of nanostructures of a certain size.

* Fax: 7–095–9393731;E-mail: dnepr@scond.phys.msu.su

Ultrafast Processes in Spectroscopy, edited by Svelto et al.
Plenum Press, New York, 1996

NONLINEAR OPTICAL ABSORPTION AT DISCRETE FREQUENCIES HAS BEEN OBSERVED IN GaAs QWrs

The molten GaAs was injected into the empty nanometer radius channels (nanotubes) of transparent chrysotile asbestos matrix [6] . The average diameter of the obtained QWrs $\approx$ 6 nm. The registered bleaching of the samples at discrete frequencies has been attributed to the saturation of optical transitions between the highest valence subband - the lowest conduction subband, and the holes of split off by spin orbit coupling valence subband - the lowest conduction subband. The values of the energies of these transitions allowed to estimate the evarage diameter of QWrs. the transition energies between the space quantization levels in the corresponding valence bands and the conduction bands were culculated within the effective mass approximation combined with the assumption of an infinitely deep cylindrical potential well and negligible role of Coulomb interaction between carriers. The estimated average diameter of QWrs (6 nm) coinsides with that measured indipendantly [6] using electron microscope. For some parts of the samples (the spot sizes of the focused pump and probing beams were smaller than the sample's size) aditional bleaching band shifted to the lower energy side have been observed only at the moment of excitation. It may be attributed to the saturation of the lowest transition in QWrs of greater cross section that are formed in the space between the crysotile asbestos nanotubes. The decrease of the relaxation time of this band as compared with other bleaching bands may be explained by the stronger influence of nonradiative surface recombination in QWrs with greater specific surface. Strong ($\chi^{(3)} \approx 3.10^{-8}$ e.s.u.) and fast (relaxation time $\leq$50 ps, T=300K) third order optical nonlinearity has been determined for GaAs QWrs in crysotile asbestos matrix.

POROUS SILICON ATTRACTS RESEARCHER'S ATTENTION BECAUSE OF ITS EFFICIENT VISIBLE LUMINESCENCE

Some researchers attribute the enhanced luminescence of porous silicon to the quantum confinement (optical transitions in nanostructures). However, till now there is no evidence of a discrete energy spectrum of porous silicon determined by quantum confinement in nanocrystals or wires with nanometer lateral dimensions. The linear absorption spectra of porous silicon are shifted to the high-energy side compared with the original bulk material. But unfortunately there are no peculiarities in these spectra determined by discrete optical transitions. This may be explained by significant inhomogeneous broadening of optical transitions due to the size dispersion and different shapes of nanostructures.

The method of picosecond laser saturation spectroscopy has been applied to measure the time-resolved nonlinear absorption and to define the energy spectrum of porous silicon. The nonlinear absorption is less dependent on surface properties than luminescence.

The registered increase of transmission at discrete frequencies in plane-parallel platelets of porous silicon detached from the silicon substrate has been attributed [3,4] to a saturation of optical transitions in nanostructures. In evaluating the energies of optical transitions in porous silicon two types of nanostructures have been considered: 1. wires that are long parallelepipeds or cylinders oriented along [100] axis and 2. nanocrystals (quantum dots) of spherical shape. The estimated lateral dimensions (or diameters) of the wires are 2.6 - 4.5 nm, the average radius of quantum dots is about 2 nm. The results of independent measurements using electron and tunnel microscopy have confirmed the existence of both QWrs and QDs of corresponding size in different parts of porous silicon samples.

For the used porous silicon samples $\chi^{(3)} \approx 10^{-8}$ e.s.u., $\tau \approx 40$–50 ps, (T=300K).

The growths of bleaching at $\Delta t \leq 10$ ps (Δt is the delay time between the ultrashort pumping and probing pulses) have been registered at the frequency of the lowest optical transitions of QDs and QWrs in the cooled (80K) samples of porous silicon. This growth of bleaching can be explained by the enlargement of population of the lowest energy levels in quantum wires and dots due to the relaxation of the carriers from the upper energy levels. The relaxation process is relatively slow compared with the intraband relaxation in the bulk silicon. Such slowed-down relaxation is one of the reasons of poor radiative efficiency of quantum dots and wires — the luminescence properties don't reach the improvement predicted from the concentration of the spectral oscillator strength.

In summary, strong and fast optical nonlinearity have been observed in quantum wires of GaAs. The saturation of optical transitions between the energy levels of size quantization in quantum wires at high density of photoexcited carriers is the physical process leading to this nonlinearity.

The method of picosecond laser saturation spectroscopy allowed to observe the saturation of discrete optical transitions in porous silicon and to determine the energy spectra of size quantization in nanostructures of two types: quasi-zero-dimensional (quantum dots) and quasi-one-dimensional (quantum wires). The quantum confined origin of optical transitions in porous silicon is confirmed by the slowed-down energy relaxation of photoexcited carriers compared with intraband relaxation in the bulk silicon.

ACKNOWLEDGMENT

The research described in this publication was made possible in part by Grants No. Ph.M5D000 and M5D300 from the International Science Foundation , and by Grant 1–034 of the russian program "Physics of Solid Nanostructures".

REFERENCES

1. N. Gushina, V.Dneprovskii, V.Poborchii et al, JETP Lett. 61 (6), 491 (1995)
2. V. Dneprovskii, N. Gushina, O.Pavlov, V.Poborchii, I.Salamatina, E.Zhukov, Phys. Lett. A204, 59 (1995)
3. V. Dneprovskii, A.Ejov et al, phys. stat. sol.(b),188, 297 (1995)
4. V. Dneprovskii, N.Gushina, D.Okorokov, V.Karavanskii, E.Dovidenko, Superlattices and Microstructures 17 (1), 41 (1995)
5. S. Schmitt-Rink, D.A.B. Miller, D.S. Chemla, Phys. Rev. B35,8113 (1987)
6. V.V. Poborchii et al., Superlattices and Microstructures,16,133 (1994)

130

CONTRIBUTIONS OF THE QUANTUM-WELL AND THE SURFACE ELECTRON STATES TO THE PICOSECOND OPTICAL RESPONSE OF THE CdSe NANOCRYSTALS FORMED IN THIN-FILM CaF$_2$-MATRIX

O. Goncharova and S. Tikhomirov

Institute of Molecular and Atomic Physics
Belarus Academy of Sciences, Minsk, Belarus

Nanometer-sized II-VI semiconductor microcrystallites (nanocrystals — NCs) dispersed in thin-film dielectrics are very interesting subject. Such media are suitable for investigating the effect of the microstructure (including the size, the concentration, the arranging types of NCs and the various matrice materials) on the quantum confinement of the electron–hole motion. Additionally, thin-film quasi-zero-dimensional (QZD) media have the potential for applications in areas such as nonlinear optical microdevices and fast all-optical switches[1].

A lot of studies have been devoted to the QZD materials, mainly on the optical properties of NCs formed in amorphous matrices [2]. In this paper, we report on the spectral region, as well as the magnitude and the speed limits of the picosecond optical response in CdSe+CaF$_2$ -sample. We focus our attention on the mechanisms of the bleaching decay near the QW-level position ($\lambda_0 \approx 630$ nm) (QW-bleaching), bleaching decay in the spectral region near band edge ($\lambda_0 \approx 725$ nm), as well as transient absorption decay in the region below band edge ($\lambda_0 \approx 950$ nm).

Test NC-sample was prepared using special multicomponent target. The prepared sample has the microstructure near the one of semiconductor-doped glasses, but the matrix is microcrystalline CaF$_2$ -thin film as well as the concentration of NCs with average size ≈ 10 nm is around 20%.

The time-resolved transmission T(Δt) were measured on a thin (≤ 0.3 μm) CdSe+CaF$_2$ film using a pump-probe scheme with delay time, Δt, scanned up to 1.3 ns. Experimental setup has been described in detail elsewhere.[3] Briefly, 3-ps-pulses of the second harmonic of the mode-locked Nd–glass laser ($\lambda_{exc} = 528$ nm) excited the sample with the 0.5-Hz repetition rate, the delayed pulses of 3-ps spectral continuum were used to probe the transmission spectra at room temperature. The probe wavelength (λ_0) is selected with interference filters over a spectral region from 450 nm to 950 nm. The inten-

Ultrafast Processes in Spectroscopy, edited by Svelto et al.
Plenum Press, New York, 1996

sity of the pump pulse can be varied over nearly two orders of magnitude. The absorption coefficient $\Delta\alpha(\Delta t)$ is deduced from the measured optical density change. The decay components of the nonlinear response were obtained using reasonable multiexponential fitting to pump-probe data.

Evolution of the QW-bleaching decays with growth of excitation powers I_{exc} is shown in Fig. 1. At the relatively low pump powers (up to 10 MW/cm^2) the intraband energy relaxation time (time of carriers transitions from higher energy levels to the lower one seen as initial changes of the amplitude of the QW-bleaching) in test-sample is about 10 ps. Such distinction of the intraband relaxation time compared with one in the bulk semiconductor be explained by the quantization of the energy spectrum of carriers. The following resolved times of the QW-bleaching relaxation are about 60 and 900 ps. At 10 MW/cm^2, more than 80% of the initially generated population decays in less than 60 ps and only 12% of one have 900-ps life times. Shortening of the time of optical nonlinearity relaxation with respect to the time of interband recombination of electrons and holes in host semiconductor ($\approx$1–2 ns) we attribute to the existence of the channel of high-speed carriers recombination in the near surface region 'NC—matrix.'

At the pump powers from 10 to 20 MW/cm^2 we observed additional shortening of the time of QW-bleaching relaxation. For example, at $I_{exc} \approx$ 20 MW/cm^2 QW-bleaching has the relaxation times about 6, 40 and 350 ps. Such behaviour may be the consequence of the increase in the rate of electron–phonon scattering resulting sharp increase in the concentration of excited electron–hole pairs which confined in NCs. At $I_{exc} \geq$ 100 MW/cm^2 the appearance of subpicosecond component in the QW-bleaching decay was detected. We attribute this fast decay to secondary or Auger recombination. As the result of these both effects, at $I_{exc} \approx$ 120 MW/cm^2 more than 80% of the optical nonlinearity relax with time shorter than 6 ps and the other part of one relax with $\tau_{relax} \approx$ 30 ps.

The appearance of subpicosecond component in the QW-bleaching decay was accompanied by the evidence of long-lived induced absorption. This effect is attributed to intra-band (free-carrier) absorption. As a result of last effect, at $I_{exc} \approx$ 240 MW/cm^2 a slow response with a time constant of longer than 2 ns (about 7%) was also observed. At the same time, about 30% of the optical nonlinearity at this intensity relax with time more shorter than $\tau_{relax} \approx$1 ps as well as about 50% and 10% of one have 6-ps and 30-ps decay components respectively.

Time evolution of the near band edge bleaching is shown in Fig. 2. The resolved times of the bleaching relaxation are about 60 and 900 ps. We can see from Fig. 1–2, that the amplitude of bleaching in the vicinity of the band edge of test NC-sample saturates with growth of pump power and increases in the spectral region near the QW-level. For instance, at $I_{exc} \approx$ 200 MW/cm^2 the amplitude of bleaching decreases from $\Delta\alpha \approx$ – 6.1 × 10^4 cm^{-1} at wavelength λ_0 = 630 nm to $\Delta\alpha \approx$ – 1.3 × 10^4 cm^{-1} at λ_0 = 725 nm. To the contrast, in bulk semiconductors or NC-samples with densely packed NCs the increase of bleaching has been observed near the fundamental band edge.

Our observation in the spectral region below band edge (Fig. 2,c,d) is consistent with conclusions drawn by others for the case of extremely fast surface electron trapping. The observed long time offset may actually decay through nonradiative electron–hole recombination in the near surface region 'NC—matrix' outside of the NC and contain some contributions from electrons ejected into matrix volume. The short-lived increase of absorption observed at 950 nm right after excitation is attributed to filling of intrinsic surface states inside of NCs.[4] The double exponential decay (1–2 and 15–35 ps) suggests that the electrons trapped inside of NC disappear through two different pathways. The amplitude of both decays are sensitive to the pump power used, especially the 1–2-ps decay.

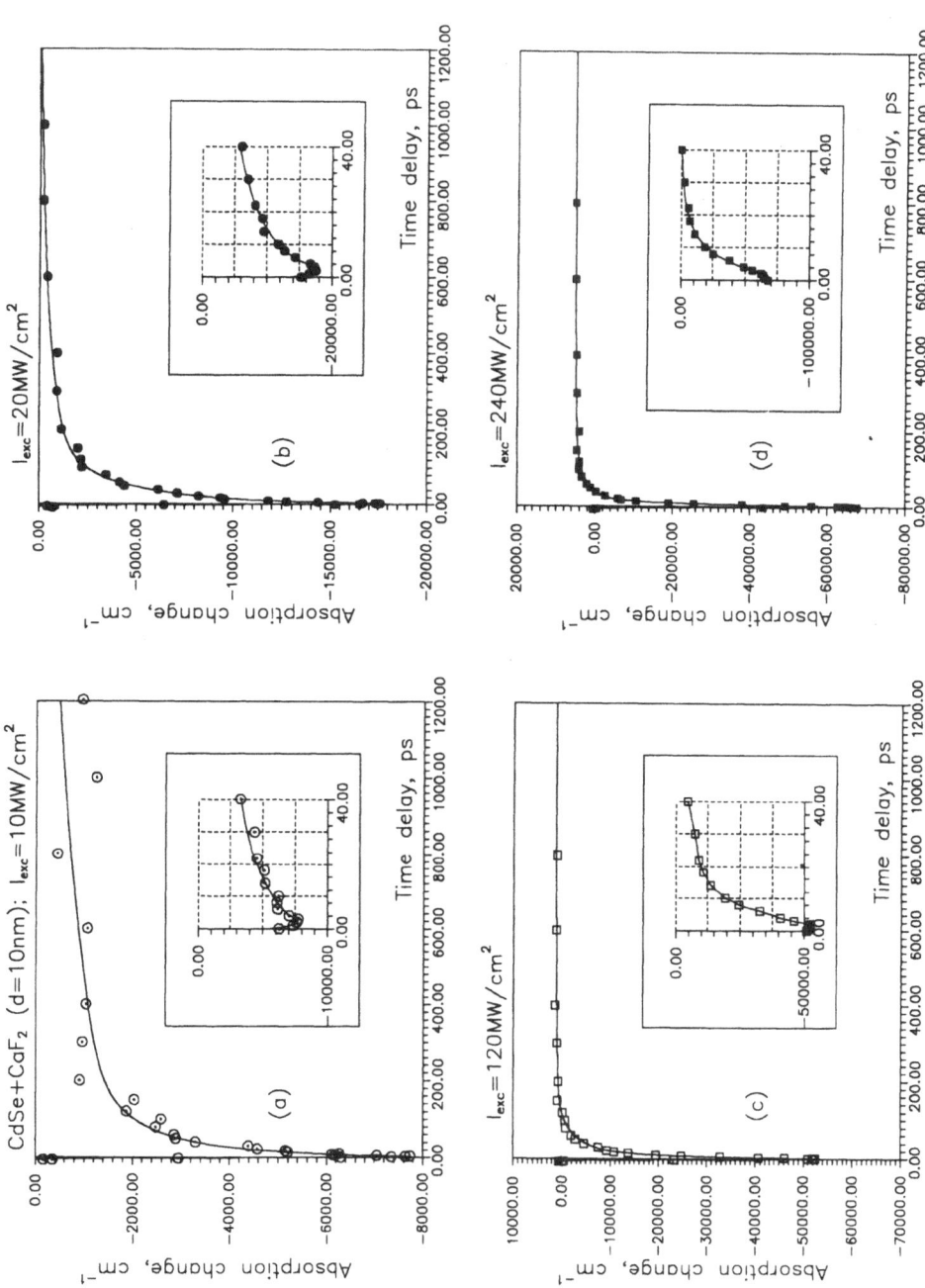

Figure 1. Absorption change decay dynamics of the CdSe+CaF$_2$-sample at the probe wavelength $\lambda_0 = 630$ nm as a function of the delay time. $I_{exc} \approx 10$ (a) (hollow circles), 20 (b) (filled circles), 120 (c) (boxes) and 200–240 MW/cm^2 (d) (filled boxes). The solid lines give the best multiexponential fit.

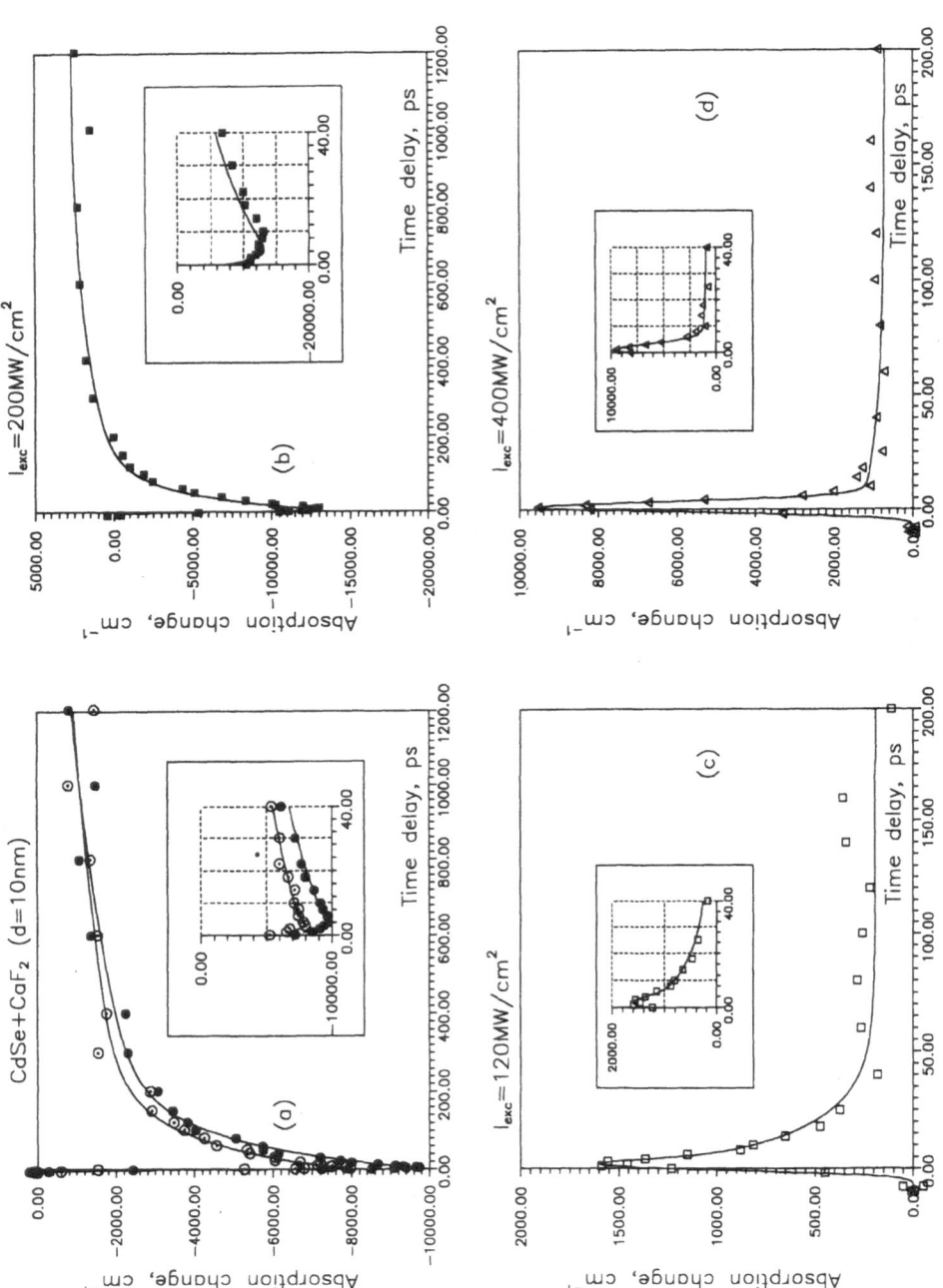

Figure 2. Absorption change decay dynamics of the CdSe+CaF$_2$-sample at the probe wavelengths $\lambda_0 = 725$ (a,b), and 950 (c,d) as a function of the delay time. $I_{exc} \approx 10$ (hollow circles), 20 (filled circles), 120 (boxes), 200–240 (filled boxes), and 400 MW/cm^2 (triangles). The solid lines give the best multiexponential fit.

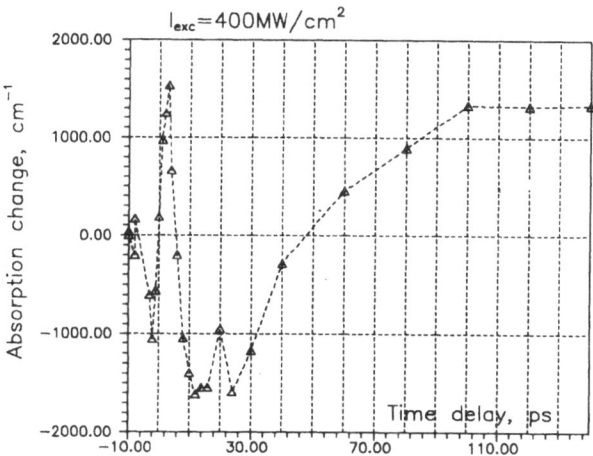

Figure 3. Absorption change decay dynamics of the CdSe+CaF$_2$-sample at the probe wavelength λ_0 = 775 nm as a function of the delay time. I$_{exc}$ ≈ 400 MW/cm^2 (triangles). The dashed line only connects the experimental data.

When the power is varied from 6 μJ/pulse to 15 μJ/pulse, the amplitude of the 1–2-ps decay component increase drastically, while the 15–35-ps decay component decreases but is still weakly visible. It is considered that this tendency as well as the appearance of subpicosecond component in the QW-bleaching decay are the result of concentration annigilation of trapped electrons with 15–35-ps relaxation times. High pump power leads to a high concentrations of free and trapped carriers and thus to decreased average tunneling distance to secondary recombination. Auger recombination is likely to be more important for near surface region due to its small size in compare with one of NCs.

It is also interesting to note that the competition between saturated band-edge bleaching and short-lived increase of absorption results in complex dynamics of the nonlinear response observed at 775 nm (Fig. 3).

In conclusion, for CdSe-NCs embedded in microcrystalline CaF$_2$ -film we detected both the fast (6–10 ps) bleaching and the appearance of the subpicosecond (≤ 1ps) component at higher pump power near the QW-level position, saturated bleaching near the band-edge, short-lived (1–15 ps) transient absorption observed near the middle of the forbidden gap of initial CdSe, as well as, for what we believe to be the first time, the competition between two last effects.

REFERENCES

1. O. V. Goncharova, in: *New Materials for Thin-Film Electronic-Technological Components*, ed. V. A. Labunov (Nauka i Tekhnika, Minsk, 1994).
2. O. V. Goncharova and S. A. Tikhomirov, Quantum Electronics, **25**, 357 (1995).
3. N. A. Lysak, S. V. Melnichuk, S. A. Tikhomirov, and G. B. Tolstorozhev, J. Appl. Spectroscopy, **47**, 832 (1988).
4. O. V. Goncharova, F. V. Karpushko, and G. V. Sinitsyn, Sov. Phys.- Tech. Phys. **28**, 1142 (1983).

131

COMBINATION OF SCANNING TUNNELING MICROSCOPY AND ULTRAFAST LASER SPECTROSCOPY

G. Gerber,[1] J. -Y. Grand,[2] R. Möller,[2] and W. Pfeiffer[1]

[1] Physikalisches Institut
University of Wuerzburg
Wuerzburg, Germany
[2] Physikalisches Institut
University of Stuttgart
Stuttgart, Germany

In recent years the application of ultrafast laser spectroscopy in solid state physics has revealed a tremendous amount of information on hot carrier distributions in bulk metals, semiconductors and insulators. Especially the study of relaxation and recombination processes in semiconductors and semiconductor heterostructures is of major importance for optoelectronic applications. In heterostructures it is possible to observe coherent wave packet dynamics over several picoseconds. This indicates that incoherent processes, usually dominant in bulk materials can be suppressed by careful sample design. A major breakthrough in these bulk experiments was achieved due to the elimination of surface effects by the use of heterostructures acting as confinement for the photogenerated carriers. Beside the fact that carrier relaxation and recombination via the surface is an extremely fast and efficient process little is known about the basic mechanisms of surface induced relaxation processes. Time resolved investigations of electronic dynamics at surfaces open therefore a new and exciting field for fundamental and applied research.

In order to gain access to the carrier dynamics at surfaces a combination of ultrafast laser spectroscopy and a surface sensitive method is required. Since "real" surfaces show a rather arbitrary behavior it is necessary to perform experiments under ultra high vacuum (UHV) conditions. This allows the preparation of well defined surfaces and to achieve reproducible results. During the last decade the rapidly growing field of surface physics has established several experimental methods that in principle can be combined with ultrashort optical excitation and allow the study of ultrafast electronic processes. Photoelectron spectroscopy (PS) as well as surface induced second harmonic generation (SHG) or sum frequency generation (SFG) have already been successfully combined with ultrafast laser spectroscopy. R. Haight and coworkers for instance have used time resolved PS to investigate the scattering rate of bulk carriers into Ge (111) surface states[1]. The combina-

Ultrafast Processes in Spectroscopy, edited by Svelto et al.
Plenum Press, New York, 1996

tion of ultrafast laser spectroscopy and SHG was recently used to study the desorption and vibrational excitation of adsorbates[2]. Whereas both methods reveal information about the dynamic properties of the surface it is not possible to gain a spatially resolved signal. This makes the interpretation of the signals on a microscopic level rather indirect and complex. Obviously there is demand for a surface sensitive method with both spatial and temporal resolution, giving to some extend complementary information to the above mentioned methods.

Since the invention of the scanning tunneling microscope by Binning and Rohrer in 1982 this method has become a standard tool in surface physics[3]. It has been shown that scanning tunneling microscopy (STM) in combination with tunneling spectroscopy reveals atomic scale information on surface topography as well as electronic surface states and defects. Unfortunately the STM based direct investigation of spatially resolved dynamic electronic processes at the surface is limited by the bandwidth of the high gain amplifier used to detect the tunneling current and the limited scan speed. S. Weiss et al have shown that this limitation could be overcome by optical gating of the tunneling current using a fast optoelectronic switch[4]. In their original work and in a recent work by R. Greoneveld et al.[5] using the same approach it is shown that this method allows to investigate the propagation of optically induced electronic transients with a time resolution of a few ps. However, this time resolution is not yet sufficient to investigate surface induced carrier relaxation that occurs on a sub-ps time scale. Further improvement of time resolution following this approach is limited by the capacitance of the tunneling tip itself.

In order to overcome this fundamental limitation and in order to obtain a time resolution only limited by the length of the optical pulse we follow another approach[6]. The basic concept is shown in Figure 1. A sequence of fs-light pulses is directed between tip and sample. Any nonlinear behavior of the tunneling contact is then reflected in the tunneling current and could thus be measured as a function of the delay time between optical pump and probe pulses. In this concept the time resolution is no longer limited by the electronic response of the whole tip but is only governed by the interaction of the light pulses with the microscopic tunneling contact. With this approach we hope to obtain a femtosecond time resolved atomic topography of the surface. In the following we present first experimental results on the combination of STM and ultrafast laser spectroscopy based on the concept shown in Figure 1.

In order to guaranty reproducible and stable tunneling conditions, which are crucial for our approach, all experiments were performed under UHV conditions. GaAs (110) surfaces were prepared in situ by cleavage of (100) oriented crystals. In all experiments we

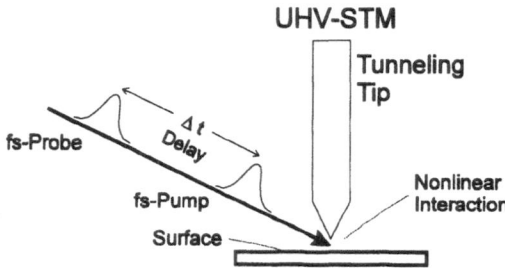

Figure 1. Schematic drawing of our approach to achieve ultrafast time resolution in scanning tunneling microscopy.

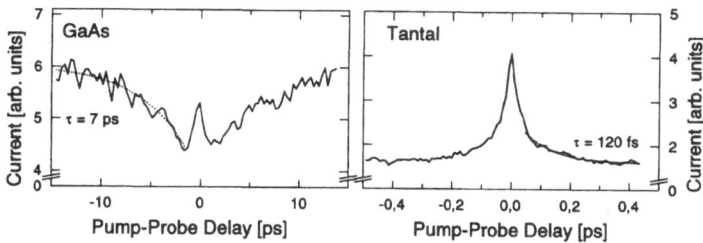

Figure 2. Pump-probe spectra from tantalum (a) and GaAs (b) surfaces. The tip was retracted (tip-sample distance about 1000 nm).

have used electrochemically etched tungsten tips. In situ field emission was used to further improve the contact quality. Both the direct output of an Ti-sapphire laser (740 nm, 50 fs, 80 MHz repetition rate) and amplified pulses (60 fs, up to 20 μJ per pulse, 50 Hz repetition rate) were used in our experiments. A Michelson interferometer is used to produce pump and probe pulses with variable delay and subsequently the laser pulses are focused between tip and substrate using a focusing lens outside the UHV chamber.

In a first run the low repetition rate amplified pulses were used to excite the tunneling gap. An laser fluence of 5 mJcm^{-2} is estimated based upon the pulse energy and the focal length of the collimator lens. In all experiments the distance control feedback loop of the scanning tunneling microscope was locked. The current picked up by the tunneling tip was recorded by boxcar averaging with the window set on the detected current pulse.

Typical results of femtosecond time resolved measurements are shown in Figure 2 for a Tantalum and a GaAs (110) surface. In both cases clear pump-probe effects are visible. With coincident pump and probe pulses the detected current from the tantalum surface is more than doubled. The transient signal can be described by an exponential decay with a time constant of 120 fs. We connect this decay time to the relaxation of hot substrate electrons. The GaAs surface shows a different behavior. With coincident pump and probe pulses a local maximum in current is observed. With increasing delay between pump and probe pulses the current first decreases and than increases again.

From the different pump-probe response of the Tantalum and the GaAs surface we conclude that the recorded current is dominated by processes in the substrate, although the incoming laser beam illuminates both, tip and sample. In the experiments it was observed, that the current picked up by the tunneling tip does not depend on the distance between tip and substrate. The spectra shown in Figure 2 were recorded with retracted tip (tip sample distance about 1 μm). From this, it is obvious that carrier tunneling processes contribute only to a minor extend to the total signal under the chosen conditions.

In the following only the GaAs results are discussed. The most obvious explanation for the observed distance independent current is the generation of photoelectrons by a multiphoton process. In order to test this hypothesis the current was measured in dependence of the pulse intensity (Fig. 3). At low intensities a slope of 4 is seen in the log-log plot of the experimental data. This experimental evidence is in agreement with the interpretation, that a four-photon process generates photoelectrons that are subsequently picked up by the tip. The electron affinity in GaAs is about 4 eV. Together with the band gap energy (1.42 eV) one can estimate that at least four photons of 740 nm (1.68 eV) are required to produce one free electron by photoemission. Above a laser fluence of about 0.3

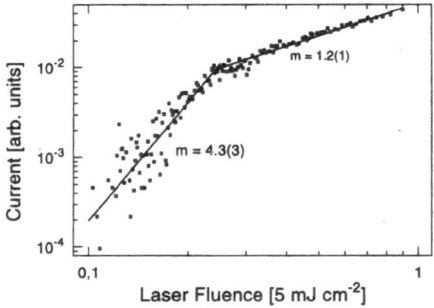

Figure 3. Current picked up by the retracted tunneling tip as function of laser fluence.

mJcm^{-2} the current exhibits a reduced slope with increasing laser intensity, most likely due to saturation effects.

Additionally we performed "photoelectron spectroscopy" with the scanning tunneling microscope. The application of a variable tip bias voltage allows to monitor the energy distribution of the electrons (Fig 4). At negative tip bias voltage the electrons are decelerated and do not reach the tip, whereas the current increases at positive tip bias voltage. The derivative of this signal is a measure for the energy of the electrons picked up by the tip (Fig. 4b). The maximum at -0.6 V suggests that indeed photoelectrons are emitted with an excess energy of about 0.6 eV. This is also in agreement with a fourth order multiphoton process, since 4 photons of 740 nm amount to a total energy of 6.72 eV.

Since a further reduction of laser pulse energy at the low repetition rate of 50 Hz leads to a vanishing small signal we have tested another detection scheme for the tunneling current. Whereas the excitation frequency (50 Hz) in the above mentioned experiments was well below the bandwidth of the STM electronics (50 kHz), we now use the

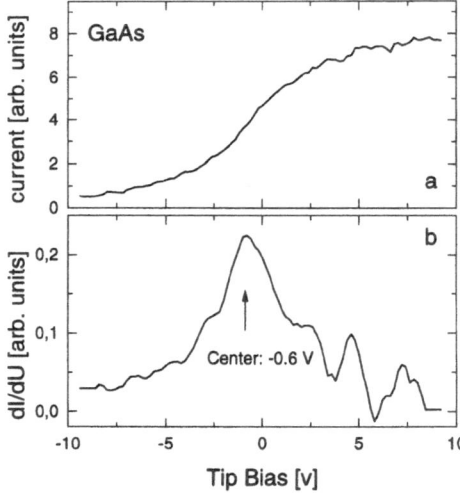

Figure 4. a) Current as function of tip bias voltage. The laser fluence for this experiment was 0.3 mJ cm-2. b) Derivative of curve a). The peak at -0.6 V corresponds to the energy of the generated photoelectrons.

direct output of the Ti-sapphire laser. The high repetition rate (80 MHz) of femtosecond laser pulses acts as a quasi cw illumination of the gap, i. e. no direct response on a single ultrashort optical pulse can be detected in the tunneling signal. Lock-In detection is used to measure the laser induced signal in the tunneling current. However with 80 MHz repetition rate the average illumination intensity is now much higher and thermal expansion of the tunneling tip can no longer be neglected. Up to now we did not yet achieve the primary goal to identify an ultrafast tunneling current component in our experiments.

Summarizing our preliminary experimental results we can draw the following conclusions: Identification of an ultrafast tunneling current component by direct illumination of the tunneling gap is a most challenging task. This is mostly due to the fact that the experimental results depend crucially on the actual gap conditions, i.e. the atomic tip geometry. Reproducible gap conditions are only accessible under UHV conditions and after the appropriate tip preparation. It is well known that the tunneling current and especially the current-voltage characteristics carry information about carrier recombination[7] and surface states[8]. Accordingly we expect that further refinement of tip preparation will finally allow us to identify an ultrafast component in the tunneling current. An additional problem, however, arises from thermal expansion of the tunneling tip leading to a strong linear response in the current signal. Excitation with high energy pulses at low repetition rate reduces the problem of thermal tip expansion but leads to the undesired generation of photoelectrons as shown in our experiments. Grafström et al. have shown that the influence of thermal tip expansion is decreasing with increasing modulation frequency[9], indicating a possible way to circumvent the problem of thermal expansion. Certainly, more and refined experiments which are underway are needed to detect a tunneling current in combination with fs time resolution.

ACKNOWLEDGMENT

We gratefully acknowledge the contributions of V. Gerstner, F. Sattler, A. Thon and S. Vogler to the presented work and the financial support by the "Deutsche Forschungsgemeinschaft."

REFERENCES

1. R. Haight, and M. Baeumler, Surf. Sci. **287/288**, 482 (1993).
2. J. A. Prybyla, H.W. Tom, and G.D. Aumiller, Phys. Rev. Lett. **68**, 503 (1992) and P. Guyot-Sionnest, Phys. Rev. Lett. **66**, 1489 (1991).
3. H. Rohrer, Surf. Sci. **299/300**, 956 (1994).
4. S. Weiss, D.F. Ogletree, D. Botkin, M. Salmeron, and D. S. Chemla, Appl. Phys. Lett. **63**, 2567 (1993).
5. R.H.M Groeneveld, Th. Rasing, L.M.F. Kaufmann, E. Smallbrugge, J.H. Wolter, M.R. Melloch, and H. van Kempen, J. Vac. Sci. Technol. **B 14**, (1996).
6. G. Gerber, F. Sattler, S. Vogler, J.Y. Grand, P. Leiderer, and R. Möller, Springer Series in Chemical Physics **60**, 149 (1994).
7. R.J. Hamers, and K. Markert, Phys. Rev. Lett. **64**, 1051 (1990).
8. J. A. Stroscio, and R.M. Feenstra, Methods of Experiemental Physics **27**, 95 (1993).
9. S. Grafström, J. Kowalski, R. Neumann, O. Probst, and M. Wörtge, J. Vac. Sci. Technol. **B9**, 568 (1991).

132

ULTRASENSITIVE PHASE MEASUREMENTS WITH FEMTOSECOND RING LASERS

Scott Diddams, Briggs Atherton, and Jean-Claude Diels

Department of Physics and Astronomy
University of New Mexico
Albuquerque, NM

1. INTRODUCTION

The decoupling in frequency of the two counterpropagating pulses in a femtosecond ring laser makes such a laser a sensitive probe of phase differences between the counterpropagating pulses. The phase difference per round trip is manifest as a beat frequency between the counterpropagating pulses. Phase differences between the counter propagating pulses of 10^{-6} have been measured, corresponding to a cavity length difference of 10^{-12} m. In this work, we will first discuss mechanisms involved in the decoupling of the counterpropagating pulses, and then we will show how the laser can be applied to ultrafast intracavity electro-optic sampling.

2. FREQUENCY LOCKING AND UNLOCKING IN A FEMTOSECOND RING LASER

The absence of frequency coupling between the counterpropagating fields of the femtosecond ring dye laser shown in Fig. 1 is primarily attributed to the pulses meeting at only two points inside the cavity.[1,2] This small overlap minimizes injection locking via the radiation backscattered from the various optical elements. One might expect the meeting of the pulses in the strongly scattering absorber jet to result in coupling between the waves, but no measurable coupling occurs. The coupling through the scattering is averaged out because of the transverse motion of the dye and solvent molecules through the focus of the beam.[2]

However, frequency locking does occur between the counterpropagating pulses when a stationary piece of anti-reflection coated is introduced at the pulse crossing point opposite the absorber. As would be expected, the frequency locking is dependent on the position of the scattering surface relative to the pulse crossing point as shown in Fig. 2. When the scattering surface is moved through the pulse crossing point, lock-in is observed over a range

Ultrafast Processes in Spectroscopy
Edited by Svelto et al., Plenum Press, New York, 1996

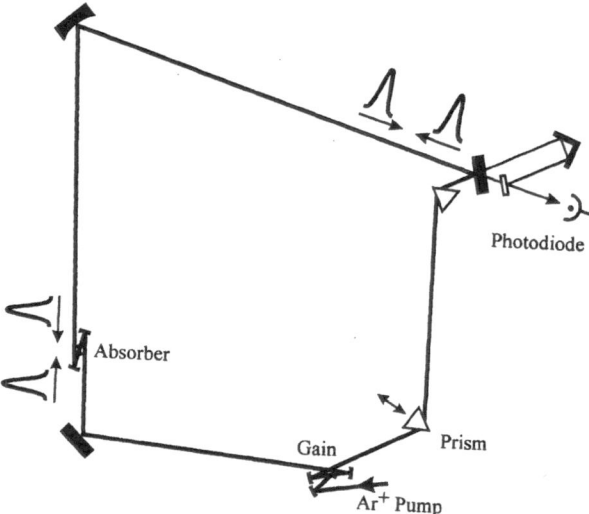

Figure 1. The femtosecond ring dye laser. Frequency coupling between the two counterpropagating pulses occurs when the pulses overlap near a scattering surface as shown

of about 200 μm, with a anomalous return of the beat frequency near the center of the locking band. This return of the beat frequency is peculiar, for one would expect the center of the dead-band to correspond to the point of maximum overlap of the counterpropagating pulses at the glass/air interface, and consequently, maximum coupling. Conditions under which the frequency locking is eliminated even with a scattering surface located at the pulse crossing point are plausible. This can be seen from a simple analysis of the differential equation governing the time evolution of the phase difference ψ between the counterpropagating fields. This is the measured beat frequency and has the explicit form:

$$\frac{\dot{\psi}}{2\pi} = \nu_{\text{bias}} - \frac{cr(\tau)}{2\pi P} \left\langle \frac{\mathcal{E}_1}{\mathcal{E}_2} \sin(\psi - \epsilon_1) + \frac{\mathcal{E}_2}{\mathcal{E}_1} \sin(\psi + \epsilon_2) \right\rangle. \tag{1}$$

In the above equation c is the speed of light, P is the cavity perimeter, $\mathcal{E}_{1,2}$ are the counterpropagating fields, and ϵ is the backscattering phase shift. The term ν_{bias} represents the bias beat frequency, which in this case is introduced by unequal saturation of the two pulses in the gain. The backscattering amplitude coefficient $r(\tau)$ has the form of a correlation between the counterpropagating pulses at the scatterer, where $\tau = 2d/c$ (d being the distance between the pulse crossing point and the scatterer). The angled braces $\langle \cdots \rangle$ indicated a time averaging over the duration of the pulse.

Equation (1) explains the major features of Fig. 2. When the scattering surface is far from the pulse crossing point, the effective backscattering $r(\tau)$ goes to zero. As the scattering surface is moved closer to the pulse crossing point, the scattered pulse in each direction will travel slightly ahead or behind the proper cavity pulse. The small amount of backscattered radiation then acts to frequency lock the counterpropagating fields. In this case Eq. (1) describes a phase-locked oscillator.[3] When the scattering surface is positioned exactly at the pulse crossing point, the scattering is symmetric provided the ratios $r(\tau)\mathcal{E}_1/\mathcal{E}_2$ and $r(\tau)\mathcal{E}_2/\mathcal{E}_1$ are equal. Furthermore if the backscattering phases are such that $\epsilon_1 = \pi$ and

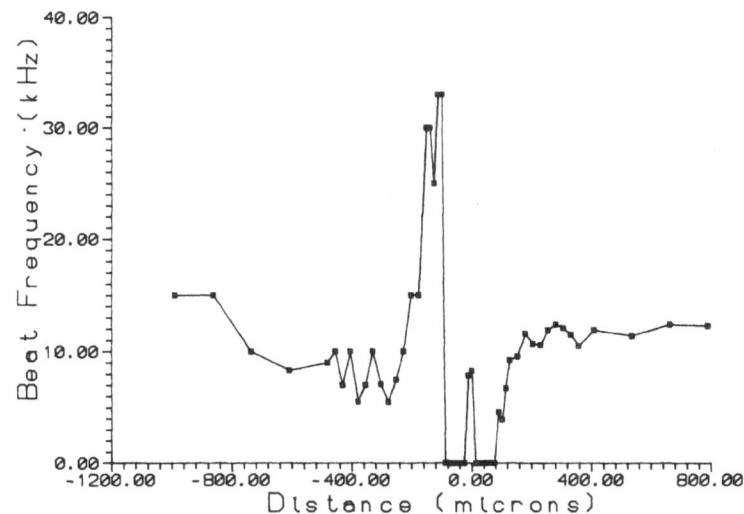

Figure 2. Beat frequency between the counterpropagating pulses of the ring laser as a function of the position of a scattering surface inside the cavity. The zero is the pulse crossing point.

$\epsilon_2 = 0$, then the two sine terms of Eq. (1) cancel, leaving once again only the bias response of the laser gyroscope. The result is the peak in the center of the locking band shown in Fig. 2. The instability of the cavity length does not permit us to look for structure which should exist on a wavelength scale under the central peak. It is however interesting that the width of this central peak is on the order of $50\,\mu$m, which is approximately the width of the correlation of two 100 fs pulses.

3. APPLICATION TO ULTRAFAST INTRACAVITY E-O SAMPLING

As a proof-of-principle experiment, we present the data of Fig. 3. A Pockels cell oriented as a phase modulator is inserted in the cavity of Fig. 1 at a point opposite the gain jet. The scattering piece of glass discussed above is removed, and the minimum resolvable beat frequency of 100 Hz corresponds to an optical path length difference $(\Delta n)d$ between the counterpropagating pulses of 10^{-12} m, or an index difference of 10^{-10} over a 1 cm length. An avalanche photodiode detects the pulse train in one direction from the laser and applies an electrical pulse to the Pockels cell. Appropriate optical delay ensures temporal coincidence of the electrical pulse and only *one* of the cavity pulses at the electrooptic crystal each round trip. Because the other cavity pulse always reaches the Pockels cell when no voltage is applied, the two counterpropagating pulses experience different indices of refraction in the cell. By varying the optical delay and recording the beat frequency, one can map out the temporal response of the photodetector/electrooptic crystal combination as shown in Fig. 3.

The temporal response in this case is about 300 ps, limited by the detector, cable, and the capacitance of the electrooptic crystal. The intrinsic resolution of the method however

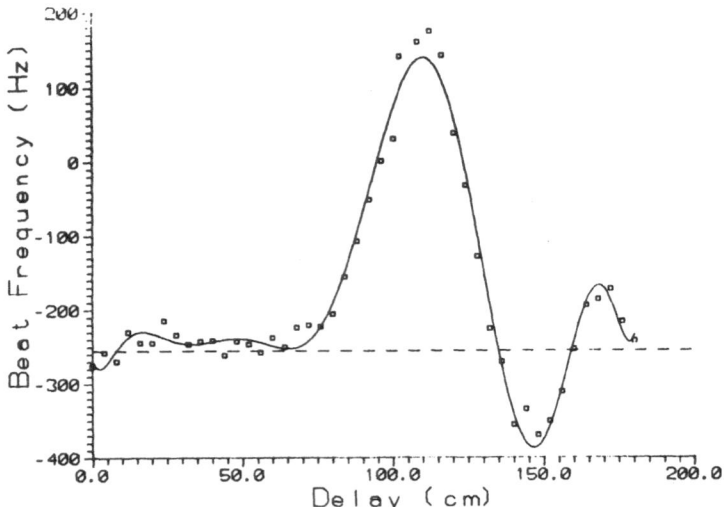

Figure 3. Measurement of an electrical signal coming from a fast photodiode via the beat frequency. The squares are the data points and the solid line is a polynomial fit.

is in the femtosecond regime, limited ultimately by the pulse duration and the transit time through the electrooptic medium. With thin samples, this arrangement could be implemented as a fast and sensitive tool to study the intrinsic response of semiconductors (by measuring directly the change of index due to the generated carriers), or photodetector/modulator combinations.

REFERENCES

1. M. L. Dennis, J.-C. M. Diels, and M. Lai, Opt. Lett. **16**, 529 (1991).
2. M. Lai, J.-C. Diels, and M. L. Dennis, Opt. Lett. **17**, 1535 (1992).
3. F. Aronowitz and R. J. Collins, J. Appl. Phys. **41**, 130 (1970).

133

FEMTOSECOND INTERFEROMETRIC CORRELATIONS BETWEEN COHERENT POLARIZATION WAVES

A Novel Approach to Dephasing and Quantum Beats

M. Gurioli, S. Ceccherini, F. Bogani, and M. Colocci

Dipartimento di Fisica-Unità INFM and LENS
Largo E.Fermi 2, 50125 Firenze, Italy

The coherent optical properties of material systems are commonly investigated by means of non-linear experimental techniques, generally referred to as four wave mixing processes[1] allowing, inter alias the direct measurement of the dephasing time T_2,[2] and the detection of quantum beats.[3]

In this contribution we show that the coherent dynamics can be studied with interferometric accuracy by means of a linear time-resolved correlation technique where the detected signal is either the coherent emission from the sample or the absorbed energy (more precisely any physical quantity related to it such as the incoherent photoluminescence (PL)). The experimental apparatus for this kind of measurements is rather simple. The beam of a tunable ultrafast source is sent in a Michelson-type interferometer providing at the output two collinear pulse trains delayed by t_D . The two pulses have the same intensity and polarization and are resonant with a radiative transition of the medium; then the incident electric field can be written as $E(t)=[A(t)+A(t-t_D)exp(i\omega_L t_D)]exp(-i\omega_L t)$. Each pulse induces a coherent polarization wave into the material system and solving the optical Bloch equations[1] one has, neglecting the non-linear terms, $P(t)=P_o(t)+P_o(t-t_D)$, where P_o is the polarization induced by a single pulse:

$$P_o(t) = \exp(-i\omega_o t - \frac{t}{T_2})\int_{-\infty}^{t} dt'\, A(t')\exp(i(\omega_o - \omega_L)t' + \frac{t'}{T_2}) \tag{1}$$

The two polarization waves interfere between them and from the measurement of the coherently scattered light and/or the absorbed energy as a function of the delay time t_D information on the coherent dynamics can be extracted.

Ultrafast Processes in Spectroscopy, edited by Svelto et al.
Plenum Press, New York, 1996

The energy absorbed by the material system W_A is obviously proportional to the integral over the time of the real part of $E \cdot dP^*/dt \approx i\omega_o E \cdot P^*$, and for Gaussian pulses it becomes:

$$W_A \propto 2\,erfc\left(\frac{\sigma}{T_2}\right) + erfc\left(\frac{\sigma}{T_2} - \frac{t_D}{2\sigma}\right)e^{(i\omega_o - 1/T_2)t_D} + erfc\left(\frac{\sigma}{T_2} + \frac{t_D}{2\sigma}\right)e^{-(i\omega_o + 1/T_2)t_D}$$

(2)

where $8\sqrt{\ln(2)}\sigma$ is the full width at half maximum of the Gaussian pulses and $erfc$ is the complementary error function [5]. In the limit $\sigma << T_2$ Eq.2 becomes

$$W_A = \omega_0 |\mu A|^2 \left(1 + \cos(\omega_o t_D)e^{-|t_D|/T_2}\right)/\hbar$$

(3)

We see that, as a consequence of the interference, W_A shows a rapid oscillation at the optical frequency ω_o which decays with the dephasing time T_2; the same conclusions hold for Eq.2 as the $erfc$ terms give only a small correction at very short times. It is also worth noting that in presence of an inhomogeneously broadened band one has to convolve Eq.2–3 with the distribution of frequencies, resulting in a decay of the interference pattern with the inverse of the inhomogeneous broadening.

At the same time, by integrating the optical Bloch equations, the energy of the coherent elastic emission in the whole solid angle is given by

$$W_S = \frac{2\omega_o^4}{3c^3} \int_{-\infty}^{+\infty} dt |P(t)|^2 = \frac{T_2}{T_{2,rad}} W_A$$

(4)

where $T_{2,rad}$ is the radiative dephasing time. This relation, which essentially reflects the optical theorem, shows that the coherent emission and the absorption have the same dependence on the time delay t_D. Therefore the same information can be obtained by monitoring the coherent emission W_S (i.e. transmission, reflection or Rayleigh scattering (RS)[4]) and/or any incoherent signal proportional to the absorbed energy W_A (i.e. PL or photo current) as a function of t_D. The resulting interferograms contain the information on the transition frequency ω_o and on the coherence status of the material system.

In Fig.1a we show the experimental interferogram obtained with a resonant excitation (120 fs pulse duration) and detection of both the heavy (HH) and light (LH) hole exciton transitions in a 180 Å GaAs/Al$_{0.3}$Ga$_{0.7}$As single quantum well (SQW). In addition to the fast interferometric oscillations (due to fluctuations of the interferometer arms) and to the phase relaxation we observe a modulation of the signal due to the quantum beating between the two transitions.

The smoothed interference amplitude is reported in a logarithmic scale in Fig.1b; the continuous line is the best-fit to the experimental data (dots). We get a dephasing time of 2.1 and 1.8 ps for the HH and LH exciton respectively and a period of the quantum beats of 0.79 ps with a value of 0.8 for the ratio between the LH and HH oscillator strengths, in good agreement with those evaluated by the cw photoluminescence excitation spectra. Similar measurements have been performed on a series of GaAs/Al$_{0.3}$Ga$_{0.7}$As SQWs heterostructures. Typical examples are reported in Fig.1c for a 120 Å SQW where the effects of the excitation density on the dephasing time are also shown. It is worth stressing that the linear correlation technique allows to measure the coherent dynamics down to extremely low powers.

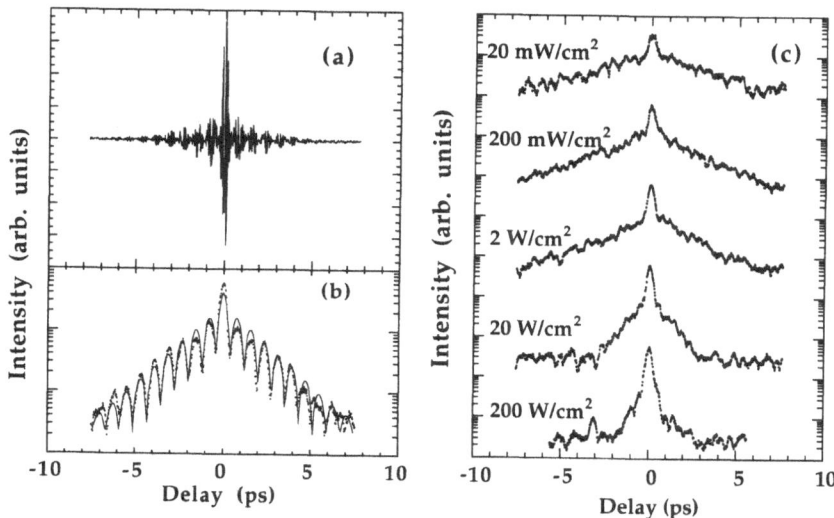

Figure 1. Examples of the interferometric correlation technique. a) Resonant excitation and detection at the HH and LH exciton transitions in a 180 Å GaAs/Al$_{0.3}$Ga$_{0.7}$As SQW. b) Interference amplitude of Fig.1a reported in a logarithmic scale. c) Interferograms of a 120 Å GaAs/Al$_{0.3}$Ga$_{0.7}$As single quantum well (SQW) for different excitation powers.

In addition the correlation technique can give an answer to one of the most important and long lasting problem in analyzing time resolved experiments,[4,5] that is the discrimination between the coherent (RS) or incoherent light emission (PL) following a resonant excitation. In fact in a RS interferogram the interference of the Fourier components of the coherent polarization is directly monitored by detecting the corresponding coherent emission and therefore the measurement strongly depends on the spectral band-pass of the detection devices.[4] In other words the spectral bandwidth must be much larger than the inverse of the dephasing time in order to avoid instrumental distortion effects on the interferogram. On the contrary, the PL interferogram comes from the interference between the polarization waves during the absorption process and the spectral bandwidth used in the detection of the incoherent emission process, which follows the coherent transient, does not introduce any distortion. Therefore spectral filtering of the emitted signal after resonant excitation can be used for separating the coherent (RS) and incoherent contributions (PL).

Let us conclude with few experimental examples concerning this feature of the interferometric correlation technique. In order to illustrate the method we have reported in Fig.2a the interferogram amplitudes of the elastic light scattering (i.e. a coherent emission) from a rough metal surface; in this case the system response is instantaneous and the interferogram with a large spectral filter gives the autocorrelation of the pulse. Reducing the spectral band-pass the interferograms show dramatic changes reflecting the Fourier transform of the instrumental response function. A completely different behaviour is found when detecting the incoherent photoluminescence (for example for non-resonant detection); the interferograms do not indeed depend on the spectral filter. Finally we report in Fig.2b the results obtained on a GaAs bulk structure exciting resonantly at the fundamental excitonic transition (818.5 nm) and detecting the emitted signal at the same

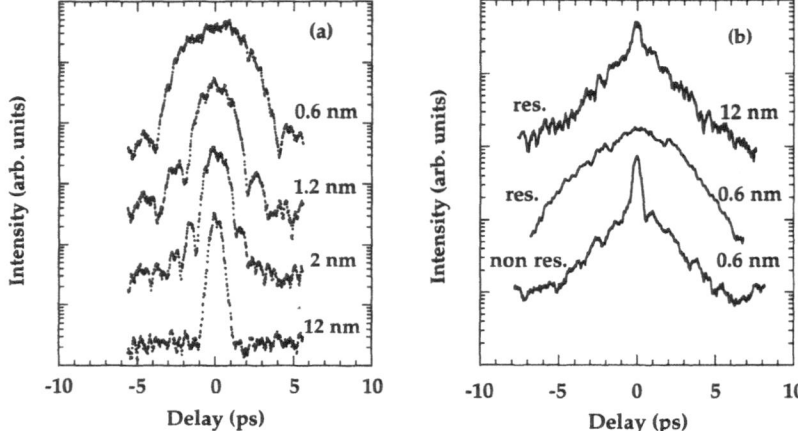

Figure 2. Effects of the spectral filtering on the interferograms. a) Coherent emission from a rough metal surface for different band-pass filters; the corresponding values are reported. b) Interferogram obtained exciting the GaAs exciton transition (818.5 nm) with resonant (818.5 nm) and non resonant (835 nm) detection for different spectral filters. Note that the interferogram for non resonant emission does not depend on filtering.

wavelength, for two different band-pass filters of 12 nm and 0.6 nm, and the incoherent PL out of resonance. The interferogram with the 12 nm filter shows an exponential decay on which is superimposed a coherent spike due to elastic light scattering from the sample surface; from a fit to the data we find a dephasing time of 1.45ps. Reducing the spectral width of the filter we observe a distortion of the resonantly detected interferogram clearly denoting the coherent nature of the emitted signal. For comparison, the interferogram obtained detecting the out of resonance PL emission (835 nm, with a filter bandwidth of 0.6 nm) from a 15 Å GaAs/In$_{0.18}$Ga$_{0.82}$As after resonant excitation of the GaAs exciton is also reported. Note that the corresponding interferogram with the 12 nm filter is, as expected, almost identical.

REFERENCES

1. M. Wegener, D. S. Chemla, S. Schmitt-Rink, and W. Schäfer, Phys. Rev. **A42**, 5675 (1990)
2. L. Schultheis, J. Kuhl, A. Honold, and C. W. Tu, Phys. Rev. Lett. **57**, 1797 (1986)
3. K. Leo, E. O. Gobël, T. C. Damen, J. Shah, S. Schmitt-Rink, W. Schäfer, J. F. Müller, Köhler, and P. Ganser, Phys. Rev. **B44**, 5726 (1991)
4. H. Stolz in *Time-Resolved Light scattering from Excitons*, (Springer-Verlag, Berlin,1994)
5. A.Vinattieri, J. Shah, T. C. Damen, D. S. Kim, L. N. Pfeiffer, M.Z. Maialle, and, L.J. Sham, Phys. Rev. **B 50**, 10868 (1994).

134

MEASUREMENT OF THE INTENSITY AND PHASE OF ULTRASHORT PULSES USING FREQUENCY-RESOLVED OPTICAL GATING

Rick Trebino,[1] Kenneth W. DeLong,[1] David N. Fittinghoff,[1]
John N. Sweetser,[1] Marco Krumbuegel,[1] Jason Bowie,[1] Greg Taft,[2] Andy
Rundquist,[2] Margaret M. Murnane,[2] and Henry C. Kapteyn[2]

[1] Combustion Research Facility
Sandia National Labs
Livermore, California 94551
[2] Department of Physics
Washington State University
Pullman, Washington 99164

Recently, we developed a technique for measuring the full time-dependent intensity, $I(t)$, and phase, $\varphi(t)$, of the complex electric field, $E(t)$, of an ultrashort laser pulse. This technique, Frequency-Resolved Optical Gating (FROG),[1–5] has been demonstrated for pulses in the ultraviolet, visible, near-infrared, and mid-infrared. It can measure pulses from millijoules to a few picojoules (the latter using the second-harmonic-generation version). It is also routinely used to measure a single ultrashort laser pulse.

In order to do this, we first used an autocorrelation apparatus followed by a spectrometer (See Fig. 1), which directly produces a spectrogram of the pulse. This measure of the optical pulse[6–9] is analogous to a musical score for an acoustic pulse, with the pulse frequency plotted vs. time. This intuitive display of the pulse is sufficient in itself for many applications. Second, we reduced the mathematics of the problem to the two-dimensional phase-retrieval problem, a well-known, solved problem from image science.[10–12] Third, we borrowed simple algorithmic techniques[3] from this field that allow the retrieval the full intensity and phase of the ultrashort laser pulse from its spectrogram, or "FROG trace." The full pulse field is uniquely determined by the FROG trace (although in second-harmonic-generation FROG, there is ambiguity in the direction of time).

In the past couple of years, we have developed FROG into a very reliable and well-characterized technique. In addition, we have investigated many variations on the basic FROG idea. And we have also begun to apply this method to answer important fundamental questions.

Specifically, we have studied the phase-retrieval algorithm that we use in great detail,[4] showing, for example, that it reliably recovers the pulse intensity and phase even in

Ultrafast Processes in Spectroscopy, edited by Svelto et al.
Plenum Press, New York, 1996

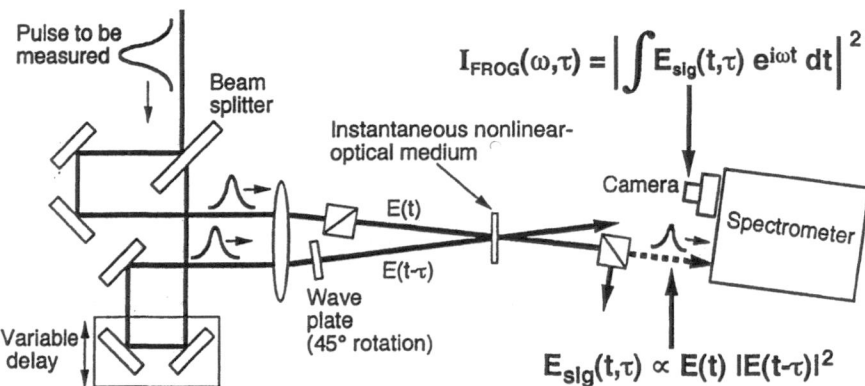

Figure 1. Multi-shot version of Frequency-Resolved Optical Gating (FROG) for measuring the intensity and phase of an arbitrary ultrashort laser pulse. In FROG, the signal pulse in an autocorrelation measurement is spectrally resolved for each delay. Here, a polarization-gate geometry is shown, in which one replica of a pulse gates out a temporal slice of another replica of the pulse, and the spectrum of the slice is measured vs. the delay between the two pulses. It is also possible to use other instantaneous nonlinear-optical interactions, and second-harmonic generation is also useful for measuring weak (e.g., picojoule) pulses.

the presence of massive noise.[4] Using standard image processing methods, we are able to retrieve a pulse with better than 1% error in the intensity and phase despite the presence of 10% additive noise in the trace.[4] This reduction in error is possible because of redundancy in the FROG trace, which contains N^2 points to determine $2N$ intensity and phase points. This redundancy has another advantage: when systematic error is present in the trace, the algorithm yields the best possible pulse for that trace, but nonetheless a pulse that fails to reproduce the experimental trace, thus informing the experimenter of the systematic error.

In addition, we are currently developing a neural-network pulse retrieval technique that operates non-iteratively, and hence, very quickly. This neural network operates by simply using the FROG trace as input and performing a series of multiplications, additions, and hyperbolic tangent evaluations to approximate the complicated function that corresponds to the FROG trace inversion algorithm. Training this neural net is time-consuming, but its use in practice requires on the order of milliseconds. So far, we have only used a limited set of pulses, but its extension to more complex pulses appears straightforward.

A variation on the FROG method has been demonstrated that can measure two pulses from a single trace. Another variation allows the use of a slowly responding nonlinear medium—in contrast to the conventional wisdom, which would imply that the medium response must be much faster than the pulse to be measured.

We have also combined FROG with spectral interferometry (SI)[17] to increase the sensitivity of pulse measurement by eight to ten order of mafnitude. Recently, we measured essentially arbitrary pulses of as little as a few zeptojoules (10^{-21} J) in energy on a multishot basis. SI allows the measurement of the phase *difference* between two pulses. So using FROG to measure a reference pulse, it is possible to then measure another pulse whose SI spectrum is available. This is very useful because SI involves no nonlinear-optical processes and so is very sensitive and can measure the phase difference between two arbitrarily weak pulses. Since a weak pulse essentially never exists without a strong pulse nearby—indeed, the strong pulse usually creates the weak pulse via some material interaction—we can use FROG to characterize the strong pulse, which then acts as a reference pulse in the SI measurement and yields the weak pulse. Because SI involves measuring the spectrum of the sum of the two pulses, we call the combination of FROG and SI: Temporal Analysis by Dispersing a Pair of Light E-fields (TADPOLE).

We have applied FROG to the problem of understanding the basis physics of the laser producing one of the shortest pulses ever created, as short as 9 fsec in duration.[5] Previous measurements of these extremely short pulses from the Ti:Sapphire oscillator have consisted of simple autocorrelations and spectra. Unfortunately, these measurements have been unsatisfying because they do not uniquely specify the pulse. Indeed, in this case, the measured pulse autocorrelation and spectrum were consistent with two different very pulse shapes that corresponded to two very different hypotheses for the physics of the limitations on the pulse width. Harvey, et al.[18] showed that a coherent-ringing phenomenon in the Ti:Sapphire yielded trailing satellite pulses that limited the pulse width, Christov, et al.[19] showed that higher-order group-velocity dispersion could yield leading, as well as trailing, satellite pulses. As both theories predicted the same autocorrelation and spectrum, it was not possible to determine the correct theory based on these limited measurements.

Because the pulse shapes predicted by the two theories are in fact quite different, their FROG traces are quite different. The traces are shown in Fig. 2. The measured FROG trace for this pulse is shown in Fig. 3, and it is clear that it is much closer in appearance to the trace of the Christov, et al. theory. Consequently, the FROG measurement of this pulse shows that higher-order dispersion is most likely the limiting factor in state-of-the-art Ti:Sapphire ultrashort-pulse oscillators.

Currently, we are working on several methods for further improving the range of this technique, especially toward weaker pulses. In the future, we look forward to applications of this versatile technique.

ACKNOWLEDGMENTS

This work was partially supported by the U.S. Department of Energy, Basic Energy Sciences, Chemical Sciences Division, by a Department of Energy Co-operative Research

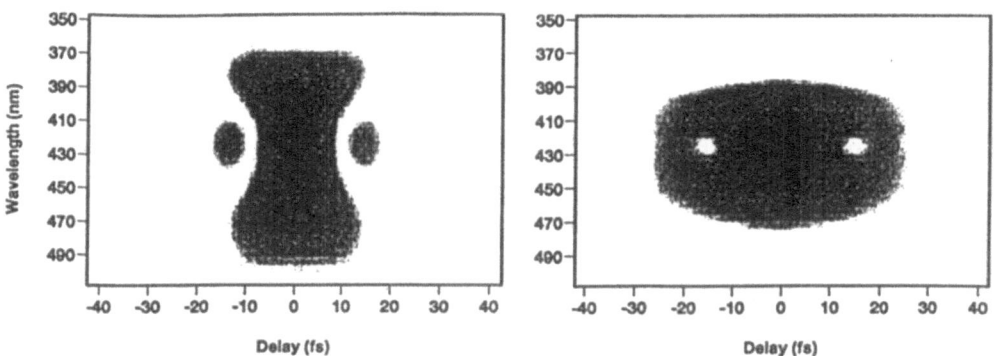

Figure 2. Theoretical FROG traces for ~9-fsec pulses for two different theories of the output of extremely short pulses from Ti:Sapphire lasers. Left: Christov, et al. Right: Harvey, et al. While these two theories yield pulses with nearly identical spectra and autocorrelations, their FROG traces differ significantly.

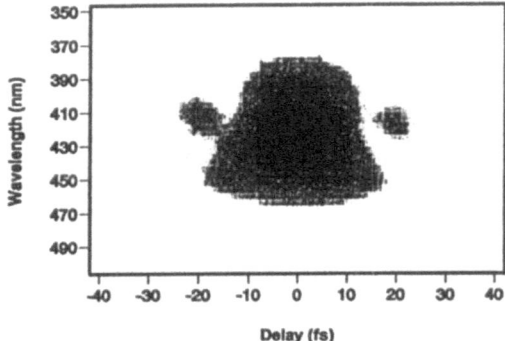

Figure 3. Experimental FROG trace for ~9-fsec pulses produced by a Ti:Sapphire laser. Note the similarity of this trace with that of with the theory of Christov, et al. (Most of the remaining discrepancy between this trace and that of the Christov, et al. trace is due to phase-matching bandwidth problems in the slightly-too-thick KDP crystal.)

And Development Agreement, and by a Sandia Laboratory-Directed Research and Development grant. M.A. Krumbügel wishes to thank Anthony E. Siegman for encouragement and the Alexander von Humboldt Foundation for support under the Feodor Lynen program.

REFERENCES

1. R. Trebino and D.J. Kane, J. Opt. Soc. Amer. A, **10**, 1101 (1993).
2. K.W. DeLong, R. Trebino, and D.J. Kane, J. Opt. Soc. Amer. B, **11**, 1595 (1994).
3. K.W. DeLong, D.N. Fittinghoff, R. Trebino, B. Kohler, and K.Wilson, Opt. Lett., **19**, 2152 (1994).
4. D.N. Fittinghoff, K.W. DeLong, R. Trebino, and C.L. Ladera, "Noise Sensitivity in Frequency-Resolved-Optical-Gating Measurements of Ultrashort laser Pulses," J. Opt. Soc. Amer. B, *in press*.
5. G. Taft, A. Rundquist, M.M. Murnane, H.C. Kapteyn, K.W. DeLong, R. Trebino, and I.P. Christov, Opt. Lett., **20**, 743 (1995).
6. W. Koenig, H. K. Dunn, and L. Y. Lacy, J. Acoust. Soc. Amer., **18**, 19 (1946).
7. S. H. Nawab, T. F. Quatieri, and J. S. Lim, IEEE Trans. Acoust. Speech Signal Process., **ASSP-31**, 986 (1983).
8. R. A. Altes, J. Acoust. Soc. Amer., **67**, 1232 (1980).
9. L. Cohen, Proc. IEEE, **77**, 941 (1989).
10. E. J. Akutowicz, Trans. Amer. Math. Soc., **83**, 234 (1956).
11. H. Stark, *Image Recovery: Theory and Application* (Academic Press, Inc., Orlando, 1987).
12. R. Barakat and G. Newsam, J. Math. Phys., **25**, 3190 (1984).
13. J. R. Fienup, J. Opt. Soc. Amer. A, **4**, 118 (1987).
14. J. H. Seldin and J. R. Fienup, J. Opt. Soc. Amer. A, **7**, 428 (1990).
15. J. R. Fienup, Appl. Opt., **21**, 2758 (1982).
16. R. G. Lane, J. Mod. Opt., **38**, 1797 (1991).
17. F. Reynaud, F. Salin, and A. Barthelemy, Opt. Lett., **14**, 275 (1989) and references therein.
18. J.D. Harvey, J.M. Dudley, P.F. Curley, C. Spielmann, and F. Krausz, Opt. Lett., **19**, 972 (1994).
19. I.P. Christov, M.M. Murnane, H.C. Kapteyn, J.P. Zhou, and C.P. Huang, Opt. Lett., **19**, 1465 (1994).

135

NEW INTERFEROMETRIC METHODS FOR GROUP-DELAY MEASUREMENT USING WHITE-LIGHT ILLUMINATION

A. P. Kovács,[1] G. Kurdi,[1] K. Osvay,[1] R. Szipöcs,[2] J. Hebling,[1] and Z. Bor[1]

[1] Department of Optics and Quantum Electronics
JATE University
H-6720 Szeged, Dóm tér 9, Hungary
Tel./Fax: +36 62 322 529
[2] Optical Coating Laboratory
Research Institute for Solid State Physics
H-1525 Budapest, P.O. Box 49, Hungary
Tel.: +36 1 169 9499, Fax: + 36 1 169 5380

Recently it has been demonstrated that in femtosecond laser systems both the intra- and extracavity group-delay dispersion (GDD) can be efficiently tailored by using specially designed dielectric mirrors, often referred as chirped mirrors[1]. It is important to know the GDD of any dielectric mirror when it is applied in a femtosecond laser system[2], since the shape of a femtosecond pulse can be strongly affected by the dispersion of the dielectric mirror. In this paper we present two different interferometric methods for group-delay (GD) measurement that can be regarded as an extension of the spectrally resolved white-light interferometry (SRWLI)[3].

In both techniques, based on a Michelson or a Fabry-Perot interferometer, a point-like white light source illuminates the interferometer containing the dielectric mirror under study (see Fig. 1 and 2). The exit plane of the interferometer is spectrally resolved by a spectrograph. Since the white-light source has a broad, continuous spectrum, the frequency-dependent GD of the dielectric mirror can be obtained from the spectrally resolved interference pattern for a wide spectral range with a single measurement. As the interference patterns are different at the two methods, therefore the way of the GD determination is different, too.

Fig. 1 shows one of our techniques, where a Michelson interferometer is used. In the reference arm a gold mirror (having negligible GDD) and in the object arm the dielectric mirror to be measured are placed[6]. When one of the mirrors is slightly tilted around the horizontal axis, horizontal interference fringes (fringes of equal thickness) are formed in the exit plane. More exactly, the intensity distribution along the Y-axis is cosinusoidal, and the argument of the cosine function is

Ultrafast Processes in Spectroscopy, edited by Svelto et al.
Plenum Press, New York, 1996

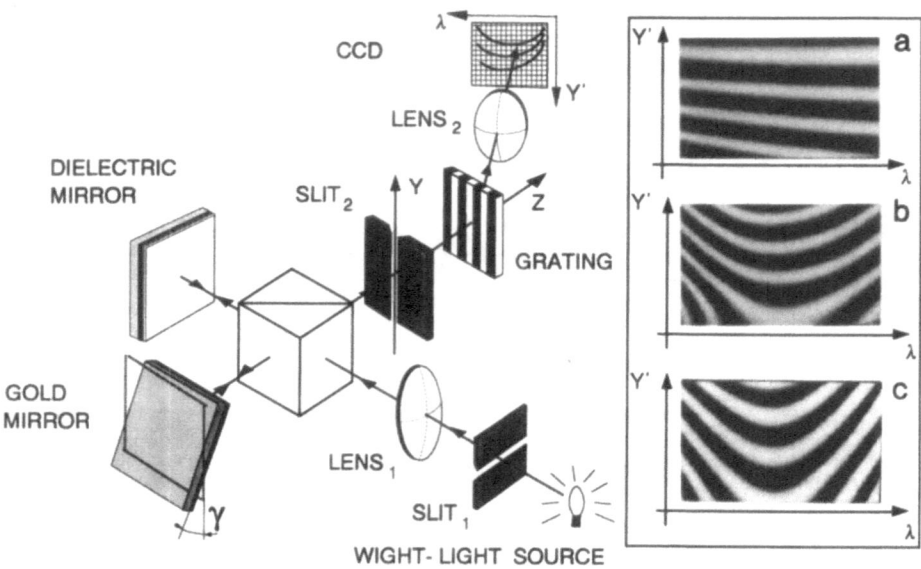

Figure 1. The SRWLI with a Michelson interferometer containing the dielectric mirror under study. Inset: spectrally resolved interference patterns recorded on a CCD camera with (a) two gold mirrors, (b) a chirped mirror designed for a sub-10-fs mirror-dispersion controlled Ti:sapphire laser[4], and (c) a chirped mirror designed for a Ti:sapphire laser amplifier[5].

$$\varphi(y,\lambda) = \varphi_D(\lambda) - \varphi_R(\lambda) + 4\pi\, y\, \gamma/\lambda, \tag{1}$$

where $\varphi_D(\lambda)$ and $\varphi_R(\lambda)$ are the phase shifts upon reflection of the dielectric and the reference mirror respectively, γ is the tilt angle and y is the vertical coordinate in the exit plane. To obtain the $\varphi_D(\lambda)$, first cosine functions are fitted for the spectrally resolved fringe pattern detected on a CCD camera at each wavelength, then it is determined from the argument of fitted functions (see Eq. 1). The group delay τ is numerically calculated by polynomial fitting of the phase shift ($\tau = \partial\varphi_D/\partial\omega$) and the GDD is similarly determined from the GD (GDD $= \partial\tau/\partial\omega$).

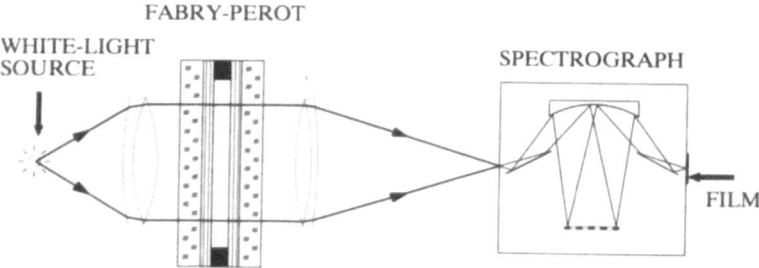

Figure 2. The SRWLI with a Fabry-Perot interferometer consisting of dielectric mirrors to be measured.

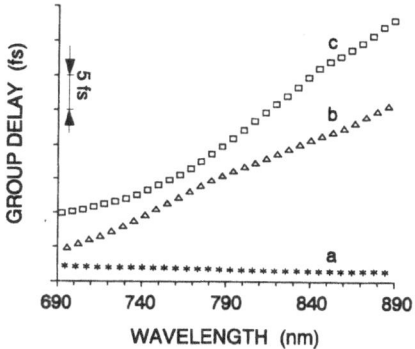

Figure 3. The group delay obtained by computer processing of the interference patterns shown in inset of Fig. 1 (every fifth point is plotted).

The resolution of the GD and GDD measurement was ±0.2 fs and ±5 fs^2 respectively. This method is well suited for rapid comparison of the dispersive properties of dielectric mirrors as well.

In the other technique a Fabry-Perot interferometer (FPI) is used. Let us take a dielectric mirror that shifts the phase of the reflected light by $\varphi(\omega)$. If such a mirror forms part of a FPI with known spacer thickness h then the condition of transmission maximum at a certain wavelength depends on $\varphi(\omega)$. From the spectral positions of transmission maxima, therefore, the phase shift and the group delay τ of the mirror can uniquely be established

$$\tau = \frac{d\varphi}{d\omega}\bigg|_{\frac{\omega_i + \omega_{i+1}}{2}} \approx \frac{\varphi_{i+1}(\omega_{i+1}) - \varphi_i(\omega_i)}{\omega_{i+1} - \omega_i} = -\frac{h}{c} + \frac{\lambda_{i+1} \cdot \lambda_i}{2c(\lambda_{i+1} - \lambda_i)} ,$$

(2)

where c is the speed of light.

Two types of mirrors have been measured. One of them is a low dispersion dielectric mirror for the visible range with R≥99.5% reflectivity and it should have constant GD

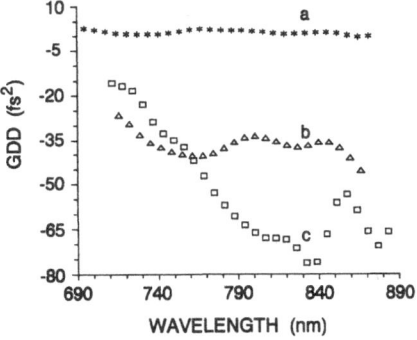

Figure 4. The group-delay dispersion obtained by numerical differentiation of the group-delay functions.

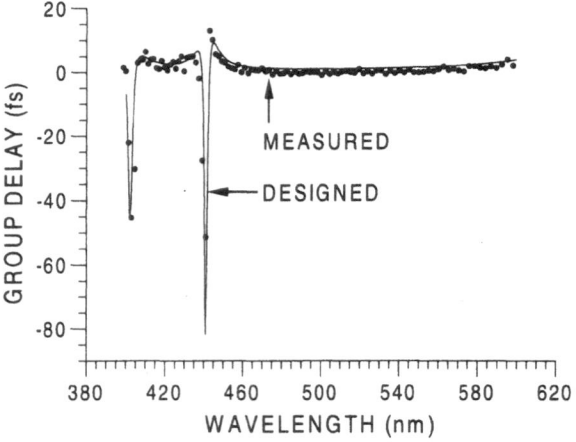

Figure 5. The measured and designed values of group delay of the low dispersion mirror.

in the spectral range of the laser operation. The other one is an intracavity mirror of an optical parametric oscillator (OPO), which has been designed to have R≥99.5% and about -120 fs^2 second order dispersion together with some third order dispersion around 1.2 μm[1]. This mirror is an improved version of the chirped mirrors used in Ref. 7.

The positions of transmission maxima of the FPI have been recorded on a photographic sheet film and evaluated by a Zeiss-made comparator. The typical readout and the spectral accuracy are ±5 μm (80 μm) and ±0.03 Å (0.5 Å), respectively. It corresponds to 0.24 fs (3.9 fs) uncertainty in determination of group delay. The numbers in bracket give a pessimistic estimation of greater error due to broader spectral stripes at the lower finesse regime[8].

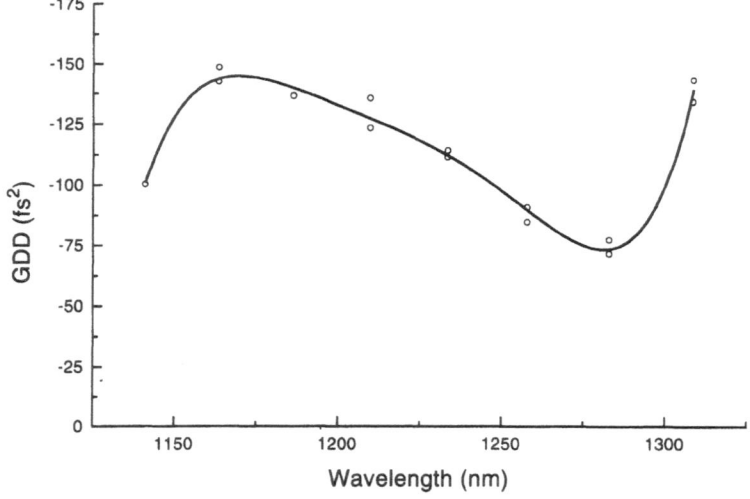

Figure 6. The measured group delay dispersion of an intracavity GDD-controlling OPO mirror.

As it is known, a FPI can be used in reflection mode as well. Thus, in the second experiment the interferometer has been built from an OPO mirror to be characterised and one semi-transparent ($R \approx 80\%$) gold mirror. Fig. 6 shows the GDD calculated from the measured wavelengths of the minima in reflection.

ACKNOWLEDGMENT

This work has been supported by the OTKA Foundation of Hungary under project Nos. F017213 and T7376.

REFERENCES

1. R. Szipöcs, K. Ferencz, Ch. Spielmann, and F. Krausz, Opt. Lett. **19,** 201 (1994).
2. W. H. Knox, Appl. Phys. **B58,** 225 (1994), and references therein.
3. C. Sainz, P. Jourdain, R. Escalona, and J. Calatroni, Opt. Comm. **110,** 381 (1994).
4. A. Stingl, M. Lenzner, Ch. Spielmann, F. Krausz, and R. Szipöcs, Opt. Lett. **20,** 602 (1995).
5. Ch. Spielmann, M. Lenzner, F. Krausz, R. Szipocs,"Compact, high-throughput expansion-compression scheme for chirped pulse amplification in the 10 fs range", Opt. Comm. in press.
6. A. P. Kovács, K. Osvay, Z. Bor, and R. Szipöcs, Opt. Lett. **20,** 788 (1995).
7. J. Hebling, E. J. Mayer, J. Kuhl, and R. Szipöcs, Opt. Lett. **20,** 788 (1995).
8. K. Osvay, G. Kurdi, J. Hebling, A. P. Kovács, Zs. Bor, and R. Szipöcs, Opt. Lett. **20,** (1995) in press.

136

MEASUREMENT AND ADJUST OF ULTRASHORT LIGHT PULSES

G. A. Sánchez and O. E. Martínez

Departamento de Física
Facultad de Ciencias Exactas y Naturales, Universidad de Buenos Aires
Ciudad Universitaria, Pabellón I
1428 Buenos Aires, Argentina

The object of this work is to describe a new method for the measuring of ultrashort light pulses in the frequency domain, which even allows real time measurements of the phase.

Its principle of operation relays in the measurement of the group delay as a function of the frequency, by means or a cross-correlation between the original pulse and a filtered version of it. The effect of filtering a narrow portion of the spectrum of a pulse is to broaden it, and to displace it in time. The time displacement is the group delay corresponding to this frequency, the magnitude to be measured. By measuring the autocorrelation of the filtered pulse with a reference one (the original ulfiltered pulse), it can be measured the group delay.

In a previous version of this method[1][2] the group delay is measured from the position of the maximum of the cross-correlation. A straightforward alternative for it is to measure the position of the zero of its derivative. This is measured with a lock-in amplifier, while the length of the non-filtered arm of the correlator is modulated dithering its mirror.

The advantage of this method is the direct retrieving of the group delay. This technique does not need to perform deconvolutions of the measured data, which often are numerically unstable and critically sensible to noise. Its mean drawback is the need to take a complete cross-correlation (or its derivative) for each group delay to be acquired. It may take excessive time, sometimes beyond the stability of the pulse generation system.

The method described in this work avoid to acquire a complete cross-correlation as a function of the time delay, measuring instead the logarithmic derivative of the cross-correlation (LDCC) for a fixed delay time. It is possible because the cross-correlation keeps the same functional shape when the frequency of the filter changes. The only changes in the cross-correlation are the possible modification of its amplitude (due to the change of the spectral amplitude), and its time displacement (due to the eventual dependence of the group delay with the frequency). The first effect is eliminated in the logarithmic deriva-

Ultrafast Processes in Spectroscopy, edited by Svelto et al.
Plenum Press, New York, 1996

tive, since it is self-normalized. Therefore, the LDCC is invariant with the frequency, except for its time displacement, which is equal to the time delay, the magnitude to be measured. Taking into account this invariance, the value of the LDCC for a given delay determines univocally the value of the frequency derivative of the phase, if the functional shape of the LDCC is known.

In our experiment the cross correlator is implemented using a diffraction grating instead the beam splitter, avoiding the phase distortion associated with this device. This divides the incoming light beam in two portions. The fist one is the reflected light, and the second one is the diffracted light. The reflected beam (the reference pulse) is reflected by a retroreflector mounted over a piezoelectric ceramic, which allows the modulation of this arm length. Then it hits on the grating for the second time, where it is reflected again. The diffracted beam pass through a lens and is reflected in a mirror. The distances between the grating and the lens, and between the lens and the mirror are both equal to the focal length of the lens. This device constitutes a zero dispersion compressor. Placing a narrow slit in front of the mirror, the filtering is performed. The light reflected by the mirror is diffracted again by the grating, when the different wavelengths present in it are recombined in a single beam. Both beams, the two times reflected, and the two times diffracted are focused over a second harmonic generator crystal, whose output is measured by a detector and constitutes the cross correlation between the reference and the filtered beams.

In order to measure the LDCC, the detector output is measured with a lock-in amplifier, whose reference signal is the oscillator that drives the retroreflector. The LDCC is measured by taking the quotient between the lock-in output and the dc component of the detector signal.

The effective filtering function is the convolution of the slit shape (usually a square function) with the image of the beam over the mirror. If the slit is very much narrower than the laser beam, then the convolution will be a gaussian function (assuming a gaussian profile in the beam). The cross-correlation will be in this case very similar to a gaussian function, and the LDCC will be a straight line. If the filtering slit is very much broaden that the laser beam, the effective filter will be a sinc function ($\sin(t)/t$), and the LDCC will not be a linear function. It can be shown that for a delay time less than five times the inverse of the spectral width of the slit, the difference between the logarithmic derivative of the sinc function and its linear approximation is less than 1%. In a real situation, the effective filtering function will be an intermediate case between the gaussian and the sinc functions, and the error will be negligible. Therefore, the calibration curve of the apparatus becomes a calibration constant, and the LDCC is proportional to the group delay.

From this result, it can be seen that a filtering slit wider than the laser beam is convenient. It avoids diffraction losses and improves the signal-to-noise ratio, due to the quantity of light available for the second harmonic generation. A very large slit limits the frequency resolution of the system. The optimal slit width depends on the phase features presents on the incoming pulse.

It must be noticed that the symmetry and the width of the reference pulse do not introduce error in the measurement. Then, it is not needed using the shortest pulse in the system as a reference, although this reference must be the same along all the experiment. A very broad reference must be avoid, because it will decrease the efficiency of the second harmonic generation, and therefore, the signal-to-noise ratio.

A critical source of error is the dispersion of the optical elements present in the filtering system. The phase due to the chromatic dispersion of the lens is added to the one of the pulse. There is also phase distortion coming from the misplacing of the lens and the mirror. This effects can be calculated, and taken into account in the measurement. If the pulse outcoming from the filtered system is used in another experiment, the dispersion can be ignored, because the phase measured is the phase of the outcoming pulse, even if it is slightly different from the phase of the incoming pulse. It will be discussed later, associated with the use of the system as a phase compensator. Another way to manage the phase distortion of the filter, is to measure it with the same apparatus. It is possible if the system is used as an interferometer. The dispersion of the filter can be retrieved from the position of the contrast maximum of the interference fringes for each position of the slit. This is a measure of the phase delay of the system as a function of the frequency. From this data, the misplacing of the optical elements can be reducing. The measurement of the remaining phase distortion allows to improve the precision of the system.

Since over the filter mirror the transversal co-ordinate is a frequency co-ordinate, it is possible to implement a phase compensation system in this place. The phase measured by the apparatus is the phase of the outcoming pulse when the slit is removed, and then it allows to adjust the phase of the pulse while measuring it.

ACKNOWLEDGMENT

This research was supported by the Consejo Nacional de Investigaciones Científicas y Técnicas (CONICET) grant PID 3336/92 and Fundación Antorchas.

Oscar E. Martínez is a member of the staff of the CONICET.

REFERENCES

1. J. A. Chilla and O. E. Martínez, Opt. Lett. **16**, 39 (1991).
2. J. A. Chilla and O. E. Martínez, IEEE J. Quantum Electron. **27**, 1228 (1991).

137

ULTRAFAST SPECTROSCOPY THROUGH DIRECT MEASUREMENT OF THE NONLINEAR POLARIZATION PHASE

L. Canioni, P. Segonds, B. Bousquet, W. Li, S. Le Boiteux, and L. Sarger

Centre de Physique Moléculaire Optique et Hertzienne
CNRS, URA No. 283
Université Bordeaux I
351 Cours de la Libération, 33405 Talence, France
Phone: (33)56846195, Fax: (33)56846970
E-mail: lsarger@frbdx11.cribx.1.u-bordeaux.fr

There is still a considerable interest in the full characterisation of ultrafast coherent transient processes occurring in semi-conductors or dye molecules. These processes directly influencing the dephasing time of an induced third-order nonlinear polarisation, are also contained in the phase of the electric field radiated by this polarisation. Several attempts have been made to observe such processes using pump-probe techniques in both the time and the frequency domains and ever relies on the amplitude decay measurements of the polarization induced by laser waves. The determination by these technics of the nonlinear properties of materials is consequently pulsewidth limited. In an other hand, the laser field optical oscillations drives the polarization oscillations and their relative phase is basically dependent on the nonlinear properties and more specially on the medium time constants (relaxation and dephasing). The direct study of this phase will lead to these response times with a limitation at the level of the optical period rather than the laser pulsewidth and will be described in this report. Moreover, since it is generally believed that optimal resolution requires pump and probe pulses as short as possible, much effort has been devoted to the generation of ultrashort light pulses -as short as 6 fs for example.[1] Beside the technical achievement, it is more fondamentally restricted as their corresponding large spectral bandwidth could not describe the degree of inhomogenous broadening in multilevel systems like dye molecules in solution for example. Recently a new method accessing the phase of the radiated electric field has been reported[2] in the frequency-domain.

We report here an alternative through the direct measurement of the phase of the third-order nonlinear polarisation induced in an isotropic medium. This new approach is based on a time resolved sampling interferometric technique[3] depicted schematically in Fig. 1.

Ultrafast Processes in Spectroscopy, edited by Svelto et al.
Plenum Press, New York, 1996

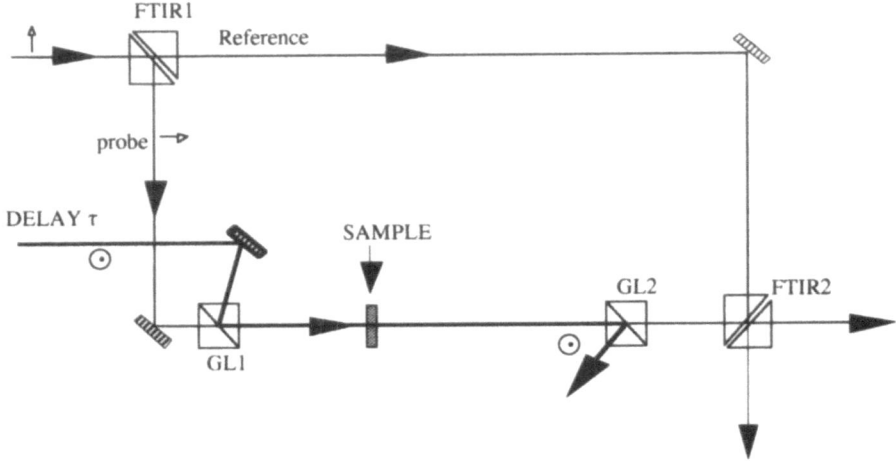

Figure 1. Experimental setup: GL-Glan prisms, FTIR-Adjustable beam splitters.

All the involved laser pulses are rather large (300 fs) and issued from the same laser source (Titane-Sapphire) at around 800 nm. The interferometric phase measurement between a reference and a probe beam measures the real part of the complex electric field induced by the third order polarisation in the medium. A time delay t between the probe and pump beams provides the basis for a sampling interferometry. Different orthogonal polarisation scheme may be used to characterise the anisotropic nature of the nonlinear process. Recording the interferometric phase as a function of the time delay τ, oscillations at twice the optical frequency modulate the second order cross-correlation function between the pump beam and the probe beam induced by classical Kerr effect. This new contribution to the signal only occurs when pump and probe beams overlap in time within the medium. and thus can be classified as coherent effects, but these oscillations are revealed as a totally new behaviour in our fully copropagating configuration. We called this signature "nonlinear fringes".

Theoretically, if one describe the orthogonally polarized copropagation of degenerate pump and probe gaussians beams in a non resonant isotropic material of susceptibility, the nonlinear polarisation writes:

$$P_x^{(3)}(t) = \frac{1}{4}\varepsilon_0 K \exp(i\omega t - ikr)\chi_{xxxx}^{electronic}\xi_{probe}(t)I_{pump}(t+\tau)(2+\exp(i2\omega\tau)) \tag{1}$$

Here the first term in braket has a simple electronic origin, is always positive and peaks at zero delay. It reproduces as a function of t the pulse temporal envelope as observed in an usual optical Kerr effect. However, coherent contributions (second term between brakets) add an oscillating term at twice the optical frequency ($2\omega\tau$). This oscillatory behaviour will fully develop in our experimental configuration where pump and probe beams strictly overlap within the nonlinear medium ($\bar{k}_{pump} = \bar{k}_{probe}$).

As an example, we presents the interferometric response for a nonresonant glass (Schott SF59) using an orthogonal polarisation scheme. The nonlinear fringes (Fig. 2) are to be compared to the well known interferometric second order autocorrelation function

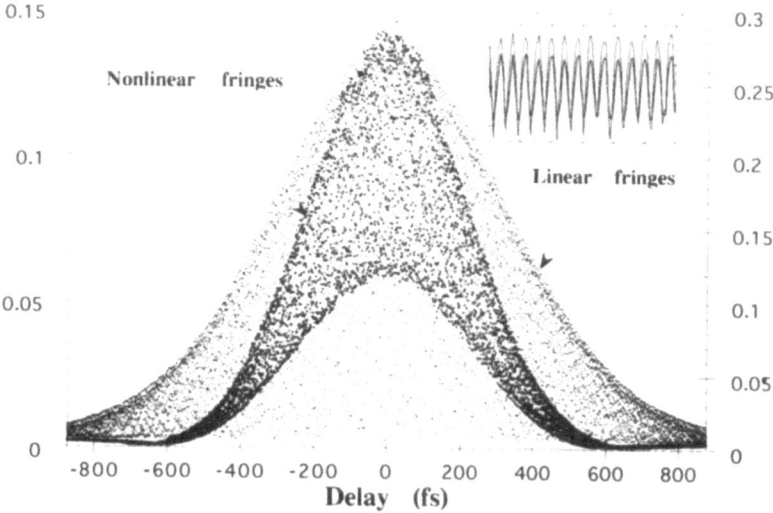

Figure 2. Interferometric response for a pure electronic non resonant process; Linear and Nonlinear fringes. Inset: Phase synchronism.

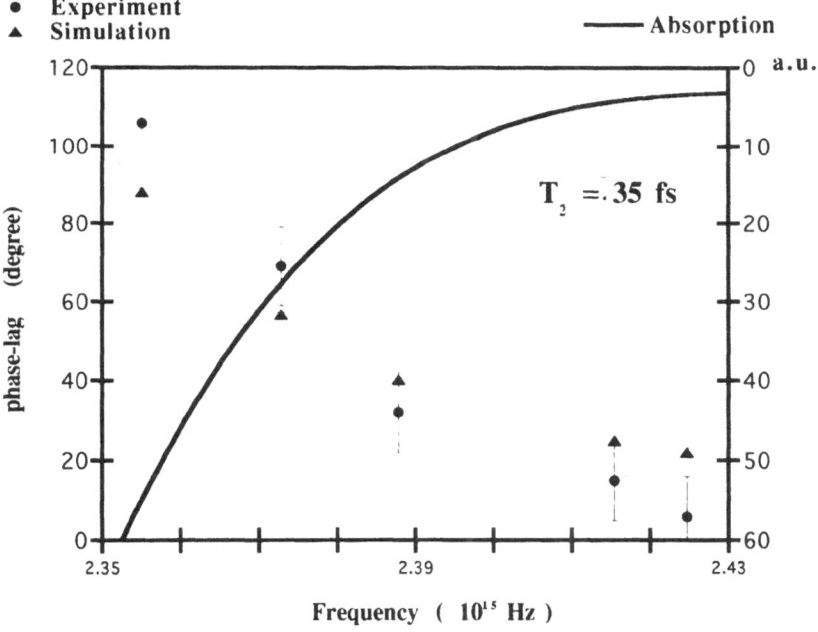

Figure 3. Phase lag versus Absorption: Triangle is the fit for a two level system.

widely used for pulse characterization. We observe in our configuration a nonlinear fringe system whose contrast of 1/2 is characteristics to the pure electronic nature of the involved process. In the same experiment, we record the interferometric phase between the reference beam and the pump beam - standard linear fringes(Fig. 2)- as a base for phase measurement. The inset of Fig. 2 display a zoom out of the simultaneous recording of the two fringes systems. Thus, it gives directly the measurement of the phase-lag of the induced third order nonlinear polarisation in the time domain with an ultimate resolution of the order of the optical period. Here, indeed, the observed lag fully confirm the expected behaviour for a system excited far above resonnance. In this technique, the use of laser pulses instead of CW beams is only required through the nonlinear process we are interested in.

To explore the potentiality of this approach to analyse the dephasing processes, we studied the 800nm transition of Nd ions in a glass matrix ($5Na_2O$-$20ZnO$-$75TeO_2$). The experimental phase of the nonlinear polarisation θ_{nl} as a function of the wavelength is presented in Fig. 3. A first simple attempt to retrieve from this data the matrix influence, we used a simple two level system to extract an reasonable equivalent T_2 of 35fs.

REFERENCES

1. J.Y Bigot, M.A. Mycek, S. Weiss, R.G. Ulbrich and D.S.Chemla, *Phys. Rev. Lett.* **70**, 3307–3310 (1993).
2. C. Froehly, A. Lacourt, J.C Vienot. *J. Opt.* **4**, 183 (1973).
3. L. Sarger, P. Segonds, L. Canioni, F. Adamietz, A. Ducasse, C. Duchesne, E. Fargin, R. Olazcuaga and G. Le Flem, *J. Opt. Soc. Am.* **B11**(6), 995–999(1994).

138

GVD-COMPENSATED PUMP-PROBE APPARATUS

S. Szatmári,[1] T. Nagy,[1] P. Simon,[2] and M. Feuerhake[2]

[1] Department of Experimental Physics
JATE University
H-6720 Szeged, Dóm tér 9., Hungary, Tel./Fax: 36/62/311–154
[2] Laser-Laboratorium Göttingen
D-37077 Göttingen, Hans-Adolf-Krebs-Weg 1., Germany

Pump-probe experiments provide the unique possibility to follow dynamics of fast processes with a temporal resolution comparable to the pulse duration of the pump and/or probe pulses used in the experiment. At present this allows a temporal resolution of several 10 fs. In most cases, where subpicosecond pulses are used, it is necessary to compress temporally both the pump and probe pulses before they are used for the experiment. On this time scale, the original pulse duration and exact synchronism between the pump and probe pulses can only be maintained for samples of submillimeter thickness, due to group velocity dispersion (GVD) of the material of the sample. This, however limits the magnitude of the change in the optical parameters of the sample which can be detected by the probe pulse. If an angular separation is applied between the pump and probe pulses for easy detection of the weak probe signal, the two pulse fronts are not completely overlapping, leading to certain loss of the temporal resolution.

It is seen from these considerations that in standard pump-probe experiments, one can have only a trade-off between temporal resolution and sensitivity by proper choice of the length of the sample.

Here we propose a novel arrangement which cancelles the GVD of the sample, therefore eliminates the depedence of the sensitivity on the achievable temporal resolution.

It is shown in Ref. [1] that a single dispersive element (eg. a grating) introduces a spatially-evolving negative chirp, which is roughly a linear function of the distance measured from the grating. This negative GVD - in a dispersion-free environment - can be used for temporal compression of the normally positively chirped pulses [1]. If this compressed pulse enters a material having a positive GVD of the same amplitude as the negative GVD of the grating, GVD-free propagation of the pulse in the material can be realized [2], preserving the compressed pulse duration during propagation (see one beam in Fig. 1). It is seen from the figure that the direction of propagation of the different spectral components

Ultrafast Processes in Spectroscopy, edited by Svelto et al.
Plenum Press, New York, 1996

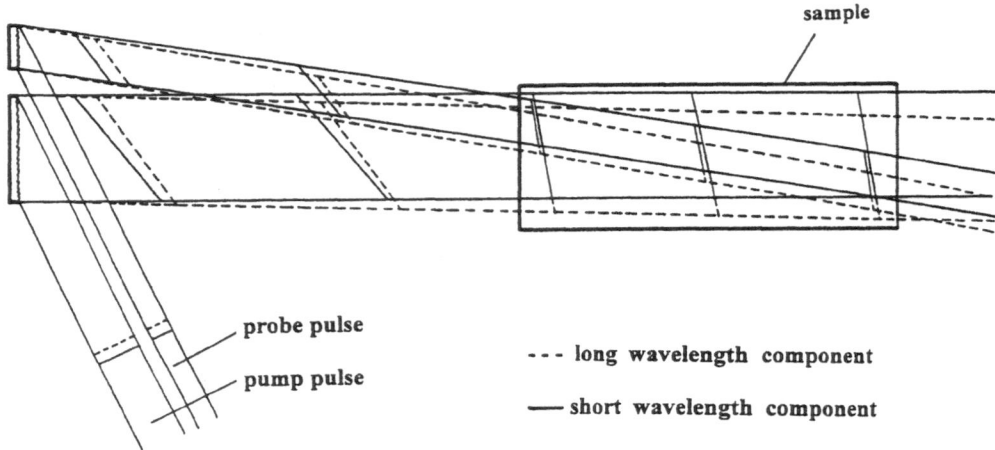

Figure 1. Schematics of the GVD compensated pump-probe arrangement.

having different phase velocities is adjusted so that the way, how the different spectral components form a tilted pulse front in the sample is independent on space.

   This consideration, however not only true for the spectral components forming one pulse, but also for different pulses of different wavelength. (In Fig. 1. an example is shown where the probe pulse has a significantly longer wavelength compared to the pump one.)

   If GVD cancellation is done in such a way both for the pump and for the probe beams, - as shown in Fig. 1 - the tilted pulse fronts of both beams are exactly parallel, their minimum pulse duration and relative delay are precisely maintained during propagation. This results in optimum temporal resolution, independently of the length of the sample.

   If the GVD of the sample is significantly different for the two pulses, separate gratings of different GVD can be used. By proper choice of the distances between the sample and the gratings, different initial chirps of the two pulses can be compensated. The relatively large angular separation of the two beams ensures easy detection of the probe signal, making this setup an ideal tool for fast dynamic studies.

REFERENCES

1. S. Szatmári, G. Kuhnle, P. Simon: Appl. Opt., **29**, 5372 (1990)
2. S. Szatmári, P. Simon, M. Feuerhake: Opt. Lett. to be published

139

CHARACTERIZATION OF ULTRAFAST MOLECULAR DYNAMICS IN LIQUIDS

Application to CS_2

L. Weiliang, L. Canioni, P. Segonds, and L. Sarger

Centre de Physique Moléculaire Optique et Hertzienne
URA 283 CNRS, Université Bordeaux I
351 cours de la Libération, 33405 Talence Cédex, France

In recent years, different kinds of time-resolved nonlinear optical techniques, such as the optical heterodyne-detection of optical Kerr effect[1,2], the optical heterodyne detected induced phase modulation[3] and the impulsive stimulated light scattering[4], have been used to study the ultrafast dynamics in molecular liquids. But because of the complex nature of the liquid dynamics involved as well as of the intrinsic limitation of the experimental approaches, the understanding of the dynamics processes, especially those taking place in subpicosecond regime, has not been satisfactorily achieved.

In this paper we present application of two complementary nonlinear techniques, i.e., a Cross-Induced Beam Deformation (CIBD) and a Sampling Mach-Zhender Interferometer (SMZI), to characterize the ultrafast molecular dynamics in CS_2. Detailed theoretical and experimental descriptions of the two techniques have been presented elsewhere[5,6]. The basic idea of the two techniques is to probe the pump-beam-induced nonlinear polarization by monitoring the phase shift of a time-delayed probe beam. In the CIBD experiment, this phase shift is measured by the far field transmission change of the probe beam through an aperture. the most advantage of the CIBD technique is the ability of measuring the sign of the third-order susceptibility. When combined with polarization selection, The technique is able to obtain symmetry information about the $\chi^{(3)}$ tensors and the pure nuclear response function corresponding to the involved ultrafast in molecular liquids, allowing their characterization. In the SMZI configuration, due to the colinear propagation of the pump and probe pulses, an oscillating contribution at twice optical frequency will add to the temporal signal envelop. The lag of this nonlinear fringe with respect to that of the linear fringe (the interference between the pump and the reference pulses) will provide information about the phase of the nonlinear polarization and thus the dephasing of the involved dynamical processes.

Fig. 1 presents the results of the CIBD experiment for liquid CS_2 and illustrates the method with which the data was analysed. curves (a) and (b) are the transmission changes

Ultrafast Processes in Spectroscopy, edited by Svelto et al.
Plenum Press, New York, 1996

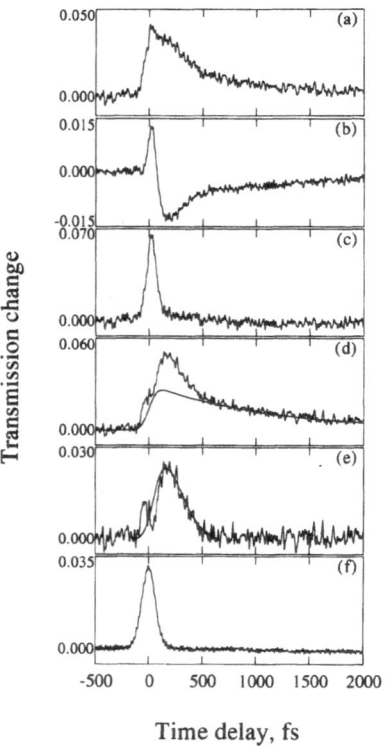

Figure 1. Results of the CIBD experiment for liquid CS_2.

as functions of the time delay for parallel and orthogonal pump-probe polarization geometries. curve (f) is the cross-correlation trace of the pump and probe pulses. From the theory of the CIBD experiment[5] we know that curves (a) and (b) in Fig. 1 measure respectively $\chi_{yyyy}^{(3)}(t)$ and $\chi_{xxyy}^{(3)}(t)$ after the coherent coupling peak. From the relative amplitudes of the two curves, we find that these two elements of $\chi^{(3)}$ obey a simple relation that

$$\chi_{yyyy}^{(3)}(t) = -2\chi_{xxyy}^{(3)}(t) \tag{1}$$

In fact, when plotting (a)+2(b) as curve (c), we see the signal vanishes except the coherent coupling peak at zero time delay. The symmetry of these two elements of $\chi^{(3)}$ reflects the anisotropic nature of the molecular dynamics involved and also implies another relation for isotropic media that

$$\chi_{yyyy}^{(3)}(t) = \frac{2}{3}\left[\chi_{xyxy}^{(3)}(t) + \chi_{xyyx}^{(3)}(t)\right] \tag{2}$$

Because $2\chi_{yyyy}^{(3)}(t)$ contributes to the coherent coupling for parallel polarization geometry while $\chi_{xyxy}^{(3)}(t) + \chi_{xyyx}^{(3)}(t)$ for orthogonal[5], plotting curve (d) such that (d)=(a)-4/3(b) will suppress the coherent coupling peak and display the pure nonlinear response function of liquid CS_2.

The temporal behavior of this nonlinear response function can be distinctly separated into a subpicosecond and a picosecond component, with the latter describing the diffusive molecular reorientation. The symmetry of the $\chi^{(3)}$ elements of the diffusive molecular reorientation obeys the simple relation as shown above. The smooth curve in Fig. 1(d) is the exponential fit to this picosecond component with 1.1ps time constant. Curve (e) in Fig. 1 shows the subpicosecond component with the picosecond term subtracted. The non-resonant electronic contribution is seen as a peak at zero time delay.

The above symmetry argument is disadvantageous to the attribution of this subpicosecond component to the interaction-induced change of molecular polarizability[1,2]. If dipole-induced-dipole (DID) intermolecular interaction is assumed, the change of one molecule dipole moment can be written as

$$\Delta\vec{\mu}^{i} = \alpha^{i} \cdot (\sum_{j \neq i} T_{ij} \cdot \vec{\mu}^{j})$$

(3)

where a^{i} is the linear molecular polarizability tensor and T_{ij} is the dipole tensor given by

$$T_{ij} = \frac{3\hat{r}_{ij}\hat{r}_{ij} - I}{r^{3}_{ij}}$$

(4)

with $\vec{r}_{ij}$ being the intermolecular distance in $\hat{r}_{ij}$ unit. Then the change of macroscopic nonlinear polarization due to this DID effect is calculated by ensemble average using proper field-dependent orientation and density distribution functions. It seems that the calculation is rather complicated and no analytical way will give the symmetry of the corresponding $\chi^{(3)}$. But when we examine the dipole tensor T_{ij}, some general results can still be accessed without being involved into the complexity of the calculation. Actually the elements of the dipole tensor can be expressed as combinations of second-rank spherical harmonics. When averaging over all intermolecular distances $\vec{r}_{ij}$ using a density distribution function, only the second-rank spherical harmonic components of the density distribution function will contribute[7]. These contributions are by no means isotropic and thus the resulted $\chi^{(3)}$ must be of different symmetry from that of isotropic media and in particular, from that experimentally observed. Thus the attribution of the subpicosecond component to the DID effect is far from being appropriate.

The above symmetry consideration instead supports the model of non-diffusive molecular libration, i.e., a rotational oscillation in the potential well of molecular interaction, whose $\chi^{(3)}$ tensor possesses the same symmetry as that of our experimental results[8]. However, the dynamics of this intermolecular libration has to be understood by a more sophisticated model than the classical oscillator used in ref. 8. Recently Tanimura and Mukamel used a multimode harmonic model to calculate the nonlinear optical response of molecular liquids[9]. We use this model with a practically homogeneously broadened under-damped mode to fit our subpicosecond signal. The best fit was obtained with frequency of 41.4 cm^{-1} and relaxation rate of 12.0ps^{-1}. This is given by the smooth curve in Fig. 1(e). Although the decay of the subpicosecond signal was well reproduced, the model failed to describe the delayed rising of the signal. Since the third-order response function is not sensitive to the inhomogeneous broadening of the mode, including inhomogeneity does not clearly improve the fitting. We feel that the dynamics of the intermolecular libration of liquid CS_2 is of subtle nature and probably the anharmonic nature of this mode has to be taken into account for its complete characterization.

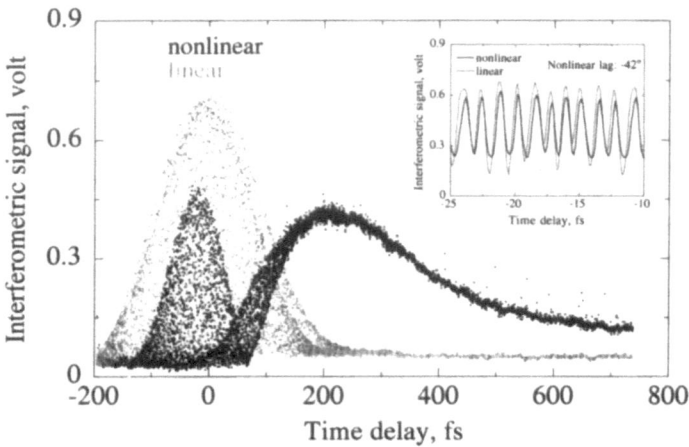

Figure 2. Time-resolved interferometric signal measured by the SMZI technique for CS_2.

As a complementary result, we show in Fig. 2 the time-resolved interferometric signal measured by the SMZI technique for CS_2. Also shown in Fig. 2 is the linear fringes which was simultaneously recorded. In this case that the polarizations of the pump and probe pulses are orthogonal, the nonlinear fringes at around zero delay consists of contributions from the non-resonant electronic excitation and from the coherent coupling between the pump and probe pulses via the positive elements ($\chi_{xyxy}^{(3)}(t) + \chi_{xyyx}^{(3)}(t)$) of the third-order nuclear nonlinear susceptibility tensor, while the nonlinear signal after the coherent coupling is the contribution from the negative element $\chi_{xxyy}^{(3)}(t)$. Because only the absolute value is recorded in the SMZI experiment, the negative signal reverses its sign. With the identification of different processes involved and the improvement of signal quality, we are able to analyse the interferometric signal in more detail. In Table 1 we summarize the parameters used to simulate the nonlinear signal. Exponential decay was assumed for each contribution.

In addition, by making a zoom on the fringes (the inset in Fig. 2), we see a lag of appproximately -42° for the nonlinear fringes with respect to the linear fringes. This lag implies that part of the contributions is complex due to near-resonant excitation. The most probable complex contribution is the subpicosecond intermolecular libration, which may

Table 1. Parameters used to simulate the experimental curve measured by the SMZI technique

Contributions	Relaxation times (fs)	Amplitudes (%)
Electronic (non-resonant)	instantaneous	Optimum: 19 { max:23 / min:18
Nuclear (picosecond reorientation)	880 ± 20	Optimum: 17 { max:20 / min:14
Nuclear (subpicosecond libration)	Optimum: 169 { max:194 / min:142	Optimum: 64 { max:65 / min:63

be Raman-excited by the frequency components within the band width of the femtosecond pulses. The ability of observing the phase of the corresponding nonlinear polarization provides the possibility of studying the dephasing of this ultrafast intermolecular motion.

REFERENCES

1. C. Kalpouzos, D. McMorrow, W. T. Lotshaw, G. A. Kenney-Wallace, Chem. Phys. Lett. **150**, 138(1988).
2. T. Hattori and T. Kobayashi, J. Chem. Phys. **94**, 3332(1991).
3. T. Kobayashi, in *Modern Nonlinear Optics*, ed. by M. Evans and S. Kielich (John Wiley & Sons, 1994).
4. S. Ruhman, A. G. Joly, B. Kohler, L. R. Williams and K. A. Nelson, Revue Phys. Appl. **22**, 1717(1987).
5. Weiliang Li, L. Sarger, L. Canioni, P. Segonds, F. Adamietz, and A. Ducasse, Opt. Commun., submitted.
6. L. Sarger, P. Segonds, L. Canioni, F. Adamietz, A. Ducasse, and C. Duchesne, E. Fargin, R. Olazcuaga, G. Le Flem, J. Am. Opt. Soc. **B11**, 995(1994).
7. D. Frenkel, in *Intermolecular Spectroscopy and Dynamical Properties of Dense Systems*, ed. by J. van Kranendonk (North-Holland, Amsterdam, 1980).
8. A. Owyoung, Ph.D. thesis, California Institute of Technology, Scientific Report No.12, 1971.
9. Y. Tanimura and S. Mukamel, J. Chem. Phys. **99**, 9496(1993).

140

Fs-TIME RESOLVED MEASUREMENTS OF NONLINEAR REFRACTION INDEX OF SEMICONDUCTOR DOPED GLASS BY DEFLECTION METHOD

Dao van Lap and Sabine Rentsch

Institute for Optics and Quantum Electronics
Friedrich-Schiller-University of Jena
Max-Wien Platz 1, D-07743 Jena, Germany

INTRODUCTION

Recently nonlinear optical properties of materials have been widely studied from the basic and practical points of view. A femtosecond pulse laser is suitable for such studies because of its high peak power and high time resolution, which allow us to distinguish several processes such as thermal and electronic processes. Their fast response and large nonlinear susceptibility make the semiconductor doped glasses promising candidates for use in all-optical signal processing devices [1]. It was shown that the nonlinear refraction index n_2 was about three times higher than in CS_2 [2]. These materials are commercially available as sharp cut-off optical filters and have also been manufactured in the form of optical fibres and waveguides. Colour filters have been used as the saturable absorbers in optical amplifiers. The size of the microcrystallites in the filter glasses may be varied by heat treatment. In commercially filter glasses semiconductor microcrystallities of 5–20nm diameter are embedded.

In this paper we report a sensitive method that enables one to separate the parallel and perpendicular parts of nonlinear refraction index on the semiconductor CdS_xSe_{1-x} doped glass by a simple measurement with femtosecond time resolution, which is based on the deflection of probe pulses. The concentration of S and Se components in glasses changes the spectral position of absorption edge to the excitation and probing laser pulses. Thereby the contributions of electrons in the different states after and by the excitation can be distinguished.

EXPERIMENTAL

The principle is depicted in Fig.1 [3]. An intense pump pulse and a weak probe pulse are irradiated on an aperture (AP). The aperture is formed by a slit of width d. The sample (S)

Ultrafast Processes in Spectroscopy, edited by Svelto et al.
Plenum Press, New York, 1996

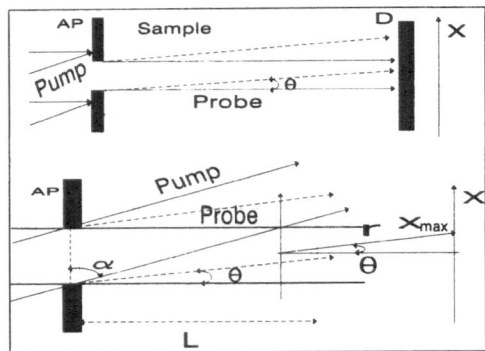

Figure 1. Scheme of beam deflection by pump beam induced refraction index.

is placed directly behind the slit. In this way we guarantee a definite, nearly rectangular spatial intensity profile and an accurate overlapping of both beams at the entrance of the sample. The probe beam enters the sample perpendicularly to the front face, whereas the pump beam crosses the sample surface at an angle a and thus induces a refraction index change in a region that provides a prism like structure for the probe beam. When the probe beam 'sees' this induced prism caused by a non-linear refraction index $\Delta n = n_2 I$, it is deflected by an angle θ.

The deflection angle can be measured by a diode array (D) as a function of the time delay between the pump and the probing pulses.

The deflection angle, defined by

$$\sin \theta = x_{max} / r$$

where x_{max} is the coordinate of the maximum of the far-field intensity, can be calculated:

$$\sin \theta \alpha(t) = \Delta n(t) \tan \alpha.$$

The output pulse ($l \cong 616nm$, duration $T \cong 75$ fs, energy $E \cong 100$ μJ) of a fs-laser system which consists of a CPM dye-laser followed by an excimer-laser pumped amplifier cascade is split by a beam splitter into a strong pump pulse and a weak probe pulse. The polarizer before the sample is oriented to pass probe light polarized at 45^0 to the plane of the pump pulse polarisation. The analyzer behind the sample makes possible the separate measurement of the probe pulse components parallel and perpendicular to the pump pulse, respectively. The pixel distance (25 μm) allowed an angle accuracy of 6×10^{-5} rad. The pump intensity amounts to about $I_p = 3 \times 10^{11}$ W/cm^2 at maximum. The deflection of the probe beam at a 2mm thick sample amounts to about more than 10^{-2} rad.

To realise various excitations conditions we used samples differing in their band-gap energy. The difference ΔE between the photon energy (2.01eV) and the respective absorption gap alters from negative to positive values within the filter series OG4, OG2, RG1 and RG2. In this way we obtain the following excitation conditions:

- OG4 (- 260 meV) far below the gap,
- OG2 (-110 meV) near below the gap,
- RG1 (26 meV) very near above the gap,
- RG2 (120 meV) above the gap.

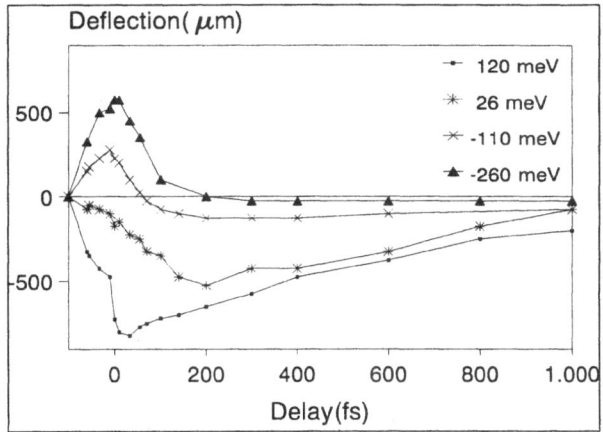

Figure 2. Change of refration index of semiconductor doped glasses (see text).

RESULTS AND DISCUSSION

The results of the $\Delta n(t)$ measurements of the doped glasses are shown in Fig.2, moreover, the transmission measured depending on time is to be seen in Fig.3.

The sample OG4 ($\Delta E = -260$meV) was measured in its transparency region. Here $\Delta n(t)$-values are detectable only during interaction of the strong laser pulse with the medium. The maximum value of Δn is $\Delta n(t=0) = 2.3 \times 10^{-4}$. From this value we get $n_2 = 4.1 \times 10^{-16}$ cm^2/W. The ratio of $\Delta n_\parallel / \Delta n_\perp = 2.8 \pm 0.4$ at intensity maximum corresponds to that of pure glass which is 2.7 ± 0.5 [3]. The transmission decreases during the pulse passing. The behaviour can be explained by degenerated four wave mixing. The diffraction of the probe beam at the grating which is induced by the pump and probe pulses causes the observed transmission decrease.

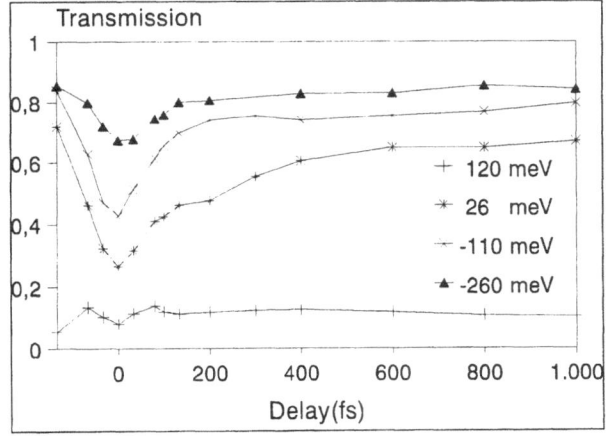

Figure 3. Time dependence of transmission at 616nm with CdS$_x$Se$_{1-x}$ doped glasses.

The samples OG2 ($\Delta E = -110meV$) and RG1 ($\Delta E = 26meV$) have band gaps nearby the laser wavelength. At OG2 at first positive and later negative Δn, but at RG1 only negative Δn were observed. The maximum of a negative value was obtained at the end of the excitation pulse where the highest number of excited carriers was achieved. Afterward Δn decays with a fast time-constant of about 1ps and a long time-constant of more than 10ps. A relaxation process after excitation was also observed at the transmission measurement.

The transmission change can be explained by gap shrinkage followed by bleaching. For the generation of free carriers at RG1 only the absorption of one photon, but OG2 one photon and additional phonons are necessary. The decay of one ps is caused by intraband relaxation.

At RG2 we measured within the absorption region. Here we observed high Δn-values and rise of transmission. The ratio $\Delta n_{\parallel}/\Delta n_{\perp}$ amounts to 1.25. To our knowledge for the first time the contribution of the hot electrons to the change of non-linear refraction index has been shown experimentally. It has been shown in [4] with the pump and probe measurement that at semiconductors CdS_xSe_{1-x} it takes about 1ps for these hot electrons to come back into thermal equilibrium. During this time, they may contribute to the non-linear refraction index.

The negative measured Δn values at CdS_xSe_{1-x} glasses are caused mainly by hot electrons, which thermalize in about 1ps.

REFERENCES

1. G. Assanto, A. Gabel, C.T. Seanton, G.I. Stegeman, C.N. Ironside, T.J. Cullen, Electron. Lett. **23**, 485 (1987)
2. J.L. Coutez and M. Kull, JOSA **B8**, 95 (1991)
3. H.-S. Albrecht, P. Heist, J, Kleinschmidt, D.V. Lap, Appl. Phys. **B57**, 193 (1993)
4. W. Rudolph, J. Puls, H. Henneberger, D.V. Lap, Phys. Stat. Sol. **B159**, 49 (1990)

141

COHERENT PROCESSES IN HIGH-T$_C$ SUPERCONDUCTORS

C. Jaekel, H. G. Roskos, M. Schröer, G. Kyas, and H. Kurz

Institut für Halbleitertechnik II
RWTH Aachen, D-52056 Aachen, Germany

The last few years have seen tremendous interest in the spectroscopy of coherent electronic and vibronic phenomena in semicondcutor materials. The investigations have led to new insight with respect to wavepacket dynamics (e.g., the existence of Bloch oscillations has been proven), phase relaxation, coupled excitations, new FIR sources etc. In this contribution, we present an extension of coherent spectroscopy to high-T$_c$ superconducting materials and device structures. Both the excitation of coherent phonons in bulk materials and the direct observation of a Josephson plasma resonance in weak link structures are discussed.

The first part of this contribution deals with the observation and generation of coherent phonons in YBa$_2$Cu$_3$O$_{7-\delta}$ (YBCO) superconductors by optical excitation with ultrashort laser pulses. Reflectivity experiments are performed in a fast-scan pump/probe setup with a Ti:sapphire laser yielding laser pulses with a pulse duration of 40–100 fs, a repetition rate of 80 MHz and wavelengths ranging from 740–820 nm. The samples are 200 nm thick YBCO films, dc-sputtered on SrTiO$_3$ substrates with their c-axis perpendicular to the substrate surface.

The weak modulations in the reflectivity transients in Fig. 1(a) and 1(c) are assigned to contributions of phonons in YBCO, which are coherently launched by the impulsive laser excitation. Fourier transformation (see Fig. 1(b) and 1(d)) yields two modes which can be attributed to A_{1g} modes of Ba (120 cm^{-1}) and of Cu(2) (150 cm^{-1}) in the CuO$_2$ plane. Coherent phonons have also been observed in non-superconducting YBCO [1], they are not a phenomenon of the superconducting compound.

In order to determine the generation mechanism for the coherent phonons in YBCO, we now discuss the temperature dependence of the phonon amplitudes. While there is no significant change of the phonon amplitudes with temperature in the non-superconducting compounds, we observe a sudden increase of the Ba-phonon amplitude just below the transition temperature T_c (see Fig. 1(d)). A closer look at the behavior below T_c reveals that the amplitude of the Ba phonons is well reproduced by the temperature dependence of the density of superconducting charge carriers n_s given by the two-fluid model with $n_s(T)=n_s(0)[1-(T/T_c)^2]$. We conclude from this behavior that the generation of the Ba phonons is connected to the optical excitation of the electronic system.

Ultrafast Processes in Spectroscopy, edited by Svelto et al.
Plenum Press, New York, 1996

633

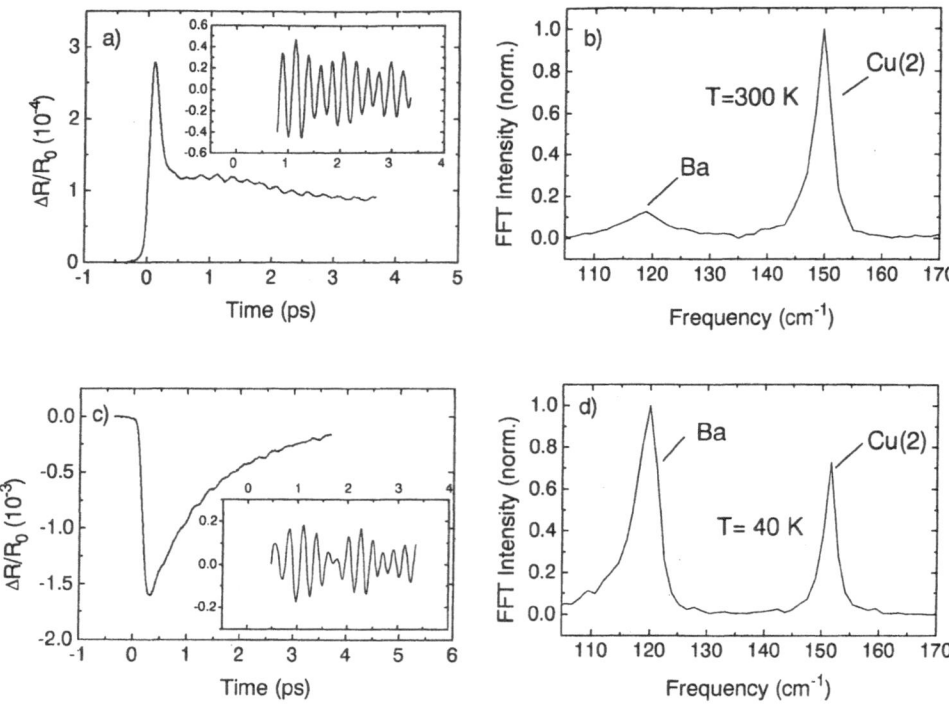

Figure 1. Temporal evolution of the reflectivity change after optical excitation of YBCO with a 60 fs laser pulse above (a) and below (c) the transition temperature. The insets in (a) and (c) show the phonon induced reflectivity changes. The Fourier transforms of these oscillations are displayed for room temperature (b) and for T=40 K (d).

It is known that the normal-conducting as well as the superconducting properties of YBCO are determined by the CuO_2-planes which are the fundamental elements of all super-conducting cuprates. Furthermore, the observed phonons come from oscillations of atoms directly neighboring the CuO_2-plane or being part of it, respectively. These observations suggest the following phonon excitation mechanism. The neighboring atoms of the planes are Y and Ba whose valences are known to be Y^{3+} and Ba^{2+} in YBCO. This difference of the valences produces an electric field extending from Y to Ba across the CuO_2-plane. The CuO_2-plane now screens that field to an extent depending on its conductivity and quantum properties. The screening can be changed suddenly by a short optical excitation resulting in a sudden displacement of the equilibrium position of the atoms. Since the unit cell of YBCO is symmetrical around the Y-atom we do not expect an oscillation of the Y-atoms but an oscillation of the Ba-atoms along the c-axis. Below T_c, the conductivity in the plane is mainly determined by the Cooper pairs. Then, the change of conductivity, i.e., of the effectiveness of screening, is due to optical Cooper-pair breaking. The more Cooper pairs are broken, the larger is the decrease of screening and therefore the displacement of the atoms. Hence, the generation mechanism of the phonons is an electronically driven one, such as the displacive excitation of coherent phonons (DECP) as introduced by Zeiger et al.[2] or the field screening mechanism discovered by Cho et al. [3,4].

In the second part, we shall discuss measurements of the frequency-dependent transmissivity of extremely thin superconductor films with a time-domain THz-spectroscopy arrangement[5,6]. We can extract both the real and imaginary part of the THz conductivity σ without using the Kramers-Kronig relation by determination of the transmission ratio of the sample and a reference sample (substrate without YBCO-film). The samples for the experiments discussed here are very thin YBCO films (d~50 nm) deposited on YSZ/Y$_2$O$_3$ buffered silicon substrates by thermal coevaporation.

YBCO on silicon is highly strained because of the severe lattice mismatch. At a critical thickness d_c, microcracks appear during the growth of the films. The critical thickness for the combination YBCO/YSZ/Y$_2$O$_3$/Si is known to be about 50 nm. Recent experimental results reveal a high strain-induced density of defects already at a thickness of 30 nm [6], which makes this system unsuitable for high-frequency applications.

In Fig. 2, the Fourier transforms of THz signals transmitted through a 50 nm thick film with microcracks are displayed for several temperatures. The general Gaussian-like shape of the spectrum is due to the antenna response. Below the transition temperature of 70 K, an absorption line is observed at a frequency that varies with temperature. This absorption is attributed to Josephson plasma oscillations of the Cooper pairs at crack-related intrinsic Josephson junctions (JJ) (or weak links) within the film. Josephson plasma resonances up to now have only been observed in a direct manner in low-temperature superconductor junctions [7]. The Josephson plasma frequency is given by [8]

$$f_{pl} = \sqrt{\frac{I_c}{2\pi\phi_0 C}} = \sqrt{\frac{j_{c,j}t}{2\pi\phi_0\varepsilon\varepsilon_0}} \, ,$$

(1)

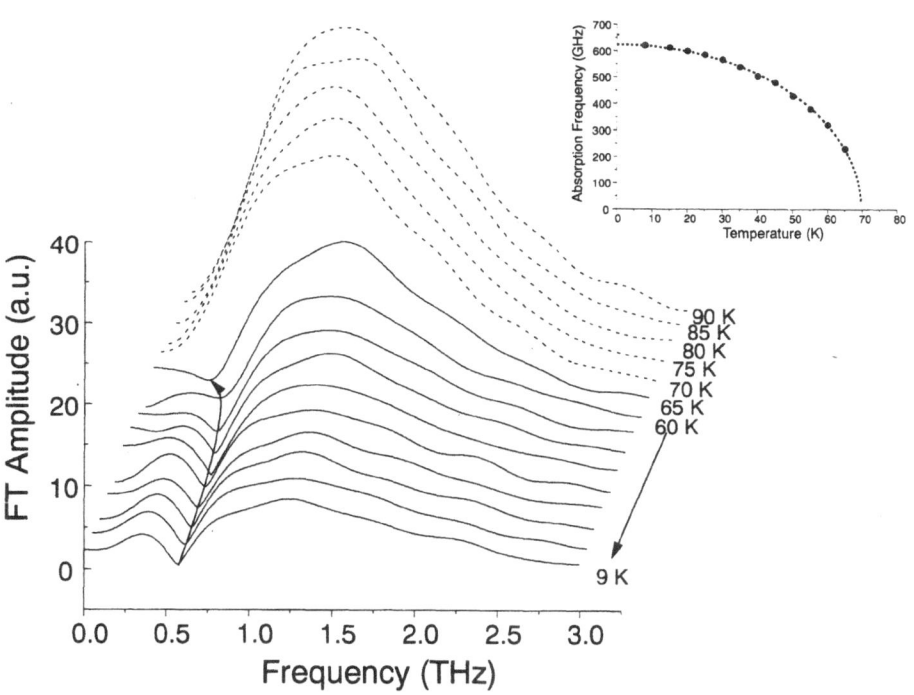

Figure 2. Frequency-domain data of THz pulses transmitted through a 50 nm YBCO film on silicon. The inset shows the peak absorption frequency as a function of temperature. The lines in the inset are fits with Eq. 1.

where $j_{c,j}$ is the critical current density of the junction, C the junction capacitance, ε the dielectric constant of the barrier, t the barrier thickness and ϕ_0 the flux quantum. Assuming a value of $j_{c,j}$ on the order of 10^3–10^5A/cm^2, an ε of 1 and values of t for the microcracks in the range from 3 to 8 nm [9], the plasma resonance frequency f_{pl} is estimated to be 200–800 GHz in good agreement with our measurements. The temperature dependence of the absorption frequency (see inset in Fig. 2) is an evidence for the absorption being a plasma resonance. According to Eq. 1, f_{pl} is expected to reflect the temperature dependence of $j_{c,j}^{1/2}$. For comparison with the experiment, we assume a two-fluid model behavior of the critical current density with $j_{c,j}=j_{c,j}(0)(1-(T/T_c)^\alpha)$. The experimental data can be fitted very well (line in Fig. 2) with Eq. 2 yielding $j_c(0)=560$ A/cm^2 with $t=8$ nm, $\alpha=2$ and $\varepsilon=1$ (i.e., vacuum tunneling).

In conclusion we have investigated coherent phenomena in high-temperature superconductors by means of optical measurement techniques. We have first studied the generation mechanism of phonons in YBCO thin films. Two phonon modes are observed that can be assigned to a Ba and a Cu(2) A_{1g} mode along the c-axis of the YBCO film. The excitation is explained by a sudden change of the screening of the crystal field present in YBCO.

In the second part, we provide evidence for the presence of intrinsic weak links within thin YBCO films on silicon. Above the critcal thickness when cracks are present within the film, an absorption line at THz frequencies is observed. It is attributed to a Josephson plasma resonance. Intrinsic JJ's may be of interest for device applications because the resonance frequency of 600 GHz is extremely high and well above the frequency range of current microelectronics. Furthermore, the results suggest applications for far-infrared radiation detectors. Because the widths of intrinsic weak links or microcracks in YBCO films on silicon are of the right order of magnitude to allow tunneling, it may be possible to produce lateral SNS/SIS-contacts without the necessity of nm-scale patterning.

ACKNOWLEDGMENT

This work is supported by the German 'Bundesministerium für Bildung und Forschung', contract No. 13N6288.

REFERENCES

1. J.M. Chwalek, C. Uher, J.F. Whitaker, G.A. Mourou and J.A. Agostinelli, Appl. Phys. Lett. **58**, 980 (1991)
2. H.J. Zeiger, J. Vidal, T.K. Cheng, E.P. Ippen, G. Dresselhaus and M.S. Dresselhaus, Phys. Rev. B **45**, 768 (1992)
3. G.C. Cho, W. Kütt and H. Kurz, Phys. Rev. Lett. **65**, 764 (1990)
4. W.A. Kütt, W. Albrecht and H. Kurz, IEEE J. Quantum Electron. **QE-28**, 2434 (1992)
5. M.C. Nuss, P.M. Mankiewich, M.L. O'Malley, E.H. Westerwick and P.B. Littlewood, Phys. Rev. Lett. **66**, 3305 (1991)
6. C. Jaekel, C. Waschke, H.G. Roskos, W. Prusseit, B. Utz, H. Kinder and H. Kurz, Appl. Phys. Lett. **64**, 3326 (1994)
7. A.J. Dahm, A. Denenstein, T.F. Finnegan, D.N. Langenberg and D.J. Scalapino, Phys. Rev. Lett. **20**, 859 (1968)
8. M. Tinkham, *Introduction to Superconductivity* (R.E. Krieger Publishing Company, Malabar, Florida, 1980)
9. W. Prusseit, S. Corsepius, M. Zwerger, P.Berberich, H. Kinder, O. Eibl, C. Jaekel, U. Breuer and H. Kurz, Physica C **201**, 249 (1992)

142

GENERATION, PROPAGATION AND DETECTION OF TERAHERTZ RADIATION FROM BIASED SEMICONDUCTOR DIPOLE ANTENNAS

P. Uhd Jepsen, R. H. Jacobsen, and S. R. Keiding

Department of Chemistry
Aarhus University
Langelandsgade 140,
Aarhus C, Denmark

The use of ultrafast lasers to generate sub-picosecond pulses of electromagnetic radiation, THz pulses, has grown into a very active research field in the last decade. For reviews of different aspects of the THz field, see the reference list.[1,2,3] The ultrafast interaction between a laserpulse and a material generates a DC-polarization, which is the source of the THz radiation. This polarization originates from either a simple flow of free carriers, or from a nonlinear χ^2-process. The resulting THz pulse consists only of a few cycles of the electric field, and consequently have a very high bandwidth. With a center-frequency of 1 THz and a bandwidth extending beyond 5 THz, the pulses have proven to be very useful in time-domain spectrometers, and has also been applied as probes of phonon dynamics in crystals and as probes of fundamental carrier transport phenomena in semiconductors. In this work we focus on a detailed description of the elements in a typical THz time-domain spectrometer, as shown in Figure 1.

We have used the simplest possible models to characterize the individual parts of the spectrometer, and by comparison between with different experiments, we show that the spectrometer is well described by the models.

To describe the THz emitter, we apply the Drude-Lorentz model of carrier transport, modified to take into account the ultrafast dynamics of the bias field. When the carrier density is high (10^{16}–10^{18} cm^{-3}), the bias field can be completely screened on a sub-picosecond time scale, and hence the dynamic screening is a very important factor in the pulse generation process.[4] In the Drude-Lorentz picture, the average velocity of the carriers is described by

$$\frac{dv(t)}{dt} = -\frac{v(t)}{\tau_s} + \frac{eE_{mol}(t)}{m^*} \tag{1}$$

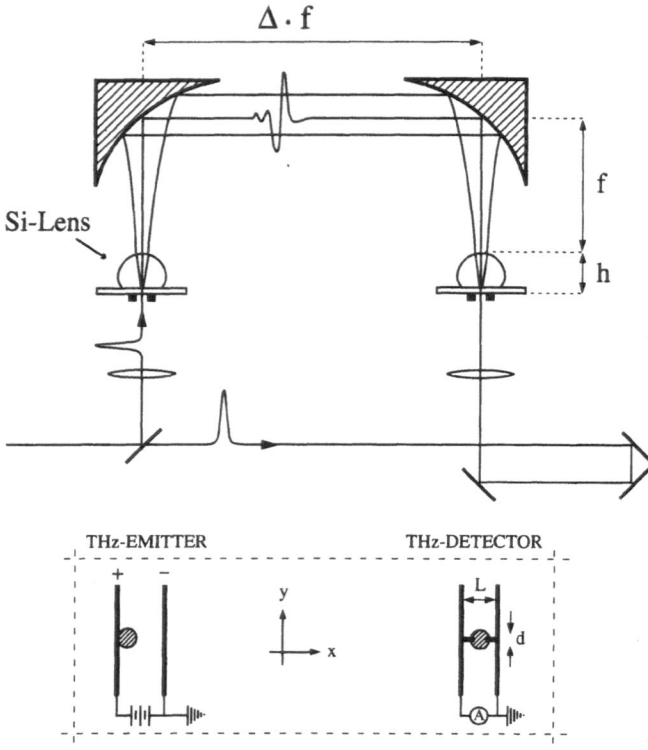

Figure 1. THz time-domain spectrometer, consisting of an emitter, two paraboloidal mirrors, and a detector. The inset shows the geometry of the emitter chip and the detector chip.

where τ_s is the scattering time, m^* is the effective mass, and $E_{mol}(t)$ is the local electric field. This term includes the important dynamic screening, since it is the sum of the bias field and the polarization field set up by the separating carriers, $E_{mol}(t) = E_{bias} - P_{sc}/\eta\epsilon$. The constant η is a geometrical factor. The polarization develops in accordance to $dP_{sc}/dt = -P_{sc}/\tau_r + j(t)$, where τ_r is the carrier recombination time, and $j(t) = n_f(t)ev(t)$ is the current density. n_f is the exponentially decaying density of free carriers, recombining with a time constant τ_r. The differential equation describing the carrier velocity can then be written as

$$\frac{d^2v}{dt^2} + \frac{1}{\tau_s}\frac{dv}{dt} + \frac{\omega_p^2}{\eta}v = \frac{eP_{sc}}{m^*\eta\epsilon\tau_r} \tag{2}$$

The radiated electric field is proportional to dv/dt, so solving the above differential equations gives us the shape of the THz pulse before the pulse-shaping performed by the optics and the detector.

The THz pulses are imaged onto the detector by the optical system illustrated in Figure 1. Due to the extreme bandwidth of the THz radiation, a frequency range covering two orders of magnitude, the optics in the beam path must be considered carefully when modelling the pulse shape at the detector. In a previous publication[5] we showed, that the radiation pattern emitted from the THz emitter can be well approximated by a Gaussian beam profile, if a truncated, spherical lens is attached to the emitter. Using this knowledge

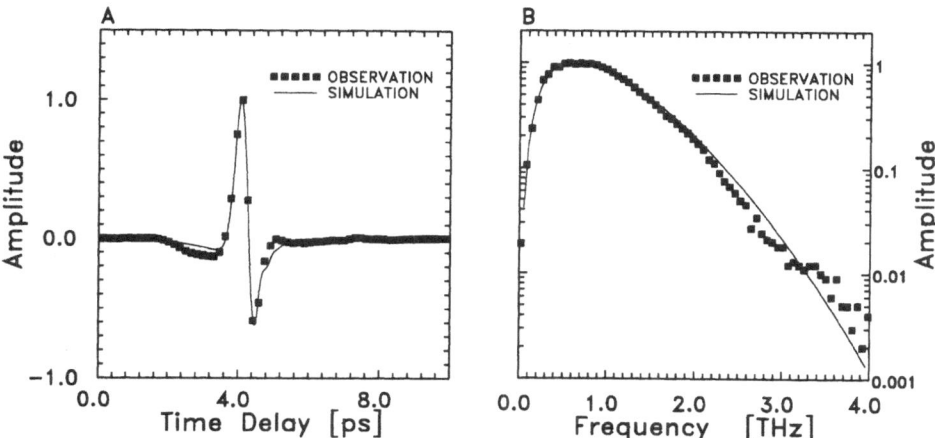

Figure 2. a) The simulated THz pulse (solid line), shown together with an experimentally measured pulse (filled squares). b) The frequency spectrum of the simulated pulse (solid line), shown together with the frequency spectrum of the experimentally measured pulse (filled squares).

allows us to apply Gaussian beam formalism to the propagation of the THz pulse from the emitter to the detector. If the paraboloidal mirrors are placed at a distance f from the emitter and detector, respectively, where f is the focal length of the mirrors, and the mirror spacing is $2 \cdot f$ (see Figure 1), the imaging of the emitter onto the detector is 1:1. After focussing through a second truncated, spherical lens, attached to the detector, the spot size of the THz beam is diffraction limited,[6] meaning that the spot size varies in inverse proportion to the frequency, and is limited in size only by the wavelength.

The spot size of the THz beam at the detector is of importance, since the detected signal is proportional to the average electric field, integrated over the active detection area. The average electric field drives a current across the detector gap, the current being the convolution of the photoconductivity induced by the probe laser pulse, and the average electric field. The high-frequency components are focussed to smaller spot sizes than the low-frequency components. Therefore, the detector response will depend on frequency. For a detector of length L (composed of a metal electrode length L_m and photoconductor length L_p) and width d, the average current flowing in the detector, as a function of frequency of the THz field, is

$$I = \sqrt{\frac{cP}{\pi \epsilon_0}} \cdot \frac{2R_L}{\rho_p L_p w_0 (n-1)\nu} \cdot \mathrm{Erf}\left(\frac{L(n-1)\pi w_0 \nu}{cR_L}\right) \cdot \mathrm{Erf}\left(\frac{d(n-1)\pi w_0 \nu}{cR_L}\right) \qquad (3)$$

where P is the power in the THz beam, R_L is the radii of the truncated, spherical lenses, ρ_p is the time-averaged resistivity of the photoconductor, and n is the index of refraction of the lenses. Equation (3) assumes a Gaussian beam $E(x,y) = E_0 \exp(-(x^2 + y^2)/w_0^2)$ in front of the emitter, where w_0 is the spot size on the surface of the emitter lens, and, consequently, also the spot size on the detector lens. Equation (3) has taken into account the optics in the beam path, and is hence specific for the described system. It is straightforward to calculate the response function for non-ideal optical systems, although analytical solutions are obtained only in special cases.

We combined all the above elements of the THz time-domain spectrometer in a simulation program, running on a standard personal computer. Figure 2 shows the result of such a simulation. Figure 2a shows the direct measured pulse, together with the simulation result, in the time domain. In Figure 2b, the pulse is shown in the frequency domain, also together with the simulation results. The agreement between the simulation and the experimentally obtained data is very good.

Three conclusions are drawn from the studies presented here. I) The THz source can be described by the Drude-Lorentz model, simply as a damped plasma oscillation of the photogenerated carriers. II) The ultrafast carrier dynamics in the emitter is of great importance, since the local electric field at high carrier densities is screened on a sub-picosecond time scale. III) The main bandwidth limiting factors in the present THz setup are the THz optics and the size of the active detector area.

The authors wish to thank M. van Exter and D. R. Grischkowsky for fruitful and interesting discussions concerning the carrier dynamics in THz emitters.

REFERENCES

1. X.-C. Chang, and D. H. Auston, J. Elec. Wav. Appl. **6**, 85 (1992).
2. D. Grischkowsky, "Nonlinear generation of sub-psec pulses of THz electromagnetic radiation by optoelectronics — Applications to time-domain spectroscopy." In *Frontiers in Nonlinear Optics*, Eds. H Walther, N. Koroteev, and M. O. Scully (Institute of Physics Publishing, Bristol and Philadelphia, 1993).
3. M. C. Nuss, P. C. M. Planken, I. Brenner, H. G. Roskos, M. S. C. Lou, and S. L. Chuang, Appl. Phys. B. **58**, 249 (1994).
4. J. E. Pedersen, V. G. Lyssenko, J. M. Hvam, P. Uhd Jepsen, S. R. Keiding, C. B. Sørensen, and P. E. Lindelof, Appl. Phys. Lett. **62**, 1265 (1993).
5. P. Uhd Jepsen, and S. R. Keiding, Opt. Lett. **20**, 807 (1995).
6. P. Uhd Jepsen, R. H. Jacobsen, and S. R. Keiding, submitted to J. Opt.Soc. Am. B, october 1995.

143

THz BLOCH OSCILLATIONS IN SUPERLATTICES AT ROOM TEMPERATURE

T. Dekorsy,[1] R. Ott,[1] P. Leisching,[1] H. J. Bakker,[1] H. Kurz[1] and K. Köhler[2]

[1] Institut für Halbleitertechnik
RWTH Aachen
52056 Aachen, Germany
[2] Fraunhoferinstitut für Angewandte Festkörperforschung
79108 Freiburg, Germany

The investigation of the dynamics of coherent wavepackets in semiconductor heterostructures has become a very attractive subject for ultrafast spectroscopy methods[1]. Up to now, most studies of optically excited quantum coherence in semiconductors are performed at low lattice temperature, since the coupling of the excited states to the lattice rapidly destroys the coherence at elevated temperatures. Especially excitonic states are destroyed on a time scale of 100 fs at room temperature, since the excitonic binding energy (4 meV for bulk GaAs) is much smaller than the thermal energy kT of 25 meV. An extensively applied technique for the time-resolved study of quantum coherence in semiconductors is four-wave mixing (FWM), where the modulation of the nonlinear interband polarization with the beat frequency immanent to the excited multi-level system is detected. At room temperature, the excitonic interband coherence detected by FWM is destroyed within 100 fs, therefore strongly aggravating the observation of quantum coherence under these conditions.

Bloch oscillations (BOs) in semiconductor superlattices are excited by the coherent superposition of several Wannier-Stark (WS) states with an ultrashort laser pulse. Their existence has been proven for the first time in FWM experiments at low lattice temperatures in GaAs/Al$_x$Ga$_{1-x}$As superlattices[2]. According to the prediction of Esaki and Tsu[3] the oscillating wavepackets are the source of tunable radiation in the THz regime[4]. We introduced a highly sensitive method for the observation of Bloch oscillations, which is based on the electro-optic detection of the birefringence induced by an oscillating dipole in the superlattice[5]. A study of the temperature dependence of the dephasing of BOs has been performed by THz emission and FWM experiments, where BOs have been observed up to 200 K[6]. Here, we report on the observation of BOs at room temperature by transmittive electro-optic sampling (TEOS)[5,7]. We show that the BOs observed at room temperature stem from charge carriers in continuum states of the conduction band. This conclusion is supported by a comparative study of BOs by FWM and TEOS under the same experimen-

Ultrafast Processes in Spectroscopy, edited by Svelto et al.
Plenum Press. New York. 1996

641

tal conditions[8] and by recent THz emission experiments[9] at low lattice temperature. This observations open the pathway towards applications of optically excited quantum coherence in semiconductor heterostructures, e.g., the application of Bloch oscillations as source of tunable THz radiation at room temperature.

We report on experiments performed with two Ti:sapphire lasers with a pulse width of either 130 fs (comparison of TEOS and FWM) or 40 fs (BOs at 300 K). FWM is performed in a standard self-diffraction geometry. The diffracted signal is proportional to the third-order interband polarization $|P^{(3)}(t)|^2$. In TEOS the change in polarization of a circularly polarized probe is measured with high sensitivity[5]. The induced birefringence is in first order linear proportional to the intraband polarization $P_z(t)$ parallel to the growth direction of the superlattice. The laser wavelength is tuned between the first heavy hole (hh) to electron -1 (e_{-1}) and electron 0 (e_0) transition of the WS ladder of the biased superlattices. We investigate two GaAs/Al$_{0.3}$Ga$_{0.7}$As superlattices with a width Δ of the first electronic miniband of 18 meV and 36 meV, respectively. The width of the miniband defines the upper limit for the tunability of electronic BOs, since for values of the applied field $F > \Delta/ed$ the electronic wavefunction gets localized in a single well (d is the superlattice period).

Figure 1 shows the comparison of TEOS and FWM signals *simultaneously* recorded at 10 K in the 18 meV sample at a reverse bias of 0.95 V applied. Distinctly different frequencies and dephasing times are observed in the two signals. In the FWM trace, the observed frequency is 2.2±0.15 THz and the dephasing time of the nonlinear interband polarization is 1 ps, while in TEOS a BO frequency of 2.6±0.05 THz and a dephasing time of the oscillatory signal of 2.4 ps is observed. We attribute the difference in the frequencies to the differences of the excitonic interband transitions observed in FWM as compared to the energy splitting of electrons in continuum states observed in TEOS[10]. Under excitation of the hh-e_{-1} and the hh-e_0 WS transition, the energy splitting of excitonic levels is smaller than eFd due to the fact that the exciton binding energy of the hh-e_{-1} transition decreases with increasing localization, while the binding energy of the hh-e_0 transition increases. The deviation from the eFd value for the given field is in accordance with numerical calculations of the exciton binding energies[8]. In TEOS, the observed oscillations are dominantly performed by electrons in continuum states, which are unaffected by the electron-hole Coulomb interaction, thus the frequency of the oscillations obeys the eFd re-

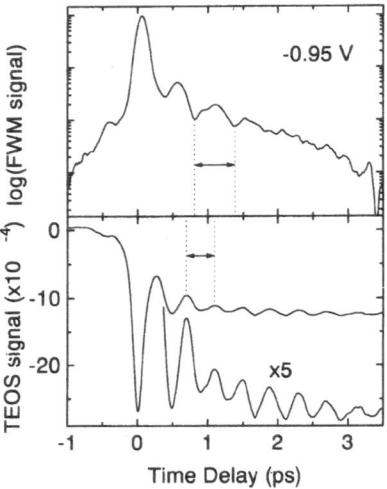

Figure 1. Simultaneously recorded FWM (upper part, logarithmic scale) and TEOS signals (lower part, linear scale) at a reverse bias of 0.95 V applied to the superlattice with 18 meV first electronic miniband width. The lattice temperature is 10 K and the excitation density is 2×10^9cm^{-2}. The arrows indicate the BO period observed in each experiment.

lation. Fig. 1 also shows that the dephasing time of the electronic continuum states is by more than a factor of two larger than the dephasing of the excitonic interband coherence. These observations combined with recent THz emission experiments, where BOs under non-resonant excitation conditions are observed[9], show that the traditional idea that the observation of optically excited coherence in semiconductors is restricted to excitonic excitations has to be expanded.

We investigate BOs at room temperature in a superlattice with a larger miniband-width (36 meV) using a different laser system with a shorter pulse duration (40 fs). We like to note that both the large miniband width and the shorter pulse width aggravate the observation of BOs due to i) the possibility of LO phonon emission in a miniband of a width of the LO phonon energy[11] and ii) due to the excitation of more than only the hh-e$_{-1}$ and the hh-e$_0$ WS transitions with a spectral pulse width of 40 meV (>Δ). Figure 2 (a) depicts the oscillatory parts of TEOS signals recorded at room temperature and Figure 2 (b) the Fourier spectra. Simultaneously performed FWM experiments reveal only an autocorrelation peak, i.e., the interband polarization is destroyed within the pulse width due to the thermionic ionization of excitons and exciton-free-carrier interaction. In the time domain data the increase of the BO frequency with the rising reverse bias is clearly visible. The dephasing time determined from the time domain data is approx. 130 fs. However, the decay of the BO amplitude is not mono-exponential. We attribute this to transient changes of the electric field in the sample after the optical excitation, thus leading to a time dependence of the splitting of the WS levels. The Fourier spectra reveal a tunability from 4.5 THz up to 8 THz. The spectra show a strong asymmetry due to the non-exponential decay discussed before. The upper frequency is close to the limit given by the miniband width and the optical phonon resonance of GaAs. When the Bloch frequency exceeds the optical phonon frequency, rapid phase destroying scattering events between the WS levels under emission of a LO phonon become possible. The lower limit is given by the requirement that the dephasing time has to be larger than half an oscillation period. The main effect leading to the rapid dephasing is attributed to the absorption of LO phonons leading to a

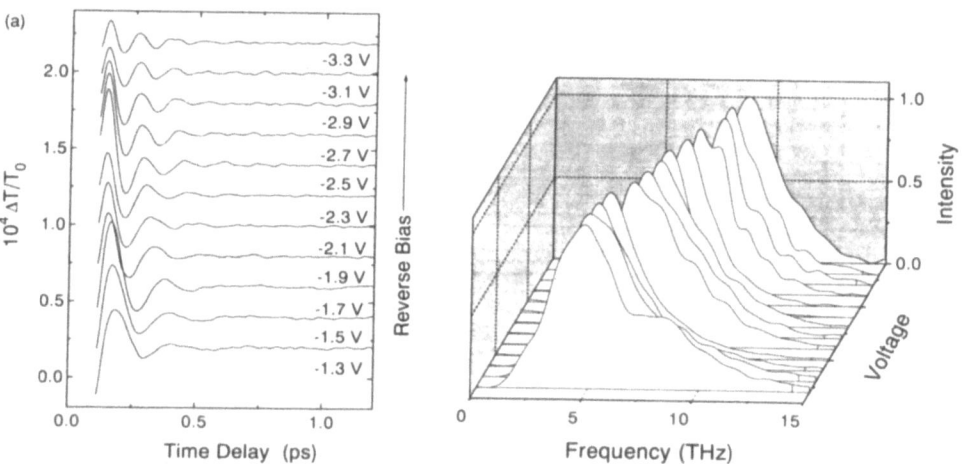

Figure 2. (a) Extracted oscillatory contributions to TEOS recorded at room temperature in a superlattice with 36 meV first electronic miniband width at different reverse bias voltages as indicated in the figure. The pulse width of the laser is 40 fs and the center wavelength 845 nm. (b) Fourier spectra of the data shown in (a).

thermalization of the excited distributions within the continuum in addition with the rapid change of the internal fields.

In a cw experiment like photocurrent measurements the minimum critical field F_c for the observation of WS transitions has to be larger than the homogeneous linewidth G_{hom} of the WS transitions divided by ed ($F_c > G_{hom}/ed$), with the relation between the linewidth and the scattering time t: $G_{hom} = 2h/t$. In cw photocurrent spectra of GaAs/AlAs and GaAs/Al$_x$Ga$_{1-x}$As superlattices several groups have observed a Wannier Stark ladder at room temperature in wide miniband superlattices (D@100 meV) with a minimum detectable WS splitting of about 30 meV[12]. In comparison, we observe a minimum BO energy of 18 meV in a 36 meV miniband sample in the time-resolved experiment, although we cannot resolve the WS ladder at room temperature in the cw photocurrent measurement due to the homogeneous broadening of the excitonic transitions. Thus one would assume that the observation of BOs in an optical technique should also be precluded. However, the experimental results show that the broadening (both homogeneous and inhomogeneous) of the intraband transitions is smaller than that of the interband transitions. We conclude that the BOs observed at 300 K in TEOS are dominantly performed by electrons in continuum states. The comparison of the room temperature data to TEOS experiments performed at the same sample at 10 K with the same laser system reveals an increase of the dephasing time by only a factor of three[7]. A stronger temperature dependence of the dephasing is expected for excitonic states, confirming our interpretation that the BOs observed are performed by electrons in continuum sates. The stability of the intraband coherence of electrons is a subject of further investigations and recent theoretical considerations[13].

ACKNOWLEDGMENT

This work is supported by the Volkswagen foundation.

REFERENCES

1. See contributions in *Coherent Optical Interactions in Semiconductors*, ed. R.T. Phillips (Plenum Press, New York, 1994).
2. J. Feldmann, K. Leo, J. Shah, D.A.B. Miler, J.E. Cunningham, S. Schmitt-Rink, T. Meier, G. von Plessen, A. Schulze, and P. Thomas, Phys. Rev. B **46**, 725 (1992); K. Leo, P. Haring Bolivar, F. Brüggemann, R. Schwedler, and K. Köhler, Solid State Comm. **84**, 943 (1992).
3. L. Esaki and R. Tsu, IBM J. Res. Dev. **14**, 61 (1970).
4. C. Waschke, H.G. Roskos, R. Schwedler, K. Leo, H. Kurz, and K. Köhler, Phys. Rev. Lett. 70, 3319 (1993).
5. T. Dekorsy, P. Leisching, K. Köhler, and H. Kurz, Phys. Rev. B **50**, 8106 (1994).
6. P. Leisching, P. Haring Bolivar, W. Beck, Y. Dhaibi, F. Brüggemann, R. Schwedler, H. Kurz, K. Leo, and K. Köhler, Phys. Rev. B **50**, 14389 (1994).
7. T. Dekorsy, R. Ott, H. Kurz, and K. Köhler, Phys. Rev. B **51**, 17275 (1995).
8. P. Leisching, T. Dekorsy, H.J. Bakker, H. Kurz, and K. Köhler, Phys. Rev. B **51**, 18015 (1995).
9. H.G. Roskos, C. Waschke, R. Schwedler, P. Leisching, Y. Dhaibi, H. Kurz, and K. Köhler, Superlatices. Microstruct. **15**, 281 (1994).
10. The *interband* dephasing time of the continuum states is below 100 fs, giving rise to the fast decaying initial peak in the FWM signal.
11. G. von Plessen, T. Meier, J. Feldmann, E.O Göbel, P. Thomas, K.W. Goosen, J.M. Kuo, and R.F. Kopf, Phys. Rev. B **49**, 14058 (1994).
12. K. Fujiwara, Jpn. J. Appl. Phys. **28**, L1718 (1989); E.E. Mendez, F. Agullo-Rueda, and J.M. Hong, Appl. Phys. Lett. **56**, 2545 (1990).
13. T. Meier, F. Rossi, P. Thomas, and S.W. Koch, Phys. Rev. Lett. **75**, 2558 (1995).

144

DETECTION OF HIGH POWER THz PULSES BY PHASE RETARDATION IN AN ELECTRO-OPTIC CRYSTAL

P. Uhd Jepsen,* M. Schall, V. Schyja, C. Winnewisser, H. Helm, and S. R. Keiding*

Fakultät für Physik
Albert-Ludwigs-Universität
Hermann-Herder-Strasse 3, Freiburg, Germany

In the last decade much research work has been done on the generation, detection and application of short-pulse THz radiation, see for an overview.[1]

With the advent of high-power fs-lasers it has become possible to generate THz pulses with high peak power, which offers the opportunity to use the THz pulse as a pump pulse, as opposed to a weak, non-interacting probe.[2,3,4] High-power THz pulses are generated by illuminating a large-aperture semiconductor emitter with a high external bias field applied across the gap. The photocarriers produced by the laser pulse are accelerated by the bias field, resulting in a large transient surface current. An electromagnetic pulse proportional to the time derivative of this current is radiated.[3,4]

In this communication we report on a novel room-temperature technique for detecting freely propagating high power THz pulses. The method is based on the electro-optic effect, which is stimulated by focusing a THz pulse into an electro-optic crystal ($LiTaO_3$). Thereby the optical properties are modified, resulting in a phase retardation (PR) of an optical probe pulse, which propagates through the crystal. The PR technique permits to determine the temporal THz pulse shape.

We illustrate the basic physics of our phase retardation experiment in Figure 1. The uniaxial $LiTaO_3$ crystal is oriented with the ordinary and extraordinary axes, with indices n_o and n_e, in the plane of the crystal surface. The THz beam and the probe beam propagate collinearly through the crystal along the z-direction (index n_o).

The polarization of the THz beam is kept parallel to the extraordinary axis. The probe beam is linearly polarized at 45 degrees with respect to the ordinary and extraordinary axes. Linearly polarized light becomes elliptically polarized due to the natural birefringence of the crystal, giving a phase change $\Delta\phi_{nat}$:

$$\Delta\phi_{nat} = \frac{\omega}{c}(n_e - n_o)L, \tag{1}$$

*Permanent address: Kemisk Institut, Århus Universitet, Langelandsgade 140, 8000 Århus C, Denmark

Ultrafast Processes in Spectroscopy
Edited by Svelto et al., Plenum Press, New York, 1996

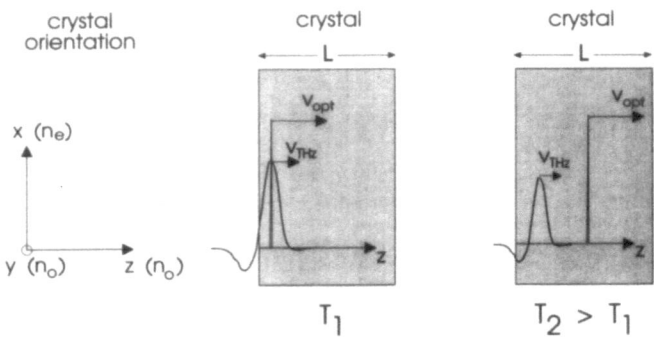

Figure 1. Principle of the phase retardation detection technique. Due to the difference in the refractive indices ($\approx$ factor 3) the probe pulse sweeps across the electrical field profile of the THz pulse inside the crystal. The length L is the crystal thickness (1 mm). The probe pulse and the THz pulse propagate collinear into the crystal, in the z-direction. The refractive indices in the directions (x, y, z) are (n_e, n_o, n_o), respectively.

where ω is the probe beam frequency and L is the length of the crystal. In the presence of an electric field (the THz field), an additional phase change, $\Delta\phi_{THz}$, will be introduced between the two electrical field components of the probe pulse,

$$\Delta\phi_{THz}(T) = \frac{1}{2}\frac{\omega}{c}(n_o^3 r_{13} - n_e^3 r_{33}) \int_0^L E_{THz}\left(\frac{z}{c}\Delta n + T\right) e^{-\alpha z} dz, \qquad (2)$$

where r_{13} and r_{33} are components of the electro-optic tensor for LiTaO$_3$.[5] The integral over the electric field E_{THz} of the THz beam accounts for a situation where the THz beam is attenuated inside the crystal (coefficient α) and where the probe pulse propagates much faster through the crystal than the THz pulse ($\Delta n = n_{probe} - n_{THz}$[6]). Therefore the probe pulse effectively sweeps over the portion of the THz pulse that lies inside the crystal. Implicit in equation (2) is the assumption that the duration of the probe pulse is much shorter than

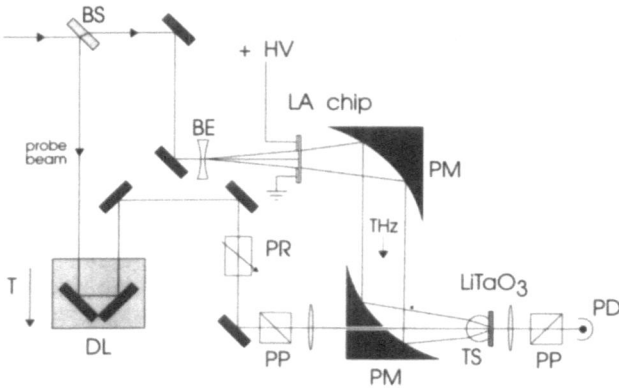

Figure 2. Schematic illustration of the setup. Legend: BS - 10% beam splitter, BE - beam expander, LA chip - large aperture emitter chip, PM - off-axis paraboloidal mirrors, DL - delay line, PR - polarization rotator, PP - polarizing prisms, TS - truncated, spherical sapphire lens and PD - photodiode.

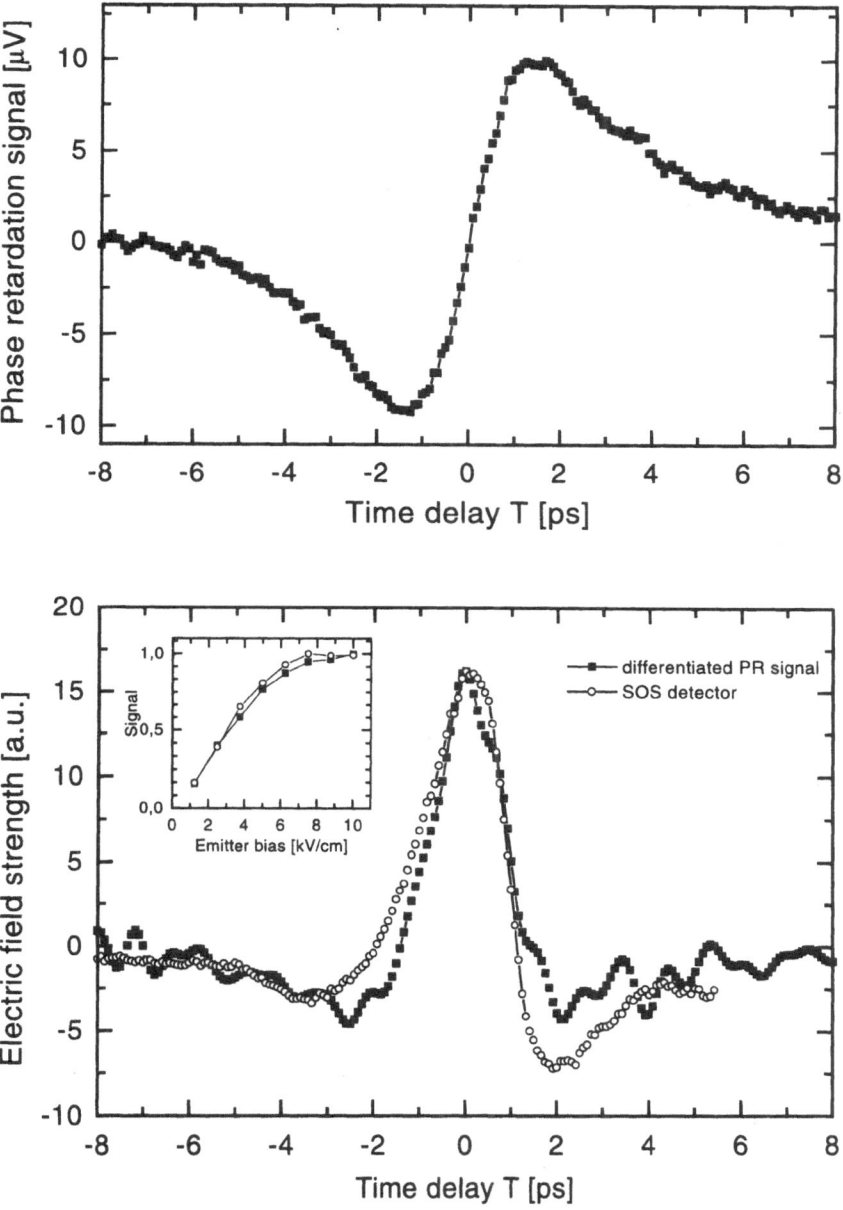

Figure 3. (a) The phase retardation signal, as a function of the time delay T. (b) The differentiated PR signal (filled squares) is shown in comparison to the silicon-on-sapphire (SOS) detector signal (open circles). The differentiated PR signal has been smoothed, and the SOS signal has been normalized with respect to the differentiated PR signal. The inset gives the THz peak emitted field strength saturation as a function of the emitter bias field, as measured by the two independent methods.

that of the THz pulse, a condition fullfilled in our experiment (140 fs vs 2 ps). T is the relative time delay between the THz pulse and the probe pulse. Converting the integration over length in equation (2) into one over time and assuming that the attenuation coefficient, α, is frequency independent, one obtains by differentiating[7]:

$$E_{THz}(T) \propto \frac{d}{dT}\Delta\phi_{THz}(T) - \frac{\alpha c}{\Delta n}\Delta\phi_{THz}(T). \tag{3}$$

Figure 2 presents a schematic drawing of the experimental setup. The laser source is a Ti:sapphire oscillator, amplified in a chirped pulse regenerative amplifier at a repetition rate of 1 kHz. The emitter consists of a wafer of high resistivity GaAs ($\rho > 10^7$ $\Omega\cdot$cm), with electrodes painted on the surface with conducting silver paint, leaving an emitter area of 1 cm^2. The THz radiation is collimated by a pair of off-axis paraboloidal mirrors, which guide the THz radiation to the crystal. A truncated, spherical sapphire lens of 5 mm radius is attached directly to the crystal to focus the THz radiation.[8]

In Figure 3b, the derivative of the PR signal is shown, together with the pulse shape measured directly with an silicon-on-sapphire (SOS) detector[9] inserted in place of the LiTaO$_3$ crystal in Figure 2. For negligible absorption the differentiated retardation signal (first term on the right hand side of equation (3)) should reveal the temporal distribution of the the THz field strength. There is a good correspondence between the SOS signal and the differentiated PR signal.

The inset in Figure 3b compares the normalized peak field strengths obtained from the SOS detection and the differentiated phase retardation signal, as functions of the field strength applied to the emitter, in the range from 1.25 kV/cm to 10 kV/cm. Both curves show a similar response to the emitter bias field, indicative of the linear response of the PR detection technique.

We have described an alternative setup for the detection of freely propagating THz pulses, and compared its performance to that of the conventional SOS detector. At present, the signal to noise ratio is lower than that of the photoconductive detection technique, but other geometries of the polarizer arrangement for the PR detection will improve the S/N ratio. While the absorption of the high frequency components of the THz pulse limits the analysis (though not the detection) of the full frequency spectrum of the THz pulse, in LiTaO$_3$, other materials may emerge with broader frequency response.

This research was supported by the Deutsche Forschungsgemeinschaft through SFB 276, TP C14. We thank Dr. G. Alber for helpful discussions.

REFERENCES

1. D. R. Dykaar, S. L. Chuang, J. Opt. Soc. Am. B **11**, 2454-2581, (1994)
2. J. T. Darrow, X.-C. Zhang, D. H. Auston, and J. D. Morse, IEEE J. Quantum Electron. **28**, 1607 (1992)
3. R. R. Jones, D. You, and P. H. Bucksbaum, Phys. Rev. Lett. **70**, 1236 (1993)
4. P. K. Benicewicz, J. P. Roberts, and A. J. Taylor, J. Opt. Soc. Am. B **11**, 2533 (1994)
5. B.E.A. Saleh and M.C. Teich, *Fundamentals of Photonics* (John Wiley & Sons, Inc., New York, 1991)
6. D. H. Auston, and M. C. Nuss, IEEE J. Quantum Electron. **24**, 184 (1988)
7. P.U. Jepsen, C. Winnewisser, M. Schall, V. Schyja, S.R. Keiding and H. Helm, submitted for publication
8. Ch. Fattinger, and D. Grischkowsky, Appl. Phys. Lett. **54**, 490 (1989)
9. D. Grischkowsky, S. Keiding, M. van Exter, and Ch. Fattinger, J. Opt. Soc. Am. B **7**, 2006 (1990)

145

NEW FIELD SENSORS FOR SUBPICOSECOND ELECTROMAGNETIC PULSES

X. -C. Zhang,[1] Q. Wu,[1] P. Campbell,[1] and L. Libelo[2]

[1] Physics Department
Rensselaer Polytechnic Institute
Troy, New York 12180–3590
[2] AMSRL-WT-NH
Army Research Laboratory
Adelphi, Maryland 20783A

In the ultrafast electronics and optoelectronics communities, especially in the subfield of applied terahertz beams, the detection of freely propagating picosecond microwave and millimeter-wave signals is primarily being carried out via photoconductive antennas and far-infrared interferometric techniques[1-8]. However, the limitation of these antenna-based detectors is the resonant behavior of their Hertzian dipole structure. This type of device structure has a resonant wavelength at the twice the dipole length. Therefore, the signal waveform, which includes the resonant detector response function, is not a simple cross-correlation of the incoming terahertz and optical gating pulses. In comparison, although far-infrared interferometric techniques provide an autocorrelation of terahertz pulses, important phase information is still lost. In most field-matter interaction applications, knowledge of the entire terahertz waveform, including both the amplitude and phase, is crucial. Thus, there is a need for the development of suitable sensing devices to support a variety of advanced scientific and technological applications.

We report the use of electro-optic field sensors for the characterization of freely-propagating terahertz pulses with subpicosecond temporal resolution. An electro-optic sensor provides a flat frequency response over an ultrawide bandwidth (DC to THz). Our work extended the conventional electro-optic sampling technique[9-11], which is primarily used for local field characterization, to free-space applications.

Fig. 1 schematically illustrates the experimental setup for free-space electro-optic sampling. When a pulsed electromagnetic radiation (terahertz beam) illuminates the electro-optic crystal, while the index of refraction is modulated via the Pockels effect. A femtosecond optical pulse probes the field-induced change in the index of refraction by passing through the crystal with a focus spot of 10 µm. To convert the field-induced ellipticity modulation into an intensity modulation, the probe pulse is analyzed by a compensator (C) and polarizer (P), then detected by a photodetector. In this experiment, a 500 µm

Ultrafast Processes in Spectroscopy, edited by Svelto et al.
Plenum Press, New York, 1996

649

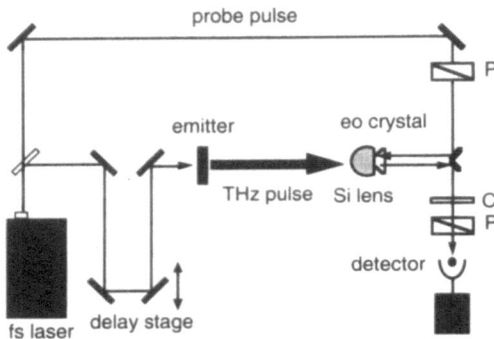

Figure 1. Experimental setup for free-space electro-optic sampling.

thick LiTaO$_3$ crystal was used as a Pockels cell with its c-axis parallel to the electric field polarization of the incoming radiation. The sensor head satisfies the phase-matching condition, which requires an angle of 69° between the terahertz beam and optical beam in LiTaO$_3$, as shown in Fig. 2. To improve detection efficiency, the terahertz beam is focused by a silicon lens.

Fig. 3 displays the near-phase-matching transient waveform. Here, the sensor crystal is 10 cm away from a biased photoconductive emitter. The risetime of the first peak is 740 femtoseconds. The minor peaks following the main peak are mainly due to multiple reflections of the electrical pulse in the photoconductive emitter (impedance mismatch along the biased electrodes), and the terahertz pulse in the sensor crystal. With a careful geometric design of the radiation emitter and a special-cut crystal, these reflections (electrical, optical and terahertz pulses) in Fig. 3 should be greatly reduced, and better temporal resolution should be achieved. Fig. 4 plots the transient waveform from an unbiased GaAs emitter with a carefully aligned probe beam path for phase-matching condition. Due the good phase-matching condition, signal-to-noise ratio is significantly improved. The FWHM of the main peak is 450 femtoseconds. The field measurement the electro-optic sensor is purely an electro-optic process. The system bandwidth is mainly limited by the dispersion of the terahertz signal and the duration of the laser pulse in the crystal, assuming it is phase-matched.

The comparison between free-space electro-optic sampling and the Hertzian dipole antenna for measurement of freely propagating quasi-optical radiation (terahertz beam) is

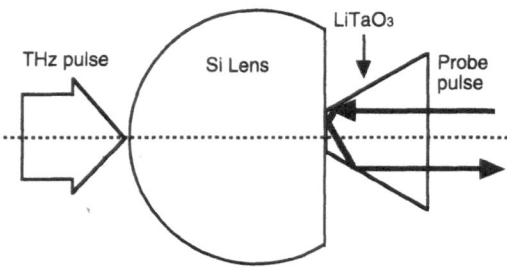

Figure 2. Phase matching electro-optic field sensor.

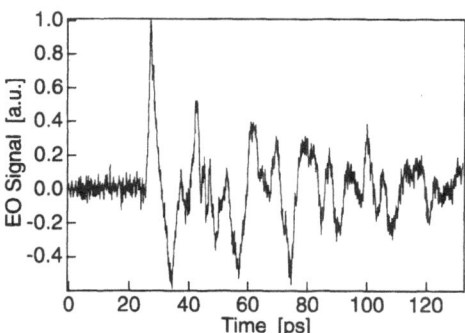

Figure 3. Near phase-matching signal measured by the electro-optic field sensor.

quite similar to the comparison between electro-optic sampling and photoconductive sampling (Auston switch) for the characterization of ultrashort electrical pulses in planar integrated circuits. Electro-optic sampling, via the Pockels effect, is attractive because it can provide a flat frequency spectrum and a true cross-correlation signal. In contrast, photoconductive sampling with a Hertzian dipole detector offers superior sensitivity. The frequency bandwidth of the two techniques should be comparable.

Our current electro-optic sensor has a minimum field sensitivity of about 22 mV/cm√Hz. The calculated minimum field sensitivity is about 5 mV/cm√ ⊦Hz. The minimum detectable voltage of the free-space electro-optic system may not be superior to some photoconductive antenna based systems. However, the bandwidth, minimal field perturbation, and true temporal cross-correlation of the free-space electro-optic system are unique. It is possible to extract the true terahertz waveform from the cross-correlation signal obtained via free-space electro-optic sampling. Specifically, because this technique eliminates the need for electrical contact with the sensor crystal, real-time terahertz imaging with an electro-optic crystal plate and a CCD camera may be more feasible than with a 2-dimensional photoconductive antenna array.

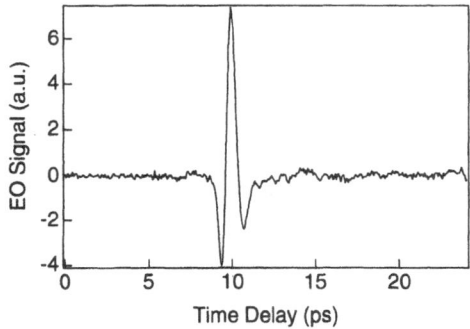

Figure 4. Phase-matching signal measured by the electro-optic field sensor.

REFERENCES

1. G. Mourou, C.V. Stancampiano and D. Blumenthal, *Appl. Phys. Lett.*, 38, 470 (1981).
2. A.P. DeFonzo, M. Jarwala and C.R. Lutz, *Appl. Phys. Lett.*, 50, 1155 (1987).
3. Ch. Fattinger and D. Grischkowsky, *Appl. Phys. Lett.*, 53, 1480 (1988); *ibid.* 54, 490 (1989).
4. M. van Exter, Ch. Fattinger and D. Grischkowsky, *Appl. Phys. Lett.*, 55, 337 (1989).
5. P.R. Smith, D.H. Auston, and M.C. Nuss, *IEEE J. Quantum Electron.*, 24, 255 (1988).
6. B.B Hu, J.T. Darrow, X.-C. Zhang, D.H. Auston and P.R. Smith, *Appl. Phys. Lett.*, 56, 886 (1990).
7. B.I. Greene, J.F. Federici, D.R. Dykaar, A.F. Levi, and L. Pfeiffer, *Opt. Lett.*, 16, 48 (1991).
8. S.E. Ralph and D. Grischkowsky, *Appl. Phys. Lett.*, 60, 1070 (1992).
9. B.H. Kolner and D.M. Bloom, *IEEE J. Quantum Electron.*, QE-22, 69 (1986).
10. J.A. Valdmanis and G.A. Mourou, *IEEE J. Quantum Electron.*, QE-22, 79 (1986).
11. D.H. Auston and M.C. Nuss, *IEEE J. Quantum Electron.*, QE-24, 184 (1988).

146

ULTRAFAST SAMPLING OF A DUAL-GATE FIELD-EFFECT TRANSISTOR

N. de B. Baynes,[1] J. Allam,[2] and J. R. A Cleaver[1]

[1] Microelectronics Research Centre
University of Cambridge, Cavendish Laboratory
Madingley Road, Cambridge CB3 0HE, United Kingdom
[2] Hitachi Cambridge Laboratory
Hitachi Europe Limited, Cavendish Laboratory
Madingley Road, Cambridge CB3 0HE, United Kingdom

Measurement of semiconductor devices at frequencies ≥ 1THz will allow direct access to the dynamics of non-equilibrium carrier transport. At present, the fastest devices such as high electron-mobility transistors operate at frequencies <1THz but significantly greater than the ≈ 100GHz bandwidth of conventional electronic measurements.[1] The reliability of extrapolating low-frequency measurements into the near-THz regime is unproven, and there is an increasing need for direct measurement methods in the 100GHz to 1THz frequency range.

Optoelectronic sampling techniques offers the possibility of measurement bandwidths approaching 1THz.[2] Illumination of photoconductive switches with ultrashort laser pulses allows the generation of sub-picosecond electromagnetic pulses, which can be propagated on coplanar transmission lines.[3] The propagating pulses can be measured using electro-optic probing, for example with an external lithium tantalate probe with a time resolution of ≈ 150fs.[4]

We have realised a state-of-the-art optoelectronic circuit for ultrafast sampling of multi-terminal devices, employing monolithic integration of the device with optimised photoconductive switches fabricated from low-temperature-grown GaAs (LT GaAs),[5-7] control of the electromagnetic modes propagating on the coplanar waveguide using microfabricated airbridges,[8] and discrimination of guided and freely-propagating modes[9] using a novel electro-optic sampling method. The device studied is a dual-gate field-effect transistor (DGFET).

DGFETs have applications in microwave circuits such as mixers[10] and also offer interesting possibilities for studying time-resolved transport in nanostructure devices. Excitation of the dual gates by a pair of time-delayed pulses corresponds to an electron time-of-flight measurement or a temporal analogue of hot-electron spectroscopy. However, such experiments place stringent requirements on the DGFET (gate lengths and

Ultrafast Processes in Spectroscopy, edited by Svelto et al.
Plenum Press, New York, 1996

separation ≪1μm, and extremely low parasitic resistance and capacitance) and on the measurement circuit (excitation pulses <1ps). Such high-speed nanostructure DGFETs have not yet been realised. In the present work, we investigate the propagation of picosecond pulses onto and through one gate in such a device, in order to demonstrate the performance of our sampling circuit.

Figure 1 shows a micrograph of a fabricated DGFET integrated with the ultrafast sampling circuit. Each of the gates is connected to the CPW at both ends, in order to study the propagation of picosecond pulses through the gate. The length of the channel between source and drain ohmic contacts was ≈ 5μm. Dual-gate FET's were fabricated with gate lengths ≈ 160 nm, and gate separations varied between ≈ 0.3 μm and 1.5 μm.

Mode-discriminating electro-optic sampling (MEOS)[9] was employed to detect the pulses incident on, reflected by and transmitted through the gates (Fig. 2). The sampling was performed at 1 mm distance from the device in order to separate in time the incident and reflected pulses. The large reflected signal is due to the impedance mismatch at the gate. A significant portion is transmitted, confirming continuity across the gate metallisation.

Figure 1. Micrograph of DGFET, showing ohmic contacts to the source and drain, dual gates connected at each end to CPW, ohmic contact to the inter-gate region and connection of all ground planes in the region of the DGFET using an airbridge.

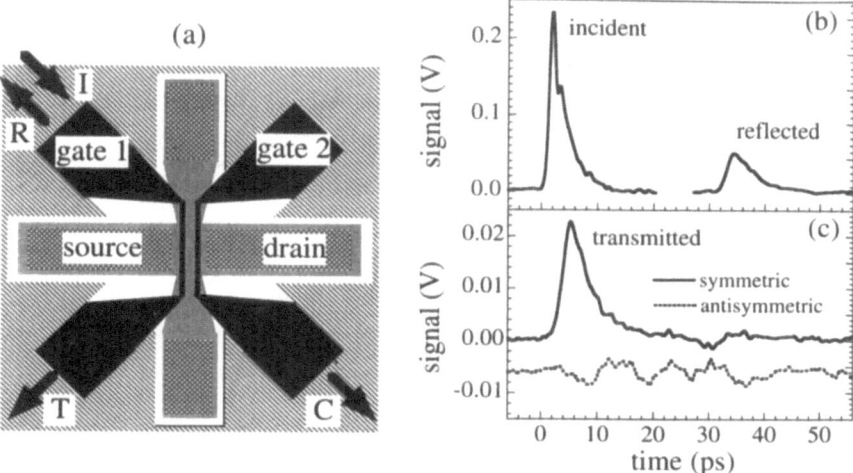

Figure 2. (a) Schematic of incident (I), reflected (R), transmitted (T) and capacitively-coupled (C) pulses associated with gate 1 of DGFET. (b) Incident and reflected pulses measured by MEOS. (c) Transmitted pulse measured by MEOS, showing both symmetric and antisymmetric signals on the CPW.

Measurement by conventional electro-optic sampling revealed features superimposed on the trailing edges of both the incident and transmitted pulses, due to freely-propagating radiation (i.e. a Terahertz beam) generated by the photoconductive switch and reflected from the substrate back surface. These reflections are discriminated from the guided mode (which is symmetric about the central conductor of the CPW) in the MEOS measurement. The transmitted antisymmetric MEOS signal (Fig. 2(c)) indicates the presence of up to 4 separate reflections. The absence of spurious features on the trailing edge of the symmetric signal allows more reliable time-windowing of incident, reflected and transmitted pulses for Fourier transformation into the frequency domain, and hence more accurate determination of S-parameters.

The Fourier transform of the incident pulse is shown in Fig. 3(a). Although the 3-dB bandwidth is only 100 GHz, the large dynamic range ensures usable spectral content at much higher frequencies. The bandwidth is limited mainly by frequency-dependent radiative losses of the CPW, due to the dielectric mismatch between the substrate (ε_r=13.1) and superstrate (ε_r=1).[3] The signal decreases smoothly for frequencies up to at least 500GHz, but at higher frequencies there are strong absorption lines, around 1.0, 1.4 and 1.7 THz, due to rotational transitions in water, which are excited by the fringing fields of the CPW.[11] Hence measurements at THz frequencies require baking of the sample to remove absorbed water, and measurement in a dry atmosphere.

The incident, reflected and transmitted pulses of Fig. 2 were Fourier-transformed, and the scattering parameters obtained by dividing the reflected (for S_{11}) or the transmitted (for S_{21}) spectrum by the incident spectrum, correcting for the different propagation distances from the device to the measurement point using a numerical simulation of the pulse propagation. The S-parameters are shown in Fig. 3 (b) at frequencies up to $\approx$300GHz. Both S_{11} and S_{21} show evenly-spaced fringes, of spacing 59GHz for S_{11} and 77GHz for S_{21}, which are absent in the incident signal. The fringes imply an etalon effect

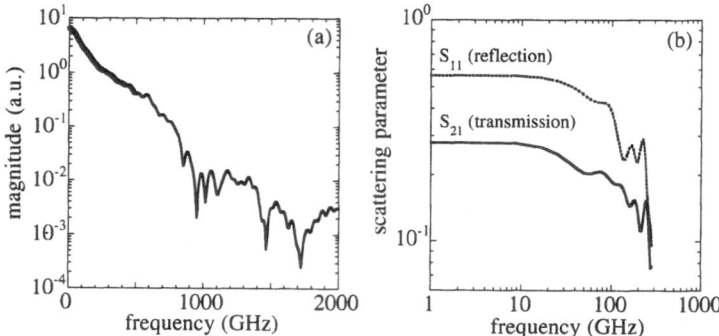

Figure 3. (a). Frequency spectrum of signal incident on the gate. (b). S-parameters for reflection (S_{11}) and transmission (S_{21}).

due to multiple reflections of the pulse with time periods of 17 and 13ps, corresponding to pulse propagation distances of ≈800μm and ≈600μm. However, their origin is not known.

The signal capacitively-coupled onto gate 2 was also measured. Comparison with the derivative of the incident pulse indicated a gate-gate capacitance of 10fF. This value is comparable to the estimated capacitance between the gates and the channel (≈7fF assuming parallel plate capacitors and ignoring fringing fields), and larger than the direct capacitance between the metal gates (≈1fF).

In conclusion, we have demonstrated the integration of a dual-gate FET with an ultrafast optoelectronic circuit for the generation and propagation of picosecond pulses. The S-parameters associated with pulse propagation in one of the gates were determined. We consider that this data represents the state-of-the-art in ultrafast sampling, in terms of bandwidth, device integration, control of propagating modes and absence of spurious features in the measured signal.

Direct measurement of transistor cut-off frequencies up to ≈300GHz appears to be possible using the above techniques. However, comparison with electrical measurements up to 100GHz is desirable to validate the data optained by optical sampling. The properties of the passive elements (e.g. the 8-way airbridge structure) require further characterisation. The measurement bandwidth can be extended by reducing radiative losses in the waveguide, by reducing the effective dielectric constant of the substrate.[3] Work is in progress to exploit these techniques for the direct measurement of 2- and 3-terminal devices in the 100GHz to 1THz frequency range.

REFERENCES

1. P. M. Smith, S.-M. J. Liu, M.-Y. Kao, P. Ho, S. C. Wang, K. H. G. Duh, S. T. Fu and P. C. Chao, IEEE Microwave Guided Wave Lett. **5**, 230 (1995).
2. M. Y. Frankel, J. F. Whittaker and G. A. Mourou, IEEE J. Quantum Electronics **QE-28**, 2313, (1992).
3. H. J. Cheng, J. F. Whitaker, T. M. Weller, and L. P. B. Katehi, IEEE Trans. Microwave Theory Tech. **42**, 2399, (1994).
4. U. D. Keil and D. R. Dykaar, Appl. Phys. Lett. **62**, 1504 (1992).
5. Y. Chen, S. Williamson, T. Brock, F. W. Smith, A. R. Calawa, Appl. Phys. Lett. **59**, 1984 (1991).
6. J. Allam, K. Ogawa, J. White, N. de B. Baynes, J. R. A. Cleaver, I. Ohbu, T. Tanoue and T. Mishima, OSA Proc. Ultrafast Electronics and Optoelectronics **14**, 197 (1993).

7. N. de B. Baynes, J. Allam, J. R. A. Cleaver, K. Ogawa, I. Ohbu, T. Mishima, Inst. Phys. Conf. Ser. **136**, 337 (1994).

8. N. de B. Baynes, J. Allam and J. R. A. Cleaver, submitted to IEEE Trans. Microwave Theory Techniques.

9. N. de B. Baynes, J. Allam and J. R. A. Cleaver, to be published in IEEE Microwave Guided Wave Lett.

10. C. Tsironis, R. Meierer and R. Stahlmann, IEEE. Trans. Microwave Theory Techniques **MTT-32**, 249 (1984).

11. M. van Exter, Ch. Fattinger and D. Grischkowsky, Optics Letters **14**, 1128 (1989).

147

ELECTRIC FIELD DYNAMICS AT A METAL–SEMICONDUCTOR INTERFACE PROBED BY TIME RESOLVED FEMTOSECOND OPTICAL SECOND HARMONIC GENERATION

W. de Jong,[1] A. F. van Etteger,[1] C. A. van 't Hof,[1] P. J. van Hall,[2] and Th. Rasing[1]

[1]Research Institute for Materials
Nijmegen, The Netherlands
[2]Department of Physics
Eindhoven University of Technology
Eindhoven, The Netherlands

The time evolution of the electric field at a Schottky barrier (SB) interface due to above bandgap excitation is directly measured using time resolved Optical Second Harmonic Generation (SHG). The results are found to agree very well with Monte Carlo simulations.

Recent time resolved Photo Luminescence (PL) experiments performed on a biased SB sample showed a PL signal with an unexpectedly long decay time.[1] This was explained to be the result of the instantaneous collapse and subsequent recovery of the SB field due to excited carriers. Possible oscillations of this field were also predicted, as have already been observed in bulk semiconductors.[2]

However PL experiments probe this interface electric field in quite an indirect way. Optical Second Harmonic Generation is a technique that has been shown to be very sensitive for symmetry breaking at interfaces.[3,4] We have recently demonstrated that SHG can also be applied to probe the (time dependence of the) electric field at semiconductor–oxide and semiconductor–metal interfaces.[5,6,7]

In this paper we will present time resolved SHG measurements on a Au/GaAs SB system as well as MC simulations which are performed on the same system. Using the experimental conditions as input and no adjustable fit parameters, an excellent agreement between these two is found.

In the depletion region near a Schottky barrier extra charge carriers can be excited by a short laser pulse. Due to the internal electric field, which can be varied with an applied bias, these charge carriers are separated and create a space charge field which is

Ultrafast Processes in Spectroscopy
Edited by Svelto et al., Plenum Press, New York, 1996

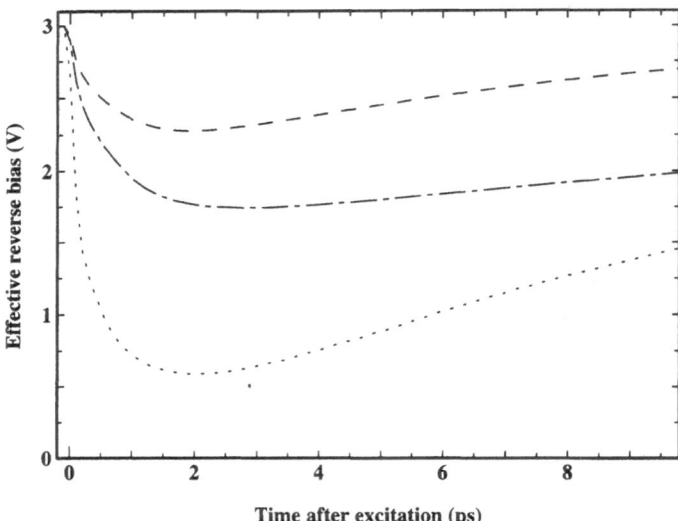

Figure 1. Time evolution of the effective bias voltage, calculated for an initial reverse bias voltage of 3 V for three different combinations of the diameter (d) of the exciting spot and the intensity (I) of the excitation. Upper curve: $d = 70~\mu$m and $I = 4$ W/cm^2, middle curve: $d = 160~\mu$m and $I = 4$ W/cm^2, lower curve: $d = 70~\mu$m and $I = 20$ W/cm^2. For $d = 70~\mu$m the regeneration time is 8 ps and for $d = 160~\mu$m it is 25 ps.

counteracting the initial field. This will cause an almost instantaneous collapse of this field. If the photo-excited charge density significantly exceeds the intrinsic one, the applied field may be quenched totally by only a small part of the excited carriers and an overshoot of the carriers can cause oscillations in the electric field.

When the excited holes reach the Au/GaAs interface they will neutralize the surface charge. The recovery of the surface charge, which also determines the recovery of the collapsed field, depends among other things on the thickness of the Au-layer and the area of the laser spot. Based on the capacitance per area of the metallic top contact (520 μF/m^2), the sizes of the excitation beam (70 μm and 160 μm) and the square resistance of the Au film (a few Ohms), we estimated the RC time constant of this process (the regeneration time) to be in the order of 10 to 30 ps.

In the, one dimensional, Monte-Carlo simulations the carriers are created using a gaussian shaped (in the time domain) excitation beam as is used in the experiments. The simulations are performed accounting for the various scattering mechanisms. Values from literature are used for all scattering probabilities. Both electron-heavy- and light-hole pairs are considered, including hole inter-band transitions. The Poisson equation is solved after every 1 fs time step, which gives the electric field conditions for the next step in the simulation. The resulting electric field is integrated to give the time evolution of the effective bias, which can be compared with the experiments. The results of the calculations are shown in Figure 1.

The Schottky barrier sample we used for our experiments, was grown on a n^+ GaAs(100) substrate. The actual barrier is formed by a 3000 Å-thick n-type GaAs (doping concentration: 10^{17} cm^{-3}) layer and a semi-transparent 100 Å thick Gold film to allow for laser excitation through the metal top contact. The sample contains a heavily doped

n-type (doping concentration: 10^{18} cm^{-3}) superlattice buffer layer (100×50 Å GaAs/50 Å Al$_{.33}$Ga$_{.67}$As) between the depletion layer and the n^+-GaAs substrate.

For the SHG measurements we used a mode-locked Titanium Sapphire laser that operated at $\lambda = 840$ nm and produced trains of 60 fs pulses at a 82 MHz repetition rate. The photon energy at this wavelength is 1.47 eV; this is about 50 meV above the bandgap of GaAs.

The SHG contributions from the bulk GaAs and the interface can be separated by looking at the azimuthal anisotropy of the SHG signal.[6] Time resolved pump probe experiments were performed at a fixed rotation angle of the sample, at a position where the electric field induced effect on the SHG-intensity is the biggest. P-polarized pump and probe beams were generated by a beam splitter. The pump beam was focussed at an incident angle of 25°. The probe beam was focussed to about 65 μm diameter on the sample at an incident angle of 45°. The probe pulses were delayed with respect to the pump pulses by a computer controlled delay stage with 0.1 μm step size. The p-component of the generated specular SHG-intensity was detected by a photon counting system. The probe and pump beams also generate a sum frequency signal, corresponding to the cross-correlation of the pump and probe pulses. This sum frequency is detected with a PM-tube lock-in system, and enables the control of the pulse width and the determination of the zero delay point between pump and probe beams. The pump-probe measurements were performed for different laser powers in the pump beam and different sizes of the pump spot.

First the SHG intensity is measured for different values of the reverse bias, between 0 V and 3 V with the pump beam blocked. The values that are found from these measurements are used as a calibration curve. The SHG intensity that is measured in the time resolved pump probe measurements is scaled in this way to a corresponding effective reverse bias value, which can be compared with the calculations. Figure 2 shows the results of the measurements after this scaling. At $t = 0$ the center of the pump pulse arrives at the sample.

In the figure it can be seen that the SHG signal and thus the effective bias rapidly decreases after the pump pulse has excited carriers in the sample. This decrease is larger for higher pump power values.

The measured time dependence of the data can be fitted with a double exponential, yielding one time constant for the collapse (τ_c) and one for the recovery (τ_r) of the electric field. This gives the following values for τ_c and τ_r for the upper, middle and lower curve respectively: (790 fs, 8 ps), (940 fs, 19 ps) and (760 fs, 64 ps). From these values it can be seen that the recovery time is considerably longer when the spot size is bigger while the excitation intensity is kept the same. This is predicted by the MC simulations as can be seen in Figure 1. Comparison of the two figures shows that for the upper and lower curve the measured decrease is equal to the calculated one. For the middle curve the measured one is bigger than calculated.

From these results we can conclude that the MC simulations describe the electric field dynamics quite well. The results show that the recovery time of the system not only depends on the excitation intensity but also on the spot size as is predicted by the simulations and demonstrated in the experiments. We are preparing experiments to measure the electric field dynamics with higher intensity pump beams to reach the regime where oscillations in the electric field are predicted by the MC simulations.

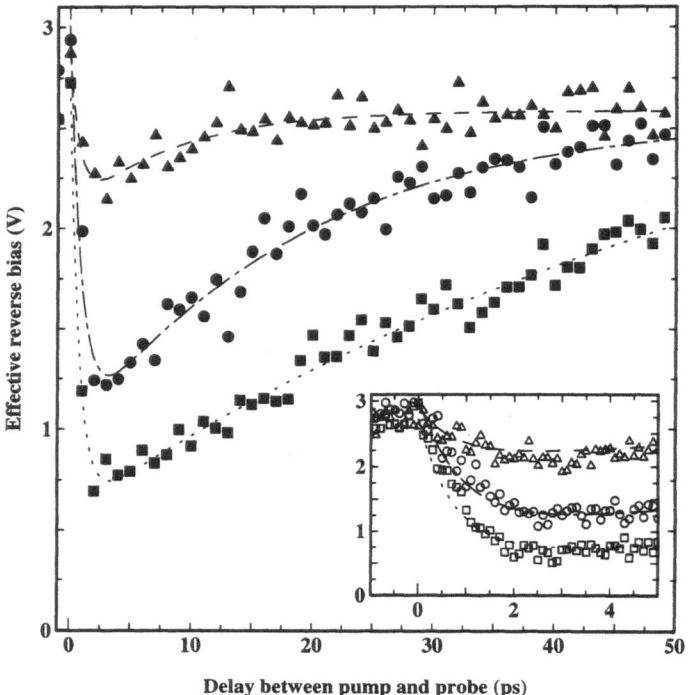

Figure 2. Measured time evolution of the effective reverse bias for different combinations of the spot diameter (d) and the intensity (I) of the exciting laser beam. The lines are an exponential collapse and recovery fit. In the upper curve $d = 70$ μm and $I = 4$ W/cm^2. In the middle one $d = 160$ μm and $I = 4$ W/cm^2. In the lower $d = 70$ μm and $I = 20$ W/cm^2. The inset shows the collapse measured with smaller time steps.

REFERENCES

1. P. C. M. Christianen, P. J. van Hall, H. J. A. Bluyssen, and J. H. Wolter, Semicond. Sci. Technol. **9**, 707 (1994).
2. W. Sha, Arthur L. Smirl, and W. F. Tseng, Phys. Rev. Lett. **74**, 4273 (1995).
3. T. F. Heinz, in *Nonlinear Surface Electromagnetic Phenomena*, ed. by H. E. Ponath and G. I. Stegeman (North Holland, Amsterdam, 1991).
4. Th. Rasing, Appl. Phys. A **59**, 531 (1994).
5. P. Godefroy, W. de Jong, C. W. van Hasselt, M. A. C. Devillers, and Th. Rasing, to be published.
6. W. de Jong, A. F. van Etteger, C. A. van 't Hof, P. J. van Hall, and Th. Rasing, Surf. Sci. **331–333**, 1372 (1995).
7. W. de Jong, A. F. van Etteger, C. A. van 't Hof, P. J. van Hall, and Th. Rasing, Surf. Sci. to appear (1996).

148

DYNAMICS OF TRAPPING, TRAP-EMPTYING, AND BREAKDOWN IN LT GaAs

J. Allam[1] and N. de B. Baynes[2]

[1] Hitachi Cambridge Laboratory
Hitachi Europe Limited, Cavendish Laboratory
Madingley Road, Cambridge CB3 0HE, United Kingdom
[2] Microelectronics Research Centre
University of Cambridge, Cavendish Laboratory
Madingley Road, Cambridge CB3 0HE, United Kingdom

Photoconductors fabricated from annealed low-temperature-grown GaAs (LT GaAs) represent the state-of-the-art in ultrafast photodetectors due to sub-ps carrier trapping, high responsivity and high breakdown voltage.[1] Much attention has been paid to measurement of the trapping time. However, the properties of the trapped charges and their influence on carrier transport has not been studied. In this paper, we investigate the dynamics of trapping, trap-emptying and breakdown in LT GaAs interdigitated photoconductors, and obtain clear evidence for field screening by trapped carriers and for avalanche breakdown. The effect of the trap-emptying time on the transport is revealed by comparison of the dynamical results with the temperature dependence of the dark current.

The 2-μm-thick layer of LT GaAs used in this study was grown at 220°C and annealed in-situ at 600°C for 10 minutes.[2] X-ray diffraction and transmission electron microscopy confirmed that the layer had good crystallinity and a uniform distribution of As precipitates[3] with average size ≈3nm and spacing ≈20nm. Ultrafast trapping in annealed LT GaAs has been correlated with the presence of these microscopic As precipitates.[3,4] Transient reflectance measurements (Fig. 1(a)) indicated a trapping time of 0.7ps.

The response time, responsivity and breakdown voltage of an interdigitated photoconductor depend on the electrode dimensions, i.e. finger width (W) and separation (S). Detectors were fabricated with W=S=0.5–8μm. The response time was measured using electro-optic sampling, and was consistent with the measured trapping time and the calculated RC time associated with the load resistance and the parasitic resistance of the interdigitated electrodes. The fastest measured response time of 1.0ps corresponds to a 3dB bandwidth of 330GHz.

The responsivity (photocurrent/incident power, $R=I_p/P$) is expected to vary inversely with the electrode separation S, for a uniform electric field and constant carrier velocity. However this relation will be modified by the non-uniform field, finite absorption length,

Ultrafast Processes in Spectroscopy, edited by Svelto et al.
Plenum Press, New York, 1996

663

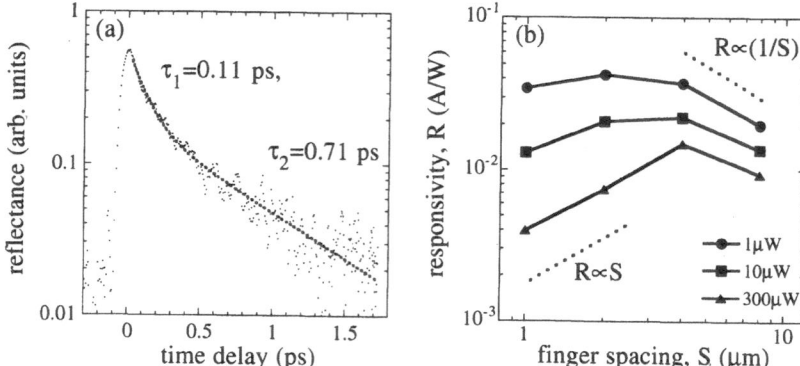

Figure 1. (a) Transient reflectance of LT GaAs. (b) Variation of responsivity R with S (1μm-8μm) and P (1μW-300μW).

field screening by photogenerated carriers, light diffraction and non-stationary transport. Klingenstein et al.[5] measured comparable responsivities for devices with S=10μm and S=1μm, whereas Chen et al.[1] found R increasing by a factor of 100 when S was decreased from 20μm to 0.2μm.

The responsivity was measured as a function of electrode separation (S=1–8μm), wavelength (410–820nm), light intensity (1μW-1mW) and electric field (5–35V/μm). Figure 1(b) shows R as a function of S for 633nm illumination and electric field of 10V/μm. At a given wavelength and electrode separation, the responsivity decreases with increasing illumination intensity. In the limit of low intensity, short wavelength, high bias and large S, the expected R∝(1/S) dependence is observed. However, in the opposite limit of high intensity, long wavelength, low bias and small S, the responsivity exhibits approximately R∝S behaviour. Correspondingly, in the photocurrent-voltage characteristic there is a transition from ohmic (I_p∝V) behaviour at high bias and high intensity, to quadratic behaviour (I_p∝V^2) at low bias and low intensity.

We associate the reduced responsivity compared to the predicted (1/S) behaviour with screening of the applied field by trapped charges. Figure 2 (a) shows the electron-hole pairs photogenerated between the contacts at time t=0. Carriers which drift into unilluminated regions of the device prior to trapping generate a dipole which opposes the applied field. The sensitivity of R to very low levels of illumination implies a long lifetime of the "frozen in" dipole. The decay time of charged traps was determined by measuring the recovery of the responsivity after illumination by a pump pulse. A novel low-frequency pump-probe experiment utilising acousto-optic modulators was used. The photocurrent recovery is shown in Fig. 2(b) for applied bias of 5V-35V across a 1μm gap. For bias <20V, reduction of the photocurrent due to field screening is observed for times as long as 50μs. Long decay times (Å1ms) have also been observed in LT GaAs photorefractive devices.[6] Contrary to previous assumptions, an LT GaAs photoconductor operating within the typical range of conditions (S≈1μm, P≈1μW, field≈10V/μm) does not fully recover between laser pulses (typically separated by ≈10ns in a high-repetition-rate laser). The differing dependence of responsivity on S reported in the literature can be explained by the different bias conditions, resulting in strong field screening in the case of Klingenstein et al.[5] but not in the case of Chen et al.[1]

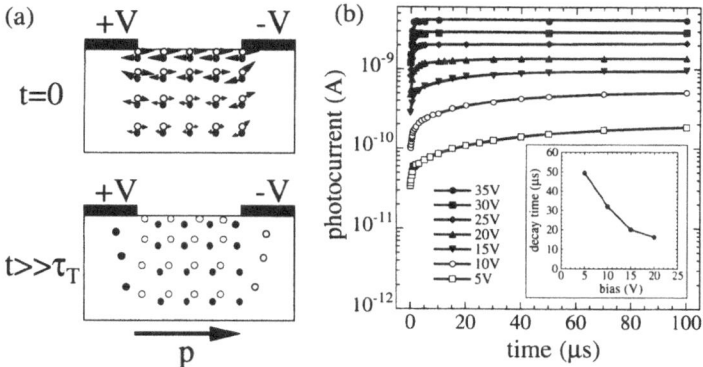

Figure 2. (a) Creation of "frozen-in" dipole due to non-uniform photocarrier generation and subsequent trapping. (b) Recovery of photocurrent after a pump pulse at time t=0. Inset: decay time of trapped charge as a function of bias.

The field screening due to trapped charge decays exponentially with time, and the decay time decreases with increasing applied bias (inset of Fig. 2(b)). Indeed, the decay rate is approximately proportional to the dark current, implying that the mechanism for the dark current is associated with trap-emptying. Figure 3(a) shows the temperature-dependence of the dark current-voltage characteristic. At fields ≤ 20V/μm, the behaviour is consistent with defect-associated hopping conduction.[7] At higher fields, the dark current increases due to thermally-assisted tunnelling of trapped carriers. This is accompanied by faster trap emptying and reduced field screening.

The maximum applied bias, and hence the maximum responsivity, is limited by breakdown. The breakdown mechanism in LT GaAs has been the subject of controversy, variously interpreted as thermionic field emission from traps[8], trap filling[9] or impact ionisation[7,10]. The breakdown field in our LT GaAs was 3.5×10^5V/cm for S=0.5–8 μm, and decreased with temperature (Fig. 3(a)), consistent with avalanche breakdown due to impact ionisation.

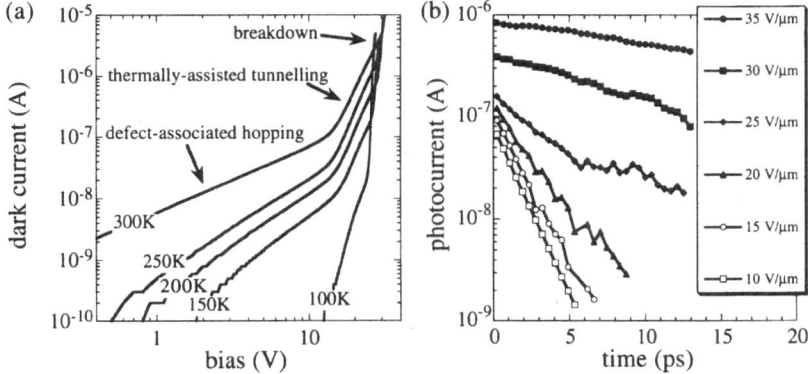

Figure 3. (a) Temperature-dependence of the dark current-voltage characteristic, for device with S=1μm. (b) Photocurrent decay as a function of applied field, for device with S=2μm.

It is remarkable that LT GaAs, which exhibits sub-picosecond trapping and very high resistivity at fields below breakdown, also exhibits a breakdown voltage which is similar to that of conventional GaAs. Clearly, it is not possible to simultaneously achieve both sub-picosecond trapping and impact ionisation, since the length scale for trapping (<0.1μm) is much less than the inverse ionisation coefficient ($1/\alpha \approx 1$μm at 3.5×10^5V/cm).[11] This is confirmed in Fig. 3(b), which shows the photocurrent decay (actually one half of a photoconductive autocorrelation curve) for a device with S=2μm, as the field is increased from 10V/μm to 35V/μm. For fields ≤20V/μm, the decay time is ≈1ps due to trapping. However for fields of 25V/μm and above, a component with decay time ≈20ps appears which dominates the response at fields close to breakdown. This corresponds to the transit time of electrons across the 2μm device, and indicates a reduced trapping rate at high fields. The responsivity increases due to generation of secondary electron-hole pairs by impact ionisation.

In conclusion, we have investigated transport in LT GaAs by means of dynamical measurements. At fields <20V/μm the trapping time is <1ps, whereas trap-emptying due to thermal emission has a time constant >10μs. The trapped charges can screen the applied field leading to reduced responsivity. At fields >20V/μm, the trap-emptying rate increases due to thermally-assisted tunnelling. In addition, the carrier trapping rate decreases, allowing carriers to reach high energy and impact ionise, causing avalanche breakdown. Such an understanding of the transport allows optimisation of the design and operation conditions of LT GaAs ultrafast photodetectors.

REFERENCES

1. Y. Chen, S. Williamson, T. Broeck, F. W. Smith and A. R. Calewa, Appl. Phys. Lett. **59**, 1984 (1991).
2. J. Allam, K. Ogawa, J. White, N. de B. Baynes, J. R. A. Cleaver, I. Ohbu, T. Tanoue and T. Mishima, OSA Proc. Ultrafast Electronics and Optoelectronics **14**, 197 (1993).
3. A. C. Warren, J. M. Woodall, J. L. Freeouf, D. Grischkowsky, D. T. McInturff, M. R. Melloch and N. Otsuka, Appl. Phys. Lett. **57**, 1331 (1990).
4. E. S. Harmon, M. R. Melloch, J. M. Woodall, D. D. Nolte, N. Otsuka and C. L. Chang, Appl. Phys. Lett. **63**, 2248 (1993).
5. M. Klingenstein, J. Kuhl, R. Nötzel, K. Ploog, J. Rosenzweig, C. Moglestue, A. Hülsmann, J. Schneider, K. Köhler, Appl. Phys. Lett. **60**, 627 (1992).
6. D. D. Nolte, M. R. Melloch, S. J. Ralph and J. M. Woodall, Appl. Phys. Lett. **61**, 3098 (1992).
7. J. P. Ibbetson, J. S. Speck, N. X. Nguyen, A. C. Gossard and U. K. Mishra, J. Electron. Mat. **22**, 1421 (1993).
8. J. K. Luo, H. Thomas, D. V. Morgan, D. Westwood and R. H. Williams, Semicond. Sci. Technol. **9**, 2199 (1994).
9. A. K. Verma, J. Tu, J. S. Smith, H. Fujioka and E. R. Webber, J. Electron. Mat. **22**, 1417 (1993).
10. N. de B. Baynes, J. Allam, J. R. A. Cleaver, K. Ogawa, I. Ohbu, T. Mishima, Inst. Phys. Conf. Ser. **136**, 337 (1994).
11. G. E. Bulman, V. M. Robbins, K. F. Brennan, K. Hess and G. E. Stillman, IEEE . Electron. Dev. Lett. **4**, 181 (1983).

INDEX